$305.95 per copy (in United States)
Price is subject to change without prior notice.

0063

RSMeans® Assemblies Cost Data

38TH ANNUAL EDITION

2013

RSMeans®
A division of Reed Construction Data, LLC
Construction Publishers & Consultants
700 Longwater Drive
Norwell, MA 02061
USA
1-800-334-3509
www.rsmeans.com

Copyright 2012 by RSMeans
A division of Reed Construction Data, LLC
All rights reserved.

Printed in the United States of America
ISSN 1524-2153
ISBN 978-1-936335-55-8

The authors, editors, and engineers of RSMeans apply diligence and judgment in locating and using reliable sources for the information published. However, RSMeans makes no express or implied warranty or guarantee in connection with the content of the information contained herein, including the accuracy, correctness, value, sufficiency, or completeness of the data, methods, and other information contained herein. RSMeans makes no express or implied warranty of merchantability or fitness for a particular purpose. RSMeans shall have no liability to any customer or third party for any loss, expense, or damage, including consequential, incidental, special, or punitive damage, including lost profits or lost revenue, caused directly or indirectly by any error or omission, or arising out of, or in connection with, the information contained herein.

No part of this publication may be reproduced, stored in a retrieval system, or transmitted in any form or by any means without prior written permission of RSMeans. The cost data contained in this publication is valuable and proprietary information of RSMeans and others and there is no transfer to you of any ownership rights in the cost data or any license granted to you to create derivative works based on or which utilize the cost data.

Senior Editor
Marilyn Phelan, AIA

Contributing Editors
Christopher Babbitt
Ted Baker
Adrian C. Charest, PE
Gary W. Christensen
David G. Drain, PE
Cheryl Elsmore
Robert J. Kuchta
Robert C. McNichols
Genevieve Medeiros
Melville J. Mossman, PE
Jeannene D. Murphy
Stephen C. Plotner
Eugene R. Spencer
Phillip R. Waier, PE

Cover Design
Wendy McBay
Paul Robertson
Wayne Engebretson

President & CEO
Iain Melville

Chief Operating Officer
Richard Remington

Chief Customer Officer
Lisa Fiondella

Vice President, Marketing & Business Development
Steven Ritchie

Vice President, Finance & Operations
Renée Pinczes

Engineering Director
Bob Mewis, CCC

Director, Product Management
Jennifer Johnson

Product Manager
Andrea Sillah

Product Marketing Manager
Ryan Cummings

Senior Director, Project & Process Management
Rodney L. Sharples

Production Manager
Michael Kokernak

Editing Supervisor
Jill Goodman

Production Supervisor
Paula Reale-Camelio

Production
Jonathan Forgit
Mary Lou Geary
Debbie Panarelli
Caroline Sullivan

Technical Support
Christopher Anderson
Gary L. Hoitt
Kathryn S. Rodriguez

This book is printed using soy-based printing ink.
This book is recyclable.

Related RSMeans Products and Services

This book is aimed primarily at commercial and industrial projects or large multi-family housing projects costing $3,500,000 and up. For civil engineering structures such as bridges, dams, highways, or the like, please refer to *RSMeans Heavy Construction Cost Data*.

The engineers at RSMeans suggest the following products and services as companion information resources to *RSMeans Assemblies Cost Data*:

Construction Cost Data Books
Building Construction Cost Data 2013
Facilities Construction Cost Data 2013
Square Foot Costs 2013

Reference Books
Building Security: Strategies & Costs
Designing & Building with the IBC
Estimating Building Costs
RSMeans Estimating Handbook
Green Building: Project Planning & Estimating
How to Estimate with RSMeans Data and CostWorks
Plan Reading & Material Takeoff
Square Foot & Assemblies Estimating Methods

Seminars and In-House Training
RSMeans CostWorks® Training
RSMeans Data for Job Order Contracting (JOC)
Square Foot & Assemblies Cost Estimating
Scheduling with MS Project for Construction Professionals
Scope of Work for Facilities Estimating
Unit Price Estimating

RSMeans on the Internet
Visit RSMeans at **www.rsmeans.com** for a one-stop portal for the most reliable and current resources available on the market. Here you can learn about our more than 20 annual cost data books, also available on CD and online versions, as well as about RSMeans seminars.

RSMeans Electronic Data
Get the information found in RSMeans cost books electronically on CD or on the Web at **MeansCostWorks.com**.

RSMeans Business Solutions
Cost engineers and analysts conduct facility life cycle and benchmark research studies and predictive cost modeling, as well as offer consultation services for conceptual estimating, real property management, and job order contracting. Research studies are designed with the application of proprietary cost and project data from Reed Construction Data and RSMeans' extensive North American databases. Analysts offer building product manufacturers qualitative and quantitative market research, as well as time/motion studies for new products, and Web-based dashboards for market opportunity analysis. Clients are from both the public and private sectors, including federal, state, and municipal agencies; corporations; institutions, construction management firms; hospitals; and associations.

RSMeans for Job Order Contracting (JOC)
Best practice job order contracting (JOC) services support the owner in the development of their program. RSMeans develops all required contracting documents, works with stakeholders, and develops qualified and experienced contractor lists. RSMeans JOCWorks™ software is the best of class in the functional tools that meet JOC specific requirements. RSMeans data is organized to meet JOC cost estimating requirements.

Construction Costs for Software Applications
Over 25 unit price and assemblies cost databases are available through a number of leading estimating and facilities management software providers (listed below). For more information see the "Other RSMeans Products" pages at the back of this publication.

RSMeansData™ is also available to federal, state, and local government agencies as multi-year, multi-seat licenses.

- 4Clicks-Solutions, LLC
- ArenaSoft Estimating
- Assetworks
- Beck Technology
- BSD – Building Systems Design, Inc.
- CMS – Construction Management Software
- Corecon Technologies, Inc.
- CorVet Systems
- Hard Dollar Corporation
- Maxwell Systems
- MC^2
- Parsons Corporation
- ProEst
- Sage Timberline Office
- UDA Technologies, Inc.
- US Cost, Inc.
- VFA – Vanderweil Facility Advisers
- WinEstimator, Inc.
- R & K Solutions
- Tririga

Table of Contents

Foreword	iv
How the Book Is Built: An Overview	v
How to Use the Book: The Details	vi
How RSMeans Assemblies Data Works	xiv
Assemblies Cost Section	1
Reference Section	441
Reference Tables	442
Square Foot Costs	544
Reference Aids	554
Estimating Forms	577
Abbreviations	583
Index	587
Other RSMeans Products and Services	606
Labor Trade Rates including Overhead & Profit	Inside Back Cover

Foreword

Our Mission

Since 1942, RSMeans has been actively engaged in construction cost publishing and consulting throughout North America.

Today, more than 70 years after RSMeans began, our primary objective remains the same: to provide you, the construction and facilities professional, with the most current and comprehensive construction cost data possible.

Whether you are a contractor, owner, architect, engineer, facilities manager, or anyone else who needs a reliable construction cost estimate, you'll find this publication to be a highly useful and necessary tool.

With the constant flow of new construction methods and materials today, it's difficult to find the time to look at and evaluate all the different construction cost possibilities. In addition, because labor and material costs keep changing, last year's cost information is not a reliable basis for today's estimate or budget.

That's why so many construction professionals turn to RSMeans. We keep track of the costs for you, along with a wide range of other key information, from city cost indexes . . . to productivity rates . . . to crew composition . . . to contractor's overhead and profit rates.

RSMeans performs these functions by collecting data from all facets of the industry and organizing it in a format that is instantly accessible to you. From the preliminary budget to the detailed unit price estimate, you'll find the data in this book useful for all phases of construction cost determination.

The Staff, the Organization, and Our Services

When you purchase one of RSMeans' publications, you are, in effect, hiring the services of a full-time staff of construction and engineering professionals.

Our thoroughly experienced and highly qualified staff works daily at collecting, analyzing, and disseminating comprehensive cost information for your needs. These staff members have years of practical construction experience and engineering training prior to joining the firm. As a result, you can count on them not only for accurate cost figures, but also for additional background reference information that will help you create a realistic estimate.

The RSMeans organization is always prepared to help you solve construction problems through its variety of data solutions, including online, CD, and print book formats, as well as cost estimating expertise available via our business solutions, training, and seminars.

Besides a full array of construction cost estimating books, RSMeans also publishes a number of other reference works for the construction industry. Subjects include construction estimating and project and business management, special topics such as green building and job order contracting, and a library of facility management references.

In addition, you can access all of our construction cost data electronically in convenient CD format or on the web. Visit **RSMeansonline.com** for more information on our 24/7 online cost data.

What's more, you can increase your knowledge and improve your construction estimating and management performance with an RSMeans construction seminar or in-house training program. These two-day seminar programs offer unparalleled opportunities for everyone in your organization to become updated on a wide variety of construction-related issues.

RSMeans is also a worldwide provider of construction cost management and analysis services for commercial and government owners.

In short, RSMeans can provide you with the tools and expertise for constructing accurate and dependable construction estimates and budgets in a variety of ways.

Robert Snow Means Established a Tradition of Quality That Continues Today

Robert Snow Means spent years building RSMeans, making certain he always delivered a quality product.

Today, at RSMeans, we do more than talk about the quality of our data and the usefulness of our books. We stand behind all of our data, from historical cost indexes to construction materials and techniques to current costs.

If you have any questions about our products or services, please call us toll-free at 1-800-334-3509. Our customer service representatives will be happy to assist you. You can also visit our website at **www.rsmeans.com**

How the Book Is Built: An Overview

A Powerful Construction Tool

You have in your hands one of the most powerful construction tools available today. A successful project is built on the foundation of an accurate and dependable estimate. This book will enable you to construct just such an estimate.

For the casual user, the book is designed to be:

- quickly and easily understood so you can get right to your estimate.
- filled with valuable information so you can understand the necessary factors that go into the cost estimate.

For the regular user, the book is designed to be:

- a handy desk reference that can be quickly referred to for key costs.
- a comprehensive, fully reliable source of current construction costs so you'll be prepared to estimate any project.
- a source book for preliminary project cost, product selections, and alternate materials and methods.

To meet all of these requirements, we have organized the book into the following clearly defined sections.

How To Use the Book: *The Details*

This section contains an in-depth explanation of how the book is arranged . . . and how you can use it to determine a reliable construction cost estimate. It includes information about how we develop our cost figures and how to completely prepare your estimate.

Assemblies Section

The cost data in this section has been organized in an "Assemblies" format. These assemblies are the functional elements of a building and are arranged according to the 7 elements of the UNIFORMAT II classification system. For a complete explanation of a typical "Assemblies" page, see "How To Use the Assemblies Cost Tables."

Reference Section

This section includes information on Reference Tables, Historical Cost Indexes, City Cost Indexes, Square Foot Costs, Reference Aids, Estimating Forms, and a listing of Abbreviations. It is visually identified by a vertical gray bar on the page edges.

Reference Tables: At the beginning of selected major classifications in the Assemblies Section are "reference numbers" shown in bold squares. These numbers refer you to related information in the Reference Section.

In this section, you'll find reference tables, explanations, and estimating information that support how we develop the assemblies data. Also included are alternate pricing methods, technical data, and estimating procedures, along with information on design and economy in construction. You'll also find helpful tips on what to expect and what to avoid when estimating and constructing your project.

It is recommended that you refer to this section if a "reference number" appears within the assembly you are estimating.

Historical Cost Indexes: These indexes provide you with data to adjust construction costs over time.

City Cost Indexes: All costs in this book are U.S. national averages. Costs may vary because of the regional economy. You can adjust the national average costs in this book to 318 major cities throughout the U.S.

Location Factors: You can adjust total project costs to over 900 locations throughout the U.S. and Canada by using the data in this section.

Square Foot Costs: This section contains costs for 59 different building types that allow you to make a rough estimate for the overall cost of a project or its major components.

Reference Aids: This section provides design criteria, code requirements, and many other reference aids useful in designing a building in various geographic locations of the country. Trade-offs by weight, material, thermal conductivity, and fire rating can be developed from this information. Design loads and capacities are based on the requirements of several nationally used building codes. In some cases, local building codes may have more stringent requirements, which would influence the final in-place cost.

Estimating Forms: Detailed estimating forms and checklists are included to be used when estimating on an "Assemblies" basis.

Abbreviations: A listing of the abbreviations used throughout this book, along with the terms they represent, is included.

Index

A comprehensive listing of all terms and subjects in this book will help you quickly find what you need.

The Scope of This Book

This book is designed to be as comprehensive and as easy to use as possible. To that end we have made certain assumptions and limited its scope in two key ways:

1. We have established material prices based on a national average.
2. We have computed labor costs based on a 30-city national average of union wage rates.

For a more detailed explanation of how the cost data is developed, see "How To Use the Book: The Details."

Project Size/Type

This book is aimed primarily at large commercial, industrial, and multi-family housing projects. Costs are primarily for new construction or major renovation of buildings rather than repairs or minor alterations.

With reasonable exercise of judgment the figures can be used for any building work.

How to Use the Book: The Details

What's Behind the Numbers? The Development of Cost Data

The staff at RSMeans continually monitors developments in the construction industry in order to ensure reliable, thorough, and up-to-date cost information. While overall construction costs may vary relative to general economic conditions, price fluctuations within the industry are dependent upon many factors. Individual price variations may, in fact, be opposite to overall economic trends. Therefore, costs are constantly tracked and complete updates are published yearly. Also, new items are frequently added in response to changes in materials and methods.

Costs— $ (U.S.)

All costs represent U.S. national averages and are given in U.S. dollars. The RSMeans City Cost Indexes can be used to adjust costs to a particular location. The City Cost Indexes for Canada can be used to adjust U.S. national averages to local costs in Canadian dollars. No exchange rate conversion is necessary.

Material Costs

The RSMeans staff contacts manufacturers, dealers, distributors, and contractors all across the U.S. and Canada to determine national average material costs. If you have access to current material costs for your specific location, you may wish to make adjustments to reflect differences from the national average. Included within material costs are fasteners for a normal installation. RSMeans engineers use manufacturers recommendations, written specifications, and/or standard construction practice for size and spacing of fasteners. Adjustments to material costs may be required for your specific application or location. Material costs do not include sales tax.

Labor Costs

Labor costs are based on the average of wage rates from 30 major U.S. cities. Rates are determined from labor union agreements or prevailing wages for construction trades for the current year. Rates, along with overhead and profit markups, are listed on the inside back cover of this book.

- If wage rates in your area vary from those used in this book, or if rate increases are expected within a given year, labor costs should be adjusted accordingly.

Labor costs reflect productivity based on actual working conditions. In addition to actual installation, these figures include time spent during a normal weekday on tasks such as material receiving and handling, mobilization at site, site movement, breaks, and cleanup. Productivity data is developed over an extended period so as not to be influenced by abnormal variations and reflects a typical average.

Equipment Costs

Equipment costs include not only rental, but also operating costs for equipment under normal use. The operating costs include parts and labor for routine servicing such as repair and replacement of pumps, filters, and worn lines. Normal operating expendables, such as fuel, lubricants, tires, and electricity (where applicable) are also included. Extraordinary operating expendables with highly variable wear patterns, such as diamond bits and blades, are excluded. These costs are included under materials. Equipment rental rates are obtained from industry sources throughout North America—contractors, suppliers, dealers, manufacturers, and distributors. Equipment costs do not include operators' wages; nor do they include the cost to move equipment to a job site (mobilization) or from a job site (demobilization).

General Conditions

Prices given in this book include the Installing Contractor's overhead and profit (O&P). General Conditions, when applicable, should also be added to the Total Cost including O&P. The costs for General Conditions are listed in the Reference Section of this book. General Conditions for the Installing Contractor may range from 0% to 10% of the Total Cost including O&P. For the General or Prime Contractor, costs for General Conditions may range from 5% to 15% of the Total Cost, including O&P, with a figure of 10% as the most typical allowance.

Overhead and Profit

All costs in this book include overhead and profit for the Installing Contractor. The material column includes a 10% profit markup. The installation column includes labor burden plus 10% profit on labor and equipment costs. The total column is the sum of the material and installation columns. To complete the assemblies estimate, one should include an additional allowance for general conditions (5%–10%) and overhead and profit for the general contractor (5%–15%).

Factors Affecting Costs

Costs can vary depending upon a number of variables. Here's how we have handled the main factors affecting costs.

Quality— The prices for materials and the workmanship upon which productivity is based represent sound construction work. They are also in line with U.S. government specifications.

Overtime—We have made no allowance for overtime. If you anticipate premium time or work beyond normal working hours, be sure to make an appropriate adjustment to your labor costs.

Productivity— The productivties used in calculating labor costs are based on working an eight-hour day in daylight hours in moderate temperatures. For work that extends beyond normal work hours or is performed under adverse conditions, productivity may decrease.

Size of Project—The size, scope of work, and type of construction project will have a significant impact on cost. Economies of scale can reduce costs for large projects. Unit costs can often run higher for small projects. Costs in this book are intended for the size and type of project as previously described in "How the Book Is Built: An Overview." Costs for projects of a significantly different size or type should be adjusted accordingly.

Location—Material prices in this book are for metropolitan areas. However, in dense urban areas, traffic and site storage limitations may increase costs. Beyond a 20-mile radius of large cities, extra trucking or transportation charges may also increase the material costs slightly. On the other hand, lower wage rates may be in effect. Be sure to consider both of these factors when preparing an estimate, particularly if the job site is located in a central city or remote rural location.

In addition, highly specialized subcontract items may require travel and per-diem expenses for mechanics.

Other Factors—
- season of year
- contractor management
- weather conditions
- local union restrictions
- building code requirements
- availability of:
  - adequate energy
  - skilled labor
  - building materials
- owner's special requirements/restrictions
- safety requirements
- environmental considerations

Unpredictable Factors— General business conditions influence "in-place" costs of all items. Substitute materials and construction methods may have to be employed. These may affect the installed cost and/or life cycle costs. Such factors may be difficult to evaluate and cannot necessarily be predicted on the basis of the job's location in a particular section of the country. Thus, where these factors apply, you may find significant but unavoidable cost variations for which you will have to apply a measure of judgment to your estimate.

Rounding of Costs

In general, all unit prices in excess of $5.00 have been rounded to make them easier to use and still maintain adequate precision of the results. The rounding rules we have chosen are in the following table.

Prices from . . .	Rounded to the nearest...
$.01 to $5.00	$.01
$5.01 to $20.00	$.05
$20.01 to $100.00	$.50
$100.01 to $300.00	$1.00
$300.01 to $1,000.00	$5.00
$1,000.01 to $10,000.00	$25.00
$10,000.01 to $50,000.00	$100.00
$50,000.01 and above	$500.00

Contingencies

The allowance for contingencies generally is included at the conceptual stage of a project to allow for design decisions that will be made as the project progresses. At the conceptual design phase an allowance of 20% is customary. During the design development process this should decrease to 15% as decisions are made. An allowance of 10% is usual during contract document developments, and 3%–5% is typical as a construction contingency.

The estimator, in addition, should consider cost escalation in his estimate. Calculate the increase in cost for the time period from the effective date of the data to the mid-point of the construction period. Use Means Historical Cost Indexes (Table J 1020-011) as a guide for year-to-year cost escalation.

Final Checklist

Estimating can be a straightforward process provided you remember the basics. Here's a checklist of some of the steps you should remember to complete before finalizing your estimate.

Did you remember to . . .

- factor in the City Cost Index for your locale?
- take into consideration which items have been marked up and by how much?
- mark up the entire estimate sufficiently for your purposes?
- read the background information on techniques and technical matters that could impact your project time span and cost?
- include all components of your project in the final estimate?
- double check your figures for accuracy?
- call RSMeans if you have any questions about your estimate or the data you've found in our publications?

Remember, RSMeans stands behind its publications. If you have any questions about your estimate . . . about the costs you've used from our books . . . or even about the technical aspects of the job that may affect your estimate, feel free to call the RSMeans editors at 1-800-334-3509.

Assemblies Estimating

The grouping of several different trades into building components or broad building elements is the "Systems" or "Assemblies" method of estimating. This method allows the estimator or designer to make quick comparisons of systems in various combinations within predetermined guidelines. Systems which are best suited to accommodate budget, code, load, insulation, fireproofing, acoustics, energy considerations, and the owner's special requirements can quickly be determined. This method can also be used to help match existing construction.

In order to understand how a System estimate is assembled, it is a good idea to compare a Unit Price estimate with a Systems estimate. In a Unit Price Estimate, each item is normally included along the guidelines of the 50-division MasterFormat of the Construction Specifications Institute. In a Systems estimate, these same items are allocated to one of seven major group elements in the UNIFORMAT II organization. Certain items that were formerly grouped into a single trade breakdown must now be allocated among two or more systems.

An example of this difference would be concrete. In a Unit Price estimate, all the concrete items on the job would be priced in the "Concrete" section of the estimate, CSI Division 03. In a Systems estimate, concrete is found in a number of locations. For instance, concrete is used in all of these systems: Division A10, Foundations; Division A20, Basement Construction; Division B10, Superstructure; and Division B20, Exterior Closure.

Conversely, other items that are listed in separate trade breakdowns in a Unit Price estimate are combined into one division in the System estimate. For example, interior partitions might include two CSI divisions: Division 06, Wood Stud Wall; and Division 09, Lath, Plaster and Paint. In the UNIFORMAT II Systems Estimate, these items are all combined in Division C, Interior Construction.

This re-allocation of the familiar items from the CSI format may, at first, seem confusing, but once the concept is understood, the resultant increase in estimating speed is well worth the initial familiarization required.

Systems or Assemblies estimating is not a substitute for Unit Price estimating. It is normally done during the earlier conceptual stage before plans have been completed or when preparing a budget. This enables the designer to bring in the project within the owner's budget. During the actual initial design process, the designer will be forced to make important decisions and "trade-offs" for each of the various systems. Some of the trade-offs can include:

a. Price of each system
b. Appearance, quality, and compatibility
c. Story height
d. Clear span
e. Complications and restrictions
f. Thermal characteristics
g. Life-cycle costs
h. Acoustical characteristics
i. Fireproofing characteristics
j. Special owner's requirements in excess of code requirements
k. Code
l. Load

Before starting a Systems Estimate, gather all the information possible pertaining to the project.

Information can be gathered from:

1. Code Requirements (see Reference Section L of this book)
2. Owner's Requirements
3. Preliminary Assumptions
4. Site Inspection and Investigation

Since the Foundation and Substructure design and price are functions of the Superstructure and the site, it is advisable to start the estimate with the Superstructure. Follow this with the Foundation and Superstructure, and then the other Systems in the sequence as applicable to your project.

Assemblies Section

Table of Contents

Table No.		Page
A SUBSTRUCTURE		**1**
A1010 Standard Foundations		
A1010 110	Strip Footings	2
A1010 210	Spread Footings	3
A1010 250	Pile Caps	5
A1010 310	Foundation Underdrain	7
A1010 320	Foundation Dampproofing	8
A1020 Special Foundations		
A1020 110	C.I.P. Concrete Piles	9
A1020 120	Precast Concrete Piles	11
A1020 130	Steel Pipe Piles	13
A1020 140	Steel H Piles	15
A1020 150	Step-Tapered Steel Piles	17
A1020 160	Treated Wood Piles	18
A1020 210	Grade Beams	20
A1020 310	Caissons	22
A1020 710	Pressure Injected Footings	23
A1030 Slab on Grade		
A1030 120	Plain & Reinforced	24
A2010 Basement Excavation		
A2010 110	Building Excav. & Backfill	26
A2020 Basement Walls		
A2020 110	Walls, Cast in Place	28
A2020 220	Subdrainage Piping	31
B SHELL		**33**
B1010 Floor Construction		
B1010 201	C.I.P. Column, Round Tied	34
B1010 202	C.I.P. Column, Round Tied - Min. Reinf.	36
B1010 203	C.I.P. Column, Square Tied	37
B1010 204	C.I.P. Column, Square Tied - Min. Reinf.	39
B1010 206	Tied, Concentric Loaded Precast Concrete Columns	40
B1010 207	Tied, Eccentric Loaded Precast Concrete Columns	41
B1010 208	Steel Columns	42
B1010 210	Wood Columns	47
B1010 213	Rectangular Precast Beams	49
B1010 214	"T" Shaped Precast Beams	51
B1010 215	"L" Shaped Precast Beams	53
B1010 216	Wood Beams	55
B1010 217	Cast in Place Slabs, One Way	56
B1010 219	C.I.P. Beam & Slab, One Way	58
B1010 220	C.I.P. Beam & Slab, Two Way	60
B1010 222	C.I.P. Flat Slab w/Drop Panels	62
B1010 223	Cast in Place Flat Plate	64
B1010 226	C.I.P. Multispan Joist Slab	65
B1010 227	Cast in Place Waffle Slab	67
B1010 229	Precast Plank, No Topping	69
B1010 230	Precast Plank, 2" Topping	70
B1010 234	Precast Double "T" Beams, No Topping	71
B1010 235	Precast Double "T" Beams, 2" Topping	72
B1010 236	Precast Beam & Plank with No Topping	74
B1010 238	Precast Beam & Plank with 2" Topping	75
B1010 239	Precast Double "T" & 2" Topping on Precast Beams	76
B1010 241	W Shape Beams & Girders	78
B1010 244	Lt. Gauge Steel Floor Systems	82
B1010 246	Deck & Joists on Brng. Walls	83
B1010 248	Steel Joists on Beam & Wall	84
B1010 250	Steel Joists, Beams & Slab on Columns	87
B1010 252	Composite Beam & C.I.P. Slab	90
B1010 254	W Shape, Composite Deck & Slab	92
B1010 256	Composite Beams, Deck & Slab	94
B1010 258	Metal Deck/Concrete Fill	96
B1010 260	Cellular Composite Deck	97
B1010 261	Wood Joist	98
B1010 264	Wood Beam & Joist	99
B1010 265	Laminated Wood Floor Beams	100
B1010 266	Wood Deck	101
B1010 710	Steel Beam Fireproofing	102
B1010 720	Steel Column Fireproofing	104
B1020 Roof Construction		
B1020 102	Wood/Flat or Pitched	106
B1020 108	Steel Joists, Beams & Deck on Columns & Walls	107
B1020 112	Steel Joists, Beams & Deck on Columns	109
B1020 116	Steel Joists & Deck on Bearing Walls	111
B1020 120	Steel Joists, Joist Girders & Deck on Columns & Walls	113
B1020 124	Steel Joists & Joist Girders on Columns	116
B1020 125	Narrow Rib, Single Span Steel Deck	119
B1020 126	Narrow Rib, Multi Span Steel Deck	119
B1020 127	Intermed. Rib, Single Span Steel Deck	119
B1020 128	Intermed. Rib, Multi Span Steel Deck	120
B1020 129	Wide Rib, Single Span Steel Deck	120
B1020 130	Wide Rib, Multi Span Steel Deck	120
B1020 131	Single Span, Steel Deck	121
B1020 132	Double Span, Steel Deck	121
B1020 134	Cellular Steel Deck, Single Span	122
B1020 135	Cellular Steel Deck, Double Span	122
B1020 136	Cellular Steel Deck, Triple Span	123
B1020 310	Canopies	123
B2010 Exterior Walls		
B2010 101	Cast in Place Concrete	124
B2010 102	Flat Precast Concrete	126
B2010 103	Fluted Window/Mullion Precast Concrete	127
B2010 104	Ribbed Precast Concrete	127
B2010 105	Precast Concrete Specialties	128
B2010 106	Tilt-Up Concrete Panel	129
B2010 109	Concrete Block Wall - Reg. Wt.	131
B2010 110	Concrete Block Wall - Lt. Wt.	132
B2010 111	Reinf. Concrete Block Wall - Reg. Wt.	133
B2010 112	Reinf. Concrete Block Wall - Lt. Wt.	134
B2010 113	Split Ribbed Block Wall - Reg. Wt.	137
B2010 114	Split Ribbed Block Wall - Lt. Wt.	138
B2010 115	Reinf. Split Rib. Block Wall - Reg. Wt.	139
B2010 116	Reinf. Split Rib. Block Wall - Lt. Wt.	140
B2010 117	Split Face Block Wall - Reg. Wt.	142
B2010 118	Split Face Blk. Wall - Lt. Wt.	143
B2010 119	Reinf. Split Face Block Wall - Reg. Wt.	145
B2010 120	Reinf. Split Face Block Wall - Lt. Wt.	146
B2010 121	Ground Face Block Wall	149
B2010 122	Reinf. Ground Face Block Wall	150
B2010 123	Misc. Block Wall	153
B2010 124	Reinf. Misc. Block Walls	155
B2010 125	Solid Brick Walls - Single Wythe	157
B2010 126	Solid Brick Walls - Double Wythe	158
B2010 127	Solid Brick Walls Reinforced	159
B2010 128	Stone Veneer	161
B2010 129	Brick Veneer/Wood Stud Backup	163
B2010 130	Brick Veneer/Metal Stud Backup	165

Table of Contents

Table No.		Page
B2010 132	Brick Face Comp. Wall - Double Wythe	169
B2010 133	Brick Face Comp. Wall - Triple Wythe	172
B2010 134	Brick Face Cavity Wall	174
B2010 135	Brick Face Cavity Wall - Insul. Backup	177
B2010 137	Block Face Cavity Wall	180
B2010 138	Block Face Cavity Wall - Insul. Backup	182
B2010 139	Block Face Composite Wall	185
B2010 140	Glass Block	189
B2010 144	Concrete Block Lintel	190
B2010 145	Concrete Block Specialties	191
B2010 146	Metal Siding Panel	192
B2010 148	Panel, Shingle & Lap Siding	194
B2010 151	Stucco Wall	196
B2010 152	E.I.F.S.	197
B2010 154	Metal Siding Support	198

B2020 Exterior Windows

B2020 102	Wood Windows	200
B2020 104	Steel Windows	201
B2020 106	Aluminum Windows	201
B2020 108	Vinyl Clad Windows	202
B2020 210	Tubular Aluminum Framing	204
B2020 220	Curtain Wall Panels	205

B2030 Exterior Doors

B2030 110	Glazed Doors, Steel or Alum.	206
B2030 210	Wood Doors	208
B2030 220	Steel Doors	209
B2030 230	Aluminum Doors	210

B3010 Roof Coverings

B3010 105	Built-Up	211
B3010 120	Single Ply Membrane	213
B3010 130	Preformed Metal Roofing	214
B3010 135	Formed Metal	135
B3010 140	Shingle & Tile	216
B3010 320	Roof Deck Rigid Insulation	217
B3010 410	Base Flashing	218
B3010 420	Roof Edges	219
B3010 430	Flashing	220
B3010 610	Gutters	221
B3010 620	Downspouts	221
B3010 630	Gravel Stop	222

B3020 Roof Openings

B3020 110	Skylights	223
B3020 210	Hatches	223

C INTERIORS — 225

C1010 Partitions

C1010 102	Concrete Block Partitions - Reg. Wt.	226
C1010 104	Concrete Block Partitions - Lt. Wt.	227
C1010 120	Tile Partitions	229
C1010 124	Drywall Partitions/Wood Stud Framing	230
C1010 126	Drywall Partitions/Metal Stud Framing	231
C1010 128	Drywall Components	233
C1010 140	Plaster Partitions Metal/Studs Framing	234
C1010 142	Plaster Partitions/Wood Stud Framing	235
C1010 144	Plaster Part. Components	237
C1010 205	Partitions	238
C1010 210	Partitions	239
C1010 230	Demountable Partitions, L.F.	239
C1010 710	Interior Glazed Opening	240

C1020 Interior Doors

C1020 102	Special Doors	243
C1020 114	Metal Door/Metal Frame	244
C1020 116	Labeled Metal Door/Metal Frames	245
C1020 120	Wood Door/Wood Frame	245
C1020 122	Wood Door/Metal Frame	246
C1020 124	Wood Fire Doors/Metal Frames	247
C1020 310	Hardware	248

C1030 Fittings

C1030 110	Toilet Partitions	249
C1030 210	Fab. Compact Units & Cubicles, Ea.	250
C1030 310	Storage Specialties, Ea.	250
C1030 510	Identifying/Visual Aid Special., Ea.	250
C1030 520	Identifying/Visual Aid Special., S.F.	250
C1030 710	Bath & Toilet Access., Ea.	250
C1030 730	Bath & Toilet Access., L.F.	251
C1030 830	Fab. Cabinets & Counters., L.F.	251
C1030 910	Other Fittings, Ea.	251

C2010 Stair Construction

C2010 110	Stairs	252

C3010 Wall Finishes

C3010 230	Paint & Covering	254
C3010 235	Paint Trim	255

C3020 Floor Finishes

C3020 410	Tile & Covering	256

C3030 Ceiling Finishes

C3030 105	Plaster Ceilings	258
C3030 110	Drywall Ceilings	259
C3030 140	Plaster Ceiling Components	260
C3030 210	Acoustical Ceilings	260
C3030 240	Acoustical Ceiling Components	261

D SERVICES — 263

D1010 Elevators & Lifts

D1010 110	Hydraulic Elevators	264
D1010 140	Traction Geared Elevators	265
D1010 150	Traction Gearless Elevators	266

D1020 Escalators/Moving Walks

D1020 110	Moving Stairs	267
D1020 210	Moving Walks	267

D1090 Other Conveying Systems

D1090 410	Miscellaneous Types	268
D1090 510	Chutes	268

D2010 Plumbing Fixtures

D2010 110	Water Closet Systems	269
D2010 120	Water Closets, Group	270
D2010 210	Urinal Systems	271
D2010 310	Lavatory Systems	272
D2010 410	Kitchen Sink Systems	273
D2010 420	Laundry Sink Systems	274
D2010 430	Laboratory Sink Systems	275
D2010 440	Service Sink Systems	276
D2010 510	Bathtub Systems	277
D2010 610	Group Wash Fountain Systems	278
D2010 710	Shower Systems	279
D2010 810	Drinking Fountain Systems	280
D2010 820	Water Cooler Systems	281
D2010 920	Two Fixture Bathroom, Two Wall Plumbing	282
D2010 922	Two Fixture Bathroom, One Wall Plumbing	282
D2010 924	Three Fixture Bathroom, One Wall Plumbing	283

Table of Contents

Table No.		Page
D2010 926	Three Fixture Bathroom, Two Wall Plumbing	283
D2010 928	Four Fixture Bathroom, Two Wall Plumbing	284
D2010 930	Four Fixture Bathroom, Three Wall Plumbing	284
D2010 932	Five Fixture Bathroom, Two Wall Plumbing	285
D2010 934	Five Fixture Bathroom, Three Wall Plumbing	285
D2010 936	Five Fixture Bathroom, One Wall Plumbing	285
D2010 951	Sample Estimate	286

D2020 Domestic Water Dist.

D2020 240	Electric Water Heaters - Comm. Systems	287
D2020 250	Gas Fired Water Heaters - Comm. Systems	288
D2020 260	Oil Fired Water Heaters - Comm. Systems	289
D2020 265	Closed Loop, External Exchange	290
D2020 270	Drainback, Immersed Exchange	292
D2020 275	Solar Draindown, Direct Collection	294
D2020 295	Closed Loop, Immersed Exchange	296

D2040 Rain Water Drainage

D2040 210	Roof Drain Systems	298

D2090 Other Plumbing Systems

D2090 810	Piping - Install. - Unit Costs	300
D2090 820	Standard Pipe Fittings - Installed - Unit Costs	302

D3010 Energy Supply

D3010 510	Fin Tube Radiation, Resi.	307
D3010 520	Fin Tube Radiation, Comm.	308
D3010 530	Terminal Unit Heaters, Comm.	309
D3010 660	Solar Swimming Pool Heater Systems	310
D3010 682	Solar Direct Gain Glazing, Door Panels	312
D3010 684	Solar Direct Gain Glazing, Window Panels	313
D3010 686	Passive Solar Indirect Gain Wall	314
D3010 688	Passive Solar Indirect Gain Panel	315
D3010 690	Passive Solar Sunspace	316

D3020 Heat Generating Systems

D3020 102	Small Heating Systems, Hydronic, Electric Boilers	317
D3020 104	Large Heating Systems, Hydronic, Electric Boilers	318
D3020 106	Boilers, Hot Water & Steam	319
D3020 108	Hydronic, Fossil Fuel Boilers, Terminal Units	320
D3020 110	Hydronic, Fossil Fuel Boilers, Fin Tube	321
D3020 310	Vent	322
D3020 315	Chimney	323

D3030 Cooling Generating Sys.

D3030 110	Chilled Water, Air Cooled	324
D3030 115	Chilled Water, Water Cooled	325
D3030 214	Heating/Cooling System	326

D3050 Terminal and Package Units

D3050 150	Rooftop Single Zone Unit Systems	327
D3050 155	Rooftop Multizone Unit	328
D3050 160	Self-Contained, Water Cooled Unit Systems	329
D3050 165	Self-Contained, Air Cooled Unit Systems	330
D3050 170	Split Systems with Air Cooled Condensing Units	331
D3050 185	Computer Room Cooling Units	332

D4010 Sprinklers

D4010 310	Dry Pipe Sprinkler Systems	333
D4010 350	Preaction Sprinkler Systems	336
D4010 370	Deluge Sprinkler Systems	339
D4010 390	Firecycle Sprinkler Systems	342
D4010 410	Wet Pipe Sprinkler Systems	345

D4020 Standpipes

D4020 310	Wet Standpipe Risers	348
D4020 330	Dry Standpipe Risers	350
D4020 410	Fire Hose Equipment	352

D4090 Other Fire Protection Sys.

D4090 910	Unit Components	353
D4090 920	FM200 Systems	353

D5010 Electrical Service/Dist.

D5010 110	High Voltage Shielded Conductors	354
D5010 120	Overhead Electric Service, 3 Phase - 4 Wire	355
D5010 130	Underground Electric Service	357
D5010 230	Feeder Installation	359
D5010 240	Switchgear	360
D5010 250	Panelboard	361

D5020 Lighting & Branch Wiring

D5020 110	Receptacle (by Wattage)	363
D5020 115	Receptacles, Floor	364
D5020 125	Receptacle & Switches, Ea.	365
D5020 130	Wall Switch by Sq. Ft.	366
D5020 135	Miscellaneous Power	367
D5020 140	Central A.C. Power (by Watt)	368
D5020 145	Motor Installation	369
D5020 155	Motor Feeder	371
D5020 160	Magnetic Starter	372
D5020 165	Safety Switches	373
D5020 170	Motor Connections	374
D5020 175	Motor & Starter	375
D5020 208	Fluorescent Fixt. (Type)	378
D5020 210	Fluorescent Fixt. (Watt.)	379
D5020 214	Incandescent Fixt. (Type)	380
D5020 216	Incandescent Fixt. (Watt.)	381
D5020 218	Fluorscent Fixt. High Bay, 8'-10' (Watt)	382
D5020 220	H.I.D. Fixt., High Bay, 8'-10' (Type)	383
D5020 222	H.I.D. Fixt., High Bay, 8'- 10' (Watt.)	384
D5020 224	H.I.D. Fixt., High Bay, 16' (Type)	385
D5020 226	H.I.D. Fixt., High Bay, 16' (Watt.)	386
D5020 228	H.I.D. Fixt., High Bay, 20' (Type)	387
D5020 230	H.I.D. Fixt., High Bay 20' (Watt.)	388
D5020 232	H.I.D. Fixt., High Bay, 30' (Type)	389
D5020 234	H.I.D. Fixt., High Bay, 30' (Watt.)	390
D5020 236	H.I.D. Fixt., Low Bay, 8'-10' (Type)	391
D5020 238	H.I.D. Fixt. Low Bay, 8'-10' (Watt.)	392
D5020 240	H.I.D. Fixt., Low Bay, 16' (Type)	393
D5020 242	H.I.D. Fixt., Low Bay, 16' (Watt.)	394

D5030 Comm. & Security

D5030 310	Telephone Systems	395
D5030 910	Comm. & Alarm Systems	396
D5030 920	Data Communications	397

D5090 Other Electrical Systems

D5090 210	Generators (by kW)	398
D5090 420	Photovoltaic Sys. Ea.	399

Table of Contents

Table No.		Page
D5090 430	Photovoltaic Sys. Grid AC/DC	400
D5090 510	Elec. Baseboard Radiation (Low Density)	401
D5090 520	Elec. Baseboard Radiation. (Med. Density)	402

E EQUIPMENT & FURNISHINGS — 403

E1010 Commercial Equipment
E1010 110	Security/Vault, Ea.	404
E1010 510	Mercantile Equipment, Ea.	404
E1010 610	Laundry/Dry Cleaning, Ea.	404

E1020 Institutional Equipment
E1020 110	Ecclesiastical Equipment, Ea.	405
E1020 130	Ecclesiastical Equipment, L.F.	405
E1020 210	Library Equipment, Ea.	405
E1020 230	Library Equipment, L.F.	405
E1020 310	Theater & Stage Equip., Ea.	405
E1020 320	Theater & Stage Equip., S.F.	405
E1020 330	Theater & Stage Equip., L.F.	405
E1020 610	Detention Equipment, Ea.	406
E1020 710	Laboratory Equipment, Ea.	406
E1020 720	Laboratory Equipment, S.F.	406
E1020 730	Laboratory Equipment, L.F.	406
E1020 810	Medical Equipment, Ea.	406

E1030 Vehicular Equipment
E1030 110	Vehicular Service Equip., Ea.	407
E1030 210	Parking Control Equip., Ea.	407
E1030 310	Loading Dock Equip., Ea.	407

E1090 Other Equipment
E1090 110	Maintenance Equip., Ea.	408
E1090 210	Solid Waste Handling Equip., Ea.	408
E1090 350	Food Service Equip., Ea.	408
E1090 360	Food Service Equip., S.F.	408
E1090 410	Residential Equip., Ea.	408
E1090 550	Darkroom Equip., Ea.	408
E1090 610	School Equipment, Ea.	409
E1090 620	School Equipment, S.F.	409
E1090 810	Athletic, Recreational & Therapeutic Equip., Ea.	409
E1090 820	Athletic, Recreational & Therapeutic Equip., S.F.	409
E1090 910	Other Equipment, Ea.	410

E2010 Fixed Furnishings
E2010 310	Window Treatment, Ea.	411
E2010 320	Window Treatment, S.F.	411
E2010 420	Fixed Floor Grilles & Mats, S.F.	411
E2010 510	Fixed Multiple Seating, Ea.	411

E2020 Moveable Furnishings
E2020 210	Furnishings, Ea.	412
E2020 220	Furniture & Access., L.F.	412

F SPECIAL CONSTRUCTION — 413

F1010 Special Structures
F1010 120	Air-Supported Struct., S.F.	414
F1010 210	Pre-Engineered Struct., Ea.	414
F1010 320	Other Special Struct., S.F.	414
F1010 330	Special Structures, Ea.	414
F1010 340	Special Structures, S.F.	414

F1020 Integrated Construction
F1020 110	Integrated Const., Ea.	415
F1020 120	Integrated Const., S.F.	415
F1020 250	Special Purpose Room, Ea.	415
F1020 260	Special Purpose Room, S.F.	415
F1020 330	Special Construction, L.F.	415

F1030 Special Const. Systems
F1030 120	Sound, Vibration & Seismic Const., S.F.	416
F1030 210	Radiation Protection, Ea.	416
F1030 220	Radiation Protection, S.F.	416
F1030 910	Other Special Construction Systems, Ea.	416

F1040 Special Facilities
F1040 210	Ice Rinks, Each	417
F1040 510	Liquid & Gas Storage Tanks, Ea.	417
F1040 910	Special Construction, Ea.	417

G BUILDING SITEWORK — 419

G1030 Site Earthwork
G1030 805	Trenching Common Earth	420
G1030 806	Trenching Loam & Sandy Clay	422
G1030 807	Trenching Sand & Gravel	424
G1030 815	Pipe Bedding	426

G2010 Roadways
G2010 210	Roadway Pavement	428
G2010 305	Curbs & Berms	429

G2020 Parking Lots
G2020 210	Parking Lots	430

G2030 Pedestrian Paving
G2030 110	Bituminous Sidewalks	431
G2030 120	Concrete Sidewalks	431
G2030 150	Brick & Tile Plazas	431

G2040 Site Development
G2040 105	Fence & Guardrails	432
G2040 810	Flag Poles	432
G2040 950	Other Site Development, Ea.	432

G3010 Water Supply
G3010 110	Water Distribution Piping	433
G3020 110	Drainage & Sewage Piping	434

G3030 Storm Sewer
G3030 210	Manholes & Catch Basins	436

G3060 Fuel Distribution
G3060 110	Gas Service Piping	438

G4020 Site Lighting
G4020 110	Site Lighting	439
G4020 210	Light Pole (Installed)	440

How RSMeans Assemblies Data Works

Assemblies estimating provides a fast and reasonably accurate way to develop construction costs. An assembly is the grouping of individual work items, with appropriate quantities, to provide a cost for a major construction component in a convenient unit of measure.

An assemblies estimate is often used during early stages of design development to compare the cost impact of various design alternatives on total building cost.

Assemblies estimates are also used as an efficient tool to verify construction estimates.

Assemblies estimates do not require a completed design or detailed drawings. Instead, they are based on the general size of the structure and other known parameters of the project. The degree of accuracy of an assemblies estimate is generally within +/- 15%.

NARRATIVE FORMAT

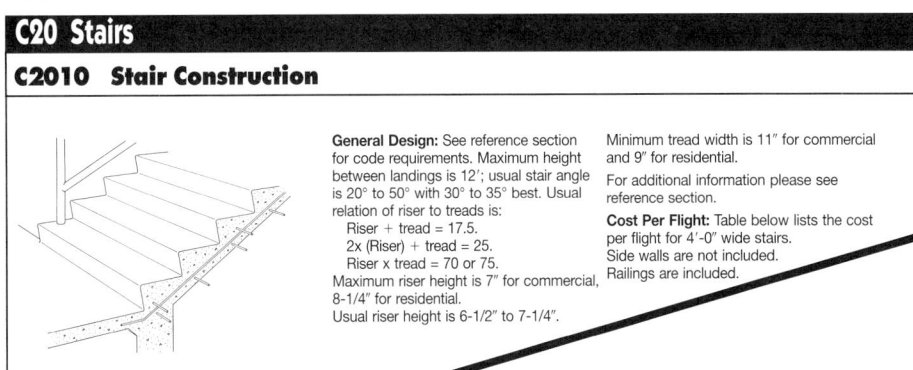

C20 Stairs
C2010 Stair Construction

General Design: See reference section for code requirements. Maximum height between landings is 12'; usual stair angle is 20° to 50° with 30° to 35° best. Usual relation of riser to treads is:
- Riser + tread = 17.5.
- 2x (Riser) + tread = 25.
- Riser x tread = 70 or 75.

Maximum riser height is 7" for commercial, 8-1/4" for residential.
Usual riser height is 6-1/2" to 7-1/4".

Minimum tread width is 11" for commercial and 9" for residential.
For additional information please see reference section.
Cost Per Flight: Table below lists the cost per flight for 4'-0" wide stairs. Side walls are not included. Railings are included.

RSMeans assemblies are identified by a **unique 12-character identifier**. The assemblies are numbered using UNIFORMAT II, ASTM Standard E1557. The first 6 characters represent this system to Level 4. The last 6 characters represent further breakdown by RSMeans in order to arrange items in understandable groups of similar tasks. Line numbers are consistent across all RSMeans publications, so a line number in any RSMeans assemblies data set will always refer to the same work.

Information is available in the Reference Section to assist the estimator with estimating procedures, alternate pricing methods, and additional technical information.

The **Reference Box** indicates the exact location of this information in the Reference Section at the back of the book. The "R" stands for "reference," and the remaining characters are the RSMeans line numbers.

RSMeans assemblies descriptions appear in two formats: narrative and table. **Narrative descriptions** are shown in a hierarchical structure to make them readable. In order to read a complete description, read up through the indents to the top of the section. Include everything that is above and to the left that is not contradicted by information below.

System Components — SYSTEM C2010 110 0560
STAIRS, C.I.P. CONCRETE WITH LANDING, 12 RISERS

System Components	QUANTITY	UNIT	MAT.	INST.	TOTAL
Concrete in place, free standing stairs not incl. safety treads	48.000	L.F.	283.20	1,917.12	2,200.32
Concrete in place, free standing stair landing	32.000	S.F.	153.28	527.36	680.64
Cast alum nosing insert, abr surface, pre-drilled, 3" wide x 4' long	12.000	Ea.	768	196.20	964.20
Industrial railing, welded, 2 rail 3'-6" high 1-1/2" pipe	18.000	L.F.	522	214.56	736.56
Wall railing with returns, steel pipe	17.000	L.F.	282.20	202.64	484.84
TOTAL			2,008.68	3,057.88	5,066.56

C2010 110 Stairs — Cost Per Flight

	Description	MAT.	INST.	TOTAL
0470	Stairs, C.I.P. concrete, w/o landing, 12 risers, w/o nosing (RC2010-100)	1,075	2,350	3,425
0480	With nosing	1,850	2,550	4,400
0550	W/landing, 12 risers, w/o nosing	1,250	2,875	4,125
0560	With nosing	2,000	3,050	5,050
0570	16 risers, w/o nosing	1,525	3,575	5,100
0580	With nosing	2,550	3,850	6,400
0590	20 risers, w/o nosing	1,800	4,325	6,125
0600	With nosing	3,075	4,650	7,725
0610	24 risers, w/o nosing	2,075	5,050	7,125
0620	With nosing	3,600	5,450	9,050
0630	Steel, grate type w/nosing & rails, 12 risers, w/o landing	5,525	1,225	6,750
0640	With landing	7,225	1,675	8,900
0660	16 risers, with landing	9,075	2,075	11,150
0680	20 risers, with landing	10,900	2,475	13,375
0700	24 risers, with landing	12,700	2,875	15,575
0710	Concrete fill metal pan & picket rail, 12 risers, w/o landing	7,200	1,225	8,425
0720	With landing	9,475	1,825	11,300
0740	16 risers, with landing	11,900	2,250	14,150
0760	20 risers, with landing	14,300	2,650	16,950
0780	24 risers, with landing	16,700	3,025	19,725
0790	Cast iron tread & pipe rail, 12 risers, w/o landing	7,200	1,225	8,425
0800	With landing	9,475	1,825	11,300
0820	16 risers, with landing	11,900	2,250	14,150

For supplemental customizable square foot estimating forms, visit: **http://www.reedconstructiondata.com/rsmeans/assemblies**

Most assemblies consist of three major elements: a graphic, the system components, and the cost data itself. The **Graphic** is a visual representation showing the typical appearance of the assembly in question, frequently accompanied by additional explanatory technical information describing the class of items. The **System Components** is a listing of the individual tasks that make up the assembly, including the quantity and unit of measure for each item, along with the cost of material and installation. The **Assemblies Data** below lists prices for other similar systems with dimensional and/or size variations.

All RSMeans assemblies costs represent the cost for the installing contractor. An allowance for profit has been added to all material, labor, and equipment rental costs. A markup for labor burdens, including workers' compensation, fixed overhead, and business overhead, is included with installation costs.

The data published in RSMeans print books represents a "national average" cost. This data should be modified to the project location using the **City Cost Indexes** or **Location Factors** tables found in the Reference Section in the back of the book.

All RSMeans assemblies data includes a typical **Unit of Measure** used for estimating that item. For instance, for continuous footings or foundation walls the unit is linear feet (L.F.). For spread footings the unit is each (Ea.). The estimator needs to take special care that the unit in the data matches the unit in the takeoff. Abbreviations and unit conversions can be found in the Reference Section.

System components are listed separately to detail what is included in the development of the total system price.

Table descriptions work in a similar fashion, except that if there is a blank in the column at a particular line number, read up to the description above in the same column.

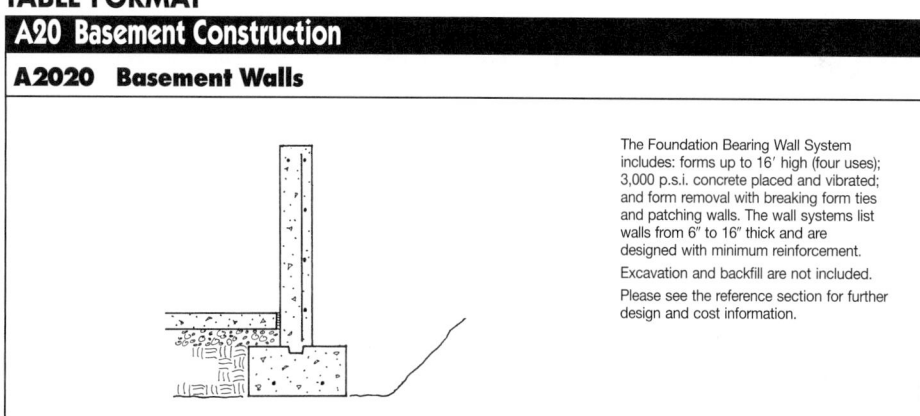

TABLE FORMAT

A20 Basement Construction
A2020 Basement Walls

The Foundation Bearing Wall System includes: forms up to 16' high (four uses); 3,000 p.s.i. concrete placed and vibrated; and form removal with breaking form ties and patching walls. The wall systems list walls from 6" to 16" thick and are designed with minimum reinforcement. Excavation and backfill are not included. Please see the reference section for further design and cost information.

System Components			COST PER L.F.		
	QUANTITY	UNIT	MAT.	INST.	TOTAL
SYSTEM A2020 110 1500					
FOUNDATION WALL, CAST IN PLACE, DIRECT CHUTE, 4' HIGH, 6" THICK					
Forms	8.000	SFCA	5.28	43.20	48.48
Reinforcing	3.300	Lb.	1.82	1.39	3.21
Unloading & sorting reinforcing	3.300	Lb.		.08	.08
Concrete, 3,000 psi	.074	C.Y.	7.92		7.92
Place concrete, direct chute	.074	C.Y.		2.28	2.28
Finish walls, break ties and patch voids, one side	4.000	S.F.	.16	3.76	3.92
TOTAL			15.18	50.71	65.89

A2020 110		Walls, Cast in Place						
	WALL HEIGHT (FT.)	PLACING METHOD	CONCRETE (C.Y. per L.F.)	REINFORCING (LBS. per L.F.)	WALL THICKNESS (IN.)	COST PER L.F.		
						MAT.	INST.	TOTAL
1500	4'	direct chute	.074	3.3	6	15.20	50.50	65.70
1520			.099	4.8	8	18.65	52	70.65
1540			.123	6.0	10	22	53	75
1560			.148	7.2	12	25	54	79
1580		RA2020 -210	.173	8.1	14	28.50	55	83.50
1600			.197	9.44	16	31.50	56	87.50
1700	4'	pumped	.074	3.3	6	15.20	51.50	66.70
1720			.099	4.8	8	18.65	54	72.65
1740			.123	6.0	10	22	55	77
1760			.148	7.2	12	25	56	81
1780			.173	8.1	14	28.50	57.50	86
1800			.197	9.44	16	31.50	58.50	90
3000	6'	direct chute	.111	4.95	6	23	76	99
3020			.149	7.20	8	28	78	106
3040			.184	9.00	10	33	79.50	112.50
3060			.222	10.8	12	38	81	119
3080			.260	12.15	14	42.50	82.50	125
3100			.300	14.39	16	48	85	133

How RSMeans Assemblies Data Works (continued)

Sample Estimate

This sample demonstrates the elements of an estimate, including a tally of the RSMeans data lines. Published assemblies costs include all markups for labor burden and profit for the installing contractor. This estimate adds a summary of the markups applied by a general contractor on the installing contractors' work. These figures represent the total cost to the owner. The RSMeans location factor is added at the bottom of the estimate to adjust the cost of the work to a specific location

Work performed: The body of the estimate shows the RSMeans data selected, including line numbers, a brief description of each item, its takeoff quantity and unit, and the total installed cost, including the installing contractor's overhead and profit.

Location Factor: RSMeans published data is based on national average costs.
If necessary, adjust the total cost of the project using a location factor from the "Location Factor" table or the "City Cost Indexes" table, found in the Reference Section. Use location factors if the work is general, covering the work of multiple trades. If the work is by a single trade (e.g. masonry) use the more specific data found in the City Cost Indexes.

To adjust costs by location factors, multiply the base cost by the factor and divide by 100.

Contingency: A factor for contingency may be added to any estimate to represent the cost of unknowns that may occur between the time that the estimate is performed and the time the project is constructed. The amount of the allowance will depend on the stage of design at which the estimate is done, and the contractor's assessment of the risk invloved.

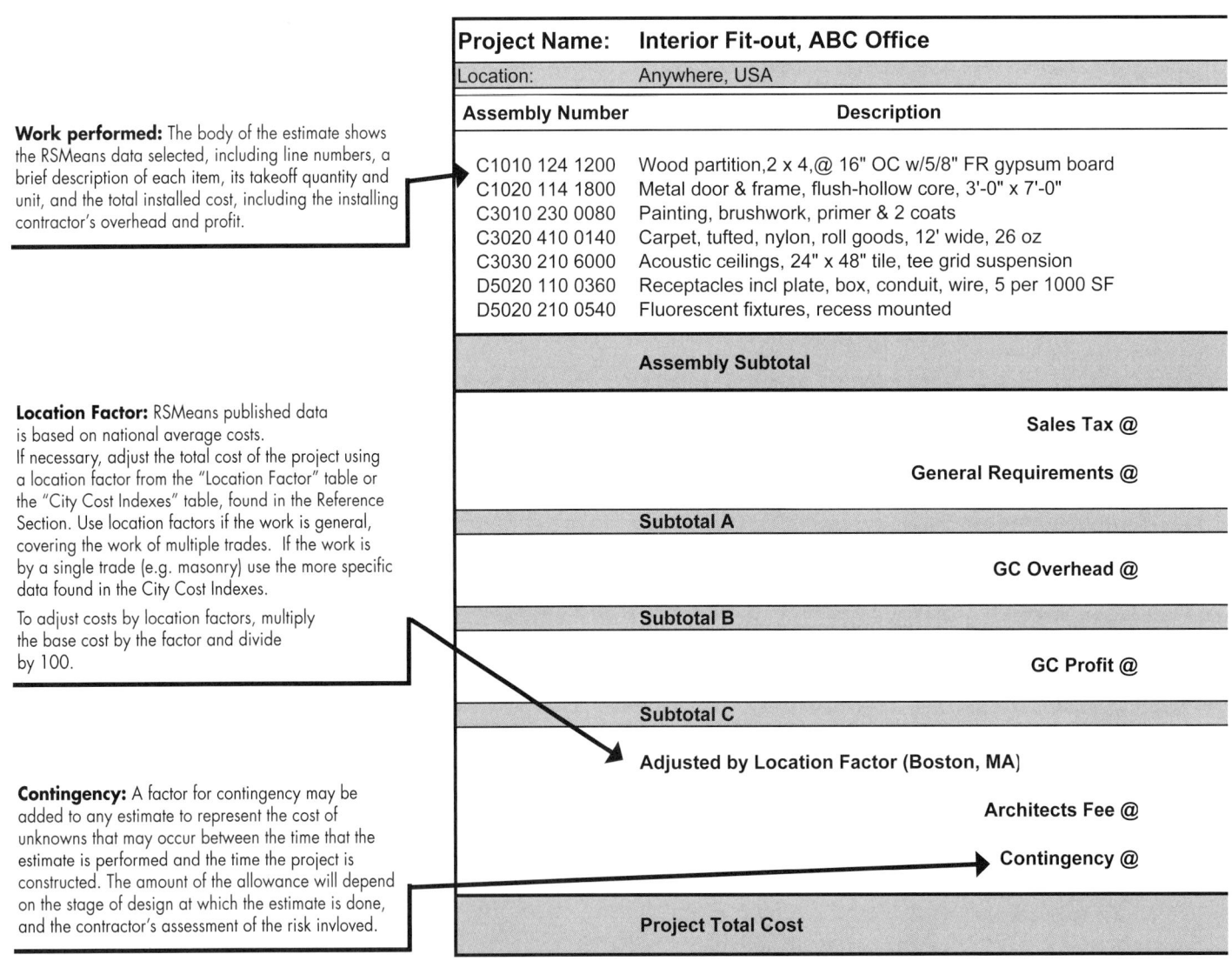

Project Name:	Interior Fit-out, ABC Office
Location:	Anywhere, USA
Assembly Number	**Description**
C1010 124 1200	Wood partition,2 x 4,@ 16" OC w/5/8" FR gypsum board
C1020 114 1800	Metal door & frame, flush-hollow core, 3'-0" x 7'-0"
C3010 230 0080	Painting, brushwork, primer & 2 coats
C3020 410 0140	Carpet, tufted, nylon, roll goods, 12' wide, 26 oz
C3030 210 6000	Acoustic ceilings, 24" x 48" tile, tee grid suspension
D5020 110 0360	Receptacles incl plate, box, conduit, wire, 5 per 1000 SF
D5020 210 0540	Fluorescent fixtures, recess mounted
Assembly Subtotal	
	Sales Tax @
	General Requirements @
Subtotal A	
	GC Overhead @
Subtotal B	
	GC Profit @
Subtotal C	
Adjusted by Location Factor (Boston, MA)	
	Architects Fee @
	Contingency @
Project Total Cost	

This estimate is based on an interactive spreadsheet.
A copy of this spreadsheet is located on the RSMeans website at
http://www.reedconstructiondata.com/rsmeans/extras/546011.
You are free to download it and adjust it to your methodology.

Date:		
Qty.	Unit	Subtotal
600.000	S.F.	$2,682.00
2.000	Ea.	$2,550.00
600.000	S.F.	$696.00
240.000	S.F.	$1,708.80
200.000	S.F.	$870.00
6.000	S.F.	$14.76
200.000	S.F.	$1,580.00
		$10,101.56
5 %		$ 252.54
7 %		$ 707.11
		$11,061.21
5 %		$ 553.06
		$11,614.27
5 %		$ 580.7
		$12,194.98
118.1		$ 14,402.27
8 %		$ 1,152.18
15 %		$ 2,160.34
		$ 17,714.80

Sales Tax: If the work is subject to state or local sales taxes, the amount must be added to the estimate. In a conceptual estimate it can be assumed that one half of the total represents material costs. Therefore, apply the sales tax rate to 50% of the assembly subtotal.

General Requirements: This item covers project-wide needs provided by the General Contractor. These items vary by project, but may include temporary facilities and utilities, security, testing, project cleanup, etc. In assemblies estimates a percentage is used, typically between 5% and 15% of project cost.

General Contractor Overhead: This entry represents the General Contractor's markup on all work to cover project administration costs.

General Contractor Profit: This entry represents the GC's profit on all work performed. The value included here can vary widely by project, and is influenced by the GC's perception of the project's financial risk and market conditions.

Architects Fee: If appropriate, add the design cost to the project estimate. These fees vary based on project complexity and size. Typical design and engineering fees can be found in the reference section.

A Substructure

No part of this publication may be reproduced, stored in a retrieval system, or transmitted in any form or by any means without prior written permission of RSMeans, a division of Reed Construction Data.

A10 Foundations

A1010 Standard Foundations

The Strip Footing System includes: excavation; hand trim; all forms needed for footing placement; forms for 2" x 6" keyway (four uses); dowels; and 3,000 p.s.i. concrete.

The footing size required varies for different soils. Soil bearing capacities are listed for 3 KSF and 6 KSF. Depths of the system range from 8" and deeper. Widths range from 16" and wider. Smaller strip footings may not require reinforcement.

Please see the reference section for further design and cost information.

System Components	QUANTITY	UNIT	COST PER L.F. MAT.	COST PER L.F. INST.	COST PER L.F. TOTAL
SYSTEM A1010 110 2500					
STRIP FOOTING, LOAD 5.1 KLF, SOIL CAP. 3 KSF, 24" WIDE X 12" DEEP, REINF.					
Trench excavation	.148	C.Y.		1.40	1.40
Hand trim	2.000	S.F.		1.84	1.84
Compacted backfill	.074	C.Y.		.28	.28
Formwork, 4 uses	2.000	S.F.	4.42	8.64	13.06
Keyway form, 4 uses	1.000	L.F.	.31	1.11	1.42
Reinforcing, fy = 60000 psi	3.000	Lb.	1.71	1.80	3.51
Dowels	2.000	Ea.	1.62	5.24	6.86
Concrete, f'c = 3000 psi	.074	C.Y.	7.92		7.92
Place concrete, direct chute	.074	C.Y.		1.71	1.71
Screed finish	2.000	S.F.		.72	.72
TOTAL			15.98	22.74	38.72

A1010 110	Strip Footings	COST PER L.F. MAT.	COST PER L.F. INST.	COST PER L.F. TOTAL
2100	Strip footing, load 2.6 KLF, soil capacity 3 KSF, 16" wide x 8" deep plain	8.20	13.65	21.85
2300	Load 3.9 KLF, soil capacity, 3 KSF, 24" wide x 8" deep, plain	10	15	25
2500	Load 5.1 KLF, soil capacity 3 KSF, 24" wide x 12" deep, reinf.	16	22.50	38.50
2700	Load 11.1 KLF, soil capacity 6 KSF, 24" wide x 12" deep, reinf.	16	22.50	38.50
2900	Load 6.8 KLF, soil capacity 3 KSF, 32" wide x 12" deep, reinf.	19.30	24.50	43.80
3100	Load 14.8 KLF, soil capacity 6 KSF, 32" wide x 12" deep, reinf.	19.30	24.50	43.80
3300	Load 9.3 KLF, soil capacity 3 KSF, 40" wide x 12" deep, reinf.	22.50	26.50	49
3500	Load 18.4 KLF, soil capacity 6 KSF, 40" wide x 12" deep, reinf.	22.50	27	49.50
3700	Load 10.1 KLF, soil capacity 3 KSF, 48" wide x 12" deep, reinf.	25	29	54
3900	Load 22.1 KLF, soil capacity 6 KSF, 48" wide x 12" deep, reinf.	26.50	30.50	57
4100	Load 11.8 KLF, soil capacity 3 KSF, 56" wide x 12" deep, reinf.	29	32	61
4300	Load 25.8 KLF, soil capacity 6 KSF, 56" wide x 12" deep, reinf.	31.50	34.50	66
4500	Load 10 KLF, soil capacity 3 KSF, 48" wide x 16" deep, reinf.	31.50	34	65.50
4700	Load 22 KLF, soil capacity 6 KSF, 48" wide, 16" deep, reinf.	32.50	34.50	67
4900	Load 11.6 KLF, soil capacity 3 KSF, 56" wide x 16" deep, reinf.	36	49	85
5100	Load 25.6 KLF, soil capacity 6 KSF, 56" wide x 16" deep, reinf.	38	51	89
5300	Load 13.3 KLF, soil capacity 3 KSF, 64" wide x 16" deep, reinf.	41	40	81
5500	Load 29.3 KLF, soil capacity 6 KSF, 64" wide x 16" deep, reinf.	44	43	87
5700	Load 15 KLF, soil capacity 3 KSF, 72" wide x 20" deep, reinf.	54.50	48	102.50
5900	Load 33 KLF, soil capacity 6 KSF, 72" wide x 20" deep, reinf.	58	51.50	109.50
6100	Load 18.3 KLF, soil capacity 3 KSF, 88" wide x 24" deep, reinf.	77	61	138
6300	Load 40.3 KLF, soil capacity 6 KSF, 88" wide x 24" deep, reinf.	83	68	151
6500	Load 20 KLF, soil capacity 3 KSF, 96" wide x 24" deep, reinf.	83.50	64.50	148
6700	Load 44 KLF, soil capacity 6 KSF, 96" wide x 24" deep, reinf.	88.50	70	158.50

A10 Foundations

A1010 Standard Foundations

The Spread Footing System includes: excavation; backfill; forms (four uses); all reinforcement; 3,000 p.s.i. concrete (chute placed); and float finish.

Footing systems are priced per individual unit. The Expanded System Listing at the bottom shows various footing sizes. It is assumed that excavation is done by a truck mounted hydraulic excavator with an operator and oiler.

Backfill is with a dozer, and compaction by air tamp. The excavation and backfill equipment is assumed to operate at 30 C.Y. per hour.

Please see the reference section for further design and cost information.

System Components	QUANTITY	UNIT	COST EACH		
			MAT.	INST.	TOTAL
SYSTEM A1010 210 7100					
SPREAD FOOTINGS, LOAD 25K, SOIL CAPACITY 3 KSF, 3' SQ X 12" DEEP					
Bulk excavation	.590	C.Y.		4.92	4.92
Hand trim	9.000	S.F.		8.28	8.28
Compacted backfill	.260	C.Y.		1	1
Formwork, 4 uses	12.000	S.F.	8.52	60.60	69.12
Reinforcing, fy = 60,000 psi	.006	Ton	6.60	7.20	13.80
Dowel or anchor bolt templates	6.000	L.F.	5.34	25.14	30.48
Concrete, f'c = 3,000 psi	.330	C.Y.	35.31		35.31
Place concrete, direct chute	.330	C.Y.		7.63	7.63
Float finish	9.000	S.F.		3.24	3.24
TOTAL			55.77	118.01	173.78

A1010 210	Spread Footings	COST EACH		
		MAT.	INST.	TOTAL
7090	Spread footings, 3000 psi concrete, chute delivered			
7100	Load 25K, soil capacity 3 KSF, 3'-0" sq. x 12" deep	56	118	174
7150	Load 50K, soil capacity 3 KSF, 4'-6" sq. x 12" deep	120	203	323
7200	Load 50K, soil capacity 6 KSF, 3'-0" sq. x 12" deep	56	118	174
7250	Load 75K, soil capacity 3 KSF, 5'-6" sq. x 13" deep	191	286	477
7300	Load 75K, soil capacity 6 KSF, 4'-0" sq. x 12" deep	97	175	272
7350	Load 100K, soil capacity 3 KSF, 6'-0" sq. x 14" deep	241	345	586
7410	Load 100K, soil capacity 6 KSF, 4'-6" sq. x 15" deep	147	239	386
7450	Load 125K, soil capacity 3 KSF, 7'-0" sq. x 17" deep	385	495	880
7500	Load 125K, soil capacity 6 KSF, 5'-0" sq. x 16" deep	190	287	477
7550	Load 150K, soil capacity 3 KSF 7'-6" sq. x 18" deep	465	575	1,040
7610	Load 150K, soil capacity 6 KSF, 5'-6" sq. x 18" deep	254	360	614
7650	Load 200K, soil capacity 3 KSF, 8'-6" sq. x 20" deep	660	760	1,420
7700	Load 200K, soil capacity 6 KSF, 6'-0" sq. x 20" deep	335	445	780
7750	Load 300K, soil capacity 3 KSF, 10'-6" sq. x 25" deep	1,225	1,250	2,475
7810	Load 300K, soil capacity 6 KSF, 7'-6" sq. x 25" deep	635	750	1,385
7850	Load 400K, soil capacity 3 KSF, 12'-6" sq. x 28" deep	1,925	1,850	3,775
7900	Load 400K, soil capacity 6 KSF, 8'-6" sq. x 27" deep	880	970	1,850
7950	Load 500K, soil capacity 3 KSF, 14'-0" sq. x 31" deep	2,700	2,400	5,100
8010	Load 500K, soil capacity 6 KSF, 9'-6" sq. x 30" deep	1,225	1,275	2,500

Note: Row 7200 references RA1010-120.

A10 Foundations

A1010 Standard Foundations

A1010 210	Spread Footings	COST EACH		
		MAT.	INST.	TOTAL
8050	Load 600K, soil capacity 3 KSF, 16'-0" sq. x 35" deep	3,925	3,300	7,225
8100	Load 600K, soil capacity 6 KSF, 10'-6" sq. x 33" deep	1,650	1,625	3,275
8150	Load 700K, soil capacity 3 KSF, 17'-0" sq. x 37" deep	4,625	3,775	8,400
8200	Load 700K, soil capacity 6 KSF, 11'-6" sq. x 36" deep	2,100	1,975	4,075
8250	Load 800K, soil capacity 3 KSF, 18'-0" sq. x 39" deep	5,475	4,375	9,850
8300	Load 800K, soil capacity 6 KSF, 12'-0" sq. x 37" deep	2,350	2,200	4,550
8350	Load 900K, soil capacity 3 KSF, 19'-0" sq. x 40" deep	6,375	5,025	11,400
8400	Load 900K, soil capacity 6 KSF, 13'-0" sq. x 39" deep	2,925	2,600	5,525
8450	Load 1000K, soil capacity 3 KSF, 20'-0" sq. x 42" deep	7,350	5,650	13,000
8500	Load 1000K, soil capacity 6 KSF, 13'-6" sq. x 41" deep	3,300	2,900	6,200
8550	Load 1200K, soil capacity 6 KSF, 15'-0" sq. x 48" deep	4,400	3,675	8,075
8600	Load 1400K, soil capacity 6 KSF, 16'-0" sq. x 47" deep	5,375	4,375	9,750
8650	Load 1600K, soil capacity 6 KSF, 18'-0" sq. x 52" deep	7,450	5,800	13,250
8700				

A10 Foundations

A1010 Standard Foundations

These pile cap systems include excavation with a truck mounted hydraulic excavator, hand trimming, compacted backfill, forms for concrete, templates for dowels or anchor bolts, reinforcing steel and concrete placed and floated.

Pile embedment is assumed as 6". Design is consistent with the Concrete Reinforcing Steel Institute Handbook f'c = 3000 psi, fy = 60,000.

Please see the reference section for further design and cost information.

System Components	QUANTITY	UNIT	COST EACH		
			MAT.	INST.	TOTAL
SYSTEM A1010 250 5100					
CAP FOR 2 PILES, 6'-6"X3'-6"X20", 15 TON PILE, 8" MIN. COL., 45K COL. LOAD					
Excavation, bulk, hyd excavator, truck mtd. 30" bucket 1/2 CY	2.890	C.Y.		24.07	24.07
Trim sides and bottom of trench, regular soil	23.000	S.F.		21.16	21.16
Dozer backfill & roller compaction	1.500	C.Y.		5.80	5.80
Forms in place pile cap, square or rectangular, 4 uses	33.000	SFCA	32.67	179.85	212.52
Templates for dowels or anchor bolts	8.000	Ea.	7.12	33.52	40.64
Reinforcing in place footings, #8 to #14	.025	Ton	27.50	17.50	45
Concrete ready mix, regular weight, 3000 psi	1.400	C.Y.	149.80		149.80
Place and vibrate concrete for pile caps, under 5 CY, direct chute	1.400	C.Y.		43.13	43.13
Float finish	23.000	S.F.		8.28	8.28
TOTAL			217.09	333.31	550.40

A1010 250			Pile Caps					
	NO. PILES	SIZE FT-IN X FT-IN X IN	PILE CAPACITY (TON)	COLUMN SIZE (IN)	COLUMN LOAD (K)	COST EACH		
						MAT.	INST.	TOTAL
5100	2	6-6x3-6x20	15	8	45	217	335	552
5150			26	40	8	155	265	410	675
5200			34	80	11	314	360	525	885
5250			37	120	14	473	385	560	945
5300	3	5-6x5-1x23	15	8	75	262	385	647
5350			28	40	10	232	291	435	726
5400			32	80	14	471	335	495	830
5450			38	120	17	709	390	565	955
5500	4	5-6x5-6x18	15	10	103	296	385	681
5550			30	40	11	308	430	550	980
5600			36	80	16	626	505	645	1,150
5650			38	120	19	945	530	675	1,205
5700	6	8-6x5-6x18	15	12	156	500	555	1,055
5750			37	40	14	458	795	825	1,620
5800			40	80	19	936	915	920	1,835
5850			45	120	24	1413	1,025	1,025	2,050
5900	8	8-6x7-9x19	15	12	205	760	755	1,515
5950			36	40	16	610	1,075	995	2,070
6000			44	80	22	1243	1,325	1,200	2,525
6050			47	120	27	1881	1,450	1,275	2,725

A10 Foundations

A1010 Standard Foundations

A1010 250 Pile Caps

	NO. PILES	SIZE FT-IN X FT-IN X IN	PILE CAPACITY (TON)	COLUMN SIZE (IN)	COLUMN LOAD (K)	COST EACH		
						MAT.	INST.	TOTAL
6100	10	11-6x7-9x21	15	14	250	1,075	920	1,995
6150			39	40	17	756	1,600	1,325	2,925
6200			47	80	25	1547	1,950	1,575	3,525
6250			49	120	31	2345	2,075	1,675	3,750
6300	12	11-6x8-6x22	15	15	316	1,375	1,125	2,500
6350			49	40	19	900	2,100	1,675	3,775
6400			52	80	27	1856	2,350	1,850	4,200
6450			55	120	34	2812	2,550	1,975	4,525
6500	14	11-6x10-9x24	15	16	345	1,750	1,375	3,125
6550			41	40	21	1056	2,275	1,725	4,000
6600			55	80	29	2155	2,975	2,175	5,150
6700	16	11-6x11-6x26	15	18	400	2,150	1,600	3,750
6750			48	40	22	1200	2,800	2,050	4,850
6800			60	80	31	2460	3,475	2,500	5,975
6900	18	13-0x11-6x28	15	20	450	2,425	1,750	4,175
6950			49	40	23	1349	3,250	2,275	5,525
7000			56	80	33	2776	3,850	2,675	6,525
7100	20	14-6x11-6x30	15	20	510	2,975	2,100	5,075
7150			52	40	24	1491	3,925	2,675	6,600

A10 Foundations

A1010 Standard Foundations

General: Footing drains can be placed either inside or outside of foundation walls depending upon the source of water to be intercepted. If the source of subsurface water is principally from grade or a subsurface stream above the bottom of the footing, outside drains should be used. For high water tables, use inside drains or both inside and outside.

The effectiveness of underdrains depends on good waterproofing. This must be carefully installed and protected during construction.

Costs below include the labor and materials for the pipe and 6" of crushed stone around pipe. Excavation and backfill are not included.

System Components	QUANTITY	UNIT	COST PER L.F.		
			MAT.	INST.	TOTAL
SYSTEM A1010 310 1000					
FOUNDATION UNDERDRAIN, OUTSIDE ONLY, PVC 4" DIAM.					
PVC pipe 4" diam. S.D.R. 35	1.000	L.F.	1.60	3.92	5.52
Crushed stone 3/4" to 1/2"	.070	C.Y.	2.38	.86	3.24
TOTAL			3.98	4.78	8.76

A1010 310	Foundation Underdrain	COST PER L.F.		
		MAT.	INST.	TOTAL
1000	Foundation underdrain, outside only, PVC, 4" diameter	3.98	4.78	8.76
1100	6" diameter	6.65	5.30	11.95
1400	Perforated HDPE, 6" diameter	4.68	2.07	6.75
1450	8" diameter	6.75	2.60	9.35
1500	12" diameter	12.90	6.30	19.20
1600	Corrugated metal, 16 ga. asphalt coated, 6" diameter	10.20	4.97	15.17
1650	8" diameter	13.10	5.30	18.40
1700	10" diameter	16.10	6.90	23
3000	Outside and inside, PVC, 4" diameter	7.95	9.55	17.50
3100	6" diameter	13.30	10.60	23.90
3400	Perforated HDPE, 6" diameter	9.35	4.13	13.48
3450	8" diameter	13.50	5.20	18.70
3500	12" diameter	26	12.55	38.55
3600	Corrugated metal, 16 ga., asphalt coated, 6" diameter	20.50	9.95	30.45
3650	8" diameter	26	10.65	36.65
3700	10" diameter	32	13.80	45.80

A10 Foundations

A1010 Standard Foundations

General: Apply foundation wall dampproofing over clean concrete giving particular attention to the joint between the wall and the footing. Use care in backfilling to prevent damage to the dampproofing.

Costs for four types of dampproofing are listed below.

System Components	QUANTITY	UNIT	COST PER L.F. MAT.	COST PER L.F. INST.	COST PER L.F. TOTAL
SYSTEM A1010 320 1000					
FOUNDATION DAMPPROOFING, BITUMINOUS, 1 COAT, 4' HIGH					
Bituminous asphalt dampproofing brushed on below grade, 1 coat	4.000	S.F.	.92	3.08	4
Labor for protection of dampproofing during backfilling	4.000	S.F.		1.31	1.31
TOTAL			.92	4.39	5.31

A1010 320	Foundation Dampproofing	MAT.	INST.	TOTAL
1000	Foundation dampproofing, bituminous, 1 coat, 4' high	.92	4.39	5.31
1400	8' high	1.84	8.75	10.59
1800	12' high	2.76	13.60	16.36
2000	2 coats, 4' high	1.84	5.45	7.29
2400	8' high	3.68	10.85	14.53
2800	12' high	5.50	16.70	22.20
3000	Asphalt with fibers, 1/16" thick, 4' high	1.36	5.45	6.81
3400	8' high	2.72	10.85	13.57
3800	12' high	4.08	16.70	20.78
4000	1/8" thick, 4' high	2.40	6.45	8.85
4400	8' high	4.80	12.95	17.75
4800	12' high	7.20	19.85	27.05
5000	Asphalt coated board and mastic, 1/4" thick, 4' high	4.08	5.85	9.93
5400	8' high	8.15	11.75	19.90
5800	12' high	12.25	18.05	30.30
6000	1/2" thick, 4' high	6.10	8.15	14.25
6400	8' high	12.25	16.30	28.55
6800	12' high	18.35	25	43.35
7000	Cementitious coating, on walls, 1/8" thick coating, 4' high	2	8.45	10.45
7400	8' high	4	16.95	20.95
7800	12' high	6	25.50	31.50
8000	Cementitious/metallic slurry, 4 coat, 1/2"thick, 2' high	.57	8.50	9.07
8400	4' high	1.14	17	18.14
8800	6' high	1.71	25.50	27.21

A10 Foundations

A1020 Special Foundations

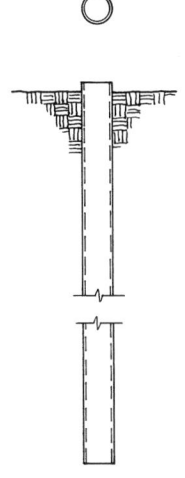

The Cast-in-Place Concrete Pile System includes: a defined number of 4,000 p.s.i. concrete piles with thin-wall, straight-sided, steel shells that have a standard steel plate driving point. An allowance for cutoffs is included.

The Expanded System Listing shows costs per cluster of piles. Clusters range from one pile to twenty piles. Loads vary from 50 Kips to 1,600 Kips. Both end-bearing and friction-type piles are shown.

Please see the reference section for cost of mobilization of the pile driving equipment and other design and cost information.

System Components	QUANTITY	UNIT	COST EACH		
			MAT.	INST.	TOTAL
SYSTEM A1020 110 2220					
CIP SHELL CONCRETE PILE, 25' LONG, 50K LOAD, END BEARING, 1 PILE					
7 Ga. shell, 12" diam.	27.000	V.L.F.	931.50	294.84	1,226.34
Steel pipe pile standard point, 12" or 14" diameter pile	1.000	Ea.	118.50	88	206.50
Pile cutoff, conc. pile with thin steel shell	1.000	Ea.		14.45	14.45
TOTAL			1,050	397.29	1,447.29

A1020 110	C.I.P. Concrete Piles	COST EACH		
		MAT.	INST.	TOTAL
2220	CIP shell concrete pile, 25' long, 50K load, end bearing, 1 pile	1,050	400	1,450
2240	100K load, end bearing, 2 pile cluster	2,100	790	2,890
2260	200K load, end bearing, 4 pile cluster	4,200	1,600	5,800
2280	400K load, end bearing, 7 pile cluster	7,350	2,775	10,125
2300	10 pile cluster	10,500	3,950	14,450
2320	800K load, end bearing, 13 pile cluster	23,000	7,275	30,275
2340	17 pile cluster	30,000	9,525	39,525
2360	1200K load, end bearing, 14 pile cluster	24,700	7,850	32,550
2380	19 pile cluster	33,600	10,700	44,300
2400	1600K load, end bearing, 19 pile cluster	33,600	10,700	44,300
2420	50' long, 50K load, end bearing, 1 pile	1,950	680	2,630
2440	Friction type, 2 pile cluster	3,725	1,375	5,100
2460	3 pile cluster	5,600	2,050	7,650
2480	100K load, end bearing, 2 pile cluster	3,900	1,375	5,275
2500	Friction type, 4 pile cluster	7,475	2,725	10,200
2520	6 pile cluster	11,200	4,075	15,275
2540	200K load, end bearing, 4 pile cluster	7,800	2,725	10,525
2560	Friction type, 8 pile cluster	14,900	5,450	20,350
2580	10 pile cluster	18,700	6,825	25,525
2600	400K load, end bearing, 7 pile cluster	13,600	4,750	18,350
2620	Friction type, 16 pile cluster	29,900	10,900	40,800
2640	19 pile cluster	35,500	13,000	48,500
2660	800K load, end bearing, 14 pile cluster	44,900	12,800	57,700
2680	20 pile cluster	64,000	18,300	82,300
2700	1200K load, end bearing, 15 pile cluster	48,100	13,800	61,900
2720	1600K load, end bearing, 20 pile cluster	64,000	18,300	82,300

RA1020-100

A10 Foundations

A1020 Special Foundations

A1020 110	C.I.P. Concrete Piles	COST EACH		
		MAT.	INST.	TOTAL
3740	75' long, 50K load, end bearing, 1 pile	2,975	1,150	4,125
3760	Friction type, 2 pile cluster	5,725	2,300	8,025
3780	3 pile cluster	8,575	3,475	12,050
3800	100K load, end bearing, 2 pile cluster	5,950	2,300	8,250
3820	Friction type, 3 pile cluster	8,575	3,475	12,050
3840	5 pile cluster	14,300	5,775	20,075
3860	200K load, end bearing, 4 pile cluster	11,900	4,625	16,525
3880	6 pile cluster	17,900	6,925	24,825
3900	Friction type, 6 pile cluster	17,200	6,925	24,125
3910	7 pile cluster	20,000	8,075	28,075
3920	400K load, end bearing, 7 pile cluster	20,900	8,075	28,975
3930	11 pile cluster	32,800	12,700	45,500
3940	Friction type, 12 pile cluster	34,300	13,900	48,200
3950	14 pile cluster	40,100	16,200	56,300
3960	800K load, end bearing, 15 pile cluster	72,500	22,600	95,100
3970	20 pile cluster	96,500	30,200	126,700
3980	1200K load, end bearing, 17 pile cluster	82,000	25,600	107,600
3990				

A10 Foundations

A1020 Special Foundations

The Precast Concrete Pile System includes: pre-stressed concrete piles; standard steel driving point; and an allowance for cutoffs.

The Expanded System Listing shows costs per cluster of piles. Clusters range from one pile to twenty piles. Loads vary from 50 Kips to 1,600 Kips. Both end-bearing and friction type piles are listed.

Please see the reference section for cost of mobilization of the pile driving equipment and other design and cost information.

System Components	QUANTITY	UNIT	COST EACH		
			MAT.	INST.	TOTAL
SYSTEM A1020 120 2220					
PRECAST CONCRETE PILE, 50' LONG, 50K LOAD, END BEARING, 1 PILE					
Precast, prestressed conc. piles, 10" square, no mobil.	53.000	V.L.F.	1,017.60	497.14	1,514.74
Steel pipe pile standard point, 8" to 10" diameter	1.000	Ea.	85	77.50	162.50
Piling special costs cutoffs concrete piles plain	1.000	Ea.		100	100
TOTAL			1,102.60	674.64	1,777.24

A1020 120	Precast Concrete Piles	COST EACH		
		MAT.	INST.	TOTAL
2220	Precast conc pile, 50' long, 50K load, end bearing, 1 pile	1,100	675	1,775
2240	Friction type, 2 pile cluster	2,825	1,400	4,225
2260	4 pile cluster	5,675	2,800	8,475
2280	100K load, end bearing, 2 pile cluster	2,200	1,350	3,550
2300	Friction type, 2 pile cluster	2,825	1,400	4,225
2320	4 pile cluster	5,675	2,800	8,475
2340	7 pile cluster	9,925	4,900	14,825
2360	200K load, end bearing, 3 pile cluster	3,300	2,025	5,325
2380	4 pile cluster	4,400	2,700	7,100
2400	Friction type, 8 pile cluster	11,300	5,575	16,875
2420	9 pile cluster	12,800	6,300	19,100
2440	14 pile cluster	19,800	9,775	29,575
2460	400K load, end bearing, 6 pile cluster	6,625	4,050	10,675
2480	8 pile cluster	8,825	5,400	14,225
2500	Friction type, 14 pile cluster	19,800	9,775	29,575
2520	16 pile cluster	22,700	11,200	33,900
2540	18 pile cluster	25,500	12,600	38,100
2560	800K load, end bearing, 12 pile cluster	14,300	8,775	23,075
2580	16 pile cluster	17,600	10,800	28,400
2600	1200K load, end bearing, 19 pile cluster	58,000	15,700	73,700
2620	20 pile cluster	61,000	16,500	77,500
2640	1600K load, end bearing, 19 pile cluster	58,000	15,700	73,700
4660	100' long, 50K load, end bearing, 1 pile	2,100	1,175	3,275
4680	Friction type, 1 pile	2,700	1,200	3,900

RA1020-100

A10 Foundations

A1020 Special Foundations

A1020 120	Precast Concrete Piles	COST EACH		
		MAT.	INST.	TOTAL
4700	2 pile cluster	5,375	2,400	7,775
4720	100K load, end bearing, 2 pile cluster	4,200	2,325	6,525
4740	Friction type, 2 pile cluster	5,375	2,400	7,775
4760	3 pile cluster	8,075	3,600	11,675
4780	4 pile cluster	10,800	4,800	15,600
4800	200K load, end bearing, 3 pile cluster	6,300	3,500	9,800
4820	4 pile cluster	8,400	4,650	13,050
4840	Friction type, 3 pile cluster	8,075	3,600	11,675
4860	5 pile cluster	13,500	5,975	19,475
4880	400K load, end bearing, 6 pile cluster	12,600	6,975	19,575
4900	8 pile cluster	16,800	9,300	26,100
4910	Friction type, 8 pile cluster	21,500	9,575	31,075
4920	10 pile cluster	26,900	12,000	38,900
4930	800K load, end bearing, 13 pile cluster	27,300	15,100	42,400
4940	16 pile cluster	33,600	18,600	52,200
4950	1200K load, end bearing, 19 pile cluster	112,500	27,200	139,700
4960	20 pile cluster	118,500	28,600	147,100
4970	1600K load, end bearing, 19 pile cluster	112,500	27,200	139,700

A10 Foundations

A1020 Special Foundations

The Steel Pipe Pile System includes: steel pipe sections filled with 4,000 p.s.i. concrete; a standard steel driving point; splices when required and an allowance for cutoffs.

The Expanded System Listing shows costs per cluster of piles. Clusters range from one pile to twenty piles. Loads vary from 50 Kips to 1,600 Kips. Both end-bearing and friction-type piles are shown.

Please see the reference section for cost of mobilization of the pile driving equipment and other design and cost information.

System Components	QUANTITY	UNIT	COST EACH		
			MAT.	INST.	TOTAL
SYSTEM A1020 130 2220 CONC. FILL STEEL PIPE PILE, 50' LONG, 50K LOAD, END BEARING, 1 PILE					
Piles, steel, pipe, conc. filled, 12" diameter	53.000	V.L.F.	1,934.50	834.75	2,769.25
Steel pipe pile, standard point, for 12" or 14" diameter pipe	1.000	Ea.	237	176	413
Pile cut off, concrete pile, thin steel shell	1.000	Ea.		14.45	14.45
TOTAL			2,171.50	1,025.20	3,196.70

A1020 130	Steel Pipe Piles	COST EACH		
		MAT.	INST.	TOTAL
2220	Conc. fill steel pipe pile, 50' long, 50K load, end bearing, 1 pile	2,175	1,025	3,200
2240	Friction type, 2 pile cluster	4,350	2,050	6,400
2250	100K load, end bearing, 2 pile cluster	4,350	2,050	6,400
2260	3 pile cluster	6,525	3,075	9,600
2300	Friction type, 4 pile cluster	8,675	4,100	12,775
2320	5 pile cluster	10,900	5,125	16,025
2340	10 pile cluster	21,700	10,300	32,000
2360	200K load, end bearing, 3 pile cluster	6,525	3,075	9,600
2380	4 pile cluster	8,675	4,100	12,775
2400	Friction type, 4 pile cluster	8,675	4,100	12,775
2420	8 pile cluster	17,400	8,200	25,600
2440	9 pile cluster	19,500	9,225	28,725
2460	400K load, end bearing, 6 pile cluster	13,000	6,150	19,150
2480	7 pile cluster	15,200	7,175	22,375
2500	Friction type, 9 pile cluster	19,500	9,225	28,725
2520	16 pile cluster	34,700	16,500	51,200
2540	19 pile cluster	41,300	19,400	60,700
2560	800K load, end bearing, 11 pile cluster	23,900	11,300	35,200
2580	14 pile cluster	30,400	14,300	44,700
2600	15 pile cluster	32,600	15,400	48,000
2620	Friction type, 17 pile cluster	36,900	17,400	54,300
2640	1200K load, end bearing, 16 pile cluster	34,700	16,500	51,200
2660	20 pile cluster	43,400	20,500	63,900
2680	1600K load, end bearing, 17 pile cluster	36,900	17,400	54,300
3700	100' long, 50K load, end bearing, 1 pile	4,200	2,025	6,225
3720	Friction type, 1 pile	4,200	2,025	6,225

A10 Foundations

A1020 Special Foundations

A1020 130	Steel Pipe Piles	COST EACH		
		MAT.	INST.	TOTAL
3740	2 pile cluster	8,400	4,075	12,475
3760	100K load, end bearing, 2 pile cluster	8,400	4,075	12,475
3780	Friction type, 2 pile cluster	8,400	4,075	12,475
3800	3 pile cluster	12,600	6,100	18,700
3820	200K load, end bearing, 3 pile cluster	12,600	6,100	18,700
3840	4 pile cluster	16,800	8,125	24,925
3860	Friction type, 3 pile cluster	12,600	6,100	18,700
3880	4 pile cluster	16,800	8,125	24,925
3900	400K load, end bearing, 6 pile cluster	25,200	12,200	37,400
3910	7 pile cluster	29,400	14,300	43,700
3920	Friction type, 5 pile cluster	21,000	10,200	31,200
3930	8 pile cluster	33,600	16,300	49,900
3940	800K load, end bearing, 11 pile cluster	46,300	22,400	68,700
3950	14 pile cluster	59,000	28,500	87,500
3960	15 pile cluster	63,000	30,500	93,500
3970	1200K load, end bearing, 16 pile cluster	67,500	32,500	100,000
3980	20 pile cluster	84,000	40,700	124,700
3990	1600K load, end bearing, 17 pile cluster	71,500	34,500	106,000

A10 Foundations

A1020 Special Foundations

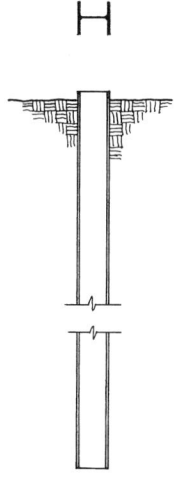

A Steel "H" Pile System includes: steel H sections; heavy duty driving point; splices where applicable and allowance for cutoffs.

The Expanded System Listing shows costs per cluster of piles. Clusters range from one pile to eighteen piles. Loads vary from 50 Kips to 2,000 Kips. All loads for Steel H Pile systems are given in terms of end bearing capacity.

Steel sections range from 10" x 10" to 14" x 14" in the Expanded System Listing. The 14" x 14" steel section is used for all H piles used in applications requiring a working load over 800 Kips.

Please see the reference section for cost of mobilization of the pile driving equipment and other design and cost information.

System Components	QUANTITY	UNIT	COST EACH		
			MAT.	INST.	TOTAL
SYSTEM A1020 140 2220					
STEEL H PILES, 50' LONG, 100K LOAD, END BEARING, 1 PILE					
Steel H piles 10" x 10", 42 #/L.F.	53.000	V.L.F.	1,272	568.16	1,840.16
Heavy duty point, 10"	1.000	Ea.	207	179	386
Pile cut off, steel pipe or H piles	1.000	Ea.		29	29
TOTAL			1,479	776.16	2,255.16

A1020 140	Steel H Piles	COST EACH		
		MAT.	INST.	TOTAL
2220	Steel H piles, 50' long, 100K load, end bearing, 1 pile	1,475	775	2,250
2260	2 pile cluster	2,950	1,575	4,525
2280	200K load, end bearing, 2 pile cluster	2,950	1,575	4,525
2300	3 pile cluster	4,425	2,325	6,750
2320	400K load, end bearing, 3 pile cluster	4,425	2,325	6,750
2340	4 pile cluster	5,925	3,100	9,025
2360	6 pile cluster	8,875	4,675	13,550
2380	800K load, end bearing, 5 pile cluster	7,400	3,875	11,275
2400	7 pile cluster	10,400	5,450	15,850
2420	12 pile cluster	17,700	9,325	27,025
2440	1200K load, end bearing, 8 pile cluster	11,800	6,200	18,000
2460	11 pile cluster	16,300	8,550	24,850
2480	17 pile cluster	25,100	13,200	38,300
2500	1600K load, end bearing, 10 pile cluster	18,400	8,125	26,525
2520	14 pile cluster	25,800	11,400	37,200
2540	2000K load, end bearing, 12 pile cluster	22,100	9,775	31,875
2560	18 pile cluster	33,200	14,700	47,900
3580	100' long, 50K load, end bearing, 1 pile	4,875	1,775	6,650
3600	100K load, end bearing, 1 pile	4,875	1,775	6,650
3620	2 pile cluster	9,750	3,525	13,275
3640	200K load, end bearing, 2 pile cluster	9,750	3,525	13,275
3660	3 pile cluster	14,600	5,300	19,900
3680	400K load, end bearing, 3 pile cluster	14,600	5,300	19,900
3700	4 pile cluster	19,500	7,075	26,575
3720	6 pile cluster	29,200	10,600	39,800
3740	800K load, end bearing, 5 pile cluster	24,400	8,850	33,250

Reference: RA1020-100

A10 Foundations

A1020 Special Foundations

A1020 140	Steel H Piles	COST EACH		
		MAT.	INST.	TOTAL
3760	7 pile cluster	34,100	12,400	46,500
3780	12 pile cluster	58,500	21,200	79,700
3800	1200K load, end bearing, 8 pile cluster	39,000	14,100	53,100
3820	11 pile cluster	53,500	19,500	73,000
3840	17 pile cluster	83,000	30,000	113,000
3860	1600K load, end bearing, 10 pile cluster	48,700	17,700	66,400
3880	14 pile cluster	68,000	24,800	92,800
3900	2000K load, end bearing, 12 pile cluster	58,500	21,200	79,700
3920	18 pile cluster	87,500	31,800	119,300

A10 Foundations

A1020 Special Foundations

A10 Foundations

A1020 Special Foundations

The Step Tapered Steel Pile System includes: step tapered piles filled with 4,000 p.s.i. concrete. The cost for splices and pile cutoffs is included.

The Expanded System Listing shows costs per cluster of piles. Clusters range from one pile to twenty-four piles. Both end bearing piles and friction piles are listed. Loads vary from 50 Kips to 1,600 Kips.

Please see the reference section for cost of mobilization of the pile driving equipment and other design and cost information.

System Components	QUANTITY	UNIT	COST EACH		
			MAT.	INST.	TOTAL
SYSTEM A1020 150 1000					
STEEL PILE, STEP TAPERED, 50' LONG, 50K LOAD, END BEARING, 1 PILE					
Steel shell step tapered conc. filled piles, 8" tip 60 ton capacity to 60'	53.000	V.L.F.	734.05	470.11	1,204.16
Pile cutoff, steel pipe or H piles	1.000	Ea.		29	29
TOTAL			734.05	499.11	1,233.16

A1020 150	Step-Tapered Steel Piles	COST EACH		
		MAT.	INST.	TOTAL
1000	Steel pile, step tapered, 50' long, 50K load, end bearing, 1 pile	735	500	1,235
1200	Friction type, 3 pile cluster	2,200	1,500	3,700
1400	100K load, end bearing, 2 pile cluster	1,475	1,000	2,475
1600	Friction type, 4 pile cluster	2,925	2,000	4,925
1800	200K load, end bearing, 4 pile cluster	2,925	2,000	4,925
2000	Friction type, 6 pile cluster	4,400	3,000	7,400
2200	400K load, end bearing, 7 pile cluster	5,150	3,500	8,650
2400	Friction type, 10 pile cluster	7,350	5,000	12,350
2600	800K load, end bearing, 14 pile cluster	10,300	7,000	17,300
2800	Friction type, 18 pile cluster	13,200	9,000	22,200
3000	1200K load, end bearing, 16 pile cluster	12,800	8,475	21,275
3200	Friction type, 21 pile cluster	16,900	11,200	28,100
3400	1600K load, end bearing, 18 pile cluster	14,500	9,550	24,050
3600	Friction type, 24 pile cluster	19,300	12,800	32,100
5000	100' long, 50K load, end bearing, 1 pile	1,475	995	2,470
5200	Friction type, 2 pile cluster	2,950	2,000	4,950
5400	100K load, end bearing, 2 pile cluster	2,950	2,000	4,950
5600	Friction type, 3 pile cluster	4,425	3,000	7,425
5800	200K load, end bearing, 4 pile cluster	5,900	3,975	9,875
6000	Friction type, 5 pile cluster	7,400	4,975	12,375
6200	400K load, end bearing, 7 pile cluster	10,300	6,975	17,275
6400	Friction type, 8 pile cluster	11,800	7,950	19,750
6600	800K load, end bearing, 15 pile cluster	22,200	14,900	37,100
6800	Friction type, 16 pile cluster	23,600	15,900	39,500
7000	1200K load, end bearing, 17 pile cluster	42,400	25,900	68,300
7200	Friction type, 19 pile cluster	31,900	19,700	51,600
7400	1600K load, end bearing, 20 pile cluster	33,600	20,700	54,300
7600	Friction type, 22 pile cluster	36,900	22,800	59,700

RA1020-100

A10 Foundations

A1020 Special Foundations

The Treated Wood Pile System includes: creosoted wood piles; a standard steel driving point; and an allowance for cutoffs.

The Expanded System Listing shows costs per cluster of piles. Clusters range from two piles to twenty piles. Loads vary from 50 Kips to 400 Kips. Both end-bearing and friction type piles are listed.

Please see the reference section for cost of mobilization of the pile driving equipment and other design and cost information.

System Components	QUANTITY	UNIT	COST EACH		
			MAT.	INST.	TOTAL
SYSTEM A1020 160 2220					
WOOD PILES, 25' LONG, 50K LOAD, END BEARING, 3 PILE CLUSTER					
Wood piles, treated, 12" butt, 8" tip, up to 30' long	81.000	V.L.F.	1,701	849.69	2,550.69
Point for driving wood piles	3.000	Ea.	82.50	82.50	165
Pile cutoff, wood piles	3.000	Ea.		43.35	43.35
TOTAL			1,783.50	975.54	2,759.04

A1020 160	Treated Wood Piles		COST EACH		
			MAT.	INST.	TOTAL
2220	Wood piles, 25' long, 50K load, end bearing, 3 pile cluster		1,775	975	2,750
2240	Friction type, 3 pile cluster		1,700	895	2,595
2260	5 pile cluster	RA1020	2,825	1,500	4,325
2280	100K load, end bearing, 4 pile cluster	-100	2,375	1,300	3,675
2300	5 pile cluster		2,975	1,625	4,600
2320	6 pile cluster		3,575	1,950	5,525
2340	Friction type, 5 pile cluster		2,825	1,500	4,325
2360	6 pile cluster		3,400	1,800	5,200
2380	10 pile cluster		5,675	2,975	8,650
2400	200K load, end bearing, 8 pile cluster		4,750	2,600	7,350
2420	10 pile cluster		5,950	3,250	9,200
2440	12 pile cluster		7,125	3,900	11,025
2460	Friction type, 10 pile cluster		5,675	2,975	8,650
2480	400K load, end bearing, 16 pile cluster		9,500	5,200	14,700
2500	20 pile cluster		11,900	6,500	18,400
4520	50' long, 50K load, end bearing, 3 pile cluster		3,800	1,450	5,250
4540	4 pile cluster		5,050	1,950	7,000
4560	Friction type, 2 pile cluster		2,450	880	3,330
4580	3 pile cluster		3,675	1,325	5,000
4600	100K load, end bearing, 5 pile cluster		6,325	2,400	8,725
4620	8 pile cluster		10,100	3,850	13,950
4640	Friction type, 3 pile cluster		3,675	1,325	5,000
4660	5 pile cluster		6,100	2,200	8,300
4680	200K load, end bearing, 9 pile cluster		11,400	4,350	15,750
4700	10 pile cluster		12,600	4,825	17,425
4720	15 pile cluster		18,900	7,225	26,125
4740	Friction type, 5 pile cluster		6,100	2,200	8,300
4760	6 pile cluster		7,325	2,650	9,975

A10 Foundations

A1020 Special Foundations

A1020 160	Treated Wood Piles	COST EACH		
		MAT.	INST.	TOTAL
4780	10 pile cluster	12,200	4,400	16,600
4800	400K load, end bearing, 18 pile cluster	22,700	8,700	31,400
4820	20 pile cluster	25,300	9,675	34,975
4840	Friction type, 9 pile cluster	11,000	3,975	14,975
4860	10 pile cluster	12,200	4,400	16,600
9000	Add for boot for driving tip, each pile	27.50	20.50	48

A10 Foundations

A1020 Special Foundations

The Grade Beam System includes: excavation with a truck mounted backhoe; hand trim; backfill; forms (four uses); reinforcing steel; and 3,000 p.s.i. concrete placed from chute.

Superimposed loads vary in the listing from 1 Kip per linear foot (KLF) and above. In the Expanded System Listing, the span of the beams varies from 15' to 40'. Depth varies from 28" to 52". Width varies from 12" and wider.

Please see the reference section for further design and cost information.

System Components	QUANTITY	UNIT	COST PER L.F.		
			MAT.	INST.	TOTAL
SYSTEM A1020 210 2220					
GRADE BEAM, 15' SPAN, 28" DEEP, 12" WIDE, 8 KLF LOAD					
Excavation, trench, hydraulic backhoe, 3/8 CY bucket	.260	C.Y.		2.46	2.46
Trim sides and bottom of trench, regular soil	2.000	S.F.		1.84	1.84
Backfill, by hand, compaction in 6" layers, using vibrating plate	.170	C.Y.		1.35	1.35
Forms in place, grade beam, 4 uses	4.700	SFCA	4.70	25.15	29.85
Reinforcing in place, beams & girders, #8 to #14	.019	Ton	20.90	17.77	38.67
Concrete ready mix, regular weight, 3000 psi	.090	C.Y.	9.63		9.63
Place and vibrate conc. for grade beam, direct chute	.090	C.Y.		1.67	1.67
TOTAL			35.23	50.24	85.47

A1020 210	Grade Beams		COST PER L.F.		
			MAT.	INST.	TOTAL
2220	Grade beam, 15' span, 28" deep, 12" wide, 8 KLF load		35	50.50	85.50
2240	14" wide, 12 KLF load		36.50	50.50	87
2260	40" deep, 12" wide, 16 KLF load	RA1020	36	62.50	98.50
2280	20 KLF load	-230	41.50	67.50	109
2300	52" deep, 12" wide, 30 KLF load		48	84.50	132.50
2320	40 KLF load		59	94	153
2340	50 KLF load		70	103	173
3360	20' span, 28" deep, 12" wide, 2 KLF load		22	39	61
3380	16" wide, 4 KLF load		29	43.50	72.50
3400	40" deep, 12" wide, 8 KLF load		36	62.50	98.50
3420	12 KLF load		46	70	116
3440	14" wide, 16 KLF load		56	78	134
3460	52" deep, 12" wide, 20 KLF load		60	95	155
3480	14" wide, 30 KLF load		83	112	195
3500	20" wide, 40 KLF load		97	118	215
3520	24" wide, 50 KLF load		120	135	255
4540	30' span, 28" deep, 12" wide, 1 KLF load		23	40	63
4560	14" wide, 2 KLF load		37	55	92
4580	40" deep, 12" wide, 4 KLF load		43.50	66	109.50
4600	18" wide, 8 KLF load		62	80.50	142.50
4620	52" deep, 14" wide, 12 KLF load		79.50	110	189.50
4640	20" wide, 16 KLF load		97	119	216
4660	24" wide, 20 KLF load		120	136	256
4680	36" wide, 30 KLF load		172	170	342
4700	48" wide, 40 KLF load		227	206	433
5720	40' span, 40" deep, 12" wide, 1 KLF load		31.50	59	90.50

A10 Foundations

A1020 Special Foundations

A1020 210	Grade Beams	COST PER L.F.		
		MAT.	INST.	TOTAL
5740	2 KLF load	39.50	65.50	105
5760	52" deep, 12" wide, 4 KLF load	59	94	153
5780	20" wide, 8 KLF load	93.50	116	209.50
5800	28" wide, 12 KLF load	129	141	270
5820	38" wide, 16 KLF load	177	172	349
5840	46" wide, 20 KLF load	231	211	442

A10 Foundations

A1020 Special Foundations

Caisson Systems are listed for three applications: stable ground, wet ground and soft rock. Concrete used is 3,000 p.s.i. placed from chute. Included are a bell at the bottom of the caisson shaft (if applicable) along with required excavation and disposal of excess excavated material up to two miles from job site.

The Expanded System lists cost per caisson. End-bearing loads vary from 200 Kips to 3,200 Kips. The dimensions of the caissons range from 2' x 50' to 7' x 200'.

Please see the reference section for further design and cost information.

System Components	QUANTITY	UNIT	COST EACH MAT.	COST EACH INST.	COST EACH TOTAL
SYSTEM A1020 310 2200					
CAISSON, STABLE GROUND, 3000 PSI CONC., 10 KSF BRNG, 200K LOAD, 2'X50'					
Caissons, drilled, to 50', 24" shaft diameter, .116 C.Y./L.F.	50.000	V.L.F.	777.50	1,470	2,247.50
Reinforcing in place, columns, #3 to #7	.060	Ton	66	100.50	166.50
4' bell diameter, 24" shaft, 0.444 C.Y.	1.000	Ea.	47.50	280	327.50
Load & haul excess excavation, 2 miles	6.240	C.Y.		42.43	42.43
TOTAL			891	1,892.93	2,783.93

A1020 310	Caissons	MAT.	INST.	TOTAL
2200	Caisson, stable ground, 3000 PSI conc, 10 KSF brng, 200K load, 2'-0"x50'-0	890	1,900	2,790
2400	400K load, 2'-6"x50'-0"	1,475	3,025	4,500
2600	800K load, 3'-0"x100'-0"	4,025	7,150	11,175
2800	1200K load, 4'-0"x100'-0"	6,975	9,050	16,025
3000	1600K load, 5'-0"x150'-0"	15,600	14,100	29,700
3200	2400K load, 6'-0"x150'-0"	22,900	18,400	41,300
3400	3200K load, 7'-0"x200'-0"	41,100	26,800	67,900
5000	Wet ground, 3000 PSI conc., 10 KSF brng, 200K load, 2'-0"x50'-0"	775	2,700	3,475
5200	400K load, 2'-6"x50'-0"	1,300	4,625	5,925
5400	800K load, 3'-0"x100'-0"	3,500	12,400	15,900
5600	1200K load, 4'-0"x100'-0"	6,025	18,400	24,425
5800	1600K load, 5'-0"x150'-0"	13,400	39,700	53,100
6000	2400K load, 6'-0"x150'-0"	19,700	49,000	68,700
6200	3200K load, 7'-0"x200'-0"	35,300	78,500	113,800
7800	Soft rock, 3000 PSI conc., 10 KSF brng, 200K load, 2'-0"x50'-0"	775	14,600	15,375
8000	400K load, 2'-6"x50'-0"	1,300	23,600	24,900
8200	800K load, 3'-0"x100'-0"	3,500	62,000	65,500
8400	1200K load, 4'-0"x100'-0"	6,025	90,500	96,525
8600	1600K load, 5'-0"x150'-0"	13,400	186,000	199,400
8800	2400K load, 6'-0"x150'-0"	19,700	221,500	241,200
9000	3200K load, 7'-0"x200'-0"	35,300	352,500	387,800

RA1020-200

A10 Foundations

A1020 Special Foundations

Pressure Injected Piles are usually uncased up to 25' and cased over 25' depending on soil conditions.

These costs include excavation and hauling of excess materials; steel casing over 25'; reinforcement; 3,000 p.s.i. concrete; plus mobilization and demobilization of equipment for a distance of up to fifty miles to and from the job site.

The Expanded System lists cost per cluster of piles. Clusters range from one pile to eight piles. End-bearing loads range from 50 Kips to 1,600 Kips.

Please see the reference section for further design and cost information.

System Components	QUANTITY	UNIT	COST EACH		
			MAT.	INST.	TOTAL
SYSTEM A1020 710 4200					
PRESSURE INJECTED FOOTING, END BEARING, 50' LONG, 50K LOAD, 1 PILE					
Pressure injected footings, cased, 30-60 ton cap., 12" diameter	50.000	V.L.F.	820	1,390	2,210
Pile cutoff, concrete pile with thin steel shell	1.000	Ea.		14.45	14.45
TOTAL			820	1,404.45	2,224.45

A1020 710	Pressure Injected Footings	COST EACH		
		MAT.	INST.	TOTAL
2200	Pressure injected footing, end bearing, 25' long, 50K load, 1 pile	380	980	1,360
2400	100K load, 1 pile	380	980	1,360
2600	2 pile cluster	765	1,950	2,715
2800	200K load, 2 pile cluster	765	1,950	2,715
3200	400K load, 4 pile cluster	1,525	3,925	5,450
3400	7 pile cluster	2,675	6,875	9,550
3800	1200K load, 6 pile cluster	3,225	7,475	10,700
4000	1600K load, 7 pile cluster	3,775	8,750	12,525
4200	50' long, 50K load, 1 pile	820	1,400	2,220
4400	100K load, 1 pile	1,425	1,400	2,825
4600	2 pile cluster	2,850	2,825	5,675
4800	200K load, 2 pile cluster	2,850	2,825	5,675
5000	4 pile cluster	5,700	5,625	11,325
5200	400K load, 4 pile cluster	5,700	5,625	11,325
5400	8 pile cluster	11,400	11,200	22,600
5600	800K load, 7 pile cluster	9,975	9,825	19,800
5800	1200K load, 6 pile cluster	9,150	8,425	17,575
6000	1600K load, 7 pile cluster	10,700	9,825	20,525

A10 Foundations

A1030 Slab on Grade

There are four types of Slab on Grade Systems listed: Non-industrial, Light industrial, Industrial and Heavy industrial. Each type is listed two ways: reinforced and non-reinforced. A Slab on Grade system includes three passes with a grader; 6" of compacted gravel fill; polyethylene vapor barrier; 3500 p.s.i. concrete placed by chute; bituminous fibre expansion joint; all necessary edge forms (4 uses); steel trowel finish; and sprayed-on membrane curing compound.

The Expanded System Listing shows costs on a per square foot basis. Thicknesses of the slabs range from 4" and above. Non-industrial applications are for foot traffic only with negligible abrasion. Light industrial applications are for pneumatic wheels and light abrasion. Industrial applications are for solid rubber wheels and moderate abrasion. Heavy industrial applications are for steel wheels and severe abrasion. All slabs are either shown unreinforced or reinforced with welded wire fabric.

System Components	QUANTITY	UNIT	COST PER S.F.		
			MAT.	INST.	TOTAL
SYSTEM A1030 120 2220					
SLAB ON GRADE, 4" THICK, NON INDUSTRIAL, NON REINFORCED					
Fine grade, 3 passes with grader and roller	.110	S.Y.		.50	.50
Gravel under floor slab, 6" deep, compacted	1.000	S.F.	.46	.30	.76
Polyethylene vapor barrier, standard, .006" thick	1.000	S.F.	.04	.15	.19
Concrete ready mix, regular weight, 3500 psi	.012	C.Y.	1.31		1.31
Place and vibrate concrete for slab on grade, 4" thick, direct chute	.012	C.Y.		.30	.30
Expansion joint, premolded bituminous fiber, 1/2" x 6"	.100	L.F.	.10	.33	.43
Edge forms in place for slab on grade to 6" high, 4 uses	.030	L.F.	.01	.10	.11
Cure with sprayed membrane curing compound	1.000	S.F.	.09	.09	.18
Finishing floor, monolithic steel trowel	1.000	S.F.		.88	.88
TOTAL			2.01	2.65	4.66

A1030 120	Plain & Reinforced		COST PER S.F.		
			MAT.	INST.	TOTAL
2220	Slab on grade, 4" thick, non industrial, non reinforced		2.01	2.65	4.66
2240	Reinforced		2.17	3.01	5.18
2260	Light industrial, non reinforced	RA1030 -200	2.59	3.25	5.84
2280	Reinforced		2.75	3.61	6.36
2300	Industrial, non reinforced		3.25	6.75	10
2320	Reinforced		3.41	7.15	10.56
3340	5" thick, non industrial, non reinforced		2.34	2.73	5.07
3360	Reinforced		2.50	3.09	5.59
3380	Light industrial, non reinforced		2.93	3.33	6.26
3400	Reinforced		3.09	3.69	6.78
3420	Heavy industrial, non reinforced		4.22	8.05	12.27
3440	Reinforced		4.32	8.50	12.82
4460	6" thick, non industrial, non reinforced		2.77	2.67	5.44
4480	Reinforced		3	3.16	6.16
4500	Light industrial, non reinforced		3.38	3.27	6.65
4520	Reinforced		3.74	3.93	7.67
4540	Heavy industrial, non reinforced		4.68	8.15	12.83
4560	Reinforced		4.91	8.65	13.56
5580	7" thick, non industrial, non reinforced		3.12	2.76	5.88
5600	Reinforced		3.42	3.28	6.70
5620	Light industrial, non reinforced		3.73	3.36	7.09
5640	Reinforced		4.03	3.88	7.91
5660	Heavy industrial, non reinforced		5	8.05	13.05
5680	Reinforced		5.25	8.55	13.80

A10 Foundations

A1030 Slab on Grade

A1030 120	Plain & Reinforced	COST PER S.F.		
		MAT.	INST.	TOTAL
6700	8" thick, non industrial, non reinforced	3.45	2.81	6.26
6720	Reinforced	3.70	3.25	6.95
6740	Light industrial, non reinforced	4.07	3.41	7.48
6760	Reinforced	4.32	3.85	8.17
6780	Heavy industrial, non reinforced	5.40	8.15	13.55
6800	Reinforced	5.75	8.60	14.35

A20 Basement Construction

A2010 Basement Excavation

Pricing Assumptions: Two-thirds of excavation is by 2-1/2 C.Y. wheel mounted front end loader and one-third by 1-1/2 C.Y. hydraulic excavator.

Two-mile round trip haul by 12 C.Y. tandem trucks is included for excavation wasted and storage of suitable fill from excavated soil. For excavation in clay, all is wasted and the cost of suitable backfill with two-mile haul is included.

Sand and gravel assumes 15% swell and compaction; common earth assumes 25% swell and 15% compaction; clay assumes 40% swell and 15% compaction (non-clay).

In general, the following items are accounted for in the costs in the table below.
1. Excavation for building or other structure to depth and extent indicated.
2. Backfill compacted in place.
3. Haul of excavated waste.
4. Replacement of unsuitable material with bank run gravel.

Note: Additional excavation and fill beyond this line of general excavation for the building (as required for isolated spread footings, strip footings, etc.) are included in the cost of the appropriate component systems.

System Components	QUANTITY	UNIT	COST PER S.F.		
			MAT.	INST.	TOTAL
SYSTEM A2010 110 2280					
EXCAVATE & FILL, 1000 S.F., 8' DEEP, SAND, ON SITE STORAGE					
Excavating bulk shovel, 1.5 C.Y. bucket, 150 cy/hr	.262	C.Y.		.49	.49
Excavation, front end loader, 2-1/2 C.Y.	.523	C.Y.		1.26	1.26
Haul earth, 12 C.Y. dump truck	.341	L.C.Y.		2.29	2.29
Backfill, dozer bulk push 300', including compaction	.562	C.Y.		2.17	2.17
TOTAL				6.21	6.21

A2010 110	Building Excavation & Backfill	COST PER S.F.		
		MAT.	INST.	TOTAL
2220	Excav & fill, 1000 S.F. 4' sand, gravel, or common earth, on site storage		1.02	1.02
2240	Off site storage		1.55	1.55
2260	Clay excavation, bank run gravel borrow for backfill	2.65	2.39	5.04
2280	8' deep, sand, gravel, or common earth, on site storage		6.20	6.20
2300	Off site storage		13.75	13.75
2320	Clay excavation, bank run gravel borrow for backfill	10.75	11.20	21.95
2340	16' deep, sand, gravel, or common earth, on site storage		17	17
2350	Off site storage		34	34
2360	Clay excavation, bank run gravel borrow for backfill	30	29	59
3380	4000 S.F., 4' deep, sand, gravel, or common earth, on site storage		.55	.55
3400	Off site storage		1.09	1.09
3420	Clay excavation, bank run gravel borrow for backfill	1.35	1.20	2.55
3440	8' deep, sand, gravel, or common earth, on site storage		4.39	4.39
3460	Off site storage		7.70	7.70
3480	Clay excavation, bank run gravel borrow for backfill	4.90	6.70	11.60
3500	16' deep, sand, gravel, or common earth, on site storage		10.55	10.55
3520	Off site storage		21	21
3540	Clay, excavation, bank run gravel borrow for backfill	13.10	15.95	29.05
4560	10,000 S.F., 4' deep, sand gravel, or common earth, on site storage		.32	.32
4580	Off site storage		.65	.65
4600	Clay excavation, bank run gravel borrow for backfill	.82	.73	1.55
4620	8' deep, sand, gravel, or common earth, on site storage		3.81	3.81
4640	Off site storage		5.80	5.80
4660	Clay excavation, bank run gravel borrow for backfill	2.97	5.25	8.22
4680	16' deep, sand, gravel, or common earth, on site storage		8.65	8.65
4700	Off site storage		14.75	14.75
4720	Clay excavation, bank run gravel borrow for backfill	7.85	11.95	19.80

A20 Basement Construction

A2010 Basement Excavation

A2010 110	Building Excavation & Backfill	MAT.	INST.	TOTAL
5740	30,000 S.F., 4' deep, sand, gravel, or common earth, on site storage		.18	.18
5760	Off site storage		.38	.38
5780	Clay excavation, bank run gravel borrow for backfill	.48	.43	.91
5800	8' deep, sand & gravel, or common earth, on site storage		3.42	3.42
5820	Off site storage		4.53	4.53
5840	Clay excavation, bank run gravel borrow for backfill	1.67	4.22	5.89
5860	16' deep, sand, gravel, or common earth, on site storage		7.40	7.40
5880	Off site storage		10.75	10.75
5900	Clay excavation, bank run gravel borrow for backfill	4.37	9.30	13.67
6910	100,000 S.F., 4' deep, sand, gravel, or common earth, on site storage		.10	.10
6920	Off site storage		.22	.22
6930	Clay excavation, bank run gravel borrow for backfill	.24	.22	.46
6940	8' deep, sand, gravel, or common earth, on site storage		3.21	3.21
6950	Off site storage		3.80	3.80
6960	Clay excavation, bank run gravel borrow for backfill	.90	3.66	4.56
6970	16' deep, sand, gravel, or common earth, on site storage		6.75	6.75
6980	Off site storage		8.50	8.50
6990	Clay excavation, bank run gravel borrow for backfill	2.36	7.70	10.06

A20 Basement Construction

A2020 Basement Walls

The Foundation Bearing Wall System includes: forms up to 16' high (four uses); 3,000 p.s.i. concrete placed and vibrated; and form removal with breaking form ties and patching walls. The wall systems list walls from 6" to 16" thick and are designed with minimum reinforcement.

Excavation and backfill are not included.

Please see the reference section for further design and cost information.

System Components	QUANTITY	UNIT	COST PER L.F.		
			MAT.	INST.	TOTAL
SYSTEM A2020 110 1500					
FOUNDATION WALL, CAST IN PLACE, DIRECT CHUTE, 4' HIGH, 6" THICK					
Formwork	8.000	SFCA	5.28	43.20	48.48
Reinforcing	3.300	Lb.	1.82	1.39	3.21
Unloading & sorting reinforcing	3.300	Lb.		.08	.08
Concrete, 3,000 psi	.074	C.Y.	7.92		7.92
Place concrete, direct chute	.074	C.Y.		2.28	2.28
Finish walls, break ties and patch voids, one side	4.000	S.F.	.16	3.76	3.92
TOTAL			15.18	50.71	65.89

A2020 110	Walls, Cast in Place							
	WALL HEIGHT (FT.)	PLACING METHOD	CONCRETE (C.Y. per L.F.)	REINFORCING (LBS. per L.F.)	WALL THICKNESS (IN.)	COST PER L.F.		
						MAT.	INST.	TOTAL
1500	4'	direct chute	.074	3.3	6	15.20	50.50	65.70
1520			.099	4.8	8	18.65	52	70.65
1540			.123	6.0	10	22	53	75
1560			.148	7.2	12	25	54	79
1580	RA2020 -210		.173	8.1	14	28.50	55	83.50
1600			.197	9.44	16	31.50	56	87.50
1700	4'	pumped	.074	3.3	6	15.20	51.50	66.70
1720			.099	4.8	8	18.65	54	72.65
1740			.123	6.0	10	22	55	77
1760			.148	7.2	12	25	56	81
1780			.173	8.1	14	28.50	57.50	86
1800			.197	9.44	16	31.50	58.50	90
3000	6'	direct chute	.111	4.95	6	23	76	99
3020			.149	7.20	8	28	78	106
3040			.184	9.00	10	33	79.50	112.50
3060			.222	10.8	12	38	81	119
3080			.260	12.15	14	42.50	82.50	125
3100			.300	14.39	16	48	85	133

A20 Basement Construction

A2020 Basement Walls

A2020 110 Walls, Cast in Place

	WALL HEIGHT (FT.)	PLACING METHOD	CONCRETE (C.Y. per L.F.)	REINFORCING (LBS. per L.F.)	WALL THICKNESS (IN.)	COST PER L.F.		
						MAT.	INST.	TOTAL
3200	6'	pumped	.111	4.95	6	23	78	101
3220			.149	7.20	8	28	80.50	108.50
3240			.184	9.00	10	33	82	115
3260			.222	10.8	12	38	85	123
3280			.260	12.15	14	42.50	86	128.50
3300			.300	14.39	16	48	88	136
5000	8'	direct chute	.148	6.6	6	30.50	101	131.50
5020			.199	9.6	8	37.50	104	141.50
5040			.250	12	10	44	106	150
5060			.296	14.39	12	50.50	108	158.50
5080			.347	16.19	14	52.50	109	161.50
5100			.394	19.19	16	63.50	113	176.50
5200	8'	pumped	.148	6.6	6	30.50	103	133.50
5220			.199	9.6	8	37.50	108	145.50
5240			.250	12	10	44	110	154
5260			.296	14.39	12	50.50	112	162.50
5280			.347	16.19	14	52.50	113	165.50
5300			.394	19.19	16	63.50	118	181.50
6020	10'	direct chute	.248	12	8	46.50	130	176.50
6040			.307	14.99	10	54.50	132	186.50
6060			.370	17.99	12	63	135	198
6080			.433	20.24	14	71	138	209
6100			.493	23.99	16	79.50	141	220.50
6220	10'	pumped	.248	12	8	46.50	134	180.50
6240			.307	14.99	10	54.50	136	190.50
6260			.370	17.99	12	63	141	204
6280			.433	20.24	14	71	143	214
6300			.493	23.99	16	79.50	147	226.50
7220	12'	pumped	.298	14.39	8	56	161	217
7240			.369	17.99	10	65.50	164	229.50
7260			.444	21.59	12	75.50	170	245.50
7280			.52	24.29	14	85.50	172	257.50
7300			.591	28.79	16	95.50	176	271.50
7420	12'	crane & bucket	.298	14.39	8	56	168	224
7440			.369	17.99	10	65.50	172	237.50
7460			.444	21.59	12	75.50	178	253.50
7480			.52	24.29	14	85.50	182	267.50
7500			.591	28.79	16	95.50	190	285.50
8220	14'	pumped	.347	16.79	8	65.50	188	253.50
8240			.43	20.99	10	76.50	191	267.50
8260			.519	25.19	12	88.50	197	285.50
8280			.607	28.33	14	99.50	200	299.50
8300			.69	33.59	16	111	206	317
8420	14'	crane & bucket	.347	16.79	8	65.50	196	261.50
8440			.43	20.99	10	76.50	200	276.50
8460			.519	25.19	12	88.50	208	296.50
8480			.607	28.33	14	99.50	213	312.50
8500			.69	33.59	16	111	220	331
9220	16'	pumped	.397	19.19	8	75	214	289
9240			.492	23.99	10	87.50	219	306.50
9260			.593	28.79	12	101	226	327
9280			.693	32.39	14	114	229	343
9300			.788	38.38	16	127	235	362

A20 Basement Construction

A2020 Basement Walls

A2020 110			Walls, Cast in Place					
	WALL HEIGHT (FT.)	PLACING METHOD	CONCRETE (C.Y. per L.F.)	REINFORCING (LBS. per L.F.)	WALL THICKNESS (IN.)	COST PER L.F.		
						MAT.	INST.	TOTAL
9420	16'	crane & bucket	.397	19.19	8	75	224	299
9440			.492	23.99	10	87.50	229	316.50
9460			.593	28.79	12	101	237	338
9480			.693	32.39	14	114	243	357
9500			.788	38.38	16	127	251	378

A20 Basement Construction

A2020 Basement Walls

A2020 220	Subdrainage Piping	COST PER L.F.		
		MAT.	INST.	TOTAL
2000	Piping, excavation & backfill excluded, PVC, perforated			
2110	3" diameter	1.60	3.92	5.52
2130	4" diameter	1.60	3.92	5.52
2140	5" diameter	3.59	4.20	7.79
2150	6" diameter	3.59	4.20	7.79
3000	Metal alum. or steel, perforated asphalt coated			
3150	6" diameter	7.15	3.87	11.02
3160	8" diameter	9.35	3.97	13.32
3170	10" diameter	11.70	5.30	17
3180	12" diameter	13.10	6.75	19.85
3220	18" diameter	19.65	9.35	29
4000	Corrugated HDPE tubing			
4130	4" diameter	.59	.73	1.32
4150	6" diameter	1.62	.97	2.59
4160	8" diameter	3	1.25	4.25
4180	12" diameter	7.45	4.32	11.77
4200	15" diameter	8.80	4.90	13.70
4220	18" diameter	14.05	6.95	21

B Shell

B10 Superstructure

B1010 Floor Construction

CONCRETE COLUMNS

General: It is desirable for purposes of consistency and simplicity to maintain constant column sizes throughout the building height. To do this, concrete strength may be varied (higher strength concrete at lower stories and lower strength concrete at upper stories), as well as varying the amount of reinforcing.

The first portion of the table provides probable minimum column sizes with related costs and weights per lineal foot of story height for bottom level columns.

The second portion of the table provides costs by column size for top level columns with minimum code reinforcement. Probable maximum loads for these columns are also given.

How to Use Table:

1. Enter the second portion (minimum reinforcing) of the table with the minimum allowable column size from the selected cast in place floor system.

   If the total load on the column does not exceed the allowable working load shown, use the cost per L.F. multiplied by the length of columns required to obtain the column cost.

2. If the total load on the column exceeds the allowable working load shown in the second portion of the table, enter the first portion of the table with the total load on the column and the minimum allowable column size from the selected cast in place floor system.

Select a cost per L.F. for bottom level columns by total load or minimum allowable column size.

Select a cost per L.F. for top level columns using the column size required for bottom level columns from the second portion of the table.

$$\frac{\text{Btm.} + \text{Top Col. Costs/L.F.}}{2} = \text{Avg. Col. Cost/L.F.}$$

Column Cost = Average Col. Cost/L.F. × Length of Cols. Required.

See reference section to determine total loads.

Design and Pricing Assumptions:
Normal wt. concrete, f'c = 4 or 6 KSI, placed by pump.
Steel, fy = 60 KSI, spliced every other level.
Minimum design eccentricity of 0.1t. Assumed load level depth is 8" (weights prorated to full story basis).
Gravity loads only (no frame or lateral loads included).

Please see the reference section for further design and cost information.

System Components	QUANTITY	UNIT	COST PER V.L.F.		
			MAT.	INST.	TOTAL
SYSTEM B1010 201 1050					
ROUND TIED COLUMNS, 4 KSI CONCRETE, 100K MAX. LOAD, 10' STORY, 12" SIZE					
Forms in place, columns, round fiber tube, 12" diam, 1 use	1.000	L.F.	2.86	14	16.86
Reinforcing in place, column ties	1.393	Lb.	4.38	6.46	10.84
Concrete ready mix, regular weight, 4000 psi	.029	C.Y.	3.25		3.25
Placing concrete, incl. vibrating, 12" sq./round columns, pumped	.029	C.Y.		2.22	2.22
Finish, burlap rub w/grout	3.140	S.F.	.13	3.55	3.68
TOTAL			10.62	26.23	36.85

B1010 201			C.I.P. Column - Round Tied					
	LOAD (KIPS)	STORY HEIGHT (FT.)	COLUMN SIZE (IN.)	COLUMN WEIGHT (P.L.F.)	CONCRETE STRENGTH (PSI)	COST PER V.L.F.		
						MAT.	INST.	TOTAL
1050	100	10	12	110	4000	10.60	26.50	37.10
1060		12	12	111	4000	10.85	26.50	37.35
1070	RB1010-112	14	12	112	4000	11.05	27	38.05
1080	150	10	12	110	4000	11.70	28	39.70
1090		12	12	111	4000	11.95	28.50	40.45
1100		14	14	153	4000	14.75	30.50	45.25
1120	200	10	14	150	4000	15.05	31	46.05
1140		12	14	152	4000	15.35	31.50	46.85

B10 Superstructure

B1010 Floor Construction

B1010 201 — C.I.P. Column - Round Tied

	LOAD (KIPS)	STORY HEIGHT (FT.)	COLUMN SIZE (IN.)	COLUMN WEIGHT (P.L.F.)	CONCRETE STRENGTH (PSI)	COST PER V.L.F. MAT.	COST PER V.L.F. INST.	COST PER V.L.F. TOTAL
1160	200	14	14	153	4000	15.65	31.50	47.15
1180	300	10	16	194	4000	18.65	34	52.65
1190		12	18	250	4000	25.50	38.50	64
1200		14	18	252	4000	26	39.50	65.50
1220	400	10	20	306	4000	29.50	44	73.50
1230		12	20	310	4000	30	44.50	74.50
1260		14	20	313	4000	30.50	45.50	76
1280	500	10	22	368	4000	36	49.50	85.50
1300		12	22	372	4000	37	50.50	87.50
1325		14	22	375	4000	37.50	51.50	89
1350	600	10	24	439	4000	41	55	96
1375		12	24	445	4000	42	56	98
1400		14	24	448	4000	43	57.50	100.50
1420	700	10	26	517	4000	51	62	113
1430		12	26	524	4000	52	63.50	115.50
1450		14	26	528	4000	53	65	118
1460	800	10	28	596	4000	57	68	125
1480		12	28	604	4000	58	70	128
1490		14	28	609	4000	59	71.50	130.50
1500	900	10	28	596	4000	60.50	73.50	134
1510		12	28	604	4000	61.50	75	136.50
1520		14	28	609	4000	63	77	140
1530	1000	10	30	687	4000	66	76	142
1540		12	30	695	4000	67	78	145
1620		14	30	701	4000	68.50	80	148.50
1640	100	10	12	110	6000	11.30	26.50	37.80
1660		12	12	111	6000	11.50	26.50	38
1680		14	12	112	6000	11.75	27	38.75
1700	150	10	12	110	6000	12.15	27.50	39.65
1710		12	12	111	6000	12.40	28	40.40
1720		14	12	112	6000	12.60	28.50	41.10
1730	200	10	12	110	6000	12.85	28.50	41.35
1740		12	12	111	6000	13.05	29	42.05
1760		14	12	112	6000	13.25	29	42.25
1780	300	10	14	150	6000	16.25	31.50	47.75
1790		12	14	152	6000	16.55	31.50	48.05
1800		14	16	153	6000	19.15	33.50	52.65
1810	400	10	16	194	6000	20.50	35	55.50
1820		12	16	196	6000	21	35.50	56.50
1830		14	18	252	6000	27	38.50	65.50
1850	500	10	18	247	6000	28	40	68
1870		12	18	250	6000	28.50	41	69.50
1880		14	20	252	6000	31.50	44	75.50
1890	600	10	20	306	6000	32.50	45.50	78
1900		12	20	310	6000	33	46.50	79.50
1905		14	20	313	6000	34	47.50	81.50
1910	700	10	22	368	6000	38.50	49.50	88
1915		12	22	372	6000	39	50.50	89.50
1920		14	22	375	6000	40	51.50	91.50
1925	800	10	24	439	6000	44	55	99
1930		12	24	445	6000	44.50	56	100.50
1935		14	24	448	6000	45.50	57.50	103

B10 Superstructure

B1010 Floor Construction

B1010 201 | C.I.P. Column - Round Tied

	LOAD (KIPS)	STORY HEIGHT (FT.)	COLUMN SIZE (IN.)	COLUMN WEIGHT (P.L.F.)	CONCRETE STRENGTH (PSI)	COST PER V.L.F. MAT.	COST PER V.L.F. INST.	COST PER V.L.F. TOTAL
1940	900	10	24	439	6000	46.50	58.50	105
1945		12	24	445	6000	47.50	60	107.50
1970	1000	10	26	517	6000	55.50	63.50	119
1980		12	26	524	6000	56.50	65	121.50
1995		14	26	528	6000	57.50	66.50	124

B1010 202 | C.I.P. Columns, Round Tied - Minimum Reinforcing

	LOAD (KIPS)	STORY HEIGHT (FT.)	COLUMN SIZE (IN.)	COLUMN WEIGHT (P.L.F.)	CONCRETE STRENGTH (PSI)	COST PER V.L.F. MAT.	COST PER V.L.F. INST.	COST PER V.L.F. TOTAL
2500	100	10-14	12	107	4000	9.55	24.50	34.05
2510	200	10-14	16	190	4000	16.70	31	47.70
2520	400	10-14	20	295	4000	26.50	39.50	66
2530	600	10-14	24	425	4000	37	48.50	85.50
2540	800	10-14	28	580	4000	51	59.50	110.50
2550	1100	10-14	32	755	4000	65.50	69	134.50
2560	1400	10-14	36	960	4000	78	81	159
2570								

B10 Superstructure

B1010 Floor Construction

CONCRETE COLUMNS

General: It is desirable for purposes of consistency and simplicity to maintain constant column sizes throughout the building height. To do this, concrete strength may be varied (higher strength concrete at lower stories and lower strength concrete at upper stories), as well as varying the amount of reinforcing.

The first portion of the table provides probable minimum column sizes with related costs and weights per lineal foot of story height for bottom level columns.

The second portion of the table provides costs by column size for top level columns with minimum code reinforcement. Probable maximum loads for these columns are also given.

How to Use Table:

1. Enter the second portion (minimum reinforcing) of the table with the minimum allowable column size from the selected cast in place floor system.

   If the total load on the column does not exceed the allowable working load shown, use the cost per L.F. multiplied by the length of columns required to obtain the column cost.

2. If the total load on the column exceeds the allowable working load shown in the second portion of the table, enter the first portion of the table with the total load on the column and the minimum allowable column size from the selected cast in place floor system.

   Select a cost per L.F. for bottom level columns by total load or minimum allowable column size.

   Select a cost per L.F. for top level columns using the column size required for bottom level columns from the second portion of the table.

$$\frac{\text{Btm. + Top Col. Costs/L.F.}}{2} = \text{Avg. Col. Cost/L.F.}$$

Column Cost = Average Col. Cost/L.F. x Length of Cols. Required.

See reference section in back of book to determine total loads.

Design and Pricing Assumptions:

Normal wt. concrete, f'c = 4 or 6 KSI, placed by pump.
Steel, fy = 60 KSI, spliced every other level.
Minimum design eccentricity of 0.1t.
Assumed load level depth is 8" (weights prorated to full story basis).
Gravity loads only (no frame or lateral loads included).

Please see the reference section for further design and cost information.

System Components	QUANTITY	UNIT	COST PER V.L.F.		
			MAT.	INST.	TOTAL
SYSTEM B1010 203 0640					
SQUARE COLUMNS, 100K LOAD, 10' STORY, 10" SQUARE					
Forms in place, columns, plywood, 10" x 10", 4 uses	3.323	SFCA	3.27	31	34.27
Chamfer strip, wood, 3/4" wide	4.000	L.F.	1.04	4.20	5.24
Reinforcing in place, column ties	1.405	Lb.	3.88	5.71	9.59
Concrete ready mix, regular weight, 4000 psi	.026	C.Y.	2.91		2.91
Placing concrete, incl. vibrating, 12" sq./round columns, pumped	.026	C.Y.		1.98	1.98
Finish, break ties, patch voids, burlap rub w/grout	3.323	S.F.	.13	3.77	3.90
TOTAL			11.23	46.66	57.89

B1010 203			C.I.P. Column, Square Tied				
	LOAD (KIPS)	STORY HEIGHT (FT.)	COLUMN SIZE (IN.)	COLUMN WEIGHT (P.L.F.)	CONCRETE STRENGTH (PSI)	COST PER V.L.F.		
						MAT.	INST.	TOTAL
0640	100	10	10	96	4000	11.25	47	58.25
0680	RB1010 -112	12	10	97	4000	11.40	47	58.40
0700		14	12	142	4000	14.85	57	71.85
0710								

B10 Superstructure

B1010 Floor Construction

B1010 203 C.I.P. Column, Square Tied

	LOAD (KIPS)	STORY HEIGHT (FT.)	COLUMN SIZE (IN.)	COLUMN WEIGHT (P.L.F.)	CONCRETE STRENGTH (PSI)	COST PER V.L.F.		
						MAT.	INST.	TOTAL
0740	150	10	10	96	4000	13.15	49.50	62.65
0780		12	12	142	4000	15.40	58	73.40
0800		14	12	143	4000	15.70	58	73.70
0840	200	10	12	140	4000	16.25	59	75.25
0860		12	12	142	4000	16.50	59.50	76
0900		14	14	196	4000	19.05	66	85.05
0920	300	10	14	192	4000	20	67.50	87.50
0960		12	14	194	4000	20.50	68	88.50
0980		14	16	253	4000	23.50	75.50	99
1020	400	10	16	248	4000	25	77.50	102.50
1060		12	16	251	4000	25.50	78.50	104
1080		14	16	253	4000	26	79	105
1200	500	10	18	315	4000	29.50	86.50	116
1250		12	20	394	4000	35.50	98.50	134
1300		14	20	397	4000	36	100	136
1350	600	10	20	388	4000	37	101	138
1400		12	20	394	4000	38	102	140
1600		14	20	397	4000	38.50	103	141.50
1900	700	10	20	388	4000	42.50	111	153.50
2100		12	22	474	4000	42.50	112	154.50
2300		14	22	478	4000	43.50	113	156.50
2600	800	10	22	388	4000	44.50	115	159.50
2900		12	22	474	4000	45.50	116	161.50
3200		14	22	478	4000	46.50	117	163.50
3400	900	10	24	560	4000	50	125	175
3800		12	24	567	4000	51	127	178
4000		14	24	571	4000	52	128	180
4250	1000	10	24	560	4000	54.50	132	186.50
4500		12	26	667	4000	57.50	138	195.50
4750		14	26	673	4000	58.50	140	198.50
5600	100	10	10	96	6000	11.80	46.50	58.30
5800		12	10	97	6000	12	47	59
6000		14	12	142	6000	15.70	57	72.70
6200	150	10	10	96	6000	13.75	49.50	63.25
6400		12	12	98	6000	16.25	58	74.25
6600		14	12	143	6000	16.55	58	74.55
6800	200	10	12	140	6000	17.10	59	76.10
7000		12	12	142	6000	17.40	59.50	76.90
7100		14	14	196	6000	20	66	86
7300	300	10	14	192	6000	21	67	88
7500		12	14	194	6000	21.50	67.50	89
7600		14	14	196	6000	21.50	68	89.50
7700	400	10	14	192	6000	22.50	69.50	92
7800		12	14	194	6000	23	70	93
7900		14	16	253	6000	25.50	75.50	101
8000	500	10	16	248	6000	26.50	77.50	104
8050		12	16	251	6000	27	78.50	105.50
8100		14	16	253	6000	27.50	79	106.50
8200	600	10	18	315	6000	31	87.50	118.50
8300		12	18	319	6000	31.50	88.50	120
8400		14	18	321	6000	32	89	121

B10 Superstructure

B1010 Floor Construction

B1010 203 C.I.P. Column, Square Tied

	LOAD (KIPS)	STORY HEIGHT (FT.)	COLUMN SIZE (IN.)	COLUMN WEIGHT (P.L.F.)	CONCRETE STRENGTH (PSI)	COST PER V.L.F.		
						MAT.	INST.	TOTAL
8500	700	10	18	315	6000	32.50	90	122.50
8600		12	18	319	6000	33.50	91	124.50
8700		14	18	321	6000	34	92	126
8800	800	10	20	388	6000	38	98.50	136.50
8900		12	20	394	6000	38.50	100	138.50
9000		14	20	397	6000	39.50	101	140.50
9100	900	10	20	388	6000	41	103	144
9300		12	20	394	6000	41.50	104	145.50
9600		14	20	397	6000	42.50	106	148.50
9800	1000	10	22	469	6000	46.50	113	159.50
9840		12	22	474	6000	47.50	115	162.50
9900		14	22	478	6000	48.50	116	164.50

B1010 204 C.I.P. Column, Square Tied-Minimum Reinforcing

	LOAD (KIPS)	STORY HEIGHT (FT.)	COLUMN SIZE (IN.)	COLUMN WEIGHT (P.L.F.)	CONCRETE STRENGTH (PSI)	COST PER V.L.F.		
						MAT.	INST.	TOTAL
9913	150	10-14	12	135	4000	13.45	55	68.45
9918	300	10-14	16	240	4000	21	72	93
9924	500	10-14	20	375	4000	31.50	93	124.50
9930	700	10-14	24	540	4000	43	115	158
9936	1000	10-14	28	740	4000	56.50	141	197.50
9942	1400	10-14	32	965	4000	70	159	229
9948	1800	10-14	36	1220	4000	87	185	272
9954	2300	10-14	40	1505	4000	106	214	320

B10 Superstructure

B1010 Floor Construction

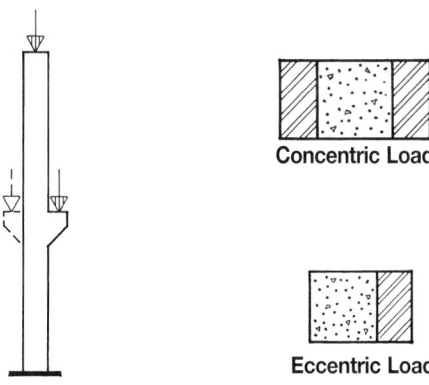

General: Data presented here is for plant produced members transported 50 miles to 100 miles to the site and erected.

Design and pricing assumptions:
Normal wt. concrete, f'c = 5 KSI

Main reinforcement, fy = 60 KSI
Ties, fy = 40 KSI

Minimum design eccentricity, 0.1t.

Concrete encased structural steel haunches are assumed where practical; otherwise galvanized rebar haunches are assumed.

Base plates are integral with columns.

Foundation anchor bolts, nuts and washers are included in price.

System Components	QUANTITY	UNIT	COST PER V.L.F.		
			MAT.	INST.	TOTAL
SYSTEM B1010 206 0560 PRECAST TIED COLUMN, CONCENTRIC LOADING, 100 K MAX. TWO STORY - 10' PER STORY, 5 KSI CONCRETE, 12"X12"					
Precast column, two story-10'/story, 5 KSI conc., 12"x12"	1.000	Ea.	158		158
Anchor bolts	.050	Set	1.70	1.63	3.33
Steel bearing plates; top, bottom, haunches	3.250	Lb.	5.10		5.10
Erection crew	.013	Hr.		12.25	12.25
TOTAL			164.80	13.88	178.68

B1010 206 — Tied, Concentric Loaded Precast Concrete Columns

	LOAD (KIPS)	STORY HEIGHT (FT.)	COLUMN SIZE (IN.)	COLUMN WEIGHT (P.L.F.)	LOAD LEVELS	COST PER V.L.F.		
						MAT.	INST.	TOTAL
0560	100	10	12x12	164	2	165	13.85	178.85
0570		12	12x12	162	2	160	11.55	171.55
0580		14	12x12	161	2	159	11.35	170.35
0590	150	10	12x12	166	3	159	11.25	170.25
0600		12	12x12	169	3	159	10.10	169.10
0610		14	12x12	162	3	159	9.95	168.95
0620	200	10	12x12	168	4	159	12.05	171.05
0630		12	12x12	170	4	159	10.90	169.90
0640		14	12x12	220	4	159	10.80	169.80
0680	300	10	14x14	225	3	216	11.25	227.25
0690		12	14x14	225	3	216	10.10	226.10
0700		14	14x14	250	3	216	9.95	225.95
0710	400	10	16x16	255	4	189	12.05	201.05
0720		12	16x16	295	4	189	10.90	199.90
0750		14	16x16	305	4	152	10.80	162.80
0790	450	10	16x16	320	3	189	11.25	200.25
0800		12	16x16	315	3	189	10.10	199.10
0810		14	16x16	330	3	189	9.95	198.95
0820	600	10	18x18	405	4	239	12.05	251.05
0830		12	18x18	395	4	238	10.90	248.90
0840		14	18x18	410	4	238	10.80	248.80
0910	800	10	20x20	495	4	218	12.05	230.05
0920		12	20x20	505	4	217	10.90	227.90
0930		14	20x20	510	4	217	10.80	227.80

B10 Superstructure

B1010 Floor Construction

B1010 206 Tied, Concentric Loaded Precast Concrete Columns

	LOAD (KIPS)	STORY HEIGHT (FT.)	COLUMN SIZE (IN.)	COLUMN WEIGHT (P.L.F.)	LOAD LEVELS	COST PER V.L.F.		
						MAT.	INST.	TOTAL
0970	900	10	22x22	625	3	264	11.25	275.25
0980		12	22x22	610	3	263	10.10	273.10
0990		14	22x22	605	3	263	9.95	272.95

B1010 207 Tied, Eccentric Loaded Precast Concrete Columns

	LOAD (KIPS)	STORY HEIGHT (FT.)	COLUMN SIZE (IN.)	COLUMN WEIGHT (P.L.F.)	LOAD LEVELS	COST PER V.L.F.		
						MAT.	INST.	TOTAL
1130	100	10	12x12	161	2	157	13.85	170.85
1140		12	12x12	159	2	157	11.55	168.55
1150		14	12x12	159	2	156	11.35	167.35
1160	150	10	12x12	161	3	154	11.25	165.25
1170		12	12x12	160	3	154	10.10	164.10
1180		14	12x12	159	3	154	9.95	163.95
1190	200	10	12x12	161	4	152	12.05	164.05
1200		12	12x12	160	4	152	10.90	162.90
1210		14	12x12	177	4	152	10.80	162.80
1250	300	10	12x12	185	3	209	11.25	220.25
1260		12	14x14	215	3	209	10.10	219.10
1270		14	14x14	215	3	209	9.95	218.95
1280	400	10	14x14	235	4	207	12.05	219.05
1290		12	16x16	285	4	207	10.90	217.90
1300		14	16x16	295	4	208	10.80	218.80
1360	450	10	14x14	245	3	184	11.25	195.25
1370		12	16x16	285	3	184	10.10	194.10
1380		14	16x16	290	3	184	9.95	193.95
1390	600	10	18x18	385	4	230	12.05	242.05
1400		12	18x18	380	4	230	10.90	240.90
1410		14	18x18	375	4	229	10.80	239.80
1480	800	10	20x20	490	4	209	12.05	221.05
1490		12	20x20	480	4	209	10.90	219.90
1500		14	20x20	475	4	210	10.80	220.80

B10 Superstructure

B1010 Floor Construction

(A) Wide Flange
(B) Pipe
(C) Pipe, Concrete Filled
(D) Square Tube
(E) Square Tube Concrete Filled
(F) Rectangular Tube
(G) Rectangular Tube, Concrete Filled

General: The following pages provide data for seven types of steel columns: wide flange, round pipe, round pipe concrete filled, square tube, square tube concrete filled, rectangular tube and rectangular tube concrete filled.

Design Assumptions: Loads are concentric; wide flange and round pipe bearing capacity is for 36 KSI steel. Square and rectangular tubing bearing capacity is for 46 KSI steel.

The effective length factor K=1.1 is used for determining column values in the tables. K=1.1 is within a frequently used range for pinned connections with cross bracing.

How To Use Tables:
a. Steel columns usually extend through two or more stories to minimize splices. Determine floors with splices.
b. Enter Table No. below with load to column at the splice. Use the unsupported height.
c. Determine the column type desired by price or design.

Cost:
a. Multiply number of columns at the desired level by the total height of the column by the cost/VLF.
b. Repeat the above for all tiers.

Please see the reference section for further design and cost information.

B1010 208 — Steel Columns

	LOAD (KIPS)	UNSUPPORTED HEIGHT (FT.)	WEIGHT (P.L.F.)	SIZE (IN.)	TYPE	COST PER V.L.F. MAT.	COST PER V.L.F. INST.	COST PER V.L.F. TOTAL
1000	25	10	13	4	A	22.50	10.65	33.15
1020			7.58	3	B	13.10	10.65	23.75
1040	RB1010-130		15	3-1/2	C	16	10.65	26.65
1060			6.87	3	D	11.90	10.65	22.55
1080			15	3	E	15.45	10.65	26.10
1100			8.15	4x3	F	14.10	10.65	24.75
1120			20	4x3	G	18.45	10.65	29.10
1200		16	16	5	A	25.50	8	33.50
1220			10.79	4	B	17.30	8	25.30
1240			36	5-1/2	C	24	8	32
1260			11.97	5	D	19.15	8	27.15
1280			36	5	E	25.50	8	33.50
1300			11.97	6x4	F	19.15	8	27.15
1320			64	8x6	G	37	8	45
1400		20	20	6	A	30.50	8	38.50
1420			14.62	5	B	22	8	30
1440			49	6-5/8	C	29.50	8	37.50
1460			11.97	5	D	18.15	8	26.15
1480			49	6	E	30	8	38
1500			14.53	7x5	F	22	8	30
1520			64	8x6	G	35	8	43
1600	50	10	16	5	A	27.50	10.65	38.15
1620			14.62	5	B	25.50	10.65	36.15
1640			24	4-1/2	C	19.05	10.65	29.70
1660			12.21	4	D	21	10.65	31.65
1680			25	4	E	21.50	10.65	32.15
1700			11.97	6x4	F	20.50	10.65	31.15
1720			28	6x3	G	24.50	10.65	35.15

B10 Superstructure

B1010 Floor Construction

B1010 208 Steel Columns

	LOAD (KIPS)	UNSUPPORTED HEIGHT (FT.)	WEIGHT (P.L.F.)	SIZE (IN.)	TYPE	COST PER V.L.F.		
						MAT.	INST.	TOTAL
1800	50	16	24	8	A	38.50	8	46.50
1820			18.97	6	B	30.50	8	38.50
1840			36	5-1/2	C	24	8	32
1860			14.63	6	D	23.50	8	31.50
1880			36	5	E	25.50	8	33.50
1900			14.53	7x5	F	23.50	8	31.50
1920			64	8x6	G	37	8	45
1940								
2000		20	28	8	A	42.50	8	50.50
2020			18.97	6	B	29	8	37
2040			49	6-5/8	C	29.50	8	37.50
2060			19.02	6	D	29	8	37
2080			49	6	E	30	8	38
2100			22.42	8x6	F	34	8	42
2120			64	8x6	G	35	8	43
2200	75	10	20	6	A	34.50	10.65	45.15
2220			18.97	6	B	33	10.65	43.65
2240			36	4-1/2	C	48	10.65	58.65
2260			14.53	6	D	25	10.65	35.65
2280			28	4	E	21.50	10.65	32.15
2300			14.33	7x5	F	25	10.65	35.65
2320			35	6x4	G	27.50	10.65	38.15
2400		16	31	8	A	49.50	8	57.50
2420			28.55	8	B	45.50	8	53.50
2440			49	6-5/8	C	31.50	8	39.50
2460			17.08	7	D	27.50	8	35.50
2480			36	5	E	25.50	8	33.50
2500			23.34	7x5	F	37.50	8	45.50
2520			64	8x6	G	37	8	45
2600		20	31	8	A	47	8	55
2620			28.55	8	B	43.50	8	51.50
2640			81	8-5/8	C	45	8	53
2660			22.42	7	D	34	8	42
2680			49	6	E	30	8	38
2700			22.42	8x6	F	34	8	42
2720			64	8x6	G	35	8	43
2800	100	10	24	8	A	41.50	10.65	52.15
2820			28.57	6	B	49.50	10.65	60.15
2840			35	4-1/2	C	48	10.65	58.65
2860			17.08	7	D	29.50	10.65	40.15
2880			36	5	E	27.50	10.65	38.15
2900			19.02	7x5	F	33	10.65	43.65
2920			46	8x4	G	33.50	10.65	44.15
3000		16	31	8	A	49.50	8	57.50
3020			28.55	8	B	45.50	8	53.50
3040			56	6-5/8	C	46.50	8	54.50
3060			22.42	7	D	36	8	44
3080			49	6	E	31.50	8	39.50
3100			22.42	8x6	F	36	8	44
3120			64	8x6	G	37	8	45

B10 Superstructure

B1010 Floor Construction

B1010 208 Steel Columns

	LOAD (KIPS)	UNSUPPORTED HEIGHT (FT.)	WEIGHT (P.L.F.)	SIZE (IN.)	TYPE	COST PER V.L.F.		
						MAT.	INST.	TOTAL
3200	100	20	40	8	A	60.50	8	68.50
3220			28.55	8	B	43.50	8	51.50
3240			81	8-5/8	C	45	8	53
3260			25.82	8	D	39	8	47
3280			66	7	E	35.50	8	43.50
3300			27.59	8x6	F	42	8	50
3320			70	8x6	G	50.50	8	58.50
3400	125	10	31	8	A	53.50	10.65	64.15
3420			28.57	6	B	49.50	10.65	60.15
3440			81	8	C	51	10.65	61.65
3460			22.42	7	D	39	10.65	49.65
3480			49	6	E	34	10.65	44.65
3500			22.42	8x6	F	39	10.65	49.65
3520			64	8x6	G	40	10.65	50.65
3600		16	40	8	A	64	8	72
3620			28.55	8	B	45.50	8	53.50
3640			81	8	C	47.50	8	55.50
3660			25.82	8	D	41.50	8	49.50
3680			66	7	E	37	8	45
3700			27.59	8x6	F	44	8	52
3720			64	8x6	G	37	8	45
3800		20	48	8	A	73	8	81
3820			40.48	10	B	61.50	8	69.50
3840			81	8	C	45	8	53
3860			25.82	8	D	39	8	47
3880			66	7	E	35.50	8	43.50
3900			37.59	10x6	F	57	8	65
3920			60	8x6	G	50.50	8	58.50
4000	150	10	35	8	A	60.50	10.65	71.15
4020			40.48	10	B	70	10.65	80.65
4040			81	8-5/8	C	51	10.65	61.65
4060			25.82	8	D	44.50	10.65	55.15
4080			66	7	E	40	10.65	50.65
4100			27.48	7x5	F	47.50	10.65	58.15
4120			64	8x6	G	40	10.65	50.65
4200		16	45	10	A	72	8	80
4220			40.48	10	B	65	8	73
4240			81	8-5/8	C	47.50	8	55.50
4260			31.84	8	D	51	8	59
4280			66	7	E	37	8	45
4300			37.69	10x6	F	60.50	8	68.50
4320			70	8x6	G	53.50	8	61.50
4400		20	49	10	A	74.50	8	82.50
4420			40.48	10	B	61.50	8	69.50
4440			123	10-3/4	C	64	8	72
4460			31.84	8	D	48.50	8	56.50
4480			82	8	E	41	8	49
4500			37.69	10x6	F	57	8	65
4520			86	10x6	G	50	8	58

B10 Superstructure

B1010 Floor Construction

B1010 208 Steel Columns

	LOAD (KIPS)	UNSUPPORTED HEIGHT (FT.)	WEIGHT (P.L.F.)	SIZE (IN.)	TYPE	COST PER V.L.F.		
						MAT.	INST.	TOTAL
4600	200	10	45	10	A	78	10.65	88.65
4620			40.48	10	B	70	10.65	80.65
4640			81	8-5/8	C	51	10.65	61.65
4660			31.84	8	D	55	10.65	65.65
4680			82	8	E	46.50	10.65	57.15
4700			37.69	10x6	F	65	10.65	75.65
4720			70	8x6	G	57.50	10.65	68.15
4800		16	49	10	A	78.50	8	86.50
4820			49.56	12	B	79.50	8	87.50
4840			123	10-3/4	C	67	8	75
4860			37.60	8	D	60	8	68
4880			90	8	E	62	8	70
4900			42.79	12x6	F	68.50	8	76.50
4920			85	10x6	G	62	8	70
5000		20	58	12	A	88	8	96
5020			49.56	12	B	75	8	83
5040			123	10-3/4	C	64	8	72
5060			40.35	10	D	61	8	69
5080			90	8	E	58.50	8	66.50
5100			47.90	12x8	F	72.50	8	80.50
5120			93	10x6	G	75.50	8	83.50
5200	300	10	61	14	A	106	10.65	116.65
5220			65.42	12	B	113	10.65	123.65
5240			169	12-3/4	C	89	10.65	99.65
5260			47.90	10	D	83	10.65	93.65
5280			90	8	E	66.50	10.65	77.15
5300			47.90	12x8	F	83	10.65	93.65
5320			86	10x6	G	86	10.65	96.65
5400		16	72	12	A	115	8	123
5420			65.42	12	B	105	8	113
5440			169	12-3/4	C	83	8	91
5460			58.10	12	D	93	8	101
5480			135	10	E	79.50	8	87.50
5500			58.10	14x10	F	93	8	101
5600		20	79	12	A	120	8	128
5620			65.42	12	B	99	8	107
5640			169	12-3/4	C	78.50	8	86.50
5660			58.10	12	D	88	8	96
5680			135	10	E	75	8	83
5700			58.10	14x10	F	88	8	96
5800	400	10	79	12	A	137	10.65	147.65
5840			178	12-3/4	C	117	10.65	127.65
5860			68.31	14	D	118	10.65	128.65
5880			135	10	E	85.50	10.65	96.15
5900			62.46	14x10	F	108	10.65	118.65
6000		16	87	12	A	139	8	147
6040			178	12-3/4	C	108	8	116
6060			68.31	14	D	109	8	117
6080			145	10	E	103	8	111
6100			76.07	14x10	F	122	8	130

B10 Superstructure

B1010 Floor Construction

B1010 208 Steel Columns

	LOAD (KIPS)	UNSUPPORTED HEIGHT (FT.)	WEIGHT (P.L.F.)	SIZE (IN.)	TYPE	COST PER V.L.F.		
						MAT.	INST.	TOTAL
6200	400	20	90	14	A	136	8	144
6240			178	12-3/4	C	102	8	110
6260			68.31	14	D	104	8	112
6280			145	10	E	97	8	105
6300			76.07	14x10	F	115	8	123
6400	500	10	99	14	A	171	10.65	181.65
6460			76.07	12	D	132	10.65	142.65
6480			145	10	E	111	10.65	121.65
6500			76.07	14x10	F	132	10.65	142.65
6600		16	109	14	A	175	8	183
6660			89.68	14	D	144	8	152
6700			89.68	16x12	F	144	8	152
6800		20	120	12	A	182	8	190
6860			89.68	14	D	136	8	144
6900			89.68	16x12	F	136	8	144
7000	600	10	120	12	A	208	10.65	218.65
7060			89.68	14	D	155	10.65	165.65
7100			89.68	16x12	F	155	10.65	165.65
7200		16	132	14	A	211	8	219
7260			103.30	16	D	165	8	173
7400		20	132	14	A	200	8	208
7460			103.30	16	D	157	8	165
7600	700	10	136	12	A	235	10.65	245.65
7660			103.30	16	D	179	10.65	189.65
7800		16	145	14	A	232	8	240
7860			103.30	16	D	165	8	173
8000		20	145	14	A	220	8	228
8060			103.30	16	D	157	8	165
8200	800	10	145	14	A	251	10.65	261.65
8300		16	159	14	A	255	8	263
8400		20	176	14	A	267	8	275
8800	900	10	159	14	A	275	10.65	285.65
8900		16	176	14	A	282	8	290
9000		20	193	14	A	293	8	301
9100	1000	10	176	14	A	305	10.65	315.65
9200		16	193	14	A	310	8	318
9300		20	211	14	A	320	8	328

B10 Superstructure

B1010 Floor Construction

Description: Table below lists costs of columns per S.F. of bay for wood columns of various sizes and unsupported heights and the maximum allowable total load per S.F. by bay size.

Design Assumptions: Columns are concentrically loaded and are not subject to bending.

Fiber stress (f) is 1200 psi maximum.

Modulus of elasticity is 1,760,000. Use table to factor load capacity figures for modulus of elasticity other than 1,760,000.

The cost of columns per S.F. of exterior bay is proportional to the area supported. For exterior bays, multiply the costs below by two. For corner bays, multiply the cost by four.

Modulus of Elasticity	Factor
1,210,000 psi	0.69
1,320,000 psi	0.75
1,430,000 psi	0.81
1,540,000 psi	0.87
1,650,000 psi	0.94
1,760,000 psi	1.00

B1010 210 — Wood Columns

	NOMINAL COLUMN SIZE (IN.)	BAY SIZE (FT.)	UNSUPPORTED HEIGHT (FT.)	MATERIAL (BF per M.S.F.)	TOTAL LOAD (P.S.F.)	COST PER S.F. MAT.	COST PER S.F. INST.	COST PER S.F. TOTAL
1000	4 x 4	10 x 8	8	133	100	.18	.25	.43
1050			10	167	60	.23	.31	.54
1100			12	200	40	.27	.37	.64
1150			14	233	30	.31	.43	.74
1200		10 x 10	8	106	80	.14	.20	.34
1250			10	133	50	.18	.25	.43
1300			12	160	30	.22	.30	.52
1350			14	187	20	.25	.35	.60
1400		10 x 15	8	71	50	.10	.13	.23
1450			10	88	30	.12	.16	.28
1500			12	107	20	.14	.20	.34
1550			14	124	15	.17	.23	.40
1600		15 x 15	8	47	30	.06	.09	.15
1650			10	59	15	.08	.11	.19
1800		15 x 20	8	35	15	.05	.06	.11
1900								
2000	6 x 6	10 x 15	8	160	230	.23	.27	.50
2050			10	200	210	.29	.34	.63
2100			12	240	140	.34	.41	.75
2150			14	280	100	.40	.48	.88
2200		15 x 15	8	107	150	.15	.18	.33
2250			10	133	140	.19	.23	.42
2300			12	160	90	.23	.27	.50
2350			14	187	60	.27	.32	.59
2400		15 x 20	8	80	110	.11	.14	.25
2450			10	100	100	.14	.17	.31
2500			12	120	70	.17	.20	.37
2550			14	140	50	.20	.24	.44
2600		20 x 20	8	60	80	.09	.10	.19
2650			10	75	70	.11	.13	.24
2700			12	90	50	.13	.15	.28
2750			14	105	30	.15	.18	.33
2800		20 x 25	8	48	60	.07	.08	.15
2850			10	60	50	.09	.10	.19
2900			12	72	40	.10	.12	.22
2950			14	84	20	.12	.14	.26

B10 Superstructure

B1010 Floor Construction

B1010 210 Wood Columns

	NOMINAL COLUMN SIZE (IN.)	BAY SIZE (FT.)	UNSUPPORTED HEIGHT (FT.)	MATERIAL (BF per M.S.F.)	TOTAL LOAD (P.S.F.)	COST PER S.F.		
						MAT.	INST.	TOTAL
3000	6 x 6	25 x 25	8	38	50	.05	.06	.11
3050			10	48	40	.07	.08	.15
3100			12	58	30	.08	.10	.18
3150			14	67	20	.10	.11	.21
3400	8 x 8	20 x 20	8	107	160	.16	.17	.33
3450			10	133	160	.20	.21	.41
3500			12	160	160	.24	.25	.49
3550			14	187	140	.25	.35	.60
3600		20 x 25	8	85	130	.11	.16	.27
3650			10	107	130	.14	.20	.34
3700			12	128	130	.20	.19	.39
3750			14	149	110	.24	.22	.46
3800		25 x 25	8	68	100	.11	.10	.21
3850			10	85	100	.14	.13	.27
3900			12	102	100	.16	.15	.31
3950			14	119	90	.19	.18	.37
4200	10 x 10	20 x 25	8	133	210	.21	.20	.41
4250			10	167	210	.27	.25	.52
4300			12	200	210	.32	.30	.62
4350			14	233	210	.37	.34	.71
4400		25 x 25	8	107	160	.17	.16	.33
4450			10	133	160	.21	.20	.41
4500			12	160	160	.26	.24	.50
4550			14	187	160	.30	.28	.58
4700	12 x 12	20 x 25	8	192	310	.32	.26	.58
4750			10	240	310	.40	.33	.73
4800			12	288	310	.48	.40	.88
4850			14	336	310	.56	.46	1.02
4900		25 x 25	8	154	240	.26	.21	.47
4950			10	192	240	.32	.26	.58
5000			12	230	240	.39	.32	.71
5050			14	269	240	.45	.37	.82

B10 Superstructure

B1010 Floor Construction

General: Beams priced in the following table are plant produced prestressed members transported to the site and erected.

Pricing Assumptions: Prices are based upon 10,000 S.F. to 20,000 S.F. projects and 50 mile to 100 mile transport.

Normal steel for connections is included in price. Deduct 20% from prices for Southern states. Add 10% to prices for Western states.

Design Assumptions: Normal weight concrete to 150 lbs./C.F. f'c = 5 KSI. Prestressing Steel: 250 KSI straight strand. Non-Prestressing Steel: fy = 60 KSI.

B1010 213					Rectangular Precast Beams			
	SPAN (FT.)	SUPERIMPOSED LOAD (K.L.F.)	SIZE W X D (IN.)	BEAM WEIGHT (P.L.F.)	TOTAL LOAD (K.L.F.)	COST PER L.F.		
						MAT.	INST.	TOTAL
2200	15	2.32	12x16	200	2.52	154	19.45	173.45
2250		3.80	12x20	250	4.05	160	19.45	179.45
2300		5.60	12x24	300	5.90	165	20	185
2350		7.65	12x28	350	8.00	171	20	191
2400		5.85	18x20	375	6.73	171	20	191
2450		8.26	18x24	450	8.71	176	22	198
2500		11.39	18x28	525	11.91	187	22	209
2550		18.90	18x36	675	19.58	198	22	220
2600		10.78	24x24	600	11.38	193	22	215
2700		15.12	24x28	700	15.82	204	22	226
2750		25.23	24x36	900	26.13	220	23	243
2800	20	1.22	12x16	200	1.44	154	14.60	168.60
2850		2.03	12x20	250	2.28	160	15.15	175.15
2900		3.02	12x24	300	3.32	165	15.15	180.15
2950		4.15	12x28	350	4.50	171	15.15	186.15
3000		6.85	12x36	450	7.30	178	16.35	194.35
3050		3.13	18x20	375	3.50	171	15.15	186.15
3100		4.45	18x24	450	4.90	176	16.35	192.35
3150		6.18	18x28	525	6.70	187	16.35	203.35
3200		10.33	18x36	675	11.00	198	17	215
3400		15.48	18x44	825	16.30	215	18.60	233.60
3450		21.63	18x52	975	22.60	226	18.60	244.60
3500		5.80	24x24	600	6.40	193	16.35	209.35
3600		8.20	24x28	700	8.90	204	17	221
3700		13.80	24x36	900	14.70	220	18.60	238.60
3800		20.70	24x44	1100	21.80	237	20.50	257.50
3850		29.20	24x52	1300	30.50	259	25.50	284.50
3900	25	1.82	12x24	300	2.12	165	12.15	177.15
4000		2.53	12x28	350	2.88	171	13.10	184.10
4050		5.18	12x36	450	5.63	178	13.10	191.10
4100		6.55	12x44	550	7.10	187	13.10	200.10
4150		9.08	12x52	650	9.73	198	13.60	211.60
4200		1.86	18x20	375	2.24	171	13.10	184.10
4250		2.69	18x24	450	3.14	176	13.10	189.10
4300		3.76	18x28	525	4.29	187	13.10	200.10
4350		6.37	18x36	675	7.05	198	14.85	212.85
4400		9.60	18x44	872	10.43	215	16.35	231.35
4450		13.49	18x52	975	14.47	226	16.35	242.35

B10 Superstructure

B1010 Floor Construction

B1010 213 — Rectangular Precast Beams

	SPAN (FT.)	SUPERIMPOSED LOAD (K.L.F.)	SIZE W X D (IN.)	BEAM WEIGHT (P.L.F.)	TOTAL LOAD (K.L.F.)	COST PER L.F.		
						MAT.	INST.	TOTAL
4500	25	18.40	18x60	1125	19.53	242	18.15	260.15
4600		3.50	24x24	600	4.10	193	14.85	207.85
4700		5.00	24x28	700	5.70	204	14.85	218.85
4800		8.50	24x36	900	9.40	220	16.35	236.35
4900		12.85	24x44	1100	13.95	237	18.15	255.15
5000		18.22	24x52	1300	19.52	259	20.50	279.50
5050		24.48	24x60	1500	26.00	275	23.50	298.50
5100	30	1.65	12x28	350	2.00	171	11.35	182.35
5150		2.79	12x36	450	3.24	178	11.35	189.35
5200		4.38	12x44	550	4.93	187	12.40	199.40
5250		6.10	12x52	650	6.75	198	12.40	210.40
5300		8.36	12x60	750	9.11	209	13.60	222.60
5400		1.72	18x24	450	2.17	176	11.35	187.35
5450		2.45	18x28	525	2.98	187	12.40	199.40
5750		4.21	18x36	675	4.89	198	12.40	210.40
6000		6.42	18x44	825	7.25	215	13.60	228.60
6250		9.07	18x52	975	10.05	226	15.15	241.15
6500		12.43	18x60	1125	13.56	242	17	259
6750		2.24	24x24	600	2.84	193	12.40	205.40
7000		3.26	24x28	700	3.96	204	13.60	217.60
7100		5.63	24x36	900	6.53	220	15.15	235.15
7200	35	8.59	24x44	1100	9.69	237	16.70	253.70
7300		12.25	24x52	1300	13.55	259	16.70	275.70
7400		16.54	24x60	1500	18.04	275	16.70	291.70
7500	40	2.23	12x44	550	2.78	187	10.20	197.20
7600		3.15	12x52	650	3.80	198	10.20	208.20
7700		4.38	12x60	750	5.13	209	11.35	220.35
7800		2.08	18x36	675	2.76	198	11.35	209.35
7900		3.25	18x44	825	4.08	215	11.35	226.35
8000		4.68	18x52	975	5.66	226	12.80	238.80
8100		6.50	18x60	1175	7.63	242	12.80	254.80
8200		2.78	24x36	900	3.60	220	15.75	235.75
8300		4.35	24x44	1100	5.45	237	15.75	252.75
8400		6.35	24x52	1300	7.65	259	17	276
8500		8.65	24x60	1500	10.15	275	17	292
8600	45	2.35	12x52	650	3.00	198	10.10	208.10
8800		3.29	12x60	750	4.04	209	10.10	219.10
9000		2.39	18x44	825	3.22	215	13	228
9020		3.49	18x52	975	4.47	226	13	239
9200		4.90	18x60	1125	6.03	242	15.15	257.15
9250		3.21	24x44	1100	4.31	237	16.55	253.55
9300		4.72	24x52	1300	6.02	259	16.55	275.55
9500		6.52	24x60	1500	8.02	275	16.55	291.55
9600	50	2.64	18x52	975	3.62	226	13.60	239.60
9700		3.75	18x60	1125	4.88	242	16.35	258.35
9750		3.58	24x52	1300	4.88	259	16.35	275.35
9800		5.00	24x60	1500	6.50	275	18.15	293.15
9950	55	3.86	24x60	1500	5.36	275	16.55	291.55

B10 Superstructure

B1010 Floor Construction

General: Beams priced in the following table are plant produced prestressed members transported to the site and erected.

Pricing Assumptions: Prices are based upon 10,000 S.F. to 20,000 S.F. projects and 50 mile to 100 mile transport.

Normal steel for connections is included in price. Deduct 20% from prices for Southern states. Add 10% to prices for Western states.

Design Assumptions: Normal weight concrete to 150 lbs./C.F. f'c = 5 KSI. Prestressing Steel: 250 KSI straight strand. Non-Prestressing Steel: fy = 60 KSI.

B1010 214		"T" Shaped Precast Beams					
	SPAN (FT.)	SUPERIMPOSED LOAD (K.L.F.)	SIZE W X D (IN.)	BEAM WEIGHT (P.L.F.)	TOTAL LOAD (K.L.F.)	COST PER L.F.		
						MAT.	INST.	TOTAL
2300	15	2.8	12x16	260	3.06	176	20	196
2350		4.33	12x20	355	4.69	187	20	207
2400		6.17	12x24	445	6.62	198	22	220
2500		8.37	12x28	515	8.89	209	22	231
2550		13.54	12x36	680	14.22	231	22	253
2600		8.83	18x24	595	9.43	215	22	237
2650		12.11	18x28	690	12.8	226	22	248
2700		19.72	18x36	905	24.63	253	23	276
2800		11.52	24x24	745	12.27	226	22	248
2900		15.85	24x28	865	16.72	242	23	265
3000		26.07	24x36	1130	27.20	275	25	300
3100	20	1.46	12x16	260	1.72	176	15.15	191.15
3200		2.28	12x20	355	2.64	187	15.15	202.15
3300		3.28	12x24	445	3.73	198	16.35	214.35
3400		4.49	12x28	515	5.00	209	16.35	225.35
3500		7.32	12x36	680	8.00	231	17	248
3600		11.26	12x44	840	12.10	242	18.60	260.60
3700		4.70	18x24	595	5.30	215	16.35	231.35
3800		6.51	18x28	690	7.20	226	17	243
3900		10.7	18x36	905	11.61	253	18.60	271.60
4300		16.19	18x44	1115	17.31	281	20.50	301.50
4400		22.77	18x52	1330	24.10	305	20.50	325.50
4500		6.15	24x24	745	6.90	226	17	243
4600		8.54	24x28	865	9.41	242	17	259
4700		14.17	24x36	1130	15.30	275	20.50	295.50
4800		21.41	24x44	1390	22.80	305	25.50	330.50
4900		30.25	24x52	1655	31.91	335	25.50	360.50

B10 Superstructure

B1010 Floor Construction

B1010 214 — "T" Shaped Precast Beams

	SPAN (FT.)	SUPERIMPOSED LOAD (K.L.F.)	SIZE W X D (IN.)	BEAM WEIGHT (P.L.F.)	TOTAL LOAD (K.L.F.)	COST PER L.F. MAT.	COST PER L.F. INST.	COST PER L.F. TOTAL
5000	25	2.68	12x28	515	3.2	209	13.10	222.10
5050		4.44	12x36	680	5.12	231	14.85	245.85
5100		6.90	12x44	840	7.74	242	14.85	256.85
5200		9.75	12x52	1005	10.76	286	15.60	301.60
5300		13.43	12x60	1165	14.60	335	16.35	351.35
5350		3.92	18x28	690	4.61	226	14.85	240.85
5400		6.52	18x36	905	7.43	253	15.60	268.60
5500		9.96	18x44	1115	11.08	281	17.20	298.20
5600		14.09	18x52	1330	15.42	305	18.15	323.15
5650		19.39	18x60	1540	20.93	330	20.50	350.50
5700		3.67	24x24	745	4.42	226	14.85	240.85
5750		5.15	24x28	865	6.02	242	16.35	258.35
5800		8.66	24x36	1130	9.79	275	18.15	293.15
5850		13.20	24x44	1390	14.59	305	20.50	325.50
5900		18.76	24x52	1655	20.42	335	22	357
5950		25.35	24x60	1916	27.27	365	23.50	388.50
6000	30	2.88	12x36	680	3.56	231	12.40	243.40
6100		4.54	12x44	840	5.38	242	13.60	255.60
6200		6.46	12x52	1005	7.47	286	13.60	299.60
6250		8.97	12x60	1165	10.14	335	15.15	350.15
6300		4.25	18x36	905	5.16	253	13.60	266.60
6350		6.57	18x44	1115	7.69	281	15.15	296.15
6400		9.38	18x52	1330	10.71	305	17	322
6500		13.00	18x60	1540	14.54	330	19.45	349.45
6700		3.31	24x28	865	4.18	242	15.15	257.15
6750		5.67	24x36	1130	6.80	275	17	292
6800		8.74	24x44	1390	10.13	305	18.15	323.15
6850		12.52	24x52	1655	14.18	335	19.45	354.45
6900		17.00	24x60	1215	18.92	365	21	386
7000	35	3.11	12x44	840	3.95	242	11.65	253.65
7100		4.48	12x52	1005	5.49	286	12.30	298.30
7200		6.28	12x60	1165	7.45	335	13.75	348.75
7300		4.53	18x44	1115	5.65	281	13.75	294.75
7500		6.54	18x52	1330	7.87	305	15.60	320.60
7600		9.14	18x60	1540	10.68	330	16.70	346.70
7700		3.87	24x36	1130	5.00	275	16.70	291.70
7800		6.05	24x44	1390	7.44	305	17.95	322.95
7900		8.76	24x52	1655	10.42	335	17.95	352.95
8000		12.00	24x60	1915	13.92	365	19.45	384.45
8100	40	3.19	12x52	1005	4.2	286	11.35	297.35
8200		4.53	12x60	1165	5.7	335	12.80	347.80
8300		4.70	18x52	1330	6.03	305	14.60	319.60
8400		6.64	18x60	1540	8.18	330	14.60	344.60
8600		4.31	24x44	1390	5.7	305	17	322
8700		6.32	24x52	1655	7.98	335	18.60	353.60
8800		8.74	24x60	1915	10.66	365	18.60	383.60
8900	45	3.34	12x60	1165	4.51	335	13	348
9000		4.92	18x60	1540	6.46	330	15.15	345.15
9250		4.64	24x52	1655	6.30	335	18.15	353.15
9900		6.5	24x60	1915	8.42	365	18.15	383.15
9950	50	4.9	24x60	1915	6.82	365	20.50	385.50

B10 Superstructure

B1010 Floor Construction

General: Beams priced in the following table are plant produced prestressed members transported to the site and erected.

Pricing Assumptions: Prices are based upon 10,000 S.F. to 20,000 S.F. projects and 50 mile to 100 mile transport.

Normal steel for connections is included in price. Deduct 20% from prices for Southern states. Add 10% to prices for Western states.

Design Assumptions: Normal weight concrete to 150 lbs./C.F. f'c = 5 KSI. Prestressing Steel: 250 KSI straight strand. Non-Prestressing Steel: fy = 60 KSI.

B1010 215	SPAN (FT.)	SUPERIMPOSED LOAD (K.L.F.)	SIZE W X D (IN.)	BEAM WEIGHT (P.L.F.)	TOTAL LOAD (K.L.F.)	COST PER L.F.		
						MAT.	INST.	TOTAL
2250	15	2.58	12x16	230	2.81	165	20	185
2300		4.10	12x20	300	4.40	165	20	185
2400		5.92	12x24	370	6.29	182	20	202
2450		8.09	12x28	435	8.53	193	20	213
2500		12.95	12x36	565	13.52	204	22	226
2600		8.55	18x24	520	9.07	198	22	220
2650		11.83	18x28	610	12.44	209	22	231
2700		19.30	18x36	790	20.09	226	22	248
2750		11.24	24x24	670	11.91	209	22	231
2800		15.40	24x28	780	16.18	220	23	243
2850		25.65	24x36	1015	26.67	248	23.50	271.50
2900	20	2.18	12x20	300	2.48	165	15.15	180.15
2950		3.17	12x24	370	3.54	182	15.75	197.75
3000		4.37	12x28	435	4.81	193	15.75	208.75
3100		7.04	12x36	565	7.60	204	17	221
3150		10.80	12x44	695	11.50	220	17.80	237.80
3200		15.08	12x52	825	15.91	237	18.60	255.60
3250		4.58	18x24	520	5.10	198	15.75	213.75
3300		6.39	18x28	610	7.00	209	16.35	225.35
3350		10.51	18x36	790	11.30	226	17	243
3400		15.73	18x44	970	16.70	248	18.60	266.60
3550		22.15	18x52	1150	23.30	264	19.45	283.45
3600		6.03	24x24	370	6.40	209	16.35	225.35
3650		8.32	24x28	780	9.10	220	18.60	238.60
3700		13.98	24x36	1015	15.00	248	19.45	267.45
3800		21.05	24x44	1245	22.30	270	23	293
3900		29.73	24x52	1475	31.21	297	25.50	322.50

"L" Shaped Precast Beams

B10 Superstructure

B1010 Floor Construction

B1010 215 "L" Shaped Precast Beams

	SPAN (FT.)	SUPERIMPOSED LOAD (K.L.F.)	SIZE W X D (IN.)	BEAM WEIGHT (P.L.F.)	TOTAL LOAD (K.L.F.)	COST PER L.F.		
						MAT.	INST.	TOTAL
4000	25	2.64	12x28	435	3.08	193	13.10	206.10
4100		4.30	12x36	565	4.87	204	13.60	217.60
4200		6.65	12x44	695	7.35	220	14.20	234.20
4250		9.35	12x52	875	10.18	237	14.85	251.85
4300		12.81	12x60	950	13.76	253	15.60	268.60
4350		2.74	18x24	570	3.76	198	13.60	211.60
4400		3.67	18x28	610	4.28	209	14.20	223.20
4450		6.44	18x36	790	7.23	226	15.60	241.60
4500		9.72	18x44	970	10.69	248	17.20	265.20
4600		13.76	18x52	1150	14.91	264	17.20	281.20
4700		18.33	18x60	1330	20.16	286	19.20	305.20
4800		3.62	24x24	370	3.99	209	14.85	223.85
4900		5.04	24x28	780	5.82	220	15.60	235.60
5000		8.58	24x36	1015	9.60	248	16.35	264.35
5100		13.00	24x44	1245	14.25	270	19.20	289.20
5200		18.50	24x52	1475	19.98	297	22	319
5300	30	2.80	12x36	565	3.37	204	11.85	215.85
5400		4.42	12x44	695	5.12	220	13	233
5500		6.24	12x52	875	7.07	237	13	250
5600		8.60	12x60	950	9.55	253	14.35	267.35
5700		2.50	18x28	610	3.11	209	13	222
5800		4.23	18x36	790	5.02	226	13	239
5900		6.45	18x44	970	7.42	248	14.35	262.35
6000		9.20	18x52	1150	10.35	264	16	280
6100		12.67	18x60	1330	14.00	286	18.15	304.15
6200		3.26	24x28	780	4.04	220	14.35	234.35
6300		5.65	24x36	1015	6.67	248	16	264
6400		8.66	24x44	1245	9.90	270	17	287
6500	35	3.00	12x44	695	3.70	220	11.65	231.65
6600		4.27	12x52	825	5.20	237	11.65	248.65
6700		6.00	12x60	950	6.95	253	13	266
6800		2.90	18x36	790	3.69	226	11.65	237.65
6900		4.48	18x44	970	5.45	248	13	261
7000		6.45	18x52	1150	7.60	264	13.75	277.75
7100		8.95	18x60	1330	10.28	286	14.60	300.60
7200		3.88	24x36	1015	4.90	248	16.70	264.70
7300		6.03	24x44	1245	7.28	270	17.95	287.95
7400		8.71	24x52	1475	10.19	297	17.95	314.95
7500	40	3.15	12x52	825	3.98	237	10.75	247.75
7600		4.43	12x60	950	5.38	253	12	265
7700		3.20	18x44	970	4.17	248	12.80	260.80
7800		4.68	18x52	1150	5.83	264	13.60	277.60
7900		6.55	18x60	1330	7.88	286	13.60	299.60
8000		4.43	24x44	1245	5.68	270	18.60	288.60
8100		6.33	24x52	1475	7.81	297	18.60	315.60
8200	45	3.30	12x60	950	4.25	253	11.35	264.35
8300		3.45	18x52	1150	4.60	264	14	278
8500		4.44	18x60	1330	5.77	286	16.55	302.55
8750		3.16	24x44	1245	4.41	270	18.15	288.15
9000		4.68	24x52	1475	6.16	297	18.15	315.15
9900	50	3.71	18x60	1330	5.04	286	16.35	302.35
9950		3.82	24x52	1475	5.30	297	18.15	315.15

B10 Superstructure

B1010 Floor Construction

Description: Table below lists the cost per L.F. and total uniform load allowable for various size beams at various spans.

Design Assumptions: Fiber strength (f) is 1,000 PSI. Maximum deflection does not exceed 1/360 the span of the beam. Modulus of elasticity (E) is 1,100,000 psi. Anything less will result in excessive deflection in the longer members so that spans must be reduced or member size increased.

The total loads are in pounds per foot.

The span is in feet and is the unsupported clear span.

The costs are based on net quantities and do not include any allowance for waste or overlap.

The member sizes are from 2" x 6" to 4" x 12" and include one, two and three pieces of each member size to arrive at the costs, spans and loading. If more than a three-piece beam is required, its cost may be calculated directly. A four-piece 2" x 6" would cost two times as much as a two-piece 2" x 6" beam. The maximum span and total load cannot be doubled, however.

B1010 216 — Wood Beams

	BEAM SIZE (IN.)	TOTAL LOAD (P.L.F.) 6' SPAN	TOTAL LOAD (P.L.F.) 10' SPAN	TOTAL LOAD (P.L.F.) 14' SPAN	TOTAL LOAD (P.L.F.) 18' SPAN	COST PER L.F. MAT.	COST PER L.F. INST.	COST PER L.F. TOTAL
1000	2 - 2 x 6	238				1.23	4.25	5.48
1050	3 - 2 x 6	357	RB1010-240			1.85	6.40	8.25
1100	2 - 2 x 8	314				1.73	4.93	6.66
1150	3 - 2 x 8	471	232			2.60	7.40	10
1200	2 - 2 x 10	400	240			2.68	5.05	7.73
1250	3 - 2 x 10	600	360			4.02	7.55	11.57
1300	2 - 2 x 12	487	292	208		2.92	5.50	8.42
1350	3 - 2 x 12	731	438	313		4.38	8.25	12.63
1400	2 - 2 x 14	574	344	246		3.80	5.70	9.50
1450	3 - 2 x 14	861	516	369	243	5.70	8.60	14.30
1550	2 - 3 x 6	397				2.99	4.73	7.72
1600	3 - 3 x 6	595				4.48	7.10	11.58
1650	1 - 3 x 8	261				2.45	2.75	5.20
1700	2 - 3 x 8	523	258			4.90	5.50	10.40
1750	3 - 3 x 8	785	388			7.35	8.25	15.60
1800	1 - 3 x 10	334	200			3.06	3.06	6.12
1850	2 - 3 x 10	668	400			6.15	6.15	12.30
1900	3 - 3 x 10	1,002	600	293		9.20	9.20	18.40
1950	1 - 3 x 12	406	243			3.68	3.30	6.98
2000	2 - 3 x 12	812	487	348		7.35	6.60	13.95
2050	3 - 3 x 12	1,218	731	522	248	11.05	9.90	20.95
2100	1 - 4 x 6	278				2.65	2.75	5.40
2150	2 - 4 x 6	556				5.30	5.50	10.80
2200	1 - 4 x 8	366				3.73	3.27	7
2250	2 - 4 x 8	733	362			7.45	6.55	14
2300	1 - 4 x 10	467	280			4.67	3.67	8.34
2350	2 - 4 x 10	935	560			9.35	7.35	16.70
2400	1 - 4 x 12	568	341			5.60	4	9.60
2450	2 - 4 x 12	1,137	682		232	11.20	8	19.20

B10 Superstructure

B1010 Floor Construction

General: Solid concrete slabs of uniform depth reinforced for flexure in one direction and for temperature and shrinkage in the other direction.

Design and Pricing Assumptions:
Concrete f'c = 3 KSI normal weight placed by pump.
Reinforcement fy = 60 KSI.
Deflection ≤ span/360.
Forms, four uses hung from steel beams plus edge forms.
Steel trowel finish for finish floor and cure.

System Components	QUANTITY	UNIT	COST PER S.F.		
			MAT.	INST.	TOTAL
SYSTEM B1010 217 2000					
6'-0" SINGLE SPAN, 4" SLAB DEPTH, 40 PSF SUPERIMPOSED LOAD					
Forms in place, floor slab forms hung from steel beams, 4 uses	1.000	S.F.	1.85	5.70	7.55
Edge forms to 6" high on elevated slab, 4 uses	.080	L.F.	.07	1.40	1.47
Reinforcing in place, elevated slabs #4 to #7	1.030	Lb.	.59	.44	1.03
Concrete, ready mix, regular weight, 3000 psi	.330	C.F.	1.32		1.32
Place and vibrate concrete, elevated slab less than 6", pumped	.330	C.F.		.52	.52
Finishing floor, monolithic steel trowel finish for finish floor	1.000	S.F.		.88	.88
Curing with sprayed membrane curing compound	1.000	S.F.	.09	.09	.18
TOTAL			3.92	9.03	12.95

B1010 217	Cast in Place Slabs, One Way							
	SLAB DESIGN & SPAN (FT.)	SUPERIMPOSED LOAD (P.S.F.)	THICKNESS (IN.)	TOTAL LOAD (P.S.F.)		COST PER S.F.		
						MAT.	INST.	TOTAL
2000	Single 6	40	4	90	RB1010 -010	3.92	9.05	12.97
2100		75	4	125		3.92	9.05	12.97
2200		125	4	175	RB1010 -100	3.92	9.05	12.97
2300		200	4	250		4.12	9.20	13.32
2500	Single 8	40	4	90		3.87	8	11.87
2600		75	4	125		3.90	8.70	12.60
2700		125	4-1/2	181		4.07	8.75	12.82
2800		200	5	262		4.53	9.05	13.58
3000	Single 10	40	4	90		4.05	8.60	12.65
3100		75	4	125		4.05	8.60	12.65
3200		125	5	188		4.50	8.80	13.30
3300		200	7-1/2	293		6	9.90	15.90
3500	Single 15	40	5-1/2	90		4.86	8.70	13.56
3600		75	6-1/2	156		5.40	9.15	14.55
3700		125	7-1/2	219		5.90	9.40	15.30
3800		200	8-1/2	306		6.40	9.65	16.05
4000	Single 20	40	7-1/2	115		5.85	9.20	15.05
4100		75	9	200		6.40	9.15	15.55
4200		125	10	250		7.05	9.55	16.60
4300		200	10	324		7.60	9.95	17.55

B10 Superstructure

B1010 Floor Construction

B1010 217 — Cast in Place Slabs, One Way

	SLAB DESIGN & SPAN (FT.)	SUPERIMPOSED LOAD (P.S.F.)	THICKNESS (IN.)	TOTAL LOAD (P.S.F.)		COST PER S.F.		
						MAT.	INST.	TOTAL
4500	Multi 6	40	4	90		3.98	8.55	12.53
4600		75	4	125		3.98	8.55	12.53
4700		125	4	175		3.98	8.55	12.53
4800		200	4	250		3.98	8.55	12.53
5000	Multi 8	40	4	90		3.98	8.35	12.33
5100		75	4	125		3.98	8.35	12.33
5200		125	4	175		3.98	8.35	12.33
5300		200	4	250		3.98	8.35	12.33
5500	Multi 10	40	4	90		3.98	8.20	12.18
5600		75	4	125		3.98	8.25	12.23
5700		125	4	175		4.14	8.40	12.54
5800		200	5	263		4.61	8.75	13.36
6600	Multi 15	40	5	103		4.58	8.50	13.08
6800		75	5	138		4.80	8.70	13.50
7000		125	6-1/2	206		5.35	8.95	14.30
7100		200	6-1/2	281		5.55	9.05	14.60
7500	Multi 20	40	5-1/2	109		5.30	8.80	14.10
7600		75	6-1/2	156		5.75	9.10	14.85
7800		125	9	238		7.20	9.90	17.10
8000		200	9	313		7.20	9.90	17.10

B10 Superstructure

B1010 Floor Construction

General: Solid concrete one-way slab cast monolithically with reinforced concrete support beams and girders.

Design and Pricing Assumptions:
Concrete f'c = 3 KSI, normal weight, placed by concrete pump.
Reinforcement, fy = 60 KSI.
Forms, four use.
Finish, steel trowel.
Curing, spray on membrane.
Based on 4 bay x 4 bay structure.

System Components	QUANTITY	UNIT	COST PER S.F.		
			MAT.	INST.	TOTAL
SYSTEM B1010 219 3000					
BM. & SLAB ONE WAY 15' X 15' BAY, 40 PSF S.LOAD, 12" MIN. COL.					
Forms in place, flat plate to 15' high, 4 uses	.858	S.F.	1.09	4.93	6.02
Forms in place, exterior spandrel, 12" wide, 4 uses	.142	SFCA	.15	1.48	1.63
Forms in place, interior beam, 12" wide, 4 uses	.306	SFCA	.38	2.62	3
Reinforcing in place, elevated slabs #4 to #7	1.600	Lb.	.91	.69	1.60
Concrete, ready mix, regular weight, 3000 psi	.410	C.F.	1.62		1.62
Place and vibrate concrete, elevated slab less than 6", pump	.410	C.F.		.63	.63
Finish floor, monolithic steel trowel finish for finish floor	1.000	S.F.		.88	.88
Cure with sprayed membrane curing compound	.010	C.S.F.	.09	.09	.18
TOTAL			4.24	11.32	15.56

B1010 219		Cast in Place Beam & Slab, One Way						
	BAY SIZE (FT.)	SUPERIMPOSED LOAD (P.S.F.)	MINIMUM COL. SIZE (IN.)	SLAB THICKNESS (IN.)	TOTAL LOAD (P.S.F.)	COST PER S.F.		
						MAT.	INST.	TOTAL
3000	15x15	40	12	4	120	4.24	11.30	15.54
3100		75	12	4	138	4.33	11.40	15.73
3200	RB1010-010	125	12	4	188	4.46	11.50	15.96
3300		200	14	4	266	4.76	11.85	16.61
3600	15x20	40	12	4	102	4.36	11.20	15.56
3700		75	12	4	140	4.57	11.55	16.12
3800	RB1010-100	125	14	4	192	4.83	11.95	16.78
3900		200	16	4	272	5.40	12.75	18.15
4200	20x20	40	12	5	115	4.79	10.95	15.74
4300		75	14	5	154	5.20	11.80	17
4400		125	16	5	206	5.40	12.40	17.80
4500		200	18	5	287	6.15	13.30	19.45
5000	20x25	40	12	5-1/2	121	4.97	11	15.97
5100		75	14	5-1/2	160	5.50	11.90	17.40
5200		125	16	5-1/2	215	5.90	12.60	18.50
5300		200	18	5-1/2	294	6.45	13.50	19.95
5500	25x25	40	12	6	129	5.25	10.80	16.05
5600		75	16	6	171	5.75	11.55	17.30
5700		125	18	6	227	6.75	13.25	20
5800		200	2	6	300	7.50	14.20	21.70
6500	25x30	40	14	6-1/2	132	5.35	11	16.35
6600		75	16	6-1/2	172	5.85	11.65	17.50
6700		125	18	6-1/2	231	6.90	13.20	20.10
6800		200	20	6-1/2	312	7.55	14.30	21.85

B10 Superstructure

B1010 Floor Construction

B1010 219 Cast in Place Beam & Slab, One Way

	BAY SIZE (FT.)	SUPERIMPOSED LOAD (P.S.F.)	MINIMUM COL. SIZE (IN.)	SLAB THICKNESS (IN.)	TOTAL LOAD (P.S.F.)	COST PER S.F.		
						MAT.	INST.	TOTAL
7000	30x30	40	14	7-1/2	150	6.25	11.80	18.05
7100		75	18	7-1/2	191	6.95	12.45	19.40
7300		125	20	7-1/2	245	7.40	13.25	20.65
7400		200	24	7-1/2	328	8.25	14.65	22.90
7500	30x35	40	16	8	158	6.60	12.20	18.80
7600		75	18	8	196	7.05	12.60	19.65
7700		125	22	8	254	7.95	13.95	21.90
7800		200	26	8	332	8.70	14.45	23.15
8000	35x35	40	16	9	169	7.40	12.55	19.95
8200		75	20	9	213	8	13.70	21.70
8400		125	24	9	272	8.85	14.20	23.05
8600		200	26	9	355	9.80	15.20	25
9000	35x40	40	18	9	174	7.60	12.80	20.40
9300		75	22	9	214	8.25	13.75	22
9400		125	26	9	273	9	14.35	23.35
9600		200	30	9	355	9.90	15.25	25.15

B10 Superstructure

B1010 Floor Construction

General: Solid concrete two-way slab cast monolithically with reinforced concrete support beams and girders.

Design and Pricing Assumptions:
Concrete f'c = 3 KSI, normal weight, placed by concrete pump.
Reinforcement, fy = 60 KSI.
Forms, four use.
Finish, steel trowel.
Curing, spray on membrane.
Based on 4 bay x 4 bay structure.

System Components	QUANTITY	UNIT	COST PER S.F.		
			MAT.	INST.	TOTAL
SYSTEM B1010 220 2000					
15' X 15' BAY, 40 PSF S. LOAD, 12" MIN. COL					
Forms in place, flat plate to 15' high, 4 uses	.894	S.F.	1.14	5.14	6.28
Forms in place, exterior spandrel, 12" wide, 4 uses	.142	SFCA	.15	1.48	1.63
Forms in place, interior beam, 12" wide, 4 uses	.178	SFCA	.22	1.52	1.74
Reinforcing in place, elevated slabs #4 to #7	1.598	Lb.	.91	.69	1.60
Concrete, ready mix, regular weight, 3000 psi	.437	C.F.	1.73		1.73
Place and vibrate concrete, elevated slab less than 6", pump	.437	C.F.		.68	.68
Finish floor, monolithic steel trowel finish for finish floor	1.000	S.F.		.88	.88
Cure with sprayed membrane curing compound	.010	C.S.F.	.09	.09	.18
TOTAL			4.24	10.48	14.72

B1010 220		Cast in Place Beam & Slab, Two Way					
	BAY SIZE (FT.)	SUPERIMPOSED LOAD (P.S.F.)	MINIMUM COL. SIZE (IN.)	SLAB THICKNESS (IN.)	TOTAL LOAD (P.S.F.)	COST PER S.F.		
						MAT.	INST.	TOTAL
2000	15x15	40	12	4-1/2	106	4.24	10.45	14.69
2200		75	12	4-1/2	143	4.37	10.60	14.97
2250	RB1010-010	125	12	4-1/2	195	4.72	10.85	15.57
2300		200	14	4-1/2	274	5.15	11.55	16.70
2400	15 x 20	40	12	5-1/2	120	4.84	10.40	15.24
2600		75	12	5-1/2	159	5.30	11.15	16.45
2800	RB1010-100	125	14	5-1/2	213	5.75	12.05	17.80
2900		200	16	5-1/2	294	6.30	12.85	19.15
3100	20 x 20	40	12	6	132	5.40	10.60	16
3300		75	14	6	167	5.75	11.60	17.35
3400		125	16	6	220	5.95	12	17.95
3600		200	18	6	301	6.60	12.85	19.45
4000	20 x 25	40	12	7	141	5.80	11.05	16.85
4300		75	14	7	181	6.65	12.10	18.75
4500		125	16	7	236	6.80	12.40	19.20
4700		200	18	7	317	7.50	13.40	20.90
5100	25 x 25	40	12	7-1/2	149	6.05	11.20	17.25
5200		75	16	7-1/2	185	6.60	12	18.60
5300		125	18	7-1/2	250	7.20	13	20.20
5400		200	20	7-1/2	332	8.65	14.05	22.70
6000	25 x 30	40	14	8-1/2	165	6.75	11.20	17.95
6200		75	16	8-1/2	201	7.25	12.20	19.45
6400		125	18	8-1/2	259	7.95	13.15	21.10

B10 Superstructure

B1010 Floor Construction

B1010 220 Cast in Place Beam & Slab, Two Way

	BAY SIZE (FT.)	SUPERIMPOSED LOAD (P.S.F.)	MINIMUM COL. SIZE (IN.)	SLAB THICKNESS (IN.)	TOTAL LOAD (P.S.F.)	COST PER S.F.		
						MAT.	INST.	TOTAL
6600	30 x 30	200	20	8-1/2	341	9	13.95	22.95
7000		40	14	9	170	7.35	12	19.35
7100		75	18	9	212	8.05	12.90	20.95
7300		125	20	9	267	8.70	13.35	22.05
7400		200	24	9	351	9.50	14.10	23.60
7600	30 x 35	40	16	10	188	8.10	12.70	20.80
7700		75	18	10	225	8.60	13.15	21.75
8000		125	22	10	282	9.60	14.15	23.75
8200		200	26	10	371	10.45	14.60	25.05
8500	35 x 35	40	16	10-1/2	193	8.70	12.95	21.65
8600		75	20	10-1/2	233	9.15	13.40	22.55
9000		125	24	10-1/2	287	10.25	14.40	24.65
9100		200	26	10-1/2	370	10.60	15.30	25.90
9300	35 x 40	40	18	11-1/2	208	9.40	13.35	22.75
9500		75	22	11-1/2	247	9.90	13.70	23.60
9800		125	24	11-1/2	302	10.55	14.20	24.75
9900		200	26	11-1/2	383	11.40	15.15	26.55

B10 Superstructure

B1010 Floor Construction

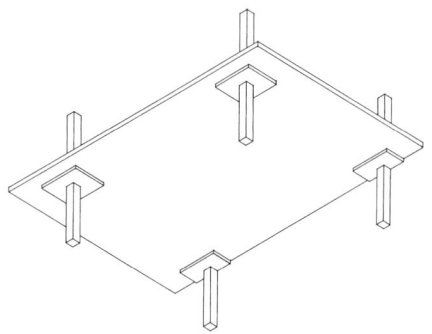

General: Flat Slab: Solid uniform depth concrete two-way slabs with drop panels at columns and no column capitals.

Design and Pricing Assumptions:
Concrete f'c = 3 KSI, placed by concrete pump.
Reinforcement, fy = 60 KSI.
Forms, four use.
Finish, steel trowel.
Curing, spray on membrane.
Based on 4 bay x 4 bay structure.

System Components	QUANTITY	UNIT	COST PER S.F.		
			MAT.	INST.	TOTAL
SYSTEM B1010 222 1700					
15' X 15' BAY, 40 PSF S. LOAD, 12" MIN. COL., 6" SLAB, 1-1/2" DROP, 117 PSF					
Forms in place, flat slab with drop panels, to 15' high, 4 uses	.993	S.F.	1.44	5.91	7.35
Forms in place, exterior spandrel, 12" wide, 4 uses	.034	SFCA	.04	.35	.39
Reinforcing in place, elevated slabs #4 to #7	1.588	Lb.	.91	.68	1.59
Concrete, ready mix, regular weight, 3000 psi	.513	C.F.	2.03		2.03
Place and vibrate concrete, elevated slab, 6" to 10" pump	.513	C.F.		.67	.67
Finish floor, monolithic steel trowel finish for finish floor	1.000	S.F.		.88	.88
Cure with sprayed membrane curing compound	.010	C.S.F.	.09	.09	.18
TOTAL			4.51	8.58	13.09

B1010 222 — Cast in Place Flat Slab with Drop Panels

	BAY SIZE (FT.)	SUPERIMPOSED LOAD (P.S.F.)	MINIMUM COL. SIZE (IN.)	SLAB & DROP (IN.)	TOTAL LOAD (P.S.F.)	COST PER S.F.		
						MAT.	INST.	TOTAL
1700	15 x 15	40	12	6 - 1-1/2	117	4.51	8.55	13.06
1720		75	12	6 - 2-1/2	153	4.61	8.65	13.26
1760	RB1010-010	125	14	6 - 3-1/2	205	4.86	8.80	13.66
1780		200	16	6 - 4-1/2	281	5.15	9	14.15
1840	15 x 20	40	12	6-1/2 - 2	124	4.81	8.70	13.51
1860		75	14	6-1/2 - 4	162	5.05	8.90	13.95
1880	RB1010-100	125	16	6-1/2 - 5	213	5.40	9.10	14.50
1900		200	18	6-1/2 - 6	293	5.55	9.25	14.80
1960	20 x 20	40	12	7 - 3	132	5.10	8.85	13.95
1980		75	16	7 - 4	168	5.40	9.10	14.50
2000		125	18	7 - 6	221	6.05	9.40	15.45
2100		200	20	8 - 6-1/2	309	6.20	9.50	15.70
2300	20 x 25	40	12	8 - 5	147	5.75	9.15	14.90
2400		75	18	8 - 6-1/2	184	6.20	9.50	15.70
2600		125	20	8 - 8	236	6.80	9.85	16.65
2800		200	22	8-1/2 - 8-1/2	323	7.10	10.15	17.25
3200	25 x 25	40	12	8-1/2 - 5-1/2	154	6	9.30	15.30
3400		75	18	8-1/2 - 7	191	6.35	9.55	15.90
4000		125	20	8-1/2 - 8-1/2	243	6.90	9.95	16.85
4400		200	24	9 - 8-1/2	329	7.25	10.20	17.45
5000	25 x 30	40	14	9-1/2 - 7	168	6.60	9.60	16.20
5200		75	18	9-1/2 - 7	203	7.05	10	17.05
5600		125	22	9-1/2 - 8	256	7.40	10.20	17.60
5800		200	24	10 - 10	342	7.90	10.50	18.40

B10 Superstructure

B1010 Floor Construction

B1010 222 — Cast in Place Flat Slab with Drop Panels

	BAY SIZE (FT.)	SUPERIMPOSED LOAD (P.S.F.)	MINIMUM COL. SIZE (IN.)	SLAB & DROP (IN.)	TOTAL LOAD (P.S.F.)	COST PER S.F.		
						MAT.	INST.	TOTAL
6400	30 x 30	40	14	10-1/2 - 7-1/2	182	7.15	9.85	17
6600		75	18	10-1/2 - 7-1/2	217	7.65	10.25	17.90
6800		125	22	10-1/2 - 9	269	8.05	10.50	18.55
7000		200	26	11 - 11	359	8.60	10.95	19.55
7400	30 x 35	40	16	11-1/2 - 9	196	7.80	10.25	18.05
7900		75	20	11-1/2 - 9	231	8.40	10.65	19.05
8000		125	24	11-1/2 - 11	284	8.75	10.95	19.70
9000	35 x 35	40	16	12 - 9	202	8.05	10.35	18.40
9400		75	20	12 - 11	240	8.70	10.80	19.50
9600		125	24	12 - 11	290	9	11	20

B10 Superstructure

B1010 Floor Construction

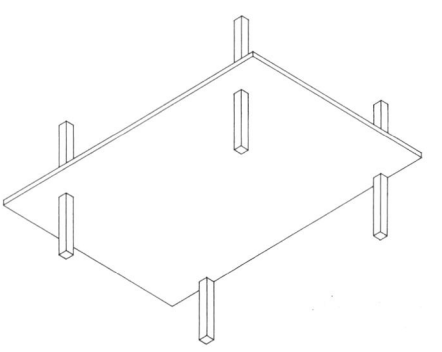

General: Flat Plates: Solid uniform depth concrete two-way slab without drops or interior beams. Primary design limit is shear at columns.

Design and Pricing Assumptions:
Concrete f'c to 4 KSI, placed by concrete pump.
Reinforcement, fy = 60 KSI.
Forms, four use.
Finish, steel trowel.
Curing, spray on membrane.
Based on 4 bay x 4 bay structure.

System Components	QUANTITY	UNIT	COST PER S.F.		
			MAT.	INST.	TOTAL
SYSTEM B1010 223 2000					
15' X 15' BAY, 40 PSF S. LOAD, 12" MIN. COL.					
Forms in place, flat plate to 15' high, 4 uses	.992	S.F.	1.26	5.70	6.96
Edge forms to 6" high on elevated slab, 4 uses	.065	L.F.	.01	.27	.28
Reinforcing in place, elevated slabs #4 to #7	1.706	Lb.	.97	.73	1.70
Concrete, ready mix, regular weight, 3000 psi	.459	C.F.	1.82		1.82
Place and vibrate concrete, elevated slab less than 6", pump	.459	C.F.		.70	.70
Finish floor, monolithic steel trowel finish for finish floor	1.000	S.F.		.88	.88
Cure with sprayed membrane curing compound	.010	C.S.F.	.09	.09	.18
TOTAL			4.15	8.37	12.52

B1010 223			Cast in Place Flat Plate				
	BAY SIZE (FT.)	SUPERIMPOSED LOAD (P.S.F.)	MINIMUM COL. SIZE (IN.)	SLAB THICKNESS (IN.)	TOTAL LOAD (P.S.F.)	COST PER S.F.		
						MAT.	INST.	TOTAL
2000	15 x 15	40	12	5-1/2	109	4.15	8.35	12.50
2200		75	14	5-1/2	144	4.18	8.35	12.53
2400	RB1010-010	125	20	5-1/2	194	4.38	8.45	12.83
2600		175	22	5-1/2	244	4.49	8.50	12.99
3000	15 x 20	40	14	7	127	4.81	8.50	13.31
3400		75	16	7-1/2	169	5.15	8.65	13.80
3600	RB1010-100	125	22	8-1/2	231	5.70	8.90	14.60
3800		175	24	8-1/2	281	5.75	8.90	14.65
4200	20 x 20	40	16	7	127	4.83	8.50	13.33
4400		75	20	7-1/2	175	5.20	8.65	13.85
4600		125	24	8-1/2	231	5.70	8.85	14.55
5000		175	24	8-1/2	281	5.75	8.90	14.65
5600	20 x 25	40	18	8-1/2	146	5.65	8.90	14.55
6000		75	20	9	188	5.85	8.95	14.80
6400		125	26	9-1/2	244	6.35	9.25	15.60
6600		175	30	10	300	6.60	9.35	15.95
7000	25 x 25	40	20	9	152	5.85	8.95	14.80
7400		75	24	9-1/2	194	6.20	9.20	15.40
7600		125	30	10	250	6.65	9.40	16.05
8000								

B10 Superstructure

B1010 Floor Construction

General: Combination of thin concrete slab and monolithic ribs at uniform spacing to reduce dead weight and increase rigidity. The ribs (or joists) are arranged parallel in one direction between supports.

Square end joists simplify forming. Tapered ends can increase span or provide for heavy load.

Costs for multiple span joists are provided in this section. Single span joist costs are not provided here.

Design and Pricing Assumptions:
Concrete f'c = 4 KSI, normal weight placed by concrete pump.
Reinforcement, fy = 60 KSI.
Forms, four use.
  4-1/2" slab.
  30" pans, sq. ends (except for shear req.).
  6" rib thickness.
  Distribution ribs as required.
Finish, steel trowel.
Curing, spray on membrane.
Based on 4 bay x 4 bay structure.

System Components	QUANTITY	UNIT	COST PER S.F.		
			MAT.	INST.	TOTAL
SYSTEM B1010 226 2000					
15' X 15' BAY, 40 PSF S. LOAD, 12" MIN. COLUMN					
Forms in place, floor slab, with 1-way joist pans, 4 use	.905	S.F.	3.06	5.84	8.90
Forms in place, exterior spandrel, 12" wide, 4 uses	.170	SFCA	.18	1.77	1.95
Forms in place, interior beam, 12" wide, 4 uses	.095	SFCA	.12	.81	.93
Edge forms, 7"-12" high on elevated slab, 4 uses	.010	L.F.	.01	.07	.08
Reinforcing in place, elevated slabs #4 to #7	.628	Lb.	.36	.27	.63
Concrete, ready mix, regular weight, 4000 psi	.555	C.F.	2.30		2.30
Place and vibrate concrete, elevated slab, 6" to 10" pump	.555	C.F.		.73	.73
Finish floor, monolithic steel trowel finish for finish floor	1.000	S.F.		.88	.88
Cure with sprayed membrane curing compound	.010	S.F.	.09	.09	.18
TOTAL			6.12	10.46	16.58

B1010 226		Cast in Place Multispan Joist Slab						
	BAY SIZE (FT.)	SUPERIMPOSED LOAD (P.S.F.)	MINIMUM COL. SIZE (IN.)	RIB DEPTH (IN.)	TOTAL LOAD (P.S.F.)	COST PER S.F.		
						MAT.	INST.	TOTAL
2000	15 x 15	40	12	8	115	6.10	10.45	16.55
2100	RB1010 -010	75	12	8	150	6.15	10.45	16.60
2200		125	12	8	200	6.30	10.55	16.85
2300		200	14	8	275	6.45	10.95	17.40
2600	15 x 20	40	12	8	115	6.25	10.40	16.65
2800	RB1010 -100	75	12	8	150	6.35	11.05	17.40
3000		125	14	8	200	6.60	11.20	17.80
3300		200	16	8	275	6.90	11.40	18.30
3600	20 x 20	40	12	10	120	6.40	10.30	16.70
3900		75	14	10	155	6.65	10.90	17.55
4000		125	16	10	205	6.70	11.10	17.80
4100		200	18	10	280	7.05	11.60	18.65
4300	20 x 25	40	12	10	120	6.35	10.45	16.80
4400		75	14	10	155	6.65	11	17.65
4500		125	16	10	205	7.05	11.55	18.60
4600		200	18	12	280	7.35	12.10	19.45
4700	25 x 25	40	12	12	125	6.45	10.20	16.65
4800		75	16	12	160	6.85	10.75	17.60
4900		125	18	12	210	7.60	11.80	19.40
5000		200	20	14	291	8	12.05	20.05

B10 Superstructure

B1010 Floor Construction

B1010 226 — Cast in Place Multispan Joist Slab

	BAY SIZE (FT.)	SUPERIMPOSED LOAD (P.S.F.)	MINIMUM COL. SIZE (IN.)	RIB DEPTH (IN.)	TOTAL LOAD (P.S.F.)	COST PER S.F.		
						MAT.	INST.	TOTAL
5400	25 x 30	40	14	12	125	6.75	10.65	17.40
5600		75	16	12	160	7	11	18
5800		125	18	12	210	7.45	11.75	19.20
6000		200	20	14	291	8.05	12.25	20.30
6200	30 x 30	40	14	14	131	7.10	10.75	17.85
6400		75	18	14	166	7.30	11.10	18.40
6600		125	20	14	216	7.75	11.70	19.45
6700		200	24	16	297	8.30	12.15	20.45
6900	30 x 35	40	16	14	131	7.25	11.25	18.50
7000		75	18	14	166	7.45	11.30	18.75
7100		125	22	14	216	7.60	11.90	19.50
7200		200	26	16	297	8.40	12.55	20.95
7400	35 x 35	40	16	16	137	7.50	11.10	18.60
7500		75	20	16	172	7.90	11.55	19.45
7600		125	24	16	222	7.95	11.55	19.50
7700		200	26	20	309	8.65	12.30	20.95
8000	35 x 40	40	18	16	137	7.70	11.55	19.25
8100		75	22	16	172	8.10	12	20.10
8300		125	26	16	222	8.15	11.85	20
8400		200	30	20	309	8.90	12.45	21.35
8750	40 x 40	40	18	20	149	8.25	11.50	19.75
8800		75	24	20	184	8.55	11.75	20.30
8900		125	26	20	234	8.90	12.20	21.10
9100	40 x 45	40	20	20	149	8.65	11.85	20.50
9500		75	24	20	184	8.70	11.95	20.65
9800		125	28	20	234	9	12.45	21.45

B10 Superstructure

B1010 Floor Construction

General: Waffle slabs are basically flat slabs with hollowed out domes on bottom side to reduce weight. Solid concrete heads at columns function as drops without increasing depth. The concrete ribs function as two-way right angle joist.

Joists are formed with standard sized domes. Thin slabs cover domes and are usually reinforced with welded wire fabric. Ribs have bottom steel and may have stirrups for shear.

Design and Pricing Assumptions:
Concrete f'c = 4 KSI, normal weight placed by concrete pump.
Reinforcement, fy = 60 KSI.
Forms, four use.
  4-1/2" slab.
  30" x 30" voids.
  6" wide ribs.
  (ribs @ 36" O.C.).
  Rib depth filler beams as required.
  Solid concrete heads at columns.
Finish, steel trowel.
Curing, spray on membrane.
Based on 4 bay x 4 bay structure.

System Components	QUANTITY	UNIT	COST PER S.F.		
			MAT.	INST.	TOTAL
SYSTEM B1010 227 3900					
20' X 20' BAY, 40 PSF S. LOAD, 12" MIN. COLUMN					
Forms in place, floor slab with 2-way waffle domes, 4 use	1.000	S.F.	3.52	6.85	10.37
Edge forms, 7"-12" high on elevated slab, 4 uses	.052	SFCA	.03	.34	.37
Forms in place, bulkhead for slab with keyway, 1 use, 3 piece	.010	L.F.	.02	.07	.09
Reinforcing in place, elevated slabs #4 to #7	1.580	Lb.	.90	.68	1.58
Welded wire fabric rolls, 6 x 6 - W4 x W4 (4 x 4) 58 lb./c.s.f	1.000	S.F.	.36	.47	.83
Concrete, ready mix, regular weight, 4000 psi	.690	C.F.	2.86		2.86
Place and vibrate concrete, elevated slab, over 10", pump	.690	C.F.		.95	.95
Finish floor, monolithic steel trowel finish for finish floor	1.000	S.F.		.88	.88
Cure with sprayed membrane curing compound	.010	C.S.F.	.09	.09	.18
TOTAL			7.78	10.33	18.11

B1010 227		Cast in Place Waffle Slab						
	BAY SIZE (FT.)	SUPERIMPOSED LOAD (P.S.F.)	MINIMUM COL. SIZE (IN.)	RIB DEPTH (IN.)	TOTAL LOAD (P.S.F.)	COST PER S.F.		
						MAT.	INST.	TOTAL
3900	20 x 20	40	12	8	144	7.80	10.30	18.10
4000		75	12	8	179	7.95	10.45	18.40
4100	RB1010-010	125	16	8	229	8.10	10.55	18.65
4200		200	18	8	304	8.60	10.95	19.55
4400	20 x 25	40	12	8	146	7.95	10.40	18.35
4500		75	14	8	181	8.15	10.55	18.70
4600	RB1010-100	125	16	8	231	8.35	10.70	19.05
4700		200	18	8	306	8.75	11	19.75
4900	25 x 25	40	12	10	150	8.20	10.50	18.70
5000		75	16	10	185	8.45	10.70	19.15
5300		125	18	10	235	8.70	10.90	19.60
5500		200	20	10	310	9	11.10	20.10
5700	25 x 30	40	14	10	154	8.35	10.60	18.95
5800		75	16	10	189	8.60	10.80	19.40
5900		125	18	10	239	8.85	11	19.85
6000		200	20	12	329	9.75	11.45	21.20
6400	30 x 30	40	14	12	169	8.90	10.80	19.70
6500		75	18	12	204	9.15	11	20.15
6600		125	20	12	254	9.30	11.10	20.40
6700		200	24	12	329	10.15	11.70	21.85

B10 Superstructure

B1010 Floor Construction

B1010 227 Cast in Place Waffle Slab

	BAY SIZE (FT.)	SUPERIMPOSED LOAD (P.S.F.)	MINIMUM COL. SIZE (IN.)	RIB DEPTH (IN.)	TOTAL LOAD (P.S.F.)	COST PER S.F.		
						MAT.	INST.	TOTAL
6900	30 x 35	40	16	12	169	9.10	10.90	20
7000		75	18	12	204	9.10	10.90	20
7100		125	22	12	254	9.60	11.25	20.85
7200		200	26	14	334	10.70	12	22.70
7400	35 x 35	40	16	14	174	9.65	11.20	20.85
7500		75	20	14	209	9.85	11.40	21.25
7600		125	24	14	259	10.10	11.55	21.65
7700		200	26	16	346	10.90	12.15	23.05
8000	35 x 40	40	18	14	176	9.90	11.40	21.30
8300		75	22	14	211	10.25	11.70	21.95
8500		125	26	16	271	10.80	11.95	22.75
8750		200	30	20	372	11.85	12.60	24.45
9200	40 x 40	40	18	14	176	10.25	11.70	21.95
9400		75	24	14	211	10.70	12	22.70
9500		125	26	16	271	10.95	12.10	23.05
9700	40 x 45	40	20	16	186	10.65	11.85	22.50
9800		75	24	16	221	11.10	12.15	23.25
9900		125	28	16	271	11.30	12.30	23.60

B10 Superstructure

B1010 Floor Construction

General: Units priced here are for plant produced prestressed members, transported to site and erected.

Normal weight concrete is most frequently used. Lightweight concrete may be used to reduce dead weight.

Structural topping is sometimes used on floors: insulating concrete or rigid insulation on roofs.

Camber and deflection may limit use by depth considerations.

Prices are based upon 10,000 S.F. to 20,000 S.F. projects, and 50 mile to 100 mile transport.

Concrete is f'c = 5 KSI and Steel is fy = 250 or 300 KSI

Note: Deduct from prices 20% for Southern states. Add to prices 10% for Western states.

Description of Table: Enter table at span and load. Most economical sections will generally consist of normal weight concrete without topping. If acceptable, note this price, depth and weight. For topping and/or lightweight concrete, note appropriate data.

Generally used on masonry and concrete bearing or reinforced concrete and steel framed structures.

The solid 4" slabs are used for light loads and short spans. The 6" to 12" thick hollow core units are used for longer spans and heavier loads. Cores may carry utilities.

Topping is used structurally for loads or rigidity and architecturally to level or slope surface.

Camber and deflection and change in direction of spans must be considered (door openings, etc.), especially untopped.

System Components	QUANTITY	UNIT	COST PER S.F.		
			MAT.	INST.	TOTAL
SYSTEM B1010 230 2000					
10' SPAN, 40 LBS S.F. WORKING LOAD, 2" TOPPING					
Precast prestressed concrete roof/floor slabs 4" thick, grouted	1.000	S.F.	6.65	3.40	10.05
Edge forms to 6" high on elevated slab, 4 uses	.100	L.F.	.02	.42	.44
Welded wire fabric 6 x 6 - W1.4 x W1.4 (10 x 10), 21 lb/csf, 10% lap	.010	C.S.F.	.16	.36	.52
Concrete, ready mix, regular weight, 3000 psi	.170	C.F.	.67		.67
Place and vibrate concrete, elevated slab less than 6", pumped	.170	C.F.		.26	.26
Finishing floor, monolithic steel trowel finish for resilient tile	1.000	S.F.		1.15	1.15
Curing with sprayed membrane curing compound	.010	C.S.F.	.09	.09	.18
TOTAL			7.59	5.68	13.27

B1010 229 — Precast Plank with No Topping

	SPAN (FT.)	SUPERIMPOSED LOAD (P.S.F.)	TOTAL DEPTH (IN.)	DEAD LOAD (P.S.F.)	TOTAL LOAD (P.S.F.)	COST PER S.F.		
						MAT.	INST.	TOTAL
0720	10	40	4	50	90	6.65	3.40	10.05
0750		75	6	50	125	7.15	2.92	10.07
0770		100	6	50	150	7.15	2.92	10.07
0800	15	40	6	50	90	7.15	2.92	10.07
0820		75	6	50	125	7.15	2.92	10.07
0850		100	6	50	150	7.15	2.92	10.07
0875	20	40	6	50	90	7.15	2.92	10.07
0900		75	6	50	125	7.15	2.92	10.07
0920		100	6	50	150	7.15	2.92	10.07
0950	25	40	6	50	90	7.15	2.92	10.07
0970		75	8	55	130	7.50	2.56	10.06
1000		100	8	55	155	7.50	2.56	10.06
1200	30	40	8	55	95	7.50	2.56	10.06
1300		75	8	55	130	7.50	2.56	10.06
1400		100	10	70	170	8.15	2.27	10.42
1500	40	40	10	70	110	8.15	2.27	10.42
1600		75	12	70	145	8.95	2.04	10.99

Note: Cells marked RB1010-010 and RB1010-100 appear in the SUPERIMPOSED LOAD column for certain rows.

B10 Superstructure

B1010 Floor Construction

B1010 229 Precast Plank with No Topping

	SPAN (FT.)	SUPERIMPOSED LOAD (P.S.F.)	TOTAL DEPTH (IN.)	DEAD LOAD (P.S.F.)	TOTAL LOAD (P.S.F.)	COST PER S.F.		
						MAT.	INST.	TOTAL
1700	45	40	12	70	110	8.95	2.04	10.99

B1010 230 Precast Plank with 2″ Concrete Topping

	SPAN (FT.)	SUPERIMPOSED LOAD (P.S.F.)	TOTAL DEPTH (IN.)	DEAD LOAD (P.S.F.)	TOTAL LOAD (P.S.F.)	COST PER S.F.		
						MAT.	INST.	TOTAL
2000	10	40	6	75	115	7.60	5.70	13.30
2100		75	8	75	150	8.10	5.20	13.30
2200		100	8	75	175	8.10	5.20	13.30
2500	15	40	8	75	115	8.10	5.20	13.30
2600		75	8	75	150	8.10	5.20	13.30
2700		100	8	75	175	8.10	5.20	13.30
2800	20	40	8	75	115	8.10	5.20	13.30
2900		75	8	75	150	8.10	5.20	13.30
3000		100	8	75	175	8.10	5.20	13.30
3100	25	40	8	75	115	8.10	5.20	13.30
3200		75	8	75	150	8.10	5.20	13.30
3300		100	10	80	180	8.45	4.84	13.29
3400	30	40	10	80	120	8.45	4.84	13.29
3500		75	10	80	155	8.45	4.84	13.29
3600		100	10	80	180	8.45	4.84	13.29
3700	35	40	12	95	135	9.10	4.55	13.65
3800		75	12	95	170	9.10	4.55	13.65
3900		100	14	95	195	9.90	4.32	14.22
4000	40	40	12	95	135	9.10	4.55	13.65
4500		75	14	95	170	9.90	4.32	14.22
5000	45	40	14	95	135	9.90	4.32	14.22

B10 Superstructure

B1010 Floor Construction

Most widely used for moderate span floors and roofs. At shorter spans, they tend to be competitive with hollow core slabs. They are also used as wall panels.

System Components	QUANTITY	UNIT	COST PER S.F.		
			MAT.	INST.	TOTAL
SYSTEM B1010 235 6700 PRECAST, DOUBLE "T", 2" TOPPING, 30' SPAN, 30 PSF SUP. LOAD, 18" X 8'					
Double "T" beams, reg. wt, 18" x 8' w, 30' span	1.000	S.F.	9.58	1.70	11.28
Edge forms to 6" high on elevated slab, 4 uses	.050	L.F.	.01	.21	.22
Concrete, ready mix, regular weight, 3000 psi	.250	C.F.	.99		.99
Place and vibrate concrete, elevated slab less than 6", pumped	.250	C.F.		.38	.38
Finishing floor, monolithic steel trowel finish for finish floor	1.000	S.F.		.88	.88
Curing with sprayed membrane curing compound	.010	C.S.F.	.09	.09	.18
TOTAL			10.67	3.26	13.93

B1010 234 — Precast Double "T" Beams with No Topping

	SPAN (FT.)	SUPERIMPOSED LOAD (P.S.F.)	DBL. "T" SIZE D (IN.) W (FT.)	CONCRETE "T" TYPE	TOTAL LOAD (P.S.F.)	COST PER S.F.		
						MAT.	INST.	TOTAL
1500	30	30	18x8	Reg. Wt.	92	9.60	1.70	11.30
1600		40	18x8	Reg. Wt.	102	9.75	2.19	11.94
1700	RB1010-010	50	18x8	Reg. Wt	112	9.75	2.19	11.94
1800		75	18x8	Reg. Wt.	137	9.75	2.28	12.03
1900		100	18x8	Reg. Wt.	162	9.75	2.28	12.03
2000	40	30	20x8	Reg. Wt.	87	8.65	1.42	10.07
2100		40	20x8	Reg. Wt.	97	8.75	1.70	10.45
2200	RB1010-100	50	20x8	Reg. Wt.	107	8.75	1.70	10.45
2300		75	20x8	Rcg. Wt.	132	8.80	1.83	10.63
2400		100	20x8	Reg. Wt.	157	8.95	2.23	11.18
2500	50	30	24x8	Reg. Wt.	103	8.70	1.28	9.98
2600		40	24x8	Reg. Wt.	113	8.80	1.56	10.36
2700		50	24x8	Reg. Wt.	123	8.80	1.66	10.46
2800		75	24x8	Reg. Wt.	148	8.80	1.69	10.49
2900		100	24x8	Reg. Wt.	173	8.95	2.08	11.03
3000	60	30	24x8	Reg. Wt.	82	8.80	1.68	10.48
3100		40	32x10	Reg. Wt.	104	8.80	1.37	10.17
3150		50	32x10	Reg. Wt.	114	8.75	1.19	9.94
3200		75	32x10	Reg. Wt.	139	8.75	1.29	10.04
3250		100	32x10	Reg. Wt.	164	8.90	1.61	10.51
3300	70	30	32x10	Reg. Wt.	94	8.75	1.27	10.02
3350		40	32x10	Reg. Wt.	104	8.75	1.28	10.03
3400		50	32x10	Reg. Wt.	114	8.90	1.61	10.51
3450		75	32x10	Reg. Wt.	139	9	1.92	10.92
3500		100	32x10	Reg. Wt.	164	9.20	2.56	11.76
3550	80	30	32x10	Reg. Wt.	94	8.90	1.61	10.51
3600		40	32x10	Reg. Wt.	104	9.10	2.24	11.34
3900		50	32x10	Reg. Wt.	114	9.20	2.55	11.75

B10 Superstructure

B1010 Floor Construction

B1010 234 — Precast Double "T" Beams with No Topping

	SPAN (FT.)	SUPERIMPOSED LOAD (P.S.F.)	DBL. "T" SIZE D (IN.) W (FT.)	CONCRETE "T" TYPE	TOTAL LOAD (P.S.F.)	COST PER S.F.		
						MAT.	INST.	TOTAL
4300	50	30	20x8	Lt. Wt.	66	9.60	1.55	11.15
4400		40	20x8	Lt. Wt.	76	9.65	1.42	11.07
4500		50	20x8	Lt. Wt.	86	9.75	1.70	11.45
4600		75	20x8	Lt. Wt.	111	9.80	1.94	11.74
4700		100	20x8	Lt. Wt.	136	9.95	2.35	12.30
4800	60	30	24x8	Lt. Wt.	70	9.70	1.54	11.24
4900		40	32x10	Lt. Wt.	88	9.60	1.07	10.67
5000		50	32x10	Lt. Wt.	98	9.70	1.30	11
5200		75	32x10	Lt. Wt.	123	9.75	1.49	11.24
5400		100	32x10	Lt. Wt.	148	9.85	1.81	11.66
5600	70	30	32x10	Lt. Wt.	78	9.60	1.07	10.67
5750		40	32x10	Lt. Wt.	88	9.70	1.30	11
5900		50	32x10	Lt. Wt.	98	9.75	1.49	11.24
6000		75	32x10	Lt. Wt.	123	9.85	1.80	11.65
6100		100	32x10	Lt. Wt.	148	10.05	2.44	12.49
6200	80	30	32x10	Lt. Wt.	78	9.75	1.49	11.24
6300		40	32x10	Lt. Wt.	88	9.85	1.80	11.65
6400		50	32x10	Lt. Wt.	98	9.95	2.12	12.07

B1010 235 — Precast Double "T" Beams With 2" Topping

	SPAN (FT.)	SUPERIMPOSED LOAD (P.S.F.)	DBL. "T" SIZE D (IN.) W (FT.)	CONCRETE "T" TYPE	TOTAL LOAD (P.S.F.)	COST PER S.F.		
						MAT.	INST.	TOTAL
6700	30	30	18x8	Reg. Wt.	117	10.65	3.26	13.91
6750		40	18x8	Reg. Wt.	127	10.65	3.26	13.91
6800		50	18x8	Reg. Wt.	137	10.75	3.54	14.29
6900		75	18x8	Reg. Wt.	162	10.75	3.54	14.29
7000		100	18x8	Reg. Wt.	187	10.80	3.68	14.48
7100	40	30	18x8	Reg. Wt.	120	8.35	3.11	11.46
7200		40	20x8	Reg. Wt.	130	9.75	2.98	12.73
7300		50	20x8	Reg. Wt.	140	9.85	3.26	13.11
7400		75	20x8	Reg. Wt.	165	9.90	3.39	13.29
7500		100	20x8	Reg. Wt.	190	10	3.79	13.79
7550	50	30	24x8	Reg. Wt.	120	9.85	3.12	12.97
7600		40	24x8	Reg. Wt.	130	9.90	3.22	13.12
7750		50	24x8	Reg. Wt.	140	9.90	3.25	13.15
7800		75	24x8	Reg. Wt.	165	10.05	3.64	13.69
7900		100	32x10	Reg. Wt.	189	9.90	2.98	12.88
8000	60	30	32x10	Reg. Wt.	118	9.75	2.53	12.28
8100		40	32x10	Reg. Wt.	129	9.85	2.75	12.60
8200		50	32x10	Reg. Wt.	139	9.90	2.98	12.88
8300		75	32x10	Reg. Wt.	164	9.95	3.17	13.12
8350		100	32x10	Reg. Wt.	189	10.05	3.49	13.54
8400	70	30	32x10	Reg. Wt.	119	9.85	2.84	12.69
8450		40	32x10	Reg. Wt.	129	9.95	3.17	13.12
8500		50	32x10	Reg. Wt.	139	10.05	3.48	13.53
8550		75	32x10	Reg. Wt.	164	10.30	4.12	14.42
8600	80	30	32x10	Reg. Wt.	119	10.30	4.11	14.41
8800	50	30	20x8	Lt. Wt.	105	10.75	3.24	13.99
8850		40	24x8	Lt. Wt.	121	10.80	3.39	14.19
8900		50	24x8	Lt. Wt.	131	10.90	3.63	14.53
8950		75	24x8	Lt. Wt.	156	10.90	3.63	14.53
9000		100	24x8	Lt. Wt.	181	11	4.04	15.04

B10 Superstructure

B1010 Floor Construction

B1010 235 Precast Double "T" Beams With 2" Topping

	SPAN (FT.)	SUPERIMPOSED LOAD (P.S.F.)	DBL. "T" SIZE D (IN.) W (FT.)	CONCRETE "T" TYPE	TOTAL LOAD (P.S.F.)	COST PER S.F.		
						MAT.	INST.	TOTAL
9200	60	30	32x10	Lt. Wt.	103	10.70	2.63	13.33
9300		40	32x10	Lt. Wt.	113	10.75	2.86	13.61
9350		50	32x10	Lt. Wt.	123	10.75	2.86	13.61
9400		75	32x10	Lt. Wt.	148	10.85	3.05	13.90
9450		100	32x10	Lt. Wt.	173	10.95	3.37	14.32
9500	70	30	32x10	Lt. Wt.	103	10.85	3.05	13.90
9550		40	32x10	Lt. Wt.	113	10.85	3.05	13.90
9600		50	32x10	Lt. Wt.	123	10.95	3.36	14.31
9650		75	32x10	Lt. Wt.	148	11.05	3.69	14.74
9700	80	30	32x10	Lt. Wt.	103	10.95	3.36	14.31
9800		40	32x10	Lt. Wt.	113	11.05	3.68	14.73
9900		50	32x10	Lt. Wt.	123	11.15	3.99	15.14

B10 Superstructure

B1010 Floor Construction

General: Units priced here are for plant produced prestressed members transported to the site and erected.

System has precast prestressed concrete beams and precast, prestressed hollow core slabs spanning the longer direction when applicable.

Camber and deflection must be considered when using untopped hollow core slabs.

Design and Pricing Assumptions:
Prices are based on 10,000 S.F. to 20,000 S.F. projects and 50 mile to 100 mile transport.

Concrete is f'c 5 KSI and prestressing steel is fy = 250 or 300 KSI.

System Components	QUANTITY	UNIT	COST PER S.F.		
			MAT.	INST.	TOTAL
SYSTEM B1010 236 5200					
20' X 20' BAY, 6" DEPTH, 40 PSF S. LOAD, 110 PSF TOTAL LOAD					
Precast concrete beam, T-shaped, 20' span, 12" x 20"	.038	L.F.	8.98	.72	9.70
Precast concrete beam, L-shaped, 20' span, 12" x 20"	.025	L.F.	5.28	.48	5.76
Precast prestressed concrete roof/floor slabs 6" deep, grouted	1.000	S.F.	7.15	2.92	10.07
TOTAL			21.41	4.12	25.53

B1010 236			Precast Beam & Plank with No Topping					
	BAY SIZE (FT.)	SUPERIMPOSED LOAD (P.S.F.)	PLANK THICKNESS (IN.)	TOTAL DEPTH (IN.)	TOTAL LOAD (P.S.F.)	COST PER S.F.		
						MAT.	INST.	TOTAL
5200	20x20	40	6	20	110	21.50	4.12	25.62
5400		75	6	24	148	16.65	3.71	20.36
5600		100	6	24	173	22	4.19	26.19
5800	20x25	40	8	24	113	19.50	3.60	23.10
6000		75	8	24	149	19.50	3.60	23.10
6100		100	8	36	183	19.80	3.53	23.33
6200	25x25	40	8	28	118	20.50	3.41	23.91
6400		75	8	36	158	20.50	3.41	23.91
6500		100	8	36	183	20.50	3.41	23.91
7000	25x30	40	8	36	110	18.70	3.31	22.01
7200		75	10	36	159	19.60	3.02	22.62
7400		100	12	36	188	20	2.74	22.74
7600	30x30	40	8	36	121	18.95	3.20	22.15
8000		75	10	44	140	18.85	3.21	22.06
8250		100	12	52	206	21.50	2.71	24.21
8500	30x35	40	12	44	135	18.90	2.61	21.51
8750		75	12	52	176	19.85	2.62	22.47
9000	35x35	40	12	52	141	19.85	2.55	22.40
9250		75	12	60	181	20.50	2.55	23.05
9500	35x40	40	12	52	137	19.85	2.55	22.40
9750	40x40	40	12	60	141	20.50	2.52	23.02

B10 Superstructure

B1010 Floor Construction

General: Beams and hollow core slabs priced here are for plant produced prestressed members transported to the site and erected.

The 2" structural topping is applied after the beams and hollow core slabs are in place and is reinforced with W.W.F.

Design and Pricing Assumptions:
Prices are based on 10,000 S.F. to 20,000 S.F. projects and 50 mile to 100 mile transport.

Concrete for prestressed members is f'c 5 KSI.

Concrete for topping is f'c 3000 PSI and placed by pump.

Prestressing steel is fy = 250 or 300 KSI.

W.W.F. is 6 x 6 – W1.4 x W1.4 (10 x 10).

System Components	QUANTITY	UNIT	COST PER S.F.		
			MAT.	INST.	TOTAL
SYSTEM B1010 238 4300					
20' X 20' BAY, 6" PLANK, 40 PSF S. LOAD, 135 PSF TOTAL LOAD					
Precast concrete beam, T-shaped, 20' span, 12" x 20"	.038	L.F.	8.79	.71	9.50
Precast concrete beam, L-shaped, 20' span, 12" x 20"	.025	L.F.	5.28	.48	5.76
Precast prestressed concrete roof/floor slabs 6" deep, grouted	1.000	S.F.	7.15	2.92	10.07
Edge forms to 6" high on elevated slab, 4 uses	.050	L.F.	.01	.21	.22
Forms in place, bulkhead for slab with keyway, 1 use, 2 piece	.013	L.F.	.03	.08	.11
Welded wire fabric rolls, 6 x 6 - W1.4 x W1.4 (10 x 10), 21 lb/csf	.010	C.S.F.	.16	.36	.52
Concrete, ready mix, regular weight, 3000 psi	.170	C.F.	.67		.67
Place and vibrate concrete, elevated slab less than 6", pump	.170	C.F.		.26	.26
Finish floor, monolithic steel trowel finish for finish floor	1.000	S.F.		.88	.88
Cure with sprayed membrane curing compound	.010	C.S.F.	.09	.09	.18
TOTAL			22.18	5.99	28.17

B1010 238		Precast Beam & Plank with 2" Topping						
	BAY SIZE (FT.)	SUPERIMPOSED LOAD (P.S.F.)	PLANK THICKNESS (IN.)	TOTAL DEPTH (IN.)	TOTAL LOAD (P.S.F.)	COST PER S.F.		
						MAT.	INST.	TOTAL
4300	20x20	40	6	22	135	22	6	28
4400	RB1010-010	75	6	24	173	23	6.05	29.05
4500		100	6	28	200	23	6.05	29.05
4600	20x25	40	6	26	134	21	5.90	26.90
5000	RB1010-100	75	8	30	177	21	5.45	26.45
5200		100	8	30	202	21	5.45	26.45
5400	25x25	40	6	38	143	21.50	5.60	27.10
5600		75	8	38	183	16.90	5.30	22.20
6000		100	8	46	216	16.70	4.89	21.59
6200	25x30	40	8	38	144	19.45	5.10	24.55
6400		75	10	46	200	20	4.77	24.77
6600		100	10	46	225	20.50	4.78	25.28
7000	30x30	40	8	46	150	20	5	25
7200		75	10	54	181	21	5	26
7600		100	10	54	231	21	5	26
7800	30x35	40	10	54	166	20	4.61	24.61
8000		75	12	54	200	20.50	4.36	24.86
8200	35x35	40	10	62	170	20.50	4.55	25.05
9300		75	12	62	206	22.50	4.35	26.85
9500	35x40	40	12	62	167	21.50	4.29	25.79
9600	40x40	40	12	62	173	21	4.26	25.26

B10 Superstructure

B1010 Floor Construction

General: Beams and double tees priced here are for plant produced prestressed members transported to the site and erected.

The 2" structural topping is applied after the beams and double tees are in place and is reinforced with W.W.F.

Design and Pricing Assumptions:
Prices are based on 10,000 S.F. to 20,000 S.F. projects and 50 mile to 100 mile transport.

Concrete for prestressed members is f'c 5 KSI.

Concrete for topping is f'c 3000 PSI and placed by pump.

Prestressing steel is fy = 250 or 300 KSI.

W.W.F. is 6 x 6 – W1.4 x W1.4 (10x10).

System Components	QUANTITY	UNIT	COST PER S.F.		
			MAT.	INST.	TOTAL
SYSTEM B1010 239 3000					
25' X 30' BAY, 38" DEPTH, 130 PSF T.L., 2" TOPPING					
Precast concrete beam, T-shaped, 25' span, 12" x 36"	.025	L.F.	7.16	.47	7.63
Precast concrete beam, L-shaped, 25' span, 12" x 28"	.017	L.F.	4.05	.28	4.33
Double T, standard weight, 16" x 8' w, 25' span	.989	S.F.	8.96	1.68	10.64
Edge forms to 6" high on elevated slab, 4 uses	.037	L.F.	.01	.16	.17
Forms in place, bulkhead for slab with keyway, 1 use, 2 piece	.010	L.F.	.02	.06	.08
Welded wire fabric rolls, 6 x 6 – W1.4 x W1.4 (10 x 10), 21 lb/csf	1.000	S.F.	.16	.36	.52
Concrete, ready mix, regular weight, 3000 psi	.170	C.F.	.67		.67
Place and vibrate concrete, elevated slab less than 6", pumped	.170	C.F.		.26	.26
Finishing floor, monolithic steel trowel finish for finish floor	1.000	S.F.		.88	.88
Curing with sprayed membrane curing compound	.010	S.F.	.09	.09	.18
TOTAL			21.12	4.24	25.36

B1010 239		Precast Double "T" & 2" Topping on Precast Beams						
	BAY SIZE (FT.)	SUPERIMPOSED LOAD (P.S.F.)	DEPTH (IN.)		TOTAL LOAD (P.S.F.)	COST PER S.F.		
						MAT.	INST.	TOTAL
3000	25x30	40	38		130	21	4.24	25.24
3100	RB1010-100	75	38		168	21	4.24	25.24
3300		100	46		196	21	4.20	25.20
3600	30x30	40	46		150	21.50	4.10	25.60
3750		75	46		174	21.50	4.10	25.60
4000		100	54		203	22.50	4.09	26.59
4100	30x40	40	46		136	18.85	3.68	22.53
4300		75	54		173	19.40	3.66	23.06
4400		100	62		204	20	3.68	23.68
4600	30x50	40	54		138	17.65	3.41	21.06
4800		75	54		181	17.45	3.46	20.91
5000		100	54		219	17.45	3.15	20.60
5200	30x60	40	62		151	17.15	3.08	20.23
5400		75	62		192	16.75	3.13	19.88
5600		100	62		215	16.90	3.13	20.03
5800	35x40	40	54		139	19.65	3.62	23.27
6000		75	62		179	19.80	3.49	23.29
6250		100	62		212	19.95	3.62	23.57
6500	35x50	40	62		142	18.20	3.40	21.60
6750		75	62		186	18.25	3.51	21.76
7300		100	62		231	18.35	3.20	21.55

B10 Superstructure

B1010 Floor Construction

B1010 239 Precast Double "T" & 2" Topping on Precast Beams

	BAY SIZE (FT.)	SUPERIMPOSED LOAD (P.S.F.)	DEPTH (IN.)		TOTAL LOAD (P.S.F.)	COST PER S.F.		
						MAT.	INST.	TOTAL
7600	35x60	40	54		154	17.65	2.96	20.61
7750		75	54		179	17.75	2.97	20.72
8000		100	62		224	18.20	3.02	21.22
8250	40x40	40	62		145	21.50	3.63	25.13
8400		75	62		187	20.50	3.67	24.17
8750		100	62		223	21.50	3.74	25.24
9000	40x50	40	62		151	19	3.42	22.42
9300		75	62		193	18.55	2.54	21.09
9800	40x60	40	62		164	21	2.45	23.45

B10 Superstructure

B1010 Floor Construction

General: The following table is based upon structural W shape beam and girder framing. Non-composite action is assumed between beams and decking. Deck costs not included.

The deck spans the short direction. The steel beams and girders are fireproofed with sprayed fiber fireproofing.

Design and Pricing Assumptions: Structural steel is A36, with high strength A325 bolts.

Fireproofing is sprayed fiber (non-asbestos).

Total load includes steel, deck & live load.

Spandrels are assumed the same as interior beams and girders to allow for exterior wall loads and bracing or moment connections. No columns included in price.

See Tables B1010 258 and B1020 128 for metal deck costs.

System Components	QUANTITY	UNIT	COST PER S.F.		
			MAT.	INST.	TOTAL
SYSTEM B1010 241 1350 15′ X 20′ BAY, 40 P.S.F. L.L., 12″ DEPTH, .535 P.S.F. FIREPROOF, 50 PSF T.LOAD					
Structural steel	3.200	Lb.	4.67	1.38	6.05
Spray mineral fiber/cement for fire proof., 1″ thick on beams	.535	S.F.	.31	.54	.85
TOTAL			4.98	1.92	6.90

B1010 241 — W Shape Beams & Girders

	BAY SIZE (FT.) BEAM X GIRD	SUPERIMPOSED LOAD (P.S.F.)	STEEL FRAMING DEPTH (IN.)	FIREPROOFING (S.F. PER S.F.)	TOTAL LOAD (P.S.F.)	COST PER S.F.		
						MAT.	INST.	TOTAL
1350	15x20	40	12	.535	50	4.98	1.92	6.90
1400		40	16	.65	90	6.50	2.47	8.97
1450		75	18	.694	125	8.60	3.10	11.70
1500		125	24	.796	175	11.85	4.31	16.16
1550		200	24	.89	263	13.40	3.96	17.36
1600	20x15	40	14	.659	50	5.05	2.05	7.10
1650		40	14	.69	90	6.80	2.60	9.40
1700		75	14	.806	125	8.35	3.14	11.49
1800		125	16	.86	175	9.85	3.75	13.60
1900		200	18	1.00	250	11.65	3.51	15.16
2000	20x20	40	12	.55	50	5.60	2.10	7.70
2050		40	14	.579	90	7.65	2.74	10.39
2100		75	16	.672	125	9.15	3.26	12.41
2150		125	16	.714	175	10.90	3.96	14.86
2200		200	24	.841	263	13.65	3.97	17.62
2300	20x20	40	14	.67	50	5.65	2.22	7.87
2400		40	14	.718	90	7.70	2.87	10.57
2500		75	18	.751	125	8.90	3.25	12.15
2550		125	21	.879	175	12.20	4.49	16.69
2600		200	21	.976	250	14.60	4.33	18.93
2650	20x20	40	14	.746	50	5.70	2.29	7.99
2700		40	14	.839	90	7.80	2.99	10.79
2750		75	18	.894	125	9.85	3.66	13.51
2800		125	21	.959	175	13.10	4.84	17.94
2850		200	21	1.10	250	16.10	4.79	20.89
2900	20x25	40	16	.53	50	6.15	2.25	8.40
2950		40	18	.621	96	9.70	3.39	13.09
3000		75	18	.651	131	11.20	3.85	15.05
3050		125	24	.77	200	14.75	5.20	19.95

B10 Superstructure

B1010 Floor Construction

B1010 241 — W Shape Beams & Girders

	BAY SIZE (FT.) BEAM X GIRD	SUPERIMPOSED LOAD (P.S.F.)	STEEL FRAMING DEPTH (IN.)	FIREPROOFING (S.F. PER S.F.)	TOTAL LOAD (P.S.F.)	COST PER S.F. MAT.	COST PER S.F. INST.	COST PER S.F. TOTAL
3100	20 x 25	200	27	.855	275	16.80	4.75	21.55
3300	20x25	40	14	.608	50	6.20	2.33	8.53
3350		40	21	.751	90	8.60	3.16	11.76
3400		75	24	.793	125	10.70	3.81	14.51
3450		125	24	.846	175	12.75	4.63	17.38
3500		200	24	.947	256	16	4.63	20.63
3550	20x25	40	14	.72	50	7.15	2.70	9.85
3600		40	21	.802	90	8.65	3.21	11.86
3650		75	24	.924	125	11.05	4.03	15.08
3700		125	24	.964	175	13.40	4.94	18.34
3750		200	27	1.09	250	16.95	4.98	21.93
3800	25x20	40	12	.512	50	6.15	2.24	8.39
3850		40	16	.653	90	8.25	2.99	11.24
3900		75	18	.726	125	10.65	3.74	14.39
4000		125	21	.827	175	13.35	4.79	18.14
4100		200	24	.928	250	16.25	4.67	20.92
4200	25x20	40	12	.65	50	6.20	2.38	8.58
4300		40	18	.702	90	9.15	3.29	12.44
4400		75	21	.829	125	10.70	3.84	14.54
4500		125	24	.914	175	13.10	4.79	17.89
4600		200	24	1.015	250	16.05	4.70	20.75
4700	25x20	40	14	.769	50	6.60	2.59	9.19
4800		40	16	.938	90	9.90	3.70	13.60
4900		75	18	.969	125	11.95	4.33	16.28
5000		125	24	1.136	175	16.70	6.10	22.80
5100		200	24	1.239	250	21.50	6.20	27.70
5200	25x25	40	18	.486	50	6.70	2.39	9.09
5300		40	18	.592	96	10	3.43	13.43
5400		75	21	.668	131	12.05	4.12	16.17
5450		125	24	.738	191	15.90	5.50	21.40
5500		200	30	.861	272	18.50	5.15	23.65
5550	25x25	40	18	.597	50	6.50	2.41	8.91
5600		40	18	.704	90	10.35	3.63	13.98
5650		75	21	.777	125	11.85	4.14	15.99
5700		125	24	.865	175	15.10	5.40	20.50
5750		200	27	.96	250	18.30	5.20	23.50
5800	25x25	40	18	.71	50	7.15	2.69	9.84
5850		40	21	.767	90	10.35	3.70	14.05
5900		75	24	.887	125	12.50	4.43	16.93
5950		125	24	.972	175	15.75	5.65	21.40
6000		200	30	1.10	250	19.25	5.55	24.80
6050	25x30	40	24	.547	50	8.50	2.95	11.45
6100		40	24	.629	103	12.05	4.07	16.12
6150		75	30	.726	138	14.45	4.85	19.30
6200		125	30	.751	206	17.10	5.90	23
6250		200	33	.868	281	20.50	5.65	26.15
6300	25x30	40	21	.568	50	7.65	2.73	10.38
6350		40	21	.694	90	10.35	3.62	13.97
6400		75	24	.776	125	13.30	4.57	17.87
6450		125	30	.904	175	16	5.70	21.70
6500		200	33	1.008	263	19.15	5.45	24.60

B10 Superstructure

B1010 Floor Construction

B1010 241 W Shape Beams & Girders

	BAY SIZE (FT.) BEAM X GIRD	SUPERIMPOSED LOAD (P.S.F.)	STEEL FRAMING DEPTH (IN.)	FIREPROOFING (S.F. PER S.F.)	TOTAL LOAD (P.S.F.)	COST PER S.F. MAT.	INST.	TOTAL
6550	25x30	40	16	.632	50	7.95	2.88	10.83
6600		40	21	.76	90	10.95	3.87	14.82
6650		75	24	.857	125	13.05	4.56	17.61
6700		125	30	.983	175	16.35	5.85	22.20
6750		200	33	1.11	250	20.50	5.90	26.40
6800	30x25	40	16	.532	50	7.30	2.59	9.89
6850		40	21	.672	96	11.20	3.87	15.07
6900		75	24	.702	131	13.25	4.49	17.74
6950		125	27	1.020	175	17.25	5.95	23.20
7000		200	30	1.160	250	22	7.45	29.45
7100	30x25	40	18	.569	50	7.65	2.73	10.38
7150		40	24	.740	90	10.65	3.75	14.40
7200		75	24	.787	125	13.30	4.58	17.88
7300		125	24	.874	175	16.55	5.85	22.40
7400		200	30	1.013	250	20.50	5.70	26.20
7450	30x25	40	16	.637	50	7.95	2.88	10.83
7500		40	24	.839	90	11.30	4.03	15.33
7550		75	24	.919	125	13.65	4.80	18.45
7600		125	27	1.02	175	17.25	6.15	23.40
7650		200	30	1.160	250	21.50	6.15	27.65
7700	30x30	40	21	.52	50	8.20	2.85	11.05
7750		40	24	.629	103	12.60	4.25	16.85
7800		75	30	.715	138	15	5	20
7850		125	36	.822	206	19.75	6.75	26.50
7900		200	36	.878	281	22	6	28
7950	30x30	40	24	.619	50	8.55	3.02	11.57
8000		40	24	.706	90	11.50	3.97	15.47
8020		75	27	.818	125	13.60	4.69	18.29
8040		125	30	.910	175	17.45	6.15	23.60
8060		200	33	.999	263	21.50	5.95	27.45
8080	30x30	40	18	.631	50	9.15	3.21	12.36
8100		40	24	.805	90	12.45	4.34	16.79
8120		75	27	.899	125	14.85	5.10	19.95
8150		125	30	1.010	175	18.40	6.50	24.90
8200		200	36	1.148	250	22	6.20	28.20
8250	30x35	40	21	.508	50	9.35	3.18	12.53
8300		40	24	.651	109	13.80	4.62	18.42
8350		75	33	.732	150	16.75	5.55	22.30
8400		125	36	.802	225	21	7.10	28.10
8450		200	36	.888	300	27.50	7.30	34.80
8500	30x35	40	24	.554	50	8.20	2.89	11.09
8520		40	24	.655	90	12.05	4.11	16.16
8540		75	30	.751	125	15.05	5.05	20.10
8600		125	33	.845	175	18.30	6.35	24.65
8650		200	36	.936	263	24	6.55	30.55
8700	30x35	40	21	.644	50	8.85	3.14	11.99
8720		40	24	.733	90	12.70	4.36	17.06
8740		75	30	.833	125	15.95	5.40	21.35
8760		125	36	.941	175	18.35	6.45	24.80
8780		200	36	1.03	250	24.50	6.65	31.15

B10 Superstructure

B1010 Floor Construction

B1010 241 — W Shape Beams & Girders

	BAY SIZE (FT.) BEAM X GIRD	SUPERIMPOSED LOAD (P.S.F.)	STEEL FRAMING DEPTH (IN.)	FIREPROOFING (S.F. PER S.F.)	TOTAL LOAD (P.S.F.)	COST PER S.F.		
						MAT.	INST.	TOTAL
8800	35x30	40	24	.540	50	9.05	3.12	12.17
8850		40	30	.670	103	13.55	4.55	18.10
8900		75	33	.748	138	16.50	5.50	22
8950		125	36	.824	206	20.50	6.95	27.45
8980		200	36	.874	281	24.50	6.60	31.10
9000	35x30	40	24	.619	50	8.85	3.11	11.96
9050		40	24	.754	90	13	4.46	17.46
9100		75	27	.844	125	15.40	5.25	20.65
9200		125	30	.856	175	18.90	6.50	25.40
9250		200	33	.953	263	21	5.80	26.80
9300	35x30	40	24	.705	50	9.45	3.38	12.83
9350		40	24	.833	90	13.35	4.62	17.97
9400		75	30	.963	125	16.35	5.60	21.95
9450		125	33	1.078	175	20.50	7.20	27.70
9500		200	36	1.172	250	24.50	6.90	31.40
9550	35x35	40	27	.560	50	9.65	3.33	12.98
9600		40	36	.706	109	17.65	5.80	23.45
9650		75	36	.750	150	18.25	6	24.25
9820		125	36	.797	225	24.50	8.20	32.70
9840		200	36	.914	300	29	7.70	36.70
9860	35x35	40	24	.580	50	8.80	3.08	11.88
9880		40	30	.705	90	13.55	4.58	18.13
9890		75	33	.794	125	16.50	5.55	22.05
9900		125	36	.878	175	20.50	7	27.50
9920		200	36	.950	263	25	6.80	31.80
9930	35x35	40	24	.689	50	9.45	3.37	12.82
9940		40	30	.787	90	13.90	4.76	18.66
9960		75	33	.871	125	17.75	5.95	23.70
9970		125	36	.949	175	19.55	6.80	26.35
9980		200	36	1.060	250	27	7.35	34.35

B10 Superstructure

B1010 Floor Construction

Description: Table below lists costs for light gauge CEE or PUNCHED DOUBLE joists to suit the span and loading with the minimum thickness subfloor required by the joist spacing.

Design Assumptions:
Maximum live load deflection is 1/360 of the clear span.
Maximum total load deflection is 1/240 of the clear span.
Bending strength is 20,000 psi.

8% allowance has been added to framing quantities for overlaps, double joists at openings under partitions, etc.; 5% added to glued & nailed subfloor for waste.

Maximum span is in feet and is the unsupported clear span.

System Components	QUANTITY	UNIT	COST PER S.F.		
			MAT.	INST.	TOTAL
SYSTEM B1010 244 1500					
LIGHT GAUGE STL & PLYWOOD FLR SYS, 40 PSF S.LOAD, 15' SPAN, 8" DEPTH, 16" O.C.					
Light gauge steel, 8" deep, 16" O.C., 12 ga.	1.870	Lb.	1.87	1.32	3.19
Subfloor plywood CDX 5/8"	1.050	S.F.	.84	.86	1.70
TOTAL			2.71	2.18	4.89

B1010 244 — Light Gauge Steel Floor Systems

	SPAN (FT.)	SUPERIMPOSED LOAD (P.S.F.)	FRAMING DEPTH (IN.)	FRAMING SPAC. (IN.)	TOTAL LOAD (P.S.F.)	COST PER S.F.		
						MAT.	INST.	TOTAL
1500	15	40	8	16	54	2.71	2.18	4.89
1550			8	24	54	2.73	2.23	4.96
1600		65	10	16	80	3.50	2.66	6.16
1650			10	24	80	3.36	2.59	5.95
1700		75	10	16	90	3.50	2.66	6.16
1750			10	24	90	3.36	2.59	5.95
1800		100	10	16	116	4.55	3.38	7.93
1850			10	24	116	3.36	2.59	5.95
1900		125	10	16	141	4.55	3.38	7.93
1950			10	24	141	3.36	2.59	5.95
2500	20	40	8	16	55	3.11	2.47	5.58
2550			8	24	55	2.73	2.23	4.96
2600		65	8	16	80	3.60	2.82	6.42
2650			10	24	80	3.37	2.61	5.98
2700		75	10	16	90	3.31	2.52	5.83
2750			12	24	90	3.79	2.27	6.06
2800		100	10	16	115	3.31	2.52	5.83
2850			10	24	116	4.35	2.94	7.29
2900		125	12	16	142	5.20	2.86	8.06
2950			12	24	141	4.73	3.12	7.85
3500	25	40	10	16	55	3.50	2.66	6.16
3550			10	24	55	3.36	2.59	5.95
3600		65	10	16	81	4.55	3.38	7.93
3650			12	24	81	4.21	2.46	6.67
3700		75	12	16	92	5.20	2.86	8.06
3750			12	24	91	4.73	3.12	7.85
3800		100	12	16	117	5.85	3.16	9.01
3850		125	12	16	143	6.65	4.17	10.82
4500	30	40	12	16	57	5.20	2.86	8.06
4550			12	24	56	4.19	2.45	6.64
4600		65	12	16	82	5.80	3.15	8.95
4650		75	12	16	92	5.80	3.15	8.95

B10 Superstructure

B1010 Floor Construction

Description: Table below lists cost per S.F. for a floor system on bearing walls using open web steel joists, galvanized steel slab form and 2-1/2" concrete slab reinforced with welded wire fabric.

Design and Pricing Assumptions:
Concrete f'c = 3 KSI placed by pump.
WWF 6 x 6 – W1.4 x W1.4 (10 x 10)
Joists are spaced as shown.
Slab form is 28 gauge galvanized.
Joists costs include appropriate bridging. Deflection is limited to 1/360 of the span. Screeds and steel trowel finish.

Design Loads	Min.	Max.
Joists	3.0 PSF	7.6 PSF
Slab Form	1.0	1.0
2-1/2" Concrete	27.0	27.0
Ceiling	3.0	3.0
Misc.	9.0	9.4
	43.0 PSF	48.0 PSF

System Components	QUANTITY	UNIT	COST PER S.F.		
			MAT.	INST.	TOTAL
SYSTEM B1010 246 1050 SPAN 20' S.LOAD 40 PSF, JOIST SPACING 2'-0" O.C., 14-1/2" DEPTH					
Open web joists	2.800	Lb.	2.46	1.03	3.49
Slab form, galvanized steel 9/16" deep, 28 gauge	1.050	S.F.	2.13	1.14	3.27
Welded wire fabric rolls, 6 x 6 - W1.4 x W1.4 (10 x 10), 21 lb/csf	1.000	S.F.	.16	.36	.52
Concrete, ready mix, regular weight, 3000 psi	.210	C.F.	.83		.83
Place and vibrate concrete, elevated slab less than 6", pumped	.210	C.F.		.32	.32
Finishing floor, monolithic steel trowel finish for finish floor	1.000	S.F.		.88	.88
Curing with sprayed membrane curing compound	.010	C.S.F.	.09	.09	.18
TOTAL			5.67	3.82	9.49

B1010 246	Deck & Joists on Bearing Walls							
	SPAN (FT.)	SUPERIMPOSED LOAD (P.S.F.)	JOIST SPACING FT. - IN.	DEPTH (IN.)	TOTAL LOAD (P.S.F.)	COST PER S.F.		
						MAT.	INST.	TOTAL
1050	20	40	2-0	14-1/2	83	5.65	3.82	9.47
1070		65	2-0	16-1/2	109	6.15	4.06	10.21
1100		75	2-0	16-1/2	119	6.15	4.06	10.21
1120		100	2-0	18-1/2	145	6.20	4.07	10.27
1150		125	1-9	18-1/2	170	6.75	4.32	11.07
1170	25	40	2-0	18-1/2	84	6.20	4.07	10.27
1200		65	2-0	20-1/2	109	6.40	4.16	10.56
1220		75	2-0	20-1/2	119	6.80	4.34	11.14
1250		100	2-0	22-1/2	145	6.95	4.39	11.34
1270		125	1-9	22-1/2	170	7.90	4.79	12.69
1300	30	40	2-0	22-1/2	84	6.95	3.95	10.90
1320		65	2-0	24-1/2	110	7.55	4.12	11.67
1350		75	2-0	26-1/2	121	7.80	4.20	12
1370		100	2-0	26-1/2	146	8.35	4.35	12.70
1400		125	2-0	24-1/2	172	9.80	4.79	14.59
1420	35	40	2-0	26-1/2	85	9.25	4.62	13.87
1450		65	2-0	28-1/2	111	9.55	4.72	14.27
1470		75	2-0	28-1/2	121	9.55	4.72	14.27
1500		100	1-11	28-1/2	147	10.40	4.97	15.37
1520		125	1-8	28-1/2	172	11.45	5.30	16.75
1550	5/8" gyp. fireproof.							
1560	On metal furring, add					.76	2.85	3.61

B10 Superstructure

B1010 Floor Construction

Table below lists costs for a floor system on exterior bearing walls and interior columns and beams using open web steel joists, galvanized steel slab form, 2-1/2" concrete slab reinforced with welded wire fabric.

Design and Pricing Assumptions:
Structural Steel is A36.
Concrete f'c = 3 KSI placed by pump.
WWF 6 x 6 – W1.4 x W1.4 (10 x 10)
Columns are 12' high.
Building is 4 bays long by 4 bays wide.
Joists are 2' O.C. ± and span the long direction of the bay.
Joists at columns have bottom chords extended and are connected to columns.

Slab form is 28 gauge galvanized. Column costs in table are for columns to support 1 floor plus roof loading in a 2-story building; however, column costs are from ground floor to 2nd floor only. Joist costs include appropriate bridging. Deflection is limited to 1/360 of the span. Screeds and steel trowel finish.

Design Loads	Min.	Max.
S.S & Joists	4.4 PSF	11.5 PSF
Slab Form	1.0	1.0
2-1/2" Concrete	27.0	27.0
Ceiling	3.0	3.0
Misc.	7.6	5.5
	43.0 PSF	48.0 PSF

System Components			COST PER S.F.		
	QUANTITY	UNIT	MAT.	INST.	TOTAL
SYSTEM B1010 248 1200					
15'X20' BAY, W.F. STEEL, STEEL JOISTS, SLAB FORM, CONCRETE SLAB					
Struct. steel, bolted-apts., nursing homes, etc., st. bearing, 1-2 stories	1.248	Lb.	1.34	.39	1.73
Open web joists	3.140	Lb.	2.76	1.16	3.92
Slab form, galvanized steel 9/16" deep, 28 gauge	1.020	S.F.	1.45	.77	2.22
Welded wire fabric 6x6 - W1.4 x W1.4 (10 x 10), 21 lb/CSF roll, 10% lap	1.000	S.F.	.16	.36	.52
Concrete, ready mix, regular weight, 3000 psi	.210	C.F.	.83		.83
Place and vibrate concrete, elevated slab less than 6", pumped	.210	C.F.		.32	.32
Finishing floor, monolithic steel trowel finish for finish floor	1.000	S.F.		.88	.88
Curing with sprayed membrane curing compound	.010	C.S.F.	.09	.09	.18
TOTAL			6.63	3.97	10.60

B1010 248 — Steel Joists on Beam & Wall

	BAY SIZE (FT.)	SUPERIMPOSED LOAD (P.S.F.)	DEPTH (IN.)	TOTAL LOAD (P.S.F.)	COLUMN ADD	COST PER S.F.		
						MAT.	INST.	TOTAL
1200	15x20 (RB1010-100)	40	17	83		6.65	3.97	10.62
1210					columns	.44	.13	.57
1220	15x20	65	19	108		6.85	4.06	10.91
1230					columns	.44	.13	.57
1250	15x20	75	19	119		7.25	4.24	11.49
1260					columns	.53	.15	.68
1270	15x20	100	19	144		7.65	4.38	12.03
1280					columns	.53	.15	.68
1300	15x20	125	19	170		7.90	4.45	12.35
1310					columns	.71	.20	.91
1350	20x20	40	19	83		6.85	4.05	10.90
1360					columns	.40	.12	.52
1370	20x20	65	23	109		7.60	4.33	11.93
1380					columns	.53	.15	.68
1400	20x20	75	23	119		7.90	4.41	12.31
1410					columns	.53	.15	.68
1420	20x20	100	23	144		8.15	4.50	12.65
1430					columns	.53	.15	.68
1450	20x20	125	23	170		9.10	4.84	13.94
1460					columns	.64	.19	.83

B10 Superstructure

B1010 Floor Construction

B1010 248 Steel Joists on Beam & Wall

	BAY SIZE (FT.)	SUPERIMPOSED LOAD (P.S.F.)	DEPTH (IN.)	TOTAL LOAD (P.S.F.)	COLUMN ADD	COST PER S.F.		
						MAT.	INST.	TOTAL
1500	20x25	40	23	84		7.20	4.20	11.40
1510					columns	.43	.13	.56
1520	20x25	65	26	110		7.80	4.41	12.21
1530					columns	.43	.13	.56
1550	20x25	75	26	120		8.25	4.60	12.85
1560					columns	.51	.15	.66
1570	20x25	100	26	145		8.70	4.76	13.46
1580					columns	.51	.15	.66
1670	20x25	125	29	170		9.90	5.20	15.10
1680					columns	.59	.18	.77
1720	25x25	40	23	84		7.80	4.37	12.17
1730					columns	.41	.12	.53
1750	25x25	65	29	110		8.20	4.52	12.72
1760					columns	.41	.12	.53
1770	25x25	75	26	120		9.15	4.86	14.01
1780					columns	.47	.14	.61
1800	25x25	100	29	145		10.30	5.30	15.60
1810					columns	.47	.14	.61
1820	25x25	125	29	170		10.85	5.50	16.35
1830					columns	.53	.15	.68
1870	25x30	40	29	84		9.10	4.41	13.51
1880					columns	.40	.12	.52
1900	25x30	65	29	110		9.15	4.42	13.57
1910					columns	.40	.12	.52
1920	25x30	75	29	120		9.65	4.58	14.23
1930					columns	.44	.13	.57
1950	25x30	100	29	145		10.50	4.82	15.32
1960					columns	.44	.13	.57
1970	25x30	125	32	170		12	5.25	17.25
1980					columns	.49	.14	.63
2020	30x30	40	29	84		9	4.38	13.38
2030					columns	.37	.10	.47
2050	30x30	65	29	110		10.20	4.73	14.93
2060					columns	.37	.10	.47
2070	30x30	75	32	120		10.50	4.82	15.32
2080					columns	.42	.13	.55
2100	30x30	100	35	145		11.60	5.15	16.75
2110					columns	.49	.14	.63
2120	30x30	125	35	172		13.45	5.70	19.15
2130					columns	.59	.18	.77
2170	30x35	40	29	85		10.15	4.72	14.87
2180					columns	.35	.10	.45
2200	30x35	65	29	111		12.15	5.30	17.45
2210					columns	.40	.12	.52
2220	30x35	75	32	121		12.10	5.30	17.40
2230					columns	.41	.13	.54
2250	30x35	100	35	148		12.35	5.35	17.70
2260					columns	.51	.15	.66
2270	30x35	125	38	173		13.85	5.80	19.65
2280					columns	.51	.15	.66
2320	35x35	40	32	85		10.50	4.82	15.32
2330					columns	.36	.10	.46

B10 Superstructure

B1010 Floor Construction

B1010 248 Steel Joists on Beam & Wall

	BAY SIZE (FT.)	SUPERIMPOSED LOAD (P.S.F.)	DEPTH (IN.)	TOTAL LOAD (P.S.F.)	COLUMN ADD	COST PER S.F.		
						MAT.	INST.	TOTAL
2350	35x35	65	35	111		12.80	5.50	18.30
2360					columns	.43	.13	.56
2370	35x35	75	35	121		13.10	5.60	18.70
2380					columns	.43	.13	.56
2400	35x35	100	38	148		13.65	5.75	19.40
2410					columns	.53	.15	.68
2420	35x35	125	41	173		15.50	6.30	21.80
2430					columns	.54	.16	.70
2460	5/8 gyp. fireproof.							
2475	On metal furring, add					.76	2.85	3.61

B10 Superstructure

B1010 Floor Construction

Table below lists costs for a floor system on steel columns and beams using open web steel joists, galvanized steel slab form, and 2-1/2" concrete slab reinforced with welded wire fabric.

Design and Pricing Assumptions:
Structural Steel is A36.
Concrete f'c = 3 KSI placed by pump.
WWF 6 x 6 – W1.4 x W1.4 (10 x 10)
Columns are 12' high.
Building is 4 bays long by 4 bays wide.
Joists are 2' O.C. ± and span the long direction of the bay.

Joists at columns have bottom chords extended and are connected to columns.

Slab form is 28 gauge galvanized. Column costs in table are for columns to support 1 floor plus roof loading in a 2-story building; however, column costs are from ground floor to 2nd floor only. Joist costs include appropriate bridging. Deflection is limited to 1/360 of the span. Screeds and steel trowel finish.

Design Loads	Min.	Max.
S.S. & Joists	6.3 PSF	15.3 PSF
Slab Form	1.0	1.0
2-1/2" Concrete	27.0	27.0
Ceiling	3.0	3.0
Misc.	5.7	1.7
	43.0 PSF	48.0 PSF

System Components	QUANTITY	UNIT	COST PER S.F.		
			MAT.	INST.	TOTAL
SYSTEM B1010 250 2350					
15' X 20'BAY 40 PSF S. LOAD, 17" DEPTH, 83 PSF TOTAL LOAD					
Struct. steel, bolted-apts., nursing homes, etc., st. bearing, 1-2 stories	1.974	Lb.	2.82	.83	3.65
Open web joists	3.140	Lb.	2.76	1.16	3.92
Slab form, galvanized steel 9/16" deep, 28 gauge	1.020	S.F.	1.45	.77	2.22
Welded wire fabric rolls, 6 x 6 - W1.4 x W1.4 (10 x 10), 21 lb/csf	1.000	S.F.	.16	.36	.52
Concrete, ready mix, regular weight, 3000 psi	.210	C.F.	.83		.83
Place and vibrate concrete, elevated slab less than 6", pumped	.210	C.F.		.32	.32
Finishing floor, monolithic steel trowel finish for finish floor	1.000	S.F.		.88	.88
Curing with sprayed membrane curing compound	.010	S.F.	.09	.09	.18
TOTAL			8.11	4.41	12.52

B1010 250 — Steel Joists, Beams & Slab on Columns

	BAY SIZE (FT.)	SUPERIMPOSED LOAD (P.S.F.)	DEPTH (IN.)	TOTAL LOAD (P.S.F.)	COLUMN ADD	COST PER S.F.		
						MAT.	INST.	TOTAL
2350	15x20 RB1010-100	40	17	83		8.10	4.41	12.51
2400					column	1.35	.40	1.75
2450	15x20	65	19	108		9.05	4.68	13.73
2500					column	1.35	.40	1.75
2550	15x20	75	19	119		9.35	4.83	14.18
2600					column	1.48	.43	1.91
2650	15x20	100	19	144		9.95	5.05	15
2700					column	1.48	.43	1.91
2750	15x20	125	19	170		10.50	5.15	15.65
2800					column	1.97	.57	2.54
2850	20x20	40	19	83		8.85	4.61	13.46
2900					column	1.11	.33	1.44
2950	20x20	65	23	109		9.75	4.93	14.68
3000					column	1.48	.43	1.91
3100	20x20	75	26	119		10.35	5.10	15.45
3200					column	1.48	.43	1.91
3400	20x20	100	23	144		10.75	5.25	16
3450					column	1.48	.43	1.91
3500	20x20	125	23	170		11.95	5.65	17.60
3600					column	1.77	.52	2.29

B10 Superstructure

B1010 Floor Construction

B1010 250 Steel Joists, Beams & Slab on Columns

	BAY SIZE (FT.)	SUPERIMPOSED LOAD (P.S.F.)	DEPTH (IN.)	TOTAL LOAD (P.S.F.)	COLUMN ADD	COST PER S.F.		
						MAT.	INST.	TOTAL
3700	20x25	40	44	83		9.50	4.84	14.34
3800					column	1.18	.34	1.52
3900	20x25	65	26	110		10.40	5.15	15.55
4000					column	1.18	.34	1.52
4100	20x25	75	26	120		10.85	5.30	16.15
4200					column	1.42	.42	1.84
4300	20x25	100	26	145		11.50	5.55	17.05
4400					column	1.42	.42	1.84
4500	20x25	125	29	170		12.80	6.05	18.85
4600					column	1.65	.48	2.13
4700	25x25	40	23	84		10.30	5.05	15.35
4800					column	1.13	.33	1.46
4900	25x25	65	29	110		10.90	5.25	16.15
5000					column	1.13	.33	1.46
5100	25x25	75	26	120		12.05	5.70	17.75
5200					column	1.32	.38	1.70
5300	25x25	100	29	145		13.40	6.20	19.60
5400					column	1.32	.38	1.70
5500	25x25	125	32	170		14.15	6.45	20.60
5600					column	1.46	.43	1.89
5700	25x30	40	29	84		11.80	5.15	16.95
5800					column	1.10	.32	1.42
5900	25x30	65	29	110		12.30	5.30	17.60
6000					column	1.10	.32	1.42
6050	25x30	75	29	120		12.65	5.45	18.10
6100					column	1.22	.36	1.58
6150	25x30	100	29	145		13.75	5.75	19.50
6200					column	1.22	.36	1.58
6250	25x30	125	32	170		15.45	6.25	21.70
6300					column	1.40	.41	1.81
6350	30x30	40	29	84		11.85	5.15	17
6400					column	1.02	.29	1.31
6500	30x30	65	29	110		13.45	5.65	19.10
6600					column	1.02	.29	1.31
6700	30x30	75	32	120		13.75	5.75	19.50
6800					column	1.17	.34	1.51
6900	30x30	100	35	145		15.25	6.15	21.40
7000					column	1.36	.40	1.76
7100	30x30	125	35	172		17.45	6.80	24.25
7200					column	1.52	.45	1.97
7300	30x35	40	29	85		13.30	5.60	18.90
7400					column	.87	.25	1.12
7500	30x35	65	29	111		15.55	6.30	21.85
7600					column	1.12	.33	1.45
7700	30x35	75	32	121		15.55	6.30	21.85
7800					column	1.15	.33	1.48
7900	30x35	100	35	148		16.05	6.40	22.45
8000					column	1.40	.41	1.81
8100	30x35	125	38	173		17.80	6.95	24.75
8200					column	1.43	.42	1.85
8300	35x35	40	32	85		13.70	5.75	19.45
8400					column	1	.29	1.29

B10 Superstructure

B1010 Floor Construction

B1010 250 Steel Joists, Beams & Slab on Columns

	BAY SIZE (FT.)	SUPERIMPOSED LOAD (P.S.F.)	DEPTH (IN.)	TOTAL LOAD (P.S.F.)	COLUMN ADD	COST PER S.F. MAT.	COST PER S.F. INST.	COST PER S.F. TOTAL
8500	35x35	65	35	111		16.35	6.50	22.85
8600					column	1.20	.36	1.56
9300	35x35	75	38	121		16.80	6.60	23.40
9400					column	1.20	.36	1.56
9500	35x35	100	38	148		18.10	7.05	25.15
9600					column	1.48	.43	1.91
9750	35x35	125	41	173		18.70	7.20	25.90
9800					column	1.51	.45	1.96
9810	5/8 gyp. fireproof.							
9815	On metal furring, add					.76	2.85	3.61

B10 Superstructure

B1010 Floor Construction

General: Composite construction of W shape flange beams and concrete slabs is most efficiently used when loads are heavy and spans are moderately long. It is stiffer with less deflection than non-composite construction of similar depth and spans.

In practice, composite construction is typically shallower in depth than non-composite would be.

Design Assumptions:
Steel, fy = 36 KSI
Beams unshored during construction
Deflection limited to span/360
Shear connectors, welded studs
Concrete, f'c = 3 KSI

System Components	QUANTITY	UNIT	COST PER S.F.		
			MAT.	INST.	TOTAL
SYSTEM B1010 252 3800					
20 X 25 BAY, 40 PSF S. LOAD, 4" THICK SLAB, 20" TOTAL DEPTH					
Structural steel	3.820	Lb.	5.58	1.64	7.22
Welded shear connectors 3/4" diameter 3-3/8" long	.140	Ea.	.08	.29	.37
Forms in place, floor slab forms hung from steel beams, 4 uses	1.000	S.F.	1.85	5.70	7.55
Edge forms to 6" high on elevated slab, 4 uses	.050	L.F.	.01	.21	.22
Reinforcing in place, elevated slabs #4 to #7	1.190	Lb.	.68	.51	1.19
Concrete, ready mix, regular weight, 3000 psi	.330	C.F.	1.31		1.31
Place and vibrate concrete, elevated slab less than 6", pumped	.330	C.F.		.51	.51
Finishing floor, monolithic steel trowel finish for resilient tile	1.000	S.F.		1.15	1.15
Curing with sprayed membrane curing compound	.010	S.F.	.09	.09	.18
Spray mineral fiber/cement for fire proof, 1" thick on beams	.520	S.F.	.30	.52	.82
TOTAL			9.90	10.62	20.52

B1010 252	Composite Beam & Cast in Place Slab							
	BAY SIZE (FT.)	SUPERIMPOSED LOAD (P.S.F.)	SLAB THICKNESS (IN.)	TOTAL DEPTH (FT.-IN.)	TOTAL LOAD (P.S.F.)	COST PER S.F.		
						MAT.	INST.	TOTAL
3800	20x25	40	4	1 - 8	94	9.90	10.60	20.50
3900		75	4	1 - 8	130	11.60	11.30	22.90
4000	RB1010-100	125	4	1 - 10	181	13.55	12.05	25.60
4100		200	5	2 - 2	272	17.70	13.65	31.35
4200	25x25	40	4-1/2	1 - 8-1/2	99	10.35	10.75	21.10
4300		75	4-1/2	1 - 10-1/2	136	12.55	11.65	24.20
4400		125	5-1/2	2 - 0-1/2	200	14.90	12.55	27.45
4500		200	5-1/2	2 - 2-1/2	278	18.60	13.95	32.55
4600	25x30	40	4-1/2	1 - 8-1/2	100	11.60	11.20	22.80
4700		75	4-1/2	1 - 10-1/2	136	13.70	12	25.70
4800		125	5-1/2	2 - 2-1/2	202	16.80	13.20	30
4900		200	5-1/2	2 - 5-1/2	279	20	14.45	34.45
5000	30x30	40	4	1 - 8	95	11.65	11.25	22.90
5200		75	4	2 - 1	131	13.55	11.85	25.40
5400		125	4	2 - 4	183	16.85	13.10	29.95
5600		200	5	2 - 10	274	21.50	14.85	36.35

B10 Superstructure

B1010 Floor Construction

B1010 252 Composite Beam & Cast in Place Slab

	BAY SIZE (FT.)	SUPERIMPOSED LOAD (P.S.F.)	SLAB THICKNESS (IN.)	TOTAL DEPTH (FT.-IN.)	TOTAL LOAD (P.S.F.)	COST PER S.F.		
						MAT.	INST.	TOTAL
5800	30x35	40	4	2 - 1	95	12.35	11.45	23.80
6000		75	4	2 - 4	139	15	12.35	27.35
6250		125	4	2 - 4	185	19.60	13.95	33.55
6500		200	5	2 - 11	276	24	15.60	39.60
7000	35x30	40	4	2 - 1	95	12.70	11.55	24.25
7200		75	4	2 - 4	139	15.20	12.45	27.65
7400		125	4	2 - 4	185	19.55	13.95	33.50
7600		200	5	2 - 11	276	24	15.60	39.60
8000	35x35	40	4	2 - 1	96	13.05	11.65	24.70
8250		75	4	2 - 4	133	15.60	12.60	28.20
8500		125	4	2 - 10	185	18.80	13.75	32.55
8750		200	5	3 - 5	276	24.50	15.75	40.25
9000	35x40	40	4	2 - 4	97	14.55	12.15	26.70
9250		75	4	2 - 4	134	17.10	12.95	30.05
9500		125	4	2 - 7	186	21	14.35	35.35
9750		200	5	3 - 5	278	27	16.55	43.55

B10 Superstructure

B1010 Floor Construction

Description: Table below lists costs per S.F. for floors using steel beams and girders, composite steel deck, concrete slab reinforced with W.W.F. and sprayed fiber fireproofing (non-asbestos) on the steel beams and girders and on the steel deck.

Design and Pricing Assumptions:
Structural Steel is A36, high strength bolted.

Composite steel deck varies from 2"–20 gauge to 3"–16 gauge galvanized.
WWF 6 x 6 – W1.4 x W1.4 (10 x 10)
Concrete f'c = 3 KSI.
Steel trowel finish and cure.

Spandrels are assumed the same weight as interior beams and girders to allow for exterior wall loads and bracing or moment connections.

System Components	QUANTITY	UNIT	COST PER S.F.		
			MAT.	INST.	TOTAL
SYSTEM B1010 254 0540					
W SHAPE BEAMS & DECK, FIREPROOFED, 15' X 20', 5" SLAB, 40 PSF LOAD					
Structural steel	4.470	Lb.	6.01	1.77	7.78
Metal decking, non-cellular composite, galv 3" deep, 20 gauge	1.050	S.F.	2.55	1.06	3.61
Sheet metal edge closure form, 12", w/2 bends, 18 ga, galv	.058	L.F.	.25	.15	.40
Welded wire fabric 6 x 6 - W1.4 x W1.4 (10 x 10), 21 lb/CSF roll, 10% lap	1.000	S.F.	.16	.36	.52
Concrete ready mix, regular weight, 3000 psi	.011	C.Y.	1.18		1.18
Place and vibrate concrete, elevated slab less than 6", pumped	.011	C.Y.		.36	.36
Finishing floor, monolithic steel trowel finish for finish floor	1.000	S.F.		.88	.88
Curing with sprayed membrane curing compound	.010	C.S.F.	.09	.09	.18
Sprayed mineral fiber/cement for fireproof, 1" thick on decks	1.000	S.F.	.87	1.21	2.08
Sprayed mineral fiber/cement for fireproof, 1" thick on beams	.615	S.F.	.36	.62	.98
TOTAL			11.47	6.50	17.97

B1010 254 — W Shape, Composite Deck, & Slab

	BAY SIZE (FT.) BEAM X GIRD	SUPERIMPOSED LOAD (P.S.F.)	SLAB THICKNESS (IN.)	TOTAL DEPTH (FT.-IN.)	TOTAL LOAD (P.S.F.)	COST PER S.F.		
						MAT.	INST.	TOTAL
0540	15x20	40	5	1-7	89	11.45	6.50	17.95
0560		75	5	1-9	125	12.65	6.90	19.55
0580	RB1010-100	125	5	1-11	176	14.85	7.55	22.40
0600		200	5	2-2	254	18.30	8.65	26.95
0620	20x15	40	4	1-4	89	11.60	6.65	18.25
0640		75	4	1-4	119	12.95	6.95	19.90
0660		125	4	1-6	170	14.55	7.40	21.95
0680		200	4	2-1	247	17.10	8.30	25.40
0700	20x20	40	5	1-9	90	12.50	6.80	19.30
0720		75	5	1-9	126	14.20	7.35	21.55
0740		125	5	1-11	177	16.15	7.95	24.10
0760		200	5	2-5	255	19.55	8.95	28.50
0780	20x25	40	5	1-9	90	12.70	6.75	19.45
0800		75	5	1-9	127	15.95	7.80	23.75
0820		125	5	1-11	180	19.10	8.50	27.60
0840		200	5	2-5	256	21.50	9.60	31.10
0860	25x20	40	5	1-11	90	12.95	6.90	19.85
0880		75	5	2-5	127	15.15	7.65	22.80
0900		125	5	2-5	178	17.20	8.35	25.55
0920		200	5	2-5	257	22.50	9.90	32.40
0940	25x25	40	5	1-11	91	14.45	7.25	21.70
0960		75	5	2-5	178	16.70	8	24.70
0980		125	5	2-5	181	20.50	8.90	29.40
1000		200	5-1/2	2-8-1/2	263	23.50	10.30	33.80

B10 Superstructure

B1010 Floor Construction

B1010 254 — W Shape, Composite Deck, & Slab

	BAY SIZE (FT.) BEAM X GIRD	SUPERIMPOSED LOAD (P.S.F.)	SLAB THICKNESS (IN.)	TOTAL DEPTH (FT.-IN.)	TOTAL LOAD (P.S.F.)	COST PER S.F.		
						MAT.	INST.	TOTAL
1400	25x30	40	5	2-5	91	14.35	7.45	21.80
1500		75	5	2-5	128	17.55	8.30	25.85
1600		125	5	2-8	180	20	9.20	29.20
1700		200	5	2-11	259	25	10.60	35.60
1800	30x25	40	5	2-5	92	15.35	7.75	23.10
1900		75	5	2-5	129	18.15	8.60	26.75
2000		125	5	2-8	181	20.50	9.50	30
2100		200	5-1/2	2-11	200	25.50	10.95	36.45
2200	30x30	40	5	2-2	92	16	7.90	23.90
2300		75	5	2-5	129	18.75	8.80	27.55
2400		125	5	2-11	182	22.50	9.90	32.40
2500		200	5	3-2	263	30.50	12.40	42.90
2600	30x35	40	5	2-5	94	17.45	8.40	25.85
2700		75	5	2-11	131	20.50	9.25	29.75
2800		125	5	3-2	183	24	10.35	34.35
2900		200	5-1/2	3-5-1/2	268	29.50	11.95	41.45
3400	35x30	40	5	2-5	93	16.85	8.25	25.10
3500		75	5	2-8	130	20.50	9.30	29.80
3600		125	5	2-11	183	24	10.50	34.50
3700		200	5	3-5	262	29.50	12.20	41.70
3800	35x35	40	5	2-8	94	17.95	8.35	26.30
3900		75	5	2-11	131	21	9.30	30.30
4000		125	5	3-5	184	25	10.65	35.65
4100		200	5-1/2	3-5-1/2	270	32.50	12.60	45.10
4200	35x40	40	5	2-11	94	18.75	8.80	27.55
4300		75	5	3-2	131	22	9.80	31.80
4400		125	5	3-5	184	26	11.10	37.10
4500		200	5	3-5-1/2	264	33	13.30	46.30

B10 Superstructure

B1010 Floor Construction

Description: Table below lists costs ($/S.F.) for a floor system using composite steel beams with welded shear studs, composite steel deck, and light weight concrete slab reinforced with W.W.F. Price includes sprayed fiber fireproofing on steel beams.

Design and Pricing Assumptions:
 Structural steel is A36, high strength bolted.
 Composite steel deck varies from 22 gauge to 16 gauge, galvanized.

Shear Studs are 3/4".
W.W.F., 6 x 6 – W1.4 x W1.4 (10 x 10)
Concrete f'c = 3 KSI, lightweight.
Steel trowel finish and cure.
Fireproofing is sprayed fiber (non-asbestos).

Spandrels are assumed the same as interior beams and girders to allow for exterior wall loads and bracing or moment connections.

System Components	QUANTITY	UNIT	COST PER S.F.		
			MAT.	INST.	TOTAL
SYSTEM B1010 256 2400					
20 X 25 BAY, 40 PSF S. LOAD, 5-1/2" SLAB, 17-1/2" TOTAL THICKNESS					
Structural steel	4.320	Lb.	6.31	1.86	8.17
Welded shear connectors 3/4" diameter 4-7/8" long	.163	Ea.	.12	.35	.47
Metal decking, non-cellular composite, galv. 3" deep, 22 gauge	1.050	S.F.	2.29	1	3.29
Sheet metal edge closure form, 12", w/2 bends, 18 ga, galv	.045	L.F.	.20	.11	.31
Welded wire fabric rolls, 6 x 6 - W1.4 x W1.4 (10 x 10), 21 lb/csf	1.000	S.F.	.16	.36	.52
Concrete ready mix, light weight, 3,000 PSI	.333	C.F.	2.40		2.40
Place and vibrate concrete, elevated slab less than 6", pumped	.333	C.F.		.52	.52
Finishing floor, monolithic steel trowel finish for finish floor	1.000	S.F.		.88	.88
Curing with sprayed membrane curing compound	.010	C.S.F.	.09	.09	.18
Shores, erect and strip vertical to 10' high	.020	Ea.		.42	.42
Sprayed mineral fiber/cement for fireproof, 1" thick on beams	.483	S.F.	.28	.49	.77
TOTAL			11.85	6.08	17.93

B1010 256		Composite Beams, Deck & Slab						
	BAY SIZE (FT.)	SUPERIMPOSED LOAD (P.S.F.)	SLAB THICKNESS (IN.)	TOTAL DEPTH (FT.-IN.)	TOTAL LOAD (P.S.F.)	COST PER S.F.		
						MAT.	INST.	TOTAL
2400	20x25	40	5-1/2	1 - 5-1/2	80	11.85	6.10	17.95
2500		75	5-1/2	1 - 9-1/2	115	12.35	6.10	18.45
2750	RB1010-100	125	5-1/2	1 - 9-1/2	167	15.20	7.10	22.30
2900		200	6-1/4	1 - 11-1/2	251	17.10	7.70	24.80
3000	25x25	40	5-1/2	1 - 9-1/2	82	11.65	5.80	17.45
3100		75	5-1/2	1 - 11-1/2	118	13	5.90	18.90
3200		125	5-1/2	2 - 2-1/2	169	13.60	6.35	19.95
3300		200	6-1/4	2 - 6-1/4	252	18.45	7.45	25.90
3400	25x30	40	5-1/2	1 - 11-1/2	83	11.90	5.75	17.65
3600		75	5-1/2	1 - 11-1/2	119	12.85	5.85	18.70
3900		125	5-1/2	1 - 11-1/2	170	15	6.60	21.60
4000		200	6-1/4	2 - 6-1/4	252	18.55	7.45	26
4200	30x30	40	5-1/2	1 - 11-1/2	81	11.95	5.90	17.85
4400		75	5-1/2	2 - 2-1/2	116	12.95	6.15	19.10
4500		125	5-1/2	2 - 5-1/2	168	15.80	6.95	22.75
4700		200	6-1/4	2 - 9-1/4	252	18.95	8.05	27
4900	30x35	40	5-1/2	2 - 2-1/2	82	12.55	6.10	18.65
5100		75	5-1/2	2 - 5-1/2	117	13.75	6.25	20
5300		125	5-1/2	2 - 5-1/2	169	16.30	7.10	23.40
5500		200	6-1/4	2 - 9-1/4	254	19.15	8.05	27.20
5750	35x35	40	5-1/2	2 - 5-1/2	84	13.40	6.15	19.55
6000		75	5-1/2	2 - 5-1/2	121	15.35	6.60	21.95

B10 Superstructure

B1010 Floor Construction

B1010 256 Composite Beams, Deck & Slab

	BAY SIZE (FT.)	SUPERIMPOSED LOAD (P.S.F.)	SLAB THICKNESS (IN.)	TOTAL DEPTH (FT.-IN.)	TOTAL LOAD (P.S.F.)	COST PER S.F.		
						MAT.	INST.	TOTAL
7000	35 x 35	125	5-1/2	2 - 8-1/2	170	18.15	7.55	25.70
7200		200	5-1/2	2 - 11-1/2	254	21	8.35	29.35
7400	35x40	40	5-1/2	2 - 5-1/2	85	14.85	6.60	21.45
7600		75	5-1/2	2 - 5-1/2	121	16.15	6.85	23
8000		125	5-1/2	2 - 5-1/2	171	18.60	7.65	26.25
9000		200	5-1/2	2 - 11-1/2	255	22.50	8.70	31.20

B10 Superstructure

B1010 Floor Construction

How to Use Tables: Enter any table at superimposed load and support spacing for deck.

Cellular decking tends to be stiffer and has the potential for utility integration with deck structure.

System Components	QUANTITY	UNIT	COST PER S.F.		
			MAT.	INST.	TOTAL
SYSTEM B1010 258 0900					
SPAN 6', LOAD 125 PSF, DECK 1-1/2"-22 GA., 4" SLAB					
Metal decking, open, galv., 1-1/2" deep, 22 ga., over 50 sq.	1.050	S.F.	1.61	.63	2.24
Welded wire fabric rolls, 6 x 6 - W1.4 x W1.4 (10 x 10), 21 lb/csf	1.000	S.F.	.16	.36	.52
Edge forms to 6" high on elevated slab, 4 uses	.050	L.F.	.01	.21	.22
Concrete ready mix, regular weight, 3000 psi	.253	C.F.	.96		.96
Place and vibrate concrete, elevated slab less than 6", pumped	.253	C.F.		.30	.30
Finishing floor, monolithic steel trowel finish for finish floor	1.000	S.F.		.88	.88
Curing with sprayed membrane curing compound	.010	C.S.F.	.09	.09	.18
TOTAL			2.83	2.47	5.30

B1010 258 — Metal Deck/Concrete Fill

	SUPERIMPOSED LOAD (P.S.F.)	DECK SPAN (FT.)	DECK GAGE DEPTH	SLAB THICKNESS (IN.)	TOTAL LOAD (P.S.F.)	COST PER S.F.		
						MAT.	INST.	TOTAL
0900	125	6	22 1-1/2	4	164	2.83	2.47	5.30
0920		7	20 1-1/2	4	164	3.11	2.62	5.73
0950		8	20 1-1/2	4	165	3.11	2.62	5.73
0970		9	18 1-1/2	4	165	3.66	2.62	6.28
1000		10	18 2	4	165	3.56	2.76	6.32
1020		11	18 3	5	169	4.40	3	7.40
1050	150	6	22 1-1/2	4	189	2.85	2.51	5.36
1070		7	22 1-1/2	4	189	2.85	2.51	5.36
1100		8	20 1-1/2	4	190	3.11	2.62	5.73
1120		9	18 1-1/2	4	190	3.66	2.62	6.28
1150		10	18 3	5	190	4.40	3	7.40
1170		11	18 3	5-1/2	194	4.40	3	7.40
1200	200	6	22 1-1/2	4	239	2.85	2.51	5.36
1220		7	22 1-1/2	4	239	2.85	2.51	5.36
1250		8	20 1-1/2	4	239	3.11	2.62	5.73
1270		9	18 2	4	241	4.21	2.82	7.03
1300		10	16 2	4	240	4.95	2.88	7.83
1320	250	6	22 1-1/2	4	289	2.85	2.51	5.36
1350		7	22 1-1/2	4	289	2.85	2.51	5.36
1370		8	18 2	4	290	4.21	2.82	7.03
1400		9	16 2	4	290	4.95	2.88	7.83

B10 Superstructure

B1010 Floor Construction

B1010 260 Cellular Composite Deck

	SUPERIMPOSED LOAD (P.S.F.)	DECK SPAN (FT.)	DECK & PLATE GAUGE	SLAB THICKNESS (IN.)	TOTAL LOAD (P.S.F.)	COST PER S.F.		
						MAT.	INST.	TOTAL
1450	150	10	20 20	5	195	10.90	4.22	15.12
1470		11	20 20	5	195	10.90	4.22	15.12
1500		12	20 20	5	195	10.90	4.22	15.12
1520		13	20 20	5	195	10.90	4.22	15.12
1570	200	10	20 20	5	250	10.90	4.22	15.12
1650		11	20 20	5	250	10.90	4.22	15.12
1670		12	20 20	5	250	10.90	4.22	15.12
1700		13	18 20	5	250	12.95	4.26	17.21

B10 Superstructure

B1010 Floor Construction

Description: Table below lists the S.F. costs for wood joists and a minimum thickness plywood subfloor.

Design Assumptions: 10% allowance has been added to framing quantities for overlaps, waste, double joists at openings or under partitions, etc. 5% added to subfloor for waste.

System Components	QUANTITY	UNIT	COST PER S.F. MAT.	COST PER S.F. INST.	COST PER S.F. TOTAL
SYSTEM B1010 261 2500					
WOOD JOISTS 2" X 6", 12" O.C.					
Framing joists, fir, 2"x6"	1.100	B.F.	.68	.97	1.65
Subfloor plywood CDX 1/2"	1.050	S.F.	.67	.78	1.45
TOTAL			1.35	1.75	3.10

B1010 261	Wood Joist	COST PER S.F. MAT.	COST PER S.F. INST.	COST PER S.F. TOTAL
2500	Wood joists, 2"x6", 12" O.C.	1.35	1.75	3.10
2550	16" O.C.	1.18	1.51	2.69
2600	24" O.C.	1.23	1.42	2.65
2900	2"x8", 12" O.C.	1.62	1.89	3.51
2950	16" O.C.	1.39	1.62	3.01
3000	24" O.C.	1.37	1.49	2.86
3300	2"x10", 12" O.C.	1.99	2.15	4.14
3350	16" O.C.	1.66	1.81	3.47
3400	24" O.C.	1.54	1.61	3.15
3700	2"x12", 12" O.C.	2.28	2.17	4.45
3750	16" O.C.	1.87	1.82	3.69
3800	24" O.C.	1.69	1.62	3.31
4100	2"x14", 12" O.C.	2.76	2.37	5.13
4150	16" O.C.	2.24	1.98	4.22
4200	24" O.C.	1.94	1.73	3.67
4500	3"x6", 12" O.C.	2.31	2.09	4.40
4550	16" O.C.	1.90	1.77	3.67
4600	24" O.C.	1.72	1.59	3.31
4900	3"x8", 12" O.C.	3.37	2.06	5.43
4950	16" O.C.	2.69	1.74	4.43
5000	24" O.C.	2.24	1.57	3.81
5300	3"x10", 12" O.C.	4.04	2.33	6.37
5350	16" O.C.	3.21	1.95	5.16
5400	24" O.C.	2.58	1.71	4.29
5700	3"x12", 12" O.C.	4.04	2.03	6.07
5750	16" O.C.	3.71	2.31	6.02
5800	24" O.C.	2.91	1.94	4.85
6100	4"x6", 12" O.C.	3.37	2.13	5.50
6150	16" O.C.	2.69	1.79	4.48
6200	24" O.C.	2.24	1.61	3.85

B10 Superstructure

B1010 Floor Construction

Description: Table lists the S.F. costs, total load, and member sizes, for various bay sizes and loading conditions.

Design Assumptions: Dead load = girder, beams, and joist weight plus 3/4" plywood floor.

Maximum deflection is 1/360 of the clear span.

Lumber is stress grade f(w) = 1,800 PSI

System Components	QUANTITY	UNIT	COST PER S.F. MAT.	COST PER S.F. INST.	COST PER S.F. TOTAL
SYSTEM B1010 264 2000					
15' X 15' BAY, S. LOAD 40 P.S.F.					
Beams and girders, structural grade, 8" x 12"	.730	B.F.	1.71	.23	1.94
Framing joists, fir 4" x 12"	.660	B.F.	.92	.41	1.33
Framing joists, 2" x 6"	.840	B.F.	.52	.74	1.26
Beam to girder saddles	.510	Lb.	3.54	.73	4.27
Column caps	.510	Lb.	.95	.10	1.05
Drilling, bolt holes	.510	Lb.		.35	.35
Machine bolts	.510	Lb.	.15	.16	.31
Joist hangers 18 ga.	.213	Ea.	.30	.71	1.01
Subfloor plywood CDX 3/4"	1.050	S.F.	.89	.93	1.82
TOTAL			8.98	4.36	13.34

B1010 264 Wood Beam & Joist

	BAY SIZE (FT.)	SUPERIMPOSED LOAD (P.S.F.)	GIRDER BEAM (IN.)	JOISTS (IN.)	TOTAL LOAD (P.S.F.)	MAT.	INST.	TOTAL
2000	15x15	40	8 x 12 4 x 12	2 x 6 @ 16	53	9	4.36	13.36
2050	RB1010-100	75	8 x 16 4 x 16	2 x 8 @ 16	90	11.90	4.77	16.67
2100		125	12 x 16 6 x 16	2 x 8 @ 12	144	18.45	6.10	24.55
2150		200	14 x 22 12 x 16	2 x 10 @ 12	227	37	9.35	46.35
2500	15x20	40	10 x 16 8 x 12	2 x 6 @ 16	58	11.60	4.45	16.05
2550		75	12 x 14 8 x 14	2 x 8 @ 16	96	15	5.15	20.15
2600		125	10 x 18 12 x 14	2 x 8 @ 12	152	22	6.90	28.90
2650		200	14 x 20 14 x 16	2 x 10 @ 12	234	32	8.50	40.50
3000	20x20	40	10 x 14 10 x 12	2 x 8 @ 16	63	11.05	4.39	15.44
3050		75	12 x 16 8 x 16	2 x 10 @ 16	102	15.80	4.92	20.72
3100		125	14 x 22 12 x 16	2 x 10 @ 12	163	31	7.80	38.80

B10 Superstructure

B1010 Floor Construction

Description: Table below lists the S.F. costs, total load, and member sizes, for various bay sizes and loading conditions.

Design Assumptions: Dead load = girder, beams, and joist weight plus 3/4" plywood floor.

Maximum deflection is 1/360 of the clear span.

Lumber is stress grade f(w) = 2,500 PSI

System Components	QUANTITY	UNIT	COST PER S.F.		
			MAT.	INST.	TOTAL
SYSTEM B1010 265 2000					
15' X 15' BAY, 40 PSF L.L., JOISTS 2 X 6, @ 16" O.C.					
Laminated framing, straight beams	1.000	B.F.	3.23	.99	4.22
Framing joists, 2"x6"	.840	B.F.	.52	.74	1.26
Saddles	3.375	Lb.	3.54	.73	4.27
Column caps	3.375	Lb.	.95	.10	1.05
Drilling, bolt holes	3.375	Lb.		.33	.33
Machine bolts	3.375	Lb.	.13	.13	.26
Joist & beam hangers 18 ga.	.213	Ea.	.30	.71	1.01
Subfloor plywood CDX 3/4"	1.050	S.F.	.89	.93	1.82
TOTAL			9.56	4.66	14.22

B1010 265 — Laminated Wood Floor Beams

	BAY SIZE (FT.)	SUPERIMPOSED LOAD (P.S.F.)	GIRDER (IN.)	BEAM (IN.)	TOTAL LOAD (P.S.F.)	COST PER S.F.		
						MAT.	INST.	TOTAL
2000	15x15	40	5 x 12-3/8	5 x 8-1/4	54	9.55	4.66	14.21
2050		75	5 x 16-1/2	5 x 11	92	12.30	5.40	17.70
2100	RB1010-100	125	8-5/8 x 15-1/8	8-5/8 x 11	147	18.25	7.55	25.80
2150		200	8-5/8 x 22	8-5/8 x 16-1/2	128	34	11	45
2500	15x20	40	5 x 13-3/4	5 x 11	55	9.60	4.61	14.21
2550		75	5 x 17-3/8	5 x 15-1/8	92	13.90	5.60	19.50
2600		125	8-5/8 x 17-7/8	8-5/8 x 13-3/4	148	21.50	8.10	29.60
2650		200	8-5/8 x 22	8-5/8 x 17-7/8	226	30	10	40
3000	20x20	40	5 x 16-1/2	5 x 13-3/4	55	9.95	4.52	14.47
3050		75	5 x 20-5/8	5 x 16-1/2	92	13.80	5.50	19.30
3100		125	8-5/8 x 22	8-5/8 x 15-1/8	148	21	7.55	28.55
3150		200	8-5/8 x 28-7/8	8-5/8 x 20-5/8	229	31.50	10.10	41.60

B10 Superstructure

B1010 Floor Construction

Description: Table below lists the S.F. costs and maximum spans for commonly used wood decking materials for various loading conditions.

Design Assumptions: Total applied load is the superimposed load plus dead load.

Maximum deflection is 1/180 of the clear span which is not suitable if plaster ceilings will be supported by the roof or floor.

Modulus of elasticity (E) and fiber strength (f) are as shown in the table to the right.

No allowance for waste has been included.

Supporting beams or purlins are not included in costs below.

Decking Material Characteristics

Deck Material	Modulus of Elasticity	Fiber Strength
Cedar	1,100,000	1000 psi
Douglas fir	1,760,000	1200 psi
Hemlock	1,600,000	1200 psi
White spruce	1,320,000	1200 psi

B1010 266 — Wood Deck

	NOMINAL THK (IN.)	TYPE WOOD	MAXIMUM SPAN 40 P.S.F. LOAD	MAXIMUM SPAN 100 P.S.F.	MAXIMUM SPAN 250 P.S.F. LOAD	MAT.	INST.	TOTAL
1000	2	cedar	6.5	5	3.5	5.10	3.16	8.26
1050		douglas fir	8	6	4	2.46	3.16	5.62
1100		hemlock	7.5	5.5	4	2.51	3.16	5.67
1150		white spruce	7	5	3.5	1.60	3.16	4.76
1200	3	cedar	11	8	6	7.65	3.46	11.11
1250		douglas fir	13	10	7	3.70	3.46	7.16
1300		hemlock	12	9	7	3.76	3.46	7.22
1350		white spruce	11	8	6	2.40	3.46	5.86
1400	4	cedar	15	11	8	10.25	4.43	14.68
1450		douglas fir	18	13	10	4.93	4.43	9.36
1500		hemlock	17	13	9	5	4.43	9.43
1550		white spruce	16	12	8	3.20	4.43	7.63
1600	6	cedar	23	17	13	15.35	5.55	20.90
1650		douglas fir	24+	21	15	7.40	5.55	12.95
1700		hemlock	24+	20	15	7.55	5.55	13.10
1750		white spruce	24	18	13	4.80	5.55	10.35

B10 Superstructure

B1010 Floor Construction

The table below lists fireproofing costs for steel beams by type, beam size, thickness and fire rating. Weights listed are for the fireproofing material only.

System Components	QUANTITY	UNIT	COST PER L.F. MAT.	COST PER L.F. INST.	COST PER L.F. TOTAL
SYSTEM B1010 710 1300					
FIREPROOFING, 5/8" F.R. GYP. BOARD, 12" X 4" BEAM, 2" THICK, 2 HR. F.R.					
1-1/4" x 1-1/4", galvanized	.020	C.L.F.	.36	3.16	3.52
L bead for drywall, galvanized	.020	C.L.F.	.44	3.68	4.12
Furring, beams & columns, 3/4" galv. channels, 24" O.C.	2.330	S.F.	.54	5.87	6.41
Drywall on beam, no finish, 2 layers at 5/8" thick	3.000	S.F.	2.25	11.07	13.32
Drywall, taping and finishing joints, add	3.000	S.F.	.15	1.65	1.80
TOTAL			3.74	25.43	29.17

B1010 710 Steel Beam Fireproofing

	ENCASEMENT SYSTEM	BEAM SIZE (IN.)	THICKNESS (IN.)	FIRE RATING (HRS.)	WEIGHT (P.L.F.)	MAT.	INST.	TOTAL
0400	Concrete	12x4	1	1	77	6.20	27	33.20
0450	3000 PSI		1-1/2	2	93	7.20	29.50	36.70
0500			2	3	121	8.05	32	40.05
0550		14x5	1	1	100	8.10	33	41.10
0600			1-1/2	2	122	9.35	37	46.35
0650			2	3	142	10.55	40.50	51.05
0700		16x7	1	1	147	9.40	35	44.40
0750			1-1/2	2	169	10.10	36.50	46.60
0800			2	3	195	11.40	40	51.40
0850		18x7-1/2	1	1	172	10.90	40.50	51.40
0900			1-1/2	2	196	12.30	44.50	56.80
0950			2	3	225	13.70	49	62.70
1000		24x9	1	1	264	14.55	49.50	64.05
1050			1-1/2	2	295	15.80	52	67.80
1100			2	3	328	17.05	55	72.05
1150		30x10-1/2	1	1	366	19.05	62	81.05
1200			1-1/2	2	404	20.50	66	86.50
1250			2	3	449	22	69.50	91.50
1300	5/8" fire rated	12x4	2	2	15	3.74	25.50	29.24
1350	Gypsum board		2-5/8	3	24	5.15	31.50	36.65
1400		14x5	2	2	17	4	26.50	30.50
1450			2-5/8	3	27	5.65	33.50	39.15
1500		16x7	2	2	20	4.49	29.50	33.99
1550			2-5/8	3	31	5.55	29	34.55
1600		18x7-1/2	2	2	22	4.81	31.50	36.31
1650			2-5/8	3	34	6.75	40	46.75

B10 Superstructure

B1010 Floor Construction

B1010 710 Steel Beam Fireproofing

	ENCASEMENT SYSTEM	BEAM SIZE (IN.)	THICKNESS (IN.)	FIRE RATING (HRS.)	WEIGHT (P.L.F.)	COST PER L.F. MAT.	COST PER L.F. INST.	COST PER L.F. TOTAL
1700	5/8" fire rated	24x9	2	2	27	5.95	39	44.95
1750	Gypsum board		2-5/8	3	42	8.35	49	57.35
1800		30x10-1/2	2	2	33	7	45.50	52.50
1850			2-5/8	3	51	9.80	57.50	67.30
1900	Gypsum	12x4	1-1/8	3	18	5.25	19.40	24.65
1950	Plaster on	14x5	1-1/8	3	21	5.95	22	27.95
2000	Metal lath	16x7	1-1/8	3	25	6.90	25	31.90
2050		18x7-1/2	1-1/8	3	32	7.90	27.50	35.40
2100		24x9	1-1/8	3	35	9.70	34	43.70
2150		30x10-1/2	1-1/8	3	44	11.65	41	52.65
2200	Perlite plaster	12x4	1-1/8	2	16	4.99	22	26.99
2250	On metal lath		1-1/4	3	20	5.30	23	28.30
2300			1-1/2	4	22	5.70	24	29.70
2350		14x5	1-1/8	2	18	6.55	26.50	33.05
2400			1-1/4	3	23	6.90	27.50	34.40
2450			1-1/2	4	25	7.70	29.50	37.20
2500		16x7	1-1/8	2	21	6.05	27	33.05
2550			1-1/4	3	26	6.40	28	34.40
2600			1-1/2	4	29	6.75	29.50	36.25
2650		18x7-1/2	1-1/8	2	26	7.55	31.50	39.05
2700			1-1/4	3	33	7.90	32.50	40.40
2750			1-1/2	4	36	8.65	34.50	43.15
2800		24x9	1-1/8	2	30	7.80	35.50	43.30
2850			1-1/4	3	38	8.10	36.50	44.60
2900			1-1/2	4	41	8.50	38	46.50
2950		30x10-1/2	1-1/8	2	37	10.10	44	54.10
3000			1-1/4	3	46	10.50	45	55.50
3050			1-1/2	4	51	11.25	47	58.25
3100	Sprayed Fiber	12x4	5/8	1	12	1.26	2.19	3.45
3150	(non asbestos)		1-1/8	2	21	2.26	3.94	6.20
3200			1-1/4	3	22	2.52	4.38	6.90
3250		14x5	5/8	1	14	1.26	2.19	3.45
3300			1-1/8	2	26	2.26	3.94	6.20
3350			1-1/4	3	29	2.52	4.38	6.90
3400		16x7	5/8	1	18	1.59	2.77	4.36
3450			1-1/8	2	32	2.86	4.98	7.84
3500			1-1/4	3	36	3.18	5.55	8.73
3550		18x7-1/2	5/8	1	20	1.76	3.06	4.82
3600			1-1/8	2	35	3.16	5.50	8.66
3650			1-1/4	3	39	3.49	6.10	9.59
3700		24x9	5/8	1	25	2.22	3.86	6.08
3750			1-1/8	2	45	4.02	7	11.02
3800			1-1/4	3	50	4.47	7.75	12.22
3850		30x10-1/2	5/8	1	31	2.69	4.68	7.37
3900			1-1/8	2	55	4.84	8.45	13.29
3950			1-1/4	3	61	5.40	9.40	14.80
4000	On Decking	Flat	1		6	.58	.63	1.21
4050	Per S.F.	Corrugated	1		7	.87	1.21	2.08
4100		Fluted	1		7	.87	1.21	2.08

B10 Superstructure

B1010 Floor Construction

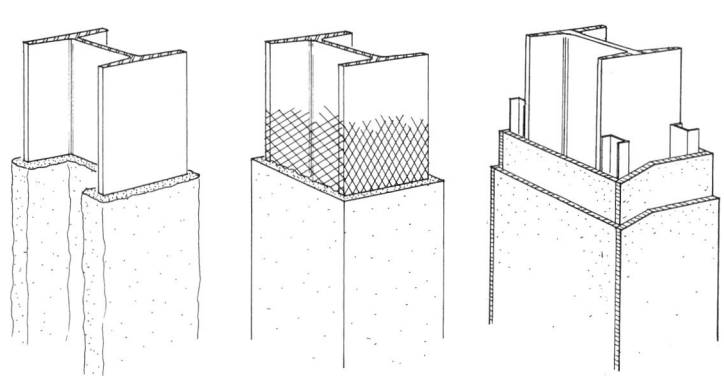

Listed below are costs per V.L.F. for fireproofing by material, column size, thickness and fire rating. Weights listed are for the fireproofing material only.

System Components	QUANTITY	UNIT	COST PER V.L.F.		
			MAT.	INST.	TOTAL
SYSTEM B1010 720 3000					
CONCRETE FIREPROOFING, 8" STEEL COLUMN, 1" THICK, 1 HR. FIRE RATING					
Forms in place, columns, plywood, 4 uses	3.330	SFCA	2.93	27.97	30.90
Welded wire fabric, 2 x 2 #14 galv. 21 lb./C.S.F., column wrap	2.700	S.F.	1.24	5.24	6.48
Concrete ready mix, regular weight, 3000 psi	.621	C.F.	2.46		2.46
Place and vibrate concrete, 12" sq./round columns, pumped	.621	C.F.		1.76	1.76
TOTAL			6.63	34.97	41.60

B1010 720			Steel Column Fireproofing				
	ENCASEMENT SYSTEM	COLUMN SIZE (IN.)	THICKNESS (IN.)	FIRE RATING (HRS.)	WEIGHT (P.L.F.)	COST PER V.L.F.		
						MAT.	INST.	TOTAL
3000	Concrete	8	1	1	110	6.65	35	41.65
3050			1-1/2	2	133	7.60	39	46.60
3100			2	3	145	8.55	42	50.55
3150		10	1	1	145	8.65	43.50	52.15
3200			1-1/2	2	168	9.60	46.50	56.10
3250			2	3	196	10.65	49.50	60.15
3300		14	1	1	258	10.90	51	61.90
3350			1-1/2	2	294	11.85	54	65.85
3400			2	3	325	13	58	71
3450	Gypsum board	8	1/2	2	8	3.25	21.50	24.75
3500	1/2" fire rated	10	1/2	2	11	3.40	22.50	25.90
3550	1 layer	14	1/2	2	18	3.48	23	26.48
3600	Gypsum board	8	1	3	14	4.33	27.50	31.83
3650	1/2" fire rated	10	1	3	17	4.61	29	33.61
3700	2 layers	14	1	3	22	4.76	30	34.76
3750	Gypsum board	8	1-1/2	3	23	5.60	34.50	40.10
3800	1/2" fire rated	10	1-1/2	3	27	6.25	38	44.25
3850	3 layers	14	1-1/2	3	35	6.90	42	48.90
3900	Sprayed fiber	8	1-1/2	2	6.3	3.93	6.85	10.78
3950	Direct application		2	3	8.3	5.40	9.45	14.85
4000			2-1/2	4	10.4	7	12.20	19.20
4050		10	1-1/2	2	7.9	4.75	8.25	13
4100			2	3	10.5	6.50	11.30	17.80
4150			2-1/2	4	13.1	8.40	14.60	23

B10 Superstructure

B1010 Floor Construction

B1010 720 Steel Column Fireproofing

	ENCASEMENT SYSTEM	COLUMN SIZE (IN.)	THICKNESS (IN.)	FIRE RATING (HRS.)	WEIGHT (P.L.F.)	COST PER V.L.F.		
						MAT.	INST.	TOTAL
4200	Sprayed fiber	14	1-1/2	2	10.8	5.90	10.25	16.15
4250	Direct application		2	3	14.5	8.05	14.05	22.10
4300			2-1/2	4	18	10.30	17.95	28.25
4350	3/4" gypsum plaster	8	3/4	1	23	6.70	27.50	34.20
4400	On metal lath	10	3/4	1	28	7.70	31	38.70
4450		14	3/4	1	38	9.35	39	48.35
4500	Perlite plaster	8	1	2	18	7.60	30.50	38.10
4550	On metal lath		1-3/8	3	23	8.30	34	42.30
4600			1-3/4	4	35	9.85	41	50.85
4650	Perlite plaster	10	1	2	21	8.75	35.50	44.25
4700			1-3/8	3	27	9.90	40.50	50.40
4750			1-3/4	4	41	11.15	46.50	57.65
4800		14	1	2	29	11	44.50	55.50
4850			1-3/8	3	35	12.40	51	63.40
4900			1-3/4	4	53	12.85	53	65.85
4950	1/2 gypsum plaster	8	7/8	1	13	6.10	23	29.10
5000	On 3/8" gypsum lath	10	7/8	1	16	7.30	27	34.30
5050		14	7/8	1	21	8.95	33	41.95
5100	5/8" gypsum plaster	8	1	1-1/2	20	6.15	25	31.15
5150	On 3/8" gypsum lath	10	1	1-1/2	24	7.30	29.50	36.80
5200		14	1	1-1/2	33	8.95	36.50	45.45
5250	1" perlite plaster	8	1-3/8	2	23	6.05	26	32.05
5300	On 3/8" gypsum lath	10	1-3/8	2	28	6.90	30	36.90
5350		14	1-3/8	2	37	8.60	38	46.60
5400	1-3/8" perlite plaster	8	1-3/4	3	27	7.50	29.50	37
5450	On 3/8" gypsum lath	10	1-3/4	3	33	8.60	34	42.60
5500		14	1-3/4	3	43	10.85	43	53.85
5550	Concrete masonry	8	4-3/4	4	126	10.30	32	42.30
5600	Units 4" thick	10	4-3/4	4	166	12.90	40	52.90
5650	75% solid	14	4-3/4	4	262	15.45	48	63.45

B10 Superstructure

B1020 Roof Construction

The table below lists prices per S.F. for roof rafters and sheathing by nominal size and spacing. Sheathing is 5/16" CDX for 12" and 16" spacing and 3/8" CDX for 24" spacing.

Factors for Converting Inclined to Horizontal

Roof Slope	Approx. Angle	Factor	Roof Slope	Approx. Angle	Factor
Flat	0°	1.000	12 in 12	45.0°	1.414
1 in 12	4.8°	1.003	13 in 12	47.3°	1.474
2 in 12	9.5°	1.014	14 in 12	49.4°	1.537
3 in 12	14.0°	1.031	15 in 12	51.3°	1.601
4 in 12	18.4°	1.054	16 in 12	53.1°	1.667
5 in 12	22.6°	1.083	17 in 12	54.8°	1.734
6 in 12	26.6°	1.118	18 in 12	56.3°	1.803
7 in 12	30.3°	1.158	19 in 12	57.7°	1.873
8 in 12	33.7°	1.202	20 in 12	59.0°	1.943
9 in 12	36.9°	1.250	21 in 12	60.3°	2.015
10 in 12	39.8°	1.302	22 in 12	61.4°	2.088
11 in 12	42.5°	1.357	23 in 12	62.4°	2.162

System Components	QUANTITY	UNIT	COST PER S.F.		
			MAT.	INST.	TOTAL
SYSTEM B1020 102 2500					
RAFTER 2" X 4", 12" O.C.					
Framing joists, fir 2"x 4"	.730	B.F.	.43	.97	1.40
Sheathing plywood on roof CDX 5/16"	1.050	S.F.	.63	.72	1.35
TOTAL			1.06	1.69	2.75

B1020 102	Wood/Flat or Pitched	COST PER S.F.		
		MAT.	INST.	TOTAL
2500	Flat rafter, 2"x4", 12" O.C.	1.06	1.69	2.75
2550	16" O.C.	.96	1.45	2.41
2600	24" O.C.	.81	1.26	2.07
2900	2"x6", 12" O.C.	1.31	1.69	3
2950	16" O.C.	1.14	1.45	2.59
3000	24" O.C.	.93	1.26	2.19
3300	2"x8", 12" O.C.	1.58	1.83	3.41
3350	16" O.C.	1.35	1.56	2.91
3400	24" O.C.	1.07	1.33	2.40
3700	2"x10", 12" O.C.	1.95	2.09	4.04
3750	16" O.C.	1.62	1.75	3.37
3800	24" O.C.	1.24	1.45	2.69
4100	2"x12", 12" O.C.	2.24	2.11	4.35
4150	16" O.C.	1.83	1.76	3.59
4200	24" O.C.	1.39	1.46	2.85
4500	2"x14", 12" O.C.	2.72	3.08	5.80
4550	16" O.C.	2.20	2.50	4.70
4600	24" O.C.	1.64	1.96	3.60
4900	3"x6", 12" O.C.	2.27	1.79	4.06
4950	16" O.C.	1.86	1.53	3.39
5000	24" O.C.	1.42	1.31	2.73
5300	3"x8", 12" O.C.	3.33	2	5.33
5350	16" O.C.	2.65	1.68	4.33
5400	24" O.C.	1.94	1.41	3.35
5700	3"x10", 12" O.C.	4	2.27	6.27
5750	16" O.C.	3.17	1.89	5.06
5800	24" O.C.	2.28	1.55	3.83
6100	3"x12", 12" O.C.	4.67	2.75	7.42
6150	16" O.C.	3.67	2.25	5.92
6200	24" O.C.	2.61	1.78	4.39

B10 Superstructure

B1020 Roof Construction

Table below lists the cost per S.F. for a roof system with steel columns, beams, and deck using open web steel joists and 1-1/2" galvanized metal deck. Perimeter of system is supported on bearing walls.

Design and Pricing Assumptions:
Columns are 18' high.
Joists are 5'-0" O.C. and span the long direction of the bay.
Joists at columns have bottom chords extended and are connected to columns. Column costs are not included but are listed separately per S.F. of floor.

Roof deck is 1-1/2", 22 gauge galvanized steel. Joist cost includes appropriate bridging. Deflection is limited to 1/240 of the span. Fireproofing is not included.

Costs/S.F. are based on a building 4 bays long and 4 bays wide.

Design Loads	Min.	Max.
Joists & Beams	3 PSF	5 PSF
Deck	2	2
Insulation	3	3
Roofing	6	6
Misc.	6	6
Total Dead Load	20 PSF	22 PSF

System Components	QUANTITY	UNIT	COST PER S.F.		
			MAT.	INST.	TOTAL
SYSTEM B1020 108 1200 METAL DECK & STEEL JOISTS,15'X20' BAY,20 PSF S. LOAD,16" DP,40 PSF T.LOAD					
Structural steel	.488	Lb.	.70	.16	.86
Open web joists	1.022	Lb.	.90	.38	1.28
Metal decking, open, galvanized, 1-1/2" deep, 22 gauge	1.050	S.F.	1.61	.63	2.24
TOTAL			3.21	1.17	4.38

B1020 108		Steel Joists, Beams & Deck on Columns & Walls						
	BAY SIZE (FT.)	SUPERIMPOSED LOAD (P.S.F.)	DEPTH (IN.)	TOTAL LOAD (P.S.F.)	COLUMN ADD	COST PER S.F.		
						MAT.	INST.	TOTAL
1200	15x20	20	16	40		3.21	1.17	4.38
1300					columns	.80	.19	.99
1400		30	16	50		3.50	1.26	4.76
1500					columns	.80	.19	.99
1600		40	18	60		3.79	1.35	5.14
1700					columns	.80	.19	.99
1800	20x20	20	16	40		3.83	1.32	5.15
1900					columns	.60	.14	.74
2000		30	18	50		3.83	1.32	5.15
2100					columns	.60	.14	.74
2200		40	18	60		4.29	1.48	5.77
2300					columns	.60	.14	.74
2400	20x25	20	18	40		3.80	1.36	5.16
2500					columns	.48	.11	.59
2600		30	18	50		3.99	1.40	5.39
2700					columns	.48	.11	.59
2800		40	20	60		4.24	1.51	5.75
2900					columns	.64	.15	.79
3000	25x25	20	18	40		4.13	1.43	5.56
3100					columns	.38	.09	.47
3200		30	22	50		4.43	1.52	5.95
3300					columns	.51	.12	.63
3400		40	20	60		4.92	1.68	6.60
3500					columns	.51	.12	.63

B10 Superstructure

B1020 Roof Construction

B1020 108 Steel Joists, Beams & Deck on Columns & Walls

	BAY SIZE (FT.)	SUPERIMPOSED LOAD (P.S.F.)	DEPTH (IN.)	TOTAL LOAD (P.S.F.)	COLUMN ADD	COST PER S.F.		
						MAT.	INST.	TOTAL
3600	25x30	20	22	40		4.38	1.38	5.76
3700					columns	.43	.10	.53
3800		30	20	50		4.96	1.54	6.50
3900					columns	.43	.10	.53
4000		40	25	60		5.15	1.58	6.73
4100					columns	.51	.12	.63
4200	30x30	20	25	42		4.82	1.48	6.30
4300					columns	.35	.09	.44
4400		30	22	52		5.25	1.61	6.86
4500					columns	.43	.10	.53
4600		40	28	62		5.50	1.66	7.16
4700					columns	.43	.10	.53
4800	30x35	20	22	42		5	1.54	6.54
4900					columns	.36	.09	.45
5000		30	28	52		5.30	1.61	6.91
5100					columns	.36	.09	.45
5200		40	25	62		5.70	1.72	7.42
5300					columns	.43	.10	.53
5400	35x35	20	28	42		5	1.55	6.55
5500					columns	.31	.08	.39
5600		30	25	52		6.10	1.82	7.92
5700					columns	.36	.09	.45
5800		40	28	62		6.20	1.85	8.05
5900					columns	.40	.10	.50

B10 Superstructure

B1020 Roof Construction

Description: Table below lists the cost per S.F. for a roof system with steel columns, beams, and deck, using open web steel joists and 1-1/2" galvanized metal deck.

Design and Pricing Assumptions:
Columns are 18' high.
Building is 4 bays long by 4 bays wide.
Joists are 5'-0" O.C. and span the long direction of the bay.
Joists at columns have bottom chords extended and are connected to columns. Column costs are not included but are listed separately per S.F. of floor.

Roof deck is 1-1/2", 22 gauge galvanized steel. Joist cost includes appropriate bridging. Deflection is limited to 1/240 of the span. Fireproofing is not included.

Design Loads	Min.	Max.
Joists & Beams	3 PSF	5 PSF
Deck	2	2
Insulation	3	3
Roofing	6	6
Misc.	6	6
Total Dead Load	20 PSF	22 PSF

System Components	QUANTITY	UNIT	COST PER S.F. MAT.	COST PER S.F. INST.	COST PER S.F. TOTAL
SYSTEM B1020 112 1100					
METAL DECK AND JOISTS, 15' X 20' BAY, 20 PSF S. LOAD					
Structural steel	.954	Lb.	1.36	.33	1.69
Open web joists	1.260	Lb.	1.11	.47	1.58
Metal decking, open, galvanized, 1-1/2" deep, 22 gauge	1.050	S.F.	1.61	.63	2.24
TOTAL			4.08	1.43	5.51

B1020 112	Steel Joists, Beams, & Deck on Columns						
	BAY SIZE (FT.)	SUPERIMPOSED LOAD (P.S.F.)	DEPTH (IN.)	TOTAL LOAD (P.S.F.)	COLUMN ADD	MAT.	INST.	TOTAL
1100	15x20	20	16	40		4.08	1.43	5.51
1200					columns	2.21	.53	2.74
1300		30	16	50		4.49	1.52	6.01
1400					columns	2.21	.53	2.74
1500		40	18	60		4.56	1.58	6.14
1600					columns	2.21	.53	2.74
1700	20x20	20	16	40		4.42	1.48	5.90
1800					columns	1.66	.39	2.05
1900		30	18	50		4.88	1.61	6.49
2000					columns	1.66	.39	2.05
2100		40	18	60		5.45	1.78	7.23
2200					columns	1.66	.39	2.05
2300	20x25	20	18	40		4.65	1.59	6.24
2400					columns	1.33	.31	1.64
2500		30	18	50		4.97	1.67	6.64
2600					columns	1.33	.31	1.64
2700		40	20	60		5.25	1.77	7.02
2800					columns	1.77	.42	2.19
2900	25x25	20	18	40		5.45	1.80	7.25
3000					columns	1.06	.25	1.31
3100		30	22	50		5.65	1.84	7.49
3200					columns	1.42	.34	1.76
3300		40	20	60		6.35	2.04	8.39
3400					columns	1.42	.34	1.76

B10 Superstructure

B1020 Roof Construction

B1020 112 — Steel Joists, Beams, & Deck on Columns

	BAY SIZE (FT.)	SUPERIMPOSED LOAD (P.S.F.)	DEPTH (IN.)	TOTAL LOAD (P.S.F.)	COLUMN ADD	COST PER S.F.		
						MAT.	INST.	TOTAL
3500	25x30	20	22	40		5.35	1.61	6.96
3600					columns	1.18	.28	1.46
3700		30	20	50		6.20	1.85	8.05
3800					columns	1.18	.28	1.46
3900		40	25	60		6.50	1.91	8.41
4000					columns	1.42	.34	1.76
4100	30x30	20	25	42		6.10	1.79	7.89
4200					columns	.98	.24	1.22
4300		30	22	52		6.70	1.96	8.66
4400					columns	1.18	.28	1.46
4500		40	28	62		7.05	2.04	9.09
4600					columns	1.18	.28	1.46
4700	30x35	20	22	42		6.25	1.85	8.10
4800					columns	1.01	.24	1.25
4900		30	28	52		6.65	1.95	8.60
5000					columns	1.01	.24	1.25
5100		40	25	62		7.30	2.11	9.41
5200					columns	1.18	.28	1.46
5300	35x35	20	28	42		6.65	1.94	8.59
5400					columns	.87	.20	1.07
5500		30	25	52		7.45	2.14	9.59
5600					columns	1.01	.24	1.25
5700		40	28	62		8.05	2.29	10.34
5800					columns	1.12	.26	1.38

B10 Superstructure

B1020 Roof Construction

Description: Table below lists cost per S.F. for a roof system using open web steel joists and 1-1/2" galvanized metal deck. The system is assumed supported on bearing walls or other suitable support. Costs for the supports are not included.

Design and Pricing Assumptions:
 Joists are 5'-0" O.C.
 Roof deck is 1-1/2", 22 gauge galvanized.

System Components	QUANTITY	UNIT	COST PER S.F. MAT.	INST.	TOTAL
SYSTEM B1020 116 1100					
METAL DECK AND JOISTS, 20' SPAN, 20 PSF S. LOAD					
Open web joists, horiz. bridging T.L. lot, to 30' span	1.114	Lb.	.98	.41	1.39
Metal decking, open type, galv 1-1/2" deep	1.050	S.F.	1.61	.63	2.24
TOTAL			2.59	1.04	3.63

B1020 116 Steel Joists & Deck on Bearing Walls

	BAY SIZE (FT.)	SUPERIMPOSED LOAD (P.S.F.)	DEPTH (IN.)	TOTAL LOAD (P.S.F.)		MAT.	INST.	TOTAL
1100	20	20	13-1/2	40		2.59	1.04	3.63
1200		30	15-1/2	50		2.65	1.06	3.71
1300		40	15-1/2	60		2.83	1.14	3.97
1400	25	20	17-1/2	40		2.84	1.15	3.99
1500		30	17-1/2	50		3.06	1.24	4.30
1600		40	19-1/2	60		3.11	1.26	4.37
1700	30	20	19-1/2	40		3.10	1.26	4.36
1800		30	21-1/2	50		3.17	1.28	4.45
1900		40	23-1/2	60		3.41	1.38	4.79
2000	35	20	23-1/2	40		3.38	1.16	4.54
2100		30	25-1/2	50		3.49	1.19	4.68
2200		40	25-1/2	60		3.71	1.25	4.96
2300	40	20	25-1/2	41		3.76	1.27	5.03
2400		30	25-1/2	51		3.99	1.34	5.33
2500		40	25-1/2	61		4.13	1.38	5.51
2600	45	20	27-1/2	41		4.53	1.50	6.03
2700		30	31-1/2	51		4.79	1.58	6.37
2800		40	31-1/2	61		5.05	1.66	6.71
2900	50	20	29-1/2	42		5.10	1.67	6.77
3000		30	31-1/2	52		5.55	1.81	7.36
3100		40	31-1/2	62		5.90	1.92	7.82
3200	60	20	37-1/2	42		6	2.15	8.15
3300		30	37-1/2	52		6.65	2.38	9.03
3400		40	37-1/2	62		6.65	2.38	9.03
3500	70	20	41-1/2	42		6.65	2.37	9.02
3600		30	41-1/2	52		7.10	2.53	9.63
3700		40	41-1/2	64		8.60	3.06	11.66

B10 Superstructure

B1020 Roof Construction

B1020 116 Steel Joists & Deck on Bearing Walls

	BAY SIZE (FT.)	SUPERIMPOSED LOAD (P.S.F.)	DEPTH (IN.)	TOTAL LOAD (P.S.F.)		COST PER S.F. MAT.	COST PER S.F. INST.	COST PER S.F. TOTAL
3800	80	20	45-1/2	44		8.35	2.97	11.32
3900		30	45-1/2	54		8.35	2.97	11.32
4000		40	45-1/2	64		9.25	3.28	12.53
4100	90	20	53-1/2	44		7.80	2.67	10.47
4200		30	53-1/2	54		8.25	2.83	11.08
4300		40	53-1/2	65		9.85	3.35	13.20
4400	100	20	57-1/2	44		8.25	2.82	11.07
4500		30	57-1/2	54		9.85	3.35	13.20
4600		40	57-1/2	65		10.85	3.69	14.54
4700	125	20	69-1/2	44		9.85	3.34	13.19
4800		30	69-1/2	56		11.45	3.87	15.32
4900		40	69-1/2	67		13	4.40	17.40

B10 Superstructure

B1020 Roof Construction

Description: Table below lists costs for a roof system supported on exterior bearing walls and interior columns. Costs include bracing, joist girders, open web steel joists and 1-1/2" galvanized metal deck.

Design and Pricing Assumptions:
  Columns are 18' high.
  Joists are 5'-0" O.C.
Joist girders and joists have bottom chords connected to columns. Roof deck is 1-1/2", 22 gauge galvanized steel. Costs include bridging and bracing. Deflection is limited to 1/240 of the span.

Fireproofing is not included.
Costs/S.F. are based on a building 4 bays long and 4 bays wide.
Costs for bearing walls are not included.
Column costs are not included but are listed separately per S.F. of floor.

System Components	QUANTITY	UNIT	COST PER S.F. MAT.	COST PER S.F. INST.	COST PER S.F. TOTAL
SYSTEM B1020 120 2050					
30' X 30' BAY SIZE, 20 PSF SUPERIMPOSED LOAD					
Joist girders, 40-ton job lots, shop primer	.475	Lb.	.38	.14	.52
Open web joists, K series, horiz bridging, 30' to 50' spans	.067	Lb.	1.23	.37	1.60
Cross bracing, rods, shop fabricated, 1" diameter	.062	Lb.	.11	.14	.25
Metal decking, open, galv., 1-1/2" deep, 22 ga., over 50 sq.	1.050	S.F.	1.61	.63	2.24
TOTAL			3.33	1.28	4.61

B1020 120 Steel Joists, Joist Girders & Deck on Columns & Walls

	BAY SIZE (FT.) GIRD X JOISTS	SUPERIMPOSED LOAD (P.S.F.)	DEPTH (IN.)	TOTAL LOAD (P.S.F.)	COLUMN ADD	MAT.	INST.	TOTAL
2050	30x30	20	17-1/2	40		3.33	1.28	4.61
2100					columns	.42	.10	.52
2150		30	17-1/2	50		3.60	1.37	4.97
2200					columns	.42	.10	.52
2250		40	21-1/2	60		3.79	1.45	5.24
2300					columns	.42	.10	.52
2350	30x35	20	32-1/2	40		3.79	1.44	5.23
2400					columns	.36	.09	.45
2450		30	36-1/2	50		3.96	1.50	5.46
2500					columns	.36	.09	.45
2550		40	36-1/2	60		4.24	1.59	5.83
2600					columns	.42	.10	.52
2650	35x30	20	36-1/2	40		3.50	1.37	4.87
2700					columns	.36	.09	.45
2750		30	36-1/2	50		3.87	1.49	5.36
2800					columns	.36	.09	.45
2850		40	36-1/2	60		5.15	1.88	7.03
2900					columns	.42	.10	.52
3000	35x35	20	36-1/2	40		4.13	1.78	5.91
3050					columns	.31	.08	.39
3100		30	36-1/2	50		4.26	1.83	6.09
3150					columns	.31	.08	.39
3200		40	36-1/2	60		4.64	1.95	6.59
3250					columns	.40	.10	.50

B10 Superstructure

B1020 Roof Construction

B1020 120 Steel Joists, Joist Girders & Deck on Columns & Walls

	BAY SIZE (FT.) GIRD X JOISTS	SUPERIMPOSED LOAD (P.S.F.)	DEPTH (IN.)	TOTAL LOAD (P.S.F.)	COLUMN ADD	COST PER S.F.		
						MAT.	INST.	TOTAL
3300	35x40	20	36-1/2	40		4.20	1.62	5.82
3350					columns	.32	.08	.40
3400		30	36-1/2	50		4.33	1.66	5.99
3450					columns	.35	.09	.44
3500		40	36-1/2	60		4.71	1.79	6.50
3550					columns	.35	.09	.44
3600	40x35	20	40-1/2	40		4.05	1.57	5.62
3650					columns	.32	.08	.40
3700		30	40-1/2	50		4.21	1.62	5.83
3750					columns	.35	.09	.44
3800		40	40-1/2	60		5.15	1.93	7.08
3850					columns	.35	.09	.44
3900	40x40	20	40-1/2	41		4.87	1.81	6.68
3950					columns	.31	.08	.39
4000		30	40-1/2	51		5.05	1.87	6.92
4050					columns	.31	.08	.39
4100		40	40-1/2	61		5.45	1.98	7.43
4150					columns	.31	.08	.39
4200	40x45	20	40-1/2	41		4.97	1.91	6.88
4250					columns	.27	.06	.33
4300		30	40-1/2	51		5.60	2.10	7.70
4350					columns	.27	.06	.33
4400		40	40-1/2	61		6.20	2.31	8.51
4450					columns	.31	.08	.39
4500	45x40	20	52-1/2	41		4.91	1.89	6.80
4550					columns	.27	.06	.33
4600		30	52-1/2	51		4.59	1.96	6.55
4650					columns	.31	.08	.39
4700		40	52-1/2	61		5.60	2.13	7.73
4750					columns	.31	.08	.39
4800	45x45	20	52-1/2	41		4.67	1.83	6.50
4850					columns	.24	.06	.30
4900		30	52-1/2	51		5.85	2.19	8.04
4950					columns	.27	.06	.33
5000		40	52-1/2	61		6	2.24	8.24
5050					columns	.31	.08	.39
5100	45x50	20	52-1/2	41		5.50	2.06	7.56
5150					columns	.22	.05	.27
5200		30	52-1/2	51		6.25	2.30	8.55
5250					columns	.25	.06	.31
5300		40	52-1/2	61		6.70	2.44	9.14
5350					columns	.32	.08	.40
5400	50x45	20	56-1/2	41		5.30	2.01	7.31
5450					columns	.22	.05	.27
5500		30	56-1/2	51		6	2.23	8.23
5550					columns	.25	.06	.31
5600		40	56-1/2	61		6.85	2.50	9.35
5650					columns	.30	.07	.37

B10 Superstructure

B1020 Roof Construction

B1020 120 Steel Joists, Joist Girders & Deck on Columns & Walls

	BAY SIZE (FT.) GIRD X JOISTS	SUPERIMPOSED LOAD (P.S.F.)	DEPTH (IN.)	TOTAL LOAD (P.S.F.)	COLUMN ADD	COST PER S.F.		
						MAT.	INST.	TOTAL
5700	50x50	20	56-1/2	42		5.95	2.24	8.19
5750					columns	.22	.05	.27
5800		30	56-1/2	53		6.45	2.41	8.86
5850					columns	.29	.06	.35
5900		40	59	64		6.85	2.73	9.58
5950					columns	.31	.08	.39
6000	55x50	20	60-1/2	43		6.05	2.29	8.34
6050					columns	.38	.09	.47
6100		30	64-1/2	54		6.95	2.57	9.52
6150					columns	.47	.11	.58
6200		40	67	65		7.10	2.81	9.91
6250					columns	.47	.11	.58
6300	60x50	20	62-1/2	43		6	2.28	8.28
6350					columns	.35	.09	.44
6400		30	68-1/2	54		6.95	2.57	9.52
6450					columns	.43	.10	.53
6500		40	71	65		7.05	2.82	9.87
6550					columns	.47	.11	.58

B10 Superstructure

B1020 Roof Construction

Description: Table below lists the cost per S.F. for a roof system supported on columns. Costs include joist girders, open web steel joists and 1-1/2" galvanized metal deck.

Design and Pricing Assumptions:
 Columns are 18' high.
 Joists are 5'-0" O.C.
Joist girders and joists have bottom chords connected to columns. Roof deck is 1-1/2", 22 gauge galvanized steel. Costs include bridging and bracing. Deflection is limited to 1/240 of the span. Fireproofing is not included.

Costs/S.F. are based on a building 4 bays long and 4 bays wide.

Costs for columns are not included, but are listed separately per S.F. of area.

System Components	QUANTITY	UNIT	COST PER S.F.		
			MAT.	INST.	TOTAL
SYSTEM B1020 124 2000 30' X 30' BAY SIZE, 20 PSF S. LOAD, 17-1/2" DEPTH, 40 PSF T. LOAD					
Joist girders, 40-ton job lots, shop primer	.758	l.b.	.61	.23	.84
Open web joists, spans to 30' average	1.580	Lb.	1.40	.58	1.98
Cross bracing, rods, shop fabricated, 1" diameter	.120	Lb.	.21	.28	.49
Metal decking, open, galv., 1-1/2" deep, 22 ga., over 50 sq.	1.050	S.F.	1.61	.63	2.24
TOTAL			3.83	1.72	5.55

B1020 124 Steel Joists & Joist Girders on Columns

	BAY SIZE (FT.) GIRD X JOISTS	SUPERIMPOSED LOAD (P.S.F.)	DEPTH (IN.)	TOTAL LOAD (P.S.F.)	COLUMN ADD	COST PER S.F.		
						MAT.	INST.	TOTAL
2000	30x30	20	17-1/2	40		3.83	1.72	5.55
2050					columns	1.18	.28	1.46
2100		30	17-1/2	50		4.16	1.86	6.02
2150					columns	1.18	.28	1.46
2200		40	21-1/2	60		4.26	1.90	6.16
2250					columns	1.18	.28	1.46
2300	30x35	20	32-1/2	40		4.07	1.61	5.68
2350					columns	1.01	.24	1.25
2400		30	36-1/2	50		4.29	1.69	5.98
2450					columns	1.01	.24	1.25
2500		40	36-1/2	60		4.58	1.78	6.36
2550					columns	1.18	.28	1.46
2600	35x30	20	36-1/2	40		3.94	1.74	5.68
2650					columns	1.01	.24	1.25
2700		30	36-1/2	50		4.36	1.91	6.27
2750					columns	1.01	.24	1.25
2800		40	36-1/2	60		4.44	1.94	6.38
2850					columns	1.18	.28	1.46
2900	35x35	20	36-1/2	40		4.42	1.95	6.37
2950					columns	.86	.20	1.06
3000		30	36-1/2	50		4.58	1.99	6.57
3050					columns	1.01	.24	1.25
3100		40	36-1/2	60		4.99	2.14	7.13
3150					columns	1.12	.27	1.39

B10 Superstructure

B1020 Roof Construction

B1020 124 Steel Joists & Joist Girders on Columns

	BAY SIZE (FT.) GIRD X JOISTS	SUPERIMPOSED LOAD (P.S.F.)	DEPTH (IN.)	TOTAL LOAD (P.S.F.)	COLUMN ADD	COST PER S.F. MAT.	COST PER S.F. INST.	COST PER S.F. TOTAL
3200	35x40	20	36-1/2	40		5.30	2.17	7.47
3250					columns	.88	.21	1.09
3300		30	36-1/2	50		5.45	2.23	7.68
3350					columns	.98	.23	1.21
3400		40	36-1/2	60		5.95	2.39	8.34
3450					columns	.98	.23	1.21
3500	40x35	20	40-1/2	40		4.68	1.99	6.67
3550					columns	.88	.21	1.09
3600		30	40-1/2	50		4.80	2.04	6.84
3650					columns	.98	.23	1.21
3700		40	40-1/2	60		5.10	2.16	7.26
3750					columns	.98	.23	1.21
3800	40x40	20	40-1/2	41		5.35	2.18	7.53
3850					columns	.85	.20	1.05
3900		30	40-1/2	51		5.55	2.25	7.80
3950					columns	.85	.20	1.05
4000		40	40-1/2	61		6	2.40	8.40
4050					columns	.85	.20	1.05
4100	40x45	20	40-1/2	41		5.75	2.48	8.23
4150					columns	.76	.18	.94
4200		30	40-1/2	51		6.25	2.63	8.88
4250					columns	.76	.18	.94
4300		40	40-1/2	61		7	2.85	9.85
4350					columns	.86	.20	1.06
4400	45x40	20	52-1/2	41		5.55	2.42	7.97
4450					columns	.76	.18	.94
4500		30	52-1/2	51		5.80	2.50	8.30
4550					columns	.76	.18	.94
4600		40	52-1/2	61		6.45	2.71	9.16
4650					columns	.86	.20	1.06
4700	45x45	20	52-1/2	41		5.85	2.46	8.31
4750					columns	.67	.16	.83
4800		30	52-1/2	51		6.60	2.70	9.30
4850					columns	.76	.18	.94
4900		40	52-1/2	61		6.75	2.76	9.51
4950					columns	.87	.20	1.07
5000	45x50	20	52-1/2	41		6.10	2.50	8.60
5050					columns	.61	.14	.75
5100		30	52-1/2	51		7	2.79	9.79
5150					columns	.69	.16	.85
5200		40	52-1/2	61		7.45	2.94	10.39
5250					columns	.78	.19	.97
5300	50x45	20	56-1/2	41		5.90	2.45	8.35
5350					columns	.61	.14	.75
5400		30	56-1/2	51		6.70	2.71	9.41
5450					columns	.69	.16	.85
5500		40	56-1/2	61		7.65	3	10.65
5550					columns	.78	.19	.97

B10 Superstructure

B1020 Roof Construction

B1020 124 Steel Joists & Joist Girders on Columns

	BAY SIZE (FT.) GIRD X JOISTS	SUPERIMPOSED LOAD (P.S.F.)	DEPTH (IN.)	TOTAL LOAD (P.S.F.)	COLUMN ADD	COST PER S.F.		
						MAT.	INST.	TOTAL
5600	50x50	20	56-1/2	42		6.60	2.69	9.29
5650					columns	.62	.15	.77
5700		30	56-1/2	53		7.25	2.89	10.14
5750					columns	.79	.19	.98
5800		40	59	64		5.95	2.57	8.52
5850					columns	.86	.20	1.06
5900	55x50	20	60-1/2	43		6.65	2.71	9.36
5950					columns	.64	.15	.79
6000		30	64-1/2	54		7.65	3.02	10.67
6050					columns	.79	.19	.98
6100		40	67	65		7.70	3.25	10.95
6150					columns	.79	.19	.98
6200	60x50	20	62-1/2	43		6.65	2.70	9.35
6250					columns	.59	.14	.73
6300		30	68-1/2	54		7.70	3.03	10.73
6350					columns	.72	.18	.90
6400		40	71	65		7.80	3.27	11.07
6450					columns	.78	.19	.97

B10 Superstructure

B1020 Roof Construction

General: Steel roof deck has the principal advantage of lightweight and speed of erection. It is covered by varying thicknesses of rigid insulation and roofing material.

Design Assumptions: Maximum spans shown in tables are based on construction and maintenance loads as specified by the Steel Deck Institute. Deflection is limited to 1/240 of the span.

Note: Costs shown in tables are based on projects over 5000 S.F. Under 5000 S.F. multiply costs by 1.16. For over four stories or congested site add 50% to the installation costs.

B1020 125	Narrow Rib, Single Span Steel Deck							
	TOTAL LOAD (P.S.F.)	MAX. SPAN (FT. - IN.)	DECK GAUGE	WEIGHT (P.S.F.)	DEPTH (IN.)	COST PER S.F.		
						MAT.	INST.	TOTAL
1250	85	3-10	22	1.80	1-1/2	1.53	.60	2.13
1300	66	4-10	20	2.15	1-1/2	1.78	.71	2.49
1350	60	5-11	18	2.85	1-1/2	2.30	.71	3.01
1360								

B1020 126	Narrow Rib, Multi Span Steel Deck							
	TOTAL LOAD (P.S.F.)	MAX. SPAN (FT.-IN.)	DECK GAUGE	WEIGHT (P.S.F.)	DEPTH (IN.)	COST PER S.F.		
						MAT.	INST.	TOTAL
1500	69	4-9	22	1.80	1-1/2	1.53	.60	2.13
1550	55	5-11	20	2.15	1-1/2	1.78	.71	2.49
1600	55	6-11	18	2.85	1-1/2	2.30	.71	3.01
1610								

B1020 127	Intermediate Rib, Single Span Steel Deck							
	TOTAL LOAD (P.S.F.)	MAX. SPAN (FT.-IN.)	DECK GAUGE	WEIGHT (P.S.F.)	DEPTH (IN.)	COST PER S.F.		
						MAT.	INST.	TOTAL
1750	72	4-6	22	1.75	1-1/2	1.53	.60	2.13
1800	65	5-3	20	2.10	1-1/2	1.78	.71	2.49
1850	64	6-2	18	2.80	1-1/2	2.30	.71	3.01
1860								

B10 Superstructure

B1020 Roof Construction

B1020 128 — Intermediate Rib, Multi Span Steel Deck

	TOTAL LOAD (P.S.F.)	MAX. SPAN (FT.-IN.)	DECK GAUGE	WEIGHT (P.S.F.)	DEPTH (IN.)	COST PER S.F.		
						MAT.	INST.	TOTAL
2000	60	5-6	22	1.75	1-1/2	1.53	.60	2.13
2050	57	6-3	20	2.10	1-1/2	1.78	.71	2.49
2100	57	7-4	18	2.80	1-1/2	2.30	.71	3.01
2110								

B1020 129 — Wide Rib, Single Span Steel Deck

	TOTAL LOAD (P.S.F.)	MAX. SPAN (FT.-IN.)	DECK GAUGE	WEIGHT (P.S.F.)	DEPTH (IN.)	COST PER S.F.		
						MAT.	INST.	TOTAL
2250	66	5-6	22	1.80	1-1/2	1.53	.60	2.13
2300	56	6-3	20	2.15	1-1/2	1.78	.71	2.49
2350	46	7-6	18	2.85	1-1/2	2.30	.71	3.01
2375								

B1020 130 — Wide Rib, Multi Span Steel Deck

	TOTAL LOAD (P.S.F.)	MAX. SPAN (FT.-IN.)	DECK GAUGE	WEIGHT (P.S.F.)	DEPTH (IN.)	COST PER S.F.		
						MAT.	INST.	TOTAL
2500	76	6-6	22	1.80	1-1/2	1.53	.60	2.13
2550	64	7-5	20	2.15	1-1/2	1.78	.71	2.49
2600	55	8-10	18	2.80	1-1/2	2.30	.71	3.01
2610								

B10 Superstructure

B1020 Roof Construction

Design & Pricing Assumptions:
Costs in table below are based upon 5000 S.F. or more.
Deflection is limited to 1/240 of the span.

B1020 131			Single Span, Steel Deck					
	TOTAL LOAD (P.S.F.)	DECK SUPPORT SPACING (FT.)	DEPTH (IN.)	DECK GAUGE	WEIGHT (P.S.F.)	COST PER S.F.		
						MAT.	INST.	TOTAL
0850	30	10	3	20	2.5	3.37	.90	4.27
0900		12	3	18	3.0	4.36	.95	5.31
0950		14	3	16	4.0	5.75	1.01	6.76
1000		16	4.5	20	3.0	3.77	1.13	4.90
1050		18	4.5	18	4.0	4.98	1.23	6.21
1100		20	4.5	16	5.0	6.50	1.30	7.80
1200		24	6.0	16	6.0	7.45	1.58	9.03
1300		28	7.5	16	7.0	8.85	1.92	10.77
1350	40	10	3.0	20	2.5	3.37	.90	4.27
1400		12	3.0	18	3.0	4.36	.95	5.31
1450		14	3.0	16	4.0	5.75	1.01	6.76
1500		16	4.5	20	3.0	3.77	1.13	4.90
1550		18	4.5	18	4.0	4.98	1.23	6.21
1600		20	6.0	16	6.0	7.45	1.58	9.03
1700		24	7.5	16	7.0	8.85	1.92	10.77
1750	50	10	3.0	20	2.5	3.37	.90	4.27
1800		12	3.0	18	3.0	4.36	.95	5.31
1850		14	3.0	16	4.0	5.75	1.01	6.76
1900		16	4.5	18	4.0	4.98	1.23	6.21
1950		18	4.5	16	5.0	6.50	1.30	7.80
2000		20	6.0	16	6.0	5.95	1.52	7.47
2100		24	7.5	16	7.0	8.85	1.92	10.77

B1020 132			Double Span, Steel Deck					
	TOTAL LOAD (P.S.F.)	DECK SUPPORT SPACING (FT.)	DEPTH (IN.)	DECK GAUGE	WEIGHT (P.S.F.)	COST PER S.F.		
						MAT.	INST.	TOTAL
2250	30	10	3	20	2.5	3.37	.90	4.27
2300		12	3	20	2.5	3.37	.90	4.27
2350		14	3	18	3.0	4.36	.95	5.31
2400		16	3	16	4.0	5.75	1.01	6.76
2450	40	10	3	20	2.5	3.37	.90	4.27
2500		12	3	20	2.5	3.37	.90	4.27
2550		14	3	18	3.0	4.36	.95	5.31
2600		16	3	16	4.0	5.75	1.01	6.76
2650	50	10	3	20	2.5	3.37	.90	4.27
2700		12	3	18	3.0	4.36	.95	5.31
2750		14	3	16	4.0	5.75	1.01	6.76
2800		16	4.5	20	4.0	3.77	1.13	4.90

B10 Superstructure

B1020 Roof Construction

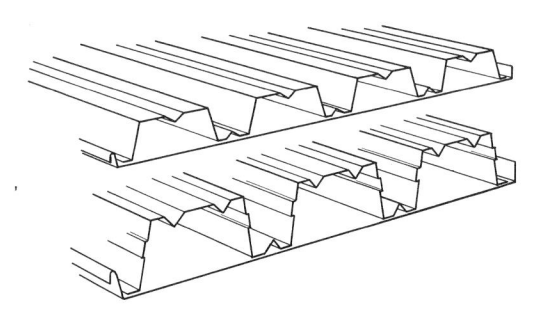

Design & Pricing Assumptions:
Costs in table below are based upon 15,000 S.F. or more.
No utility outlets or modifications included in cost.
Deflection is limited to 1/240 of the span.

B1020 134 — Cellular Steel Deck, Single Span

	TOTAL LOAD (P.S.F.)	DECK SUPPORT SPACING (FT.)	DEPTH (IN.)	WEIGHT (P.S.F.)	COST PER S.F.		
					MAT.	INST.	TOTAL
1150	30	10	1-5/8	4	9.35	2.10	11.45
1200		12	1-5/8	5	9.75	2.19	11.94
1250		14	3	5	11.15	2.26	13.41
1300		16	3	5	11.10	2.36	13.46
1350		18	3	5	11.10	2.36	13.46
1400		20	3	6	12.50	2.48	14.98
1450	40	8	1-5/8	4	9.35	2.10	11.45
1500		10	1-5/8	5	9.75	2.19	11.94
1550		12	3	5	11.10	2.36	13.46
1600		14	3	5	11.10	2.36	13.46
1650		16	3	5	11.10	2.36	13.46
1700		18	3	6	12.50	2.48	14.98
1750	50	8	1-5/8	4	9.35	2.10	11.45
1800		10	1-5/8	5	9.75	2.19	11.94
1850		12	3	4	11.15	2.26	13.41
1900		14	3	5	11.10	2.36	13.46
1950		16	3	6	12.50	2.48	14.98
1960							

B1020 135 — Cellular Steel Deck, Double Span

	TOTAL LOAD (P.S.F.)	DECK SUPPORT SPACING (FT.)	DEPTH (IN.)	WEIGHT (P.S.F.)	COST PER S.F.		
					MAT.	INST.	TOTAL
2150	30	10	1-5/8	4	9.35	2.10	11.45
2200		12	1-5/8	5	9.75	2.19	11.94
2250		14	1-5/8	5	9.75	2.19	11.94
2300		16	3	5	11.10	2.36	13.46
2350		18	3	5	11.10	2.36	13.46
2400		20	3	6	12.50	2.48	14.98
2450	40	8	1-5/8	4	9.35	2.10	11.45
2500		10	1-5/8	5	9.75	2.19	11.94
2550		12	1-5/8	5	9.75	2.19	11.94
2600		14	3	5	11.10	2.36	13.46
2650		16	3	5	11.10	2.36	13.46
2700		18	3	6	12.50	2.48	14.98

B10 Superstructure

B1020 Roof Construction

B1020 135 — Cellular Steel Deck, Double Span

	TOTAL LOAD (P.S.F.)	DECK SUPPORT SPACING (FT.)	DEPTH (IN.)	WEIGHT (P.S.F.)		COST PER S.F.		
						MAT.	INST.	TOTAL
2750	50	8	1-5/8	4		9.35	2.10	11.45
2800		10	1-5/8	5		9.75	2.19	11.94
2850		12	1-5/8	5		9.75	2.19	11.94
2900		14	3	5		11.10	2.36	13.46
2950		16	3	6		12.50	2.48	14.98
2960								

B1020 136 — Cellular Steel Deck, Triple Span

	TOTAL LOAD (P.S.F.)	DECK SUPPORT SPACING (FT.)	DEPTH (IN.)	WEIGHT (P.S.F.)		COST PER S.F.		
						MAT.	INST.	TOTAL
3200	30	10	1-5/8	4		9.35	2.10	11.45
3210								
3250	40	8	1-5/8	4		9.35	2.10	11.45
3300		10	1-5/8	4		9.35	2.10	11.45
3350		12	1-5/8	5		9.75	2.19	11.94
3400	50	8	1-5/8	4		9.35	2.10	11.45
3450		10	1-5/8	5		9.75	2.19	11.94
3500		12	1-5/8	5		9.75	2.19	11.94

B1020 310 — Canopies

	Canopies	COST PER S.F.		
		MAT.	INST.	TOTAL
0100	Canopies, wall hung, prefinished aluminum, 8' x 10'	30	21.50	51.50
0110	12' x 40'	23	7.80	30.80
0120	Canopies, wall hung, canvas awnings, with frame and lettering	83.50	12.30	95.80
0130	Freestanding (walkways), vinyl coated steel, 12' wide	25.50	9.05	34.55

B20 Exterior Enclosure

B2010 Exterior Walls

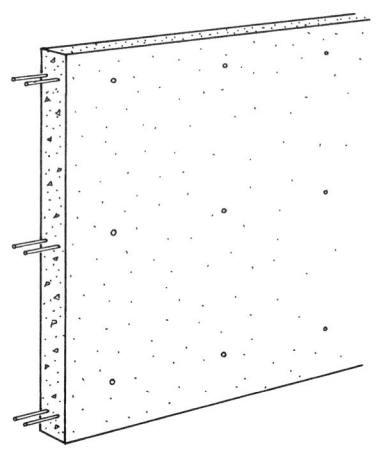

The table below describes a concrete wall system for exterior closure. There are several types of wall finishes priced from plain finish to a finish with 3/4" rustication strip.

Design Assumptions:
Conc. f'c = 3000 to 5000 psi
Reinf. fy = 60,000 psi

System Components	QUANTITY	UNIT	COST PER S.F.		
			MAT.	INST.	TOTAL
SYSTEM B2010 101 2100					
CONC. WALL, REINFORCED, 8' HIGH, 6" THICK, PLAIN FINISH, 3,000 PSI					
Forms in place, wall, job built plyform to 8' high, 4 uses	2.000	SFCA	2.12	12.80	14.92
Reinforcing in place, walls, #3 to #7	.752	Lb.	.43	.32	.75
Concrete ready mix, regular weight, 3000 psi	.018	C.Y.	1.93		1.93
Place and vibrate concrete, walls 6" thick, pump	.018	C.Y.		.83	.83
Finish wall, break ties, patch voids	2.000	S.F.	.08	1.88	1.96
TOTAL			4.56	15.83	20.39

B2010 101	Cast In Place Concrete	COST PER S.F.		
		MAT.	INST.	TOTAL
2100	Conc wall reinforced, 8' high, 6" thick, plain finish, 3000 PSI	4.56	15.85	20.41
2200	4000 PSI	4.65	15.85	20.50
2300	5000 PSI	4.77	15.85	20.62
2400	Rub concrete 1 side, 3000 PSI	4.56	18.75	23.31
2500	4000 PSI	4.65	18.75	23.40
2600	5000 PSI	4.77	18.75	23.52
2700	Aged wood liner, 3000 PSI	5.75	18	23.75
2800	4000 PSI	5.85	18	23.85
2900	5000 PSI	5.95	18	23.95
3000	Sand blast light 1 side, 3000 PSI	5.40	18.05	23.45
3100	4000 PSI	5.45	18.05	23.50
3300	5000 PSI	5.60	18.05	23.65
3400	Sand blast heavy 1 side, 3000 PSI	6.20	22.50	28.70
3500	4000 PSI	6.30	22.50	28.80
3600	5000 PSI	6.40	22.50	28.90
3700	3/4" bevel rustication strip, 3000 PSI	4.73	16.80	21.53
3800	4000 PSI	4.82	16.80	21.62
3900	5000 PSI	4.94	16.80	21.74
4000	8" thick, plain finish, 3000 PSI	5.40	16.25	21.65
4100	4000 PSI	5.50	16.25	21.75
4200	5000 PSI	5.65	16.25	21.90
4300	Rub concrete 1 side, 3000 PSI	5.40	19.15	24.55
4400	4000 PSI	5.50	19.15	24.65
4500	5000 PSI	5.65	19.15	24.80
4550	8" thick, aged wood liner, 3000 PSI	6.55	18.40	24.95
4600	4000 PSI	6.70	18.40	25.10

B20 Exterior Enclosure

B2010 Exterior Walls

B2010 101	Cast In Place Concrete	COST PER S.F.		
		MAT.	INST.	TOTAL
4700	5000 PSI	6.85	18.40	25.25
4750	Sand blast light 1 side, 3000 PSI	6.20	18.45	24.65
4800	4000 PSI	6.30	18.45	24.75
4900	5000 PSI	6.50	18.45	24.95
5000	Sand blast heavy 1 side, 3000 PSI	7	22.50	29.50
5100	4000 PSI	7.15	22.50	29.65
5200	5000 PSI	7.30	22.50	29.80
5300	3/4" bevel rustication strip, 3000 PSI	5.55	17.20	22.75
5400	4000 PSI	5.65	17.20	22.85
5500	5000 PSI	5.85	17.20	23.05
5600	10" thick, plain finish, 3000 PSI	6.15	16.60	22.75
5700	4000 PSI	6.30	16.60	22.90
5800	5000 PSI	6.55	16.60	23.15
5900	Rub concrete 1 side, 3000 PSI	6.15	19.50	25.65
6000	4000 PSI	6.30	19.50	25.80
6100	5000 PSI	6.55	19.50	26.05
6200	Aged wood liner, 3000 PSI	7.35	18.80	26.15
6300	4000 PSI	7.50	18.80	26.30
6400	5000 PSI	7.70	18.80	26.50
6500	Sand blast light 1 side, 3000 PSI	7	18.80	25.80
6600	4000 PSI	7.15	18.80	25.95
6700	5000 PSI	7.35	18.80	26.15
6800	Sand blast heavy 1 side, 3000 PSI	7.80	23	30.80
6900	4000 PSI	7.95	23	30.95
7000	5000 PSI	8.15	23	31.15
7100	3/4" bevel rustication strip, 3000 PSI	6.35	17.55	23.90
7200	4000 PSI	6.50	17.55	24.05
7300	5000 PSI	6.70	17.55	24.25
7400	12" thick, plain finish, 3000 PSI	7.10	17.05	24.15
7500	4000 PSI	7.30	17.05	24.35
7600	5000 PSI	7.55	17.05	24.60
7700	Rub concrete 1 side, 3000 PSI	7.10	19.95	27.05
7800	4000 PSI	7.30	19.95	27.25
7900	5000 PSI	7.55	19.95	27.50
8000	Aged wood liner, 3000 PSI	8.30	19.25	27.55
8100	4000 PSI	8.45	19.25	27.70
8200	5000 PSI	8.75	19.25	28
8300	Sand blast light 1 side, 3000 PSI	7.95	19.25	27.20
8400	4000 PSI	8.10	19.25	27.35
8500	5000 PSI	8.35	19.25	27.60
8600	Sand blast heavy 1 side, 3000 PSI	8.75	23.50	32.25
8700	4000 PSI	8.95	23.50	32.45
8800	5000 PSI	9.20	23.50	32.70
8900	3/4" bevel rustication strip, 3000 PSI	7.30	18	25.30
9000	4000 PSI	7.45	18	25.45
9500	5000 PSI	7.70	18	25.70

B20 Exterior Enclosure

B2010 Exterior Walls

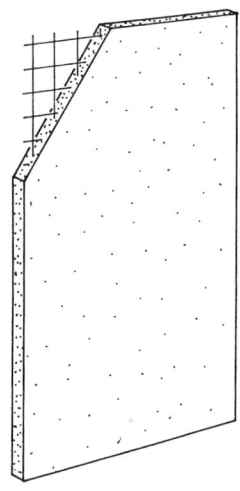

Precast concrete wall panels are either solid or insulated with plain, colored or textured finishes. Transportation is an important cost factor. Prices below are based on delivery within fifty miles of a plant. Engineering data is available from fabricators to assist with construction details. Usual minimum job size for economical use of panels is about 5000 S.F. Small jobs can double the prices below. For large, highly repetitive jobs, deduct up to 15% from the prices below.

B2010 102			Flat Precast Concrete					
	THICKNESS (IN.)	PANEL SIZE (FT.)	FINISHES	RIGID INSULATION (IN)	TYPE	COST PER S.F.		
						MAT.	INST.	TOTAL
3000	4	5x18	smooth gray	none	low rise	12.95	7.80	20.75
3050		6x18				10.80	6.55	17.35
3100		8x20				21	3.16	24.16
3150		12x20				19.90	2.97	22.87
3200	6	5x18	smooth gray	2	low rise	13.85	8.35	22.20
3250		6x18				11.75	7.05	18.80
3300		8x20				22.50	3.81	26.31
3350		12x20				20	3.49	23.49
3400	8	5x18	smooth gray	2	low rise	29	4.79	33.79
3450		6x18				27.50	4.57	32.07
3500		8x20				25	4.21	29.21
3550		12x20				22.50	3.86	26.36
3600	4	4x8	white face	none	low rise	48.50	4.52	53.02
3650		8x8				36.50	3.38	39.88
3700		10x10				32	2.97	34.97
3750		20x10				29	2.69	31.69
3800	5	4x8	white face	none	low rise	49.50	4.61	54.11
3850		8x8				37.50	3.48	40.98
3900		10x10				33.50	3.08	36.58
3950		20x20				30.50	2.82	33.32
4000	6	4x8	white face	none	low rise	51.50	4.77	56.27
4050		8x8				39	3.62	42.62
4100		10x10				34.50	3.20	37.70
4150		20x10				31.50	2.93	34.43
4200	6	4x8	white face	2	low rise	52.50	5.40	57.90
4250		8x8				40	4.24	44.24
4300		10x10				35.50	3.82	39.32
4350		20x10				31.50	2.93	34.43
4400	7	4x8	white face	none	low rise	53	4.89	57.89
4450		8x8				40.50	3.76	44.26
4500		10x10				36.50	3.37	39.87
4550		20x10				33.50	3.08	36.58
4600	7	4x8	white face	2	low rise	54	5.50	59.50
4650		8x8				41.50	4.38	45.88
4700		10x10				37.50	3.99	41.49

B20 Exterior Enclosure

B2010 Exterior Walls

B2010 102 Flat Precast Concrete

	THICKNESS (IN.)	PANEL SIZE (FT.)	FINISHES	RIGID INSULATION (IN)	TYPE	COST PER S.F. MAT.	COST PER S.F. INST.	COST PER S.F. TOTAL
4750	7	20x10				34.50	3.70	38.20
4800	8	4x8	white face	none	low rise	54	5	59
4850		8x8				41.50	3.85	45.35
4900		10x10				37.50	3.46	40.96
4950		20x10				34.50	3.19	37.69
5000	8	4x8	white face	2	low rise	55	5.60	60.60
5050		8x8				42.50	4.47	46.97
5100		10x10				38.50	4.08	42.58
5150		20x10				35.50	3.81	39.31

B2010 103 Fluted Window or Mullion Precast Concrete

	THICKNESS (IN.)	PANEL SIZE (FT.)	FINISHES	RIGID INSULATION (IN)	TYPE	COST PER S.F. MAT.	COST PER S.F. INST.	COST PER S.F. TOTAL
5200	4	4x8	smooth gray	none	high rise	29.50	17.85	47.35
5250		8x8				21	12.80	33.80
5300		10x10				40	6.05	46.05
5350		20x10				35.50	5.30	40.80
5400	5	4x8	smooth gray	none	high rise	30	18.20	48.20
5450		8x8				22	13.25	35.25
5500		10x10				42	6.30	48.30
5550		20x10				37	5.55	42.55
5600	6	4x8	smooth gray	none	high rise	31	18.65	49.65
5650		8x8				22.50	13.65	36.15
5700		10x10				43	6.45	49.45
5750		20x10				38.50	5.75	44.25
5800	6	4x8	smooth gray	2	high rise	32	19.25	51.25
5850		8x8				23.50	14.30	37.80
5900		10x10				44	7.05	51.05
5950		20x10				39.50	6.35	45.85
6000	7	4x8	smooth gray	none	high rise	31.50	19.05	50.55
6050		8x8				23.50	14.05	37.55
6100		10x10				45	6.75	51.75
6150		20x10				39.50	5.90	45.40
6200	7	4x8	smooth gray	2	high rise	32.50	19.65	52.15
6250		8x8				24.50	14.65	39.15
6300		10x10				46	7.35	53.35
6350		20x10				40.50	6.55	47.05
6400	8	4x8	smooth gray	none	high rise	32	19.35	51.35
6450		8x8				24	14.40	38.40
6500		10x10				46.50	6.95	53.45
6550		20x10				41.50	6.20	47.70
6600	8	4x8	smooth gray	2	high rise	33	20	53
6650		8x8				25	15.05	40.05
6700		10x10				47.50	7.60	55.10
6750		20x10				42.50	6.85	49.35

B2010 104 Ribbed Precast Concrete

	THICKNESS (IN.)	PANEL SIZE (FT.)	FINISHES	RIGID INSULATION (IN.)	TYPE	COST PER S.F. MAT.	COST PER S.F. INST.	COST PER S.F. TOTAL
6800	4	4x8	aggregate	none	high rise	34.50	17.85	52.35
6850		8x8				25.50	13.05	38.55
6900		10x10				43	4.74	47.74

B20 Exterior Enclosure

B2010 Exterior Walls

B2010 104 Ribbed Precast Concrete

	THICKNESS (IN.)	PANEL SIZE (FT.)	FINISHES	RIGID INSULATION (IN.)	TYPE	COST PER S.F.		
						MAT.	INST.	TOTAL
6950	4	20x10	aggregate	none	high rise	38	4.18	42.18
7000	5	4x8	aggregate	none	high rise	35	18.20	53.20
7050		8x8				26	13.40	39.40
7100		10x10				45	4.91	49.91
7150		20x10				40	4.39	44.39
7200	6	4x8	aggregate	none	high rise	36	18.55	54.55
7250		8x8				26.50	13.80	40.30
7300		10x10				46.50	5.05	51.55
7350		20x10				41.50	4.55	46.05
7400	6	4x8	aggregate	2	high rise	37	19.20	56.20
7450		8x8				28	14.40	42.40
7500		10x10				47.50	5.70	53.20
7550		20x10				42.50	5.15	47.65
7600	7	4x8	aggregate	none	high rise	36.50	19	55.50
7650		8x8				27.50	14.15	41.65
7700		10x10				47.50	5.25	52.75
7750		20x10				42.50	4.68	47.18
7800	7	4x8	aggregate	2	high rise	38	19.60	57.60
7850		8x8				28.50	14.75	43.25
7900		10x10				49	5.85	54.85
7950		20x10				43.50	5.30	48.80
8000	8	4x8	aggregate	none	high rise	37.50	19.25	56.75
8050		8x8				28	14.55	42.55
8100		10x10				49.50	5.40	54.90
8150		20x10				44.50	4.90	49.40
8200	8	4x8	aggregate	2	high rise	38.50	19.90	58.40
8250		8x8				29.50	15.20	44.70
8300		10x10				50.50	6.05	56.55
8350		20x10				45.50	5.50	51

B2010 105 Precast Concrete Specialties

	TYPE	SIZE				COST PER L.F.		
						MAT.	INST.	TOTAL
8400	Coping, precast	6" wide				20	13.05	33.05
8450	Stock units	10" wide				21.50	14	35.50
8460		12" wide				23.50	15.05	38.55
8480		14" wide				25	16.30	41.30
8500	Window sills	6" wide				13.50	14	27.50
8550	Precast	10" wide				17.60	16.30	33.90
8600		14" wide				17.60	19.55	37.15
8610								

B20 Exterior Enclosure

B2010 Exterior Walls

The advantage of tilt up construction is in the low cost of forms and placing of concrete and reinforcing. Tilt up has been used for several types of buildings, including warehouses, stores, offices, and schools. The panels are cast in forms on the ground, or floor slab. Most jobs use 5-1/2" thick solid reinforced concrete panels.

Design Assumptions:
Conc. f'c = 3000 psi
Reinf. fy = 60,000

System Components	QUANTITY	UNIT	COST EACH		
			MAT.	INST.	TOTAL
SYSTEM B2010 106 3200					
TILT-UP PANELS, 20' X 25', BROOM FINISH, 5-1/2" THICK, 3000 PSI					
Apply liquid bond release agent	500.000	S.F.	10	65	75
Edge forms in place for slab on grade	120.000	L.F.	40.80	418.80	459.60
Reinforcing in place	.350	Ton	385	385	770
Footings, form braces, steel	1.000	Set	476		476
Slab lifting inserts	1.000	Set	64		64
Framing, less than 4" angles	1.000	Set	81.70	836	917.70
Concrete, ready mix, regular weight, 3000 psi	8.550	C.Y.	914.85		914.85
Place and vibrate concrete for slab on grade, 4" thick, direct chute	8.550	C.Y.		215.12	215.12
Finish floor, monolithic broom finish	500.000	S.F.		390	390
Cure with curing compound, sprayed	500.000	S.F.	44.25	46	90.25
Erection crew	.058	Day		667	667
TOTAL	500.000		2,016.60	3,022.92	5,039.52
COST PER S.F.			4.03	6.05	10.08

B2010 106	Tilt-Up Concrete Panel	COST PER S.F.		
		MAT.	INST.	TOTAL
3200	Tilt up conc panels, broom finish, 5-1/2" thick, 3000 PSI	4.05	6.05	10.10
3250	5000 PSI	4.12	5.95	10.07
3300	6" thick, 3000 PSI	4.44	6.20	10.64
3350	5000 PSI	4.55	6.10	10.65
3400	7-1/2" thick, 3000 PSI	5.65	6.45	12.10
3450	5000 PSI	5.80	6.35	12.15
3500	8" thick, 3000 PSI	6.05	6.60	12.65
3550	5000 PSI	6.25	6.50	12.75
3700	Steel trowel finish, 5-1/2" thick, 3000 PSI	4.03	6.15	10.18
3750	5000 PSI	4.12	6.05	10.17
3800	6" thick, 3000 PSI	4.44	6.30	10.74
3850	5000 PSI	4.55	6.20	10.75
3900	7-1/2" thick, 3000 PSI	5.65	6.50	12.15
3950	5000 PSI	5.80	6.40	12.20
4000	8" thick, 3000 PSI	6.05	6.65	12.70
4050	5000 PSI	6.25	6.55	12.80
4200	Exp. aggregate finish, 5-1/2" thick, 3000 PSI	4.41	6.20	10.61
4250	5000 PSI	4.50	6.10	10.60
4300	6" thick, 3000 PSI	4.82	6.35	11.17
4350	5000 PSI	4.93	6.25	11.18
4400	7-1/2" thick, 3000 PSI	6	6.60	12.60
4450	5000 PSI	6.20	6.45	12.65

B20 Exterior Enclosure

B2010 Exterior Walls

B2010 106	Tilt-Up Concrete Panel	COST PER S.F.		
		MAT.	INST.	TOTAL
4500	8" thick, 3000 PSI	6.45	6.75	13.20
4550	5000 PSI	6.65	6.60	13.25
4600	Exposed aggregate & vert. rustication 5-1/2" thick, 3000 PSI	6.65	7.65	14.30
4650	5000 PSI	6.75	7.55	14.30
4700	6" thick, 3000 PSI	7.05	7.80	14.85
4750	5000 PSI	7.15	7.70	14.85
4800	7-1/2" thick, 3000 PSI	8.25	8	16.25
4850	5000 PSI	8.45	7.90	16.35
4900	8" thick, 3000 PSI	8.70	8.15	16.85
4950	5000 PSI	8.85	8.05	16.90
5000	Vertical rib & light sandblast, 5-1/2" thick, 3000 PSI	6.65	10	16.65
5050	5000 PSI	6.75	9.90	16.65
5100	6" thick, 3000 PSI	7.10	10.15	17.25
5150	5000 PSI	7.20	10.05	17.25
5200	7-1/2" thick, 3000 PSI	8.30	10.40	18.70
5250	5000 PSI	8.45	10.30	18.75
5300	8" thick, 3000 PSI	8.70	10.55	19.25
5350	5000 PSI	8.90	10.45	19.35
6000	Broom finish w/2" polystyrene insulation, 6" thick, 3000 PSI	3.84	7.45	11.29
6050	5000 PSI	3.99	7.45	11.44
6100	Broom finish 2" fiberplank insulation, 6" thick, 3000 PSI	4.33	7.35	11.68
6150	5000 PSI	4.48	7.35	11.83
6200	Exposed aggregate w/2" polystyrene insulation, 6" thick, 3000 PSI	4.11	7.45	11.56
6250	5000 PSI	4.26	7.45	11.71
6300	Exposed aggregate 2" fiberplank insulation, 6" thick, 3000 PSI	4.60	7.35	11.95
6350	5000 PSI	4.75	7.35	12.10

B20 Exterior Enclosure

B2010 Exterior Walls

Exterior concrete block walls are defined in the following terms; structural reinforcement, weight, percent solid, size, strength and insulation. Within each of these categories, two to four variations are shown. No costs are included for brick shelf or relieving angles.

System Components	QUANTITY	UNIT	COST PER S.F.		
			MAT.	INST.	TOTAL
SYSTEM B2010 109 1400					
UNREINFORCED CONCRETE BLOCK WALL, 8″ X 8″ X 16″, PERLITE CORE FILL					
Concrete block wall, 8″ thick	1.000	S.F.	2.50	6.65	9.15
Perlite insulation	1.000	S.F.	1.70	.41	2.11
Horizontal joint reinforcing, alternate courses	.800	S.F.	.16	.17	.33
Control joint	.050	L.F.	.07	.07	.14
TOTAL			4.43	7.30	11.73

B2010 109		Concrete Block Wall - Regular Weight						
	TYPE	SIZE (IN.)	STRENGTH (P.S.I.)	CORE FILL		COST PER S.F.		
						MAT.	INST.	TOTAL
1200	Hollow	4x8x16	2,000	none		1.85	6	7.85
1250			4,500	none		2.25	6	8.25
1300	RB2010 -200	6x8x16	2,000	perlite		3.70	6.80	10.50
1310				styrofoam		3.80	6.45	10.25
1340				none		2.55	6.45	9
1350			4,500	perlite		3.92	6.80	10.72
1360				styrofoam		4.02	6.45	10.47
1390				none		2.77	6.45	9.22
1400		8x8x16	2,000	perlite		4.43	7.30	11.73
1410				styrofoam		3.98	6.90	10.88
1440				none		2.73	6.90	9.63
1450			4,500	perlite		5.10	7.30	12.40
1460				styrofoam		4.63	6.90	11.53
1490				none		3.38	6.90	10.28
1500		12x8x16	2,000	perlite		6.85	9.75	16.60
1510				styrofoam		6.25	8.95	15.20
1540				none		4.07	8.95	13.02
1550			4,500	perlite		7.25	9.75	17
1560				styrofoam		6.65	8.95	15.60
1590				none		4.47	8.95	13.42
2000	75% solid	4x8x16	2,000	none		2.23	6.10	8.33
2050			4,500	none		2.76	6.10	8.86

B20 Exterior Enclosure

B2010 Exterior Walls

B2010 109 Concrete Block Wall - Regular Weight

	TYPE	SIZE (IN.)	STRENGTH (P.S.I.)	CORE FILL		COST PER S.F.		
						MAT.	INST.	TOTAL
2100	75% solid	6x8x16	2,000	perlite		3.67	6.75	10.42
2140				none		3.09	6.55	9.64
2150			4,500	perlite		3.96	6.75	10.71
2190				none		3.38	6.55	9.93
2200		8x8x16	2,000	perlite		4.13	7.20	11.33
2240				none		3.28	7	10.28
2250			4,500	perlite		4.98	7.20	12.18
2290				none		4.13	7	11.13
2300		12x8x16	2,000	perlite		6.30	9.45	15.75
2340				none		4.93	9.05	13.98
2350			4,500	perlite		6.85	9.45	16.30
2390				none		5.45	9.05	14.50
2500	Solid	4x8x16	2,000	none		2.11	6.20	8.31
2550			4,500	none		3.09	6.15	9.24
2600		6x8x16	2,000	none		2.15	6.70	8.85
2650			4,500	none		3.79	6.60	10.39
2700		8x8x16	2,000	none		3.88	7.20	11.08
2750			4,500	none		4.63	7.10	11.73
2800		12x8x16	2,000	none		5.80	9.35	15.15
2850			4,500	none		6.15	9.20	15.35

B2010 110 Concrete Block Wall - Lightweight

	TYPE	SIZE (IN.)	WEIGHT (P.C.F.)	CORE FILL		COST PER S.F.		
						MAT.	INST.	TOTAL
3100	Hollow	8x4x16	105	perlite		4.27	6.90	11.17
3110				styrofoam		3.82	6.50	10.32
3140				none		2.57	6.50	9.07
3150			85	perlite		5.55	6.75	12.30
3160				styrofoam		5.10	6.35	11.45
3190				none		3.84	6.35	10.19
3200		4x8x16	105	none		2	5.90	7.90
3250			85	none		2.48	5.75	8.23
3300		6x8x16	105	perlite		3.96	6.65	10.61
3310				styrofoam		4.06	6.30	10.36
3340				none		2.81	6.30	9.11
3350			85	perlite		4.60	6.50	11.10
3360				styrofoam		4.70	6.15	10.85
3390				none		3.45	6.15	9.60
3400		8x8x16	105	perlite		5.10	7.15	12.25
3410				styrofoam		4.63	6.75	11.38
3440				none		3.38	6.75	10.13
3450			85	perlite		5.35	7	12.35
3460				styrofoam		4.90	6.60	11.50
3490				none		3.65	6.60	10.25
3500		12x8x16	105	perlite		7	9.50	16.50
3510				styrofoam		6.45	8.70	15.15
3540				none		4.23	8.70	12.93
3550			85	perlite		8.30	9.25	17.55
3560				styrofoam		7.70	8.45	16.15
3590				none		5.50	8.45	13.95
3600		4x8x24	105	none		1.52	6.60	8.12
3650			85	none		3.54	5.50	9.04
3690	For stacked bond add						.40	.40

B20 Exterior Enclosure

B2010 Exterior Walls

B2010 110 Concrete Block Wall - Lightweight

	TYPE	SIZE (IN.)	WEIGHT (P.C.F.)	CORE FILL		COST PER S.F. MAT.	INST.	TOTAL
3700	Hollow	6x8x24	105	perlite		3.24	7.20	10.44
3710				styrofoam		3.34	6.85	10.19
3740				none		2.09	6.85	8.94
3750			85	perlite		6.20	6.10	12.30
3760				styrofoam		6.30	5.75	12.05
3790				none		5.05	5.75	10.80
3800		8x8x24	105	perlite		4.24	7.70	11.94
3810				styrofoam		3.79	7.30	11.09
3840				none		2.54	7.30	9.84
3850			85	perlite		7.80	6.55	14.35
3860				styrofoam		7.35	6.15	13.50
3890				none		6.10	6.15	12.25
3900		12x8x24	105	perlite		5.95	9.15	15.10
3910				styrofoam		5.40	8.35	13.75
3940				none		3.19	8.35	11.54
3950			85	perlite		10.40	8.95	19.35
3960				styrofoam		9.80	8.15	17.95
3990				none		7.60	8.15	15.75
4000	75% solid	4x8x16	105	none		2.88	5.95	8.83
4050			85	none		3.05	5.80	8.85
4100		6x8x16	105	perlite		4.68	6.55	11.23
4140				none		4.10	6.35	10.45
4150			85	perlite		4.85	6.40	11.25
4190				none		4.27	6.20	10.47
4200		8x8x16	105	perlite		5.75	7.05	12.80
4240				none		4.92	6.85	11.77
4250			85	perlite		5.35	6.85	12.20
4290				none		4.48	6.65	11.13
4300		12x8x16	105	perlite		7.55	9.20	16.75
4340				none		6.15	8.80	14.95
4350			85	perlite		8.20	8.95	17.15
4390				none		6.80	8.55	15.35
4500	Solid	4x8x16	105	none		1.97	6.15	8.12
4550			85	none		4.02	5.90	9.92
4600		6x8x16	105	none		3.10	6.60	9.70
4650			85	none		3.93	6.30	10.23
4700		8x8x16	105	none		3.92	7.10	11.02
4750			85	none		7.60	6.75	14.35
4800		12x8x16	105	none		4.50	9.20	13.70
4850			85	none		11.65	8.70	20.35
4900	For stacked bond, add						.40	.40

B2010 111 Reinforced Concrete Block Wall - Regular Weight

	TYPE	SIZE (IN.)	STRENGTH (P.S.I.)	VERT. REINF & GROUT SPACING		COST PER S.F. MAT.	INST.	TOTAL
5200	Hollow	4x8x16	2,000	#4 @ 48"		2	6.65	8.65
5250			4,500	#4 @ 48"		2.40	6.65	9.05
5300		6x8x16	2,000	#4 @ 48"		2.81	7.15	9.96
5330				#5 @ 32"		3.02	7.40	10.42
5340				#5 @ 16"		3.49	8.35	11.84
5350			4,500	#4 @ 28"		3.03	7.15	10.18
5380				#5 @ 32"		3.24	7.40	10.64
5390				#5 @ 16"		3.71	8.35	12.06

B20 Exterior Enclosure

B2010 Exterior Walls

B2010 111 Reinforced Concrete Block Wall - Regular Weight

	TYPE	SIZE (IN.)	STRENGTH (P.S.I.)	VERT. REINF & GROUT SPACING		COST PER S.F. MAT.	INST.	TOTAL
5400	Hollow	8x8x16	2,000	#4 @ 48"		3.11	7.60	10.71
5430				#5 @ 32"		3.28	8	11.28
5440				#5 @ 16"		3.82	9.15	12.97
5450		8x8x16	4,500	#4 @ 48"		3.69	7.70	11.39
5480				#5 @ 32"		3.93	8	11.93
5490				#5 @ 16"		4.47	9.15	13.62
5500		12x8x16	2,000	#4 @ 48"		4.51	9.80	14.31
5530				#5 @ 32"		4.82	10.10	14.92
5540				#5 @ 16"		5.55	11.30	16.85
5550			4,500	#4 @ 48"		4.91	9.80	14.71
5580				#5 @ 32"		5.20	10.10	15.30
5590				#5 @ 16"		5.95	11.30	17.25
6100	75% solid	6x8x16	2,000	#4 @ 48"		3.23	7.10	10.33
6130				#5 @ 32"		3.39	7.35	10.74
6140				#5 @ 16"		3.68	8.10	11.78
6150			4,500	#4 @ 48"		3.52	7.10	10.62
6180				#5 @ 32"		3.68	7.35	11.03
6190				#5 @ 16"		3.97	8.10	12.07
6200		8x8x16	2,000	#4 @ 48"		3.43	7.65	11.08
6230				#5 @ 32"		3.60	7.90	11.50
6240				#5 @ 16"		3.91	8.85	12.76
6250			4,500	#4 @ 48"		4.28	7.65	11.93
6280				#5 @ 32"		4.45	7.90	12.35
6290				#5 @ 16"		4.76	8.85	13.61
6300		12x8x16	2,000	#4 @ 48"		5.20	9.75	14.95
6330				#5 @ 32"		5.40	10	15.40
6340				#5 @ 16"		5.85	10.95	16.80
6350			4,500	#4 @ 48"		5.70	9.75	15.45
6380				#4 @ 32"		5.95	10	15.95
6390				#5 @ 16"		6.40	10.95	17.35
6500	Solid-double Wythe	2-4x8x16	2,000	#4 @ 48" E.W.		5.10	14.05	19.15
6530				#5 @ 16" E.W.		5.80	14.95	20.75
6550			4,500	#4 @ 48" E.W.		7.05	13.95	21
6580				#5 @ 16" E.W.		7.80	14.85	22.65
6600		2-6x8x16	2,000	#4 @ 48" E.W.		5.10	15.10	20.20
6630				#5 @ 16" E.W.		5.85	16	21.85
6650			4,000	#4 @ 48" E.W.		8.40	14.90	23.30
6680				#5 @ 16" E.W.		9.15	15.80	24.95

B2010 112 Reinforced Concrete Block Wall - Lightweight

	TYPE	SIZE (IN.)	WEIGHT (P.C.F.)	VERT REINF. & GROUT SPACING		COST PER S.F. MAT.	INST.	TOTAL
7100	Hollow	8x4x16	105	#4 @ 48"		2.88	7.30	10.18
7130				#5 @ 32"		3.12	7.60	10.72
7140				#5 @ 16"		3.66	8.75	12.41
7150			85	#4 @ 48"		4.15	7.15	11.30
7180				#5 @ 32"		4.39	7.45	11.84
7190				#5 @ 16"		4.93	8.60	13.53
7200		4x8x16	105	#4 @ 48"		2.15	6.55	8.70
7250			85	#4 @ 48"		2.63	6.40	9.03

B20 Exterior Enclosure

B2010 Exterior Walls

B2010 112 Reinforced Concrete Block Wall - Lightweight

	TYPE	SIZE (IN.)	WEIGHT (P.C.F.)	VERT REINF. & GROUT SPACING		COST PER S.F.		
						MAT.	INST.	TOTAL
7300	Hollow	6x8x16	105	#4 @ 48"		3.07	7	10.07
7330				#5 @ 32"		3.28	7.25	10.53
7340				#5 @ 16"		3.75	8.20	11.95
7350			85	#4 @ 48"		3.71	6.85	10.56
7380				#5 @ 32"		3.92	7.10	11.02
7390				#5 @ 16"		4.39	8.05	12.44
7400		8x8x16	105	#4 @ 48"		3.69	7.55	11.24
7430				#5 @ 32"		3.93	7.85	11.78
7440				#5 @ 16"		4.47	9	13.47
7450		8x8x16	85	#4 @ 48"		3.96	7.40	11.36
7480				#5 @ 32"		4.20	7.70	11.90
7490				#5 @ 16"		4.74	8.85	13.59
7510		12x8x16	105	#4 @ 48"		4.67	9.55	14.22
7530				#5 @ 32"		4.98	9.85	14.83
7540				#5 @ 16"		5.70	11.05	16.75
7550			85	#4 @ 48"		5.95	9.30	15.25
7580				#5 @ 32"		6.25	9.60	15.85
7590				#5 @ 16"		7	10.80	17.80
7600		4x8x24	105	#4 @ 48"		1.67	7.25	8.92
7650			85	#4 @ 48"		3.69	6.15	9.84
7700		6x8x24	105	#4 @ 48"		2.35	7.55	9.90
7730				#5 @ 32"		2.56	7.80	10.36
7740				#5 @ 16"		3.03	8.75	11.78
7750			85	#4 @ 48"		5.30	6.45	11.75
7780				#5 @ 32"		5.50	6.70	12.20
7790				#5 @ 16"		6	7.65	13.65
7800		8x8x24	105	#4 @ 48"		2.85	8.10	10.95
7840				#5 @ 16"		3.63	9.55	13.18
7850			85	#4 @ 48"		6.40	6.95	13.35
7880				#5 @ 32"		6.65	7.25	13.90
7890				#5 @ 16"		7.15	8.40	15.55
7900		12x8x24	105	#4 @ 48"		3.63	9.20	12.83
7930				#5 @ 32"		3.94	9.50	13.44
7940				#5 @ 16"		4.67	10.70	15.37
7950			85	#4 @ 48"		8.05	9	17.05
7980				#5 @ 32"		8.35	9.30	17.65
7990				#5 @ 16"		9.10	10.50	19.60
8100	75% solid	6x8x16	105	#4 @ 48"		4.24	6.90	11.14
8130				#5 @ 32"		4.40	7.15	11.55
8140				#5 @ 16"		4.69	7.90	12.59
8150			85	#4 @ 48"		4.41	6.75	11.16
8180				#5 @ 32"		4.57	7	11.57
8190				#5 @ 16"		4.86	7.75	12.61
8200		8x8x16	105	#4 @ 48"		5.05	7.50	12.55
8230				#5 @ 32"		5.25	7.75	13
8240				#5 @ 16"		5.55	8.70	14.25
8250			85	#4 @ 48"		4.63	7.30	11.93
8280				#5 @ 32"		4.80	7.55	12.35
8290				#5 @ 16"		5.10	8.50	13.60

B20 Exterior Enclosure

B2010 Exterior Walls

B2010 112 Reinforced Concrete Block Wall - Lightweight

	TYPE	SIZE (IN.)	WEIGHT (P.C.F.)	VERT REINF. & GROUT SPACING		COST PER S.F.		
						MAT.	INST.	TOTAL
8300	75% solid	12x8x16	105	#4 @ 48"		6.40	9.50	15.90
8330				#5 @ 32"		6.65	9.75	16.40
8340				#5 @ 16"		7.10	10.70	17.80
8350			85	#4 @ 48"		7.05	9.25	16.30
8380				#5 @ 32"		7.30	9.50	16.80
8390				#5 @ 16"		7.75	10.45	18.20
8500	Solid-double	2-4x8x16	105	#4 @ 48"		4.80	13.95	18.75
8530	Wythe		105	#5 @ 16"		5.55	14.85	20.40
8550			85	#4 @ 48"		8.90	13.45	22.35
8580				#5 @ 16"		9.65	14.35	24
8600		2-6x8x16	105	#4 @ 48"		7	14.90	21.90
8630				#5 @ 16"		7.75	15.80	23.55
8650			85	#4 @ 48"		8.65	14.30	22.95
8680				#5 @ 16"		9.40	15.20	24.60
8900	For stacked bond add						.40	.40
8910								

B20 Exterior Enclosure

B2010 Exterior Walls

Exterior split ribbed block walls are defined in the following terms; structural reinforcement, weight, percent solid, size, number of ribs and insulation. Within each of these categories two to four variations are shown. No costs are included for brick shelf or relieving angles. Costs include control joints every 20' and horizontal reinforcing.

System Components	QUANTITY	UNIT	COST PER S.F.		
			MAT.	INST.	TOTAL
SYSTEM B2010 113 1430					
UNREINFORCED SPLIT RIB BLOCK WALL, 8"X8"X16", 8 RIBS(HEX), PERLITE FILL					
Split ribbed block wall, 8" thick	1.000	S.F.	5.40	8.20	13.60
Perlite insulation	1.000	S.F.	1.70	.41	2.11
Horizontal joint reinforcing, alternate courses	.800	L.F.	.16	.17	.33
Control joint	.050	L.F.	.07	.07	.14
TOTAL			7.33	8.85	16.18

B2010 113		Split Ribbed Block Wall - Regular Weight						
	TYPE	SIZE (IN.)	RIBS	CORE FILL		COST PER S.F.		
						MAT.	INST.	TOTAL
1220	Hollow	4x8x16	4	none		3.86	7.45	11.31
1250			8	none		4.19	7.45	11.64
1280			16	none		4.52	7.55	12.07
1330	RB2010-200	6x8x16	4	none		6	8.25	14.25
1340				styrofoam		6.10	7.90	14
1350				none		4.86	7.90	12.76
1360			16	perlite		6.35	8.35	14.70
1370				styrofoam		6.45	8	14.45
1380				none		5.20	8	13.20
1430		8x8x16	8	perlite		7.35	8.85	16.20
1440				styrofoam		6.90	8.45	15.35
1450				none		5.65	8.45	14.10
1460			16	perlite		7.75	9	16.75
1470				styrofoam		7.30	8.60	15.90
1480				none		6.05	8.60	14.65
1530		12x8x16	8	perlite		9.35	11.75	21.10
1540				styrofoam		8.75	10.95	19.70
1550				none		6.55	10.95	17.50
1560			16	perlite		9.85	11.95	21.80
1570				styrofoam		9.25	11.15	20.40
1580				none		7.05	11.15	18.20

B20 Exterior Enclosure

B2010 Exterior Walls

B2010 113 — Split Ribbed Block Wall - Regular Weight

	TYPE	SIZE (IN.)	RIBS	CORE FILL		COST PER S.F.		
						MAT.	INST.	TOTAL
2120	75% solid	4x8x16	4	none		4.85	7.55	12.40
2150			8	none		5.30	7.55	12.85
2180			16	none		5.70	7.65	13.35
2230		6x8x16	8	perlite		6.70	8.20	14.90
2250				none		6.10	8	14.10
2260			16	perlite		7.15	8.35	15.50
2280				none		6.55	8.15	14.70
2330		8x8x16	8	perlite		7.90	8.80	16.70
2350				none		7.05	8.60	15.65
2360			16	perlite		8.45	8.90	17.35
2380				none		7.60	8.70	16.30
2430		12x8x16	8	perlite		9.55	11.55	21.10
2450				none		8.15	11.15	19.30
2460			16	perlite		10.20	11.80	22
2480				none		8.80	11.40	20.20
2520	Solid	4x8x16	4	none		5.50	7.65	13.15
2550			8	none		6.50	7.65	14.15
2580			16	none		6.50	7.75	14.25
2620		6x8x16	8	none		7.50	8.15	15.65
2650			16	none		7.50	8.25	15.75
2680		8x8x16	8	none		8.65	8.70	17.35
2720			16	none		8.65	8.85	17.50
2750		12x8x16	8	none		10	11.40	21.40
2780			16	none		10	11.60	21.60

B2010 114 — Split Ribbed Block Wall - Lightweight

	TYPE	SIZE (IN.)	RIBS	CORE FILL		COST PER S.F.		
						MAT.	INST.	TOTAL
3250	Hollow	4x8x16	8	none		5.10	7.25	12.35
3280			16	none		5.50	7.35	12.85
3330		6x8x16	8	perlite		7.05	8	15.05
3340				styrofoam		7.15	7.65	14.80
3350				none		5.90	7.65	13.55
3360			16	perlite		7.50	8.10	15.60
3370				styrofoam		7.60	7.75	15.35
3380				none		6.35	7.75	14.10
3430		8x8x16	8	perlite		8.50	8.60	17.10
3440				styrofoam		8.05	8.20	16.25
3450				none		6.80	8.20	15
3460			16	perlite		9.05	8.70	17.75
3470				styrofoam		8.60	8.30	16.90
3480				none		7.35	8.30	15.65
3530		12x8x16	8	perlite		10.70	11.40	22.10
3540				styrofoam		10.10	10.60	20.70
3550				none		7.90	10.60	18.50
3560			16	perlite		11.30	11.60	22.90
3570				styrofoam		10.70	10.80	21.50
3580				none		8.50	10.80	19.30
4150	75% solid	4x8x16	8	none		6.45	7.35	13.80
4180			16	none		7	7.45	14.45

B20 Exterior Enclosure

B2010 Exterior Walls

B2010 114 Split Ribbed Block Wall - Lightweight

	TYPE	SIZE (IN.)	RIBS	CORE FILL		COST PER S.F.		
						MAT.	INST.	TOTAL
4230	75% solid	6x8x16	8	perlite		8.05	7.95	16
4250				none		7.45	7.75	15.20
4260			16	perlite		8.65	8.10	16.75
4280				none		8.05	7.90	15.95
4330		8x8x16	8	perlite		9.45	8.50	17.95
4350				none		8.60	8.30	16.90
4360			16	perlite		10.15	8.65	18.80
4380		8x8x16	16	none		9.30	8.45	17.75
4430		12x8x16	8	perlite		11.35	11.20	22.55
4450				none		9.95	10.80	20.75
4460			16	perlite		12.15	11.35	23.50
4480				none		10.75	10.95	21.70
4550	Solid	4x8x16	8	none		7.35	7.45	14.80
4580			16	none		8	7.55	15.55
4650		6x8x16	8	none		8.45	7.90	16.35
4680			16	none		9.15	8	17.15
4750		8x8x16	8	none		9.80	8.45	18.25
4780			16	none		10.60	8.60	19.20
4850		12x8x16	8	none		11.30	10.95	22.25
4880			16	none		12.20	11.15	23.35

B2010 115 Reinforced Split Ribbed Block Wall - Regular Weight

	TYPE	SIZE (IN.)	RIBS	VERT. REINF. & GROUT SPACING		COST PER S.F.		
						MAT.	INST.	TOTAL
5200	Hollow	4x8x16	4	#4 @ 48"		4.01	8.10	12.11
5230			8	#4 @ 48"		4.34	8.10	12.44
5260			16	#4 @ 48"		4.67	8.20	12.87
5330		6x8x16	8	#4 @ 48"		5.10	8.60	13.70
5340				#5 @ 32"		5.35	8.85	14.20
5350				#5 @ 16"		5.80	9.80	15.60
5360			16	#4 @ 48"		5.50	8.70	14.20
5370				#5 @ 32"		5.70	8.95	14.65
5380				#5 @ 16"		6.15	9.90	16.05
5430		8x8x16	8	#4 @ 48"		5.95	9.25	15.20
5440				#5 @ 32"		6.20	9.55	15.75
5450				#5 @ 16"		6.70	10.70	17.40
5460			16	#4 @ 48"		6.35	9.40	15.75
5470				#5 @ 32"		6.60	9.70	16.30
5480				#5 @ 16"		7.10	10.85	17.95
5530		12x8x16	8	#4 @ 48"		7	11.80	18.80
5540				#5 @ 32"		7.30	12.10	19.40
5550				#5 @ 16"		8.05	13.30	21.35
5560			16	#4 @ 48"		7.50	12	19.50
5570				#5 @ 32"		7.80	12.30	20.10
5580				#5 @ 16"		8.55	13.50	22.05
6230	75% solid	6x8x16	8	#4 @ 48"		6.25	8.55	14.80
6240				#5 @ 32"		6.40	8.80	15.20
6250				#5 @ 16"		6.70	9.55	16.25
6260			16	#4 @ 48"		6.70	8.70	15.40
6270				#5 @ 32"		6.85	8.95	15.80
6280				#5 @ 16"		7.15	9.70	16.85

B20 Exterior Enclosure

B2010 Exterior Walls

B2010 115 Reinforced Split Ribbed Block Wall - Regular Weight

	TYPE	SIZE (IN.)	RIBS	VERT. REINF. & GROUT SPACING		COST PER S.F.		
						MAT.	INST.	TOTAL
6330	75% solid	8x8x16	8	#4 @ 48"		7.20	9.25	16.45
6340				#5 @ 32"		7.35	9.50	16.85
6350				#5 @ 16"		7.65	10.45	18.10
6360			16	#4 @ 48"		7.75	9.35	17.10
6370				#5 @ 32"		7.90	9.60	17.50
6380				#5 @ 16"		8.20	10.55	18.75
6430		12x8x16	8	#4 @ 48"		8.40	11.85	20.25
6440				#5 @ 32"		8.65	12.10	20.75
6450				#5 @ 16"		9.10	13.05	22.15
6460			16	#4 @ 48"		9.05	12.10	21.15
6470				#5 @ 32"		9.30	12.35	21.65
6480				#5 @ 16"		9.75	13.30	23.05
6500	Solid-double Wythe	2-4x8x16	4	#4 @ 48"		6.45	9.25	15.70
6520				#5 @ 16"		6.85	9.65	16.50
6530			8	#4 @ 48"		7.45	9.25	16.70
6550				#5 @ 16"		7.85	9.65	17.50
6560			16	#4 @ 48"		7.45	9.35	16.80
6580				#5 @ 16"		7.85	9.75	17.60
6630		2-6x8x16	8	#4 @ 48"		8.45	9.80	18.25
6650				#5 @ 16"		8.80	10.20	19
6660			16	#4 @ 48"		8.45	9.90	18.35
6680				#5 @ 16"		8.80	10.30	19.10

B2010 116 Reinforced Split Ribbed Block Wall - Lightweight

	TYPE	SIZE (IN.)	RIBS	VERT. REINF. & GROUT SPACING		COST PER S.F.		
						MAT.	INST.	TOTAL
7230	Hollow	4x8x16	8	#4 @ 48"		5.25	7.90	13.15
7260			16	#4 @ 48"		5.65	8	13.65
7330		6x8x16	8	#4 @ 48"		6.20	8.35	14.55
7340				#5 @ 32"		6.40	8.60	15
7350				#5 @ 16"		6.85	9.55	16.40
7360			16	#4 @ 48"		6.65	8.45	15.10
7370				#5 @ 32"		6.85	8.70	15.55
7380				#5 @ 16"		7.30	9.65	16.95
7430		8x8x16	8	#4 @ 48"		7.10	9	16.10
7440				#5 @ 32"		7.35	9.30	16.65
7450				#5 @ 16"		7.85	10.45	18.30
7460			16	#4 @ 48"		7.65	9.10	16.75
7470				#5 @ 32"		7.90	9.40	17.30
7480				#5 @ 16"		8.40	10.55	18.95
7530		12x8x16	8	#4 @ 48"		8.35	11.45	19.80
7540				#5 @ 32"		8.65	11.75	20.40
7550				#5 @ 16"		9.40	12.95	22.35
7560			16	#4 @ 48"		8.95	11.65	20.60
7570				#5 @ 32"		9.25	11.95	21.20
7580				#5 @ 16"		10	13.15	23.15
8230	75% solid	6x8x16	8	#4 @ 48"		7.60	8.30	15.90
8240				#5 @ 32"		7.75	8.55	16.30
8250				#5 @ 16"		8.05	9.30	17.35
8260			16	#4 @ 48"		8.20	8.45	16.65
8270				#5 @ 32"		8.35	8.70	17.05
8280				#5 @ 16"		8.65	9.45	18.10

B20 Exterior Enclosure

B2010 Exterior Walls

B2010 116 Reinforced Split Ribbed Block Wall - Lightweight

	TYPE	SIZE (IN.)	RIBS	VERT. REINF. & GROUT SPACING		COST PER S.F.		
						MAT.	INST.	TOTAL
8330	75% solid	8x8x16	8	#4 @ 48"		8.75	8.95	17.70
8340				#5 @ 32"		8.90	9.20	18.10
8350				#5 @ 16"		9.20	10.15	19.35
8360			16	#4 @ 48"		9.45	9.10	18.55
8370				#5 @ 32"		9.60	9.35	18.95
8380				#5 @ 16"		9.90	10.30	20.20
8430		12x8x16	8	#4 @ 48"		10.20	11.50	21.70
8440				#5 @ 32"		10.45	11.75	22.20
8450				#5 @ 16"		10.90	12.70	23.60
8460			16	#4 @ 48"		11	11.65	22.65
8470				#5 @ 32"		11.25	11.90	23.15
8480				#5 @ 16"		11.70	12.85	24.55
8530	Solid double	2-4x8x16	8	#4 @ 48" E.W		8.40	9.30	17.70
8550	Wythe			#5 @ 16" E.W.		9.15	10.20	19.35
8560			16	#4 @ 48" E.W.		9.05	9.40	18.45
8580				#5 @ 16" E.W.		9.80	10.30	20.10
8630		2-6x8x16	8	#4 @ 48" E.W.		9.50	9.80	19.30
8650				#5 @ 16" E.W.		10.25	10.70	20.95
8660			16	#4 @ 48" E.W.		10.20	9.90	20.10
8680				#5 @ 16" E.W.		10.95	10.80	21.75
8900	For stacked bond, add						.40	.40

B20 Exterior Enclosure

B2010 Exterior Walls

Exterior split face block walls are defined in the following terms; structural reinforcement, weight, percent solid, size, scores and insulation. Within each of these categories two to four variations are shown. No costs are included for brick shelf or relieving angles. Costs include control joints every 20′ and horizontal reinforcing.

System Components	QUANTITY	UNIT	COST PER S.F.		
			MAT.	INST.	TOTAL
SYSTEM B2010 117 1600					
UNREINFORCED SPLIT FACE BLOCK WALL, 8″X8″X16″, 0 SCORES, PERLITE FILL					
Split face block wall, 8″ thick	1.000	S.F.	4.49	8.45	12.94
Perlite insulation	1.000	S.F.	1.70	.41	2.11
Horizontal joint reinforcing, alternate course	.800	L.F.	.16	.17	.33
Control joint	.050	L.F.	.07	.07	.14
TOTAL			6.42	9.10	15.52

B2010 117				Split Face Block Wall - Regular Weight				
	TYPE	SIZE (IN.)	SCORES	CORE FILL		COST PER S.F.		
						MAT.	INST.	TOTAL
1200	Hollow	8x4x16	0	perlite		7.10	9.75	16.85
1210				styrofoam		6.65	9.35	16
1240				none		5.40	9.35	14.75
1250	RB2010 -200		1	perlite		7.45	9.75	17.20
1260				styrofoam		7	9.35	16.35
1290				none		5.75	9.35	15.10
1300		12x4x16	0	perlite		9.35	11.10	20.45
1310				styrofoam		8.75	10.30	19.05
1340				none		6.55	10.30	16.85
1350			1	perlite		9.85	11.10	20.95
1360				styrofoam		9.25	10.30	19.55
1390				none		7.05	10.30	17.35
1400		4x8x16	0	none		3.53	7.35	10.88
1450			1	none		3.82	7.45	11.27
1500		6x8x16	0	perlite		5.25	8.50	13.75
1510				styrofoam		5.35	8.15	13.50
1540				none		4.09	8.15	12.24
1550			1	perlite		5.60	8.60	14.20
1560				styrofoam		5.70	8.25	13.95
1590				none		4.43	8.25	12.68

B20 Exterior Enclosure

B2010 Exterior Walls

B2010 117 Split Face Block Wall - Regular Weight

	TYPE	SIZE (IN.)	SCORES	CORE FILL		MAT.	INST.	TOTAL
					COST PER S.F.			
1600	Hollow	8x8x16	0	perlite		6.40	9.10	15.50
1610				styrofoam		5.95	8.70	14.65
1640				none		4.72	8.70	13.42
1650		8x8x16	1	perlite		6.80	9.25	16.05
1660				styrofoam		6.35	8.85	15.20
1690				none		5.10	8.85	13.95
1700		12x8x16	0	perlite		8.45	11.95	20.40
1710				styrofoam		7.85	11.15	19
1740				none		5.65	11.15	16.80
1750			1	perlite		8.90	12.20	21.10
1760				styrofoam		8.30	11.40	19.70
1790				none		6.10	11.40	17.50
1800	75% solid	8x4x16	0	perlite		7.40	9.70	17.10
1840				none		6.55	9.50	16.05
1850			1	perlite		7.90	9.70	17.60
1890				none		7.05	9.50	16.55
1900		12x4x16	0	perlite		9.35	10.90	20.25
1940				none		7.95	10.50	18.45
1950			1	perlite		9.90	10.90	20.80
1990				none		8.50	10.50	19
2000		4x8x16	0	none		4.41	7.45	11.86
2050			1	none		4.80	7.55	12.35
2100		6x8x16	0	perlite		5.70	9.85	15.55
2140				none		5.10	9.65	14.75
2150			1	perlite		6.10	8.60	14.70
2190				none		5.50	8.40	13.90
2200		8x8x16	0	perlite		6.75	9.05	15.80
2240				none		5.90	8.85	14.75
2250			1	perlite		7.25	9.20	16.45
2290				none		6.40	9	15.40
2300		12x8x16	0	perlite		8.40	11.80	20.20
2340				none		7	11.40	18.40
2350			1	perlite		9	12	21
2390				none		7.60	11.60	19.20
2400	Solid	8x4x16	0	none		7.30	9.70	17
2450			1	none		7.90	9.70	17.60
2500		12x4x16	0	none		8.85	10.70	19.55
2550			1	none		9.50	12.55	22.05
2600		4x8x16	0	none		5	7.55	12.55
2650			1	none		5.45	7.65	13.10
2700		6x8x16	0	none		5.75	8.40	14.15
2750			1	none		6.25	8.55	14.80
2800		8x8x16	0	none		6.65	9	15.65
2850			1	none		7.25	9.20	16.45
2900		12x8x16	0	none		7.90	11.60	19.50
2950			1	none		8.60	11.80	20.40

B2010 118 Split Face Block Wall - Lightweight

	TYPE	SIZE (IN.)	SCORES	CORE FILL		MAT.	INST.	TOTAL
					COST PER S.F.			
3200	Hollow	8x4x16	0	perlite		8.05	10.95	19
3210				styrofoam		7.60	10.55	18.15
3240				none		6.35	10.55	16.90

B20 Exterior Enclosure

B2010 Exterior Walls

B2010 118 Split Face Block Wall - Lightweight

	TYPE	SIZE (IN.)	SCORES	CORE FILL		COST PER S.F.		
						MAT.	INST.	TOTAL
3250	Hollow	8x4x16	1	perlite		8.50	9.40	17.90
3260				styrofoam		8.05	9	17.05
3290				none		6.80	9	15.80
3300		12x4x16	0	perlite		10.50	10.70	21.20
3310				styrofoam		9.90	9.90	19.80
3340				none		7.70	9.90	17.60
3350			1	perlite		11.05	12.40	23.45
3360				styrofoam		10.45	11.60	22.05
3390				none		8.25	11.60	19.85
3400		4x8x16	0	none		4.27	7.15	11.42
3450			1	none		4.64	7.25	11.89
3500		6x8x16	0	perlite		6.10	8.25	14.35
3510				styrofoam		6.20	7.90	14.10
3540				none		4.93	7.90	12.83
3550			1	perlite		6.50	8.35	14.85
3560				styrofoam		6.60	8	14.60
3590				none		5.35	8	13.35
3600		8x8x16	0	perlite		7.40	8.85	16.25
3610				styrofoam		6.95	8.45	15.40
3640				none		5.70	8.45	14.15
3650			1	perlite		7.85	9	16.85
3660				styrofoam		7.40	8.60	16
3690				none		6.15	8.60	14.75
3700		12x8x16	0	perlite		9.60	11.60	21.20
3710				styrofoam		9	10.80	19.80
3740				none		6.80	10.80	17.60
3750			1	perlite		10.15	11.75	21.90
3760				styrofoam		9.55	10.95	20.50
3790				none		7.35	10.95	18.30
3800	75% solid	8x4x16	0	perlite		7.90	9.70	17.60
3840				none		7.05	9.50	16.55
3850			1	perlite		9.25	9.40	18.65
3890				none		8.40	9.20	17.60
3900		12x4x16	0	perlite		9.90	10.90	20.80
3940				none		8.50	10.50	19
3950			1	perlite		11.55	12.20	23.75
3990				none		10.15	11.80	21.95
4000		4x8x16	0	none		5.40	7.25	12.65
4050			1	none		5.85	7.35	13.20
4100		6x8x16	0	perlite		6.75	8.20	14.95
4140				none		6.15	8	14.15
4150			1	perlite		7.30	8.35	15.65
4190				none		6.70	8.15	14.85
4200		8x8x16	0	perlite		8	8.80	16.80
4240				none		7.15	8.60	15.75
4250			1	perlite		8.60	8.90	17.50
4290				none		7.75	8.70	16.45
4300		12x8x16	0	perlite		9.90	11.35	21.25
4340				none		8.50	10.95	19.45
4350			1	perlite		10.60	11.55	22.15
4390				none		9.20	11.15	20.35

B20 Exterior Enclosure

B2010 Exterior Walls

B2010 118 — Split Face Block Wall - Lightweight

	TYPE	SIZE (IN.)	SCORES	CORE FILL	MAT.	INST.	TOTAL
4400	Solid	8x4x16	0	none	8.75	9.35	18.10
4450			1	none	9.45	9.35	18.80
4500		12x4x16	0	none	10.55	10.30	20.85
4550			1	none	11.40	12.05	23.45
4600		4x8x16	0	none	6.10	7.35	13.45
4650			1	none	6.65	7.45	14.10
4700		6x8x16	0	none	7	8.15	15.15
4750			1	none	7.65	8.25	15.90
4800		8x8x16	0	none	8.10	8.70	16.80
4850			1	none	8.80	8.85	17.65
4900		12x8x16	0	none	9.60	11.15	20.75
4950			1	none	10.45	11.40	21.85

B2010 119 — Reinforced Split Face Block Wall - Regular Weight

	TYPE	SIZE (IN.)	SCORES	VERT. REINF. & GROUT SPACING	MAT.	INST.	TOTAL
5200	Hollow	8x4x16	0	#4 @ 48"	5.70	10.15	15.85
5210				#5 @ 32"	5.95	10.45	16.40
5240				#5 @ 16"	6.45	11.60	18.05
5250			1	#4 @ 48"	6.05	10.15	16.20
5260				#5 @ 32"	6.30	10.45	16.75
5290				#5 @ 16"	6.80	11.60	18.40
5300		12x4x16	0	#4 @ 48"	7	11.15	18.15
5310				#5 @ 32"	7.30	11.45	18.75
5340				#5 @ 16"	8.05	12.65	20.70
5350			1	#4 @ 48"	7.50	11.15	18.65
5360				#5 @ 32"	7.80	11.45	19.25
5390				#5 @ 16"	8.55	12.65	21.20
5400		4x8x16	0	#4 @ 48"	3.68	8	11.68
5450			1	#4 @ 48"	3.97	8.10	12.07
5500		6x8x16	0	#4 @ 48"	4.35	8.85	13.20
5510				#5 @ 32"	4.56	9.10	13.66
5540				#5 @ 16"	5.05	10.05	15.10
5550			1	#4 @ 48"	4.69	8.95	13.64
5560				#5 @ 32"	4.90	9.20	14.10
5590				#5 @ 16"	5.35	10.15	15.50
5600		8x8x16	0	#4 @ 48"	5.05	9.50	14.55
5610				#5 @ 32"	5.25	9.80	15.05
5640				#5 @ 16"	5.45	10.40	15.85
5650			1	#4 @ 48"	8.85	13.10	21.95
5660				#5 @ 32"	5.65	9.95	15.60
5690				#5 @ 16"	6.20	11.10	17.30
5700		12x8x16	0	#4 @ 48"	6.10	12	18.10
5710				#5 @ 32"	6.40	12.30	18.70
5740				#5 @ 16"	7.15	13.50	20.65
5750			1	#4 @ 48"	6.55	12.25	18.80
5760				#5 @ 32"	6.85	12.55	19.40
5790				#5 @ 16"	7.60	13.75	21.35

B20 Exterior Enclosure

B2010 Exterior Walls

B2010 119 Reinforced Split Face Block Wall - Regular Weight

	TYPE	SIZE (IN.)	SCORES	VERT. REINF. & GROUT SPACING		COST PER S.F.		
						MAT.	INST.	TOTAL
6000	75% solid	8x4x16	0	#4 @ 48"		6.70	10.15	16.85
6010				#5 @ 32"		6.85	10.40	17.25
6040				#5 @ 16"		7.15	11.35	18.50
6050			1	#4 @ 48"		7.20	10.15	17.35
6060				#5 @ 32"		7.35	10.40	17.75
6090				#5 @ 16"		7.65	11.35	19
6100		12x4x16	0	#4 @ 48"		8.20	11.20	19.40
6110				#5 @ 32"		8.45	11.45	19.90
6140				#5 @ 16"		8.85	12.20	21.05
6150			1	#4 @ 48"		8.75	11.20	19.95
6160				#5 @ 32"		9	11.45	20.45
6190				#5 @ 16"		9.45	12.40	21.85
6200		6x8x16	0	#4 @ 48"		5.25	10.20	15.45
6210				#5 @ 32"		5.40	10.45	15.85
6240				#5 @ 16"		5.70	11.20	16.90
6250			1	#4 @ 48"		5.65	8.95	14.60
6260				#5 @ 32"		5.80	9.20	15
6290				#5 @ 16"		6.10	9.95	16.05
6300		8x8x16	0	#4 @ 48"		6.05	9.50	15.55
6310				#5 @ 32"		6.20	9.75	15.95
6340				#5 @ 16"		6.50	10.70	17.20
6350			1	#4 @ 48"		6.55	9.65	16.20
6360				#5 @ 32"		6.70	9.90	16.60
6390				#5 @ 16"		7	10.85	17.85
6400		12x8x16	0	#4 @ 48"		7.25	12.10	19.35
6410				#5 @ 32"		7.50	12.35	19.85
6440				#5 @ 16"		7.95	13.30	21.25
6450			1	#4 @ 48"		7.85	12.30	20.15
6460				#5 @ 32"		8.10	12.55	20.65
6490				#5 @ 16"		8.55	13.50	22.05
6700	Solid-double Wythe	2-4x8x16	0	#4 @ 48" E.W.		10.90	16.75	27.65
6710				#5 @ 32" E.W.		11.15	16.90	28.05
6740				#5 @ 16" E.W.		11.60	17.65	29.25
6750			1	#4 @ 48" E.W.		11.75	16.95	28.70
6760				#5 @ 32" E.W.		12.05	17.10	29.15
6790				#5 @ 16" E.W.		12.50	17.85	30.35
6800		2-6x8x16	0	#4 @ 48" E.W.		12.35	18.50	30.85
6810				#5 @ 32" E.W.		12.60	18.65	31.25
6840				#5 @ 16" E.W.		13.10	19.40	32.50
6850			1	#4 @ 48" E.W.		13.35	18.80	32.15
6860				#5 @ 32" E.W.		13.60	18.95	32.55
6890				#5 @ 16" E.W.		14.10	19.70	33.80

B2010 120 Reinforced Split Face Block Wall - Lightweight

	TYPE	SIZE (IN.)	SCORES	VERT. REINF. & GROUT SPACING		COST PER S.F.		
						MAT.	INST.	TOTAL
7200	Hollow	8x4x16	0	#4 @ 48"		6.65	11.35	18
7210				#5 @ 32"		6.90	11.65	18.55
7240				#5 @ 16"		7.40	12.80	20.20
7250			1	#4 @ 48"		7.10	9.80	16.90
7260				#5 @ 32"		7.35	10.10	17.45
7290				#5 @ 16"		7.85	11.25	19.10

B20 Exterior Enclosure

B2010 Exterior Walls

B2010 120 Reinforced Split Face Block Wall - Lightweight

	TYPE	SIZE (IN.)	SCORES	VERT. REINF. & GROUT SPACING		COST PER S.F.		
						MAT.	INST.	TOTAL
7300	Hollow	12x4x16	0	#4 @ 48"		8.15	10.75	18.90
7340				#5 @ 16"		9.20	12.25	21.45
7350			1	#4 @ 48"		8.70	12.45	21.15
7360				#5 @ 32"		9	12.75	21.75
7390				#5 @ 16"		9.75	13.95	23.70
7400		4x8x16	0	#4 @ 48"		4.42	7.80	12.22
7450			1	#4 @ 48"		4.79	7.90	12.69
7500		6x8x16	0	#4 @ 48"		5.20	8.60	13.80
7510				#5 @ 32"		5.40	8.85	14.25
7540				#5 @ 16"		5.85	9.80	15.65
7550			1	#4 @ 48"		5.65	8.70	14.35
7560				#5 @ 32"		5.85	8.95	14.80
7590				#5 @ 16"		6.30	9.90	16.20
7600		8x8x16	0	#4 @ 48"		6	9.25	15.25
7610				#5 @ 32"		6.25	9.55	15.80
7640				#5 @ 16"		6.75	10.70	17.45
7650			1	#4 @ 48"		6.45	9.40	15.85
7660				#5 @ 32"		6.70	9.70	16.40
7690				#5 @ 16"		7.20	10.85	18.05
7700		12x8x16	0	#4 @ 48"		7.25	11.65	18.90
7710				#5 @ 32"		7.55	11.95	19.50
7740		12x8x16	0	#5 @ 16"		8.30	13.15	21.45
7750			1	#4 @ 48"		7.80	11.80	19.60
7760				#5 @ 16"		8.10	12.10	20.20
7790				#5 @ 16"		8.85	13.30	22.15
8000	75% solid	8x4x16	0	#4 @ 48"		7.20	10.15	17.35
8010				#5 @ 32"		7.35	10.40	17.75
8040				#5 @ 16"		7.65	11.35	19
8050			1	#4 @ 48"		8.55	9.85	18.40
8060				#5 @ 32"		8.70	10.10	18.80
8090				#5 @ 16"		9	11.05	20.05
8100		12x4x16	0	#4 @ 48"		8.75	11.20	19.95
8110				#5 @ 32"		9	11.45	20.45
8140				#5 @ 16"		9.45	12.40	21.85
8150			1	#4 @ 48"		10.40	12.50	22.90
8160				#5 @ 32"		10.65	12.75	23.40
8190				#5 @ 16"		11.10	13.70	24.80
8200		6x8x16	0	#4 @ 48"		6.30	8.55	14.85
8210				#5 @ 32"		6.45	8.80	15.25
8240				#5 @ 16"		6.75	9.55	16.30
8250			1	#4 @ 48"		6.85	8.70	15.55
8260				#5 @ 32"		7	8.95	15.95
8290				#5 @ 16"		7.30	9.70	17
8300		8x8x16	0	#4 @ 48"		7.30	9.25	16.55
8310				#5 @ 32"		7.45	9.50	16.95
8340				#5 @ 16"		7.75	10.45	18.20
8350			1	#4 @ 48"		7.90	9.35	17.25
8360				#5 @ 32"		8.05	9.60	17.65
8390				#5 @ 16"		8.35	10.55	18.90

B20 Exterior Enclosure

B2010 Exterior Walls

B2010 120 Reinforced Split Face Block Wall - Lightweight

	TYPE	SIZE (IN.)	SCORES	VERT. REINF. & GROUT SPACING		COST PER S.F.		
						MAT.	INST.	TOTAL
8400	75% solid	12x8x16	0	#4 @ 48"		8.75	11.65	20.40
8410				#5 @ 32"		9	11.90	20.90
8440				#5 @ 16"		9.45	12.85	22.30
8450			1	#4 @ 48"		9.45	11.85	21.30
8460				#5 @ 32"		9.70	12.10	21.80
8490				#5 @ 16"		10.15	13.05	23.20
8700	Solid-double	2-4x8x16	0	#4 @ 48" E.W.		13.05	16.35	29.40
8710	Wythe			#5 @ 32" E.W.		13.35	16.50	29.85
8740				#5 @ 16" E.W.		13.80	17.25	31.05
8750			1	#4 @ 48" E.W.		14.15	16.55	30.70
8760				#5 @ 32" E.W.		14.45	16.70	31.15
8790				#5 @ 16" E.W.		14.90	17.45	32.35
8800		2-6x8x16	0	#4 @ 48" E.W.		14.85	18	32.85
8810				#5 @ 32" E.W.		15.10	18.15	33.25
8840				#5 @ 16" E.W.		15.60	18.90	34.50
8850			1	#4 @ 48" E.W.		16.15	18.20	34.35
8860				#5 @ 32" E.W.		16.40	18.35	34.75
8890				#5 @ 16" E.W.		16.90	19.10	36

B20 Exterior Enclosure

B2010 Exterior Walls

Exterior ground face block walls are defined in the following terms; structural reinforcement, weight, percent solid, size, scores and insulation. Within each of these categories two to four variations are shown. No costs are included for brick shelf or relieving angles. Costs include control joints every 20' and horizontal reinforcing.

System Components	QUANTITY	UNIT	COST PER S.F.		
			MAT.	INST.	TOTAL
SYSTEM B2010 121 1600					
UNREINF. GROUND FACE BLOCK WALL, 8"X8"X16", 0 SCORES, PERLITE FILL					
Ground face block wall, 8" thick	1.000	S.F.	8	8.60	16.60
Perlite insulation	1.000	S.F.	1.70	.41	2.11
Horizontal joint reinforcing, alternate course	.800	L.F.	.16	.17	.33
Control joint	.050	L.F.	.07	.07	.14
TOTAL			9.93	9.25	19.18

B2010 121				Ground Face Block Wall				
	TYPE	SIZE (IN.)	SCORES	CORE FILL		COST PER S.F.		
						MAT.	INST.	TOTAL
1200	Hollow	4x8x16	0	none		6.45	7.45	13.90
1250			1	none		7.05	7.55	14.60
1300			2 to 5	none		7.65	7.65	15.30
1400	RB2010-200	6x8x16	0	perlite		8.40	8.60	17
1410				styrofoam		8.50	8.25	16.75
1440				none		7.25	8.25	15.50
1450			1	perlite		9.05	8.75	17.80
1460				styrofoam		9.15	8.40	17.55
1490				none		7.90	8.40	16.30
1500			2 to 5	perlite		9.70	8.90	18.60
1510				styrofoam		9.80	8.55	18.35
1540				none		8.55	8.55	17.10
1600		8x8x16	0	perlite		9.95	9.25	19.20
1610				styrofoam		9.50	8.85	18.35
1640				none		8.25	8.85	17.10
1650			1	perlite		10.65	9.40	20.05
1660				styrofoam		10.20	9	19.20
1690				none		8.95	9	17.95
1700			2 to 5	perlite		11.40	9.60	21
1710				styrofoam		10.95	9.20	20.15
1740				none		9.70	9.20	18.90

B20 Exterior Enclosure

B2010 Exterior Walls

B2010 121 Ground Face Block Wall

	TYPE	SIZE (IN.)	SCORES	CORE FILL		MAT.	INST.	TOTAL
						COST PER S.F.		
1800	Hollow	12x8x16	0	perlite		15.20	12.20	27.40
1810				styrofoam		14.60	11.40	26
1840				none		12.40	11.40	23.80
1850			1	perlite		16.30	12.40	28.70
1860				styrofoam		15.70	11.60	27.30
1890				none		13.50	11.60	25.10
1900			2 to 5	perlite		17.45	12.60	30.05
1910				styrofoam		16.85	11.80	28.65
1940				none		14.65	11.80	26.45
2200	75% solid	4x8x16	0	none		8.25	7.55	15.80
2250			1	none		9	7.65	16.65
2300			2 to 5	none		9.75	7.75	17.50
2400		6x8x16	0	perlite		9.80	8.60	18.40
2440				none		9.20	8.40	17.60
2450			1	perlite		10.65	8.75	19.40
2490				none		10.05	8.55	18.60
2500			2 to 5	perlite		11.50	8.85	20.35
2540				none		10.90	8.65	19.55
2600		8x8x16	0	perlite		11.25	9.20	20.45
2640				none		10.40	9	19.40
2650			1	perlite		12.25	9.40	21.65
2690				none		11.40	9.20	20.60
2700			2 to 5	perlite		13.20	9.55	22.75
2740				none		12.35	9.35	21.70
2800		12x8x16	0	perlite		17.20	12	29.20
2840				none		15.80	11.60	27.40
2850			1	perlite		18.65	12.20	30.85
2890				none		17.25	11.80	29.05
2900			2 to 5	perlite		20	12.45	32.45
2940				none		18.70	12.05	30.75
3200	Solid	4x8x16	0	none		9.40	7.65	17.05
3250			1	none		10.30	7.75	18.05
3300			2 to 5	none		11.15	7.90	19.05
3400		6x8x16	0	none		10.50	8.55	19.05
3450			1	none		11.50	8.65	20.15
3500			2 to 5	none		12.45	8.80	21.25
3600		8x8x16	0	none		11.90	9.20	21.10
3650			1	none		13	9.35	22.35
3700			2 to 5	none		14.10	9.50	23.60
3800		12x8x16	0	none		18.05	11.80	29.85
3850			1	none		19.70	12.05	31.75
3900			2 to 5	none		21	12.30	33.30

B2010 122 Reinforced Ground Face Block Wall

	TYPE	SIZE (IN.)	SCORES	VERT. REINF. & GROUT SPACING		MAT.	INST.	TOTAL
						COST PER S.F.		
5200	Hollow	4x8x16	0	#4 @ 48"		6.60	8.10	14.70
5250			1	#4 @ 48"		7.20	8.20	15.40
5300			2 to 5	#4 @ 48"		7.80	8.30	16.10
5400		6x8x16	0	#4 @ 48"		7.55	8.95	16.50
5420				#5 @ 32"		7.75	9.20	16.95
5430				#5 @ 16"		8.20	10.15	18.35

B20 Exterior Enclosure

B2010 Exterior Walls

B2010 122 Reinforced Ground Face Block Wall

	TYPE	SIZE (IN.)	SCORES	VERT. REINF. & GROUT SPACING		COST PER S.F.		
						MAT.	INST.	TOTAL
5450	Hollow	6x8x16	1	#4 @ 48"		8.20	9.10	17.30
5470				#5 @ 32"		8.40	9.35	17.75
5480				#5 @ 16"		8.85	10.30	19.15
5500			2 to 5	#4 @ 48"		8.85	9.25	18.10
5520				#5 @ 32"		9.05	9.50	18.55
5530				#5 @ 16"		9.50	10.45	19.95
5600	Hollow	8x8x16	0	#4 @ 48"		8.55	9.65	18.20
5620				#5 @ 32"		8.80	9.95	18.75
5630				#5 @ 16"		9.30	11.10	20.40
5650			1	#4 @ 48"		9.25	9.80	19.05
5670				#5 @ 32"		9.50	10.10	19.60
5680				#5 @ 16"		10	11.25	21.25
5700			2 to 5	#4 @ 48"		10	10	20
5720				#5 @ 32"		10.25	10.30	20.55
5730				#5 @ 16"		10.75	11.45	22.20
5800		12x8x16	0	#4 @ 48"		12.85	12.25	25.10
5820				#5 @ 32"		13.15	12.55	25.70
5830				#5 @ 16"		13.90	13.75	27.65
5850			1	#4 @ 48"		13.95	12.45	26.40
5870				#5 @ 32"		14.25	12.75	27
5880				#5 @ 16"		15	13.95	28.95
5900			2 to 5	#4 @ 48"		15.10	12.65	27.75
5920				#5 @ 32"		15.40	12.95	28.35
5930				#5 @ 16"		16.15	14.15	30.30
6200	75% solid	6x8x16	0	#4 @ 48"		9.40	8.85	18.25
6220				#5 @ 32"		9.60	9.05	18.65
6230				#5 @ 16"		9.90	9.70	19.60
6250			1	#4 @ 48"		10.25	9	19.25
6270				#5 @ 32"		10.45	9.20	19.65
6280				#5 @ 16"		10.75	9.85	20.60
6300			2 to 5	#4 @ 48"		11.10	9.10	20.20
6330				#5 @ 16"		11.60	9.95	21.55
6400		8x8x16	0	#4 @ 48"		10.60	9.50	20.10
6420				#5 @ 32"		10.80	9.75	20.55
6430				#5 @ 16"		11.15	10.45	21.60
6450			1	#4 @ 48"		11.60	9.70	21.30
6470				#5 @ 32"		11.80	9.95	21.75
6480				#5 @ 16"		12.15	10.65	22.80
6500			2 to 5	#4 @ 48"		12.55	9.85	22.40
6520				#5 @ 32"		12.75	10.10	22.85
6530				#5 @ 16"		13.10	10.80	23.90
6600		12x8x16	0	#4 @ 48"		16.10	12.10	28.20
6620				#5 @ 32"		16.30	12.40	28.70
6630				#5 @ 16"		16.80	13.10	29.90
6650			1	#4 @ 48"		17.55	12.30	29.85
6670				#5 @ 32"		17.75	12.60	30.35
6680				#5 @ 16"		18.25	13.30	31.55
6700			2 to 5	#4 @ 48"		19	12.55	31.55
6720				#5 @ 32"		19.20	12.85	32.05
6730				#5 @ 16"		19.70	13.55	33.25

B20 Exterior Enclosure

B2010 Exterior Walls

B2010 122 Reinforced Ground Face Block Wall

	TYPE	SIZE (IN.)	SCORES	VERT. REINF. & GROUT SPACING		COST PER S.F.		
						MAT.	INST.	TOTAL
6800	Solid-double	2-4x8x16	0	#4 @ 48" E.W.		19.55	16.80	36.35
6830	Wythe			#5 @ 16" E.W.		20	17.50	37.50
6850			1	#4 @ 48" E.W.		21.50	17	38.50
6880				#5 @ 16" E.W.		22	17.70	39.70
6900			2 to 5	#4 @ 48" E.W.		23	17.30	40.30
6930				#5 @ 16" E.W.		23.50	18	41.50
7000		2-6x8x16	0	#4 @ 48" E.W.		21.50	18.60	40.10
7030				#5 @ 16" E.W.		22.50	19.30	41.80
7050			1	#4 @ 48" E.W.		23.50	18.80	42.30
7080				#5 @ 16" E.W.		24.50	19.50	44
7100			2 to 5	#4 @ 48" E.W.		25.50	19.10	44.60
7130				#5 @ 16" E.W.		26.50	19.80	46.30

B20 Exterior Enclosure

B2010 Exterior Walls

Exterior miscellaneous block walls are defined in the following terms: structural reinforcement, finish texture, percent solid, size, weight and insulation. Within each of these categories two to four variations are shown. No costs are included for brick shelf or relieving angles. Costs include control joints every 20' and horizontal reinforcing.

System Components	QUANTITY	UNIT	COST PER S.F.		
			MAT.	INST.	TOTAL
SYSTEM B2010 123 2700					
UNREINFORCED HEXAGONAL BLOCK WALL, 8" X 8" X 16", 125 PCF, PERLITE FILL					
Hexagonal block partition, 8" thick	1.000	S.F.	4.30	8.60	12.90
Perlite insulation	1.000	S.F.	1.70	.41	2.11
Horizontal joint reinf., alternate courses	.800	L.F.	.16	.17	.33
Control joint	.050	L.F.	.07	.07	.14
TOTAL			6.23	9.25	15.48

B2010 123			Miscellaneous Block Wall				
	TYPE	SIZE (IN.)	WEIGHT (P.C.F.)	CORE FILL		COST PER S.F.	
						MAT.	INST.	TOTAL
1100	Deep groove-hollow	4x8x16	125	none		2.78	7.45	10.23
1150			105	none		3.34	7.25	10.59
1200	RB2010 -200	6x8x16	125	perlite		5.20	8.25	13.45
1210				styrofoam		5.30	7.90	13.20
1240				none		4.05	7.90	11.95
1250			105	perlite		6.05	7.60	13.65
1260				styrofoam		6.15	7.25	13.40
1290				none		4.88	7.25	12.13
1300		8x8x16	125	perlite		6.35	9	15.35
1310				styrofoam		5.90	8.60	14.50
1340				none		4.65	8.60	13.25
1350			105	perlite		7.30	8.70	16
1360				styrofoam		6.85	8.30	15.15
1390				none		5.60	8.30	13.90
1400	Deep groove-75% solid	4x8x16	125	none		3.45	7.55	11
1450			105	none		4.17	7.35	11.52
1500		6x8x16	125	perlite		5.60	8.20	13.80
1540				none		5.05	8	13.05
1550			105	perlite		6.70	7.95	14.65
1590				none		6.10	7.75	13.85

B20 Exterior Enclosure

B2010 Exterior Walls

B2010 123 Miscellaneous Block Wall

	TYPE	SIZE (IN.)	WEIGHT (P.C.F.)	CORE FILL		COST PER S.F.		
						MAT.	INST.	TOTAL
1600	Deep groove	8x8x16	125	perlite		6.65	8.90	15.55
1640	75% solid			none		5.80	8.70	14.50
1650			105	perlite		7.90	8.65	16.55
1690				none		7.05	8.45	15.50
1700	Deep groove-solid	4x8x16	125	none		3.89	7.65	11.54
1750			105	none		4.72	7.45	12.17
1800		6x8x16	125	none		5.70	8.15	13.85
1850			105	none		6.95	7.90	14.85
1900		8x8x16	125	none		6.55	8.85	15.40
1950			105	none		9.40	8.60	18
2100	Fluted	4x8x16	125	none		4.47	7.45	11.92
2150			105	none		5.65	7.25	12.90
2200		6x8x16	125	perlite		7.70	8.25	15.95
2210				styrofoam		7.80	7.90	15.70
2240				none		6.55	7.90	14.45
2300		8x8x16	125	perlite		7.10	9	16.10
2310				styrofoam		6.65	8.60	15.25
2340				none		5.40	8.60	14
2500	Hex-hollow	4x8x16	125	none		3.65	7.45	11.10
2550			105	none		4.57	7.25	11.82
2600		6x8x16	125	perlite		5.25	8.60	13.85
2610				styrofoam		5.35	8.25	13.60
2640				none		4.10	8.25	12.35
2650			105	perlite		6.25	8.35	14.60
2660				styrofoam		6.35	8	14.35
2690				none		5.10	8	13.10
2700		8x8x16	125	perlite		6.25	9.25	15.50
2710				styrofoam		5.80	8.85	14.65
2740				none		4.53	8.85	13.38
2750			105	perlite		7.35	9	16.35
2760				styrofoam		6.90	8.60	15.50
2790				none		5.65	8.60	14.25
2800	Hex-solid	4x8x16	125	none		4.53	7.65	12.18
2850			105	none		5.70	7.45	13.15
3100	Slump block	4x4x16	125	none		4.08	6.15	10.23
3200		6x4x16	125	perlite		6.70	6.65	13.35
3210				styrofoam		6.80	6.30	13.10
3240				none		5.55	6.30	11.85
3300		8x4x16	125	perlite		7.55	6.95	14.50
3310				styrofoam		7.10	6.55	13.65
3340				none		5.85	6.55	12.40
3400		12x4x16	125	perlite		14.25	8.65	22.90
3410				styrofoam		13.60	7.85	21.45
3440				none		11.45	7.85	19.30
3600		6x6x16	125	perlite		6.35	6.85	13.20
3610				styrofoam		6.45	6.50	12.95
3640				none		5.20	6.50	11.70
3700		8x6x16	125	perlite		9.25	7.15	16.40
3710				styrofoam		8.80	6.75	15.55
3740				none		7.55	6.75	14.30
3800		12x6x16	125	perlite		14.75	9.25	24
3810				styrofoam		14.10	8.45	22.55
3840				none		11.95	8.45	20.40

B20 Exterior Enclosure

B2010 Exterior Walls

B2010 124 Reinforced Misc. Block Walls

	TYPE	SIZE (IN.)	WEIGHT (P.C.F.)	VERT. REINF. & GROUT SPACING		COST PER S.F.		
						MAT.	INST.	TOTAL
5100	Deep groove-hollow	4x8x16	125	#4 @ 48"		2.93	8.10	11.03
5150			105	#4 @ 48"		3.49	7.90	11.39
5200		6x8x16	125	#4 @ 48"		4.31	8.60	12.91
5220				#5 @ 32"		4.52	8.85	13.37
5230				#5 @ 16"		4.99	9.80	14.79
5250			105	#4 @ 48"		5.15	7.95	13.10
5270				#5 @ 32"		5.35	8.20	13.55
5280				#5 @ 16"		5.80	9.15	14.95
5300		8x8x16	125	#4 @ 48"		4.96	9.40	14.36
5320				#5 @ 32"		5.20	9.70	14.90
5330				#5 @ 16"		5.75	10.85	16.60
5350			105	#4 @ 48"		5.90	9.10	15
5370				#5 @ 32"		6.15	9.40	15.55
5380				#5 @ 16"		6.65	10.55	17.20
5500	Deep groove-75% solid	6x8x16	125	#4 @ 48"		5.20	8.45	13.65
5520				#5 @ 32"		5.40	8.65	14.05
5530				#5 @ 16"		5.75	9.30	15.05
5550			105	#4 @ 48"		6.30	8.20	14.50
5570				#5 @ 32"		6.50	8.55	15.05
5580				#5 @ 16"		6.80	9.05	15.85
5600		8x8x16	125	#4 @ 48"		6	9.20	15.20
5620				#5 @ 32"		6.20	9.45	15.65
5630				#5 @ 16"		6.55	10.15	16.70
5650			105	#4 @ 48"		7.25	8.95	16.20
5670				#5 @ 32"		7.45	9.20	16.65
5680				#5 @ 16"		7.80	9.90	17.70
5700	Deep groove-solid Double wythe	2-4x8x16	125	#4 @ 48" E.W.		8.50	16.80	25.30
5730				#5 @ 16" E.W.		9.25	17.60	26.85
5750			105	#4 @ 48" E.W.		10.20	16.40	26.60
5780				#5 @ 16" E.W.		10.90	17.20	28.10
5800		2-6x8x16	125	#4 @ 48" E.W.		12.15	17.80	29.95
5830				#5 @ 16" E.W.		12.90	18.60	31.50
5850			105	#4 @ 48" E.W.		14.65	17.30	31.95
5880				#5 @ 16" E.W.		15.40	18.10	33.50
6100	Fluted	4x8x16	125	#4 @ 48"		4.62	8.10	12.72
6150			105	#4 @ 48"		5.80	7.90	13.70
6200		6x8x16	125	#4 @ 48"		6.85	8.60	15.45
6220				#5 @ 32"		7.05	8.85	15.90
6230				#5 @ 16"		7.50	9.80	17.30
6300		8x8x16	125	#4 @ 48"		5.70	9.40	15.10
6320				#5 @ 32"		5.95	9.70	15.65
6330				#5 @ 16"		6.45	10.85	17.30
6500	Hex-hollow	4x8x16	125	#4 @ 48"		3.80	8.10	11.90
6550			105	#4 @ 48"		4.72	7.90	12.62
6620				#5 @ 32"		4.57	9.20	13.77
6630				#5 @ 16"		5.05	10.15	15.20
6650		6x8x16	105	#4 @ 48"		5.40	8.70	14.10
6670				#5 @ 32"		5.60	8.95	14.55
6680				#5 @ 16"		6.05	9.90	15.95
6700		8x8x16	125	#4 @ 48"		4.84	9.65	14.49
6720				#5 @ 32"		5.10	9.95	15.05
6730				#5 @ 16"		5.60	11.10	16.70

B20 Exterior Enclosure

B2010 Exterior Walls

B2010 124 Reinforced Misc. Block Walls

	TYPE	SIZE (IN.)	WEIGHT (P.C.F.)	VERT. REINF. & GROUT SPACING		COST PER S.F.		
						MAT.	INST.	TOTAL
6750	Hex-hollow	8x8x16	105	#4 @ 48"		5.95	9.40	15.35
6770				#5 @ 32"		6.20	9.70	15.90
6800	Hex-solid	2-4x8x16	125	#4 @ 48" E.W.		9.75	16.75	26.50
6830	Double wythe			#5 @ 16" E.W.		10.55	17.60	28.15
6850			105	#4 @ 48" E.W.		12.05	16.35	28.40
6880				#5 @ 16" E.W.		12.90	17.20	30.10
7100	Slump block	4x4x16	125	#4 @ 48"		4.23	6.80	11.03
7200		6x4x16	125	#4 @ 48"		5.85	7	12.85
7220				#5 @ 32"		6.05	7.55	13.60
7230				#5 @ 16"		6.50	8.20	14.70
7300		8x4x16	125	#4 @ 48"		6.15	7.35	13.50
7320				#5 @ 32"		6.40	7.65	14.05
7330				#5 @ 16"		6.90	8.80	15.70
7400		12x4x16	125	#4 @ 48"		11.90	8.70	20.60
7420				#5 @ 32"		12.20	9	21.20
7430				#5 @ 16"		12.95	10.20	23.15
7600		6x6x16	125	#4 @ 48"		5.45	7.20	12.65
7620				#5 @ 32"		5.65	7.45	13.10
7630				#5 @ 16"		6.15	8.40	14.55
7700		8x6x16	125	#4 @ 48"		7.85	7.55	15.40
7720				#5 @ 32"		8.10	7.85	15.95
7730				#5 @ 16"		8.60	9	17.60
7800		12x6x16	125	#4 @ 48"		12.40	9.30	21.70
7820				#5 @ 32"		12.70	9.60	22.30
7830				#5 @ 16"		13.45	10.80	24.25

B20 Exterior Enclosure

B2010 Exterior Walls

Exterior solid brick walls are defined in the following terms; structural reinforcement, type of brick, thickness, and bond. Sixteen different types of face bricks are presented, with single wythes shown in four different bonds. Six types of reinforced single wythe walls are also given, and twice that many for reinforced double wythe. Shelf angles are included in system components. These walls do not include ties and as such are not tied to a backup wall.

System Components	QUANTITY	UNIT	COST PER S.F.		
			MAT.	INST.	TOTAL
SYSTEM B2010 125 1000					
BRICK WALL, COMMON, SINGLE WYTHE, 4" THICK, RUNNING BOND					
Common brick wall, running bond	1.000	S.F.	3.09	10.85	13.94
Wash brick	1.000	S.F.	.07	.97	1.04
Control joint, backer rod	.100	L.F.		.12	.12
Control joint, sealant	.100	L.F.	.03	.24	.27
Shelf angle	1.000	Lb.	1.10	.99	2.09
Flashing	.100	S.F.	.12	.36	.48
TOTAL			4.41	13.53	17.94

B2010 125			Solid Brick Walls - Single Wythe					
	TYPE	THICKNESS (IN.)	BOND			COST PER S.F.		
						MAT.	INST.	TOTAL
1000	Common	4	running			4.41	13.55	17.96
1010			common			5.25	15.50	20.75
1050			Flemish			5.80	18.85	24.65
1100			English			6.50	19.95	26.45
1150	Standard	4	running			5.70	14.05	19.75
1160			common			6.70	16.20	22.90
1200			Flemish			7.40	19.35	26.75
1250			English			8.30	20.50	28.80
1300	Glazed	4	running			19.65	14.60	34.25
1310			common			23.50	17	40.50
1350			Flemish			26.50	20.50	47
1400			English			29.50	22	51.50
1450	Engineer	4	running			5.30	12.30	17.60
1460			common			6.25	14.05	20.30
1500			Flemish			6.85	17	23.85
1550			English			7.70	17.85	25.55
1600	Economy	4	running			8.05	10.75	18.80
1610			common			9.50	12.30	21.80
1650			Flemish			10.50	14.60	25.10
1700			English			21	15.50	36.50
1750	Double	4	running			12.70	8.80	21.50
1760			common			15.10	9.95	25.05
1800			Flemish			16.65	11.80	28.45
1850			English			18.75	12.50	31.25

B20 Exterior Enclosure

B2010 Exterior Walls

B2010 125 Solid Brick Walls - Single Wythe

	TYPE	THICKNESS (IN.)	BOND			COST PER S.F.		
						MAT.	INST.	TOTAL
1900	Fire	4-1/2	running			12.60	12.30	24.90
1910			common			15	14.05	29.05
1950			Flemish			16.65	17	33.65
2000			English			18.70	17.85	36.55
2050	King	3-1/2	running			4.30	11.05	15.35
2060			common			5	12.70	17.70
2100			Flemish			6	15.20	21.20
2150			English			6.65	15.85	22.50
2200	Roman	4	running			9.40	12.70	22.10
2210			common			11.25	14.60	25.85
2250			Flemish			12.40	17.40	29.80
2300			English			14	18.85	32.85
2350	Norman	4	running			6.50	10.50	17
2360			common			7.65	11.95	19.60
2400			Flemish			17.50	14.35	31.85
2450			English			9.50	15.20	24.70
2500	Norwegian	4	running			6.40	9.35	15.75
2510			common			7.55	10.65	18.20
2550			Flemish			8.35	12.70	21.05
2600			English			9.35	13.35	22.70
2650	Utility	4	running			6.50	8.25	14.75
2660			common			7.60	9.30	16.90
2700			Flemish			8.40	11.05	19.45
2750			English			9.40	11.65	21.05
2800	Triple	4	running			4.67	7.45	12.12
2810			common			5.40	8.40	13.80
2850			Flemish			5.90	9.85	15.75
2900			English			6.50	10.25	16.75
2950	SCR	6	running			8.50	10.75	19.25
2960			common			10.10	12.30	22.40
3000			Flemish			11.25	14.60	25.85
3050			English			12.70	15.50	28.20
3100	Norwegian	6	running			7.60	9.55	17.15
3110			common			8.95	10.90	19.85
3150			Flemish			9.90	12.90	22.80
3200			English			11.10	13.55	24.65
3250	Jumbo	6	running			6.35	8.45	14.80
3260			common			7.50	9.55	17.05
3300			Flemish			8.30	11.30	19.60
3350			English			9.35	11.95	21.30

B2010 126 Solid Brick Walls - Double Wythe

	TYPE	THICKNESS (IN.)	COLLAR JOINT THICKNESS (IN.)			COST PER S.F.		
						MAT.	INST.	TOTAL
4100	Common	8	3/4			8.35	23	31.35
4150	Standard	8	1/2			10.75	23.50	34.25
4200	Glazed	8	1/2			38.50	24.50	63
4250	Engineer	8	1/2			10	21	31
4300	Economy	8	3/4			15.50	17.85	33.35
4350	Double	8	3/4			15.40	14.40	29.80
4400	Fire	8	3/4			24.50	21	45.50
4450	King	7	3/4			7.95	18.25	26.20

B20 Exterior Enclosure

B2010 Exterior Walls

B2010 126 Solid Brick Walls - Double Wythe

	TYPE	THICKNESS (IN.)	COLLAR JOINT THICKNESS (IN.)			COST PER S.F.		
						MAT.	INST.	TOTAL
4500	Roman	8	1			18.25	21.50	39.75
4550	Norman	8	3/4			12.40	17.45	29.85
4600	Norwegian	8	3/4			12.25	15.45	27.70
4650	Utility	8	3/4			12.35	13.35	25.70
4700	Triple	8	3/4			8.70	12	20.70
4750	SCR	12	3/4			16.30	17.90	34.20
4800	Norwegian	12	3/4			14.50	15.80	30.30
4850	Jumbo	12	3/4			12.05	13.80	25.85

B2010 127 Solid Brick Walls, Reinforced

	TYPE	THICKNESS (IN.)	WYTHE	REINF. & SPACING		COST PER S.F.		
						MAT.	INST.	TOTAL
7200	Common	4"	1	#4 @ 48"vert		4.61	14.05	18.66
7220				#5 @ 32"vert		4.76	14.25	19.01
7230				#5 @ 16"vert		5.05	14.90	19.95
7300	Utility	4"	1	#4 @ 48"vert		6.70	8.75	15.45
7320				#5 @ 32"vert		6.85	8.95	15.80
7330				#5 @ 16"vert		7.15	9.60	16.75
7400	SCR	6"	1	#4 @ 48"vert		8.75	11.25	20
7420				#5 @ 32"vert		8.95	11.45	20.40
7430				#5 @ 16"vert		9.30	12.10	21.40
7500	Jumbo	6"	1	#4 @ 48"vert		6.55	8.90	15.45
7520				#5 @ 32"vert		6.70	9.05	15.75
7530				#5 @ 16"vert		6.95	9.65	16.60
7600		8"	1	#4 @ 48"vert		8.20	9.15	17.35
7620				#5 @ 32"vert		8.40	9.40	17.80
7630				#5 @ 16"vert		8.75	10.15	18.90
7700	King	9"	2	#4 @ 48" E.W.		8.35	19.15	27.50
7720				#5 @ 32" E.W.		8.60	19.30	27.90
7730				#5 @ 16" E.W.		9.10	19.85	28.95
7800	Common	10"	2	#4 @ 48" E.W.		8.75	23.50	32.25
7820				#5 @ 32" E.W.		9	23.50	32.50
7830				#5 @ 16" E.W.		9.50	24.50	34
7900	Standard	10"	2	#4 @ 48" E.W.		11.15	24.50	35.65
7920				#5 @ 32" E.W.		11.40	24.50	35.90
7930				#5 @ 16" E.W.		11.90	25	36.90
8000	Engineer	10"	2	#4 @ 48" E.W.		10.40	21.50	31.90
8020				#5 @ 32" E.W.		10.65	21.50	32.15
8030				#5 @ 16" E.W.		11.15	22.50	33.65
8100	Economy	10"	2	#4 @ 48" E.W.		15.90	18.75	34.65
8120				#5 @ 32" E.W.		16.15	18.90	35.05
8130				#5 @ 16" E.W.		16.65	19.45	36.10
8200	Double	10"	2	#4 @ 48" E.W.		15.80	15.30	31.10
8220				#5 @ 32" E.W.		16.05	15.45	31.50
8230				#5 @ 16" E.W.		16.55	16	32.55
8300	Fire	10"	2	#4 @ 48" E.W.		25	21.50	46.50
8320				#5 @ 32" E.W.		25	21.50	46.50
8330				#5 @ 16" E.W.		25.50	22.50	48
8400	Roman	10"	2	#4 @ 48" E.W.		18.65	22	40.65
8420				#5 @ 32" E.W.		18.90	22.50	41.40
8430				#5 @ 16" E.W.		19.40	23	42.40

B20 Exterior Enclosure

B2010 Exterior Walls

B2010 127 Solid Brick Walls, Reinforced

	TYPE	THICKNESS (IN.)	WYTHE	REINF. & SPACING		COST PER S.F.		
						MAT.	INST.	TOTAL
8500	Norman	10"	2	#4 @ 48" E.W.		12.80	18.35	31.15
8520				#5 @ 32" E.W.		13.05	18.50	31.55
8530				#5 @ 16" E.W.		13.55	19.05	32.60
8600	Norwegian	10"	2	#4 @ 48" E.W.		12.65	16.35	29
8620				#5 @ 32" E.W.		12.90	16.50	29.40
8630				#5 @ 16" E.W.		13.40	17.05	30.45
8700	Utility	10"	2	#4 @ 48" E.W.		12.75	14.25	27
8720				#5 @ 32" E.W.		13	14.40	27.40
8730				#5 @ 16" E.W.		13.50	14.95	28.45
8800	Triple	10"	2	#4 @ 48" E.W.		9.10	12.90	22
8820				#5 @ 32" E.W.		9.35	13.05	22.40
8830				#5 @ 16" E.W.		9.85	13.60	23.45
8900	SCR	14"	2	#4 @ 48" E.W.		16.75	18.75	35.50
8920				#5 @ 32" E.W.		17	18.90	35.90
8930				#5 @ 16" E.W.		17.50	19.45	36.95
9000	Jumbo	14"	2	#4 @ 48" E.W.		12.50	14.65	27.15
9030				#5 @ 16" E.W.		13.25	15.35	28.60
9100	Oversized, 6" x 16"	4"	1	#5@12" vert		17.90	9.20	27.10
9110		8"	1	#5@12" vert		15.80	8.45	24.25
9120	Oversized, 8" x 16"	4"	1	#5@12" vert		18.35	10.40	28.75
9130		8"	1	#5@12" vert		16.75	9	25.75

B20 Exterior Enclosure

B2010 Exterior Walls

The table below lists costs per S.F. for stone veneer walls on various backup using different stone. Typical components for a system are shown in the component block below.

System Components	QUANTITY	UNIT	COST PER S.F.		
			MAT.	INST.	TOTAL
SYSTEM B2010 128 2000 **ASHLAR STONE VENEER, 4", 2"X4" STUD, 16" O.C. BACK UP, 8' HIGH**					
Sawn face, split joints, low priced stone	1.000	S.F.	12.35	17.85	30.20
Framing, 2x4 studs 8' high 16" O.C.	.920	B.F.	.45	1.38	1.83
Wall ties for stone veneer, galv. corrg 7/8" x 7", 22 gauge	.700	Ea.	.10	.36	.46
Asphalt felt sheathing paper, 15 lb.	1.000	S.F.	.06	.15	.21
Sheathing plywood on wall CDX 1/2"	1.000	S.F.	.64	.98	1.62
Fiberglass insulation, 3-1/2", cr-11	1.000	S.F.	.41	.41	.82
Flashing, copper, paperback 1 side, 3 oz	.125	S.F.	.27	.20	.47
TOTAL			14.28	21.33	35.61

B2010 128	Stone Veneer	COST PER S.F.		
		MAT.	INST.	TOTAL
2000	Ashlar veneer, 4", 2"x4" stud backup, 16" O.C., 8' high, low priced stone	14.30	21.50	35.80
2100	Metal stud backup, 8' high, 16" O.C.	15.10	21.50	36.60
2150	24" O.C.	14.75	21	35.75
2200	Conc. block backup, 4" thick	15.60	25.50	41.10
2300	6" thick	16.25	25.50	41.75
2350	8" thick	16.45	26	42.45
2400	10" thick	16.95	27.50	44.45
2500	12" thick	17.80	29.50	47.30
3100	High priced stone, wood stud backup, 10' high, 16" O.C.	21.50	24.50	46
3200	Metal stud backup, 10' high, 16" O.C.	22	25	47
3250	24" O.C.	21.50	24.50	46
3300	Conc. block backup, 10' high, 4" thick	22.50	28.50	51
3350	6" thick	23	28.50	51.50
3400	8" thick	23.50	29.50	53
3450	10" thick	24	30.50	54.50
3500	12" thick	24.50	32.50	57
4000	Indiana limestone 2" thick, sawn finish, wood stud backup, 10' high, 16" O.C.	40	13.15	53.15
4100	Metal stud backup, 10' high, 16" O.C.	40.50	13.45	53.95
4150	24" O.C.	40.50	12.95	53.45
4200	Conc. block backup, 4" thick	41	17.20	58.20
4250	6" thick	42	17.45	59.45
4300	8" thick	42	18	60

B20 Exterior Enclosure

B2010 Exterior Walls

B2010 128	Stone Veneer	COST PER S.F.		
		MAT.	INST.	TOTAL
4350	10" thick	42.50	19.30	61.80
4400	12" thick	43.50	21	64.50
4450	2" thick, smooth finish, wood stud backup, 8' high, 16" O.C.	40	13.15	53.15
4550	Metal stud backup, 8' high, 16" O.C.	40.50	13.25	53.75
4600	24" O.C.	40.50	12.75	53.25
4650	Conc. block backup, 4" thick	41	17	58
4700	6" thick	42	17.25	59.25
4750	8" thick	42	17.80	59.80
4800	10" thick	42.50	19.30	61.80
4850	12" thick	43.50	21	64.50
5350	4" thick, smooth finish, wood stud backup, 8' high, 16" O.C.	46.50	13.15	59.65
5450	Metal stud backup, 8' high, 16" O.C.	47	13.45	60.45
5500	24" O.C.	47	12.95	59.95
5550	Conc. block backup, 4" thick	47.50	17.20	64.70
5600	6" thick	48.50	17.45	65.95
5650	8" thick	48.50	18	66.50
5700	10" thick	49	19.30	68.30
5750	12" thick	50	21	71
6000	Granite, gray or pink, 2" thick, wood stud backup, 8' high, 16" O.C.	24	26.50	50.50
6100	Metal studs, 8' high, 16" O.C.	24.50	27	51.50
6150	24" O.C.	24.50	26.50	51
6200	Conc. block backup, 4" thick	25	30.50	55.50
6250	6" thick	26	31	57
6300	8" thick	26	31.50	57.50
6350	10" thick	26.50	32.50	59
6400	12" thick	27.50	34.50	62
6900	4" thick, wood stud backup, 8' high, 16" O.C.	38.50	30	68.50
7000	Metal studs, 8' high, 16" O.C.	39	30.50	69.50
7050	24" O.C.	39	30	69
7100	Conc. block backup, 4" thick	39.50	34	73.50
7150	6" thick	40.50	34.50	75
7200	8" thick	40.50	35	75.50
7250	10" thick	41	36	77
7300	12" thick	42	38	80

B20 Exterior Enclosure

B2010 Exterior Walls

Exterior brick veneer/stud backup walls are defined in the following terms: type of brick and studs, stud spacing and bond. All systems include a back-up wall, a control joint every 20', a brick shelf every 12' of height, ties to the backup and the necessary dampproofing, flashing and insulation.

System Components	QUANTITY	UNIT	COST PER S.F.		
			MAT.	INST.	TOTAL
SYSTEM B2010 129 1100					
STANDARD BRICK VENEER, 2"X4" STUD BACKUP @ 16"O.C., RUNNING BOND					
Standard brick wall, 4" thick, running bond	1.000	S.F.	4.37	11.35	15.72
Wash smooth brick	1.000	S.F.	.07	.97	1.04
Joint backer rod	.100	L.F.		.12	.12
Sealant	.100	L.F.	.03	.24	.27
Wall ties, corrugated, 7/8" x 7", 22 gauge	.003	Ea.	.04	.15	.19
Shelf angle	1.000	Lb.	1.10	.99	2.09
Wood stud partition, backup, 2" x 4" @ 16" O.C.	1.000	S.F.	.42	1.11	1.53
Sheathing, plywood, CDX, 1/2"	1.000	S.F.	.64	.79	1.43
Building paper, asphalt felt, 15 lb.	1.000	S.F.	.06	.15	.21
Fiberglass insulation, batts, 3-1/2" thick paper backing	1.000	S.F.	.41	.41	.82
Flashing, copper, paperbacked	.100	S.F.	.12	.36	.48
TOTAL			7.26	16.64	23.90

B2010 129	Brick Veneer/Wood Stud Backup						
	FACE BRICK	STUD BACKUP	STUD SPACING (IN.)	BOND	FACE	COST PER S.F.		
						MAT.	INST.	TOTAL
1100	Standard	2x4-wood	16	running		7.25	16.65	23.90
1120				common		8.30	18.80	27.10
1140				Flemish		9	22	31
1160				English		9.90	23	32.90
1400		2x6-wood	16	running		7.50	16.75	24.25
1420				common		8.50	18.90	27.40
1440				Flemish		9.20	22	31.20
1460				English		10.10	23.50	33.60
1500			24	running		7.35	16.50	23.85
1520				common		8.35	18.65	27
1540				Flemish		9.05	22	31.05
1560				English		9.95	23	32.95
1700	Glazed	2x4-wood	16	running		21	17.20	38.20
1720				common		25	19.60	44.60
1740				Flemish		28	23	51
1760				English		31	24.50	55.50

B20 Exterior Enclosure

B2010 Exterior Walls

B2010 129 Brick Veneer/Wood Stud Backup

	FACE BRICK	STUD BACKUP	STUD SPACING (IN.)	BOND	FACE	COST PER S.F.		
						MAT.	INST.	TOTAL
2000	Glazed	2x6-wood	16	running		21.50	17.30	38.80
2020				common		25	19.70	44.70
2040				Flemish		28	23.50	51.50
2060				English		31	24.50	55.50
2100			24	running		21.50	17.05	38.55
2120				common		25	19.45	44.45
2140				Flemish		28	23	51
2160				English		31	24.50	55.50
2300	Engineer	2x4-wood	16	running		6.90	14.90	21.80
2320				common		7.80	16.65	24.45
2340				Flemish		8.45	19.60	28.05
2360				English		9.30	20.50	29.80
2600		2x6-wood	16	running		7.10	15	22.10
2620				common		8.05	16.75	24.80
2640				Flemish		8.65	19.70	28.35
2660				English		9.50	20.50	30
2700			24	running		6.95	14.75	21.70
2720				common		7.85	16.50	24.35
2740				Flcmish		8.50	19.45	27.95
2760				English		9.35	20.50	29.85
2900	Roman	2x4-wood	16	running		11	15.30	26.30
2920				common		12.85	17.20	30.05
2940				Flemish		14	20	34
2960				English		15.60	21.50	37.10
3200		2x6-wood	16	running		11.20	15.40	26.60
3220				common		13.05	17.30	30.35
3240				Flemish		14.20	20	34.20
3260				English		15.80	21.50	37.30
3300			24	running		11.05	15.15	26.20
3320				common		12.90	17.05	29.95
3340				Flemish		14.05	19.85	33.90
3360				English		15.65	21.50	37.15
3500	Norman	2x4-wood	16	running		8.10	13.10	21.20
3520				common		9.25	14.55	23.80
3540				Flemish		19.10	16.95	36.05
3560				English		11.10	17.80	28.90
3800		2x6-wood	16	running		8.30	13.20	21.50
3820				common		9.45	14.65	24.10
3840				Flemish		18.75	16.35	35.10
3860				English		11.30	17.90	29.20
3900			24	running		8.15	12.95	21.10
3920				common		9.30	14.40	23.70
3940				Flemish		19.15	16.80	35.95
3960				English		11.15	17.65	28.80
4100	Norwegian	2x4-wood	16	running		8	11.95	19.95
4120				common		9.15	13.25	22.40
4140				Flemish		9.95	15.30	25.25
4160				English		10.95	15.95	26.90
4400		2x6-wood	16	running		8.20	12.05	20.25
4420				common		9.35	13.35	22.70
4440				Flemish		10.15	15.40	25.55
4460				English		11.15	16.05	27.20

B20 Exterior Enclosure

B2010 Exterior Walls

B2010 129 — Brick Veneer/Wood Stud Backup

	FACE BRICK	STUD BACKUP	STUD SPACING (IN.)	BOND	FACE	COST PER S.F. MAT.	COST PER S.F. INST.	COST PER S.F. TOTAL
4500	Norwegian	2x6-wood	24	running		8.05	11.80	19.85
4520				common		9.20	13.10	22.30
4540				Flemish		10	15.15	25.15
4560				English		11	15.80	26.80
4600	Oversized, 4" x 2-1/4" x 16"	2x4-wood	16	running	plain face	10	11.75	21.75
4610					1 to 3 slot face	10.50	12.25	22.75
4620					4 to 7 slot face	10.95	12.60	23.55
4630	Oversized, 4" x 2-3/4" x 16"	2x4-wood	16	running	plain face	8.70	11.35	20.05
4640					1 to 3 slot face	9.15	11.80	20.95
4650					4 to 7 slot face	9.50	12.15	21.65
4660	Oversized, 4" x 4" x 16"	2x4-wood	16	running	plain face	8.75	10.75	19.50
4670					1 to 3 slot face	9.20	11.15	20.35
4680					4 to 7 slot face	9.55	11.50	21.05
4690	Oversized, 4" x 8" x 16"	2x4-wood	16	running	plain face	9.65	10	19.65
4700					1 to 3 slot face	10.15	10.35	20.50
4710					4 to 7 slot face	10.55	10.60	21.15
4720	Oversized, 4" x 2-1/4" x 16"	2x6-wood	16	running	plain face	10.20	11.85	22.05
4730					1 to 3 slot face	10.75	12.35	23.10
4740					4 to 7 slot face	11.20	12.75	23.95
4750	Oversized, 4" x 2-3/4" x 16"	2x6-wood	16	running	plain face	8.90	11.45	20.35
4760					1 to 3 slot face	9.35	11.90	21.25
4770					4 to 7 slot face	9.70	12.30	22
4780	Oversized, 4" x 4" x 16"	2x6-wood	16	running	plain face	8.95	10.85	19.80
4790					1 to 3 slot face	9.40	11.25	20.65
4800					4 to 7 slot face	9.75	11.60	21.35
4810	Oversized, 4" x 8" x 16"	2x6-wood	16	running	plain face	9.85	10.10	19.95
4820					1 to 3 slot face	10.40	10.45	20.85
4830					4 to 7 slot face	10.80	10.75	21.55

B2010 130 — Brick Veneer/Metal Stud Backup

	FACE BRICK	STUD BACKUP	STUD SPACING (IN.)	BOND	FACE	COST PER S.F. MAT.	COST PER S.F. INST.	COST PER S.F. TOTAL
5100	Standard	25ga.x6"NLB	24	running		6.85	16.25	23.10
5120				common		7.90	18.40	26.30
5140				Flemish		8.25	21	29.25
5160				English		9.50	23	32.50
5200		20ga.x3-5/8"NLB	16	running		7	16.95	23.95
5220				common		8.05	19.10	27.15
5240				Flemish		8.75	22.50	31.25
5260				English		9.65	23.50	33.15
5300			24	running		6.85	16.45	23.30
5320				common		7.90	18.60	26.50
5340				Flemish		8.60	22	30.60
5360				English		9.50	23	32.50
5400		16ga.x3-5/8"LB	16	running		7.75	17.20	24.95
5420				common		8.75	19.35	28.10
5440				Flemish		9.45	22.50	31.95
5460				English		10.35	23.50	33.85
5500			24	running		7.40	16.70	24.10
5520				common		8.40	18.85	27.25
5540				Flemish		9.10	22	31.10
5560				English		10	23	33

B20 Exterior Enclosure

B2010 Exterior Walls

B2010 130 — Brick Veneer/Metal Stud Backup

	FACE BRICK	STUD BACKUP	STUD SPACING (IN.)	BOND	FACE	COST PER S.F.		
						MAT.	INST.	TOTAL
5700	Glazed	25ga.x6"NLB	24	running		21	16.80	37.80
5720				common		24.50	19.20	43.70
5740				Flemish		27.50	23	50.50
5760				English		30.50	24	54.50
5800		20ga.x3-5/8"NLB	24	running		21	17	38
5820				common		24.50	19.40	43.90
5840				Flemish		27.50	23	50.50
5860				English		30.50	24.50	55
6000	Glazed	16ga.x3-5/8"LB	16	running		21.50	17.75	39.25
6020				common		25.50	20	45.50
6040				Flemish		28.50	23.50	52
6060				English		31.50	25	56.50
6100			24	running		21.50	17.25	38.75
6120				common		25	19.65	44.65
6140				Flemish		28	23	51
6160				English		31	24.50	55.50
6300	Engineer	25ga.x6"NLB	24	running		6.50	14.50	21
6320				common		7.40	16.25	23.65
6340				Flemish		8.05	19.20	27.25
6360				English		8.90	20	28.90
6400		20ga.x3-5/8"NLB	16	running		6.65	15.20	21.85
6420				common		7.55	16.95	24.50
6440				Flemish		8.20	19.90	28.10
6460				English		9.05	21	30.05
6500		20ga.x3-5/8"NLB	24	running		6.15	13.85	20
6520				common		7.40	16.45	23.85
6540				Flemish		8.05	19.40	27.45
6560				English		8.90	20.50	29.40
6600		16ga.x3-5/8"LB	16	running		7.35	15.45	22.80
6620				common		8.25	17.20	25.45
6640				Flemish		8.90	20	28.90
6660				English		9.75	21	30.75
6700			24	running		7	14.95	21.95
6720				common		7.90	16.70	24.60
6740				Flemish		8.55	19.65	28.20
6760				English		9.40	20.50	29.90
6900	Roman	25ga.x6"NLB	24	running		10.60	14.90	25.50
6920				common		12.45	16.75	29.20
6940				Flemish		13.60	19.60	33.20
6960				English		15.20	21	36.20
7000		20ga.x3-5/8"NLB	16	running		10.75	15.60	26.35
7020				common		12.60	17.50	30.10
7040				Flemish		13.75	20.50	34.25
7060				English		15.35	22	37.35
7100			24	running		10.60	15.10	25.70
7120				common		12.45	17	29.45
7140				Flemish		13.60	19.80	33.40
7160				English		15.20	21.50	36.70
7200		16ga.x3-5/8"LB	16	running		11.45	15.85	27.30
7220				common		13.30	17.75	31.05
7240				Flemish		14.45	20.50	34.95
7260				English		16.05	22	38.05

B20 Exterior Enclosure

B2010 Exterior Walls

B2010 130 Brick Veneer/Metal Stud Backup

	FACE BRICK	STUD BACKUP	STUD SPACING (IN.)	BOND	FACE	COST PER S.F.		
						MAT.	INST.	TOTAL
7300	Roman		24	running		11.10	15.35	26.45
7320				common		12.95	17.25	30.20
7340				Flemish		14.10	20	34.10
7360				English		15.70	21.50	37.20
7500	Norman	25ga.x6"NLB	24	running		7.70	12.70	20.40
7520				common		8.85	14.15	23
7540				Flemish		18.70	16.55	35.25
7560				English		10.70	17.40	28.10
7600		20ga.x3-5/8"NLB	24	running		7.70	12.90	20.60
7620				common		8.85	14.35	23.20
7640				Flemish		18.70	16.75	35.45
7660				English		10.70	17.60	28.30
7800		16ga.x3-5/8"LB	16	running		8.55	13.65	22.20
7820				common		9.70	15.10	24.80
7840				Flemish		19.55	17.50	37.05
7860				English		11.55	18.35	29.90
7900			24	running		8.20	13.15	21.35
7920				common		9.35	14.60	23.95
7940				Flemish		19.20	17	36.20
7960				English		11.20	17.85	29.05
8100	Norwegian	25ga.x6"NLB	24	running		7.60	11.55	19.15
8120				common		8.75	12.85	21.60
8140				Flemish		9.55	14.90	24.45
8160				English		10.55	15.55	26.10
8200		20ga.x3-5/8"NLB	16	running		7.75	12.25	20
8220				common		8.90	13.55	22.45
8240				Flemish		9.70	15.60	25.30
8260				English		10.70	16.25	26.95
8300			24	running		7.60	11.75	19.35
8320				common		8.75	13.05	21.80
8340				Flemish		9.55	15.10	24.65
8360				English		10.55	15.75	26.30
8400		16ga.x3-5/8"LB	16	running		8.45	12.50	20.95
8420				common		9.60	13.80	23.40
8440				Flemish		10.40	15.85	26.25
8460				English		11.40	16.50	27.90
8500			24	running		8.10	12	20.10
8520				common		9.25	13.30	22.55
8540				Flemish		10.05	15.35	25.40
8560				English		11.05	16	27.05
8600	Oversized, 4" x 2-1/4" x 16"	25 ga. x 6" NLB	24	running	plain face	9.95	11.40	21.35
8610					1 to 3 slot face	10.50	11.90	22.40
8620					4 to 7 slot face	10.95	12.25	23.20
8630	Oversized, 4" x 2-3/4" x 16"	25 ga. x 6" NLB	24	running	plain face	8.65	11	19.65
8640					1 to 3 slot face	9.10	11.45	20.55
8650					4 to 7 slot face	9.45	11.80	21.25
8660	Oversized, 4" x 4" x 16"	25 ga. x 6" NLB	24	running	plain face	8.70	10.40	19.10
8670					1 to 3 slot face	9.15	10.80	19.95
8680					4 to 7 slot face	9.50	11.15	20.65
8690	Oversized, 4" x 8" x 16"	25 ga x 6" NLB	24	running	plain face	9.60	9.65	19.25
8700					1 to 3 slot face	10.15	10	20.15
8710					4 to 7 slot face	10.55	10.25	20.80

B20 Exterior Enclosure

B2010 Exterior Walls

B2010 130 — Brick Veneer/Metal Stud Backup

	FACE BRICK	STUD BACKUP	STUD SPACING (IN.)	BOND	FACE	COST PER S.F.		
						MAT.	INST.	TOTAL
8720	Oversized, 4" x 2-1/4" x 16"	20 ga. x 3-5/8" NLB	24	running	plain face	9.95	11.55	21.50
8730					1 to 3 slot face	10.50	12.05	22.55
8740					4 to 7 slot face	10.95	12.45	23.40
8750	Oversized, 4" x 2-3/4" x 16"	20 ga. x 3-5/8" NLB	24	running	plain face	8.65	11.15	19.80
8760					1 to 3 slot face	9.10	11.65	20.75
8770					4 to 7 slot face	9.45	12	21.45
8780	Oversized, 4" x 4" x 16"	20 ga. x 3-5/8" NLB	24	running	plain face	8.70	10.55	19.25
8790					1 to 3 slot face	9.15	11	20.15
8800					4 to 7 slot face	9.50	11.30	20.80
8810	Oversized, 4" x 8" x 16"	20 ga x 3-5/8" NLB	24	running	plain face	9.60	9.80	19.40
8820					1 to 3 slot face	10.15	10.15	20.30
8830					4 to 7 slot face	10.55	10.45	21

B20 Exterior Enclosure

B2010 Exterior Walls

Exterior brick face composite walls are defined in the following terms: type of face brick and backup masonry, thickness of backup masonry and insulation. A special section is included on triple wythe construction at the back. Seven types of face brick are shown with various thicknesses of seven types of backup. All systems include a brick shelf, ties to the backup and necessary dampproofing, flashing, and control joints every 20'.

System Components	QUANTITY	UNIT	COST PER S.F.		
			MAT.	INST.	TOTAL
SYSTEM B2010 132 1120					
COMPOSITE WALL, STANDARD BRICK FACE, 6" C.M.U. BACKUP, PERLITE FILL					
Face brick veneer, standard, running bond	1.000	S.F.	4.37	11.35	15.72
Wash brick	1.000	S.F.	.07	.97	1.04
Concrete block backup, 6" thick	1.000	S.F.	2.33	6.25	8.58
Wall ties	.300	Ea.	.09	.15	.24
Perlite insulation, poured	1.000	S.F.	1.15	.33	1.48
Flashing, aluminum	.100	S.F.	.12	.36	.48
Shelf angle	1.000	Lb.	1.10	.99	2.09
Control joint	.050	L.F.	.07	.07	.14
Backer rod	.100	L.F.		.12	.12
Sealant	.100	L.F.	.03	.24	.27
Collar joint	1.000	S.F.	.40	.62	1.02
TOTAL			9.73	21.45	31.18

B2010 132		Brick Face Composite Wall - Double Wythe					
	FACE BRICK	BACKUP MASONRY	BACKUP THICKNESS (IN.)	BACKUP CORE FILL		COST PER S.F.	
						MAT.	INST.	TOTAL
1000	Standard	common brick	4	none		9.35	25.50	34.85
1040		SCR brick	6	none		13.45	23	36.45
1080		conc. block	4	none		7.90	20.50	28.40
1120			6	perlite		9.75	21.50	31.25
1160				styrofoam		9.85	21	30.85
1200			8	perlite		10.45	22	32.45
1240				styrofoam		10	21.50	31.50
1280		L.W. block	4	none		8.05	20.50	28.55
1320			6	perlite		10	21.50	31.50
1360				styrofoam		10.10	21	31.10
1400			8	perlite		11.10	21.50	32.60
1440				styrofoam		10.65	21.50	32.15
1520		glazed block	4	none		16.65	22	38.65
1560			6	perlite		18.65	22.50	41.15
1600				styrofoam		18.75	22.50	41.25
1640			8	perlite		19.60	23.50	43.10
1680				styrofoam		19.15	23	42.15

B20 Exterior Enclosure

B2010 Exterior Walls

B2010 132 Brick Face Composite Wall - Double Wythe

	FACE BRICK	BACKUP MASONRY	BACKUP THICKNESS (IN.)	BACKUP CORE FILL		COST PER S.F.		
						MAT.	INST.	TOTAL
1720	Standard	clay tile	4	none		12.45	19.90	32.35
1760			6	none		14.65	20.50	35.15
1800			8	none		16.70	21	37.70
1840		glazed tile	4	none		19.25	26	45.25
1880								
2000	Glazed	common brick	4	none		23.50	26	49.50
2040		SCR brick	6	none		27.50	23.50	51
2080		conc. block	4	none		22	21	43
2120			6	perlite		23.50	22	45.50
2160				styrofoam		24	21.50	45.50
2200			8	perlite		24.50	22.50	47
2240				styrofoam		24	22	46
2280		L.W. block	4	none		22	21	43
2320			6	perlite		24	22	46
2360				styrofoam		24	21.50	45.50
2400			8	perlite		25	22.50	47.50
2440				styrofoam		24.50	22	46.50
2520		glazed block	4	none		30.50	22.50	53
2560			6	perlite		32.50	23.50	56
2600				styrofoam		32.50	23	55.50
2640			8	perlite		33.50	24	57.50
2680				styrofoam		33	23.50	56.50
2720		clay tile	4	none		26.50	20.50	47
2760			6	none		28.50	21	49.50
2800			8	none		30.50	21.50	52
2840		glazed tile	4	none		33	26.50	59.50
2880								
3000	Engineer	common brick	4	none		8.95	24	32.95
3040		SCR brick	6	none		13.05	21	34.05
3080		conc. block	4	none		7.50	18.95	26.45
3120			6	perlite		9.35	19.70	29.05
3160				styrofoam		9.45	19.40	28.85
3200			8	perlite		10.05	20	30.05
3240				styrofoam		9.60	19.80	29.40
3280		L.W. block	4	none		5.95	13.40	19.35
3320			6	perlite		9.60	19.55	29.15
3360				styrofoam		9.70	19.25	28.95
3400			8	perlite		10.70	20	30.70
3440				styrofoam		10.25	19.65	29.90
3520		glazed block	4	none		16.25	20.50	36.75
3560			6	perlite		18.25	21	39.25
3600				styrofoam		18.35	20.50	38.85
3640			8	perlite		19.20	21.50	40.70
3680				styrofoam		18.75	21	39.75
3720		clay tile	4	none		12.05	18.15	30.20
3760			6	none		14.25	18.70	32.95
3800			8	none		16.30	19.40	35.70
3840		glazed tile	4	none		18.85	24.50	43.35
4000	Roman	common brick	4	none		13.05	24.50	37.55
4040		SCR brick	6	none		17.20	21.50	38.70

B20 Exterior Enclosure

B2010 Exterior Walls

B2010 132 Brick Face Composite Wall - Double Wythe

	FACE BRICK	BACKUP MASONRY	BACKUP THICKNESS (IN.)	BACKUP CORE FILL		COST PER S.F.		
						MAT.	INST.	TOTAL
4080	Roman	conc. block	4	none		11.65	19.35	31
4120			6	perlite		13.45	20	33.45
4160				styrofoam		13.55	19.80	33.35
4200			8	perlite		14.20	20.50	34.70
4240				styrofoam		13.75	20	33.75
4280		L.W. block	4	none		11.80	19.25	31.05
4320			6	perlite		13.70	19.95	33.65
4360				styrofoam		13.80	19.65	33.45
4400			8	perlite		14.85	20.50	35.35
4440				styrofoam		14.40	20	34.40
4520		glazed block	4	none		20.50	20.50	41
4560			6	perlite		22.50	21.50	44
4600				styrofoam		22.50	21	43.50
4640			8	perlite		23.50	22	45.50
4680				styrofoam		23	21.50	44.50
4720		claytile	4	none		16.20	18.55	34.75
4760			6	none		18.40	19.10	37.50
4800			8	none		20.50	19.80	40.30
4840		glazed tile	4	none		23	25	48
5000	Norman	common brick	4	none		10.15	22	32.15
5040		SCR brick	6	none		14.30	19.40	33.70
5080		conc. block	4	none		8.75	17.15	25.90
5120			6	perlite		10.55	17.90	28.45
5160				styrofoam		10.65	17.60	28.25
5200			8	perlite		11.30	18.40	29.70
5240				styrofoam		10.85	18	28.85
5280		L.W. block	4	none		8.90	17.05	25.95
5320			6	perlite		10.85	17.85	28.70
5360				styrofoam		10.90	17.45	28.35
5400			8	perlite		11.95	18.25	30.20
5440				styrofoam		11.50	17.85	29.35
5520		glazed block	4	nonc		17.50	18.60	36.10
5560			6	perlite		19.50	19.20	38.70
5600				styrofoam		19.60	18.90	38.50
5640			8	perlite		20.50	19.80	40.30
5680				styrofoam		20	19.40	39.40
5720		clay tile	4	none		13.30	16.35	29.65
5760			6	none		15.50	16.90	32.40
5800			8	none		17.55	17.60	35.15
5840		glazed tile	4	none		20	22.50	42.50
6000	Norwegian	common brick	4	none		10.05	21	31.05
6040		SCR brick	6	none		14.20	18.25	32.45
6080		conc. block	4	none		8.65	16	24.65
6120			6	perlite		10.45	16.75	27.20
6160				styrofoam		10.55	16.45	27
6200			8	perlite		11.20	17.25	28.45
6240				styrofoam		10.75	16.85	27.60
6280		L.W. block	4	none		8.80	15.90	24.70
6320			6	perlite		10.70	16.60	27.30
6360				styrofoam		10.80	16.30	27.10
6400			8	perlite		11.85	17.10	28.95
6440				styrofoam		11.40	16.70	28.10

B20 Exterior Enclosure

B2010 Exterior Walls

B2010 132 — Brick Face Composite Wall - Double Wythe

	FACE BRICK	BACKUP MASONRY	BACKUP THICKNESS (IN.)	BACKUP CORE FILL		COST PER S.F. MAT.	COST PER S.F. INST.	COST PER S.F. TOTAL
6520	Norwegian	glazed block	4	none		17.40	17.45	34.85
6560			6	perlite		19.40	18.05	37.45
6600				styrofoam		19.50	17.75	37.25
6640			8	perlite		20.50	18.65	39.15
6680				styrofoam		19.90	18.25	38.15
6720		clay tile	4	none		13.20	15.20	28.40
6760			6	none		15.40	15.75	31.15
6800			8	none		17.45	16.45	33.90
6840		glazed tile	4	none		20	21.50	41.50
6880								
7000	Utility	common brick	4	none		10.15	19.95	30.10
7040		SCR brick	6	none		14.30	17.15	31.45
7080		conc. block	4	none		8.75	14.90	23.65
7120			6	perlite		10.55	15.65	26.20
7160				styrofoam		10.65	15.35	26
7200			8	perlite		11.30	16.15	27.45
7240				styrofoam		10.85	15.75	26.60
7280		L.W. block	4	none		8.90	14.80	23.70
7320			6	perlite		10.80	15.50	26.30
7360				styrofoam		10.90	15.20	26.10
7400			8	perlite		11.95	16	27.95
7440				styrofoam		11.50	15.60	27.10
7520		glazed block	4	none		17.50	16.35	33.85
7560			6	perlite		19.50	16.95	36.45
7600				styrofoam		19.60	16.65	36.25
7640			8	perlite		20.50	17.55	38.05
7680				styrofoam		20	17.15	37.15
7720		clay tile	4	none		13.30	14.10	27.40
7760			6	none		15.50	14.65	30.15
7800			8	none		17.55	15.35	32.90
7840		glazed tile	4	none		20	20.50	40.50
7880								

B2010 133 — Brick Face Composite Wall - Triple Wythe

	FACE BRICK	MIDDLE WYTHE	INSIDE MASONRY	TOTAL THICKNESS (IN.)		COST PER S.F. MAT.	COST PER S.F. INST.	COST PER S.F. TOTAL
8000	Standard	common brick	standard brick	12		14.20	37	51.20
8100		4" conc. brick	standard brick	12		13.60	38	51.60
8120		4" conc. brick	common brick	12		12.30	37.50	49.80
8200	Glazed	common brick	standard brick	12		28	37.50	65.50
8300		4" conc. brick	standard brick	12		27.50	38.50	66
8320		4" conc. brick	glazed brick	12		26.50	38	64.50
8400	Engineer	common brick	standard brick	12		13.80	35	48.80
8500		4" conc. brick	standard brick	12		13.20	36	49.20
8520		4" conc. brick	engineer brick	12		11.90	35.50	47.40
8600	Roman	common brick	standard brick	12		17.95	35.50	53.45
8700		4" conc. brick	standard brick	12		17.30	36.50	53.80
8720		4" conc. brick	Roman brick	12		16.05	36	52.05
8800	Norman	common brick	standard brick	12		15.05	33	48.05
8900		4" conc. brick	standard brick	12		14.40	34.50	48.90
8920		4" conc. brick	Norman brick	12		13.15	34	47.15

B20 Exterior Enclosure

B2010 Exterior Walls

B2010 133 Brick Face Composite Wall - Triple Wythe

	FACE BRICK	MIDDLE WYTHE	INSIDE MASONRY	TOTAL THICKNESS (IN.)		COST PER S.F.		
						MAT.	INST.	TOTAL
9000	Norwegian	common brick	standard brick	12		14.95	32	46.95
9100		4" conc. brick	standard brick	12		14.30	33	47.30
9120		4" conc. brick	Norwegian brick	12		13.05	32.50	45.55
9200	Utility	common brick	standard brick	12		15.05	31	46.05
9300		4" conc. brick	standard brick	12		14.25	32	46.25
9320		4" conc. brick	utility brick	12		13.15	31.50	44.65

B20 Exterior Enclosure

B2010 Exterior Walls

Exterior brick face cavity walls are defined in the following terms: cavity treatment, type of face brick, backup masonry, total thickness and insulation. Seven types of face brick are shown with various types of backup. All systems include a brick shelf, ties to the backups and necessary dampproofing, flashing, and control joints every 20'.

System Components	QUANTITY	UNIT	COST PER S.F. MAT.	COST PER S.F. INST.	COST PER S.F. TOTAL
SYSTEM B2010 134 1000					
CAVITY WALL, STANDARD BRICK FACE, COMMON BRICK BACKUP, POLYSTYRENE					
Face brick veneer, standard, running bond	1.000	S.F.	4.37	11.35	15.72
Wash brick	1.000	S.F.	.07	.97	1.04
Common brick wall backup, 4" thick	1.000	S.F.	3.09	10.85	13.94
Wall ties	.300	L.F.	.13	.15	.28
Polystyrene insulation board, 1" thick	1.000	S.F.	.26	.69	.95
Flashing, aluminum	.100	S.F.	.12	.36	.48
Shelf angle	1.000	Lb.	1.10	.99	2.09
Control joint	.050	L.F.	.07	.07	.14
Backer rod	.100	L.F.		.12	.12
Sealant	.100	L.F.	.03	.24	.27
TOTAL			9.24	25.79	35.03

B2010 134	Brick Face Cavity Wall						
	FACE BRICK	BACKUP MASONRY	TOTAL THICKNESS (IN.)	CAVITY INSULATION		COST PER S.F.	
						MAT.	INST.	TOTAL
1000	Standard	4" common brick	10	polystyrene		9.25	26	35.25
1020				none		9	25	34
1040		6" SCR brick	12	polystyrene		13.40	23	36.40
1060				none		13.10	22.50	35.60
1080		4" conc. block	10	polystyrene		7.80	20.50	28.30
1100				none		7.55	20	27.55
1120		6" conc. block	12	polystyrene		8.50	21	29.50
1140				none		8.25	20.50	28.75
1160		4" L.W. block	10	polystyrene		7.95	20.50	28.45
1180				none		7.70	19.95	27.65
1200		6" L.W. block	12	polystyrene		8.75	21	29.75
1220				none		8.50	20.50	29
1240		4" glazed block	10	polystyrene		16.55	22	38.55
1260				none		16.30	21.50	37.80
1280		6" glazed block	12	polystyrene		16.60	22	38.60
1300				none		16.30	21.50	37.80

B20 Exterior Enclosure

B2010 Exterior Walls

B2010 134 Brick Face Cavity Wall

	FACE BRICK	BACKUP MASONRY	TOTAL THICKNESS (IN.)	CAVITY INSULATION		COST PER S.F. MAT.	COST PER S.F. INST.	COST PER S.F. TOTAL
1320	Standard	4" clay tile	10	polystyrene		12.35	19.95	32.30
1340				none		12.10	19.25	31.35
1360		4" glazed tile	10	polystyrene		19.15	26.50	45.65
1380				none		18.90	25.50	44.40
1500	Glazed	4" common brick	10	polystyrene		23	26.50	49.50
1520				none		23	25.50	48.50
1540		6" SCR brick	12	polystyrene		27.50	23.50	51
1560				none		27	23	50
1580		4" conc. block	10	polystyrene		22	21.50	43.50
1600				none		21.50	20.50	42
1620		6" conc. block	12	polystyrene		22.50	21.50	44
1640				none		22	21	43
1660		4" L.W. block	10	polystyrene		22	21	43
1680				none		21.50	20.50	42
1700		6" L.W. block	12	polystyrene		23	21.50	44.50
1720				none		22.50	21	43.50
1740		4" glazed block	10	polystyrene		30.50	22.50	53
1760				none		30.50	22	52.50
1780		6" glazed block	12	polystyrene		31.50	23	54.50
1800				none		31	22.50	53.50
1820		4" clay tile	10	polystyrene		26.50	20.50	47
1840				none		26	19.80	45.80
1860		4" glazed tile	10	polystyrene		28.50	21	49.50
1880				none		28.50	20.50	49
2000	Engineer	4" common brick	10	polystyrene		8.85	24	32.85
2020				none		8.60	23.50	32.10
2040		6" SCR brick	12	polystyrene		13	21	34
2060				none		12.75	20.50	33.25
2080		4" conc. block	10	polystyrene		7.40	19	26.40
2100				none		7.15	18.30	25.45
2120		6" conc. block	12	polystyrene		8.15	19.45	27.60
2140				none		7.85	18.75	26.60
2160		4" L.W. block	10	polystyrene		7.55	18.90	26.45
2180				none		7.30	18.20	25.50
2200		6" L.W. block	12	polystyrene		8.40	19.30	27.70
2220				none		8.15	18.60	26.75
2240		4" glazed block	10	polystyrene		16.15	20.50	36.65
2260				none		15.90	19.75	35.65
2280		6" glazed block	12	polystyrene		17.05	20.50	37.55
2300				none		16.80	20	36.80
2320		4" clay tile	10	polystyrene		11.95	18.20	30.15
2340				none		11.70	17.50	29.20
2360		4" glazed tile	10	polystyrene		18.75	24.50	43.25
2380				none		18.50	24	42.50
2500	Roman	4" common brick	10	polystyrene		12.95	24.50	37.45
2520				none		12.70	24	36.70
2540		6" SCR brick	12	polystyrene		17.10	21.50	38.60
2560				none		16.85	21	37.85

B20 Exterior Enclosure

B2010 Exterior Walls

B2010 134 Brick Face Cavity Wall

	FACE BRICK	BACKUP MASONRY	TOTAL THICKNESS (IN.)	CAVITY INSULATION		COST PER S.F.		
						MAT.	INST.	TOTAL
2580	Roman	4" conc. block	10	polystyrene		11.55	19.40	30.95
2600				none		11.25	18.70	29.95
2620		6" conc. block	12	polystyrene		12.25	19.85	32.10
2640				none		12	19.15	31.15
2660		4" L.W. block	10	polystyrene		11.70	19.30	31
2680				none		11.40	18.60	30
2700		6" L.W. block	12	polystyrene		12.50	19.70	32.20
2720				none		12.25	19	31.25
2740		4" glazed block	10	polystyrene		20.50	21	41.50
2760				none		20	20	40
2780		6" glazed block	12	polystyrene		21	21	42
2800				none		21	20.50	41.50
2820		4" clay tile	10	polystyrene		16.10	18.60	34.70
2840				none		15.80	17.90	33.70
2860		4" glazed tile	10	polystyrene		23	25	48
2880				none		22.50	24.50	47
3000	Norman	4" common brick	10	polystyrene		10.05	22	32.05
3020				none		9.80	21.50	31.30
3040		6" SCR brick	12	polystyrene		14.20	19.45	33.65
3060				none		13.95	18.75	32.70
3080		4" conc. block	10	polystyrene		8.65	17.20	25.85
3100				none		8.35	16.50	24.85
3120		6" conc. block	12	polystyrene		9.35	17.65	27
3140				none		9.10	16.95	26.05
3160		4" L.W. block	10	polystyrene		8.80	17.10	25.90
3180				none		8.50	16.40	24.90
3200		6" L.W. block	12	polystyrene		9.60	17.50	27.10
3220				none		9.35	16.80	26.15
3240		4" glazed block	10	polystyrene		17.40	18.65	36.05
3260				none		17.10	17.95	35.05
3320		4" clay tile	10	polystyrene		13.20	16.40	29.60
3340				none		12.90	15.70	28.60
3360		4" glazed tile	10	polystyrene		20	22.50	42.50
3380				none		19.70	22	41.70
3500	Norwegian	4" common brick	10	polystyrene		9.95	21	30.95
3520				none		9.70	20.50	30.20
3540		6" SCR brick	12	polystyrene		14.10	18.30	32.40
3560				none		13.85	17.60	31.45
3580		4" conc. block	10	polystyrene		8.55	16.05	24.60
3600				none		8.25	15.35	23.60
3620		6" conc. block	12	polystyrene		9.25	16.50	25.75
3640				none		9	15.80	24.80

B20 Exterior Enclosure

B2010 Exterior Walls

B2010 134 Brick Face Cavity Wall

	FACE BRICK	BACKUP MASONRY	TOTAL THICKNESS (IN.)	CAVITY INSULATION		COST PER S.F.		
						MAT.	INST.	TOTAL
3660	Norwegian	4" L.W. block	10	polystyrene		8.70	15.95	24.65
3680				none		8.40	15.25	23.65
3700		6" L.W. block	12	polystyrene		9.50	16.35	25.85
3720				none		9.25	15.65	24.90
3740		4" glazed block	10	polystyrene		17.30	17.50	34.80
3760				none		17	16.80	33.80
3780		6" glazed block	12	polystyrene		18.15	17.80	35.95
3800				none		17.90	17.10	35
3820		4" clay tile	10	polystyrene		13.10	15.25	28.35
3840				none		12.80	14.55	27.35
3860		4" glazed tile	10	polystyrene		19.90	21.50	41.40
3880				none		19.60	21	40.60
4000	Utility	4" common brick	10	polystyrene		10.05	20	30.05
4020				none		9.80	19.30	29.10
4040		6" SCR brick	12	polystyrene		14.20	17.20	31.40
4060				none		13.95	16.50	30.45
4080		4" conc. block	10	polystyrene		8.65	14.95	23.60
4100				none		8.35	14.25	22.60
4120		6" conc. block	12	polystyrene		9.35	15.40	24.75
4140				none		9.10	14.70	23.80
4160		4" L.W. block	10	polystyrene		8.80	14.85	23.65
4180				none		8.50	14.15	22.65
4200		6" L.W. block	12	polystyrene		9.60	15.25	24.85
4220				none		9.35	14.55	23.90
4240		4" glazed block	10	polystyrene		17.40	16.40	33.80
4260				none		17.10	15.70	32.80
4280		6" glazed block	12	polystyrene		18.25	16.70	34.95
4300				none		18	16	34
4320		4" clay tile	10	polystyrene		13.20	14.15	27.35
4340				none		12.90	13.45	26.35
4360		4" glazed tile	10	polystyrene		20	20.50	40.50
4380				none		19.70	19.80	39.50

B2010 135 Brick Face Cavity Wall - Insulated Backup

	FACE BRICK	BACKUP MASONRY	TOTAL THICKNESS (IN.)	BACKUP CORE FILL		COST PER S.F.		
						MAT.	INST.	TOTAL
5100	Standard	6" conc. block	10	perlite		9.35	21	30.35
5120				styrofoam		9.45	20.50	29.95
5140		8" conc. block	12	perlite		10.10	21.50	31.60
5160				styrofoam		9.65	21	30.65
5180		6" L.W. block	10	perlite		9.65	20.50	30.15
5200				styrofoam		9.75	20.50	30.25
5220		8" L.W. block	12	perlite		10.75	21	31.75
5240				styrofoam		10.30	21	31.30
5260		6" glazed block	10	perlite		18.30	22	40.30
5280				styrofoam		18.40	22	40.40
5300		8" glazed block	12	perlite		19.25	22.50	41.75
5320				styrofoam		18.80	22.50	41.30
5340		6" clay tile	10	none		14.30	19.80	34.10
5360		8" clay tile	12	none		16.35	20.50	36.85
5600	Glazed	6" conc. block	10	perlite		23.50	21.50	45
5620				styrofoam		23.50	21	44.50

B20 Exterior Enclosure

B2010 Exterior Walls

B2010 135 Brick Face Cavity Wall - Insulated Backup

	FACE BRICK	BACKUP MASONRY	TOTAL THICKNESS (IN.)	BACKUP CORE FILL		COST PER S.F.		
						MAT.	INST.	TOTAL
5640	Glazed	8" conc. block	12	perlite		24	22	46
5660				styrofoam		23.50	21.50	45
5680		6" L.W. block	10	perlite		23.50	21	44.50
5700				styrofoam		23.50	21	44.50
5720		8" L.W. block	12	perlite		25	21.50	46.50
5740				styrofoam		24.50	21.50	46
5760		6" glazed block	10	perlite		32.50	22.50	55
5780				styrofoam		32.50	22.50	55
5800		8" glazed block	12	perlite		33.50	23.50	57
5820				styrofoam		33	23	56
5840		6" clay tile	10	none		28.50	20.50	49
5860		8" clay tile	8	none		30.50	21	51.50
6100	Engineer	6" conc. block	10	perlite		9	19.10	28.10
6120				styrofoam		9.10	18.75	27.85
6140		8" conc. block	12	perlite		9.75	19.55	29.30
6160				styrofoam		9.30	19.15	28.45
6180		6" L.W. block	10	perlite		9.25	18.95	28.20
6200				styrofoam		9.35	18.60	27.95
6220		8" L.W. block	12	perlite		10.40	19.40	29.80
6240				styrofoam		9.95	19	28.95
6260		6" glazed block	10	perlite		17.90	20.50	38.40
6280				styrofoam		18	20	38
6300		8" glazed block	12	perlite		18.90	21	39.90
6320				styrofoam		18.45	20.50	38.95
6340		6" clay tile	10	none		13.90	18.05	31.95
6360		8" clay tile	12	none		16	18.75	34.75
6600	Roman	6" conc. block	10	perlite		13.10	19.50	32.60
6620				styrofoam		13.20	19.15	32.35
6640		8" conc. block	12	perlite		13.85	19.95	33.80
6660				styrofoam		13.40	19.55	32.95
6680		6" L.W. block	10	perlite		13.35	19.35	32.70
6700				styrofoam		13.45	19	32.45
6720		8" L.W. block	12	perlite		14.50	19.80	34.30
6740				styrofoam		14.05	19.40	33.45
6760		6" glazed block	10	perlite		22	21	43
6780				styrofoam		22	20.50	42.50
6800		8" glazed block	12	perlite		23	21.50	44.50
6820				styrofoam		22.50	21	43.50
6840		6" clay tile	10	none		18	18.45	36.45
6860		8" clay tile	12	none		20	19.15	39.15
7100	Norman	6" conc. block	10	perlite		10.20	17.30	27.50
7120				styrofoam		10.30	16.95	27.25
7140		8" conc. block	12	perlite		10.95	17.75	28.70
7160				styrofoam		10.50	17.35	27.85
7180		6" L.W. block	10	perlite		10.45	17.15	27.60
7200				styrofoam		10.55	16.80	27.35
7220		8" L.W. block	12	perlite		11.60	17.60	29.20
7240				styrofoam		11.15	17.20	28.35
7260		6" glazed block	10	perlite		19.10	18.60	37.70
7280				styrofoam		19.20	18.25	37.45
7300		8" glazed block	12	perlite		20	19.15	39.15
7320				styrofoam		19.65	18.75	38.40

B20 Exterior Enclosure

B2010 Exterior Walls

B2010 135 Brick Face Cavity Wall - Insulated Backup

	FACE BRICK	BACKUP MASONRY	TOTAL THICKNESS (IN.)	BACKUP CORE FILL		COST PER S.F.		
						MAT.	INST.	TOTAL
7340	Norwegian	6" clay tile	10	none		15.10	16.25	31.35
7360		8" clay tile	12	none		17.20	16.95	34.15
7600		6" conc. block	10	perlite		10.10	16.15	26.25
7620				styrofoam		10.20	15.80	26
7640		8" conc. block	12	perlite		10.85	16.60	27.45
7660				styrofoam		10.40	16.20	26.60
7680		6" L.W. block	10	perlite		10.35	16	26.35
7700				styrofoam		10.45	15.65	26.10
7720		8" L.W. block	12	perlite		11.50	16.45	27.95
7740				styrofoam		11.05	16.05	27.10
7760		6" glazed block	10	perlite		19	17.45	36.45
7780				styrofoam		19.10	17.10	36.20
7800		8" glazed block	12	perlite		20	18	38
7820				styrofoam		19.55	17.60	37.15
7840		6" clay tile	10	none		15	15.10	30.10
7860		8" clay tile	12	none		17.10	15.80	32.90
8100	Utility	6" conc. block	10	perlite		10.20	15.05	25.25
8120				styrofoam		10.30	14.70	25
8140		8" conc. block	12	perlite		10.95	15.50	26.45
8160				styrofoam		10.50	15.10	25.60
8180		6" L.W. block	10	perlite		10.45	14.90	25.35
8200				styrofoam		10.55	14.55	25.10
8220		8" L.W. block	12	perlite		11.60	15.35	26.95
8240				styrofoam		11.15	14.95	26.10
8260		6" glazed block	10	perlite		19.10	16.35	35.45
8280				styrofoam		19.20	16	35.20
8300		8" glazed block	12	perlite		20	16.90	36.90
8320				styrofoam		19.65	16.50	36.15
8340		6" clay tile	10	none		15.10	14	29.10
8360		8" clay tile	12	none		17.20	14.70	31.90

B20 Exterior Enclosure

B2010 Exterior Walls

Exterior block face cavity walls are defined in the following terms: cavity treatment, type of face block, backup masonry, total thickness and insulation. Multiple types of face block are shown with various types of backup. All systems include a brick shelf and necessary dampproofing, flashing, and control joints every 20'.

System Components	QUANTITY	UNIT	COST PER S.F.		
			MAT.	INST.	TOTAL
SYSTEM B2010 137 1600					
CAVITY WALL, FLUTED BLOCK FACE, 4" C.M.U. BACKUP, POLYSTYRENE BOARD					
Fluted block partition, 4" thick	1.000	S.F.	3.69	7.45	11.14
Conc. block wall backup, 4" thick	1.000	S.F.	1.65	5.80	7.45
Horizontal joint reinforcing	.800	L.F.	.14	.22	.36
Polystyrene insulation board, 1" thick	1.000	S.F.	.26	.69	.95
Flashing, aluminum	.100	S.F.	.12	.36	.48
Shelf angle	1.000	Lb.	1.10	.99	2.09
Control joint	.050	L.F.	.07	.07	.14
Backer rod	.100	L.F.		.12	.12
Sealant	.100	L.F.	.03	.24	.27
TOTAL			7.06	15.94	23

B2010 137			Block Face Cavity Wall				
	FACE BLOCK	BACKUP MASONRY	TOTAL THICKNESS (IN.)	CAVITY INSULATION		COST PER S.F.	
						MAT.	INST.	TOTAL
1000	Deep groove	4" conc. block	10	polystyrene		7.05	15.95	23
1050	Reg. wt.			none		6.80	15.25	22.05
1060		6" conc. block	12	polystyrene		7.75	16.40	24.15
1110				none		7.50	15.70	23.20
1120		4" L.W. block	10	polystyrene		7.20	15.85	23.05
1170				none		6.95	15.15	22.10
1180		6" L.W. block	12	polystyrene		8	16.25	24.25
1230				none		7.75	15.55	23.30
1300	Deep groove	4" conc. block	10	polystyrene		7.90	15.75	23.65
1350	L.W.			none		7.65	15.05	22.70
1360		6" conc. block	12	polystyrene		8.55	16.20	24.75
1410				none		8.30	15.50	23.80
1420		4" L.W. block	10	polystyrene		8.05	15.65	23.70
1470				none		7.80	14.95	22.75
1480		6" L.W. block	12	polystyrene		8.85	16.05	24.90
1530				none		8.55	15.35	23.90

B20 Exterior Enclosure

B2010 Exterior Walls

B2010 137 Block Face Cavity Wall

	FACE BLOCK	BACKUP MASONRY	TOTAL THICKNESS (IN.)	CAVITY INSULATION		COST PER S.F.		
						MAT.	INST.	TOTAL
1600	Fluted	4" conc. block	10	polystyrene		7.05	15.95	23
1650	Reg. wt.			none		7.40	15.05	22.45
1660		6" conc. block	12	polystyrene		8.30	16.20	24.50
1710				none		8.05	15.50	23.55
1720		4" L.W. block	10	polystyrene		7.80	15.65	23.45
1770				none		7.55	14.95	22.50
1780		6" L.W. block	12	polystyrene		8.60	16.05	24.65
1830				none		8.30	15.35	23.65
1900	Fluted	4" conc. block	10	polystyrene		8.80	15.55	24.35
1950	L.W.			none		8.55	14.85	23.40
1960		6" conc. block	12	polystyrene		9.50	16	25.50
2010				none		9.25	15.30	24.55
2020		4" L.W. block	10	polystyrene		8.95	15.45	24.40
2070				none		8.70	14.75	23.45
2080		6" L.W. block	12	polystyrene		9.75	15.85	25.60
2130				none		9.50	15.15	24.65
2200	Ground face	4" conc. block	10	polystyrene		13.45	16.05	29.50
2250	1 Score			none		13.20	15.35	28.55
2260		6" conc. block	12	polystyrene		14.15	16.50	30.65
2310				none		13.90	15.80	29.70
2320		4" L.W. block	10	polystyrene		13.60	15.95	29.55
2370				none		13.35	15.25	28.60
2380		6" L.W. block	12	polystyrene		14.40	16.35	30.75
2430				none		14.15	15.65	29.80
2500	Hexagonal	4" conc. block	10	polystyrene		6.80	15.75	22.55
2550	Reg. wt.			none		6.55	15.05	21.60
2560		6" conc. block	12	polystyrene		7.50	16.20	23.70
2610				none		7.25	15.50	22.75
2620		4" L.W. block	10	polystyrene		6.95	15.65	22.60
2670				none		6.70	14.95	21.65
2680		6" L.W. block	12	polystyrene		7.75	16.05	23.80
2730				none		7.50	15.35	22.85
2800	Hexagonal	4" conc. block	10	polystyrene		7.75	15.55	23.30
2850	L.W.			none		7.50	14.85	22.35
2860		6" conc. block	12	polystyrene		8.40	16	24.40
2910				none		8.15	15.30	23.45
2920		4" L.W. block	10	polystyrene		7.90	15.45	23.35
2970				none		7.65	14.75	22.40
2980		6" L.W. block	12	polystyrene		8.70	15.85	24.55
3030				none		8.40	15.15	23.55
3100	Slump block	4" conc. block	10	polystyrene		7.25	14.45	21.70
3150	4x16			none		7	13.75	20.75
3160		6" conc. block	12	polystyrene		7.95	14.90	22.85
3210				none		7.65	14.20	21.85
3220		4" L.W. block	10	polystyrene		7.40	14.35	21.75
3270				none		7.15	13.65	20.80
3280		6" L.W. block	12	polystyrene		8.20	14.75	22.95
3330				none		7.95	14.05	22

B20 Exterior Enclosure

B2010 Exterior Walls

B2010 137 Block Face Cavity Wall

	FACE BLOCK	BACKUP MASONRY	TOTAL THICKNESS (IN.)	CAVITY INSULATION		COST PER S.F.		
						MAT.	INST.	TOTAL
3400	Split face	4" conc. block	10	polystyrene		8.60	15.95	24.55
3450	1 Score			none		8.35	15.25	23.60
3460	Reg. wt.	6" conc. block	12	polystyrene		9.30	16.40	25.70
3510				none		9.05	15.70	24.75
3520		4" L.W. block	10	polystyrene		8.75	15.85	24.60
3570				none		8.50	15.15	23.65
3580		6" L.W. block	12	polystyrene		9.55	16.25	25.80
3630				none		9.30	15.55	24.85
3700	Split face	4" conc. block	10	polystyrene		9.80	15.75	25.55
3750	1 Score			none		9.55	15.05	24.60
3760	L.W.	6" conc. block	12	polystyrene		10.50	16.20	26.70
3810				none		10.25	15.50	25.75
3820		4" L.W. block	10	polystyrene		9.95	15.65	25.60
3870				none		9.70	14.95	24.65
3880		6" L.W. block	12	polystyrene		10.75	16.05	26.80
3930				none		10.50	15.35	25.85
4000	Split rib	4" conc. block	10	polystyrene		9.65	15.95	25.60
4050	8 Rib			none		9.40	15.25	24.65
4060	Reg. wt.	6" conc. block	12	polystyrene		10.35	16.40	26.75
4110				none		10.10	15.70	25.80
4120		4" L.W. block	10	polystyrene		9.80	15.85	25.65
4170				none		9.55	15.15	24.70
4180		6" L.W. block	12	polystyrene		10.60	16.25	26.85
4230				none		10.35	15.55	25.90
4300	Split rib	4" conc. block	10	polystyrene		10.50	15.75	26.25
4350	8 Rib			none		10.25	15.05	25.30
4360	L.W.	6" conc. block	12	polystyrene		11.20	16.20	27.40
4410				none		10.95	15.50	26.45
4420		4" L.W. block	10	polystyrene		10.65	15.65	26.30
4470				none		10.40	14.95	25.35
4480		6" L.W. block	12	polystyrene		11.45	16.05	27.50
4530				none		11.20	15.35	26.55

B2010 138 Block Face Cavity Wall - Insulated Backup

	FACE BLOCK	BACKUP MASONRY	TOTAL THICKNESS (IN.)	BACKUP CORE INSULATION		COST PER S.F.		
						MAT.	INST.	TOTAL
5010	Deep groove	6" conc. block	10	perlite		8.65	16.05	24.70
5050	Reg. wt.			styrofoam		8.75	15.70	24.45
5060		8" conc. block	12	perlite		9.35	16.50	25.85
5110				styrofoam		8.90	16.10	25
5120		6" L.W. block	10	perlite		8.90	15.90	24.80
5170				styrofoam		9	15.55	24.55
5180		8" L.W. block	12	perlite		10	16.35	26.35
5230				styrofoam		9.55	15.95	25.50
5300	Deep groove	6" conc. block	10	perlite		9.45	15.85	25.30
5350	L.W.			styrofoam		9.55	15.50	25.05
5360		8" conc. block	12	perlite		10.20	16.30	26.50
5410				styrofoam		9.75	15.90	25.65
5420		6" L.W. block	10	perlite		9.70	15.70	25.40
5470				styrofoam		9.80	15.35	25.15
5480		8" L.W. block	12	perlite		10.85	16.15	27
5530				styrofoam		10.40	15.75	26.15

B20 Exterior Enclosure

B2010 Exterior Walls

B2010 138   Block Face Cavity Wall - Insulated Backup

	FACE BLOCK	BACKUP MASONRY	TOTAL THICKNESS (IN.)	BACKUP CORE INSULATION		COST PER S.F.		
						MAT.	INST.	TOTAL
5600	Fluted	6" conc. block	10	perlite		9.20	15.85	25.05
5650	Reg.			styrofoam		9.30	15.50	24.80
5660		8" conc. block	12	perlite		9.95	16.30	26.25
5710				styrofoam		9.50	15.90	25.40
5720		6" L.W. block	10	perlite		9.45	15.70	25.15
5770				styrofoam		9.55	15.35	24.90
5780		8" L.W. block	12	perlite		10.60	16.15	26.75
5830				styrofoam		10.15	15.75	25.90
5900	Fluted	6 conc. block	10	perlite		10.40	15.65	26.05
5950	L.W.			styrofoam		10.50	15.30	25.80
5960		8" conc. block	12	perlite		11.10	16.10	27.20
6010				styrofoam		10.65	15.70	26.35
6020		6" L.W. block	10	perlite		10.65	15.50	26.15
6070				styrofoam		10.75	15.15	25.90
6080		8" L.W. block	12	perlite		11.75	15.95	27.70
6130				styrofoam		11.30	15.55	26.85
6200	Ground face	6" conc. block	10	perlite		15.05	16.15	31.20
6250	1 Score			styrofoam		15.15	15.80	30.95
6260		8" conc. block	12	perlite		15.75	16.60	32.35
6310				styrofoam		15.30	16.20	31.50
6320		6" L.W. block	10	perlite		15.30	16	31.30
6370				styrofoam		15.40	15.65	31.05
6380		8" L.W. block	12	perlite		16.40	16.45	32.85
6430				styrofoam		17	19.25	36.25
6500	Hexagonal	6" conc. block	10	perlite		8.40	15.85	24.25
6550	Reg. wt.			styrofoam		8.50	15.50	24
6560		8" conc. block	12	perlite		9.10	16.30	25.40
6610				styrofoam		8.65	15.90	24.55
6620		6" L.W. block	10	perlite		8.65	15.70	24.35
6670				styrofoam		8.75	15.35	24.10
6680		8" L.W. block	12	perlite		9.75	16.15	25.90
6730				styrofoam		9.30	15.75	25.05
6800	Hexagonal	6" conc. block	10	perlite		9.30	15.65	24.95
6850	L.W.			styrofoam		9.40	15.30	24.70
6860		8" conc. block	12	perlite		10.05	16.10	26.15
6910				styrofoam		9.60	15.70	25.30
6920		6" L.W. block	10	perlite		9.55	15.50	25.05
6970				styrofoam		9.65	15.15	24.80
6980		8" L.W. block	12	perlite		10.70	15.95	26.65
7030				styrofoam		10.25	15.55	25.80
7100	Slump block	6" conc. block	10	perlite		8.80	14.55	23.35
7150	4x16			styrofoam		8.90	14.20	23.10
7160		8" conc. block	12	perlite		9.55	15	24.55
7210				styrofoam		9.10	14.60	23.70
7220		6" L.W. block	10	perlite		9.10	14.40	23.50
7270				styrofoam		9.20	14.05	23.25
7280		8" L.W. block	12	perlite		10.20	14.85	25.05
7330				styrofoam		9.75	14.45	24.20

B20 Exterior Enclosure

B2010 Exterior Walls

B2010 138 Block Face Cavity Wall - Insulated Backup

	FACE BLOCK	BACKUP MASONRY	TOTAL THICKNESS (IN.)	BACKUP CORE INSULATION		COST PER S.F.		
						MAT.	INST.	TOTAL
7400	Split face	6" conc. block	10	perlite		10.20	16.05	26.25
7450	1 Score			styrofoam		10.30	15.70	26
7460	Reg. wt.	8" conc. block	12	perlite		10.90	16.50	27.40
7510				styrofoam		10.45	16.10	26.55
7520		6" L.W. block	10	perlite		10.45	15.90	26.35
7570				styrofoam		10.55	15.55	26.10
7580		8" L.W. block	12	perlite		11.55	16.35	27.90
7630				styrofoam		11.10	15.95	27.05
7700	Split rib	6" conc. block	10	perlite		11.25	16.05	27.30
7750	8 Rib			styrofoam		11.35	15.70	27.05
7760	Reg. wt.	8" conc. block	12	perlite		11.95	16.50	28.45
7810				styrofoam		11.50	16.10	27.60
7820		6" L.W. block	10	perlite		11.50	15.90	27.40
7870				styrofoam		11.60	15.55	27.15
7880		8" L.W. block	12	perlite		12.60	16.35	28.95
7930				styrofoam		12.15	15.95	28.10
8000	Split rib	6" conc. block	10	perlite		12.10	15.85	27.95
8050	8 Rib			styrofoam		12.25	15.55	27.80
8060	L.W.	8" conc. block	12	perlite		12.80	16.30	29.10
8110				styrofoam		12.35	15.90	28.25
8120		6" L.W. block	10	perlite		12.35	15.70	28.05
8170				styrofoam		12.45	15.35	27.80
8180		8" L.W. block	12	perlite		13.45	16.15	29.60
8230				styrofoam		13	15.75	28.75

B20 Exterior Enclosure

B2010 Exterior Walls

Exterior block face composite walls are defined in the following terms: type of face block and backup masonry, total thickness and insulation. All systems include shelf angles and necessary dampproofing, flashing, and control joints every 20'.

System Components	QUANTITY	UNIT	COST PER S.F.		
			MAT.	INST.	TOTAL
SYSTEM B2010 139 1000					
COMPOSITE WALL, GROOVED BLOCK FACE, 4" C.M.U. BACKUP, PERLITE FILL					
Deep groove block veneer, 4" thick	1.000	S.F.	3.69	7.45	11.14
Concrete block wall, backup, 4" thick	1.000	S.F.	1.65	5.80	7.45
Horizontal joint reinforcing	.800	L.F.	.19	.17	.36
Perlite insulation, poured	1.000	S.F.	.76	.20	.96
Flashing, aluminum	.100	S.F.	.12	.36	.48
Shelf angle	1.000	Lb.	1.10	.99	2.09
Control joint	.050	L.F.	.07	.07	.14
Backer rod	.100	L.F.		.12	.12
Sealant	.100	L.F.	.03	.24	.27
Collar joint	1.000	S.F.	.40	.62	1.02
TOTAL			8.01	16.02	24.03

B2010 139		Block Face Composite Wall					
	FACE BLOCK	BACKUP MASONRY	TOTAL THICKNESS (IN.)	BACKUP CORE INSULATION		COST PER S.F.	
						MAT.	INST.	TOTAL
1000	Deep groove	4" conc. block	8	perlite		8	16.05	24.05
1050	Reg. wt.			none		7.25	15.85	23.10
1060		6" conc. block	10	perlite		9.05	16.65	25.70
1110				styrofoam		9.15	16.35	25.50
1120		8" conc. block	12	perlite		9.80	17.15	26.95
1170				styrofoam		9.35	16.75	26.10
1200		4" L.W. block	8	perlite		8.15	15.95	24.10
1250				none		7.40	15.75	23.15
1260		6" L.W. block	10	perlite		9.30	16.50	25.80
1310				styrofoam		9.40	16.20	25.60
1320		8" L.W. block	12	perlite		10.45	17	27.45
1370				styrofoam		10	16.60	26.60

B20 Exterior Enclosure

B2010 Exterior Walls

B2010 139 Block Face Composite Wall

	FACE BLOCK	BACKUP MASONRY	TOTAL THICKNESS (IN.)	BACKUP CORE INSULATION		COST PER S.F.		
						MAT.	INST.	TOTAL
1600	Deep groove	4" conc. block	8	perlite		8.85	15.85	24.70
1650	L.W.			none		8.10	15.65	23.75
1660		6" conc. block	10	perlite		9.90	16.45	26.35
1710				styrofoam		10	16.15	26.15
1720		8" conc. block	12	perlite		10.60	16.95	27.55
1770				styrofoam		10.15	16.55	26.70
1800		4" L.W. block	8	perlite		9	15.75	24.75
1850				none		8.25	15.55	23.80
1860		6" L.W. block	10	perlite		10.15	16.30	26.45
1910				styrofoam		10.25	16	26.25
1920		8" L.W. block	12	perlite		11.25	16.80	28.05
1970				styrofoam		10.80	16.40	27.20
2200	Fluted	4" conc. block	8	perlite		8.60	15.85	24.45
2250	Reg. wt.			none		7.85	15.65	23.50
2260		6" conc. block	10	perlite		9.65	16.45	26.10
2310				styrofoam		9.75	16.15	25.90
2320		8" conc. block	12	perlite		10.35	16.95	27.30
2370				styrofoam		9.90	16.55	26.45
2400		4" L.W. block	8	perlite		8.75	15.75	24.50
2450				styrofoam		8	15.55	23.55
2460		6" L.W. block	10	perlite		9.90	16.30	26.20
2510				styrofoam		10	16	26
2520		8" L.W. block	12	perlite		11	16.80	27.80
2570				styrofoam		10.55	16.40	26.95
2860	Fluted	6" conc. block	10	perlite		10.95	16.60	27.55
2910	L.W.			styrofoam		10.90	15.95	26.85
2920		8" conc. block	12	perlite		11.55	16.75	28.30
2970				styrofoam		11.10	16.35	27.45
3060		6" L.W. block	10	perlite		11.10	16.10	27.20
3110				styrofoam		11.20	15.80	27
3120		8" L.W. block	12	perlite		12.20	16.60	28.80
3170				styrofoam		11.75	16.20	27.95
3400	Ground face	4" conc. block	8	perlite		14.40	16.15	30.55
3450	1 Score			none		13.65	15.95	29.60
3460		6" conc. block	10	perlite		15.45	16.75	32.20
3510				styrofoam		15.55	16.45	32
3520		8" conc. block	12	perlite		16.20	17.35	33.55
3570				styrofoam		15.75	16.85	32.60
3600		4" L.W. block	8	perlite		14.55	16.05	30.60
3650				none		13.80	15.85	29.65
3660		6" L.W. block	10	perlite		15.75	16.60	32.35
3710				styrofoam		15.85	16.30	32.15
3720		8" L.W. block	12	perlite		16.85	17.10	33.95
3770				styrofoam		16.40	16.70	33.10

B20 Exterior Enclosure

B2010 Exterior Walls

B2010 139 Block Face Composite Wall

	FACE BLOCK	BACKUP MASONRY	TOTAL THICKNESS (IN.)	BACKUP CORE INSULATION		COST PER S.F.		
						MAT.	INST.	TOTAL
4000	Hexagonal	4" conc. block	8	perlite		7.75	15.85	23.60
4050	Reg. wt.			none		7	15.65	22.65
4060		6" conc. block	10	perlite		8.80	16.45	25.25
4110				styrofoam		8.90	16.15	25.05
4120		8" conc. block	12	perlite		9.60	17	26.60
4170				styrofoam		9.10	16.55	25.65
4200		4" L.W. block	8	perlite		7.90	15.75	23.65
4250				none		7.15	15.55	22.70
4260		6" L.W. block	10	perlite		9.10	16.30	25.40
4310				styrofoam		9.20	16	25.20
4320		8" L.W. block	12	perlite		10.20	16.80	27
4370				styrofoam		9.75	16.40	26.15
4600	Hexagonal	4" conc. block	8	perlite		8.70	15.65	24.35
4650	L.W.			none		7.95	15.45	23.40
4660		6" conc. block	10	perlite		9.35	16.15	25.50
4710				styrofoam		9.85	15.95	25.80
4720		8" conc. block	12	perlite		10.45	16.75	27.20
4770				styrofoam		10	16.35	26.35
4800		4" L.W. block	8	perlite		8.85	15.55	24.40
4850				none		8.10	15.35	23.45
4860		6" L.W. block	10	perlite		10	16.10	26.10
4910				styrofoam		10.10	15.80	25.90
4920		8" L.W. block	12	perlite		11.10	16.60	27.70
4970				styrofoam		10.65	16.20	26.85
5200	Slump block	4" conc. block	8	perlite		8.20	14.55	22.75
5250	4x16			none		7.45	14.35	21.80
5260		6" conc. block	10	perlite		9.25	15.15	24.40
5310				styrofoam		9.35	14.85	24.20
5320		8" conc. block	12	perlite		10	15.65	25.65
5370				styrofoam		9.55	15.25	24.80
5400		4" L.W. block	8	perlite		8.35	14.45	22.80
5450				none		7.60	14.25	21.85
5460		6" L.W. block	10	perlite		9.50	15	24.50
5510				styrofoam		9.60	14.70	24.30
5520		8" L.W. block	12	perlite		10.65	15.50	26.15
5570				styrofoam		10.20	15.10	25.30
5800	Split face	4" conc. block	8	perlite		9.55	16.05	25.60
5850	1 Score			none		8.80	15.85	24.65
5860	Reg. wt.	6" conc. block	10	perlite		10.60	16.65	27.25
5910				styrofoam		10.70	16.35	27.05
5920		8" conc. block	12	perlite		11.35	17.15	28.50
5970				styrofoam		10.90	16.75	27.65
6000		4" L.W. block	8	perlite		9.70	15.95	25.65
6050				none		8.95	15.75	24.70
6060		6" L.W. block	10	perlite		10.90	16.50	27.40
6110				styrofoam		11	16.20	27.20
6120		8" L.W. block	12	perlite		12	17	29
6170				styrofoam		11.55	16.60	28.15

B20 Exterior Enclosure

B2010 Exterior Walls

B2010 139 — Block Face Composite Wall

	FACE BLOCK	BACKUP MASONRY	TOTAL THICKNESS (IN.)	BACKUP CORE INSULATION		COST PER S.F.		
						MAT.	INST.	TOTAL
6460	Split face	6" conc. block	10	perlite		11.80	16.45	28.25
6510	1 Score			styrofoam		11.90	16.15	28.05
6520	L.W.	8" conc. block	12	perlite		12.55	16.95	29.50
6570				styrofoam		12.10	16.55	28.65
6660		6" L.W. block	10	perlite		12.10	16.30	28.40
6710				styrofoam		12.05	16.20	28.25
6720		8" L.W. block	12	perlite		13.20	16.80	30
6770				styrofoam		12.60	16.60	29.20
7000	Split rib	4" conc. block	8	perlite		10.60	16.05	26.65
7050	8 Rib			none		9.85	15.85	25.70
7060	Reg. wt.	6" conc. block	10	perlite		11.65	16.65	28.30
7110				styrofoam		11.75	16.35	28.10
7120		8" conc. block	12	perlite		12.40	17.15	29.55
7170				styrofoam		11.95	16.75	28.70
7200		4" L.W. block	8	perlite		10.75	15.95	26.70
7250				none		10	15.75	25.75
7260		6" L.W. block	10	perlite		11.95	16.50	28.45
7310				styrofoam		12.05	16.20	28.25
7320		8" L.W. block	12	perlite		13.05	17	30.05
7370				styrofoam		12.60	16.60	29.20
7600	Split rib	4" conc. block	8	perlite		11.45	15.85	27.30
7650	8 Rib			none		10.70	15.65	26.35
7660	L.W.	6" conc. block	10	perlite		12.50	16.45	28.95
7710				styrofoam		12.60	16.15	28.75
7720		8" conc. block	12	perlite		13.25	16.95	30.20
7770				styrofoam		12.80	16.55	29.35
7800		4" conc. block	8	perlite		11.60	15.75	27.35
7850				none		10.85	15.55	26.40
7860		6" conc. block	10	perlite		12.80	16.30	29.10
7910				styrofoam		12.90	16	28.90
7920		8" conc. block	12	perlite		13.90	16.80	30.70
7970				styrofoam		13.45	16.40	29.85

B20 Exterior Enclosure

B2010 Exterior Walls

The table below lists costs per S.F. for glass block walls. Included in the costs are the following special accessories required for glass block walls.

Glass block accessories required for proper installation.

Wall ties: Galvanized double steel mesh full length of joint.

Fiberglass expansion joint at sides and top.

Silicone caulking: One gallon does 95 L.F.

Oakum: One lb. does 30 L.F.

Asphalt emulsion: One gallon does 600 L.F.

If block are not set in wall chase, use 2'-0" long wall anchors at 2'-0" O.C.

System Components	QUANTITY	UNIT	COST PER S.F.		
			MAT.	INST.	TOTAL
SYSTEM B2010 140 2300					
GLASS BLOCK, 4" THICK, 6" X 6" PLAIN, UNDER 1,000 S.F.					
Glass block, 4" thick, 6" x 6" plain, under 1000 S.F.	4.100	Ea.	30.50	21.50	52
Glass block, cleaning blocks after installation, both sides add	2.000	S.F.	.18	2.50	2.68
TOTAL			30.68	24	54.68

B2010 140	Glass Block	COST PER S.F.		
		MAT.	INST.	TOTAL
2300	Glass block, 4" thick, 6"x6" plain, under 1,000 S.F.	30.50	24	54.50
2400	1,000 to 5,000 S.F.	30	21	51
2500	Over 5,000 S.F.	29	19.75	48.75
2600	Solar reflective, under 1,000 S.F.	43	32.50	75.50
2700	1,000 to 5,000 S.F.	42	28.50	70.50
2800	Over 5,000 S.F.	41	26.50	67.50
3500	8"x8" plain, under 1,000 S.F.	18.20	18.10	36.30
3600	1,000 to 5,000 S.F.	17.85	15.65	33.50
3700	Over 5,000 S.F.	17.30	14.15	31.45
3800	Solar reflective, under 1,000 S.F.	25.50	24.50	50
3900	1,000 to 5,000 S.F.	25	21	46
4000	Over 5,000 S.F.	24	18.80	42.80
5000	12"x12" plain, under 1,000 S.F.	25	16.80	41.80
5100	1,000 to 5,000 S.F.	24.50	14.15	38.65
5200	Over 5,000 S.F.	24	12.90	36.90
5300	Solar reflective, under 1,000 S.F.	35	22.50	57.50
5400	1,000 to 5,000 S.F.	34.50	18.80	53.30
5600	Over 5,000 S.F.	34	17.05	51.05
5800	3" thinline, 6"x6" plain, under 1,000 S.F.	22.50	24	46.50
5900	Over 5,000 S.F.	21.50	19.75	41.25
6000	Solar reflective, under 1,000 S.F.	31.50	32.50	64
6100	Over 5,000 S.F.	30.50	26.50	57
6200	8"x8" plain, under 1,000 S.F.	12.65	18.10	30.75
6300	Over 5,000 S.F.	12.35	14.15	26.50
6400	Solar reflective, under 1,000 S.F.	17.60	24.50	42.10
6500	Over 5,000 S.F.	17.20	18.80	36

B20 Exterior Enclosure

B2010 Exterior Walls

Concrete Block Lintel

Bond Beam

Pilaster

Concrete block specialties are divided into lintels, pilasters, and bond beams. Lintels are defined by span, thickness, height and wall loading. Span refers to the clear opening but the cost includes 8" of bearing at both ends.

Bond beams and pilasters are defined by height thickness and weight of the masonry unit itself. Components for bond beams also include grout and reinforcing. For pilasters, components include four #5 reinforcing bars and type N mortar.

System Components	QUANTITY	UNIT	COST PER LINTEL		
			MAT.	INST.	TOTAL
SYSTEM B2010 144 3100					
CONCRETE BLOCK LINTEL, 6"X8", LOAD 300 LB/L.F. 3'-4" SPAN					
Lintel blocks, 6" x 8" x 8"	7.000	Ea.	9.73	22.82	32.55
Joint reinforcing, #3 & #4 steel bars, horizontal	3.120	Lb.	1.72	3.78	5.50
Grouting, 8" deep, 6" thick, .15 C.F./L.F.	.700	C.F.	3.92	5.29	9.21
Temporary shoring, lintel forms	1.000	Set	.92	12.56	13.48
Temporary shoring, wood joists	1.000	Set		22	22
Temporary shoring, vertical members	1.000	Set		40	40
Mortar, masonry cement, 1:3 mix, type N	.270	C.F.	1.21	.82	2.03
TOTAL			17.50	107.27	124.77

B2010 144		Concrete Block Lintel					
	SPAN	THICKNESS (IN.)	HEIGHT (IN.)	WALL LOADING P.L.F.		COST PER LINTEL	
						MAT.	INST.	TOTAL
3100	3'-4"	6	8	300		17.50	107	124.50
3150		6	16	1,000		28.50	116	144.50
3200		8	8	300		17.80	115	132.80
3300		8	8	1,000		19.50	119	138.50
3400		8	16	1,000		30	126	156
3450	4'-0"	6	8	300		20.50	119	139.50
3500		6	16	1,000		37.50	135	172.50
3600		8	8	300		20.50	125	145.50
3700		8	16	1,000		34.50	136	170.50

B20 Exterior Enclosure

B2010 Exterior Walls

B2010 144 Concrete Block Lintel

	SPAN	THICKNESS (IN.)	HEIGHT (IN.)	WALL LOADING P.L.F.		COST PER LINTEL		
						MAT.	INST.	TOTAL
3800	4'-8"	6	8	300		22.50	120	142.50
3900		6	16	1,000		37	133	170
4000		8	8	300		23.50	134	157.50
4100		8	16	1,000		39	146	185
4200	5'-4"	6	8	300		27.50	135	162.50
4300		6	16	1,000		42.50	140	182.50
4400		8	8	300		28.50	147	175.50
4500		8	16	1,000		44.50	159	203.50
4600	6'-0"	6	8	300		33.50	163	196.50
4700		6	16	300		45	171	216
4800		6	16	1,000		48	177	225
4900		8	8	300		31.50	178	209.50
5000		8	16	1,000		50	194	244
5100	6'-8"	6	16	300		43.50	172	215.50
5200		6	16	1,000		46.50	179	225.50
5300		8	8	300		37	185	222
5400		8	16	1,000		55	205	260
5500	7'-4"	6	16	300		48	179	227
5600		6	16	1,000		61	193	254
5700		8	8	300		46.50	201	247.50
5800		8	16	300		66	216	282
5900		8	16	1,000		66	216	282
6000	8'-0"	6	16	300		52	187	239
6100		6	16	1,000		66	202	268
6200		8	16	300		73	228	301
6500		8	16	1,000		73	228	301

B2010 145 Concrete Block Specialties

	TYPE	HEIGHT	THICKNESS	WEIGHT (P.C.F.)		COST PER L.F.		
						MAT.	INST.	TOTAL
7100	Bond beam	8	8	125		5.10	7.35	12.45
7200				105		5.65	7.30	12.95
7300			12	125		6.80	9.20	16
7400				105		7.15	9.10	16.25
7500	Pilaster	8	16	125		12.25	24.50	36.75
7600			20	125		14.75	29	43.75

B20 Exterior Enclosure

B2010 Exterior Walls

The table below lists costs for metal siding of various descriptions, not including the steel frame, or the structural steel, of a building. Costs are per S.F. including all accessories and insulation.

For steel frame support see System B2010 154.

System Components	QUANTITY	UNIT	COST PER S.F. MAT.	COST PER S.F. INST.	COST PER S.F. TOTAL
SYSTEM B2010 146 1400					
METAL SIDING ALUMINUM PANEL, CORRUGATED, .024" THICK, NATURAL					
Corrugated aluminum siding, industrial type, .024" thick	1.000	S.F.	2.33	2.81	5.14
Flashing alum mill finish .032" thick	.111	S.F.	.15	.39	.54
Tapes, sealant, p.v.c. foam adheasive, 1/16"x1"	.004	C.L.F.	.07		.07
Closure strips for aluminum siding, corrugated, .032" thick	.111	L.F.	.10	.30	.40
Pre-engineered steel building insulation, vinyl faced, 1-1/2" thick, R5	1.000	S.F.	.30	.48	.78
TOTAL			2.95	3.98	6.93

B2010 146	Metal Siding Panel	MAT.	INST.	TOTAL
1400	Metal siding aluminum panel, corrugated, .024" thick, natural	2.95	3.98	6.93
1450	Painted	3.13	3.98	7.11
1500	.032" thick, natural	3.30	3.98	7.28
1550	Painted	3.89	3.98	7.87
1600	Ribbed 4" pitch, .032" thick, natural	3.23	3.98	7.21
1650	Painted	3.93	3.98	7.91
1700	.040" thick, natural	3.72	3.98	7.70
1750	Painted	4.26	3.98	8.24
1800	.050" thick, natural	4.19	3.98	8.17
1850	Painted	4.75	3.98	8.73
1900	8" pitch panel, .032" thick, natural	3.08	3.82	6.90
1950	Painted	3.76	3.83	7.59
2000	.040" thick, natural	3.57	3.84	7.41
2050	Painted	4.09	3.88	7.97
2100	.050" thick, natural	4.04	3.86	7.90
2150	Painted	4.62	3.89	8.51
3000	Steel, corrugated or ribbed, 29 Ga., .0135" thick, galvanized	2.26	3.63	5.89
3050	Colored	3.21	3.69	6.90
3100	26 Ga., .0179" thick, galvanized	2.34	3.65	5.99
3150	Colored	3.32	3.71	7.03
3200	24 Ga., .0239" thick, galvanized	3	3.66	6.66
3250	Colored	3.74	3.72	7.46
3300	22 Ga., .0299" thick, galvanized	3.01	3.68	6.69
3350	Colored	4.53	3.74	8.27
3400	20 Ga., .0359" thick, galvanized	3.01	3.68	6.69
3450	Colored	4.82	3.95	8.77

B20 Exterior Enclosure

B2010 Exterior Walls

B2010 146	Metal Siding Panel	COST PER S.F.		
		MAT.	INST.	TOTAL
4100	Sandwich panels, factory fab., 1" polystyrene, steel core, 26 Ga., galv.	6.20	5.70	11.90
4200	Colored, 1 side	7.60	5.70	13.30
4300	2 sides	9.75	5.70	15.45
4400	2" polystyrene, steel core, 26 Ga., galvanized	7.35	5.70	13.05
4500	Colored, 1 side	8.75	5.70	14.45
4600	2 sides	10.90	5.70	16.60
4700	22 Ga., baked enamel exterior	14.40	6.05	20.45
4800	Polyvinyl chloride exterior	15.15	6.05	21.20
5100	Textured aluminum, 4' x 8' x 5/16" plywood backing, single face	4.24	3.47	7.71
5200	Double face	5.75	3.47	9.22
5300	4' x 10' x 5/16" plywood backing, single face	4.53	3.47	8

B20 Exterior Enclosure

B2010 Exterior Walls

The table below lists costs per S.F. for exterior walls with wood siding. A variety of systems are presented using both wood and metal studs at 16" and 24" O.C.

System Components	QUANTITY	UNIT	COST PER S.F. MAT.	INST.	TOTAL
SYSTEM B2010 148 1400					
2"X4" STUDS, 16" O.C., INSUL. WALL, W/ 5/8" TEXTURE 1-11 FIR PLYWOOD					
Partitions, 2" x 4" studs 8' high 16" O.C.	1.000	B.F.	.45	1.38	1.83
Sheathing plywood on wall CDX 1/2"	1.000	S.F.	.64	.98	1.62
Building paper, asphalt felt sheathing paper 15 lb	1.000	S.F.	.06	.15	.21
Fiberglass insulation batts, paper or foil back, 3-1/2", R11	1.000	S.F.	.41	.41	.82
Siding plywood texture 1-11 fir 5/8", natural	1.000	S.F.	1.38	1.64	3.02
Exterior wood stain on other than shingles, 2 coats & sealer	1.000	S.F.	.24	1.09	1.33
TOTAL			3.18	5.65	8.83

B2010 148	Panel, Shingle & Lap Siding	COST PER S.F. MAT.	INST.	TOTAL
1400	Wood siding w/2"x4"studs, 16"O.C., insul. wall, 5/8"text 1-11 fir plywood	3.18	5.65	8.83
1450	5/8" text 1-11 cedar plywood	4.47	5.45	9.92
1500	1" x 4" vert T.&G. redwood	4.59	6.50	11.09
1600	1" x 8" vert T.&G. redwood	6.30	5.70	12
1650	1" x 5" rabbetted cedar bev. siding	6.85	5.90	12.75
1700	1" x 6" cedar drop siding	6.95	5.95	12.90
1750	1" x 12" rough sawn cedar	3.16	5.70	8.86
1800	1" x 12" sawn cedar, 1" x 4" battens	6.20	5.35	11.55
1850	1" x 10" redwood shiplap siding	6.35	5.50	11.85
1900	18" no. 1 red cedar shingles, 5-1/2" exposed	4.77	7.45	12.22
1950	6" exposed	4.56	7.20	11.76
2000	6-1/2" exposed	4.34	6.95	11.29
2100	7" exposed	4.13	6.70	10.83
2150	7-1/2" exposed	3.92	6.45	10.37
3000	8" wide aluminum siding	4.21	5.05	9.26
3150	8" plain vinyl siding	2.75	5.15	7.90
3250	8" insulated vinyl siding	3.08	5.70	8.78
3400	2" x 6" studs, 16" O.C., insul. wall, w/ 5/8" text 1-11 fir plywood	3.49	5.80	9.29
3500	5/8" text 1-11 cedar plywood	4.78	5.80	10.58
3600	1" x 4" vert T.&G. redwood	4.90	6.70	11.60
3700	1" x 8" vert T.&G. redwood	6.60	5.85	12.45
3800	1" x 5" rabbetted cedar bev siding	7.15	6.05	13.20
3900	1" x 6" cedar drop siding	7.25	6.10	13.35
4000	1" x 12" rough sawn cedar	3.47	5.85	9.32
4200	1" x 12" sawn cedar, 1" x 4" battens	6.55	5.50	12.05
4500	1" x 10" redwood shiplap siding	6.65	5.65	12.30

B20 Exterior Enclosure

B2010 Exterior Walls

B2010 148	Panel, Shingle & Lap Siding	COST PER S.F.		
		MAT.	INST.	TOTAL
4550	18" no. 1 red cedar shingles, 5-1/2" exposed	5.10	7.60	12.70
4600	6" exposed	4.87	7.35	12.22
4650	6-1/2" exposed	4.65	7.10	11.75
4700	7" exposed	4.44	6.90	11.34
4750	7-1/2" exposed	4.23	6.65	10.88
4800	8" wide aluminum siding	4.52	5.25	9.77
4850	8" plain vinyl siding	3.06	5.30	8.36
4900	8" insulated vinyl siding	3.39	5.85	9.24
5000	2" x 6" studs, 24" O.C., insul. wall, 5/8" text 1-11, fir plywood	3.34	5.45	8.79
5050	5/8" text 1-11 cedar plywood	4.63	5.45	10.08
5100	1" x 4" vert T.&G. redwood	4.75	6.35	11.10
5150	1" x 8" vert T.&G. redwood	6.45	5.50	11.95
5200	1" x 5" rabbetted cedar bev siding	7	5.70	12.70
5250	1" x 6" cedar drop siding	7.10	5.75	12.85
5300	1" x 12" rough sawn cedar	3.32	5.50	8.82
5400	1" x 12" sawn cedar, 1" x 4" battens	6.40	5.15	11.55
5450	1" x 10" redwood shiplap siding	6.50	5.30	11.80
5500	18" no. 1 red cedar shingles, 5-1/2" exposed	4.93	7.25	12.18
5550	6" exposed	4.72	7.05	11.77
5650	7" exposed	4.29	6.55	10.84
5700	7-1/2" exposed	4.08	6.30	10.38
5750	8" wide aluminum siding	4.37	4.89	9.26
5800	8" plain vinyl siding	2.91	4.97	7.88
5850	8" insulated vinyl siding	3.24	5.50	8.74
5900	3-5/8" metal studs, 16 Ga., 16" O.C. insul.wall, 5/8"text 1-11 fir plywood	3.99	5.95	9.94
5950	5/8" text 1-11 cedar plywood	5.30	5.95	11.25
6000	1" x 4" vert T.&G. redwood	5.40	6.80	12.20
6050	1" x 8" vert T.&G. redwood	7.10	6	13.10
6100	1" x 5" rabbetted cedar bev siding	7.65	6.20	13.85
6150	1" x 6" cedar drop siding	7.75	6.25	14
6200	1" x 12" rough sawn cedar	3.97	6	9.97
6250	1" x 12" sawn cedar, 1" x 4" battens	7.05	5.65	12.70
6300	1" x 10" redwood shiplap siding	7.15	5.80	12.95
6350	18" no. 1 red cedar shingles, 5-1/2" exposed	5.60	7.75	13.35
6500	6" exposed	5.35	7.50	12.85
6550	6-1/2" exposed	5.15	7.25	12.40
6600	7" exposed	4.94	7	11.94
6650	7-1/2" exposed	4.94	6.85	11.79
6700	8" wide aluminum siding	5	5.35	10.35
6750	8" plain vinyl siding	3.56	5.25	8.81
6800	8" insulated vinyl siding	3.89	5.80	9.69
7000	3-5/8" metal studs, 16 Ga. 24" O.C. insul wall, 5/8" text 1-11 fir plywood	3.64	5.30	8.94
7050	5/8" text 1-11 cedar plywood	4.93	5.30	10.23
7100	1" x 4" vert T.&G. redwood	5.05	6.35	11.40
7150	1" x 8" vert T.&G. redwood	6.75	5.50	12.25
7200	1" x 5" rabbetted cedar bev siding	7.30	5.70	13
7250	1" x 6" cedar drop siding	7.40	5.75	13.15
7300	1" x 12" rough sawn cedar	3.62	5.55	9.17
7350	1" x 12" sawn cedar 1" x 4" battens	6.70	5.15	11.85
7400	1" x 10" redwood shiplap siding	6.80	5.30	12.10
7450	18" no. 1 red cedar shingles, 5-1/2" exposed	5.45	7.55	13
7500	6" exposed	5.25	7.30	12.55
7550	6-1/2" exposed	5	7.05	12.05
7600	7" exposed	4.80	6.80	11.60
7650	7-1/2" exposed	4.38	6.30	10.68
7700	8" wide aluminum siding	4.67	4.90	9.57
7750	8" plain vinyl siding	3.21	4.98	8.19
7800	8" insul. vinyl siding	3.54	5.55	9.09

B20 Exterior Enclosure

B2010 Exterior Walls

Exterior Stucco Wall

The table below lists costs for some typical stucco walls including all the components as demonstrated in the component block below. Prices are presented for backup walls using wood studs, metal studs and CMU.

System Components	QUANTITY	UNIT	COST PER S.F.		
			MAT.	INST.	TOTAL
SYSTEM B2010 151 2100 **7/8" CEMENT STUCCO, PLYWOOD SHEATHING, STUD WALL, 2" X 4", 16" O.C.**					
Framing 2x4 studs 8' high 16" O.C.	1.000	S.F.	.45	1.38	1.83
Plywood sheathing on ext. stud wall, 5/8" thick	1.000	S.F.	.80	1.05	1.85
Building paper, asphalt felt sheathing paper 15 lb	1.000	S.F.	.06	.15	.21
Stucco, 3 coats 7/8" thk, float finish, on frame construction	1.000	S.F.	.73	5.28	6.01
Fiberglass insulation batts, paper or foil back, 3-1/2", R-11	1.000	S.F.	.41	.41	.82
Paint exterior stucco, brushwork, primer & 2 coats	1.000	S.F.	.22	.93	1.15
TOTAL			2.67	9.20	11.87

B2010 151	Stucco Wall	COST PER S.F.		
		MAT.	INST.	TOTAL
2100	Cement stucco, 7/8" th., plywood sheathing, stud wall, 2" x 4", 16" O.C.	2.67	9.20	11.87
2200	24" O.C.	2.57	8.95	11.52
2300	2" x 6", 16" O.C.	2.98	9.40	12.38
2400	24" O.C.	2.83	9.05	11.88
2500	No sheathing, metal lath on stud wall, 2" x 4", 16" O.C.	1.81	8	9.81
2600	24" O.C.	1.71	7.75	9.46
2700	2" x 6", 16" O.C.	2.12	8.20	10.32
2800	24" O.C.	1.97	7.85	9.82
2900	1/2" gypsum sheathing, 3-5/8" metal studs, 16" O.C.	2.85	8.35	11.20
2950	24" O.C.	2.50	7.90	10.40
3000	Cement stucco, 5/8" th., 2 coats on std. CMU block, 8"x 16", 8" thick	3.14	11.15	14.29
3100	10" thick	4.34	11.45	15.79
3200	12" thick	4.49	13.15	17.64
3300	Std. light Wt. block 8" x 16", 8" thick	3.76	11	14.76
3400	10" thick	4.40	11.25	15.65
3500	12" thick	4.61	12.90	17.51
3600	3 coat stucco, self furring metal lath 3.4 Lb/SY, on 8" x 16", 8" thick	3.18	11.80	14.98
3700	10" thick	4.38	12.10	16.48
3800	12" thick	4.53	13.80	18.33
3900	Lt. Wt. block, 8" thick	3.80	11.65	15.45
4000	10" thick	4.44	11.90	16.34
4100	12" thick	4.65	13.55	18.20

B20 Exterior Enclosure

B2010 Exterior Walls

B2010 152	E.I.F.S.	COST PER S.F.		
		MAT.	INST.	TOTAL
5100	E.I.F.S., plywood sheathing, stud wall, 2" x 4", 16" O.C., 1" EPS	3.79	9.30	13.09
5110	2" EPS	4.05	9.30	13.35
5120	3" EPS	4.32	9.30	13.62
5130	4" EPS	4.58	9.30	13.88
5140	2" x 6", 16" O.C., 1" EPS	4.10	9.45	13.55
5150	2" EPS	4.36	9.45	13.81
5160	3" EPS	4.63	9.45	14.08
5170	4" EPS	4.89	9.45	14.34
5180	Cement board sheathing, 3-5/8" metal studs, 16" O.C., 1" EPS	4.55	11.50	16.05
5190	2" EPS	4.81	11.50	16.31
5200	3" EPS	5.10	11.50	16.60
5210	4" EPS	6.10	11.50	17.60
5220	6" metal studs, 16" O.C., 1" EPS	5	11.55	16.55
5230	2" EPS	5.25	11.55	16.80
5240	3" EPS	5.55	11.55	17.10
5250	4" EPS	6.55	11.55	18.10
5260	CMU block, 8" x 8" x 16", 1" EPS	4.81	13.20	18.01
5270	2" EPS	5.05	13.20	18.25
5280	3" EPS	5.35	13.20	18.55
5290	4" EPS	5.60	13.20	18.80
5300	8" x 10" x 16", 1" EPS	6	13.50	19.50
5310	2" EPS	6.25	13.50	19.75
5320	3" EPS	6.55	13.50	20.05
5330	4" EPS	6.80	13.50	20.30
5340	8" x 12" x 16", 1" EPS	6.15	15.20	21.35
5350	2" EPS	6.40	15.20	21.60
5360	3" EPS	6.70	15.20	21.90
5370	4" EPS	6.95	15.20	22.15

B2010 153	Oversized Brick Curtain Wall	COST PER S.F.		
		MAT.	INST.	TOTAL
1000	Oversized brk curtain wall, 6"x4"x16", incl rebar, insul, bond beam, plain	47.50	16.30	63.80
1010	1 to 3 slots in face	54.50	18.40	72.90
1020	4 to 7 slots in face	59.50	19.85	79.35
1030	8" x 4" x 16", incl rebar, insul & bond beam, plain face	54.50	18.90	73.40
1040	1 to 3 slots in face	62.50	21.50	84
1050	4 to 7 slots in face	67.50	23	90.50
1060	6" x 8" x 16", incl rebar, insul & bond beam, plain face	45	14.65	59.65
1070	1 to 3 slots in face	51.50	16.50	68
1080	4 to 7 slots in face	55.50	17.75	73.25
1090	8" x 8" x 16", incl rebar, insul & bond beam, plain face	65.50	15.85	81.35
1100	1 to 3 slots in face	75	17.90	92.90
1110	4 to 7 slots in face	81.50	19.30	100.80

B20 Exterior Enclosure

B2010 Exterior Walls

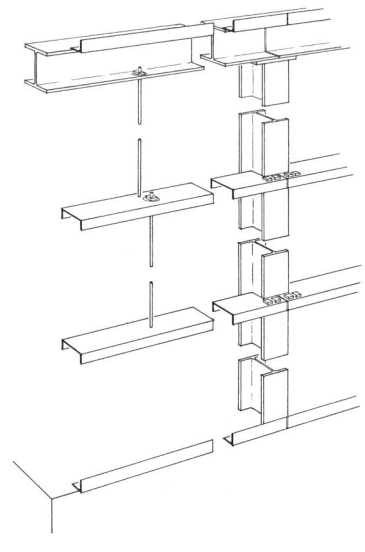

Description: The table below lists costs, $/S.F., for channel girts with sag rods and connector angles top and bottom for various column spacings, building heights and wind loads. Additive costs are shown for wind columns.

How to Use this Table: Add the cost of girts, sag rods, and angles to the framing costs of steel buildings clad in metal or composition siding. If the column spacing is in excess of the column spacing shown, use intermediate wind columns. Additive costs are shown under "Wind Columns".

System Components			COST PER S.F.		
	QUANTITY	UNIT	MAT.	INST.	TOTAL
SYSTEM B2010 154 3000					
METAL SIDING SUPPORT, 18' HIGH, 20 PSF WIND LOAD, 20' SPACING					
Structural steel	1.417	Lb.	2.03	.48	2.51
Lightgage framing, angles less than 4"	.400	Lb.	.34	3.52	3.86
TOTAL			2.37	4	6.37
SYSTEM B2010 154 3100					
WIND COLUMNS					
Structural steel	.856	S.F.	1.22	.29	1.51

B2010 154			Metal Siding Support				
	BLDG. HEIGHT (FT.)	WIND LOAD (P.S.F.)	COL. SPACING (FT.)		INTERMEDIATE COLUMNS	COST PER S.F.	
						MAT.	INST.	TOTAL
3000	18	20	20			2.37	4	6.37
3100					wind cols.	1.22	.29	1.51
3200	RB2010-430	20	25			2.59	4.05	6.64
3300					wind cols.	.98	.23	1.21
3400		20	30			2.86	4.12	6.98
3500					wind cols.	.82	.19	1.01
3600		20	35			3.15	4.19	7.34
3700					wind cols.	.70	.16	.86
3800		30	20			2.62	4.06	6.68
3900					wind cols.	1.22	.29	1.51
4000		30	25			2.86	4.12	6.98
4100					wind cols.	.98	.23	1.21
4200		30	30			3.16	4.19	7.35
4300					wind cols.	1.11	.26	1.37
4400		30	35			4.20	4.44	8.64
4500					wind cols.	1.10	.26	1.36
4600	30	20	20			2.30	2.81	5.11
4700					wind cols.	1.73	.41	2.14
4800		20	25			2.59	2.88	5.47
4900					wind cols.	1.38	.33	1.71

B20 Exterior Enclosure

B2010 Exterior Walls

B2010 154 Metal Siding Support

	BLDG. HEIGHT (FT.)	WIND LOAD (P.S.F.)	COL. SPACING (FT.)		INTERMEDIATE COLUMNS	COST PER S.F.		
						MAT.	INST.	TOTAL
5000	30	20	30			2.91	2.96	5.87
5100					wind cols.	1.36	.33	1.69
5200		20	35			3.25	3.04	6.29
5300					wind cols.	1.35	.32	1.67
5400		30	20			2.62	2.89	5.51
5500					wind cols.	2.04	.49	2.53
5600		30	25			2.90	2.95	5.85
5700					wind cols.	1.95	.47	2.42
5800		30	30			3.27	3.04	6.31
5900					wind cols.	1.83	.44	2.27
6000		30	35			4.52	3.34	7.86
6100					wind cols.	1.80	.43	2.23

B20 Exterior Enclosure

B2020 Exterior Windows

The table below lists window systems by material, type and size. Prices between sizes listed can be interpolated with reasonable accuracy. Prices include frame, hardware, and casing as illustrated in the component block below.

System Components	QUANTITY	UNIT	COST PER UNIT		
			MAT.	INST.	TOTAL
SYSTEM B2020 102 3000					
WOOD, DOUBLE HUNG, STD. GLASS, 2'-8" X 4'-6"					
Framing rough opening, header w/jacks	14.680	B.F.	9.54	28.04	37.58
Casing, stock pine 11/16" x 2-1/2"	16.000	L.F.	27.04	35.36	62.40
Stool cap, pine, 11/16" x 3-1/2"	5.000	L.F.	14.60	13.85	28.45
Residential wood window, double hung, 2'-8" x 4'-6" standard glazed	1.000	Ea.	198	163	361
TOTAL			249.18	240.25	489.43

B2020 102 — Wood Windows

	MATERIAL	TYPE	GLAZING	SIZE	DETAIL	MAT.	INST.	TOTAL
3000	Wood	double hung	std. glass	2'-8" x 4'-6"		249	240	489
3050				3'-0" x 5'-6"		325	275	600
3100			insul. glass	2'-8" x 4'-6"		264	240	504
3150				3'-0" x 5'-6"		355	275	630
3200		sliding	std. glass	3'-4" x 2'-7"		330	200	530
3250				4'-4" x 3'-3"		345	219	564
3300				5'-4" x 6'-0"		435	263	698
3350			insul. glass	3'-4" x 2'-7"		400	234	634
3400				4'-4" x 3'-3"		415	254	669
3450				5'-4" x 6'-0"		525	300	825
3500		awning	std. glass	2'-10" x 1'-9"		231	120	351
3600				4'-4" x 2'-8"		370	116	486
3700			insul. glass	2'-10" x 1'-9"		282	138	420
3800				4'-4" x 2'-8"		450	127	577
3900		casement	std. glass	1'-10" x 3'-2"	1 lite	335	158	493
3950				4'-2" x 4'-2"	2 lite	595	210	805
4000				5'-11" x 5'-2"	3 lite	900	270	1,170
4050				7'-11" x 6'-3"	4 lite	1,275	325	1,600
4100				9'-11" x 6'-3"	5 lite	1,650	375	2,025
4150			insul. glass	1'-10" x 3'-2"	1 lite	335	158	493
4200				4'-2" x 4'-2"	2 lite	610	210	820
4250				5'-11" x 5'-2"	3 lite	945	270	1,215
4300				7'-11" x 6'-3"	4 lite	1,350	325	1,675
4350				9'-11" x 6'-3"	5 lite	1,750	375	2,125

B20 Exterior Enclosure

B2020 Exterior Windows

B2020 102 Wood Windows

	MATERIAL	TYPE	GLAZING	SIZE	DETAIL	COST PER UNIT MAT.	COST PER UNIT INST.	COST PER UNIT TOTAL
4400	Wood	picture	std. glass	4'-6" x 4'-6"		470	271	741
4450				5'-8" x 4'-6"		530	300	830
4500		picture	insul. glass	4'-6" x 4'-6"		570	315	885
4550				5'-8" x 4'-6"		650	350	1,000
4600		fixed bay	std. glass	8' x 5'		1,625	500	2,125
4650				9'-9" x 5'-4"		1,175	700	1,875
4700			insul. glass	8' x 5'		2,150	500	2,650
4750				9'-9" x 5'-4"		1,275	700	1,975
4800		casement bay	std. glass	8' x 5'		1,625	575	2,200
4850			insul. glass	8' x 5'		1,825	575	2,400
4900		vert. bay	std. glass	8' x 5'		1,825	575	2,400
4950			insul. glass	8' x 5'		1,925	575	2,500

B2020 104 Steel Windows

	MATERIAL	TYPE	GLAZING	SIZE	DETAIL	COST PER UNIT MAT.	COST PER UNIT INST.	COST PER UNIT TOTAL
5000	Steel	double hung	1/4" tempered	2'-8" x 4'-6"		880	191	1,071
5050				3'-4" x 5'-6"		1,350	291	1,641
5100			insul. glass	2'-8" x 4'-6"		895	218	1,113
5150				3'-4" x 5'-6"		1,375	335	1,710
5200		horz. pivoted	std. glass	2' x 2'		269	63.50	332.50
5250				3' x 3'		605	143	748
5300				4' x 4'		1,075	254	1,329
5350				6' x 4'		1,625	380	2,005
5400			insul. glass	2' x 2'		274	73	347
5450				3' x 3'		615	164	779
5500				4' x 4'		1,100	291	1,391
5550				6' x 4'		1,650	435	2,085
5600		picture window	std. glass	3' x 3'		380	143	523
5650				6' x 4'		1,025	380	1,405
5700			insul. glass	3' x 3'		390	164	554
5750				6' x 4'		1,050	435	1,485
5800		industrial security	std. glass	2'-9" x 4'-1"		810	178	988
5850				4'-1" x 5'-5"		1,600	350	1,950
5900			insul. glass	2'-9" x 4'-1"		825	204	1,029
5950				4'-1" x 5'-5"		1,625	400	2,025
6000		comm. projected	std. glass	3'-9" x 5'-5"		1,325	325	1,650
6050				6'-9" x 4'-1"		1,800	440	2,240
6100			insul. glass	3'-9" x 5'-5"		1,350	370	1,720
6150				6'-9" x 4'-1"		1,825	500	2,325
6200		casement	std. glass	4'-2" x 4'-2"	2 lite	1,175	276	1,451
6250			insul. glass	4'-2" x 4'-2"		1,200	315	1,515
6300			std. glass	5'-11" x 5'-2"	3 lite	2,475	485	2,960
6350			insul. glass	5'-11" x 5'-2"		2,500	555	3,055

B2020 106 Aluminum Windows

	MATERIAL	TYPE	GLAZING	SIZE	DETAIL	COST PER UNIT MAT.	COST PER UNIT INST.	COST PER UNIT TOTAL
6400	Aluminum	double hung	insul. glass	2'-0" x 4'-0"		385	143	528
6450				4'-5" x 5'-3"		435	179	614
6500			insul. glass	3'-1" x 3'-2"		460	172	632
6550				4'-5" x 5'-3"		520	215	735

B20 Exterior Enclosure

B2020 Exterior Windows

B2020 106 — Aluminum Windows

	MATERIAL	TYPE	GLAZING	SIZE	DETAIL	COST PER UNIT MAT.	COST PER UNIT INST.	COST PER UNIT TOTAL
6600	Aluminum	sliding	std. glass	3' x 2'		239	143	382
6650				5' x 3'		365	159	524
6700				8' x 4'		385	238	623
6750				9' x 5'		585	355	940
6800			insul. glass	3' x 2'		255	143	398
6850				5' x 3'		425	159	584
6900				8' x 4'		620	238	858
6950				9' x 5'		935	355	1,290
7000		single hung	std. glass	2' x 3'		226	143	369
7050				2'-8" x 6'-8"		400	179	579
7100				3'-4" x 5'-0"		325	159	484
7150			insul. glass	2' x 3'		274	143	417
7200				2'-8" x 6'-8"		515	179	694
7250				3'-4" x 5'		360	159	519
7300		double hung	std. glass	2' x 3'		310	95.50	405.50
7350				2'-8" x 6'-8"		920	283	1,203
7400				3'-4" x 5'		860	265	1,125
7450			insul. glass	2' x 3'		320	109	429
7500				2'-8" x 6'-8"		945	325	1,270
7550				3'-4" x 5'-0"		885	305	1,190
7600		casement	std. glass	3'-1" x 3'-2"		273	155	428
7650				4'-5" x 5'-3"		660	375	1,035
7700			insul. glass	3'-1" x 3'-2"		287	177	464
7750				4'-5" x 5'-3"		695	430	1,125
7800		hinged swing	std. glass	3' x 4'		570	191	761
7850				4' x 5'		955	320	1,275
7900			insul. glass	3' x 4'		590	218	808
7950				4' x 5'		980	365	1,345
8200		picture unit	std. glass	2'-0" x 3'-0"		175	95.50	270.50
8250				2'-8" x 6'-8"		520	283	803
8300				3'-4" x 5'-0"		485	265	750
8350			insul. glass	2'-0" x 3'-0"		183	109	292
8400				2'-8" x 6'-8"		545	325	870
8450				3'-4" x 5'-0"		510	305	815
8500		awning type	std. glass	3'-0" x 3'-0"	2 lite	460	102	562
8550				3'-0" x 4'-0"	3 lite	540	143	683
8600				3'-0" x 5'-4"	4 lite	650	143	793
8650				4'-0" x 5'-4"	4 lite	715	159	874
8700			insul. glass	3'-0" x 3'-0"	2 lite	490	102	592
8750				3'-0" x 4'-0"	3 lite	620	143	763
8800				3'-0" x 5'-4"	4 lite	765	143	908
8850				4'-0" x 5'-4"	4 lite	855	159	1,014

B2020 108 — Vinyl Clad Windows

	MATERIAL	TYPE	GLAZING	SIZE	DETAIL	COST PER UNIT MAT.	COST PER UNIT INST.	COST PER UNIT TOTAL
1000	Vinyl clad	casement	insul. glass	1'-4" x 4'-0"		410	128	538
1010				2'-0" x 3'-0"		360	128	488
1020				2'-0" x 4'-0"		410	134	544
1030				2'-0" x 5'-0"		460	141	601
1040				2'-0" x 6'-0"		480	141	621
1050				2'-4" x 4'-0"		460	141	601

B20 Exterior Enclosure

B2020 Exterior Windows

B2020 108 — Vinyl Clad Windows

	MATERIAL	TYPE	GLAZING	SIZE	DETAIL	COST PER UNIT		
						MAT.	INST.	TOTAL
1060				2'-6" x 5'-0"		565	137	702
1070				3'-0" x 5'-0"		790	141	931
1080				4'-0" x 3'-0"		865	141	1,006
1090				4'-0" x 4'-0"		745	141	886
1100				4'-8" x 4'-0"		820	141	961
1110				4'-8" x 5'-0"		925	164	1,089
1120				4'-8" x 6'-0"		1,025	164	1,189
1130				6'-0" x 4'-0"		945	164	1,109
1140				6'-0" x 5'-0"		1,050	164	1,214
3000		double-hung	insul. glass	2'-0" x 4'-0"		440	123	563
3050				2'-0" x 5'-0"		500	123	623
3100				2'-4" x 4'-0"		455	128	583
3150				2'-4" x 4'-8"		455	128	583
3200				2'-4" x 6'-0"		515	128	643
3250				2'-6" x 4'-0"		455	128	583
3300				2'-8" x 4'-0"		680	164	844
3350				2'-8" x 5'-0"		535	141	676
3375				2'-8" x 6'-0"		545	141	686
3400				3'-0 x 3'-6"		430	128	558
3450				3'-0 x 4'-0"		495	134	629
3500				3'-0 x 4'-8"		530	134	664
3550				3'-0 x 5'-0"		545	141	686
3600				3'-0 x 6'-0"		490	141	631
3700				4'-0 x 5'-0"		680	151	831
3800				4'-0 x 6'-0"		830	151	981

B20 Exterior Enclosure

B2020 Exterior Windows

The table below lists costs per S.F of opening for framing with 1-3/4" x 4-1/2" clear anodized tubular aluminum framing. This is the type often used for 1/4" plate glass flush glazing.

For bronze finish, add 18% to material cost. For black finish, add 27% to material cost. For stainless steel, add 75% to material cost. For monumental grade, add 50% to material cost. This tube framing is usually installed by a glazing contractor.

Note: The costs below do not include the glass. For glazing, use Assembly B2020 220.

System Components	QUANTITY	UNIT	COST/S.F. OPNG. MAT.	INST.	TOTAL
SYSTEM B2020 210 1250					
ALUM FLUSH TUBE, FOR 1/4" GLASS, 5'X20' OPENING, 3 INTER. HORIZONTALS					
Flush tube frame, alum mill fin, 1-3/4"x4" open headr for 1/4" glass	.450	L.F.	6.21	5.90	12.11
Flush tube frame, alum mill fin, 1-3/4"x4" open sill for 1/4" glass	.050	L.F.	.56	.64	1.20
Flush tube frame, alum mill fin, 1-3/4"x4" closed back sill, 1/4" glass	.150	L.F.	2.83	1.85	4.68
Aluminum structural shapes, 1" to 10" members, under 1 ton	.040	Lb.	.18	.07	.25
Joints for tube frame, 90° clip type	.100	Ea.	2.70		2.70
Caulking/sealants, polysulfide, 1 or 2 part,1/2x1/4"bead 154 lf/gal	.500	L.F.	.20	.94	1.14
TOTAL			12.68	9.40	22.08

B2020 210	Tubular Aluminum Framing	COST/S.F. OPNG. MAT.	INST.	TOTAL
1100	Alum flush tube frame, for 1/4" glass, 1-3/4"x4", 5'x6'opng, no inter horiz	13.80	10.95	24.75
1150	One intermediate horizontal	18.85	13.10	31.95
1200	Two intermediate horizontals	24	15.20	39.20
1250	5' x 20' opening, three intermediate horizontals	12.70	9.40	22.10
1400	1-3/4" x 4-1/2", 5' x 6' opening, no intermediate horizontals	15.95	10.95	26.90
1450	One intermediate horizontal	21	13.10	34.10
1500	Two intermediate horizontals	26.50	15.20	41.70
1550	5' x 20' opening, three intermediate horizontals	14.30	9.40	23.70
1700	For insulating glass, 2"x4-1/2", 5'x6' opening, no intermediate horizontals	16.25	11.60	27.85
1750	One intermediate horizontal	21	13.85	34.85
1800	Two intermediate horizontals	26	16.10	42.10
1850	5' x 20' opening, three intermediate horizontals	14.15	9.95	24.10
2000	Thermal break frame, 2-1/4"x4-1/2", 5'x6'opng, no intermediate horizontals	16.90	11.75	28.65
2050	One intermediate horizontal	22.50	14.45	36.95
2100	Two intermediate horizontals	28	17.10	45.10
2150	5' x 20' opening, three intermediate horizontals	15.30	10.40	25.70

B20 Exterior Enclosure

B2020 Exterior Windows

The table below lists costs of curtain wall and spandrel panels per S.F. Costs do not include structural framing used to hang the panels.

B2020 220	Curtain Wall Panels	MAT.	INST.	TOTAL
1000	Glazing panel, insulating, 1/2" thick, 2 lites 1/8" float, clear	10.60	11.05	21.65
1100	Tinted	14.85	11.05	25.90
1200	5/8" thick units, 2 lites 3/16" float, clear	14.85	11.65	26.50
1400	1" thick units, 2 lites, 1/4" float, clear	17.80	14	31.80
1700	Light and heat reflective glass, tinted	34	12.35	46.35
2000	Plate glass, 1/4" thick, clear	6.20	8.75	14.95
2050	Tempered	9.20	8.75	17.95
2100	Tinted	9.70	8.75	18.45
2200	3/8" thick, clear	10.80	14	24.80
2250	Tempered	17.80	14	31.80
2300	Tinted	17	14	31
2400	1/2" thick, clear	18.30	19.05	37.35
2450	Tempered	26.50	19.05	45.55
2500	Tinted	30	19.05	49.05
2600	3/4" thick, clear	38.50	30	68.50
2650	Tempered	44.50	30	74.50
3000	Spandrel glass, panels, 1/4" plate glass insul w/fiberglass, 1" thick	18.25	8.75	27
3100	2" thick	21.50	8.75	30.25
3200	Galvanized steel backing, add	6.35		6.35
3300	3/8" plate glass, 1" thick	31	8.75	39.75
3400	2" thick	34.50	8.75	43.25
4000	Polycarbonate, masked, clear or colored, 1/8" thick	11	6.15	17.15
4100	3/16" thick	13.20	6.35	19.55
4200	1/4" thick	14.30	6.75	21.05
4300	3/8" thick	24.50	7	31.50
5000	Facing panel, textured alum., 4' x 8' x 5/16" plywood backing, sgl face	4.24	3.47	7.71
5100	Double face	5.75	3.47	9.22
5200	4' x 10' x 5/16" plywood backing, single face	4.53	3.47	8
5300	Double face	6.10	3.47	9.57
5400	4' x 12' x 5/16" plywood backing, single face	4.21	3.47	7.68
5500	Sandwich panel, 22 Ga. galv., both sides 2" insulation, enamel exterior	14.40	6.05	20.45
5600	Polyvinylidene fluoride exterior finish	15.15	6.05	21.20
5700	26 Ga., galv. both sides, 1" insulation, colored 1 side	7.60	5.70	13.30
5800	Colored 2 sides	9.75	5.70	15.45

B20 Exterior Enclosure

B2030 Exterior Doors

Costs are listed for exterior door systems by material, type and size. Prices between sizes listed can be interpolated with reasonable accuracy. Prices are per opening for a complete door system including frame.

B2030 110			Glazed Doors, Steel or Aluminum				
	MATERIAL	TYPE	DOORS	SPECIFICATION	OPENING	COST PER OPNG.		
						MAT.	INST.	TOTAL
5600	St. Stl. & glass	revolving	stock unit	manual oper.	6'-0" x 7'-0"	43,700	9,475	53,175
5650				auto Cntrls.	6'-10" x 7'-0"	60,000	10,100	70,100
5700	Bronze	revolving	stock unit	manual oper.	6'-10" x 7'-0"	51,500	19,100	70,600
5750				auto Cntrls.	6'-10" x 7'-0"	67,500	19,700	87,200
5800	St. Stl. & glass	balanced	standard	economy	3'-0" x 7'-0"	9,650	1,600	11,250
5850				premium	3'-0" x 7'-0"	16,900	2,100	19,000
6300	Alum. & glass	w/o transom	narrow stile	w/panic Hrdwre.	3'-0" x 7'-0"	1,725	1,050	2,775
6350				dbl. door, Hrdwre.	6'-0" x 7'-0"	2,975	1,725	4,700
6400			wide stile	hdwre.	3'-0" x 7'-0"	2,025	1,025	3,050
6450				dbl. door, Hdwre.	6'-0" x 7'-0"	3,925	2,050	5,975
6500			full vision	hdwre.	3'-0" x 7'-0"	2,550	1,625	4,175
6550				dbl. door, Hdwre.	6'-0" x 7'-0"	3,575	2,325	5,900
6600			non-standard	hdwre.	3'-0" x 7'-0"	2,200	1,025	3,225
6650				dbl. door, Hdwre.	6'-0" x 7'-0"	4,400	2,050	6,450
6700			bronze fin.	hdwre.	3'-0" x 7'-0"	1,600	1,025	2,625
6750				dbl. door, Hrdwre.	6'-0" x 7'-0"	3,225	2,050	5,275
6800			black fin.	hdwre.	3'-0" x 7'-0"	2,325	1,025	3,350
6850				dbl. door, Hdwre.	6'-0" x 7'-0"	4,650	2,050	6,700
6900		w/transom	narrow stile	hdwre.	3'-0" x 10'-0"	2,275	1,175	3,450
6950				dbl. door, Hdwre.	6'-0" x 10'-0"	3,575	2,075	5,650
7000			wide stile	hdwre.	3'-0" x 10'-0"	2,600	1,425	4,025
7050				dbl. door, Hdwre.	6'-0" x 10'-0"	3,925	2,450	6,375
7100			full vision	hdwre.	3'-0" x 10'-0"	2,875	1,575	4,450
7150				dbl. door, Hdwre.	6'-0" x 10'-0"	4,300	2,725	7,025
7200			non-standard	hdwre.	3'-0" x 10'-0"	2,275	1,100	3,375
7250				dbl. door, Hdwre.	6'-0" x 10'-0"	4,550	2,200	6,750
7300			bronze fin.	hdwre.	3'-0" x 10'-0"	1,700	1,100	2,800
7350				dbl. door, Hdwre.	6'-0" x 10'-0"	3,375	2,200	5,575
7400			black fin.	hdwre.	3'-0" x 10'-0"	2,400	1,100	3,500
7450				dbl. door, Hdwre.	6'-0" x 10'-0"	4,800	2,200	7,000
7500		revolving	stock design	minimum	6'-10" x 7'-0"	22,300	3,800	26,100
7550				average	6'-0" x 7'-0"	26,900	4,775	31,675
7600				maximum	6'-10" x 7'-0"	46,100	6,350	52,450
7650				min., automatic	6'-10" x 7'-0"	38,500	4,425	42,925

B20 Exterior Enclosure

B2030 Exterior Doors

B2030 110 Glazed Doors, Steel or Aluminum

	MATERIAL	TYPE	DOORS	SPECIFICATION	OPENING	COST PER OPNG.		
						MAT.	INST.	TOTAL
7700	Alum. & glass	revolving	stock design	avg., automatic	6'-10" x 7'-0"	43,100	5,400	48,500
7750				max., automatic	6'-10" x 7'-0"	62,500	6,975	69,475
7800		balanced	standard	economy	3'-0" x 7'-0"	7,000	1,600	8,600
7850				premium	3'-0" x 7'-0"	8,550	2,050	10,600
7900		mall front	sliding panels	alum. fin.	16'-0" x 9'-0"	3,650	805	4,455
7950					24'-0" x 9'-0"	5,175	1,500	6,675
8000				bronze fin.	16'-0" x 9'-0"	4,250	940	5,190
8050					24'-0" x 9'-0"	6,025	1,750	7,775
8100			fixed panels	alum. fin.	48'-0" x 9'-0"	9,650	1,175	10,825
8150				bronze fin.	48'-0" x 9'-0"	11,200	1,375	12,575
8200		sliding entrance	5' x 7' door	electric oper.	12'-0" x 7'-6"	9,025	1,500	10,525
8250		sliding patio	temp. glass	economy	6'-0" x 7'-0"	1,825	277	2,102
8300			temp. glass	economy	12'-0" x 7'-0"	4,275	370	4,645
8350				premium	6'-0" x 7'-0"	2,750	415	3,165
8400					12'-0" x 7'-0"	6,425	555	6,980

B20 Exterior Enclosure

B2030 Exterior Doors

Costs are listed for exterior door systems by material, type and size. Prices between sizes listed can be interpolated with reasonable accuracy. Prices are per opening for a complete door system including frame as illustrated in the component block.

System Components	QUANTITY	UNIT	COST PER OPNG.		
			MAT.	INST.	TOTAL
SYSTEM B2030 210 2500					
WOOD DOOR, SOLID CORE, SINGLE, HINGED, 3'-0" X 7'-0"					
Exterior, flush, solid core, birch, 1-3/4" x 2'-6" x 7'-0"	1.000	Ea.	190	74	264
Wood door exterior frame, pine, w/trim, 5/4 x 5-3/16" deep	1.000	Set	146.20	50.15	196.35
Hinges, full mortise-high freq, brass base, 4-1/2" x 4-1/2", US10	1.500	Pr.	116.25		116.25
Panic device for mortise locks, single door exit only	1.000	Ea.	650	138	788
Paint exterior door & frame one side 3' x 7', primer & 2 coats	1.000	Ea.	2.14	27.50	29.64
Sill, 8/4 x 8" deep, oak, 2" horns	1.000	Ea.	23.50	12.30	35.80
TOTAL			1,128.09	301.95	1,430.04

B2030 210			Wood Doors				
	MATERIAL	TYPE	DOORS	SPECIFICATION	OPENING	COST PER OPNG.		

	MATERIAL	TYPE	DOORS	SPECIFICATION	OPENING	MAT.	INST.	TOTAL
2350	Birch	solid core	single door	hinged	2'-6" x 6'-8"	1,125	300	1,425
2400					2'-6" x 7'-0"	1,125	300	1,425
2450					2'-8" x 7'-0"	1,100	300	1,400
2500					3'-0" x 7'-0"	1,125	300	1,425
2550			double door		2'-6" x 6'-8"	2,100	550	2,650
2600					2'-6" x 7'-0"	2,100	550	2,650
2650					2'-8" x 7'-0"	2,075	550	2,625
2700					3'-0" x 7'-0"	1,950	480	2,430
2750	Wood	combination	storm & screen		3'-0" x 6'-8"	335	74	409
2800					3'-0" x 7'-0"	370	81.50	451.50
2850		overhead	panels, H.D.	manual oper.	8'-0" x 8'-0"	1,125	555	1,680
2900					10'-0" x 10'-0"	1,625	615	2,240
2950					12'-0" x 12'-0"	2,225	740	2,965
3000					14'-0" x 14'-0"	3,550	850	4,400
3050					20'-0" x 16'-0"	5,850	1,700	7,550
3100				electric oper.	8'-0" x 8'-0"	2,150	830	2,980
3150					10'-0" x 10'-0"	2,650	890	3,540
3200					12'-0" x 12'-0"	3,250	1,025	4,275
3250					14'-0" x 14'-0"	4,575	1,125	5,700
3300					20'-0" x 16'-0"	7,050	2,250	9,300

B20 Exterior Enclosure

B2030 Exterior Doors

B2030 220 Steel Doors

	MATERIAL	TYPE	DOORS	SPECIFICATION	OPENING	COST PER OPNG. MAT.	INST.	TOTAL
3350	Steel 18 Ga.	hollow metal	1 door w/frame	no label	2'-6" x 7'-0"	1,600	310	1,910
3400					2'-8" x 7'-0"	1,600	310	1,910
3450					3'-0" x 7'-0"	1,600	310	1,910
3500					3'-6" x 7'-0"	1,650	325	1,975
3550			1 door w/frame	no label	4'-0" x 8'-0"	1,950	330	2,280
3600			2 doors w/frame	no label	5'-0" x 7'-0"	3,125	575	3,700
3650					5'-4" x 7'-0"	3,125	575	3,700
3700					6'-0" x 7'-0"	3,125	575	3,700
3750					7'-0" x 7'-0"	3,225	610	3,835
3800					8'-0" x 8'-0"	3,775	610	4,385
3850			1 door w/frame	"A" label	2'-6" x 7'-0"	1,925	375	2,300
3900					2'-8" x 7'-0"	1,925	380	2,305
3950					3'-0" x 7'-0"	1,925	380	2,305
4000					3'-6" x 7'-0"	2,075	390	2,465
4050					4'-0" x 8'-0"	2,275	405	2,680
4100			2 doors w/frame		5'-0" x 7'-0"	3,700	695	4,395
4150					5'-4" x 7'-0"	3,700	705	4,405
4200					6'-0" x 7'-0"	3,700	705	4,405
4250					7'-0" x 7'-0"	4,025	720	4,745
4300					8'-0" x 8'-0"	4,025	720	4,745
4350	Steel 24 Ga.	overhead	sectional	manual oper.	8'-0" x 8'-0"	890	555	1,445
4400					10'-0" x 10'-0"	1,200	615	1,815
4450					12'-0" x 12'-0"	1,450	740	2,190
4500					20'-0" x 14'-0"	4,100	1,575	5,675
4550				electric oper.	8'-0" x 8'-0"	1,925	830	2,755
4600					10'-0" x 10'-0"	2,225	890	3,115
4650					12'-0" x 12'-0"	2,475	1,025	3,500
4700					20'-0" x 14'-0"	5,300	2,125	7,425
4750	Steel		rolling	manual oper.	8'-0" x 8'-0"	1,225	895	2,120
4800					10'-0" x 10'-0"	2,025	1,025	3,050
4850					12'-0" x 12'-0"	2,050	1,200	3,250
4900					14'-0" x 14'-0"	3,325	1,775	5,100
4950					20'-0" x 12'-0"	2,325	1,600	3,925
5000					20'-0" x 16'-0"	3,825	2,375	6,200
5050				electric oper.	8'-0" x 8'-0"	2,400	1,175	3,575
5100					10'-0" x 10'-0"	3,200	1,300	4,500
5150					12'-0" x 12'-0"	3,225	1,475	4,700
5200					14'-0" x 14'-0"	4,500	2,050	6,550
5250					20'-0" x 12'-0"	3,475	1,875	5,350
5300					20'-0" x 16'-0"	4,975	2,650	7,625
5350				fire rated	10'-0" x 10'-0"	2,250	1,300	3,550
5400			rolling grille	manual oper.	10'-0" x 10'-0"	2,750	1,425	4,175
5450					15'-0" x 8'-0"	3,175	1,775	4,950
5500		vertical lift	1 door w/frame	motor operator	16'-0" x 16'-0"	23,700	5,525	29,225
5550					32'-0" x 24'-0"	51,000	3,675	54,675

B20 Exterior Enclosure

B2030 Exterior Doors

B2030 230 Aluminum Doors

	MATERIAL	TYPE	DOORS	SPECIFICATION	OPENING	COST PER OPNG.		
						MAT.	INST.	TOTAL
6000	Aluminum	combination	storm & screen	hinged	3'-0" x 6'-8"	330	79	409
6050					3'-0" x 7'-0"	365	87	452
6100		overhead	rolling grille	manual oper.	12'-0" x 12'-0"	4,400	2,500	6,900
6150				motor oper.	12'-0" x 12'-0"	5,775	2,800	8,575
6200	Alum. & Fbrgls.		heavy duty	manual oper.	12'-0" x 12'-0"	2,975	740	3,715
6250				electric oper.	12'-0" x 12'-0"	4,000	1,025	5,025

B30 Roofing

B3010 Roof Coverings

Multiple ply roofing is the most popular covering for minimum pitch roofs. Lines 1200 through 6300 list the costs of the various types, plies and weights per S.F.

System Components	QUANTITY	UNIT	COST PER S.F. MAT.	INST.	TOTAL
SYSTEM B3010 105 2500					
ASPHALT FLOOD COAT, W/GRAVEL, 4 PLY ORGANIC FELT					
Organic #30 base felt	1.000	S.F.	.11	.09	.20
Organic #15 felt, 3 plies	3.000	S.F.	.17	.27	.44
Asphalt mopping of felts	4.000	S.F.	.48	.82	1.30
Asphalt flood coat	1.000	S.F.	.31	.65	.96
Gravel aggregate, washed river stone	4.000	Lb.	.07	.15	.22
TOTAL			1.14	1.98	3.12

B3010 105	Built-Up	COST PER S.F. MAT.	INST.	TOTAL
1200	Asphalt flood coat w/gravel; not incl. insul, flash., nailers			
1300				
1400	Asphalt base sheets & 3 plies #15 asphalt felt, mopped	1.08	1.79	2.87
1500	On nailable deck	1.09	1.87	2.96
1600	4 plies #15 asphalt felt, mopped	1.45	1.96	3.41
1700	On nailable deck	1.28	2.07	3.35
1800	Coated glass base sheet, 2 plies glass (type IV), mopped	1.14	1.79	2.93
1900	For 3 plies	1.37	1.96	3.33
2000	On nailable deck	1.28	2.07	3.35
2300	4 plies glass fiber felt (type IV), mopped	1.68	1.96	3.64
2400	On nailable deck	1.51	2.07	3.58
2500	Organic base sheet & 3 plies #15 organic felt, mopped	1.14	1.98	3.12
2600	On nailable deck	1.12	2.07	3.19
2700	4 plies #15 organic felt, mopped	1.40	1.79	3.19
2750				
2800	Asphalt flood coat, smooth surface, not incl. insul, flash., nailers			
2850				
2900	Asphalt base sheet & 3 plies #15 asphalt felt, mopped	1.12	1.64	2.76
3000	On nailable deck	1.03	1.71	2.74
3100	Coated glass fiber base sheet & 2 plies glass fiber felt, mopped	1.09	1.58	2.67
3200	On nailable deck	1.03	1.64	2.67
3300	For 3 plies, mopped	1.32	1.71	3.03
3400	On nailable deck	1.23	1.79	3.02
3700	4 plies glass fiber felt (type IV), mopped	1.55	1.71	3.26
3800	On nailable deck	1.46	1.79	3.25
3900	Organic base sheet & 3 plies #15 organic felt, mopped	1.15	1.64	2.79
4000	On nailable decks	1.07	1.71	2.78
4100	4 plies #15 organic felt, mopped	1.35	1.79	3.14
4200	Coal tar pitch with gravel surfacing			
4300	4 plies #15 tarred felt, mopped	2.56	1.87	4.43
4400	3 plies glass fiber felt (type IV), mopped	2.11	2.07	4.18
4500	Coated glass fiber base sheets 2 plies glass fiber felt, mopped	2.13	2.07	4.20

B30 Roofing

B3010 Roof Coverings

B3010 105	Built-Up	COST PER S.F.		
		MAT.	INST.	TOTAL
4600	On nailable decks	1.85	2.18	4.03
4800	3 plies glass fiber felt (type IV), mopped	2.89	1.87	4.76
4900	On nailable decks	2.61	1.96	4.57
5300	Asphalt mineral surface, roll roofing			
5400	1 ply #15 organic felt, 1 ply mineral surfaced			
5500	Selvage roofing, lap 19", nailed and mopped	.80	1.46	2.26
5600	3 plies glass fiber felt (type IV), 1 ply mineral surfaced			
5700	Selvage roofing, lapped 19", mopped	1.31	1.58	2.89
5800	Coated glass fiber base sheet			
5900	2 plies glass fiber, felt (type IV), 1 ply mineral surfaced			
6000	Selvage, roofing, lapped 19", mopped	1.41	1.58	2.99
6100	On nailable deck	1.28	1.64	2.92
6200	3 plies glass fiber felt (type IV), 1 ply mineral surfaced			
6300	Selvage roofing, lapped 19", mopped	1.31	1.58	2.89

B30 Roofing

B3010 Roof Coverings

Fully Adhered

Ballasted

The systems listed below reflect only the cost for the single ply membrane.

B3010 120	Single Ply Membrane	MAT.	INST.	TOTAL
1000	CSPE (Chlorosulfonated polyethylene), 35 mils, fully adhered	2.14	.98	3.12
1100	Loosely laid with stone ballast	2.23	.50	2.73
2000	EPDM (Ethylene propylene diene monomer), 45 mils, fully adhered	1.25	.98	2.23
2100	Loosely laid with stone ballast	.92	.50	1.42
2200	Mechanically fastened with batten strips	.85	.73	1.58
3300	60 mils, fully adhered	1.41	.98	2.39
3400	Loosely laid with stone ballast	1.09	.50	1.59
3500	Mechanically fastened with batten strips	1.01	.73	1.74
4000	Modified bit., SBS modified, granule surface cap sheet, mopped, 150 mils	.72	1.96	2.68
4100	Smooth surface cap sheet, mopped, 145 mils	.86	1.87	2.73
4500	APP modified, granule surface cap sheet, torched, 180 mils	.98	1.28	2.26
4600	Smooth surface cap sheet, torched, 170 mils	.77	1.21	1.98
5000	PIB (Polyisobutylene), 100 mils, fully adhered with contact cement	2.66	.98	3.64
5100	Loosely laid with stone ballast	2.05	.50	2.55
5200	Partially adhered with adhesive	2.57	.73	3.30
5300	Hot asphalt attachment	2.46	.73	3.19
6000	Reinforced PVC, 48 mils, loose laid and ballasted with stone	1.24	.50	1.74
6100	Partially adhered with mechanical fasteners	1.17	.73	1.90
6200	Fully adhered with adhesive	1.55	.98	2.53
6300	Reinforced PVC, 60 mils, loose laid and ballasted with stone	1.27	.50	1.77
6400	Partially adhered with mechanical fasteners	1.20	.73	1.93
6500	Fully adhered with adhesive	1.58	.98	2.56

B30 Roofing

B3010 Roof Coverings

Listed below are installed prices for preformed roofing materials with applicable specifications for each material.

B3010 130	Preformed Metal Roofing	COST PER S.F.		
		MAT.	INST.	TOTAL
0200	Corrugated roofing, aluminum, mill finish, .0175" thick, .272 P.S.F.	1.09	1.81	2.90
0250	.0215" thick, .334 P.S.F.	1.43	1.81	3.24
0300	.0240" thick, .412 P.S.F.	2.04	1.81	3.85
0350	.0320" thick, .552 P.S.F.	2.63	1.81	4.44
0400	Painted, .0175" thick, .280 P.S.F.	1.58	1.81	3.39
0450	.0215" thick, .344 P.S.F.	1.76	1.81	3.57
0500	.0240" thick, .426 P.S.F.	2.48	1.81	4.29
0550	.0320" thick, .569 P.S.F.	3.37	1.81	5.18
0700	Fiberglass, 6 oz., .375 P.S.F.	2.29	2.17	4.46
0750	8 oz., .5 P.S.F.	3.91	2.17	6.08
0900	Steel, galvanized, 29 ga., .72 P.S.F.	1.98	1.98	3.96
0930	26 ga., .91 P.S.F.	2.11	2.07	4.18
0950	24 ga., 1.26 P.S.F.	2.18	2.17	4.35
0970	22 ga., 1.45 P.S.F.	3.43	2.29	5.72
1000	Colored, 26 ga., 1.08 P.S.F.	1.98	2.07	4.05
1050	24 ga., 1.43 P.S.F.	2.34	2.17	4.51

B30 Roofing

B3010 Roof Coverings

Batten Seam

Flat Seam

Standing Seam

Formed metal roofing is practical on all sloped roofs from 1/4" per foot of rise to vertical. Its use is more aesthetic than economical. Table below lists the various materials used and the weight per S.F.

System Components	QUANTITY	UNIT	COST PER S.F.		
			MAT.	INST.	TOTAL
SYSTEM B3010 135 1000					
BATTEN SEAM, FORMED COPPER ROOFING 3" MIN. SLOPE, 16 OZ., 1.2 PSF.					
Copper batten seam roofing, over 10 sq, 16 oz., 130 lb per sq	1.000	S.F.	13.25	5.90	19.15
Asphalt impregnated felt, 30 lb, 2 sq. per roll, not mopped	1.000	S.F.	.11	.09	.20
TOTAL			13.36	5.99	19.35

B3010 135	Formed Metal	COST PER S.F.		
		MAT.	INST.	TOTAL
1000	Batten seam, formed copper roofing, 3" min slope, 16 oz., 1.2 P.S.F.	13.35	6	19.35
1100	18 oz., 1.35 P.S.F.	14.85	6.60	21.45
1200	20 oz., 1.50 P.S.F.	17.35	6.60	23.95
1500	Lead, 3" min slope, 3 Lb., 3.6 P.S.F.	13.10	5.50	18.60
2000	Zinc copper alloy, 3" min slope, .020" thick, .88 P.S.F.	12.85	5.50	18.35
2050	.027" thick, 1.18 P.S.F.	16.10	5.75	21.85
2100	.032" thick, 1.41 P.S.F.	20.50	6	26.50
2150	.040" thick, 1.77 P.S.F.	25.50	6.30	31.80
3000	Flat seam, copper, 1/4" min. slope, 16 oz., 1.2 P.S.F.	9.75	5.50	15.25
3100	18 oz., 1.35 P.S.F.	10.85	5.75	16.60
3200	20 oz., 1.50 P.S.F.	12.60	6	18.60
3500	Lead, 1/4" min. slope, 3 Lb. 3.6 P.S.F.	11.10	5.10	16.20
4000	Zinc copper alloy, 1/4" min. slope, .020" thick, .84 P.S.F.	11.10	5.50	16.60
4050	.027" thick, 1.13 P.S.F.	13.85	5.75	19.60
4100	.032" thick, 1.33 P.S.F.	17.60	5.90	23.50
4150	.040" thick, 1.67 P.S.F.	22	6.30	28.30
5000	Standing seam, copper, 2-1/2" min. slope, 16 oz., 1.25 P.S.F.	10.60	5.10	15.70
5100	18 oz., 1.40 P.S.F.	11.85	5.50	17.35
5200	20 oz., 1.56 P.S.F.	13.60	6	19.60
5500	Lead, 2-1/2" min. slope, 3 Lb., 3.6 P.S.F.	12.60	5.10	17.70
6000	Zinc copper alloy, 2-1/2" min. slope, .020" thick, .87 P.S.F.	12.35	5.50	17.85
6050	.027" thick, 1.18 P.S.F.	15.35	5.75	21.10
6100	.032" thick, 1.39 P.S.F.	19.60	6	25.60
6150	.040" thick, 1.74 P.S.F.	24.50	6.30	30.80

B30 Roofing

B3010 Roof Coverings

Shingles and tiles are practical in applications where the roof slope is more than 3-1/2" per foot of rise. Table below lists the various materials and the weight per S.F.

System Components	QUANTITY	UNIT	COST PER S.F. MAT.	COST PER S.F. INST.	COST PER S.F. TOTAL
SYSTEM B3010 140 1200					
STRIP SHINGLES, 4" SLOPE, MULTI-LAYERED, CLASS A					
Class A, 260-300 lb./sq., 4 bundles/sq.	1.000	S.F.	1.69	1.47	3.16
Building paper, asphalt felt sheathing paper 15 lb	1.000	S.F.	.06	.15	.21
TOTAL			1.75	1.62	3.37

B3010 140	Shingle & Tile	MAT.	INST.	TOTAL
1095	Asphalt roofing			
1100	Strip shingles, 4" slope, inorganic class A 210-235 lb./sq.	.95	1.09	2.04
1150	Organic, class C, 235-240 lb./sq.	1.19	1.18	2.37
1200	Premium laminated multi-layered, class A, 260-300 lb./sq.	1.75	1.62	3.37
1250	Class C, 300-385 lb./sq.	2.22	1.87	4.09
1545	Metal roofing			
1550	Alum., shingles, colors, 3" min slope, .019" thick, 0.4 PSF	2.76	1.20	3.96
1600	.020" thick, 0.6 PSF	2.77	1.20	3.97
1850	Steel, colors, 3" min slope, 26 gauge, 1.0 PSF	4.02	2.45	6.47
2000	24 gauge, 1.7 PSF	4.12	2.45	6.57
2795	Slate roofing			
2800	Slate roofing, 4" min. slope, shingles, 3/16" thick, 8.0 PSF	5.35	3.05	8.40
2900	1/4" thick, 9.0 PSF	5.35	3.05	8.40
3495	Wood roofing			
3500	4" min slope, cedar shingles, 16" x 5", 5" exposure 1.6 PSF	2.82	2.38	5.20
4000	Shakes, 18", 8-1/2" exposure, 2.8 PSF	2.33	2.94	5.27
4100	24", 10" exposure, 3.5 PSF	2.63	2.38	5.01
5095	Tile roofing			
5100	Aluminum, mission, 3" min slope, .019" thick, 0.65 PSF	9.05	2.30	11.35
5200	S type, 0.48 PSF	6.25	1.93	8.18
6000	Clay, Americana, 3" minimum slope, 8 PSF	5.40	2.68	8.08
6500	Mission, 4" minimum slope, 12.2 PSF	9	2.92	11.92

B30 Roofing

B3010 Roof Coverings

B3010 320	Roof Deck Rigid Insulation		COST PER S.F.		
			MAT.	INST.	TOTAL
0100	Fiberboard low density, 1/2" thick, R1.39	RB3010 -010			
0150	1" thick R2.78		.55	.70	1.25
0300	1 1/2" thick R4.17		.84	.70	1.54
0350	2" thick R5.56		1.11	.70	1.81
0370	Fiberboard high density, 1/2" thick R1.3		.29	.61	.90
0380	1" thick R2.5		.57	.70	1.27
0390	1 1/2" thick R3.8		.86	.70	1.56
0410	Fiberglass, 3/4" thick R2.78		.62	.61	1.23
0450	15/16" thick R3.70		.81	.61	1.42
0500	1-1/16" thick R4.17		1.02	.61	1.63
0550	1-5/16" thick R5.26		1.39	.61	2
0600	2-1/16" thick R8.33		1.46	.70	2.16
0650	2 7/16" thick R10		1.69	.70	2.39
1260	Perlite, 1/2" thick R1.32		.34	.59	.93
1300	3/4" thick R2.08		.39	.70	1.09
1350	1" thick R2.78		.53	.70	1.23
1400	1 1/2" thick R4.17		.79	.70	1.49
1450	2" thick R5.56		1.06	.78	1.84
1510	Polyisocyanurate 2#/CF density, 1" thick		.55	.49	1.04
1550	1 1/2" thick		.70	.53	1.23
1600	2" thick		.87	.57	1.44
1650	2 1/2" thick		1.11	.59	1.70
1700	3" thick		1.34	.61	1.95
1750	3 1/2" thick		2.15	.61	2.76
1800	Tapered for drainage		.63	.49	1.12
1810	Expanded polystyrene, 1#/CF density, 3/4" thick R2.89		.20	.47	.67
1820	2" thick R7.69		.53	.53	1.06
1825	Extruded Polystyrene				
1830	Extruded polystyrene, 15 PSI compressive strength, 1" thick, R5		.53	.47	1
1835	2" thick R10		.68	.53	1.21
1840	3" thick R15		1.38	.61	1.99
1900	25 PSI compressive strength, 1" thick R5		.73	.47	1.20
1950	2" thick R10		1.38	.53	1.91
2000	3" thick R15		2.10	.61	2.71
2050	4" thick R20		3	.61	3.61
2150	Tapered for drainage		.64	.47	1.11
2550	40 PSI compressive strength, 1" thick R5		.56	.47	1.03
2600	2" thick R10		1.06	.53	1.59
2650	3" thick R15		1.54	.61	2.15
2700	4" thick R20		2.04	.61	2.65
2750	Tapered for drainage		.80	.49	1.29
2810	60 PSI compressive strength, 1" thick R5		.78	.48	1.26
2850	2" thick R10		1.50	.54	2.04
2900	Tapered for drainage		1.02	.49	1.51
4000	Composites with 1-1/2" polyisocyanurate				
4010	1" fiberboard		1.24	.70	1.94
4020	1" perlite		1.11	.68	1.79
4030	7/16" oriented strand board		.98	.70	1.68

B30 Roofing

B3010 Roof Coverings

The table below lists the costs for various types of base flashing per L.F.

All the systems below are also based on using a 4" x 4" diagonally cut treated wood cant and a 2" x 8" treated wood nailer fastened at 4'-0" O.C.

System Components	QUANTITY	UNIT	COST PER L.F. MAT.	INST.	TOTAL
SYSTEM B3010 410 1000					
ALUMINUM BASE, FLASHING MILL FINISH, .019" THICK					
Treated 2" x 8" wood blocking	1.333	B.F.	.94	1.70	2.64
Anchor bolts, 4' O.C.	.250	Ea.	1.47	1.53	3
Treated wood cant, 4" x 4" split	1.000	L.F.	2.09	1.58	3.67
Roof covering, splice sheet	1.000	S.F.	2.23	.50	2.73
Base flashing, .019" thick, aluminum	1.000	L.F.	1.18	3.55	4.73
Reglet .025" thick counter flashing .032" thick	1.000	L.F.	1.95	2.46	4.41
Counter flashing, aluminum	1.000	L.F.	2.43	4.34	6.77
TOTAL			12.29	15.66	27.95

B3010 410			Base Flashing					
	TYPE	DESCRIPTION	SPECIFICATION	REGLET	COUNTER FLASHING	COST PER L.F. MAT.	INST.	TOTAL
1000	Aluminum	mill finish	.019" thick	alum. .025	alum. .032	12.30	15.65	27.95
1100			.050" thick	.025	.032	13.60	15.65	29.25
1200		fabric, 2 sides	.004" thick	.025	.032	12.65	13.70	26.35
1300			.016" thick	.025	.032	12.95	13.70	26.65
1400		mastic, 2 sides	.004" thick	.025	.032	12.65	13.70	26.35
1500			.016" thick	.025	.032	13.15	13.70	26.85
1700	Copper	sheets, plain	16 oz.	copper 16 oz.	copper 16 oz.	27	16.60	43.60
1800			20 oz.	16 oz.	16 oz.	29.50	16.80	46.30
1900			24 oz.	16 oz.	16 oz.	32	17	49
2000			32 oz.	16 oz.	16 oz.	36.50	17.25	53.75
2200	Sheets, Metal	galvanized	20 Ga.	galv. 24 Ga.	galv. 24 Ga.	11.95	36.50	48.45
2300			24 Ga.	24 Ga.	24 Ga.	11.20	29.50	40.70
2400			30 Ga.	24 Ga.	24 Ga.	10.45	22	32.45
2600	Rubber	butyl	1/32" thick	galv. 24 Ga.	galv. 24 Ga.	11.55	13.95	25.50
2700			1/16" thick	24 Ga.	24 Ga.	12.30	13.95	26.25
2800		neoprene	1/16" thick	24 Ga.	24 Ga.	11.90	13.95	25.85
2900			1/8" thick	24 Ga.	24 Ga.	14.25	13.95	28.20

B30 Roofing

B3010 Roof Coverings

The table below lists the costs for various types of roof perimeter edge treatments.

All systems below include the cost per L.F. for a 2" x 8" treated wood nailer fastened at 4'-0" O.C. and a diagonally cut 4" x 6" treated wood cant.

Roof edge and base flashing are assumed to be made from the same material.

System Components	QUANTITY	UNIT	COST PER L.F.		
			MAT.	INST.	TOTAL
SYSTEM B3010 420 1000					
ALUMINUM ROOF EDGE, MILL FINISH, .050" THICK, 4" HIGH					
Treated 2" x 8" wood blocking	1.333	B.F.	.94	1.70	2.64
Anchor bolts, 4' O.C.	.250	Ea.	1.47	1.53	3
Treated wood cant	1.000	L.F.	2.09	1.58	3.67
Roof covering, splice sheet	1.000	S.F.	2.23	.50	2.73
Aluminum roof edge, .050" thick, 4" high	1.000	L.F.	6.60	4.49	11.09
TOTAL			13.33	9.80	23.13

B3010 420 Roof Edges

	EDGE TYPE	DESCRIPTION	SPECIFICATION	FACE HEIGHT		COST PER L.F.		
						MAT.	INST.	TOTAL
1000	Aluminum	mill finish	.050" thick	4"		13.35	9.80	23.15
1100				6"		13.85	10.15	24
1200				8"		14.75	10.50	25.25
1300		duranodic	.050" thick	4"		13.90	9.80	23.70
1400				6"		14.40	10.15	24.55
1500				8"		15.55	10.50	26.05
1600		painted	.050" thick	4"		14.25	9.80	24.05
1700				6"		15.65	10.15	25.80
1800				8"		15.70	10.50	26.20
2000	Copper	plain	16 oz.	4"		49.50	9.80	59.30
2100				6"		49.50	10.15	59.65
2200				8"		64	11.75	75.75
2300			20 oz.	4"		64	11.30	75.30
2400				6"		56.50	10.90	67.40
2500				8"		72.50	12.70	85.20
2700	Sheet Metal	galvanized	20 Ga.	4"		16.20	11.75	27.95
2800				6"		16.20	11.75	27.95
2900				8"		19.35	13.90	33.25
3000			24 Ga.	4"		13.35	9.80	23.15
3100				6"		13.35	9.80	23.15
3200				8"		15.55	11.30	26.85

B30 Roofing

B3010 Roof Coverings

B3010 430 Flashing

	MATERIAL	BACKING	SIDES	SPECIFICATION	QUANTITY	COST PER S.F. MAT.	COST PER S.F. INST.	COST PER S.F. TOTAL
0040	Aluminum	none		.019"		1.18	3.55	4.73
0050				.032"		1.38	3.55	4.93
0100				.040"		2.35	3.55	5.90
0150				.050"		2.48	3.55	6.03
0300		fabric	2	.004"		1.55	1.56	3.11
0350				.016"		1.84	1.56	3.40
0400		mastic		.004"		1.55	1.56	3.11
0450				.005"		1.84	1.56	3.40
0500				.016"		2.02	1.56	3.58
0510								
0700	Copper	none		16 oz.	<500 lbs.	8.35	4.48	12.83
0710					>2000 lbs.	8.35	3.32	11.67
0750				20 oz.	<500 lbs.	10.85	4.68	15.53
0760					>2000 lbs.	10.30	3.55	13.85
0800				24 oz.	<500 lbs.	13.35	4.90	18.25
0810					>2000 lbs.	12.65	3.81	16.46
0850				32 oz.	<500 lbs.	17.80	5.15	22.95
0860					>2000 lbs.	16.90	3.96	20.86
1000		paper backed	1	2 oz.		1.84	1.56	3.40
1100				3 oz.		2.17	1.56	3.73
1200			2	2 oz.		1.89	1.56	3.45
1300				5 oz.		2.96	1.56	4.52
3000	Lead	none		2.5 lb.	to 12" wide	5.50	3.81	9.31
3100					over 12"	4.39	3.81	8.20
3500	PVC black			.010"		.23	1.81	2.04
3600				.020"		.31	1.81	2.12
3700				.030"		.39	1.81	2.20
3800				.056"		.94	1.81	2.75
4000	Rubber butyl			1/32"		2.11	1.81	3.92
4100				1/16"		2.86	1.81	4.67
4200	Neoprene			1/16"		2.45	1.81	4.26
4300				1/8"		4.84	1.81	6.65
4500	Stainless steel			.015"	<500 lbs.	6.40	4.48	10.88
4600	Copper clad				>2000 lbs.	6.10	3.32	9.42
4700				.018"	<500 lbs.	7.20	5.15	12.35
4800					>2000 lbs.	6.85	3.55	10.40
5000	Plain			32 ga.		4.40	3.32	7.72
5100				28 ga.		6.20	3.32	9.52
5200				26 ga.		8	3.32	11.32
5300				24 ga.		9.80	3.32	13.12
5400	Terne coated			28 ga.		8.10	3.32	11.42
5500				26 ga.		9.10	3.32	12.42

B30 Roofing

B3010 Roof Coverings

B3010 610 Gutters

	SECTION	MATERIAL	THICKNESS	SIZE	FINISH	COST PER L.F.		
						MAT.	INST.	TOTAL
0050	Box	aluminum	.027"	5"	enameled	2.84	5.40	8.24
0100					mill	2.91	5.40	8.31
0200			.032"	5"	enameled	3.24	5.40	8.64
0300					mill	3.02	5.40	8.42
0500		copper	16 Oz.	4"	lead coated	21	5.40	26.40
0600					mill	13.85	5.40	19.25
0700				5"	lead coated	19.20	5.40	24.60
0800					mill	9.80	5.40	15.20
1000		steel galv.	28 Ga.	5"	enameled	2.22	5.40	7.62
1100					mill	2.40	5.40	7.80
1200			26 Ga.	5"	mill	2.40	5.40	7.80
1300				6"	mill	2.98	5.40	8.38
1500		stainless steel		4"	mill	7.70	5.40	13.10
1600		stainless		5"	mill	9.80	5.40	15.20
1800		vinyl		4"	colors	1.28	5.05	6.33
1900				5"	colors	1.71	5.05	6.76
2200		wood clear,		3"x4"	treated	11	5.55	16.55
2300		hemlock or fir		4"x5"	treated	14.85	5.55	20.40
3000	Half round	copper	16 Oz.	4"	lead coated	15.20	5.40	20.60
3100					mill	10.75	5.40	16.15
3200				5"	lead coated	17.50	6.20	23.70
3300					mill	9.20	5.40	14.60
3600		steel galv.	28 Ga.	5"	enameled	2.22	5.40	7.62
3700					mill	2.40	5.40	7.80
3800			26 Ga.	5"	mill	2.40	5.40	7.80
4000		stainless steel		4"	mill	7.70	5.40	13.10
4100				5"	mill	9.80	5.40	15.20
5000		vinyl		4"	white	1.19	5.05	6.24

B3010 620 Downspouts

	MATERIALS	SECTION	SIZE	FINISH	THICKNESS	COST PER V.L.F.		
						MAT.	INST.	TOTAL
0100	Aluminum	rectangular	2"x3"	embossed mill	.020"	1.07	3.42	4.49
0150				enameled	.020"	1.45	3.42	4.87
0200				enameled	.024"	2.26	3.61	5.87
0250			3"x4"	enameled	.024"	3.17	4.65	7.82
0300		round corrugated	3"	enameled	.020"	1.89	3.42	5.31
0350			4"	enameled	.025"	2.92	4.65	7.57
0500	Copper	rectangular corr.	2"x3"	mill	16 Oz.	9.85	3.42	13.27
0550				lead coated	16 Oz.	14.85	3.42	18.27
0600		smooth		mill	16 Oz.	16.65	3.42	20.07
0650				lead coated	16 Oz.	26.50	3.42	29.92
0700		rectangular corr.	3"x4"	mill	16 Oz.	12.95	4.49	17.44
0750				lead coated	16 Oz.	19.50	3.89	23.39
0800		smooth		mill	16 Oz.	18.60	4.49	23.09
0850				lead coated	16 Oz.	22.50	4.49	26.99
0900		round corrugated	2"	mill	16 Oz.	10.90	3.52	14.42
0950		smooth		lead coated	16 Oz.	24.50	3.42	27.92
1000		corrugated	3"	mill	16 Oz.	9.75	3.54	13.29
1050		smooth		lead coated	16 Oz.	24	3.42	27.42

B30 Roofing

B3010 Roof Coverings

B3010 620 Downspouts

	MATERIALS	SECTION	SIZE	FINISH	THICKNESS	COST PER V.L.F.		
						MAT.	INST.	TOTAL
1100	Copper	corrugated	4"	mill	16 Oz.	12.65	4.67	17.32
1150		smooth		lead coated	16 Oz.	24.50	4.49	28.99
1200		corrugated	5"	mill	16 Oz.	22.50	5.25	27.75
1250				lead coated	16 Oz.	26.50	5	31.50
1300	Steel	rectangular corr.	2"x3"	galvanized	28 Ga.	2.13	3.42	5.55
1350				epoxy coated	24 Ga.	2.42	3.42	5.84
1400		smooth		galvanized	28 Ga.	3.33	3.42	6.75
1450		rectangular corr.	3"x4"	galvanized	28 Ga.	2	4.49	6.49
1500				epoxy coated	24 Ga.	3.03	4.49	7.52
1550		smooth		galvanized	28 Ga.	3.37	4.49	7.86
1600		round corrugated	2"	galvanized	28 Ga.	2.29	3.42	5.71
1650			3"	galvanized	28 Ga.	2.29	3.42	5.71
1700			4"	galvanized	28 Ga.	2.87	4.49	7.36
1750			5"	galvanized	28 Ga.	4.04	5	9.04
1800				galvanized	26 Ga.	4.04	5	9.04
1850			6"	galvanized	28 Ga.	4.62	6.20	10.82
1900				galvanized	26 Ga.	6.40	6.20	12.60
2500	Stainless steel	rectangular	2"x3"	mill		41.50	3.42	44.92
2550	Tubing sch.5		3"x4"	mill		53	4.49	57.49
2600			4"x5"	mill		109	4.82	113.82
2650		round	3"	mill		41.50	3.42	44.92
2700			4"	mill		53	4.49	57.49
2750			5"	mill		109	4.82	113.82

B3010 630 Gravel Stop

	MATERIALS	SECTION	SIZE	FINISH	THICKNESS	COST PER L.F.		
						MAT.	INST.	TOTAL
5100	Aluminum	extruded	4"	mill	.050"	6.60	4.49	11.09
5200			4"	duranodic	.050"	7.15	4.49	11.64
5300			8"	mill	.050"	8	5.20	13.20
5400			8"	duranodic	.050"	8.80	5.20	14
5500			12"-2 pc.	mill	.050"	11.20	6.50	17.70
6000			12"-2 pc.	duranodic	.050"	11.05	6.50	17.55
6100	Stainless	formed	6"	mill	24 Ga.	10.80	4.82	15.62
6200			12"	mill	24 Ga.	21.50	6.50	28

B30 Roofing

B3020 Roof Openings

Roof Hatch

Smoke Hatch

Skylight

B3020 110	Skylights	COST PER S.F.		
		MAT.	INST.	TOTAL
5100	Skylights, plastic domes, insul curbs, nom. size to 10 S.F., single glaze	29	13.60	42.60
5200	Double glazing	33	16.75	49.75
5300	10 S.F. to 20 S.F., single glazing	28	5.50	33.50
5400	Double glazing	28	6.90	34.90
5500	20 S.F. to 30 S.F., single glazing	21.50	4.68	26.18
5600	Double glazing	26.50	5.50	32
5700	30 S.F. to 65 S.F., single glazing	19.70	3.56	23.26
5800	Double glazing	27	4.68	31.68
6000	Sandwich panels fiberglass, 1-9/16" thick, 2 S.F. to 10 S.F.	18.90	10.85	29.75
6100	10 S.F. to 18 S.F.	17.05	8.20	25.25
6200	2-3/4" thick, 25 S.F. to 40 S.F.	27.50	7.35	34.85
6300	40 S.F. to 70 S.F.	22.50	6.60	29.10

B3020 210	Hatches	COST PER OPNG.		
		MAT.	INST.	TOTAL
0200	Roof hatches with curb and 1" fiberglass insulation, 2'-6"x3'-0", aluminum	1,175	217	1,392
0300	Galvanized steel 165 lbs.	705	217	922
0400	Primed steel 164 lbs.	695	217	912
0500	2'-6"x4'-6" aluminum curb and cover, 150 lbs.	1,000	242	1,242
0600	Galvanized steel 220 lbs.	975	242	1,217
0650	Primed steel 218 lbs.	980	242	1,222
0800	2'x6"x8'-0" aluminum curb and cover, 260 lbs.	1,975	330	2,305
0900	Galvanized steel, 360 lbs.	1,975	330	2,305
0950	Primed steel 358 lbs.	1,575	330	1,905
1200	For plexiglass panels, add to the above	485		485
2100	Smoke hatches, unlabeled not incl. hand winch operator, 2'-6"x3', galv	855	263	1,118
2200	Plain steel, 160 lbs.	840	263	1,103
2400	2'-6"x8'-0", galvanized steel, 360 lbs.	2,150	360	2,510
2500	Plain steel, 350 lbs.	1,725	360	2,085
3000	4'-0"x8'-0", double leaf low profile, aluminum cover, 359 lb.	3,075	272	3,347
3100	Galvanized steel 475 lbs.	2,625	272	2,897
3200	High profile, aluminum cover, galvanized curb, 361 lbs.	2,950	272	3,222

C Interiors

No part of this publication may be reproduced, stored in a retrieval system, or transmitted in any form or by any means without prior written permission of RSMeans, a division of Reed Construction Data.

C10 Interior Construction

C1010 Partitions

The Concrete Block Partition Systems are defined by weight and type of block, thickness, type of finish and number of sides finished. System components include joint reinforcing on alternate courses and vertical control joints.

System Components	QUANTITY	UNIT	COST PER S.F.		
			MAT.	INST.	TOTAL
SYSTEM C1010 102 1020					
CONC. BLOCK PARTITION, 8" X 16", 4" TK., 2 CT. GYP. PLASTER, 2 SIDES					
Conc. block partition, 4" thick	1.000	S.F.	1.65	5.80	7.45
Control joint	.050	L.F.	.13	.14	.27
Horizontal joint reinforcing	.800	L.F.	.07	.07	.14
Gypsum plaster, 2 coat, on masonry	2.000	S.F.	.76	5.57	6.33
TOTAL			2.61	11.58	14.19

C1010 102		Concrete Block Partitions - Regular Weight					
	TYPE	THICKNESS (IN.)	TYPE FINISH	SIDES FINISHED		COST PER S.F.		
						MAT.	INST.	TOTAL
1000	Hollow	4	none	0		1.85	6	7.85
1010			gyp. plaster 2 coat	1		2.23	8.75	10.98
1020				2		2.61	11.60	14.21
1100			lime plaster - 2 coat	1		2.08	8.75	10.83
1150			lime portland - 2 coat	1		2.10	8.75	10.85
1200			portland - 3 coat	1		2.12	9.20	11.32
1400			5/8" drywall	1		2.40	7.90	10.30
1410				2		2.95	9.75	12.70
1500		6	none	0		2.55	6.45	9
1510			gyp. plaster 2 coat	1		2.93	9.20	12.13
1520				2		3.31	12.05	15.36
1600			lime plaster - 2 coat	1		2.78	9.20	11.98
1650			lime portland - 2 coat	1		2.80	9.20	12
1700			portland - 3 coat	1		2.82	9.65	12.47
1900			5/8" drywall	1		3.10	8.35	11.45
1910				2		3.65	10.20	13.85
2000		8	none	0		2.73	6.90	9.63
2010			gyp. plaster 2 coat	1		3.11	9.65	12.76
2020			gyp. plaster 2 coat	2		3.49	12.45	15.94
2100			lime plaster - 2 coat	1		2.96	9.65	12.61
2150			lime portland - 2 coat	1		2.98	9.65	12.63
2200			portland - 3 coat	1		3	10.10	13.10
2400			5/8" drywall	1		3.28	8.75	12.03
2410				2		3.83	10.65	14.48

C10 Interior Construction

C1010 Partitions

C1010 102 Concrete Block Partitions - Regular Weight

	TYPE	THICKNESS (IN.)	TYPE FINISH	SIDES FINISHED		COST PER S.F.		
						MAT.	INST.	TOTAL
2500	Hollow	10	none	0		3.23	7.25	10.48
2510			gyp. plaster 2 coat	1		3.61	10	13.61
2520				2		3.99	12.80	16.79
2600			lime plaster - 2 coat	1		3.46	10	13.46
2650			lime portland - 2 coat	1		3.48	10	13.48
2700			portland - 3 coat	1		3.50	10.45	13.95
2900			5/8" drywall	1		3.78	9.10	12.88
2910				2		4.33	11	15.33
3000	Solid	2	none	0		1.60	5.95	7.55
3010			gyp. plaster	1		1.98	8.70	10.68
3020				2		2.36	11.55	13.91
3100			lime plaster - 2 coat	1		1.83	8.70	10.53
3150			lime portland - 2 coat	1		1.85	8.70	10.55
3200			portland - 3 coat	1		1.87	9.15	11.02
3400			5/8" drywall	1		2.15	7.85	10
3410				2		2.70	9.70	12.40
3500		4	none	0		2.11	6.20	8.31
3510			gyp. plaster	1		2.62	9	11.62
3520				2		2.87	11.80	14.67
3600			lime plaster - 2 coat	1		2.34	8.95	11.29
3650			lime portland - 2 coat	1		2.36	8.95	11.31
3700			portland - 3 coat	1		2.38	9.40	11.78
3900			5/8" drywall	1		2.66	8.10	10.76
3910				2		3.21	9.95	13.16
4000		6	none	0		2.15	6.70	8.85
4010			gyp. plaster	1		2.53	9.45	11.98
4020				2		2.91	12.30	15.21
4100			lime plaster - 2 coat	1		2.38	9.45	11.83
4150			lime portland - 2 coat	1		2.40	9.45	11.85
4200			portland - 3 coat	1		2.42	9.90	12.32
4400			5/8" drywall	1		2.70	8.60	11.30
4410				2		3.25	10.45	13.70

C1010 104 Concrete Block Partitions - Lightweight

	TYPE	THICKNESS (IN.)	TYPE FINISH	SIDES FINISHED		COST PER S.F.		
						MAT.	INST.	TOTAL
5000	Hollow	4	none	0		2	5.90	7.90
5010			gyp. plaster	1		2.38	8.65	11.03
5020				2		2.76	11.50	14.26
5100			lime plaster - 2 coat	1		2.23	8.65	10.88
5150			lime portland - 2 coat	1		2.25	8.65	10.90
5200			portland - 3 coat	1		2.27	9.10	11.37
5400			5/8" drywall	1		2.55	7.80	10.35
5410				2		3.10	9.65	12.75
5500		6	none	0		2.81	6.30	9.11
5510			gyp. plaster	1		3.19	9.05	12.24
5520			gyp. plaster	2		3.57	11.90	15.47
5600			lime plaster - 2 coat	1		3.04	9.05	12.09
5650			lime portland - 2 coat	1		3.06	9.05	12.11
5700			portland - 3 coat	1		3.08	9.50	12.58
5900			5/8" drywall	1		3.36	8.20	11.56
5910				2		3.91	10.05	13.96

C10 Interior Construction

C1010 Partitions

C1010 104 Concrete Block Partitions - Lightweight

	TYPE	THICKNESS (IN.)	TYPE FINISH	SIDES FINISHED		COST PER S.F.		
						MAT.	INST.	TOTAL
6000	Hollow	8	none	0		3.38	6.75	10.13
6010			gyp. plaster	1		3.76	9.50	13.26
6020				2		4.14	12.30	16.44
6100			lime plaster - 2 coat	1		3.61	9.50	13.11
6150			lime portland - 2 coat	1		3.63	9.50	13.13
6200			portland - 3 coat	1		3.65	9.95	13.60
6400			5/8" drywall	1		3.93	8.60	12.53
6410				2		4.48	10.50	14.98
6500		10	none	0		4.02	7.05	11.07
6510			gyp. plaster	1		4.40	9.80	14.20
6520				2		4.78	12.60	17.38
6700			portland - 3 coat	1		4.29	10.25	14.54
6900			5/8" drywall	1		4.57	8.90	13.47
6910				2		5.10	10.80	15.90
7000	Solid	4	none	0		1.97	6.15	8.12
7010			gyp. plaster	1		2.35	8.90	11.25
7020				2		2.73	11.75	14.48
7100			lime plaster - 2 coat	1		2.20	8.90	11.10
7150			lime portland - 2 coat	1		2.22	8.90	11.12
7200			portland - 3 coat	1		2.24	9.35	11.59
7400			5/8" drywall	1		2.52	8.05	10.57
7410				2		3.07	9.90	12.97
7500		6	none	0		3.10	6.60	9.70
7510			gyp. plaster	1		3.59	9.45	13.04
7520				2		3.86	12.20	16.06
7600			lime plaster - 2 coat	1		3.33	9.35	12.68
7650			lime portland - 2 coat	1		3.35	9.35	12.70
7700			portland - 3 coat	1		3.37	9.80	13.17
7900			5/8" drywall	1		3.65	8.50	12.15
7910				2		4.20	10.35	14.55
8000		8	none	0		3.92	7.10	11.02
8010			gyp. plaster	1		4.30	9.85	14.15
8020				2		4.68	12.65	17.33
8100			lime plaster - 2 coat	1		4.15	9.85	14
8150			lime portland - 2 coat	1		4.17	9.85	14.02
8200			portland - 3 coat	1		4.19	10.30	14.49
8400			5/8" drywall	1		4.47	8.95	13.42
8410				2		5	10.85	15.85

C10 Interior Construction

C1010 Partitions

Single Wythe

Double Wythe

C1010 120	Tile Partitions	COST PER S.F.		
		MAT.	INST.	TOTAL
1000	8W series 8"x16", 4" thick wall, reinf every 2 courses, glazed 1 side	14.80	7.25	22.05
1100	Glazed 2 sides	18.30	7.70	26
1200	Glazed 2 sides, using 2 wythes of 2" thick tile	22	13.90	35.90
1300	6" thick wall, horizontal reinf every 2 courses, glazed 1 side	25.50	7.55	33.05
1400	Glazed 2 sides, each face different color, 2" and 4" tile	26	14.20	40.20
1500	8" thick wall, glazed 2 sides using 2 wythes of 4" thick tile	29.50	14.50	44
1600	10" thick wall, glazed 2 sides using 1 wythe of 4" tile & 1 wythe of 6" til	40.50	14.80	55.30
1700	Glazed 2 sides cavity wall, using 2 wythes of 4" thick tile	29.50	14.50	44
1800	12" thick wall, glazed 2 sides using 2 wythes of 6" thick tile	51	15.10	66.10
1900	Glazed 2 sides cavity wall, using 2 wythes of 4" thick tile	29.50	14.50	44
2100	6T series 5-1/3"x12" tile, 4" thick, non load bearing glazed one side,	13	11.35	24.35
2200	Glazed two sides	17.10	12.80	29.90
2300	Glazed two sides, using two wythes of 2" thick tile	19.10	22	41.10
2400	6" thick, glazed one side	19.40	11.90	31.30
2500	Glazed two sides	23	13.50	36.50
2600	Glazed two sides using 2" thick tile and 4" thick tile	22.50	22.50	45
2700	8" thick, glazed one side	25.50	13.90	39.40
2800	Glazed two sides using two wythes of 4" thick tile	26	22.50	48.50
2900	Glazed two sides using 6" thick tile and 2" thick tile	29	23	52
3000	10" thick cavity wall, glazed two sides using two wythes of 4" tile	26	22.50	48.50
3100	12" thick, glazed two sides using 4" thick tile and 8" thick tile	38.50	25.50	64
3200	2" thick facing tile, glazed one side, on 6" concrete block	13.45	13.20	26.65
3300	On 8" concrete block	13.60	13.60	27.20
3400	On 10" concrete block	14.10	13.90	28

C10 Interior Construction

C1010 Partitions

Wood Stud Framing

Metal Stud Framing

The Drywall Partitions/Stud Framing Systems are defined by type of drywall and number of layers, type and spacing of stud framing, and treatment on the opposite face. Components include taping and finishing.

Cost differences between regular and fire resistant drywall are negligible, and terminology is interchangeable. In some cases fiberglass insulation is included for additional sound deadening.

System Components	QUANTITY	UNIT	COST PER S.F.		
			MAT.	INST.	TOTAL
SYSTEM C1010 124 1250					
DRYWALL PARTITION, 5/8" F.R.1 SIDE, 5/8" REG.1 SIDE, 2"X4"STUDS, 16" O.C.					
Gypsum plasterboard, nailed/screwed to studs, 5/8" fire resistant	1.000	S.F.	.32	.55	.87
Gypsum plasterboard, nailed/screwed to studs, 5/8" regular	1.000	S.F.	.30	.55	.85
Taping and finishing joints	2.000	S.F.	.10	1.10	1.20
Framing, 2 x 4 studs @ 16" O.C., 10' high	1.000	S.F.	.42	1.11	1.53
TOTAL			1.14	3.31	4.45

C1010 124			Drywall Partitions/Wood Stud Framing					
	FACE LAYER	BASE LAYER	FRAMING	OPPOSITE FACE	INSULATION	COST PER S.F.		
						MAT.	INST.	TOTAL
1200	5/8" FR drywall	none	2 x 4, @ 16" O.C.	same	0	1.16	3.31	4.47
1250				5/8" reg. drywall	0	1.14	3.31	4.45
1300				nothing	0	.79	2.21	3
1400		1/4" SD gypsum	2 x 4 @ 16" O.C.	same	1-1/2" fiberglass	2.41	5.10	7.51
1450				5/8" FR drywall	1-1/2" fiberglass	1.99	4.47	6.46
1500				nothing	1-1/2" fiberglass	1.62	3.37	4.99
1600		resil. channels	2 x 4 @ 16", O.C.	same	1-1/2" fiberglass	1.96	6.45	8.41
1650				5/8" FR drywall	1-1/2" fiberglass	1.77	5.15	6.92
1700				nothing	1-1/2" fiberglass	1.40	4.06	5.46
1800		5/8" FR drywall	2 x 4 @ 24" O.C.	same	0	1.70	4.19	5.89
1850				5/8" FR drywall	0	1.38	3.64	5.02
1900				nothing	0	1.01	2.54	3.55
1950		5/8" FR drywall	2 x 4, 16" O.C.	same	0	1.80	4.41	6.21
1955				5/8" FR drywall	0	1.48	3.86	5.34
2000				nothing	0	1.11	2.76	3.87
2010		5/8" FR drywall	staggered, 6" plate	same	0	2.22	5.55	7.77
2015				5/8" FR drywall	0	1.90	4.98	6.88
2020				nothing	0	1.53	3.88	5.41
2200		5/8" FR drywall	2 rows-2 x 4	same	2" fiberglass	2.66	6.05	8.71
2250			16" O.C.	5/8" FR drywall	2" fiberglass	2.34	5.50	7.84
2300				nothing	2" fiberglass	1.97	4.41	6.38
2400	5/8" WR drywall	none	2 x 4, @ 16" O.C.	same	0	1.44	3.31	4.75
2450				5/8" FR drywall	0	1.30	3.31	4.61
2500				nothing	0	.93	2.21	3.14
2600		5/8" FR drywall	2 x 4, @ 24" O.C.	same	0	1.98	4.19	6.17
2650				5/8" FR drywall	0	1.52	3.64	5.16
2700				nothing	0	1.15	2.54	3.69

C10 Interior Construction

C1010 Partitions

C1010 124 — Drywall Partitions/Wood Stud Framing

	FACE LAYER	BASE LAYER	FRAMING	OPPOSITE FACE	INSULATION	COST PER S.F. MAT.	COST PER S.F. INST.	COST PER S.F. TOTAL
2800	5/8" VF drywall	none	2 x 4, @ 16" O.C.	same	0	2.34	3.57	5.91
2850				5/8" FR drywall	0	1.75	3.44	5.19
2900				nothing	0	1.38	2.34	3.72
3000		5/8" FR drywall	2 x 4, 24" O.C.	same	0	2.88	4.45	7.33
3050				5/8" FR drywall	0	1.97	3.77	5.74
3100				nothing	0	1.60	2.67	4.27
3200	1/2" reg drywall	3/8" reg drywall	2 x 4, @ 16" O.C.	same	0	1.54	4.41	5.95
3250				5/8" FR drywall	0	1.35	3.86	5.21
3300				nothing	0	.98	2.76	3.74

C1010 126 — Drywall Partitions/Metal Stud Framing

	FACE LAYER	BASE LAYER	FRAMING	OPPOSITE FACE	INSULATION	COST PER S.F. MAT.	COST PER S.F. INST.	COST PER S.F. TOTAL
5200	5/8" FR drywall	none	1-5/8" @ 24" O.C.	same	0	.94	2.93	3.87
5250				5/8" reg. drywall	0	.92	2.93	3.85
5300				nothing	0	.57	1.83	2.40
5400			3-5/8" @ 24" O.C.	same	0	1.02	2.95	3.97
5450				5/8" reg. drywall	0	1	2.95	3.95
5500				nothing	0	.65	1.85	2.50
5600		1/4" SD gypsum	1-5/8" @ 24" O.C.	same	0	1.78	4.15	5.93
5650				5/8" FR drywall	0	1.36	3.54	4.90
5700				nothing	0	.99	2.44	3.43
5800			2-1/2" @ 24" O.C.	same	0	1.83	4.16	5.99
5850				5/8" FR drywall	0	1.41	3.55	4.96
5900				nothing	0	1.04	2.45	3.49
6000		5/8" FR drywall	2-1/2" @ 16" O.C.	same	0	1.86	4.72	6.58
6050				5/8" FR drywall	0	1.54	4.17	5.71
6100				nothing	0	1.17	3.07	4.24
6200			3-5/8" @ 24" O.C.	same	0	1.66	4.05	5.71
6250				5/8"FR drywall	3-1/2" fiberglass	1.75	3.91	5.66
6300				nothing	0	.97	2.40	3.37
6400	5/8" WR drywall	none	1-5/8" @ 24" O.C.	same	0	1.22	2.93	4.15
6450				5/8" FR drywall	0	1.08	2.93	4.01
6500				nothing	0	.71	1.83	2.54
6600			3-5/8" @ 24" O.C.	same	0	1.30	2.95	4.25
6650				5/8" FR drywall	0	1.16	2.95	4.11
6700				nothing	0	.79	1.85	2.64
6800		5/8" FR drywall	2-1/2" @ 16" O.C.	same	0	2.14	4.72	6.86
6850				5/8" FR drywall	0	1.68	4.17	5.85
6900				nothing	0	1.31	3.07	4.38
7000			3-5/8" @ 24" O.C.	same	0	1.94	4.05	5.99
7050				5/8"FR drywall	3-1/2" fiberglass	1.89	3.91	5.80
7100				nothing	0	1.11	2.40	3.51
7200	5/8" VF drywall	none	1-5/8" @ 24" O.C.	same	0	2.12	3.19	5.31
7250				5/8" FR drywall	0	1.53	3.06	4.59
7300				nothing	0	1.16	1.96	3.12
7400			3-5/8" @ 24" O.C.	same	0	2.20	3.21	5.41
7450				5/8" FR drywall	0	1.61	3.08	4.69
7500				nothing	0	1.24	1.98	3.22

C10 Interior Construction

C1010 Partitions

C1010 126 Drywall Partitions/Metal Stud Framing

	FACE LAYER	BASE LAYER	FRAMING	OPPOSITE FACE	INSULATION	COST PER S.F.		
						MAT.	INST.	TOTAL
7600	5/8" VF drywall	5/8" FR drywall	2-1/2" @ 16" O.C.	same	0	3.04	4.98	8.02
7650				5/8" FR drywall	0	2.13	4.30	6.43
7700				nothing	0	1.76	3.20	4.96
7800			3-5/8" @ 24" O.C.	same	0	2.84	4.31	7.15
7850				5/8"FR drywall	3-1/2" fiberglass	2.34	4.04	6.38
7900				nothing	0	1.56	2.53	4.09

C10 Interior Construction

C1010 Partitions

C1010 128	Drywall Components	COST PER S.F.		
		MAT.	INST.	TOTAL
0140	Metal studs, 24" O.C. including track, load bearing, 18 gage, 2-1/2"	.66	1.04	1.70
0160	3-5/8"	.78	1.06	1.84
0180	4"	.70	1.08	1.78
0200	6"	1.04	1.10	2.14
0220	16 gage, 2-1/2"	.77	1.18	1.95
0240	3-5/8"	.91	1.21	2.12
0260	4"	.96	1.23	2.19
0280	6"	1.20	1.26	2.46
0300	Non load bearing, 25 gage, 1-5/8"	.20	.73	.93
0340	3-5/8"	.28	.75	1.03
0360	4"	.32	.75	1.07
0380	6"	.40	.76	1.16
0400	20 gage, 2-1/2"	.35	.92	1.27
0420	3-5/8"	.40	.94	1.34
0440	4"	.47	.94	1.41
0460	6"	.57	.95	1.52
0540	Wood studs including blocking, shoe and double top plate, 2"x4", 12" O.C.	.52	1.39	1.91
0560	16" O.C.	.42	1.11	1.53
0580	24" O.C.	.32	.89	1.21
0600	2"x6", 12" O.C.	.80	1.58	2.38
0620	16" O.C.	.65	1.23	1.88
0640	24" O.C.	.49	.96	1.45
0642	Furring one side only, steel channels, 3/4", 12" O.C.	.39	1.98	2.37
0644	16" O.C.	.35	1.76	2.11
0646	24" O.C.	.23	1.33	1.56
0647	1-1/2" , 12" O.C.	.49	2.22	2.71
0648	16" O.C.	.44	1.94	2.38
0649	24" O.C.	.29	1.53	1.82
0650	Wood strips, 1" x 3", on wood, 12" O.C.	.38	1.01	1.39
0651	16" O.C.	.29	.76	1.05
0652	On masonry, 12" O.C.	.38	1.12	1.50
0653	16" O.C.	.29	.84	1.13
0654	On concrete, 12" O.C.	.38	2.13	2.51
0655	16" O.C.	.29	1.60	1.89
0665	Gypsum board, one face only, exterior sheathing, 1/2"	.45	.98	1.43
0680	Interior, fire resistant, 1/2"	.30	.55	.85
0700	5/8"	.32	.55	.87
0720	Sound deadening board 1/4"	.42	.61	1.03
0740	Standard drywall 3/8"	.25	.55	.80
0760	1/2"	.26	.55	.81
0780	5/8"	.30	.55	.85
0800	Tongue & groove coreboard 1"	.79	2.30	3.09
0820	Water resistant, 1/2"	.40	.55	.95
0840	5/8"	.46	.55	1.01
0860	Add for the following:, foil backing	.15		.15
0880	Fiberglass insulation, 3-1/2"	.41	.41	.82
0900	6"	.48	.41	.89
0920	Rigid insulation 1"	.53	.55	1.08
0940	Resilient furring @ 16" O.C.	.26	1.74	2
0960	Taping and finishing	.05	.55	.60
0980	Texture spray	.04	.58	.62
1000	Thin coat plaster	.12	.69	.81
1050	Sound wall framing, 2x6 plates, 2x4 staggered studs, 12" O.C.	.63	1.35	1.98

C10 Interior Construction

C1010 Partitions

Plaster Partitions are defined as follows: type of plaster, type and spacing of framing, type of lath and treatment of opposite face.

Included in the system components are expansion joints. Metal studs are assumed to be non-loadbearing.

System Components	QUANTITY	UNIT	COST PER S.F.		
			MAT.	INST.	TOTAL
SYSTEM C1010 140 1000					
GYP PLASTER PART'N, 2 COATS, 2-1/2" MTL. STUD, 16" O.C., 3/8" GYP. LATH					
Gypsum plaster, 2 coats, on walls	2.000	S.F.	1.09	5.33	6.42
Gypsum lath, 3/8" thick, on metal studs	2.000	S.F.	1.03	1.38	2.41
Metal studs, 25 ga., 2-1/2" @ 16" O.C.	1.000	S.F.	.34	1.13	1.47
Expansion joint	.100	L.F.	.09	.17	.26
TOTAL			2.55	8.01	10.56

C1010 140 — Plaster Partitions/Metal Stud Framing

	TYPE	FRAMING	LATH	OPPOSITE FACE		MAT.	INST.	TOTAL
1000	2 coat gypsum	2-1/2" @ 16"O.C.	3/8" gypsum	same		2.55	8.05	10.60
1010				nothing		1.50	4.66	6.16
1100		3-1/4" @ 24"O.C.	1/2" gypsum	same		2.56	7.75	10.31
1110				nothing		1.47	4.33	5.80
1500	2 coat vermiculite	2-1/2" @ 16"O.C.	3/8" gypsum	same		2.41	8.75	11.16
1510				nothing		1.42	5	6.42
1600		3-1/4" @ 24"O.C.	1/2" gypsum	same		2.42	8.45	10.87
1610				nothing		1.39	4.68	6.07
2000	3 coat gypsum	2-1/2" @ 16"O.C.	3/8" gypsum	same		2.38	9.05	11.43
2010				nothing		1.41	5.20	6.61
2020			3.4lb. diamond	same		2.44	9.05	11.49
2030				nothing		1.44	5.20	6.64
2040			2.75lb. ribbed	same		2.07	9.05	11.12
2050				nothing		1.25	5.20	6.45
2100		3-1/4" @ 24"O.C.	1/2" gypsum	same		2.39	8.80	11.19
2110				nothing		1.38	4.86	6.24
2120			3.4lb. ribbed	same		2.47	8.80	11.27
2130				nothing		1.42	4.86	6.28
2500	3 coat lime	2-1/2" @ 16"O.C.	3.4lb. diamond	same		2.10	9.05	11.15
2510				nothing		1.27	5.20	6.47
2520			2.75lb. ribbed	same		1.73	9.05	10.78
2530				nothing		1.08	5.20	6.28
2600		3-1/4" @ 24"O.C.	3.4lb. ribbed	same		2.13	8.80	10.93
2610				nothing		1.25	4.86	6.11

C10 Interior Construction

C1010 Partitions

C1010 140 — Plaster Partitions/Metal Stud Framing

	TYPE	FRAMING	LATH	OPPOSITE FACE		COST PER S.F.		
						MAT.	INST.	TOTAL
3000	3 coat Portland	2-1/2" @ 16"O.C.	3.4lb. diamond	same		2.06	9.05	11.11
3010				nothing		1.25	5.20	6.45
3020			2.75lb. ribbed	same		1.69	9.05	10.74
3030				nothing		1.06	5.20	6.26
3100		3-1/4" @ 24"O.C.	3.4lb. ribbed	same		2.09	8.80	10.89
3110				nothing		1.23	4.86	6.09
3500	3 coat gypsum	2-1/2" @ 16"O.C.	3/8" gypsum	same		2.88	11.65	14.53
3510	W/med. Keenes			nothing		1.66	6.50	8.16
3520			3.4lb. diamond	same		2.94	11.65	14.59
3530				nothing		1.69	6.50	8.19
3540			2.75lb. ribbed	same		2.57	11.65	14.22
3550				nothing		1.50	6.50	8
3600		3-1/4" @ 24"O.C.	1/2" gypsum	same		2.89	11.35	14.24
3610				nothing		1.63	6.20	7.83
3620			3.4lb. ribbed	same		2.97	11.35	14.32
3630				nothing		1.67	6.20	7.87
4000	3 coat gypsum	2-1/2" @ 16"O.C.	3/8" gypsum	same		2.89	12.95	15.84
4010	W/hard Keenes			nothing		1.66	7.10	8.76
4020			3.4lb. diamond	same		2.95	12.95	15.90
4030				nothing		1.69	7.10	8.79
4040			2.75lb. ribbed	same		2.58	12.95	15.53
4050				nothing		1.50	7.10	8.60
4100		3-1/4" @ 24"O.C.	1/2" gypsum	same		2.90	12.65	15.55
4110				nothing		1.63	6.75	8.38
4120			3.4lb. ribbed	same		2.98	12.65	15.63
4130				nothing		1.67	6.75	8.42
4500	3 coat lime	2-1/2" @ 16"O.C.	3.4lb. diamond	same		2.07	9.05	11.12
4510	Portland			nothing		1.26	5.20	6.46
4520			2.75lb. ribbed	same		1.70	9.05	10.75
4530				nothing		1.07	5.20	6.27
4600		3-1/4" @ 24"O.C.	3.4lb. ribbed	same		2.10	8.80	10.90
4610				nothing		1.24	4.86	6.10

C1010 142 — Plaster Partitions/Wood Stud Framing

	TYPE	FRAMING	LATH	OPPOSITE FACE		COST PER S.F.		
						MAT.	INST.	TOTAL
5000	2 coat gypsum	2"x4" @ 16"O.C.	3/8" gypsum	same		2.63	7.85	10.48
5010				nothing		1.58	4.56	6.14
5100		2"x4" @ 24"O.C.	1/2" gypsum	same		2.60	7.70	10.30
5110				nothing		1.51	4.38	5.89
5500	2 coat vermiculite	2"x4" @ 16"O.C.	3/8" gypsum	same		2.49	8.55	11.04
5510				nothing		1.50	4.91	6.41
5600		2"x4" @ 24"O.C.	1/2" gypsum	same		2.46	8.40	10.86
5610				nothing		1.43	4.73	6.16

C10 Interior Construction

C1010 Partitions

C1010 142 — Plaster Partitions/Wood Stud Framing

	TYPE	FRAMING	LATH	OPPOSITE FACE		COST PER S.F. MAT.	COST PER S.F. INST.	COST PER S.F. TOTAL
6000	3 coat gypsum	2"x4" @ 16"O.C.	3/8" gypsum	same		2.46	8.90	11.36
6010				nothing		1.49	5.10	6.59
6020			3.4lb. diamond	same		2.52	8.95	11.47
6030				nothing		1.52	5.15	6.67
6040			2.75lb. ribbed	same		2.13	9	11.13
6050				nothing		1.32	5.15	6.47
6100		2"x4" @ 24"O.C.	1/2" gypsum	same		2.43	8.75	11.18
6110				nothing		1.42	4.91	6.33
6120			3.4lb. ribbed	same		2.05	8.85	10.90
6130				nothing		1.23	4.95	6.18
6500	3 coat lime	2"x4" @ 16"O.C.	3.4lb. diamond	same		2.18	8.95	11.13
6510				nothing		1.35	5.15	6.50
6520			2.75lb. ribbed	same		1.79	9	10.79
6530				nothing		1.15	5.15	6.30
6600		2"x4" @ 24"O.C.	3.4lb. ribbed	same		1.71	8.85	10.56
6610				nothing		1.06	4.95	6.01
7000	3 coat Portland	2"x4" @ 16"O.C.	3.4lb. diamond	same		2.14	8.95	11.09
7010				nothing		1.33	5.15	6.48
7020			2.75lb. ribbed	same		1.75	9	10.75
7030				nothing		1.13	5.15	6.28
7100		2"x4" @ 24"O.C.	3.4lb. ribbed	same		2.13	8.95	11.08
7110				nothing		1.27	5	6.27
7500	3 coat gypsum W/med Keenes	2"x4" @ 16"O.C.	3/8" gypsum	same		2.96	11.45	14.41
7510				nothing		1.74	6.40	8.14
7520			3.4lb. diamond	same		3.02	11.55	14.57
7530				nothing		1.77	6.45	8.22
7540			2.75lb. ribbed	same		2.63	11.60	14.23
7550				nothing		1.57	6.45	8.02
7600		2"x4" @ 24"O.C.	1/2" gypsum	same		2.93	11.35	14.28
7610				nothing		1.67	6.25	7.92
7620			3.4lb. ribbed	same		3.01	11.50	14.51
7630				nothing		1.71	6.30	8.01
8000	3 coat gypsum W/hard Keenes	2"x4" @ 16"O.C.	3/8" gypsum	same		2.97	12.75	15.72
8010				nothing		1.74	7	8.74
8020			3.4lb. diamond	same		3.03	12.85	15.88
8030				nothing		1.77	7.05	8.82
8040			2.75lb. ribbed	same		2.64	12.90	15.54
8050				nothing		1.57	7.05	8.62
8100		2"x4" @ 24"O.C.	1/2" gypsum	same		2.94	12.65	15.59
8110				nothing		1.67	6.80	8.47
8120			3.4lb. ribbed	same		3.02	12.80	15.82
8130				nothing		1.71	6.90	8.61
8500	3 coat lime Portland	2"x4" @ 16"O.C.	3.4lb. diamond	same		2.15	8.95	11.10
8510				nothing		1.34	5.15	6.49
8520			2.75lb. ribbed	same		1.76	9	10.76
8530				nothing		1.14	5.15	6.29
8600		2"x4" @ 24"O.C.	3.4lb. ribbed	same		2.14	8.95	11.09
8610				nothing		1.28	5	6.28

C10 Interior Construction

C1010 Partitions

C1010 144	Plaster Partition Components	COST PER S.F.		
		MAT.	INST.	TOTAL
0060	Metal studs, 16" O.C., including track, non load bearing, 25 gage, 1-5/8"	.34	1.13	1.47
0080	2-1/2"	.34	1.13	1.47
0100	3-1/4"	.38	1.15	1.53
0120	3-5/8"	.38	1.15	1.53
0140	4"	.43	1.16	1.59
0160	6"	.54	1.18	1.72
0180	Load bearing, 20 gage, 2-1/2"	.89	1.44	2.33
0200	3-5/8"	1.05	1.46	2.51
0220	4"	1.11	1.50	2.61
0240	6"	1.40	1.52	2.92
0260	16 gage 2-1/2"	1.05	1.63	2.68
0280	3-5/8"	1.26	1.68	2.94
0300	4"	1.32	1.70	3.02
0320	6"	1.65	1.73	3.38
0340	Wood studs, including blocking, shoe and double plate, 2"x4", 12" O.C.	.52	1.39	1.91
0360	16" O.C.	.42	1.11	1.53
0380	24" O.C.	.32	.89	1.21
0400	2"x6", 12" O.C.	.80	1.58	2.38
0420	16" O.C.	.65	1.23	1.88
0440	24" O.C.	.49	.96	1.45
0460	Furring one face only, steel channels, 3/4", 12" O.C.	.39	1.98	2.37
0480	16" O.C.	.35	1.76	2.11
0500	24" O.C.	.23	1.33	1.56
0520	1-1/2", 12" O.C.	.49	2.22	2.71
0540	16" O.C.	.44	1.94	2.38
0560	24"O.C.	.29	1.53	1.82
0580	Wood strips 1"x3", on wood., 12" O.C.	.38	1.01	1.39
0600	16"O.C.	.29	.76	1.05
0620	On masonry, 12" O.C.	.38	1.12	1.50
0640	16" O.C.	.29	.84	1.13
0660	On concrete, 12" O.C.	.38	2.13	2.51
0680	16" O.C.	.29	1.60	1.89
0700	Gypsum lath. plain or perforated, nailed to studs, 3/8" thick	.52	.61	1.13
0720	1/2" thick	.55	.65	1.20
0740	Clipped to studs, 3/8" thick	.52	.69	1.21
0760	1/2" thick	.55	.74	1.29
0780	Metal lath, diamond painted, nailed to wood studs, 2.5 lb.	.47	.61	1.08
0800	3.4 lb.	.55	.65	1.20
0820	Screwed to steel studs, 2.5 lb.	.47	.65	1.12
0840	3.4 lb.	.46	.69	1.15
0860	Rib painted, wired to steel, 2.75 lb	.36	.69	1.05
0880	3.4 lb	.59	.74	1.33
0900	4.0 lb	.69	.80	1.49
0910				
0920	Gypsum plaster, 2 coats	.42	2.67	3.09
0940	3 coats	.59	3.20	3.79
0960	Perlite or vermiculite plaster, 2 coats	.47	3.02	3.49
0980	3 coats	.77	3.79	4.56
1000	Stucco, 3 coats, 1" thick, on wood framing	.73	5.30	6.03
1020	On masonry	.28	4.15	4.43
1100	Metal base galvanized and painted 2-1/2" high	.66	1.94	2.60

C10 Interior Construction

C1010 Partitions

Folding Accordion

Folding Leaf

Movable and Borrow Lites

C1010 205	Partitions	MAT.	INST.	TOTAL
0360	Folding accordion, vinyl covered, acoustical, 3 lb. S.F., 17 ft max. hgt	30	11.05	41.05
0380	5 lb. per S.F. 27 ft max height	42	11.65	53.65
0400	5.5 lb. per S.F., 17 ft. max height	49	12.30	61.30
0420	Commercial, 1.75 lb per S.F., 8 ft. max height	27.50	4.92	32.42
0440	2.0 Lb per S.F., 17 ft. max height	28.50	7.40	35.90
0460	Industrial, 4.0 lb. per S.F. 27 ft. max height	42.50	14.75	57.25
0480	Vinyl clad wood or steel, electric operation 6 psf	60	6.90	66.90
0500	Wood, non acoustic, birch or mahogany	32	3.69	35.69
0560	Folding leaf, aluminum framed acoustical 12 ft.high., 5.5 lb per S.F., min.	42.50	18.45	60.95
0580	Maximum	51.50	37	88.50
0600	6.5 lb. per S.F., minimum	45	18.45	63.45
0620	Maximum	55.50	37	92.50
0640	Steel acoustical, 7.5 per S.F., vinyl faced, minimum	63.50	18.45	81.95
0660	Maximum	77.50	37	114.50
0665	Steel with vinyl face, economy	63.50	18.45	81.95
0670	Deluxe	77.50	37	114.50
0680	Wood acoustic type, vinyl faced to 18' high 6 psf, minimum	59.50	18.45	77.95
0700	Average	71	24.50	95.50
0720	Maximum	92	37	129
0740	Formica or hardwood faced, minimum	61.50	18.45	79.95
0760	Maximum	65.50	37	102.50
0780	Wood, low acoustical type to 12 ft. high 4.5 psf	44.50	22	66.50
0840	Demountable, trackless wall, cork finish, semi acous, 1-5/8" th, min	40.50	3.40	43.90
0860	Maximum	45	5.80	50.80
0880	Acoustic, 2" thick, minimum	38.50	3.63	42.13
0900	Maximum	59	4.92	63.92
0920	In-plant modular office system, w/prehung steel door			
0940	3" thick honeycomb core panels			
0960	12' x 12', 2 wall	13.60	.69	14.29
0970	4 wall	15.55	.92	16.47
0980	16' x 16', 2 wall	13.60	.49	14.09
0990	4 wall	9.55	.49	10.04
1000	Gypsum, demountable, 3" to 3-3/4" thick x 9' high, vinyl clad	7.05	2.55	9.60
1020	Fabric clad	17.55	2.78	20.33
1040	1.75 system, vinyl clad hardboard, paper honeycomb core panel			
1060	1-3/4" to 2-1/2" thick x 9' high	11.90	2.55	14.45
1080	Unitized gypsum panel system, 2" to 2-1/2" thick x 9' high			
1100	Vinyl clad gypsum	15.20	2.55	17.75
1120	Fabric clad gypsum	25	2.78	27.78
1140	Movable steel walls, modular system			
1160	Unitized panels, 48" wide x 9' high			
1180	Baked enamel, pre-finished	17.20	2.05	19.25
1200	Fabric clad	25	2.19	27.19

C10 Interior Construction

C1010 Partitions

C1010 210	Partitions	COST PER S.F.		
		MAT.	INST.	TOTAL
0100	Partitions, folding accordion, 1.25 lb. vinyl (residential), to 8' high	24	3.69	27.69
0110	4 lb. vinyl (industrial), to 27' high	42.50	14.75	57.25
0120	Acoustical, 3 lb. vinyl, to 17' high	30	11.05	41.05
0130	4.5 lb. vinyl, fire rated 40, 20' high	49	6.90	55.90
0485	Operable acoustic, no track, 1-5/8" thick, economy	34.50	2.95	37.45
0490	With track, 3" thick, deluxe	105	3.69	108.69

C1010 230	Demountable Partitions, L.F.	COST PER L.F.		
		MAT.	INST.	TOTAL
0100	Office-demountable gypsum, 3" to 3-3/4" thick x 9' high, vinyl clad	63.50	23	86.50
0110	Fabric clad	158	25	183
0120	1.75 system, vinyl clad hardboard, paper honeycomb			
0130	core panel, 1-3/4" to 2-1/2" thick x 9' high	107	23	130
0140	Steel, modular, unitized panels, 48" wide x 9' high			
0150	Baked enamel, pre-finished	155	18.45	173.45
0160	Portable, fiber core, 4' high, fabric face	99.50	6.30	105.80

C10 Interior Construction

C1010 Partitions

Alum. Tube Frame

Oakwood Frame

Concealed Frame Butt Glazed

Interior Glazed Openings are defined as follows: framing material, glass, size and intermediate framing members. Components for each system include gasket setting or glazing bead and typical wood blocking.

System Components	QUANTITY	UNIT	COST PER OPNG.		
			MAT.	INST.	TOTAL
SYSTEM C1010 710 1000					
GLAZED OPENING, ALUMINUM TUBE FRAME, FLUSH, 1/4″ FLOAT GLASS, 6′ X 4′					
Aluminum tube frame, flush, anodized bronze, head & jamb	16.500	L.F.	227.70	216.15	443.85
Aluminum tube frame, flush, anodized bronze, open sill	7.000	L.F.	78.40	89.60	168
Joints for tube frame, clip type	4.000	Ea.	108		108
Gasket setting, add	20.000	L.F.	125		125
Wood blocking	8.000	B.F.	4.76	24.60	29.36
Float glass, 1/4″ thick, clear, plain	24.000	S.F.	148.80	210	358.80
TOTAL			692.66	540.35	1,233.01

C1010 710		Interior Glazed Opening					
	FRAME	GLASS	OPENING-SIZE W X H	INTERMEDIATE MULLION	INTERMEDIATE HORIZONTAL	COST PER OPNG.		
						MAT.	INST.	TOTAL
1000	Aluminum flush	1/4″ float	6′x4′	0	0	695	540	1,235
1040	Tube		12′x4′	3	0	1,700	1,150	2,850
1080			4′x5′	0	0	630	465	1,095
1120			8′x5′	1	0	1,150	865	2,015
1160			12′x5′	2	0	1,700	1,250	2,950
1240		3/8″ float	9′x6′	2	0	1,900	1,450	3,350
1280			4′x8′-6″	0	1	1,225	935	2,160
1320			16′x10′	3	1	4,925	3,825	8,750
1400		1/4″ tempered	6′x4′	0	0	765	540	1,305
1440			12′x4′	3	0	1,850	1,150	3,000
1480			4′x5′	0	0	690	465	1,155
1520			8′x5′	1	0	1,275	865	2,140
1560			12′x5′	2	0	1,875	1,250	3,125
1640		3/8″ tempered	9′x6′	2	0	2,275	1,450	3,725
1680			4′x8′-6″	0	0	1,275	875	2,150
1720			16′x10′	3	0	5,350	3,600	8,950
1800		1/4″ one way mirror	6′x4′	0	0	1,025	530	1,555
1840			12′x4′	3	0	2,400	1,150	3,550

C10 Interior Construction

C1010 Partitions

C1010 710 — Interior Glazed Opening

	FRAME	GLASS	OPENING-SIZE W X H	INTERMEDIATE MULLION	INTERMEDIATE HORIZONTAL	COST PER OPNG.		
						MAT.	INST.	TOTAL
1880	Aluminum flush	1/4" one way	4'x5'	0	0	915	460	1,375
1920	Tube	mirror	8'x5'	1	0	1,725	850	2,575
1960			12'x5'	2	0	2,550	1,250	3,800
2040		3/8" laminated safety	9'x6'	2	0	2,600	1,425	4,025
2080			4'x8'-6"	0	0	1,475	860	2,335
2120			16'x10'	3	0	6,350	3,500	9,850
2200		1-3/16" bullet proof	6'x4'	0	0	3,400	1,925	5,325
2240			12'x4'	3	0	7,100	3,925	11,025
2280			4'x5'	0	0	2,900	1,625	4,525
2320			8'x5'	1	0	5,650	3,150	8,800
2360			12'x5'	2	0	8,425	4,700	13,125
2440		1" acoustical	9'x6'	2	0	3,425	1,300	4,725
2480			4'x8'-6"	0	0	2,025	785	2,810
2520			16'x10'	3	0	8,625	3,100	11,725
3000	Oakwood	1/4" float	6'x4'	0	0	390	390	780
3040			12'x4'	3	0	970	950	1,920
3080			4'x5'	0	0	335	320	655
3120			8'x5'	1	0	680	675	1,355
3160			12'x5'	2	0	1,025	1,025	2,050
3240		3/8" float	9'x6'	2	0	1,225	1,250	2,475
3280			4'x8'-6"	0	1	755	735	1,490
3320			16'x10'	3	1	3,450	3,575	7,025
3400		1/4" tempered	6'x4'	0	0	460	390	850
3440			12'x4'	3	0	1,125	950	2,075
3480			4'x5'	0	0	395	320	715
3520			8'x5'	1	0	800	675	1,475
3560			12'x5'	2	0	1,200	1,025	2,225
3640		3/8" tempered	9'x6'	2	0	1,600	1,250	2,850
3680			4'x8'-6"	0	0	895	660	1,555
3720			16'x10'	3	0	4,200	3,250	7,450
3800		1/4" one way mirror	6'x4'	0	0	730	380	1,110
3840			12'x4'	3	0	1,650	930	2,580
3880			4'x5'	0	0	620	310	930
3920			8'x5'	1	0	1,250	660	1,910
3960			12'x5'	2	0	1,875	1,000	2,875
4040		3/8" laminated safety	9'x6'	2	0	1,950	1,200	3,150
4080			4'x8'-6"	0	0	1,100	640	1,740
4120			16'x10'	3	0	5,175	3,175	8,350
4200		1-3/16" bullet proof	6'x4'	0	0	3,025	1,750	4,775
4240			12'x4'	3	0	6,250	3,675	9,925
4280			4'x5'	0	0	2,525	1,450	3,975
4360			12'x5'	2	0	7,600	4,425	12,025
4440		1" acoustical	9'x6'	2	0	2,650	1,050	3,700
4520			16'x10'	3	0	7,250	2,700	9,950

C10 Interior Construction

C1010 Partitions

C1010 710 Interior Glazed Opening

	FRAME	GLASS	OPENING-SIZE W X H	INTERMEDIATE MULLION	INTERMEDIATE HORIZONTAL	COST PER OPNG.		
						MAT.	INST.	TOTAL
5000	Concealed frame	1/2" float	6'x4'	1		670	840	1,510
5040	Butt glazed		9'x4'	2		960	1,200	2,160
5080			8'x5'	1		1,025	1,250	2,275
5120			16'x5'	3		1,950	2,375	4,325
5200		3/4" float	6'x8'	1		2,175	2,000	4,175
5240			9'x8'	2		3,175	2,875	6,050
5280			8'x10'	1		3,500	3,100	6,600
5320			16'x10'	3		6,775	5,925	12,700
5400		1/4" tempered	6'x4'	1		450	595	1,045
5440			9'x4'	2		635	830	1,465
5480			8'x5'	1		665	845	1,510
5520			16'x5'	3		1,225	1,550	2,775
5600		1/2" tempered	6'x8'	1		1,600	1,475	3,075
5640			9'x8'	2		2,300	2,075	4,375
5680			8'x10'	1		2,525	2,225	4,750
5720			16'x10'	3		4,850	4,150	9,000
5800		1/4" laminated safety	6'x4'	1		545	665	1,210
5840			9'x4'	2		775	935	1,710
5880			8'x5'	1		825	960	1,785
5920			16'x5'	3		1,550	1,775	3,325
6000		1/2" laminated safety	6'x8'	1		1,650	1,325	2,975
6040			9'x8'	2		2,375	1,875	4,250
6080			8'x10'	1		2,625	2,000	4,625
6120			16'x10'	3		5,025	3,700	8,725
6200		1-3/16" bullet proof	6'x4'	1		3,025	1,950	4,975
6240			9'x4'	2		4,475	2,875	7,350
6280			8'x5'	1		4,950	3,125	8,075
6320			16'x5'	3		9,775	6,075	15,850
6400		2" bullet proof	6'x8'	1		7,275	4,750	12,025
6440			9'x8'	2		10,800	7,000	17,800
6480			8'x10'	1		12,000	7,700	19,700
6520			16'x10'	3		23,800	15,100	38,900

C10 Interior Construction

C1020 Interior Doors

Sliding Panel-Mall Front

Rolling Overhead Steel Door

C1020 102	Special Doors	COST PER OPNG.		
		MAT.	INST.	TOTAL
3800	Sliding entrance door and system mill finish	10,300	2,650	12,950
3900	Bronze finish	11,300	2,875	14,175
4000	Black finish	11,800	2,975	14,775
4100	Sliding panel mall front, 16'x9' opening, mill finish	3,650	805	4,455
4200	Bronze finish	4,750	1,050	5,800
4300	Black finish	5,850	1,300	7,150
4400	24'x9' opening mill finish	5,175	1,500	6,675
4500	Bronze finish	6,725	1,950	8,675
4600	Black finish	8,275	2,400	10,675
4700	48'x9' opening mill finish	9,650	1,175	10,825
4800	Bronze finish	12,500	1,525	14,025
4900	Black finish	15,400	1,875	17,275
5000	Rolling overhead steel door, manual, 8' x 8' high	1,225	895	2,120
5100	10' x 10' high	2,025	1,025	3,050
5200	20' x 10' high	3,375	1,425	4,800
5300	12' x 12' high	2,050	1,200	3,250
5400	Motor operated, 8' x 8' high	2,400	1,175	3,575
5500	10' x 10' high	3,200	1,300	4,500
5600	20' x 10' high	4,550	1,700	6,250
5700	12' x 12' high	3,225	1,475	4,700
5800	Roll up grille, aluminum, manual, 10' x 10' high, mill finish	3,050	1,750	4,800
5900	Bronze anodized	4,900	1,750	6,650
6000	Motor operated, 10' x 10' high, mill finish	4,425	2,025	6,450
6100	Bronze anodized	6,275	2,025	8,300
6200	Steel, manual, 10' x 10' high	2,750	1,425	4,175
6300	15' x 8' high	3,175	1,775	4,950
6400	Motor operated, 10' x 10' high	4,125	1,700	5,825
6500	15' x 8' high	4,550	2,050	6,600

C10 Interior Construction

C1020 Interior Doors

Steel Door, Half Glass
Steel Frame

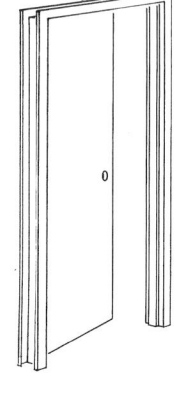
Steel Door, Flush
Steel Frame

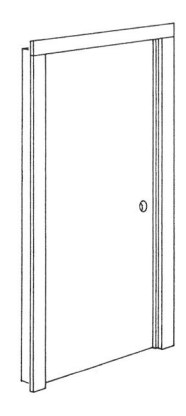
Wood Door, Flush
Wood Frame

The Metal Door/Metal Frame Systems are defined as follows: door type, design and size, frame type and depth. Included in the components for each system is painting the door and frame. No hardware has been included in the systems.

System Components	QUANTITY	UNIT	COST EACH		
			MAT.	INST.	TOTAL
SYSTEM C1020 114 1200					
STEEL DOOR, HOLLOW, 20 GA., HALF GLASS, 3'-0"X7'-0", D.W.FRAME, 4-7/8" DP					
Steel door, flush, hollow core, 1-3/4" thk, half gl, 20 Ga., 3'-0" x 7'-0"	1.000	Ea.	650	61.50	711.50
Steel frame, 16 ga., up to 4-7/8" deep, 7'-0" H, 3'-0" W, single	1.000	Ea.	180	69	249
Hinges full mortise, avg. freq., steel base, USP, 4-1/2" x 4-1/2"	1.500	Pr.	42.75		42.75
Float glass, 3/16" thick, clear, tempered	5.000	S.F.	36.50	40.25	76.75
Paint door and frame each side, primer	1.000	Ea.	6.78	93	99.78
Paint door and frame each side, 2 coats	1.000	Ea.	9.44	155	164.44
TOTAL			925.47	418.75	1,344.22

C1020 114			Metal Door/Metal Frame				
	TYPE	DESIGN	SIZE	FRAME	DEPTH	COST EACH		
						MAT.	INST.	TOTAL
1000	Flush-hollow	20 ga. full panel	3'-0" x 7'-0"	drywall K.D.	4-7/8"	715	380	1,095
1020				butt welded	8-3/4"	825	430	1,255
1160			6'-0" x 7'-0"	drywall K.D.	4-7/8"	1,275	605	1,880
1180				butt welded	8-3/4"	1,400	655	2,055
1200		20 ga. half glass	3'-0" x 7'-0"	drywall K.D.	4-7/8"	925	420	1,345
1220				butt welded	8-3/4"	1,025	470	1,495
1360			6'-0" x 7'-0"	drywall K.D.	4-7/8"	1,675	680	2,355
1380				butt welded	8-3/4"	1,825	730	2,555
1800		18 ga. full panel	3'-0" x 7'-0"	drywall K.D.	4-7/8"	895	380	1,275
1820				butt welded	8-3/4"	1,000	430	1,430
1960			6'-0" x 7'-0"	drywall K.D.	4-7/8"	1,625	605	2,230
1980				butt welded	8-3/4"	1,750	655	2,405
2000		18 ga. half glass	3'-0" x 7'-0"	drywall K.D.	4-7/8"	985	420	1,405
2020				butt welded	8-3/4"	1,100	470	1,570
2160			6'-0" x 7'-0"	drywall K.D.	4-7/8"	1,800	685	2,485
2180				butt welded	8-3/4"	1,925	735	2,660

C10 Interior Construction

C1020 Interior Doors

C1020 116 Labeled Metal Door/Metal Frames

	TYPE	DESIGN	SIZE	FRAME	DEPTH	COST EACH MAT.	COST EACH INST.	COST EACH TOTAL
6000	Hollow-1-1/2 hour	20 ga. full panel	3'-0" x 6'-8"	drywall K.D.	4-7/8"	785	385	1,170
6020				butt welded	8-3/4"	970	435	1,405
6160			6'-0" x 7'-0"	drywall K.D.	4-7/8"	1,400	620	2,020
6180				butt welded	8-3/4"	1,575	670	2,245
6200		20 ga. vision	3'-0" x 7'-0"	drywall K.D.	4-7/8"	945	405	1,350
6220				butt welded	8-3/4"	1,125	455	1,580
6360			6'-0" x 7'-0"	drywall K.D.	4-7/8"	1,725	655	2,380
6380				butt welded	8-3/4"	1,900	705	2,605
6600		18 ga. full panel	3'-0" x 7'-0"	drywall K.D.	4-7/8"	845	390	1,235
6620				butt welded	8-3/4"	1,025	440	1,465
6760			6'-0" x 7'-0"	drywall K.D.	4-7/8"	1,575	625	2,200
6780				butt welded	8-3/4"	1,600	675	2,275
6800		18 ga. vision	3'-0" x 7'-0"	drywall K.D.	4-7/8"	1,175	410	1,585
6820				butt welded	8-3/4"	1,350	460	1,810
6960			6'-0" x 7'-0"	drywall K.D.	4-7/8"	2,175	665	2,840
6980				butt welded	8-3/4"	2,350	715	3,065
7200	Hollow-3 hour	18 ga. full panel	3'-0" x 7'-0"	butt K.D.	5-3/4"	845	390	1,235
7220				butt welded	8-3/4"	1,025	440	1,465
7360			6'-0" x 7'-0"	butt K.D.	5-3/4"	1,525	625	2,150
7380				butt welded	8-3/4"	1,700	675	2,375
8000	Composite-1-1/2 hour	20 ga. full panel	3'-0" x 7'-0"	drywall K.D.	4-3/8"	1,050	390	1,440
8020				butt welded	8-3/4"	1,225	440	1,665
8160			6'-0" x 7'-0"	drywall K.D.	4-7/8"	1,900	625	2,525
8180				butt welded	8-3/4"	2,075	675	2,750
8200		20 ga. vision	3'-0" x 7'-0"	drywall K.D.	4-7/8"	1,375	410	1,785
8220				butt welded	8-3/4"	1,550	460	2,010
8360			6'-0" x 7'-0"	drywall K.D.	4-7/8"	2,550	665	3,215
8400	Mineral core-3 hour	18 ga. full panel	3'-0" x 7'-0"	drywall K.D.	5-3/4"	1,325	470	1,795
8420				butt welded	8-3/4"	1,175	445	1,620
8560			6'-0" x 7'-0"	drywall K.D.	5-3/4"	2,025	635	2,660

C1020 120 Wood Door/Wood Frame

	TYPE	FACE	SIZE	FRAME	DEPTH	COST EACH MAT.	COST EACH INST.	COST EACH TOTAL
1600	Hollow core/flush	lauan	3'-0" x 7'-0"	pine	3-5/8"	335	244	579
1620					5-3/16"	345	257	602
1760			6'-0" x 7'-0"	pine	3-5/8"	560	400	960
1780					5-3/16"	570	420	990
1800		birch	3'-0" x 7'-0"	pine	3-5/8"	320	244	564
1820					5-3/16"	335	262	597
1960			6'-0" x 7'-0"	pine	3-5/8"	515	395	910
1980					5-3/16"	525	415	940
2000		oak	3'-0" x 7'-0"	oak	3-5/8"	315	198	513
2020					5-3/16"	535	257	792
2160			6'-0" x 7'-0"	oak	3-5/8"	770	400	1,170
2180					5-3/16"	870	415	1,285
3000	Particle core/flush	lauan	3'-0" x 7'-0"	pine	3-5/8"	330	256	586
3020					5-3/16"	335	270	605
3160			6'-0" x 7'-0"	pine	3-5/8"	545	430	975
3180					5-3/16"	555	445	1,000

C10 Interior Construction

C1020 Interior Doors

C1020 120 Wood Door/Wood Frame

	TYPE	FACE	SIZE	FRAME	DEPTH	COST EACH MAT.	COST EACH INST.	COST EACH TOTAL
3200	Particle core/flush	birch	3'-0" x 7'-0"	pine	3-5/8"	330	256	586
3220					5-3/16"	335	270	605
3360			6'-0" x 7'-0"	pine	3-5/8"	545	430	975
3380					5-3/16"	555	445	1,000
3398								
3400		oak	3'-0" x 7'-0"	oak	3-5/8"	420	259	679
3420					5-3/16"	500	273	773
3560			6'-0" x 7'-0"	oak	3-5/8"	700	430	1,130
3580					5-3/16"	800	450	1,250
5000	Solid core/flush	birch	3'-0" x 7'-0"	pine	3-5/8"	415	250	665
5020					5-3/16"	425	264	689
5160			6'-0" x 7'-0"	pine	3-5/8"	720	415	1,135
5180					5-3/16"	730	435	1,165
5200		oak	3'-0" x 7'-0"	oak	3-5/8"	475	253	728
5220					5-3/16"	555	267	822
5360			6'-0" x 7'-0"	oak	3-5/8"	810	420	1,230
5380					5-3/16"	910	435	1,345
5400		M.D. overlay	3'-0" x 7'-0"	pine	3-5/8"	385	410	795

C1020 122 Wood Door/Metal Frame

	TYPE	FACE	SIZE	FRAME	DEPTH	COST EACH MAT.	COST EACH INST.	COST EACH TOTAL
1600	Hollow core/flush	lauan	3'-0" x 7'-0"	drywall K.D.	4-7/8"	390	221	611
1620				butt welded	8-3/4"	535	230	765
1760			6'-0" x 7'-0"	drywall K.D.	4-7/8"	640	375	1,015
1780				butt welded	8-3/4"	785	385	1,170
1800		birch	3'-0" x 7'-0"	drywall K.D.	4-7/8"	375	221	596
1820				butt welded	8-3/4"	520	230	750
1960			6'-0" x 7'-0"	drywall K.D.	4-7/8"	610	375	985
1980				butt welded	8-3/4"	755	385	1,140
2000		oak	3'-0" x 7'-0"	drywall K.D.	4-7/8"	475	221	696
2020				butt welded	8-3/4"	620	230	850
2160			6'-0" x 7'-0"	drywall K.D.	4-7/8"	815	375	1,190
2180				butt welded	8-3/4"	960	385	1,345
3000	Particle core/flush	lauan	3'-0" x 7'-0"	drywall K.D.	4-7/8"	390	237	627
3020				butt welded	8-3/4"	535	246	781
3160			6'-0" x 7'-0"	drywall K.D.	4-7/8"	640	405	1,045
3180				butt welded	8-3/4"	785	415	1,200
3200		birch	3'-0" x 7'-0"	drywall K.D.	4-7/8"	390	237	627
3220				butt welded	8-3/4"	535	246	781
3360			6'-0" x 7'-0"	drywall K.D.	4-7/8"	640	405	1,045
3380				butt welded	8-3/4"	785	415	1,200
3400		oak	3'-0" x 7'-0"	drywall K.D.	4-7/8"	440	237	677
3420				butt welded	8-3/4"	585	246	831
3560			6'-0" x 7'-0"	drywall K.D.	4-7/8"	745	405	1,150
3580				butt welded	8-3/4"	890	415	1,305
5000	Solid core/flush	birch	2'-8" x 6'-8"	drywall K.D.	4-7/8"	365	226	591
5020				butt welded	8-3/4"	510	235	745
5160			6'-0" x 7'-0"	drywall K.D.	4-7/8"	640	405	1,045
5180				butt welded	8-3/4"	785	415	1,200

C10 Interior Construction

C1020 Interior Doors

C1020 122 Wood Door/Metal Frame

	TYPE	FACE	SIZE	FRAME	DEPTH	MAT.	INST.	TOTAL
						\multicolumn{3}{c}{COST EACH}
5200	Solid core/flush	oak	2'-8" x 6'-8"	drywall K.D.	4-7/8"	380	226	606
5220				butt welded	8-3/4"	525	235	760
5360			6'-0" x 7'-0"	drywall K.D.	4-7/8"	665	405	1,070
5380				butt welded	8-3/4"	810	415	1,225
6000	1 hr/flush	birch	3'-0" x 7'-0"	drywall K.D.	4-7/8"	755	244	999
6020				butt welded	8-3/4"	940	253	1,193
6160			6'-0" x 7'-0"	drywall K.D.	4-7/8"	1,350	420	1,770
6180				butt welded	8-3/4"	1,525	430	1,955
6200		oak	3'-0" x 7'-0"	drywall K.D.	4-7/8"	780	244	1,024
6220				butt welded	8-3/4"	965	253	1,218
6360			6'-0" x 7'-0"	drywall K.D.	4-7/8"	1,400	420	1,820
6380				butt welded	8-3/4"	1,575	430	2,005
6400		walnut	3'-0" x 7'-0"	drywall K.D.	4-7/8"	835	244	1,079
6420				butt welded	8-3/4"	1,025	253	1,278
6560			6'-0" x 7'-0"	drywall K.D.	4-7/8"	1,525	420	1,945
6580				butt welded	8-3/4"	1,675	430	2,105
7600	1-1/2 hr/flush	birch	3'-0" x 7'-0"	drywall K.D.	4-7/8"	700	244	944
7620				butt welded	8-3/4"	885	253	1,138
7760			6'-0" x 7'-0"	drywall K.D.	4-7/8"	1,250	420	1,670
7780				butt welded	8-3/4"	1,425	430	1,855
7800		oak	3'-0" x 7'-0"	drywall K.D.	4-7/8"	775	244	1,019
7820				butt welded	8-3/4"	960	253	1,213
7960			6'-0" x 7'-0"	drywall K.D.	4-7/8"	1,400	420	1,820
7980				butt welded	8-3/4"	1,575	430	2,005
8000		walnut	3'-0" x 7'-0"	drywall K.D.	4-7/8"	790	244	1,034
8020				butt welded	8-3/4"	975	253	1,228
8160			6'-0" x 7'-0"	drywall K. D.	4 7/8"	1,425	420	1,845
8180				butt welded	8-3/4"	1,600	430	2,030

C1020 124 Wood Fire Doors/Metal Frames

	FIRE RATING (HOURS)	CORE MATERIAL	THICKNESS (IN.)	FACE MATERIAL	SIZE W(FT)XH(FT)	MAT.	INST.	TOTAL
0840	1	mineral	1-3/4"	birch	2'-8" x 6'-8"	695	231	926
0860		3 ply stile			3'-0" x 7'-0"	715	250	965
0880				oak	2'-8" x 6'-8"	730	231	961
0900					3'-0" x 7'-0"	740	250	990
0920				walnut	2'-8" x 6'-8"	750	231	981
0940					3'-0" x 7'-0"	795	250	1,045
0960				m.d. overlay	2'-8" x 6'-8"	585	395	980
0980					3'-0" x 7'-0"	675	405	1,080
1000	1-1/2	mineral	1-3/4"	birch	2'-8" x 6'-8"	585	231	816
1020		3 ply stile			3'-0" x 7'-0"	660	250	910
1040				oak	2'-8" x 6'-8"	710	231	941
1060					3'-0" x 7'-0"	735	250	985
1080				walnut	2'-8" x 6'-8"	675	231	906
1100					3'-0" x 7'-0"	750	250	1,000
1120				m.d. overlay	2'-8" x 6'-8"	630	395	1,025
1140					3'-0" x 7'-0"	685	405	1,090

C10 Interior Construction

C1020 Interior Doors

C1020 310	Hardware	COST EACH		
		MAT.	INST.	TOTAL
0060	**HINGES**			
0080				
0100	Full mortise, low frequency, steel base, 4-1/2" x 4-1/2", USP	9		9
0120	5" x 5" USP	16.25		16.25
0140	6" x 6" USP	49		49
0160	Average frequency, steel base, 4-1/2" x 4-1/2" USP	14.25		14.25
0180	5" x 5" USP	24		24
0200	6" x 6" USP	50.50		50.50
0220	High frequency, steel base, 4-1/2" x 4-1/2", USP	35		35
0240	5" x 5" USP	28		28
0260	6" x 6" USP	73		73
0280				
0300	**LOCKSETS**			
0320				
0340	Heavy duty cylindrical, passage door			
0360	Non-keyed, passage	67	46	113
0380	Privacy	78	46	124
0400	Keyed, single cylinder function	117	55.50	172.50
0420	Hotel	220	69	289
0440				
0460	For re-core cylinder, add	49.50		49.50
0480				
0500	**CLOSERS**			
0520				
0540	Rack & pinion			
0560	Adjustable backcheck, 3 way mount, all sizes, regular arm	192	92	284
0580	Hold open arm	192	92	284
0600	Fusible link	178	85	263
0620	Non-sized, regular arm	165	92	257
0640	4 way mount, non-sized, regular arm	208	92	300
0660	Hold open arm	274	92	366
0680				
0700	**PUSH, PULL**			
0720				
0740	Push plate, aluminum	14.95	46	60.95
0760	Bronze	29.50	46	75.50
0780	Pull handle, push bar, aluminum	145	50.50	195.50
0800	Bronze	186	55.50	241.50
0810				
0820	**PANIC DEVICES**			
0840				
0860	Narrow stile, rim mounted, bar, exit only	710	92	802
0880	Outside key & pull	985	111	1,096
0900	Bar and vertical rod, exit only	1,050	111	1,161
0920	Outside key & pull	1,050	138	1,188
0940	Bar and concealed rod, exit only	1,175	184	1,359
0960	Mortise, bar, exit only	650	138	788
0980	Touch bar, exit only	860	138	998
1000				
1020	**WEATHERSTRIPPING**			
1040				
1060	Interlocking, 3' x 7', zinc	48.50	184	232.50
1080	Spring type, 3' x 7', bronze	51	184	235

C10 Interior Construction

C1030 Fittings

Toilet Units

Entrance Screens

Urinal Screens

C1030 110	Toilet Partitions	COST PER UNIT		
		MAT.	INST.	TOTAL
0380	Toilet partitions, cubicles, ceiling hung, marble	1,775	500	2,275
0400	Painted metal	565	277	842
0420	Plastic laminate	605	277	882
0460	Stainless steel	1,100	277	1,377
0480	Handicap addition	475		475
0520	Floor and ceiling anchored, marble	2,250	400	2,650
0540	Painted metal	645	221	866
0560	Plastic laminate	860	221	1,081
0600	Stainless steel	1,225	221	1,446
0620	Handicap addition	345		345
0660	Floor mounted marble	1,325	335	1,660
0680	Painted metal	640	158	798
0700	Plastic laminate	535	158	693
0740	Stainless steel	1,525	158	1,683
0760	Handicap addition	345		345
0780	Juvenile deduction	45.50		45.50
0820	Floor mounted with headrail marble	1,250	335	1,585
0840	Painted metal	425	184	609
0860	Plastic laminate	825	184	1,009
0900	Stainless steel	1,150	184	1,334
0920	Handicap addition	410		410
0960	Wall hung, painted metal	700	158	858
1020	Stainless steel	1,825	158	1,983
1040	Handicap addition	410		410
1080	Entrance screens, floor mounted, 54" high, marble	830	112	942
1100	Painted metal	266	74	340
1140	Stainless steel	1,000	74	1,074
1300	Urinal screens, floor mounted, 24" wide, laminated plastic	198	138	336
1320	Marble	700	163	863
1340	Painted metal	253	138	391
1380	Stainless steel	655	138	793
1428	Wall mounted wedge type, painted metal	148	111	259
1430	Urinal screens, wall hung, plastic laminate/particle board	227	138	365
1460	Stainless steel	635	111	746
1500	Partitions, shower stall, single wall, painted steel, 2'-8" x 2'-8"	1,125	260	1,385
1510	Fiberglass, 2'-8" x 2'-8"	875	289	1,164
1520	Double wall, enameled steel, 2'-8" x 2'-8"	1,125	260	1,385
1530	Stainless steel, 2'-8" x 2'-8"	2,500	260	2,760
1540	Doors, plastic, 2' wide	158	72.50	230.50
1550	Tempered glass, chrome/brass frame-deluxe	685	650	1,335
1560	Tub enclosure, sliding panels, tempered glass, aluminum frame	450	325	775
1570	Chrome/brass frame-deluxe	1,325	435	1,760

C10 Interior Construction

C1030 Fittings

C1030 210	Fabricated Compact Units and Cubicles, EACH	COST EACH		
		MAT.	INST.	TOTAL
0500	Telephone enclosure, shelf type, wall hung, economy	1,125	221	1,346
0510	Deluxe	3,125	221	3,346
0520	Booth type, painted steel, economy	3,825	740	4,565
0530	Stainless steel, deluxe	12,700	740	13,440

C1030 310	Storage Specialties, EACH	COST EACH		
		MAT.	INST.	TOTAL
0200	Lockers, steel, single tier, 5' to 6' high, per opening, min.	171	46.50	217.50
0210	Maximum	300	54	354
0220	Two tier, minimum	87	25	112
0230	Maximum	127	32.50	159.50
0440	Parts bins, 3' wide x 7' high, 14 bins, 18" x 12" x 12"	350	87.50	437.50
0450	84 bins, 6" x 6" x 12"	945	109	1,054
0600	Shelving, metal industrial, braced, 3' wide, 1' deep	24.50	12.25	36.75
0610	2' deep	31.50	12.95	44.45
0620	Enclosed, 3' wide, 1' deep	45.50	14.30	59.80
0630	2' deep	61	16.80	77.80

C1030 510	Identifying/Visual Aid Specialties, EACH	COST EACH		
		MAT.	INST.	TOTAL
0100	Control boards, magnetic, porcelain finish, framed, 24" x 18"	219	138	357
0110	96" x 48"	1,175	221	1,396
0120	Directory boards, outdoor, black plastic, 36" x 24"	840	555	1,395
0130	36" x 36"	970	740	1,710
0140	Indoor, economy, open faced, 18" x 24"	168	158	326
0150	36" x 48"	276	184	460
0160	Building, aluminum, black felt panel, 24" x 18"	345	277	622
0170	48" x 72"	1,075	1,100	2,175
0500	Signs, interior electric exit sign, wall mounted, 6"	92	78.50	170.50
0510	Street, reflective alum., dbl. face, 4 way, w/bracket	254	37	291
0520	Letters, cast aluminum, 1/2" deep, 4" high	29	30.50	59.50
0530	1" deep, 10" high	61.50	30.50	92
0540	Plaques, cast aluminum, 20" x 30"	1,500	277	1,777
0550	Cast bronze, 36" x 48"	5,075	555	5,630

C1030 520	Identifying/Visual Aid Specialties, S.F.	COST PER S.F.		
		MAT.	INST.	TOTAL
0100	Bulletin board, cork sheets, no frame, 1/4" thick	2.28	3.82	6.10
0110	1/2" thick	4.55	3.82	8.37
0120	Aluminum frame, 1/4" thick, 3' x 5'	8.90	4.62	13.52
0130	4' x 8'	8.90	2.45	11.35
0200	Chalkboards, wall hung, alum, frame & chalktrough	14.40	2.45	16.85
0210	Wood frame & chalktrough	12.10	2.64	14.74
0220	Sliding board, one board with back panel	53.50	2.37	55.87
0230	Two boards with back panel	89	2.37	91.37
0240	Liquid chalk type, alum. frame & chalktrough	12.55	2.45	15
0250	Wood frame & chalktrough	33.50	2.45	35.95

C1030 710	Bath and Toilet Accessories, EACH	COST EACH		
		MAT.	INST.	TOTAL
0100	Specialties, bathroom accessories, st. steel, curtain rod, 5' long, 1" diam	37	42.50	79.50
0110	1-1/2" diam.	30	42.50	72.50
0120	Dispenser, towel, surface mounted	49	34.50	83.50
0130	Flush mounted with waste receptacle	315	55.50	370.50
0140	Grab bar, 1-1/4" diam., 12" long	32	23	55
0150	1-1/2" diam. 36" long	39.50	27.50	67

C10 Interior Construction

C1030 Fittings

C1030 710 Bath and Toilet Accessories, EACH

		COST EACH		
		MAT.	INST.	TOTAL
0160	Mirror, framed with shelf, 18" x 24"	205	27.50	232.50
0170	72" x 24"	330	92	422
0180	Toilet tissue dispenser, surface mounted, single roll	19.60	18.45	38.05
0190	Double roll	26	23	49
0200	Towel bar, 18" long	46	24	70
0210	30" long	123	26.50	149.50
0300	Medicine cabinets, sliding mirror doors, 20" x 16" x 4-3/4", unlighted	137	79	216
0310	24" x 19" x 8-1/2", lighted	197	111	308
0320	Triple door, 30" x 32", unlighted, plywood body	350	79	429
0330	Steel body	385	79	464
0340	Oak door, wood body, beveled mirror, single door	214	79	293
0350	Double door	430	92	522

C1030 730 Bath and Toilet Accessories, L.F.

		COST PER L.F.		
		MAT.	INST.	TOTAL
0100	Partitions, hospital curtain, ceiling hung, polyester oxford cloth	23	5.40	28.40
0110	Designer oxford cloth	19.45	6.85	26.30

C1030 830 Fabricated Cabinets & Counters, L.F.

		COST PER L.F.		
		MAT.	INST.	TOTAL
0110	Household, base, hardwood, one top drawer & one door below x 12" wide	264	44.50	308.50
0120	Four drawer x 24" wide	330	49.50	379.50
0130	Wall, hardwood, 30" high with one door x 12" wide	214	50.50	264.50
0140	Two doors x 48" wide	500	60	560
0150	Counter top-laminated plastic, stock, economy	18.35	18.45	36.80
0160	Custom-square edge, 7/8" thick	16.30	41.50	57.80
0170	School, counter, wood, 32" high	251	55.50	306.50
0180	Metal, 84" high	450	74	524

C1030 910 Other Fittings, EACH

		COST EACH		
		MAT.	INST.	TOTAL
0500	Mail boxes, horizontal, rear loaded, aluminum, 5" x 6" x 15" deep	38.50	16.25	54.75
0510	Front loaded, aluminum, 10" x 12" x 15" deep	119	27.50	146.50
0520	Vertical, front loaded, aluminum, 15" x 5" x 6" deep	33	16.25	49.25
0530	Bronze, duranodic finish	44	16.25	60.25
0540	Letter slot, post office	116	69	185
0550	Mail counter, window, post office, with grille	605	277	882
0600	Partition, woven wire, wall panel, 4' x 7' high	151	44.50	195.50
0610	Wall panel with window & shelf, 5' x 8' high	475	74	549
0620	Doors, swinging w/ no transom, 3' x 7' high	305	184	489
0630	Sliding, 6' x 10' high	830	277	1,107
0700	Turnstiles, one way, 4' arm, 46" diam., manual	1,800	221	2,021
0710	Electric	2,125	920	3,045
0720	3 arm, 5'-5" diam. & 7' high, manual	6,275	1,100	7,375
0730	Electric	8,475	1,850	10,325

C20 Stairs

C2010 Stair Construction

General Design: See reference section for code requirements. Maximum height between landings is 12'; usual stair angle is 20° to 50° with 30° to 35° best. Usual relation of riser to treads is:
- Riser + tread = 17.5.
- 2x (Riser) + tread = 25.
- Riser x tread = 70 or 75.

Maximum riser height is 7" for commercial, 8-1/4" for residential.
Usual riser height is 6-1/2" to 7-1/4".

Minimum tread width is 11" for commercial and 9" for residential.

For additional information please see reference section.

Cost Per Flight: Table below lists the cost per flight for 4'-0" wide stairs.
Side walls are not included.
Railings are included.

System Components	QUANTITY	UNIT	COST PER FLIGHT MAT.	INST.	TOTAL
SYSTEM C2010 110 0560					
STAIRS, C.I.P. CONCRETE WITH LANDING, 12 RISERS					
Concrete in place, free standing stairs not incl. safety treads	48.000	L.F.	283.20	1,917.12	2,200.32
Concrete in place, free standing stair landing	32.000	S.F.	153.28	527.36	680.64
Cast alum nosing insert, abr surface, pre-drilled, 3" wide x 4' long	12.000	Ea.	768	196.20	964.20
Industrial railing, welded, 2 rail 3'-6" high 1-1/2" pipe	18.000	L.F.	522	214.56	736.56
Wall railing with returns, steel pipe	17.000	L.F.	282.20	202.64	484.84
TOTAL			2,008.68	3,057.88	5,066.56

C2010 110	Stairs	MAT.	INST.	TOTAL
0470	Stairs, C.I.P. concrete, w/o landing, 12 risers, w/o nosing	1,075	2,350	3,425
0480	With nosing	1,850	2,550	4,400
0550	W/landing, 12 risers, w/o nosing	1,250	2,875	4,125
0560	With nosing	2,000	3,050	5,050
0570	16 risers, w/o nosing	1,525	3,575	5,100
0580	With nosing	2,550	3,850	6,400
0590	20 risers, w/o nosing	1,800	4,325	6,125
0600	With nosing	3,075	4,650	7,725
0610	24 risers, w/o nosing	2,075	5,050	7,125
0620	With nosing	3,600	5,450	9,050
0630	Steel, grate type w/nosing & rails, 12 risers, w/o landing	5,525	1,225	6,750
0640	With landing	7,225	1,675	8,900
0660	16 risers, with landing	9,075	2,075	11,150
0680	20 risers, with landing	10,900	2,475	13,375
0700	24 risers, with landing	12,700	2,875	15,575
0710	Concrete fill metal pan & picket rail, 12 risers, w/o landing	7,200	1,225	8,425
0720	With landing	9,475	1,825	11,300
0740	16 risers, with landing	11,900	2,250	14,150
0760	20 risers, with landing	14,300	2,650	16,950
0780	24 risers, with landing	16,700	3,025	19,725
0790	Cast iron tread & pipe rail, 12 risers, w/o landing	7,200	1,225	8,425
0800	With landing	9,475	1,825	11,300
0820	16 risers, with landing	11,900	2,250	14,150
0840	20 risers, with landing	14,300	2,650	16,950
0860	24 risers, with landing	16,700	3,025	19,725
0870	Pan tread & flat bar rail, pre-assembled, 12 risers, w/o landing	6,050	880	6,930
0880	With landing	11,000	1,375	12,375
0900	16 risers, with landing	12,100	1,550	13,650
0920	20 risers, with landing	14,200	1,850	16,050
0940	24 risers, with landing	16,200	2,125	18,325

Note: Row 0470/0480 references RC2010-100.

C20 Stairs

C2010 Stair Construction

C2010 110	Stairs	COST PER FLIGHT		
		MAT.	INST.	TOTAL
0950	Spiral steel, industrial checkered plate 4'-6' dia., 12 risers	6,950	810	7,760
0960	16 risers	9,275	1,075	10,350
0970	20 risers	11,600	1,350	12,950
0980	24 risers	13,900	1,600	15,500
0990	Spiral steel, industrial checkered plate 6'-0" dia., 12 risers	8,400	915	9,315
1000	16 risers	11,200	1,225	12,425
1010	20 risers	14,000	1,525	15,525
1020	24 risers	16,800	1,825	18,625
1030	Aluminum, spiral, stock units, 5'-0" dia., 12 risers	7,375	810	8,185
1040	16 risers	9,850	1,075	10,925
1050	20 risers	12,300	1,350	13,650
1060	24 risers	14,800	1,600	16,400
1070	Custom 5'-0" dia., 12 risers	14,100	810	14,910
1080	16 risers	18,800	1,075	19,875
1090	20 risers	23,500	1,350	24,850
1100	24 risers	28,200	1,600	29,800
1120	Wood, prefab box type, oak treads, wood rails 3'-6" wide, 14 risers	2,125	460	2,585
1150	Prefab basement type, oak treads, wood rails 3'-0" wide, 14 risers	960	114	1,074

C30 Interior Finishes

C3010 Wall Finishes

C3010 230	Paint & Covering	COST PER S.F.		
		MAT.	INST.	TOTAL
0060	Painting, interior on plaster and drywall, brushwork, primer & 1 coat	.13	.71	.84
0080	Primer & 2 coats	.21	.95	1.16
0100	Primer & 3 coats	.29	1.16	1.45
0120	Walls & ceilings, roller work, primer & 1 coat	.13	.48	.61
0140	Primer & 2 coats	.21	.61	.82
0160	Woodwork incl. puttying, brushwork, primer & 1 coat	.13	1.03	1.16
0180	Primer & 2 coats	.21	1.37	1.58
0200	Primer & 3 coats	.29	1.86	2.15
0260	Cabinets and casework, enamel, primer & 1 coat	.14	1.16	1.30
0280	Primer & 2 coats	.21	1.43	1.64
0300	Masonry or concrete, latex, brushwork, primer & 1 coat	.31	.97	1.28
0320	Primer & 2 coats	.40	1.39	1.79
0340	Addition for block filler	.23	1.19	1.42
0380	Fireproof paints, intumescent, 1/8" thick 3/4 hour	2.33	.95	3.28
0420	7/16" thick 2 hour	6.70	3.32	10.02
0440	1-1/16" thick 3 hour	10.95	6.65	17.60
0480	Miscellaneous metal brushwork, exposed metal, primer & 1 coat	.14	1.16	1.30
0500	Gratings, primer & 1 coat	.35	1.45	1.80
0600	Pipes over 12" diameter	.81	4.64	5.45
0700	Structural steel, brushwork, light framing 300-500 S.F./Ton	.10	1.80	1.90
0720	Heavy framing 50-100 S.F./Ton	.10	.90	1
0740	Spraywork, light framing 300-500 S.F./Ton	.11	.40	.51
0800	Varnish, interior wood trim, no sanding sealer & 1 coat	.08	1.16	1.24
0820	Hardwood floor, no sanding 2 coats	.17	.25	.42
0840	Wall coatings, acrylic glazed coatings, minimum	.34	.88	1.22
0860	Maximum	.72	1.52	2.24
0880	Epoxy coatings, minimum	.44	.88	1.32
1080	Maximum	1.14	3.20	4.34
1100	High build epoxy 50 mil, minimum	.74	1.19	1.93
1120	Maximum	1.25	4.89	6.14
1140	Laminated epoxy with fiberglass minimum	.79	1.57	2.36
1160	Maximum	1.43	3.20	4.63
1180	Sprayed perlite or vermiculite 1/16" thick, minimum	.30	.16	.46
1200	Maximum	.80	.73	1.53
1260	Wall coatings, vinyl plastic, minimum	.36	.63	.99
1280	Maximum	.89	1.93	2.82
1300	Urethane on smooth surface, 2 coats, minimum	.30	.41	.71
1320	Maximum	.64	.70	1.34
1340	3 coats, minimum	.39	.55	.94
1360	Maximum	.86	.99	1.85
1380	Ceramic-like glazed coating, cementitious, minimum	.52	1.06	1.58
1400	Maximum	.88	1.35	2.23
1420	Resin base, minimum	.36	.73	1.09
1440	Maximum	.58	1.41	1.99
1460	Wall coverings, aluminum foil	1.14	1.70	2.84
1500	Vinyl backing	6.10	1.94	8.04
1520	Cork tiles, 12"x12", light or dark, 3/16" thick	6.15	1.94	8.09
1540	5/16" thick	2.53	1.98	4.51
1580	Natural, non-directional, 1/2" thick	8.70	1.94	10.64
1600	12"x36", granular, 3/16" thick	1.42	1.21	2.63
1620	1" thick	1.83	1.26	3.09
1660	5/16" thick	6.20	1.98	8.18
1661	Paneling, prefinished plywood, birch	1.51	2.63	4.14
1662	Mahogany, African	2.83	2.77	5.60
1664	Oak or cherry	3.61	2.77	6.38
1665	Rosewood	5.15	3.46	8.61
1666	Teak	3.61	2.77	6.38
1720	Gypsum based, fabric backed, minimum	.88	.58	1.46

C30 Interior Finishes

C3010 Wall Finishes

C3010 230	Paint & Covering	COST PER S.F.		
		MAT.	INST.	TOTAL
1740	Average	1.38	.65	2.03
1760	Maximum	1.54	.73	2.27
1780	Vinyl wall covering, fabric back, light weight	.94	.73	1.67
1800	Medium weight	.91	.97	1.88
1820	Heavy weight	1.42	1.07	2.49
1840	Wall paper, double roll, solid pattern, avg. workmanship	.47	.73	1.20
1860	Basic pattern, avg. workmanship	1.07	.87	1.94
1880	Basic pattern, quality workmanship	2.13	1.07	3.20
1900	Grass cloths with lining paper, minimum	.73	1.17	1.90
1920	Maximum	2.93	1.33	4.26
1940	Ceramic tile, thin set, 4-1/4" x 4-1/4"	2.43	4.58	7.01
1960	12" x 12"	3.89	5.45	9.34

C3010 235	Paint Trim	COST PER L.F.		
		MAT.	INST.	TOTAL
2040	Painting, wood trim, to 6" wide, enamel, primer & 1 coat	.14	.58	.72
2060	Primer & 2 coats	.21	.74	.95
2080	Misc. metal brushwork, ladders	.71	5.80	6.51
2120	6" to 8" dia.	.21	2.44	2.65
2140	10" to 12" dia.	.64	3.66	4.30
2160	Railings, 2" pipe	.25	2.90	3.15
2180	Handrail, single	.18	1.16	1.34
2185	Caulking & Sealants, Polyurethane, In place, 1 or 2 component, 1/2" X 1/4"	.36	1.88	2.24

C30 Interior Finishes

C3020 Floor Finishes

C3020 410	Tile & Covering	COST PER S.F.		
		MAT.	INST.	TOTAL
0060	Carpet tile, nylon, fusion bonded, 18" x 18" or 24" x 24", 24 oz.	3.55	.67	4.22
0080	35 oz.	4.38	.67	5.05
0100	42 oz.	5.65	.67	6.32
0140	Carpet, tufted, nylon, roll goods, 12' wide, 26 oz.	6.35	.77	7.12
0160	36 oz.	9.45	.77	10.22
0180	Woven, wool, 36 oz.	16	.77	16.77
0200	42 oz.	16.55	.77	17.32
0220	Padding, add to above, minimum	.60	.36	.96
0240	Maximum	1.10	.36	1.46
0260	Composition flooring, acrylic, 1/4" thick	1.79	5.35	7.14
0280	3/8" thick	2.27	6.20	8.47
0300	Epoxy, minimum	3.03	4.13	7.16
0320	Maximum	4.37	5.70	10.07
0340	Epoxy terrazzo, minimum	6.45	6	12.45
0360	Maximum	9.80	8	17.80
0380	Mastic, hot laid, 1-1/2" thick, minimum	4.48	4.05	8.53
0400	Maximum	5.75	5.35	11.10
0420	Neoprene 1/4" thick, minimum	4.41	5.10	9.51
0440	Maximum	6.05	6.45	12.50
0460	Polyacrylate with ground granite 1/4", minimum	3.83	3.80	7.63
0480	Maximum	6.90	5.80	12.70
0500	Polyester with colored quart 2 chips 1/16", minimum	3.42	2.62	6.04
0520	Maximum	5.10	4.13	9.23
0540	Polyurethane with vinyl chips, minimum	8.25	2.62	10.87
0560	Maximum	11.90	3.24	15.14
0600	Concrete topping, granolithic concrete, 1/2" thick	.34	4.47	4.81
0620	1" thick	.69	4.58	5.27
0640	2" thick	1.38	5.25	6.63
0660	Heavy duty 3/4" thick, minimum	.47	6.90	7.37
0680	Maximum	.75	8.20	8.95
0700	For colors, add to above, minimum	.73	1.60	2.33
0720	Maximum	1.22	1.76	2.98
0740	Exposed aggregate finish, minimum	.27	.81	1.08
0760	Maximum	.83	1.09	1.92
0780	Abrasives, .25 P.S.F. add to above, minimum	.54	.60	1.14
0800	Maximum	.85	.60	1.45
0820	Dust on coloring, add, minimum	.73	.39	1.12
0840	Maximum	1.22	.81	2.03
0860	Floor coloring using 0.6 psf powdered color, 1/2" integral, minimum	5.45	4.47	9.92
0880	Maximum	5.90	4.47	10.37
1020	Integral topping and finish, 1:1:2 mix, 3/16" thick	.12	2.63	2.75
1040	1/2" thick	.31	2.78	3.09
1060	3/4" thick	.47	3.11	3.58
1080	1" thick	.62	3.51	4.13
1100	Terrazzo, minimum	3.58	15.95	19.53
1120	Maximum	6.85	8	14.85
1340	Cork tile, minimum	8.25	1.54	9.79
1360	Maximum	17.55	1.54	19.09
1380	Polyethylene, in rolls, minimum	3.82	1.76	5.58
1400	Maximum	7.35	1.76	9.11
1420	Polyurethane, thermoset, minimum	5.85	4.85	10.70
1440	Maximum	7.45	9.70	17.15
1460	Rubber, sheet goods, minimum	8.40	4.04	12.44
1480	Maximum	13.50	5.40	18.90
1500	Tile, minimum	7.90	1.21	9.11
1520	Maximum	15.35	1.76	17.11
1540	Synthetic turf, minimum	6.90	2.31	9.21
1560	Maximum	14.75	2.55	17.30

C30 Interior Finishes

C3020 Floor Finishes

C3020 410	Tile & Covering	COST PER S.F.		
		MAT.	INST.	TOTAL
1580	Vinyl, composition tile, minimum	.95	.97	1.92
1600	Maximum	1.05	.97	2.02
1620	Vinyl tile, minimum	3.74	.97	4.71
1640	Maximum	4.02	.97	4.99
1660	Sheet goods, minimum	4.51	1.94	6.45
1680	Maximum	7.85	2.42	10.27
1720	Tile, ceramic natural clay	6.40	4.75	11.15
1730	Marble, synthetic 12"x12"x5/8"	6.95	14.50	21.45
1740	Porcelain type, minimum	5.50	4.75	10.25
1760	Maximum	6.15	5.60	11.75
1800	Quarry tile, mud set, minimum	7.95	6.20	14.15
1820	Maximum	9.30	7.90	17.20
1840	Thin set, deduct		1.24	1.24
1850	Tile, natural stone, marble, in mortar bed, 12" x 12" x 3/8" thick	99.50	24	123.50
1860	Terrazzo precast, minimum	5.15	6.45	11.60
1880	Maximum	10.25	6.45	16.70
1900	Non-slip, minimum	21.50	16.30	37.80
1920	Maximum	22.50	21.50	44
2020	Wood block, end grain factory type, natural finish, 2" thick	4.48	4.43	8.91
2040	Fir, vertical grain, 1"x4", no finish, minimum	2.84	2.17	5.01
2060	Maximum	3.01	2.17	5.18
2080	Prefinished white oak, prime grade, 2-1/4" wide	5.35	3.25	8.60
2100	3-1/4" wide	5.95	2.99	8.94
2120	Maple strip, sanded and finished, minimum	5.70	4.73	10.43
2140	Maximum	5.90	4.73	10.63
2160	Oak strip, sanded and finished, minimum	3.59	4.73	8.32
2180	Maximum	5.30	4.73	10.03
2200	Parquetry, sanded and finished, minimum	6.95	4.94	11.89
2220	Maximum	5.40	7.05	12.45
2260	Add for sleepers on concrete, treated, 24" O.C., 1"x2"	1.73	2.83	4.56
2280	1"x3"	1.88	2.23	4.11
2300	2"x4"	.69	1.13	1.82
2320	2"x6"	.70	.85	1.55
2340	Underlayment, plywood, 3/8" thick	.89	.74	1.63
2350	1/2" thick	1.03	.76	1.79
2360	5/8" thick	1.20	.79	1.99
2370	3/4" thick	1.36	.85	2.21
2380	Particle board, 3/8" thick	.41	.74	1.15
2390	1/2" thick	.45	.76	1.21
2400	5/8" thick	.55	.79	1.34
2410	3/4" thick	.75	.85	1.60
2420	Hardboard, 4' x 4', .215" thick	.63	.74	1.37

C30 Interior Finishes

C3030 Ceiling Finishes

2 Coats of Plaster on Gypsum Lath on Wood Furring

Fiberglass Board on Exposed Suspended Grid System

Plaster and Metal Lath on Metal Furring

System Components	QUANTITY	UNIT	COST PER S.F.		
			MAT.	INST.	TOTAL
SYSTEM C3030 105 2400					
GYPSUM PLASTER, 2 COATS, 3/8″GYP. LATH, WOOD FURRING, FRAMING					
Gypsum plaster 2 coats no lath, on ceilings	.110	S.Y.	.42	3.02	3.44
Gypsum lath plain/perforated nailed, 3/8″ thick	.110	S.Y.	.52	.61	1.13
Add for ceiling installation	.110	S.Y.		.24	.24
Furring, 1″ x 3″ wood strips on wood joists	.750	L.F.	.29	1.19	1.48
Paint, primer and one coat	1.000	S.F.	.13	.48	.61
TOTAL			1.36	5.54	6.90

C3030 105			Plaster Ceilings				
	TYPE	LATH	FURRING	SUPPORT		COST PER S.F.	
						MAT.	INST.	TOTAL
2400	2 coat gypsum	3/8″ gypsum	1″x3″ wood, 16″ O.C.	wood		1.36	5.55	6.91
2500	Painted			masonry		1.36	5.65	7.01
2600				concrete		1.36	6.35	7.71
2700	3 coat gypsum	3.4# metal	1″x3″ wood, 16″ O.C.	wood		1.56	5.95	7.51
2800	Painted			masonry		1.56	6.05	7.61
2900				concrete		1.56	6.70	8.26
3000	2 coat perlite	3/8″ gypsum	1″x3″ wood, 16″ O.C.	wood		1.41	5.80	7.21
3100	Painted			masonry		1.41	5.90	7.31
3200				concrete		1.41	6.60	8.01
3300	3 coat perlite	3.4# metal	1″x3″ wood, 16″ O.C.	wood		1.44	5.95	7.39
3400	Painted			masonry		1.44	6.05	7.49
3500				concrete		1.44	6.70	8.14
3600	2 coat gypsum	3/8″ gypsum	3/4″ CRC, 12″ O.C.	1-1/2″ CRC, 48″O.C.		1.46	6.60	8.06
3700	Painted		3/4″ CRC, 16″ O.C.	1-1/2″ CRC, 48″O.C.		1.42	5.95	7.37
3800			3/4″ CRC, 24″ O.C.	1-1/2″ CRC, 48″O.C.		1.30	5.45	6.75
3900	2 coat perlite	3/8″ gypsum	3/4″ CRC, 12″ O.C.	1-1/2″ CRC, 48″O.C		1.51	7.10	8.61
4000	Painted		3/4″ CRC, 16″ O.C.	1-1/2″ CRC, 48″O.C.		1.47	6.45	7.92
4100			3/4″ CRC, 24″ O.C.	1-1/2″ CRC, 48″O.C.		1.35	5.95	7.30
4200	3 coat gypsum	3.4# metal	3/4″ CRC, 12″ O.C.	1-1/2″ CRC, 36″ O.C.		1.85	8.65	10.50
4300	Painted		3/4″ CRC, 16″ O.C.	1-1/2″ CRC, 36″ O.C.		1.81	8	9.81
4400			3/4″ CRC, 24″ O.C.	1-1/2″ CRC, 36″ O.C.		1.69	7.50	9.19
4500	3 coat perlite	3.4# metal	3/4″ CRC, 12″ O.C.	1-1/2″ CRC,36″ O.C.		2.03	9.55	11.58
4600	Painted		3/4″ CRC, 16″ O.C.	1-1/2″ CRC, 36″ O.C.		1.99	8.95	10.94
4700			3/4″ CRC, 24″ O.C.	1-1/2″ CRC, 36″ O.C.		1.87	8.45	10.32

C30 Interior Finishes

C3030 Ceiling Finishes

C3030 110 Drywall Ceilings

	TYPE	FINISH	FURRING	SUPPORT		COST PER S.F.		
						MAT.	INST.	TOTAL
4800	1/2" F.R. drywall	painted and textured	1"x3" wood, 16" O.C.	wood		.81	3.40	4.21
4900				masonry		.81	3.51	4.32
5000				concrete		.81	4.18	4.99
5100	5/8" F.R. drywall	painted and textured	1"x3" wood, 16" O.C.	wood		.83	3.40	4.23
5200				masonry		.83	3.51	4.34
5300				concrete		.83	4.18	5.01
5400	1/2" F.R. drywall	painted and textured	7/8" resil. channels	24" O.C.		.68	3.30	3.98
5500			1"x2" wood	stud clips		.85	3.17	4.02
5600			1-5/8" metal studs	24" O.C.		.72	2.94	3.66
5700	5/8" F.R. drywall	painted and textured	1-5/8" metal studs	24" O.C.		.74	2.94	3.68
5702								

C30 Interior Finishes

C3030 Ceiling Finishes

C3030 140 — Plaster Ceiling Components

		COST PER S.F.		
		MAT.	INST.	TOTAL
0060	Plaster, gypsum incl. finish, 2 coats			
0080	3 coats	.59	3.55	4.14
0100	Perlite, incl. finish, 2 coats	.47	3.55	4.02
0120	3 coats	.77	4.44	5.21
0140	Thin coat on drywall	.12	.69	.81
0200	Lath, gypsum, 3/8" thick	.52	.85	1.37
0220	1/2" thick	.55	.89	1.44
0240	5/8" thick	.30	1.04	1.34
0260	Metal, diamond, 2.5 lb.	.47	.69	1.16
0280	3.4 lb.	.55	.74	1.29
0300	Flat rib, 2.75 lb.	.36	.69	1.05
0320	3.4 lb.	.59	.74	1.33
0440	Furring, steel channels, 3/4" galvanized, 12" O.C.	.39	2.22	2.61
0460	16" O.C.	.35	1.61	1.96
0480	24" O.C.	.23	1.11	1.34
0500	1-1/2" galvanized, 12" O.C.	.49	2.46	2.95
0520	16" O.C.	.44	1.79	2.23
0540	24" O.C.	.29	1.20	1.49
0560	Wood strips, 1"x3", on wood, 12" O.C.	.38	1.58	1.96
0580	16" O.C.	.29	1.19	1.48
0600	24" O.C.	.19	.79	.98
0620	On masonry, 12" O.C.	.38	1.73	2.11
0640	16" O.C.	.29	1.30	1.59
0660	24" O.C.	.19	.87	1.06
0680	On concrete, 12" O.C.	.38	2.63	3.01
0700	16" O.C.	.29	1.97	2.26
0720	24" O.C.	.19	1.32	1.51

C3030 210 — Acoustical Ceilings

	TYPE	TILE	GRID	SUPPORT		COST PER S.F.		
						MAT.	INST.	TOTAL
5800	5/8" fiberglass board	24" x 48"	tee	suspended		2.03	1.66	3.69
5900		24" x 24"	tee	suspended		2.23	1.82	4.05
6000	3/4" fiberglass board	24" x 48"	tee	suspended		2.66	1.69	4.35
6100		24" x 24"	tee	suspended		2.86	1.85	4.71
6500	5/8" mineral fiber	12" x 12"	1"x3" wood, 12" O.C.	wood		2.56	3.42	5.98
6600				masonry		2.56	3.57	6.13
6700				concrete		2.56	4.47	7.03
6800	3/4" mineral fiber	12" x 12"	1"x3" wood, 12" O.C.	wood		2.34	3.42	5.76
6900				masonry		2.34	3.42	5.76
7000				concrete		2.34	3.42	5.76
7100	3/4" mineral fiber on	12" x 12"	25 ga. channels	runners		2.75	4.21	6.96
7102	5/8" F.R. drywall							
7200	5/8" plastic coated	12" x 12"		adhesive backed		2.60	1.84	4.44
7201	Mineral fiber							
7202								
7300	3/4" plastic coated	12" x 12"		adhesive backed		2.38	1.84	4.22
7301	Mineral fiber							
7302								
7400	3/4" mineral fiber	12" x 12"	conceal 2" bar & channels	suspended		2.70	4.16	6.86
7401								
7402								

C30 Interior Finishes

C3030 Ceiling Finishes

C3030 240	Acoustical Ceiling Components	COST PER S.F.		
		MAT.	INST.	TOTAL
2480	Ceiling boards, eggcrate, acrylic, 1/2" x 1/2" x 1/2" cubes	2	1.11	3.11
2500	Polystyrene, 3/8" x 3/8" x 1/2" cubes	1.68	1.09	2.77
2520	1/2" x 1/2" x 1/2" cubes	2.24	1.11	3.35
2540	Fiberglass boards, plain, 5/8" thick	1.06	.89	1.95
2560	3/4" thick	1.69	.92	2.61
2580	Grass cloth faced, 3/4" thick	2.35	1.11	3.46
2600	1" thick	3.10	1.14	4.24
2620	Luminous panels, prismatic, acrylic	2.44	1.38	3.82
2640	Polystyrene	1.25	1.38	2.63
2660	Flat or ribbed, acrylic	4.25	1.38	5.63
2680	Polystyrene	2.92	1.38	4.30
2700	Drop pan, white, acrylic	6.25	1.38	7.63
2720	Polystyrene	5.20	1.38	6.58
2740	Mineral fiber boards, 5/8" thick, standard	.75	.82	1.57
2760	Plastic faced	1.16	1.38	2.54
2780	2 hour rating	1.27	.82	2.09
2800	Perforated aluminum sheets, .024 thick, corrugated painted	2.52	1.13	3.65
2820	Plain	4.16	1.11	5.27
3080	Mineral fiber, plastic coated, 12" x 12" or 12" x 24", 5/8" thick	2.18	1.84	4.02
3100	3/4" thick	1.96	1.84	3.80
3120	Fire rated, 3/4" thick, plain faced	1.49	1.84	3.33
3140	Mylar faced	1.41	1.84	3.25
3160	Add for ceiling primer	.14		.14
3180	Add for ceiling cement	.42		.42
3240	Suspension system, furring, 1" x 3" wood 12" O.C.	.38	1.58	1.96
3260	T bar suspension system, 2' x 4' grid	.74	.69	1.43
3280	2' x 2' grid	.94	.85	1.79
3300	Concealed Z bar suspension system 12" module	.86	1.06	1.92
3320	Add to above for 1-1/2" carrier channels 4' O.C.	.12	1.18	1.30
3340	Add to above for carrier channels for recessed lighting	.22	1.20	1.42

D Services

No part of this publication may be reproduced, stored in a retrieval system, or transmitted in any form or by any means without prior written permission of RSMeans, a division of Reed Construction Data.

D10 Conveying

D1010 Elevators and Lifts

The hydraulic elevator obtains its motion from the movement of liquid under pressure in the piston connected to the car bottom. These pistons can provide travel to a maximum rise of 70' and are sized for the intended load. As the rise reaches the upper limits the cost tends to exceed that of a geared electric unit.

System Components	QUANTITY	UNIT	COST EACH		
			MAT.	INST.	TOTAL
SYSTEM D1010 110 2000					
PASS. ELEV., HYDRAULIC, 2500 LB., 5 FLOORS, 100 FPM					
Passenger elevator, hydraulic, 1500 lb capacity, 2 stop, 100 FPM	1.000	Ea.	43,400	17,000	60,400
Over 10' travel height, passenger elevator, hydraulic, add	40.000	V.L.F.	32,200	9,400	41,600
Passenger elevator, hydraulic, 2500 lb. capacity over standard, add	1.000	Ea.	2,925		2,925
Over 2 stops, passenger elevator, hydraulic, add	3.000	Stop	6,225	18,900	25,125
Hall lantern	5.000	Ea.	2,775	1,065	3,840
Maintenance agreement for pass. elev. 9 months	1.000	Ea.	4,125		4,125
Position indicator at lobby	1.000	Ea.	485	53	538
TOTAL			92,135	46,418	138,553

D1010 110	Hydraulic		COST EACH		
			MAT.	INST.	TOTAL
1300	Pass. elev., 1500 lb., 2 Floors, 100 FPM	RD1010 -010	48,600	17,400	66,000
1400	5 Floors, 100 FPM		89,000	46,400	135,400
1600	2000 lb., 2 Floors, 100 FPM		50,500	17,400	67,900
1700	5 floors, 100 FPM		91,000	46,400	137,400
1900	2500 lb., 2 Floors, 100 FPM		51,500	17,400	68,900
2000	5 floors, 100 FPM		92,000	46,400	138,400
2200	3000 lb., 2 Floors, 100 FPM		53,000	17,400	70,400
2300	5 floors, 100 FPM		93,500	46,400	139,900
2500	3500 lb., 2 Floors, 100 FPM		55,500	17,400	72,900
2600	5 floors, 100 FPM		96,000	46,400	142,400
2800	4000 lb., 2 Floors, 100 FPM		57,000	17,400	74,400
2900	5 floors, 100 FPM		97,500	46,400	143,900
3100	4500 lb., 2 Floors, 100 FPM		59,500	17,400	76,900
3200	5 floors, 100 FPM		100,000	46,400	146,400
4000	Hospital elevators, 3500 lb., 2 Floors, 100 FPM		74,500	17,400	91,900
4100	5 floors, 100 FPM		124,000	46,400	170,400
4300	4000 lb., 2 Floors, 100 FPM		74,500	17,400	91,900
4400	5 floors, 100 FPM		124,000	46,400	170,400
4600	4500 lb., 2 Floors, 100 FPM		81,000	17,400	98,400
4800	5 floors, 100 FPM		131,000	46,400	177,400
4900	5000 lb., 2 Floors, 100 FPM		84,500	17,400	101,900
5000	5 floors, 100 FPM		134,000	46,400	180,400
6700	Freight elevators (Class "B"), 3000 lb., 2 Floors, 50 FPM		118,500	22,200	140,700
6800	5 floors, 100 FPM		216,000	58,000	274,000
7000	4000 lb., 2 Floors, 50 FPM		125,500	22,200	147,700
7100	5 floors, 100 FPM		223,000	58,000	281,000
7500	10,000 lb., 2 Floors, 50 FPM		155,500	22,200	177,700
7600	5 floors, 100 FPM		253,000	58,000	311,000
8100	20,000 lb., 2 Floors, 50 FPM		206,500	22,200	228,700
8200	5 Floors, 100 FPM		304,000	58,000	362,000

D10 Conveying

D1010 Elevators and Lifts

Geared traction elevators are the intermediate group both in cost and in operating areas. These are in buildings of four to fifteen floors and speed ranges from 200' to 500' per minute.

System Components	QUANTITY	UNIT	COST EACH MAT.	COST EACH INST.	COST EACH TOTAL
SYSTEM D1010 140 1600					
PASSENGER, 2500 LB., 5 FLOORS, 200 FPM					
Passenger elevator, geared, 2000 lb. capacity, 4 stop, 200 FPM	1.000	Ea.	99,000	34,000	133,000
Over 40' travel height, passenger elevator electric, add	10.000	V.L.F.	8,750	2,350	11,100
Passenger elevator, electric, 2500 lb. cap., add	1.000	Ea.	4,050		4,050
Over 4 stops, passenger elevator, electric, add	1.000	Stop	6,350	6,300	12,650
Hall lantern	5.000	Ea.	2,775	1,065	3,840
Maintenance agreement for passenger elevator, 9 months	1.000	Ea.	4,125		4,125
Position indicator at lobby	1.000	Ea.	485	53	538
TOTAL			125,535	43,768	169,303

D1010 140	Traction Geared Elevators		MAT.	INST.	TOTAL
1300	Passenger, 2000 Lb., 5 floors, 200 FPM		121,500	43,800	165,300
1500	15 floors, 350 FPM		288,500	132,500	421,000
1600	2500 Lb., 5 floors, 200 FPM	RD1010 -010	125,500	43,800	169,300
1800	15 floors, 350 FPM		292,500	132,500	425,000
1900	3000 Lb., 5 floors, 200 FPM		126,000	43,800	169,800
2100	15 floors, 350 FPM		293,000	132,500	425,500
2200	3500 Lb., 5 floors, 200 FPM		127,500	43,800	171,300
2400	15 floors, 350 FPM		294,500	132,500	427,000
2500	4000 Lb., 5 floors, 200 FPM		128,500	43,800	172,300
2700	15 floors, 350 FPM		295,500	132,500	428,000
2800	4500 Lb., 5 floors, 200 FPM		131,000	43,800	174,800
3000	15 floors, 350 FPM		298,000	132,500	430,500
3100	5000 Lb., 5 floors, 200 FPM		133,500	43,800	177,300
3300	15 floors, 350 FPM		300,500	132,500	433,000
4000	Hospital, 3500 Lb., 5 floors, 200 FPM		122,000	43,800	165,800
4200	15 floors, 350 FPM		302,000	132,500	434,500
4300	4000 Lb., 5 floors, 200 FPM		122,000	43,800	165,800
4500	15 floors, 350 FPM		302,000	132,500	434,500
4600	4500 Lb., 5 floors, 200 FPM		127,500	43,800	171,300
4800	15 floors, 350 FPM		308,000	132,500	440,500
4900	5000 Lb., 5 floors, 200 FPM		129,500	43,800	173,300
5100	15 floors, 350 FPM		310,000	132,500	442,500
6000	Freight, 4000 Lb., 5 floors, 50 FPM class 'B'		143,500	45,300	188,800
6200	15 floors, 200 FPM class 'B'		347,500	157,500	505,000
6300	8000 Lb., 5 floors, 50 FPM class 'B'		182,000	45,300	227,300
6500	15 floors, 200 FPM class 'B'		386,000	157,500	543,500
7000	10,000 Lb., 5 floors, 50 FPM class 'B'		193,000	45,300	238,300
7200	15 floors, 200 FPM class 'B'		566,500	157,500	724,000
8000	20,000 Lb., 5 floors, 50 FPM class 'B'		213,500	45,300	258,800
8200	15 floors, 200 FPM class 'B'		587,000	157,500	744,500

D10 Conveying

D1010 Elevators and Lifts

Gearless traction elevators are used in high rise situations where speeds to over 1000 FPM are required to move passengers efficiently. This type of installation is also the most costly.

System Components	QUANTITY	UNIT	COST EACH MAT.	COST EACH INST.	COST EACH TOTAL
SYSTEM D1010 150 1700					
PASSENGER, 2500 LB., 10 FLOORS, 200 FPM					
Passenger elevator, geared, 2000 lb. capacity, 4 stop 100 FPM	1.000	Ea.	99,000	34,000	133,000
Over 10' travel height, passenger elevator electric, add	60.000	V.L.F.	52,500	14,100	66,600
Passenger elevator, electric, 2500 lb. cap., add	1.000	Ea.	4,050		4,050
Over 2 stops, passenger elevator, electric, add	6.000	Stop	38,100	37,800	75,900
Gearless variable voltage machinery, overhead	1.000	Ea.	90,000	24,300	114,300
Hall lantern	9.000	Ea.	5,550	2,130	7,680
Position indicator at lobby	1.000	Ea.	485	53	538
Fireman services control	1.000	Ea.	8,975	1,425	10,400
Emergency power manual switching, add	1.000	Ea.	540	213	753
Maintenance agreement for passenger elevator, 9 months	1.000	Ea.	4,125		4,125
TOTAL			303,325	114,021	417,346

D1010 150	Traction Gearless Elevators		COST EACH MAT.	COST EACH INST.	COST EACH TOTAL
1700	Passenger, 2500 Lb., 10 floors, 200 FPM		303,500	114,000	417,500
1900	30 floors, 600 FPM		662,000	291,500	953,500
2000	3000 Lb., 10 floors, 200 FPM	RD1010 -010	304,000	114,000	418,000
2200	30 floors, 600 FPM		663,000	291,500	954,500
2300	3500 Lb., 10 floors, 200 FPM		305,500	114,000	419,500
2500	30 floors, 600 FPM		664,000	291,500	955,500
2700	50 floors, 800 FPM		988,000	468,500	1,456,500
2800	4000 Lb., 10 floors, 200 FPM		306,500	114,000	420,500
3000	30 floors, 600 FPM		665,500	291,500	957,000
3200	50 floors, 800 FPM		989,000	468,500	1,457,500
3300	4500 Lb., 10 floors, 200 FPM		309,000	114,000	423,000
3500	30 floors, 600 FPM		667,500	291,500	959,000
3700	50 floors, 800 FPM		991,500	468,500	1,460,000
3800	5000 Lb., 10 floors, 200 FPM		311,500	114,000	425,500
4000	30 floors, 600 FPM		670,000	291,500	961,500
4200	50 floors, 800 FPM		994,000	468,500	1,462,500
6000	Hospital, 3500 Lb., 10 floors, 200 FPM		306,500	114,000	420,500
6200	30 floors, 600 FPM		694,500	291,500	986,000
6400	4000 Lb., 10 floors, 200 FPM		306,500	114,000	420,500
6600	30 floors, 600 FPM		694,500	291,500	986,000
6800	4500 Lb., 10 floors, 200 FPM		312,500	114,000	426,500
7000	30 floors, 600 FPM		700,500	291,500	992,000
7200	5000 Lb., 10 floors, 200 FPM		314,000	114,000	428,000
7400	30 floors, 600 FPM		702,000	291,500	993,500

D10 Conveying

D1020 Escalators and Moving Walks

Moving stairs can be used for buildings where 600 or more people are to be carried to the second floor or beyond. Freight cannot be carried on escalators and at least one elevator must be available for this function.

Carrying capacity is 5000 to 8000 people per hour. Power requirement is 2 KW to 3 KW per hour and incline angle is 30°.

D1020 110 Moving Stairs

	TYPE	HEIGHT	WIDTH	BALUSTRADE MATERIAL		COST EACH		
						MAT.	INST.	TOTAL
0100	Escalator	10ft	32"	glass		90,500	46,900	137,400
0150				metal		96,500	46,900	143,400
0200			48"	glass		96,500	46,900	143,400
0250				metal		104,000	46,900	150,900
0300		15ft	32"	glass		95,000	55,000	150,000
0350				metal		102,000	55,000	157,000
0400			48"	glass		100,000	55,000	155,000
0450				metal		107,500	55,000	162,500
0500		20ft	32"	glass		100,000	67,000	167,000
0550				metal		107,500	67,000	174,500
0600			48"	glass		108,500	67,000	175,500
0650				metal		117,000	67,000	184,000
0700		25ft	32"	glass		108,500	82,500	191,000
0750				metal		116,500	82,500	199,000
0800			48"	glass		125,000	82,500	207,500
0850				metal		134,000	82,500	216,500

D1020 210 Moving Walks

	TYPE	STORY HEIGHT	DEGREE SLOPE	WIDTH	COST RANGE	COST PER L.F.		
						MAT.	INST.	TOTAL
1500	Ramp	10'-23'	12	3'-4"	minimum	2,525	920	3,445
1600					maximum	3,200	1,125	4,325
2000	Walk	0'	0	2'-0"	minimum	940	505	1,445
2500					maximum	1,300	740	2,040
3000				3'-4"	minimum	2,125	740	2,865
3500					maximum	2,475	860	3,335

D10 Conveying

D1090 Other Conveying Systems

D1090 410	Miscellaneous Types	COST EACH		
		MAT.	INST.	TOTAL
1050	Conveyors, Horiz. belt, ctr drive, 60 FPM, cap 40 lb./L.F., belt x 26.5'L	3,475	2,175	5,650
1100	24" belt, 42' length	5,375	2,725	8,100
1200	24" belt, 62' length	7,175	3,650	10,825
1500	Inclined belt, horiz. loading and end idler assembly, 16" belt x 27.5'L	9,350	3,650	13,000
1600	24" belt, 27.5' length	10,500	7,300	17,800
3500	Pneumatic tube system, single, 2 station-100', stock economy model, 3" dia.	7,250	11,400	18,650
3600	4" diameter	8,200	15,200	23,400
4000	Twin tube system base cost, minimum	7,025	1,825	8,850
4100	Maximum	13,900	5,450	19,350
4200	Add per L.F., of system, minimum	9.05	14.60	23.65
4300	Maximum	26.50	36.50	63
5000	Completely automatic system, 15 to 50 stations, 4" dia., cost/station	22,100	4,700	26,800
5100	50 to 144 stations, 4" diameter, cost per station	17,200	4,275	21,475
7100	Corresp. lift, Self cont., 45 lb. cap., 2 stop, elec. enam. steel doors	9,200	6,275	15,475

D1090 510	Chutes	COST EACH		
		MAT.	INST.	TOTAL
0100	Chutes, linen or refuse incl. sprinklers, galv. steel, 18" diam., per floor	1,875	370	2,245
0110	Aluminized steel, 36" diam., per floor	2,900	465	3,365
0120	Mail, 8-3/4" x 3-1/2", aluminum & glass, per floor	715	260	975
0130	Bronze or stainless, per floor	1,100	289	1,389

D20 Plumbing

D2010 Plumbing Fixtures

Systems are complete with trim seat and rough-in (supply, waste and vent) for connection to supply branches and waste mains.

One Piece Wall Hung

Supply

Waste/Vent

Floor Mount

System Components	QUANTITY	UNIT	COST EACH		
			MAT.	INST.	TOTAL
SYSTEM D2010 110 1880					
WATER CLOSET, VITREOUS CHINA					
TANK TYPE, WALL HUNG, TWO PIECE					
Water closet, tank type vit china wall hung 2 pc. w/seat supply & stop	1.000	Ea.	690	228	918
Pipe Steel galvanized, schedule 40, threaded, 2" diam.	4.000	L.F.	58.60	75.60	134.20
Pipe, CI soil, no hub, cplg 10' OC, hanger 5' OC, 4" diam.	2.000	L.F.	36.70	42	78.70
Pipe, coupling, standard coupling, CI soil, no hub, 4" diam.	2.000	Ea.	40	73	113
Copper tubing type L solder joint, hangar 10' O.C., 1/2" diam.	6.000	L.F.	27.96	49.80	77.76
Wrought copper 90° elbow for solder joints 1/2" diam.	2.000	Ea.	6.22	67	73.22
Wrought copper Tee for solder joints 1/2" diam.	1.000	Ea.	5.30	51.50	56.80
Supports/carrier, water closet, siphon jet, horiz, single, 4" waste	1.000	Ea.	720	126	846
TOTAL			1,584.78	712.90	2,297.68

D2010 110	Water Closet Systems		COST EACH		
			MAT.	INST.	TOTAL
1800	Water closet, vitreous china				
1840	Tank type, wall hung				
1880	Close coupled two piece	RD2010 -400	1,575	715	2,290
1920	Floor mount, one piece		1,525	760	2,285
1960	One piece low profile		920	760	1,680
2000	Two piece close coupled		645	760	1,405
2040	Bowl only with flush valve				
2080	Wall hung		1,900	810	2,710
2120	Floor mount		840	770	1,610
2160	Floor mount, ADA compliant with 18" high bowl		840	790	1,630

D20 Plumbing

D2010 Plumbing Fixtures

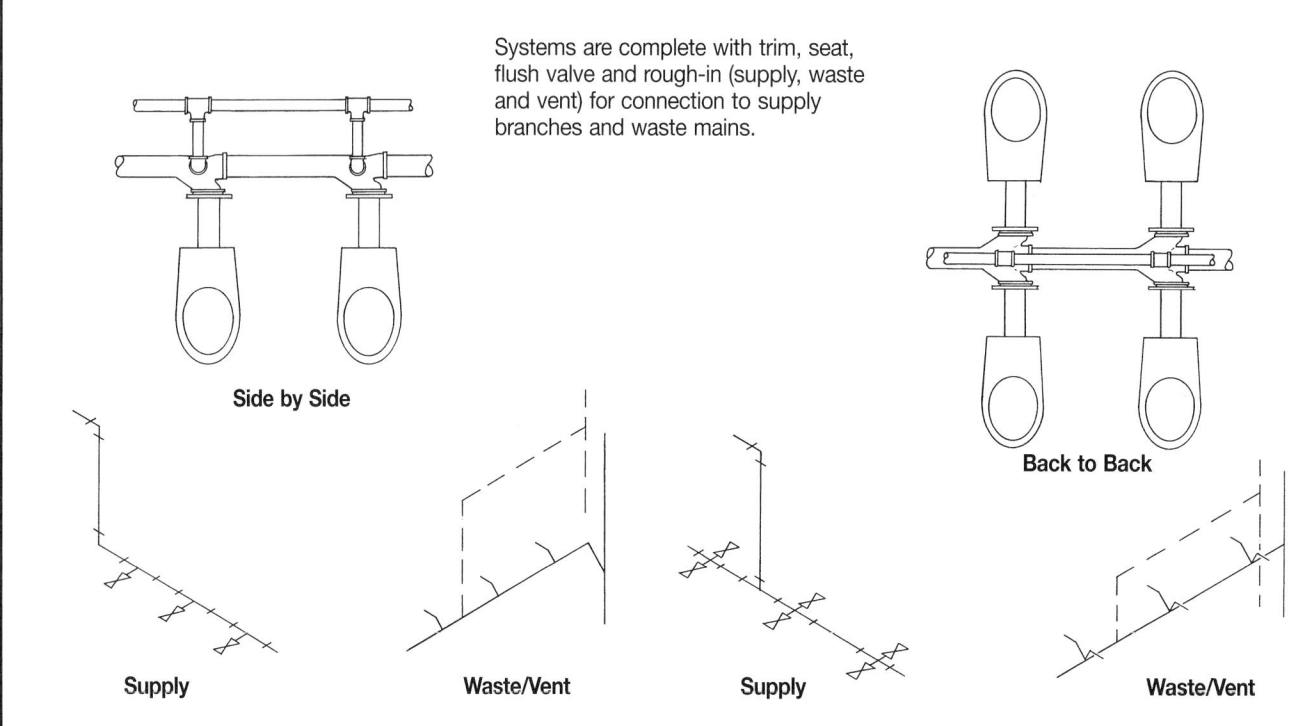

Systems are complete with trim, seat, flush valve and rough-in (supply, waste and vent) for connection to supply branches and waste mains.

System Components	QUANTITY	UNIT	COST EACH MAT.	COST EACH INST.	COST EACH TOTAL
SYSTEM D2010 120 1760					
WATER CLOSETS, BATTERY MOUNT, WALL HUNG, SIDE BY SIDE, FIRST CLOSET					
Water closet, bowl only w/flush valve, seat, wall hung	1.000	Ea.	780	209	989
Pipe, CI soil, no hub, cplg 10' OC, hanger 5' OC, 4" diam.	3.000	L.F.	55.05	63	118.05
Coupling, standard, CI, soil, no hub, 4" diam.	2.000	Ea.	40	73	113
Copper tubing, type L, solder joints, hangers 10' OC, 1" diam.	6.000	L.F.	63.30	59.40	122.70
Copper tubing, type DWV, solder joints, hangers 10'OC, 2" diam.	6.000	L.F.	138	91.80	229.80
Wrought copper 90° elbow for solder joints 1" diam.	1.000	Ea.	17.20	42	59.20
Wrought copper Tee for solder joints 1" diam.	1.000	Ea.	39.50	67	106.50
Support/carrier, siphon jet, horiz, adjustable single, 4" pipe	1.000	Ea.	720	126	846
Valve, gate, bronze, 125 lb, NRS, soldered 1" diam.	1.000	Ea.	77.50	35.50	113
Wrought copper, DWV, 90° elbow, 2" diam.	1.000	Ea.	59	67	126
TOTAL			1,989.55	833.70	2,823.25

D2010 120	Water Closets, Group		MAT.	INST.	TOTAL
1760	Water closets, battery mount, wall hung, side by side, first closet		2,000	835	2,835
1800	Each additional water closet, add		1,900	790	2,690
3000	Back to back, first pair of closets	RD2010 -400	3,450	1,100	4,550
3100	Each additional pair of closets, back to back		3,400	1,100	4,500

D20 Plumbing

D2010 Plumbing Fixtures

Systems are complete with trim, flush valve and rough-in (supply, waste and vent) for connection to supply branches and waste mains.

Stall Type

Supply   Waste/Vent

Wall Hung

System Components	QUANTITY	UNIT	COST EACH MAT.	COST EACH INST.	COST EACH TOTAL
SYSTEM D2010 210 2000					
URINAL, VITREOUS CHINA, WALL HUNG					
Urinal, wall hung, vitreous china, incl. hanger	1.000	Ea.	335	405	740
Pipe, steel, galvanized, schedule 40, threaded, 1-1/2" diam.	5.000	L.F.	54.75	75.75	130.50
Copper tubing type DWV, solder joint, hangers 10' OC, 2" diam.	3.000	L.F.	69	45.90	114.90
Combination Y & 1/8 bend for CI soil pipe, no hub, 3" diam.	1.000	Ea.	19.10		19.10
Pipe, CI, no hub, cplg. 10' OC, hanger 5' OC, 3" diam.	4.000	L.F.	56.20	75.60	131.80
Pipe coupling standard, CI soil, no hub, 3" diam.	3.000	Ea.	34.20	64	98.20
Copper tubing type L, solder joint, hanger 10' OC 3/4" diam.	5.000	L.F.	36	44.25	80.25
Wrought copper 90° elbow for solder joints 3/4" diam.	1.000	Ea.	7	35.50	42.50
Wrought copper Tee for solder joints, 3/4" diam.	1.000	Ea.	12.85	56	68.85
TOTAL			624.10	802	1,426.10

D2010 210	Urinal Systems	COST EACH MAT.	COST EACH INST.	COST EACH TOTAL
2000	Urinal, vitreous china, wall hung	625	800	1,425
2040	Stall type	1,375	960	2,335

D20 Plumbing

D2010 Plumbing Fixtures

Systems are complete with trim and rough-in (supply, waste and vent) to connect to supply branches and waste mains.

Vanity Top

Supply Waste/Vent

Wall Hung

System Components	QUANTITY	UNIT	COST EACH		
			MAT.	INST.	TOTAL
SYSTEM D2010 310 1560					
LAVATORY W/TRIM, VANITY TOP, P.E. ON C.I., 20" X 18"					
Lavatory w/trim, PE on CI, white, vanity top, 20" x 18" oval	1.000	Ea.	355	189	544
Pipe, steel, galvanized, schedule 40, threaded, 1-1/4" diam.	4.000	L.F.	37.40	54.40	91.80
Copper tubing type DWV, solder joint, hanger 10' OC 1-1/4" diam.	4.000	L.F.	53.40	44.80	98.20
Wrought copper DWV, Tee, sanitary, 1-1/4" diam.	1.000	Ea.	59.50	74.50	134
P trap w/cleanout, 20 ga., 1-1/4" diam.	1.000	Ea.	207	37.50	244.50
Copper tubing type L, solder joint, hanger 10' OC 1/2" diam.	10.000	L.F.	46.60	83	129.60
Wrought copper 90° elbow for solder joints 1/2" diam.	2.000	Ea.	6.22	67	73.22
Wrought copper Tee for solder joints, 1/2" diam.	2.000	Ea.	10.60	103	113.60
Stop, chrome, angle supply, 1/2" diam.	2.000	Ea.	21.80	61	82.80
TOTAL			797.52	714.20	1,511.72

D2010 310	Lavatory Systems	COST EACH		
		MAT.	INST.	TOTAL
1560	Lavatory w/trim, vanity top, PE on CI, 20" x 18", Vanity top by others.	800	715	1,515
1600	19" x 16" oval	630	715	1,345
1640	18" round	705	715	1,420
1680	Cultured marble, 19" x 17"	645	715	1,360
1720	25" x 19"	675	715	1,390
1760	Stainless, self-rimming, 25" x 22"	840	715	1,555
1800	17" x 22"	830	715	1,545
1840	Steel enameled, 20" x 17"	660	735	1,395
1880	19" round	630	735	1,365
1920	Vitreous china, 20" x 16"	730	750	1,480
1960	19" x 16"	735	750	1,485
2000	22" x 13"	735	750	1,485
2040	Wall hung, PE on CI, 18" x 15"	965	790	1,755
2080	19" x 17"	985	790	1,775
2120	20" x 18"	830	790	1,620
2160	Vitreous china, 18" x 15"	785	810	1,595
2200	19" x 17"	740	810	1,550
2240	24" x 20"	830	810	1,640
2300	20" x 27", handicap	1,100	875	1,975

Note: RD2010-400 (reference for line 1680)

D20 Plumbing

D2010 Plumbing Fixtures

Systems are complete with trim and rough-in (supply, waste and vent) to connect to supply branches and waste mains.

Countertop Single Bowl

Supply Waste/Vent

Countertop Double Bowl

System Components	QUANTITY	UNIT	COST EACH MAT.	COST EACH INST.	COST EACH TOTAL
SYSTEM D2010 410 1720					
KITCHEN SINK W/TRIM, COUNTERTOP, P.E. ON C.I., 24" X 21", SINGLE BOWL					
Kitchen sink, counter top, PE on CI, 1 bowl, 24" x 21" OD	1.000	Ea.	305	216	521
Pipe, steel, galvanized, schedule 40, threaded, 1-1/4" diam.	4.000	L.F.	37.40	54.40	91.80
Copper tubing, type DWV, solder, hangers 10' OC 1-1/2" diam.	6.000	L.F.	100.50	74.70	175.20
Wrought copper, DWV, Tee, sanitary, 1-1/2" diam.	1.000	Ea.	74	84	158
P trap, standard, copper, 1-1/2" diam.	1.000	Ea.	200	39.50	239.50
Copper tubing, type L, solder joints, hangers 10' OC 1/2" diam.	10.000	L.F.	46.60	83	129.60
Wrought copper 90° elbow for solder joints 1/2" diam.	2.000	Ea.	6.22	67	73.22
Wrought copper Tee for solder joints, 1/2" diam.	2.000	Ea.	10.60	103	113.60
Stop, angle supply, chrome, 1/2" CTS	2.000	Ea.	21.80	61	82.80
TOTAL			802.12	782.60	1,584.72

D2010 410	Kitchen Sink Systems	MAT.	INST.	TOTAL
1720	Kitchen sink w/trim, countertop, PE on CI, 24"x21", single bowl	800	785	1,585
1760	30" x 21" single bowl	1,100	785	1,885
1800	32" x 21" double bowl	870	845	1,715
1840	42" x 21" double bowl	1,800	855	2,655
1880	Stainless steel, 19" x 18" single bowl	1,125	785	1,910
1920	25" x 22" single bowl	1,200	785	1,985
1960	33" x 22" double bowl	1,575	845	2,420
2000	43" x 22" double bowl	1,750	855	2,605
2040	44" x 22" triple bowl	2,025	890	2,915
2080	44" x 24" corner double bowl	1,450	855	2,305
2120	Steel, enameled, 24" x 21" single bowl	1,000	785	1,785
2160	32" x 21" double bowl	1,050	845	1,895
2240	Raised deck, PE on CI, 32" x 21", dual level, double bowl	970	1,075	2,045
2280	42" x 21" dual level, triple bowl	1,850	1,175	3,025

D20 Plumbing

D2010 Plumbing Fixtures

Systems are complete with trim and rough-in (supply, waste and vent) to connect to supply branches and waste mains.

Single Compartment Sink   Supply   Waste/Vent   Double Compartment Sink

System Components	QUANTITY	UNIT	COST EACH MAT.	INST.	TOTAL
SYSTEM D2010 420 1760					
LAUNDRY SINK W/TRIM, PE ON CI, BLACK IRON FRAME					
24" X 20" OD, SINGLE COMPARTMENT					
Laundry sink PE on CI w/trim & frame, 24" x 21" OD, 1 compartment	1.000	Ea.	470	202	672
Pipe, steel, galvanized, schedule 40, threaded, 1-1/4" diam	4.000	L.F.	37.40	54.40	91.80
Copper tubing, type DWV, solder joint, hanger 10' OC 1-1/2"diam	6.000	L.F.	100.50	74.70	175.20
Wrought copper, DWV, Tee, sanitary, 1-1/2" diam	1.000	Ea.	74	84	158
P trap, standard, copper, 1-1/2" diam	1.000	Ea.	200	39.50	239.50
Copper tubing type L, solder joints, hangers 10' OC, 1/2" diam	10.000	L.F.	46.60	83	129.60
Wrought copper 90° elbow for solder joints 1/2" diam	2.000	Ea.	6.22	67	73.22
Wrought copper Tee for solder joints, 1/2" diam	2.000	Ea.	10.60	103	113.60
Stop, angle supply, 1/2" diam	2.000	Ea.	21.80	61	82.80
TOTAL			967.12	768.60	1,735.72

D2010 420	Laundry Sink Systems		COST EACH MAT.	INST.	TOTAL
1740	Laundry sink w/trim, PE on CI, black iron frame				
1760	24" x 20", single compartment		965	770	1,735
1800	24" x 23" single compartment	RD2010 -400	990	770	1,760
1840	48" x 21" double compartment		1,450	835	2,285
1920	Molded stone, on wall, 22" x 21" single compartment		705	770	1,475
1960	45"x 21" double compartment		895	835	1,730
2040	Plastic, on wall or legs, 18" x 23" single compartment		665	755	1,420
2080	20" x 24" single compartment		665	755	1,420
2120	36" x 23" double compartment		730	810	1,540
2160	40" x 24" double compartment		830	810	1,640

D20 Plumbing

D2010 Plumbing Fixtures

Corrosion resistant laboratory sink systems are complete with trim and rough-in (supply, waste and vent) to connect to supply branches and waste mains.

Laboratory Sink

Supply   Waste/Vent

Polypropylene Cup Sink

System Components	QUANTITY	UNIT	COST EACH		
			MAT.	INST.	TOTAL
SYSTEM D2010 430 1600					
LABORATORY SINK W/TRIM, STAINLESS STEEL, SINGLE BOWL					
DOUBLE DRAINBOARD, 54″ X 24″ O.D.					
Sink w/trim, stainless steel, 1 bowl, 2 drainboards 54″ x 24″ OD	1.000	Ea.	1,825	405	2,230
Pipe, polypropylene, schedule 40, acid resistant 1-1/2″ diam.	10.000	L.F.	83.50	178	261.50
Tee, sanitary, polypropylene, acid resistant, 1-1/2″ diam.	1.000	Ea.	18.05	67	85.05
P trap, polypropylene, acid resistant, 1-1/2″ diam.	1.000	Ea.	58	39.50	97.50
Copper tubing type L, solder joint, hanger 10′ O.C. 1/2″ diam.	10.000	L.F.	46.60	83	129.60
Wrought copper 90° elbow for solder joints 1/2″ diam.	2.000	Ea.	6.22	67	73.22
Wrought copper Tee for solder joints, 1/2″ diam.	2.000	Ea.	10.60	103	113.60
Stop, angle supply, chrome, 1/2″ diam.	2.000	Ea.	21.80	61	82.80
TOTAL			2,069.77	1,003.50	3,073.27

D2010 430	Laboratory Sink Systems	COST EACH		
		MAT.	INST.	TOTAL
1580	Laboratory sink w/trim, stainless steel, single bowl,			
1590	Stainless steel, single bowl,			
1600	Double drainboard, 54″ x 24″ O.D. R2010 -400	2,075	1,000	3,075
1640	Single drainboard, 47″ x 24″O.D.	1,525	1,000	2,525
1670	Stainless steel, double bowl,			
1680	70″ x 24″ O.D.	2,275	1,000	3,275
1750	Polyethylene, single bowl,			
1760	Flanged, 14-1/2″ x 14-1/2″ O.D.	495	905	1,400
1800	18-1/2″ x 18-1/2″ O.D.	595	905	1,500
1840	23-1/2″ x 20-1/2″ O.D.	670	905	1,575
1920	Polypropylene, cup sink, oval, 7″ x 4″ O.D.	370	795	1,165
1960	10″ x 4-1/2″ O.D.	395	795	1,190

D20 Plumbing

D2010 Plumbing Fixtures

Corrosion resistant laboratory sink systems are complete with trim and rough-in (supply, waste and vent) to connect to supply branches and waste mains.

Wall Hung

Supply

Waste/Vent

Corner, Floor

System Components	QUANTITY	UNIT	COST EACH		
			MAT.	INST.	TOTAL
SYSTEM D2010 440 4260					
SERVICE SINK, PE ON CI, CORNER FLOOR, 28"X28", W/RIM GUARD & TRIM					
Service sink, corner floor, PE on CI, 28" x 28", w/rim guard & trim	1.000	Ea.	1,050	275	1,325
Copper tubing type DWV, solder joint, hanger 10'OC 3" diam.	6.000	L.F.	261	126	387
Copper tubing type DWV, solder joint, hanger 10'OC 2" diam	4.000	L.F.	92	61.20	153.20
Wrought copper DWV, Tee, sanitary, 3" diam.	1.000	Ea.	355	173	528
P trap with cleanout & slip joint, copper 3" diam	1.000	Ea.	740	61	801
Copper tubing, type L, solder joints, hangers 10' OC, 1/2" diam	10.000	L.F.	46.60	83	129.60
Wrought copper 90° elbow for solder joints 1/2" diam	2.000	Ea.	6.22	67	73.22
Wrought copper Tee for solder joints, 1/2" diam	2.000	Ea.	10.60	103	113.60
Stop, angle supply, chrome, 1/2" diam	2.000	Ea.	21.80	61	82.80
TOTAL			2,583.22	1,010.20	3,593.42

D2010 440	Service Sink Systems	COST EACH		
		MAT.	INST.	TOTAL
4260	Service sink w/trim, PE on CI, corner floor, 28" x 28", w/rim guard	2,575	1,000	3,575
4300	Wall hung w/rim guard, 22" x 18"	2,775	1,175	3,950
4340	24" x 20"	2,875	1,175	4,050
4380	Vitreous china, wall hung 22" x 20"	2,600	1,175	3,775

D20 Plumbing

D2010 Plumbing Fixtures

Systems are complete with trim and rough-in (supply, waste and vent) to connect to supply branches and waste mains.

Recessed Bathtub

Supply

Waste/Vent

Corner Bathtub

System Components	QUANTITY	UNIT	COST EACH		
			MAT.	INST.	TOTAL
SYSTEM D2010 510 2000					
BATHTUB, RECESSED, PORCELAIN ENAMEL ON CAST IRON,, 48" x 42"					
Bath tub, porcelain enamel on cast iron, w/fittings, 48" x 42"	1.000	Ea.	2,825	305	3,130
Pipe, steel, galvanized, schedule 40, threaded, 1-1/4" diam.	4.000	L.F.	37.40	54.40	91.80
Pipe, CI no hub soil w/couplings 10' OC, hangers 5' OC, 4" diam.	3.000	L.F.	55.05	63	118.05
Combination Y and 1/8 bend for C.I. soil pipe, no hub, 4" pipe size	1.000	Ea.	49.50		49.50
Drum trap, 3" x 5", copper, 1-1/2" diam.	1.000	Ea.	280	42	322
Copper tubing type L, solder joints, hangers 10' OC 1/2" diam.	10.000	L.F.	46.60	83	129.60
Wrought copper 90° elbow, solder joints, 1/2" diam.	2.000	Ea.	6.22	67	73.22
Wrought copper Tee, solder joints, 1/2" diam.	2.000	Ea.	10.60	103	113.60
Stop, angle supply, 1/2" diameter	2.000	Ea.	21.80	61	82.80
Copper tubing type DWV, solder joints, hanger 10' OC 1-1/2" diam.	3.000	L.F.	50.25	37.35	87.60
Pipe coupling, standard, C.I. soil no hub, 4" pipe size	2.000	Ea.	40	73	113
TOTAL			3,422.42	888.75	4,311.17

D2010 510	Bathtub Systems		COST EACH		
			MAT.	INST.	TOTAL
2000	Bathtub, recessed, P.E. on Cl., 48" x 42"	RD2010 -400	3,425	890	4,315
2040	72" x 36"		3,525	990	4,515
2080	Mat bottom, 5' long		1,800	860	2,660
2120	5'-6" long		2,525	890	3,415
2160	Corner, 48" x 42"		3,425	860	4,285
2200	Formed steel, enameled, 4'-6" long		1,150	795	1,945

D20 Plumbing

D2010 Plumbing Fixtures

Circular Fountain

Systems are complete with trim, flush valve and rough-in (supply, waste and vent) for connection to supply branches and waste mains.

Supply

Waste/Vent

Semi-Circular Fountain

System Components	QUANTITY	UNIT	COST EACH		
			MAT.	INST.	TOTAL
SYSTEM D2010 610 1760					
GROUP WASH FOUNTAIN, PRECAST TERRAZZO					
CIRCULAR, 36" DIAMETER					
Wash fountain, group, precast terrazzo, foot control 36" diam.	1.000	Ea.	6,000	630	6,630
Copper tubing type DWV, solder joint, hanger 10' OC, 2" diam.	10.000	L.F.	230	153	383
P trap, standard, copper, 2" diam.	1.000	Ea.	310	45	355
Wrought copper, Tee, sanitary, 2" diam.	1.000	Ea.	86.50	96	182.50
Copper tubing type L, solder joint, hanger 10' OC 1/2" diam.	20.000	L.F.	93.20	166	259.20
Wrought copper 90° elbow for solder joints 1/2" diam.	3.000	Ea.	9.33	100.50	109.83
Wrought copper Tee for solder joints, 1/2" diam.	2.000	Ea.	10.60	103	113.60
TOTAL			6,739.63	1,293.50	8,033.13

D2010 610	Group Wash Fountain Systems	COST EACH		
		MAT.	INST.	TOTAL
1740	Group wash fountain, precast terrazzo			
1760	Circular, 36" diameter	6,750	1,300	8,050
1800	54" diameter	8,225	1,425	9,650
1840	Semi-circular, 36" diameter	6,000	1,300	7,300
1880	54" diameter	7,400	1,425	8,825
1960	Stainless steel, circular, 36" diameter	6,750	1,200	7,950
2000	54" diameter	8,500	1,325	9,825
2040	Semi-circular, 36" diameter	5,975	1,200	7,175
2080	54" diameter	7,350	1,325	8,675
2160	Thermoplastic, circular, 36" diameter	5,075	980	6,055
2200	54" diameter	5,775	1,125	6,900
2240	Semi-circular, 36" diameter	4,075	980	5,055
2280	54" diameter	5,600	1,125	6,725

D20 Plumbing

D2010 Plumbing Fixtures

Systems are complete with trim and rough-in (supply, waste and vent) for connection to supply branches and waste mains.

Three Wall

Supply   Waste/Vent

Corner Angle

System Components	QUANTITY	UNIT	COST EACH		
			MAT.	INST.	TOTAL
SYSTEM D2010 710 1560					
SHOWER, STALL, BAKED ENAMEL, MOLDED STONE RECEPTOR, 30" SQUARE					
Shower stall, enameled steel, molded stone receptor, 30" square	1.000	Ea.	1,275	233	1,508
Copper tubing type DWV, solder joints, hangers 10' OC, 2" diam.	6.000	L.F.	100.50	74.70	175.20
Wrought copper DWV, Tee, sanitary, 2" diam.	1.000	Ea.	74	84	158
Trap, standard, copper, 2" diam.	1.000	Ea.	200	39.50	239.50
Copper tubing type L, solder joint, hanger 10' OC 1/2" diam.	16.000	L.F.	74.56	132.80	207.36
Wrought copper 90° elbow for solder joints 1/2" diam.	3.000	Ea.	9.33	100.50	109.83
Wrought copper Tee for solder joints, 1/2" diam.	2.000	Ea.	10.60	103	113.60
Stop and waste, straightway, bronze, solder joint 1/2" diam.	2.000	Ea.	25.50	56	81.50
TOTAL			1,769.49	823.50	2,592.99

D2010 710	Shower Systems		COST EACH		
			MAT.	INST.	TOTAL
1560	Shower, stall, baked enamel, molded stone receptor, 30" square		1,775	825	2,600
1600	32" square		1,775	835	2,610
1640	Terrazzo receptor, 32" square	RD2010 -400	1,975	835	2,810
1680	36" square		2,125	845	2,970
1720	36" corner angle		2,400	845	3,245
1800	Fiberglass one piece, three walls, 32" square		1,100	810	1,910
1840	36" square		1,125	810	1,935
1880	Polypropylene, molded stone receptor, 30" square		1,200	1,200	2,400
1920	32" square		1,200	1,200	2,400
1960	Built-in head, arm, bypass, stops and handles		121	310	431
2050	Shower, stainless steel panels, handicap				
2100	w/fixed and handheld head, control valves, grab bar, and seat		6,025	3,700	9,725
2500	Shower, group with six heads, thermostatic mix valves & balancing valve		5,450	910	6,360
2520	Five heads		4,525	825	5,350

D20 Plumbing

D2010 Plumbing Fixtures

Wall Mounted, No Back

Systems are complete with trim and rough-in (supply, waste and vent) to connect to supply branches and waste mains.

Supply   Waste/Vent

Wall Mounted, Low Back

System Components	QUANTITY	UNIT	COST EACH		
			MAT.	INST.	TOTAL
SYSTEM D2010 810 1800					
DRINKING FOUNTAIN, ONE BUBBLER, WALL MOUNTED					
NON RECESSED, BRONZE, NO BACK					
Drinking fountain, wall mount, bronze, 1 bubbler	1.000	Ea.	1,150	168	1,318
Copper tubing, type L, solder joint, hanger 10' OC 3/8" diam.	5.000	L.F.	20.95	40	60.95
Stop, supply, straight, chrome, 3/8" diam.	1.000	Ea.	12.45	28	40.45
Wrought copper 90° elbow for solder joints 3/8" diam.	1.000	Ea.	9.35	30.50	39.85
Wrought copper Tee for solder joints, 3/8" diam.	1.000	Ea.	15.75	48	63.75
Copper tubing, type DWV, solder joint, hanger 10' OC 1-1/4" diam.	4.000	L.F.	53.40	44.80	98.20
P trap, standard, copper drainage, 1-1/4" diam.	1.000	Ea.	207	37.50	244.50
Wrought copper, DWV, Tee, sanitary, 1-1/4" diam.	1.000	Ea.	59.50	74.50	134
TOTAL			1,528.40	471.30	1,999.70

D2010 810	Drinking Fountain Systems		COST EACH		
			MAT.	INST.	TOTAL
1740	Drinking fountain, one bubbler, wall mounted				
1760	Non recessed				
1800	Bronze, no back	RD2010 -400	1,525	470	1,995
1840	Cast iron, enameled, low back		1,375	470	1,845
1880	Fiberglass, 12" back		2,075	470	2,545
1920	Stainless steel, no back		1,525	470	1,995
1960	Semi-recessed, poly marble		1,400	470	1,870
2040	Stainless steel		1,625	470	2,095
2080	Vitreous china		1,350	470	1,820
2120	Full recessed, poly marble		2,125	470	2,595
2200	Stainless steel		2,050	470	2,520
2240	Floor mounted, pedestal type, aluminum		2,850	640	3,490
2320	Bronze		2,500	640	3,140
2360	Stainless steel		2,300	640	2,940

D20 Plumbing

D2010 Plumbing Fixtures

Systems are complete with trim and rough-in (supply, waste and vent) for connection to supply branches and waste mains.

Wall Hung

Supply

Waste/Vent

Floor Mounted

System Components	QUANTITY	UNIT	COST EACH MAT.	COST EACH INST.	COST EACH TOTAL
SYSTEM D2010 820 1840					
WATER COOLER, ELECTRIC, SELF CONTAINED, WALL HUNG, 8.2 G.P.H.					
Water cooler, wall mounted, 8.2 GPH	1.000	Ea.	795	305	1,100
Copper tubing type DWV, solder joint, hanger 10' OC 1-1/4" diam.	4.000	L.F.	53.40	44.80	98.20
Wrought copper DWV, Tee, sanitary 1-1/4" diam.	1.000	Ea.	59.50	74.50	134
P trap, copper drainage, 1-1/4" diam.	1.000	Ea.	207	37.50	244.50
Copper tubing type L, solder joint, hanger 10' OC 3/8" diam.	5.000	L.F.	20.95	40	60.95
Wrought copper 90° elbow for solder joints 3/8" diam.	1.000	Ea.	9.35	30.50	39.85
Wrought copper Tee for solder joints, 3/8" diam.	1.000	Ea.	15.75	48	63.75
Stop and waste, straightway, bronze, solder, 3/8" diam.	1.000	Ea.	12.45	28	40.45
TOTAL			1,173.40	608.30	1,781.70

D2010 820	Water Cooler Systems		MAT.	INST.	TOTAL
1840	Water cooler, electric, wall hung, 8.2 G.P.H.		1,175	610	1,785
1880	Dual height, 14.3 G.P.H.		1,450	625	2,075
1920	Wheelchair type, 7.5 G.P.H.	RD2010 -400	1,375	610	1,985
1960	Semi recessed, 8.1 G.P.H.		1,175	610	1,785
2000	Full recessed, 8 G.P.H.		2,500	650	3,150
2040	Floor mounted, 14.3 G.P.H.		1,325	525	1,850
2080	Dual height, 14.3 G.P.H.		1,650	640	2,290
2120	Refrigerated compartment type, 1.5 G.P.H.		1,925	525	2,450

D20 Plumbing

D2010 Plumbing Fixtures

Two Fixture Bathroom Systems consisting of a lavatory, water closet, and rough-in service piping.

- Prices for plumbing and fixtures only.

*Common wall is with an adjacent bathroom.

System Components	QUANTITY	UNIT	COST EACH		
			MAT.	INST.	TOTAL
SYSTEM D2010 920 1180					
BATHROOM, LAVATORY & WATER CLOSET, 2 WALL PLUMBING, STAND ALONE					
Water closet, 2 Pc. close cpld. vit .china flr. mntd. w/seat, supply & stop	1.000	Ea.	253	228	481
Water closet, rough-in waste & vent	1.000	Set	355	395	750
Lavatory w/ftngs., wall hung, white, PE on CI, 20" x 18"	1.000	Ea.	300	151	451
Lavatory, rough-in waste & vent	1.000	Set	530	730	1,260
Copper tubing type L, solder joint, hanger 10' OC 1/2" diam.	10.000	L.F.	46.60	83	129.60
Pipe, steel, galvanized, schedule 40, threaded, 2" diam.	12.000	L.F.	175.80	226.80	402.60
Pipe, CI soil, no hub, coupling 10' OC, hanger 5' OC, 4" diam.	7.000	L.F.	130.20	154	284.20
TOTAL			1,790.60	1,967.80	3,758.40

D2010 920	Two Fixture Bathroom, Two Wall Plumbing	COST EACH		
		MAT.	INST.	TOTAL
1180	Bathroom, lavatory & water closet, 2 wall plumbing, stand alone	1,800	1,975	3,775
1200	Share common plumbing wall*	1,600	1,700	3,300

D2010 922	Two Fixture Bathroom, One Wall Plumbing	COST EACH		
		MAT.	INST.	TOTAL
2220	Bathroom, lavatory & water closet, one wall plumbing, stand alone	1,650	1,775	3,425
2240	Share common plumbing wall*	1,450	1,500	2,950

D20 Plumbing

D2010 Plumbing Fixtures

Three Fixture Bathroom Systems consisting of a lavatory, water closet, bathtub or shower and rough-in service piping.
- Prices for plumbing and fixtures only.

*Common wall is with an adjacent bathroom.

System Components	QUANTITY	UNIT	COST EACH MAT.	COST EACH INST.	COST EACH TOTAL
SYSTEM D2010 924 1170					
BATHROOM, LAVATORY, WATER CLOSET & BATHTUB					
ONE WALL PLUMBING, STAND ALONE					
Wtr closet, 2 pc close cpld vit china flr mntd w/seat supply & stop	1.000	Ea.	253	228	481
Water closet, rough-in waste & vent	1.000	Set	355	395	750
Lavatory w/ftngs, wall hung, white, PE on CI, 20" x 18"	1.000	Ea.	300	151	451
Lavatory, rough-in waste & vent	1.000	Set	530	730	1,260
Bathtub, white PE on CI, w/ftgs, mat bottom, recessed, 5' long	1.000	Ea.	1,200	275	1,475
Baths, rough-in waste and vent	1.000	Set	468	526.50	994.50
TOTAL			3,106	2,305.50	5,411.50

D2010 924		Three Fixture Bathroom, One Wall Plumbing	MAT.	INST.	TOTAL
1150	Bathroom, three fixture, one wall plumbing				
1160		Lavatory, water closet & bathtub			
1170		Stand alone	3,100	2,300	5,400
1180		Share common plumbing wall *	2,650	1,675	4,325

D2010 926		Three Fixture Bathroom, Two Wall Plumbing	MAT.	INST.	TOTAL
2130	Bathroom, three fixture, two wall plumbing				
2140		Lavatory, water closet & bathtub			
2160		Stand alone	3,125	2,325	5,450
2180		Long plumbing wall common *	2,800	1,850	4,650
3610		Lavatory, bathtub & water closet			
3620		Stand alone	3,425	2,650	6,075
3640		Long plumbing wall common *	3,175	2,400	5,575
4660		Water closet, corner bathtub & lavatory			
4680		Stand alone	4,775	2,375	7,150
4700		Long plumbing wall common *	4,275	1,775	6,050
6100		Water closet, stall shower & lavatory			
6120		Stand alone	3,650	2,650	6,300
6140		Long plumbing wall common *	3,450	2,450	5,900
7060		Lavatory, corner stall shower & water closet			
7080		Stand alone	4,000	2,350	6,350
7100		Short plumbing wall common *	3,200	1,575	4,775

D20 Plumbing

D2010 Plumbing Fixtures

Four Fixture Bathroom Systems consisting of a lavatory, water closet, bathtub, shower and rough-in service piping.

- Prices for plumbing and fixtures only.

*Common wall is with an adjacent bathroom.

System Components	QUANTITY	UNIT	COST EACH MAT.	COST EACH INST.	COST EACH TOTAL
SYSTEM D2010 928 1160					
BATHROOM, BATHTUB, WATER CLOSET, STALL SHOWER & LAVATORY					
TWO WALL PLUMBING, STAND ALONE					
Wtr closet, 2 pc close cpld vit china flr mntd w/seat supply & stop	1.000	Ea.	253	228	481
Water closet, rough-in waste & vent	1.000	Set	355	395	750
Lavatory w/ftngs, wall hung, white PE on CI, 20" x 18"	1.000	Ea.	300	151	451
Lavatory, rough-in waste & vent	1.000	Set	53	73	126
Bathtub, white PE on CI, w/ftgs, mat bottom, recessed, 5' long	1.000	Ea.	1,200	275	1,475
Baths, rough-in waste and vent	1.000	Set	520	585	1,105
Shower stall, bkd enam, molded stone receptor, door & trim 32" sq.	1.000	Ea.	1,275	242	1,517
Shower stall, rough-in supply, waste & vent	1.000	Set	655	590	1,245
TOTAL			4,611	2,539	7,150

D2010 928	Four Fixture Bathroom, Two Wall Plumbing	MAT.	INST.	TOTAL
1140	Bathroom, four fixture, two wall plumbing			
1150	Bathtub, water closet, stall shower & lavatory			
1160	Stand alone	4,600	2,550	7,150
1180	Long plumbing wall common *	4,100	1,950	6,050
2260	Bathtub, lavatory, corner stall shower & water closet			
2280	Stand alone	5,225	2,550	7,775
2320	Long plumbing wall common *	4,725	1,975	6,700
3620	Bathtub, stall shower, lavatory & water closet			
3640	Stand alone	5,100	3,200	8,300
3660	Long plumbing wall (opp. door) common *	4,575	2,600	7,175

D2010 930	Four Fixture Bathroom, Three Wall Plumbing	MAT.	INST.	TOTAL
4680	Bathroom, four fixture, three wall plumbing			
4700	Bathtub, stall shower, lavatory & water closet			
4720	Stand alone	5,925	3,525	9,450
4760	Long plumbing wall (opposite door) common *	5,750	3,250	9,000

D20 Plumbing

D2010 Plumbing Fixtures

Five Fixture Bathroom Systems consisting of two lavatories, a water closet, bathtub, shower and rough-in service piping.

- Prices for plumbing and fixtures only.

*Common wall is with an adjacent bathroom.

System Components	QUANTITY	UNIT	COST EACH MAT.	INST.	TOTAL
SYSTEM D2010 932 1360					
BATHROOM, BATHTUB, WATER CLOSET, STALL SHOWER & TWO LAVATORIES					
TWO WALL PLUMBING, STAND ALONE					
Wtr closet, 2 pc close cpld vit china flr mntd incl seat, supply & stop	1.000	Ea.	253	228	481
Water closet, rough-in waste & vent	1.000	Set	355	395	750
Lavatory w/ftngs, wall hung, white PE on CI, 20" x 18"	2.000	Ea.	600	302	902
Lavatory, rough-in waste & vent	2.000	Set	1,060	1,460	2,520
Bathtub, white PE on CI, w/ftgs, mat bottom, recessed, 5' long	1.000	Ea.	1,200	275	1,475
Baths, rough-in waste and vent	1.000	Set	520	585	1,105
Shower stall, bkd enam molded stone receptor, door & ftng, 32" sq.	1.000	Ea.	1,275	242	1,517
Shower stall, rough-in supply, waste & vent	1.000	Set	655	590	1,245
TOTAL			5,918	4,077	9,995

D2010 932	Five Fixture Bathroom, Two Wall Plumbing	COST EACH MAT.	INST.	TOTAL
1320	Bathroom, five fixture, two wall plumbing			
1340	Bathtub, water closet, stall shower & two lavatories			
1360	Stand alone	5,925	4,075	10,000
1400	One short plumbing wall common *	5,400	3,500	8,900
1500	Bathtub, two lavatories, corner stall shower & water closet			
1520	Stand alone	6,550	4,075	10,625
1540	Long plumbing wall common*	5,875	3,175	9,050

D2010 934	Five Fixture Bathroom, Three Wall Plumbing	COST EACH MAT.	INST.	TOTAL
2360	Bathroom, five fixture, three wall plumbing			
2380	Water closet, bathtub, two lavatories & stall shower			
2400	Stand alone	6,550	4,075	10,625
2440	One short plumbing wall common *	6,025	3,500	9,525

D2010 936	Five Fixture Bathroom, One Wall Plumbing	COST EACH MAT.	INST.	TOTAL
4080	Bathroom, five fixture, one wall plumbing			
4100	Bathtub, two lavatories, corner stall shower & water closet			
4120	Stand alone	6,225	3,650	9,875
4160	Share common wall *	5,250	2,350	7,600

D20 Plumbing

D2010 Plumbing Fixtures

Example of Plumbing Cost Calculations: The bathroom system includes the individual fixtures such as bathtub, lavatory, shower and water closet. These fixtures are listed below as separate items merely as a checklist.

D2010 951 Plumbing Systems 20 Unit, 2 Story Apartment Building

	FIXTURE	SYSTEM	LINE	QUANTITY	UNIT	COST EACH		
						MAT.	INST.	TOTAL
0440	Bathroom	D2010 926	3640	20	Ea.	63,500	48,000	111,500
0480	Bathtub							
0520	Booster pump[1]	not req'd.						
0560	Drinking fountain							
0600	Garbage disposal[1]	not incl.						
0660								
0680	Grease interceptor							
0720	Water heater	D2020 250	2140	1	Ea.	13,500	3,275	16,775
0760	Kitchen sink	D2010 410	1960	20	Ea.	31,600	16,900	48,500
0800	Laundry sink	D2010 420	1840	4	Ea.	5,775	3,325	9,100
0840	Lavatory							
0900								
0920	Roof drain, 1 floor	D2040 210	4200	2	Ea.	1,850	1,750	3,600
0960	Roof drain, add'l floor	D2040 210	4240	20	L.F.	370	440	810
1000	Service sink	D2010 440	4300	1	Ea.	2,775	1,175	3,950
1040	Sewage ejector[1]	not req'd.						
1080	Shower							
1100								
1160	Sump pump							
1200	Urinal							
1240	Water closet							
1320								
1360	**SUB TOTAL**					119,500	75,000	194,500
1480	Water controls	RD2010-031		10%[2]		12,000	7,475	19,475
1520	Pipe & fittings[3]	RD2010-031		30%[2]		35,900	22,500	58,400
1560	Other							
1600	Quality/complexity	RD2010-031		15%[2]		17,900	11,200	29,100
1680								
1720	**TOTAL**					185,500	116,000	301,500
1740								

[1] **Note:** Cost for items such as booster pumps, backflow preventers, sewage ejectors, water meters, etc., may be obtained from RSMeans "Plumbing Cost Data" book, or be added as an increase in the Water Controls percentage. Water Controls, Pipe and Fittings, and the Quality/Complexity factors come from Table RD2010-031.

[2] Percentage of subtotal.

[3] Long easily discernable runs of pipe would be more accurately priced from Table D2090 810. If this is done, reduce the miscellaneous percentage in proportion.

D20 Plumbing

D2020 Domestic Water Distribution

Systems below include piping and fittings within 10' of heater. Electric water heaters do not require venting.

System Components	QUANTITY	UNIT	COST EACH		
			MAT.	INST.	TOTAL
SYSTEM D2020 240 1820					
ELECTRIC WATER HEATER, COMMERCIAL, 100°F RISE					
50 GALLON TANK, 9 KW, 37 GPH					
Water heater, commercial, electric, 50 Gal, 9 KW, 37 GPH	1.000	Ea.	4,375	375	4,750
Copper tubing, type L, solder joint, hanger 10' OC, 3/4" diam	34.000	L.F.	244.80	300.90	545.70
Wrought copper 90° elbow for solder joints 3/4" diam	5.000	Ea.	35	177.50	212.50
Wrought copper Tee for solder joints, 3/4" diam	2.000	Ea.	25.70	112	137.70
Wrought copper union for soldered joints, 3/4" diam.	2.000	Ea.	93	75	168
Valve, gate, bronze, 125 lb, NRS, soldered 3/4" diam	2.000	Ea.	109	67	176
Relief valve, bronze, press & temp, self-close, 3/4" IPS	1.000	Ea.	171	24	195
Wrought copper adapter, copper tubing to male, 3/4" IPS	1.000	Ea.	10.95	39.50	50.45
TOTAL			5,064.45	1,170.90	6,235.35

D2020 240	Electric Water Heaters - Commercial Systems	COST EACH		
		MAT.	INST.	TOTAL
1800	Electric water heater, commercial, 100°F rise RD2020-100			
1820	50 gallon tank, 9 KW 37 GPH	5,075	1,175	6,250
1860	80 gal, 12 KW 49 GPH	6,800	1,450	8,250
1900	36 KW 147 GPH	9,225	1,550	10,775
1940	120 gal, 36 KW 147 GPH	9,875	1,675	11,550
1980	150 gal, 120 KW 490 GPH	28,600	1,800	30,400
2020	200 gal, 120 KW 490 GPH	30,000	1,850	31,850
2060	250 gal, 150 KW 615 GPH	33,800	2,150	35,950
2100	300 gal, 180 KW 738 GPH	43,200	2,275	45,475
2140	350 gal, 30 KW 123 GPH	27,600	2,450	30,050
2180	180 KW 738 GPH	37,700	2,450	40,150
2220	500 gal, 30 KW 123 GPH	35,600	2,900	38,500
2260	240 KW 984 GPH	60,500	2,900	63,400
2300	700 gal, 30 KW 123 GPH	29,400	3,300	32,700
2340	300 KW 1230 GPH	42,600	3,300	45,900
2380	1000 gal, 60 KW 245 GPH	37,100	4,600	41,700
2420	480 KW 1970 GPH	57,500	4,625	62,125
2460	1500 gal, 60 KW 245 GPH	75,500	5,700	81,200
2500	480 KW 1970 GPH	102,500	5,700	108,200

D20 Plumbing

D2020 Domestic Water Distribution

Units may be installed in multiples for increased capacity.

Included below is the heater with self-energizing gas controls, safety pilots, insulated jacket, hi-limit aquastat and pressure relief valve.

Installation includes piping and fittings within 10' of heater. Gas heaters require vent piping (not included in these prices).

System Components	QUANTITY	UNIT	COST EACH		
			MAT.	INST.	TOTAL
SYSTEM D2020 250 1780					
GAS FIRED WATER HEATER, COMMERCIAL, 100°F RISE					
75.5 MBH INPUT, 63 GPH					
Water heater, commercial, gas, 75.5 MBH, 63 GPH	1.000	Ea.	3,475	480	3,955
Copper tubing, type L, solder joint, hanger 10' OC, 1-1/4" diam	30.000	L.F.	451.50	348	799.50
Wrought copper 90° elbow for solder joints 1-1/4" diam	4.000	Ea.	104	180	284
Wrought copper tee for solder joints, 1-1/4" diam	2.000	Ea.	109	149	258
Wrought copper union for soldered joints, 1-1/4" diam	2.000	Ea.	278	96	374
Valve, gate, bronze, 125 lb, NRS, soldered 1-1/4" diam	2.000	Ea.	238	90	328
Relief valve, bronze, press & temp, self-close, 3/4" IPS	1.000	Ea.	171	24	195
Copper tubing, type L, solder joints, 3/4" diam	8.000	L.F.	57.60	70.80	128.40
Wrought copper 90° elbow for solder joints 3/4" diam	1.000	Ea.	7	35.50	42.50
Wrought copper, adapter, CTS to MPT, 3/4" IPS	1.000	Ea.	10.95	39.50	50.45
Pipe steel black, schedule 40, threaded, 3/4" diam	10.000	L.F.	38.10	110	148.10
Pipe, 90° elbow, malleable iron black, 150 lb threaded, 3/4" diam	2.000	Ea.	7.68	96	103.68
Pipe, union with brass seat, malleable iron black, 3/4" diam	1.000	Ea.	16.15	51.50	67.65
Valve, gas stop w/o check, brass, 3/4" IPS	1.000	Ea.	17.05	30.50	47.55
TOTAL			4,981.03	1,800.80	6,781.83

D2020 250	Gas Fired Water Heaters - Commercial Systems		COST EACH		
			MAT.	INST.	TOTAL
1760	Gas fired water heater, commercial, 100°F rise				
1780	75.5 MBH input, 63 GPH		4,975	1,800	6,775
1860	100 MBH input, 91 GPH		7,250	1,875	9,125
1980	155 MBH input, 150 GPH	RD2020-100	10,100	2,175	12,275
2060	200 MBH input, 192 GPH		11,200	2,650	13,850
2140	300 MBH input, 278 GPH		13,500	3,275	16,775
2180	390 MBH input, 374 GPH		15,600	3,325	18,925
2220	500 MBH input, 480 GPH		20,700	3,550	24,250
2260	600 MBH input, 576 GPH		24,500	3,850	28,350

D20 Plumbing

D2020 Domestic Water Distribution

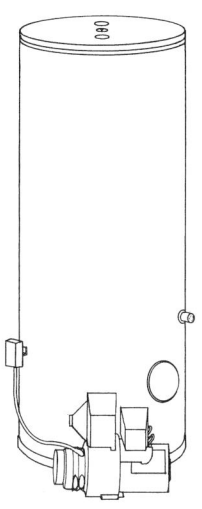

Units may be installed in multiples for increased capacity.

Included below is the heater, wired-in flame retention burners, cadmium cell primary controls, hi-limit controls, ASME pressure relief valves, draft controls, and insulated jacket.

Oil fired water heater systems include piping and fittings within 10' of heater. Oil fired heaters require vent piping (not included in these systems).

System Components	QUANTITY	UNIT	COST EACH MAT.	COST EACH INST.	COST EACH TOTAL
SYSTEM D2020 260 1820					
OIL FIRED WATER HEATER, COMMERCIAL, 100°F RISE					
140 GAL., 140 MBH INPUT, 134 GPH					
Water heater, commercial, oil, 140 gal., 140 MBH input, 134 GPH	1.000	Ea.	20,900	570	21,470
Copper tubing, type L, solder joint, hanger 10' OC, 3/4" diam.	34.000	L.F.	244.80	300.90	545.70
Wrought copper 90° elbow for solder joints 3/4" diam.	5.000	Ea.	35	177.50	212.50
Wrought copper Tee for solder joints, 3/4" diam.	2.000	Ea.	25.70	112	137.70
Wrought copper union for soldered joints, 3/4" diam.	2.000	Ea.	93	75	168
Valve, bronze, 125 lb, NRS, soldered 3/4" diam.	2.000	Ea.	109	67	176
Relief valve, bronze, press & temp, self-close, 3/4" IPS	1.000	Ea.	171	24	195
Wrought copper adapter, copper tubing to male, 3/4" IPS	1.000	Ea.	10.95	39.50	50.45
Copper tubing, type L, solder joint, hanger 10' OC, 3/8" diam.	10.000	L.F.	41.90	80	121.90
Wrought copper 90° elbow for solder joints 3/8" diam.	2.000	Ea.	18.70	61	79.70
Valve, globe, fusible, 3/8" IPS	1.000	Ea.	14.40	28.50	42.90
TOTAL			21,664.45	1,535.40	23,199.85

D2020 260	Oil Fired Water Heaters - Commercial Systems		MAT.	INST.	TOTAL
1800	Oil fired water heater, commercial, 100°F rise				
1820	140 gal., 140 MBH input, 134 GPH		21,700	1,525	23,225
1900	140 gal., 255 MBH input, 247 GPH		23,700	1,925	25,625
1940	140 gal., 270 MBH input, 259 GPH	RD2020 -100	28,900	2,175	31,075
1980	140 gal., 400 MBH input, 384 GPH		30,100	2,525	32,625
2060	140 gal., 720 MBH input, 691 GPH		32,000	2,650	34,650
2100	221 gal., 300 MBH input, 288 GPH		42,500	2,900	45,400
2140	221 gal., 600 MBH input, 576 GPH		47,100	2,925	50,025
2180	221 gal., 800 MBH input, 768 GPH		47,800	3,050	50,850
2220	201 gal., 1000 MBH input, 960 GPH		48,900	3,075	51,975
2260	201 gal., 1250 MBH input, 1200 GPH		49,800	3,150	52,950
2300	201 gal., 1500 MBH input, 1441 GPH		54,000	3,225	57,225
2340	411 gal., 600 MBH input, 576 GPH		54,500	3,300	57,800
2380	411 gal., 800 MBH input, 768 GPH		54,500	3,400	57,900
2420	411 gal., 1000 MBH input, 960 GPH		60,000	3,900	63,900
2460	411 gal., 1250 MBH input, 1200 GPH		61,000	4,025	65,025
2500	397 gal., 1500 MBH input, 1441 GPH		62,000	4,150	66,150
2540	397 gal., 1750 MBH input, 1681 GPH		66,500	4,300	70,800

D20 Plumbing

D2020 Domestic Water Distribution

In this closed-loop indirect collection system, fluid with a low freezing temperature, such as propylene glycol, transports heat from the collectors to water storage. The transfer fluid is contained in a closed-loop consisting of collectors, supply and return piping, and a remote heat exchanger. The heat exchanger transfers heat energy from the fluid in the collector loop to potable water circulated in a storage loop. A typical two-or-three panel system contains 5 to 6 gallons of heat transfer fluid.

When the collectors become approximately 20°F warmer than the storage temperature, a controller activates the circulator on the collector and storage loops. The circulators will move the fluid and potable water through the heat exchanger until heat collection no longer occurs. At that point, the system shuts down. Since the heat transfer medium is a fluid with a very low freezing temperature, there is no need for it to be drained from the system between periods of collection.

D20 Plumbing

D2020 Domestic Water Distribution

System Components		QUANTITY	UNIT	COST EACH		
				MAT.	INST.	TOTAL
	SYSTEM D2020 265 2760					
	SOLAR, CLOSED LOOP, ADD-ON HOT WATER SYS., EXTERNAL HEAT EXCHANGER					
	3/4" TUBING, TWO 3'X7' BLACK CHROME COLLECTORS					
A,B,G,L,K,M	Heat exchanger fluid-fluid pkg incl 2 circulators, expansion tank,					
	Check valve, relief valve, controller, hi temp cutoff, & 2 sensors	1.000	Ea.	820	485	1,305
C	Thermometer, 2" dial	3.000	Ea.	69	126	195
D, T	Fill & drain valve, brass, 3/4" connection	1.000	Ea.	10.05	28	38.05
E	Air vent, manual, 1/8" fitting	2.000	Ea.	31.80	42	73.80
F	Air purger	1.000	Ea.	51.50	56	107.50
H	Strainer, Y type, bronze body, 3/4" IPS	1.000	Ea.	36	36	72
I	Valve, gate, bronze, NRS, soldered 3/4" diam	6.000	Ea.	327	201	528
J	Neoprene vent flashing	2.000	Ea.	27.40	67	94.40
N-1, N	Relief valve temp & press, 150 psi 210°F self-closing 3/4" IPS	1.000	Ea.	22	22.50	44.50
O	Pipe covering, urethane, ultraviolet cover, 1" wall 3/4" diam	20.000	L.F.	60.20	121	181.20
P	Pipe covering, fiberglass, all service jacket, 1" wall, 3/4" diam	50.000	L.F.	53.50	241.50	295
Q	Collector panel solar energy blk chrome on copper, 1/8" temp glass 3'x7'	2.000	Ea.	2,250	254	2,504
	Roof clamps for solar energy collector panels	2.000	Set	5.58	34.60	40.18
R	Valve, swing check, bronze, regrinding disc, 3/4" diam	2.000	Ea.	196	67	263
S	Pressure gauge, 60 psi, 2" dial	1.000	Ea.	25.50	21	46.50
U	Valve, water tempering, bronze, sweat connections, 3/4" diam	1.000	Ea.	126	33.50	159.50
W-2, V	Tank water storage w/heating element, drain, relief valve, existing	1.000	Ea.			
	Copper tubing type L, solder joint, hanger 10' OC 3/4" diam	20.000	L.F.	144	177	321
	Copper tubing, type M, solder joint, hanger 10' OC 3/4" diam	70.000	L.F.	381.50	602	983.50
	Sensor wire, #22-2 conductor multistranded	.500	C.L.F.	7.83	31.25	39.08
	Solar energy heat transfer fluid, propylene glycol anti-freeze	6.000	Gal.	95.70	144	239.70
	Wrought copper fittings & solder, 3/4" diam	76.000	Ea.	532	2,698	3,230
	TOTAL			5,272.56	5,488.35	10,760.91

D2020 265	Solar, Closed Loop, Add-On Hot Water Systems		COST EACH		
			MAT.	INST.	TOTAL
2550	Solar, closed loop, add-on hot water system, external heat exchanger				
2700	3/4" tubing, 3 ea. 3'x7' black chrome collectors		6,400	5,625	12,025
2720	3 ea. 3'x7' flat black absorber plate collectors		5,375	5,650	11,025
2740	2 ea. 4'x9' flat black w/plastic glazing collectors	RD3010-600	4,975	5,675	10,650
2760	2 ea. 3'x7' black chrome collectors		5,275	5,500	10,775
2780	1" tubing, 4 ea 2'x9' plastic absorber & glazing collectors		6,675	6,375	13,050
2800	4 ea. 3'x7' black chrome absorber collectors		8,825	6,400	15,225
2820	4 ea. 3'x7' flat black absorber collectors		7,450	6,425	13,875

D20 Plumbing

D2020 Domestic Water Distribution

In the drainback indirect-collection system, the heat transfer fluid is distilled water contained in a loop consisting of collectors, supply and return piping, and an unpressurized holding tank. A large heat exchanger containing incoming potable water is immersed in the holding tank. When a controller activates solar collection, the distilled water is pumped through the collectors and heated and pumped back down to the holding tank. When the temperature differential between the water in the collectors and water in storage is such that collection no longer occurs, the pump turns off and gravity causes the distilled water in the collector loop to drain back to the holding tank. All the loop piping is pitched so that the water can drain out of the collectors and piping and not freeze there. As hot water is needed in the home, incoming water first flows through the holding tank with the immersed heat exchanger and is warmed and then flows through a conventional heater for any supplemental heating that is necessary.

D20 Plumbing

D2020 Domestic Water Distribution

System Components		QUANTITY	UNIT	COST EACH		
				MAT.	INST.	TOTAL
SYSTEM D2020 270 2760						
SOLAR, DRAINBACK, ADD ON, HOT WATER, IMMERSED HEAT EXCHANGER						
3/4" TUBING, THREE EA 3'X7' BLACK CHROME COLLECTOR						
A, B	Differential controller 2 sensors, thermostat, solar energy system	1.000	Ea.	625	56	681
C	Thermometer 2" dial	3.000	Ea.	69	126	195
D, T	Fill & drain valve, brass, 3/4" connection	1.000	Ea.	10.05	28	38.05
E-1	Automatic air vent 1/8" fitting	1.000	Ea.	25	21	46
H	Strainer, Y type, bronze body, 3/4" IPS	1.000	Ea.	36	36	72
I	Valve, gate, bronze, NRS, soldered 3/4" diam	2.000	Ea.	109	67	176
J	Neoprene vent flashing	2.000	Ea.	27.40	67	94.40
L	Circulator, solar heated liquid, 1/20 HP	1.000	Ea.	177	101	278
N	Relief valve temp. & press. 150 psi 210°F self-closing 3/4" IPS	1.000	Ea.	22	22.50	44.50
O	Pipe covering, urethane, ultraviolet cover, 1" wall, 3/4" diam	20.000	L.F.	60.20	121	181.20
P	Pipe covering, fiberglass, all service jacket, 1" wall, 3/4" diam	50.000	L.F.	53.50	241.50	295
Q	Collector panel solar energy blk chrome on copper, 1/8" temp glas 3'x7'	3.000	Ea.	3,375	381	3,756
	Roof clamps for solar energy collector panels	3.000	Set	8.37	51.90	60.27
R	Valve, swing check, bronze, regrinding disc, 3/4" diam	1.000	Ea.	98	33.50	131.50
U	Valve, water tempering, bronze sweat connections, 3/4" diam	1.000	Ea.	126	33.50	159.50
V	Tank, water storage w/heating element, drain, relief valve, existing	1.000	Ea.			
W	Tank, water storage immersed heat exchr elec 2"x1/2# insul 120 gal	1.000	Ea.	1,950	480	2,430
X	Valve, globe, bronze, rising stem, 3/4" diam, soldered	3.000	Ea.	417	100.50	517.50
Y	Flow control valve	1.000	Ea.	148	30.50	178.50
Z	Valve, ball, bronze, solder 3/4" diam, solar loop flow control	1.000	Ea.	23	33.50	56.50
	Copper tubing, type L, solder joint, hanger 10' OC 3/4" diam	20.000	L.F.	144	177	321
	Copper tubing, type M, solder joint, hanger 10' OC 3/4" diam	70.000	L.F.	381.50	602	983.50
	Sensor wire, #22-2 conductor, multistranded	.500	C.L.F.	7.83	31.25	39.08
	Wrought copper fittings & solder, 3/4" diam	76.000	Ea.	532	2,698	3,230
	TOTAL			8,424.85	5,539.65	13,964.50

D2020 270		Solar, Drainback, Hot Water Systems	COST EACH		
			MAT.	INST.	TOTAL
2550	Solar, drainback, hot water, immersed heat exchanger				
2560		3/8" tubing, 3 ea. 4' x 4'-4" vacuum tube collectors	7,500	4,975	12,475
2760		3/4" tubing, 3 ea 3'x7' black chrome collectors, 120 gal tank	8,425	5,550	13,975
2780		3 ea. 3'x7' flat black absorber collectors, 120 gal tank	7,400	5,550	12,950
2800		2 ea. 4'x9' flat blk w/plastic glazing collectors 120 gal tank	7,000	5,575	12,575
2840		1" tubing, 4 ea. 2'x9' plastic absorber & glazing collectors, 120 gal tank	8,825	6,275	15,100
2860		4 ea. 3'x7' black chrome absorber collectors, 120 gal tank	11,000	6,300	17,300
2880		4 ea. 3'x7' flat black absorber collectors, 120 gal tank	9,600	6,325	15,925

(Reference: RD3010-600 at row 2760)

D20 Plumbing

D2020 Domestic Water Distribution

In the draindown direct-collection system, incoming domestic water is heated in the collectors. When the controller activates solar collection, domestic water is first heated as it flows through the collectors and is then pumped to storage. When conditions are no longer suitable for heat collection, the pump shuts off and the water in the loop drains down and out of the system by means of solenoid valves and properly pitched piping.

D20 Plumbing

D2020 Domestic Water Distribution

System Components		QUANTITY	UNIT	COST EACH MAT.	COST EACH INST.	COST EACH TOTAL
	SYSTEM D2020 275 2760 SOLAR, DRAINDOWN, HOT WATER, DIRECT COLLECTION 3/4" TUBING, THREE 3'X7' BLACK CHROME COLLECTORS					
A, B	Differential controller, 2 sensors, thermostat, solar energy system	1.000	Ea.	625	56	681
A-1	Solenoid valve, solar heating loop, brass, 3/4" diam, 24 volts	3.000	Ea.	381	223.50	604.50
B-1	Solar energy sensor, freeze prevention	1.000	Ea.	26	21	47
C	Thermometer, 2" dial	3.000	Ea.	69	126	195
E-1	Vacuum relief valve, 3/4" diam	1.000	Ea.	31.50	21	52.50
F-1	Air vent, automatic, 1/8" fitting	1.000	Ea.	25	21	46
H	Strainer, Y type, bronze body, 3/4" IPS	1.000	Ea.	36	36	72
I	Valve, gate, bronze, NRS, soldered, 3/4" diam	2.000	Ea.	109	67	176
J	Vent flashing neoprene	2.000	Ea.	27.40	67	94.40
K	Circulator, solar heated liquid, 1/25 HP	1.000	Ea.	295	86.50	381.50
N	Relief valve temp & press 150 psi 210°F self-closing 3/4" IPS	1.000	Ea.	22	22.50	44.50
O	Pipe covering, urethane, ultraviolet cover, 1" wall, 3/4" diam	20.000	L.F.	60.20	121	181.20
P	Pipe covering, fiberglass, all service jacket, 1" wall, 3/4" diam	50.000	L.F.	53.50	241.50	295
	Roof clamps for solar energy collector panels	3.000	Set	8.37	51.90	60.27
Q	Collector panel solar energy blk chrome on copper, 1/8" temp glass 3'x7'	3.000	Ea.	3,375	381	3,756
R	Valve, swing check, bronze, regrinding disc, 3/4" diam, soldered	2.000	Ea.	196	67	263
T	Drain valve, brass, 3/4" connection	2.000	Ea.	20.10	56	76.10
U	Valve, water tempering, bronze, sweat connections, 3/4" diam	1.000	Ea.	126	33.50	159.50
W-2, W	Tank, water storage elec elem 2"x1/2# insul 120 gal	1.000	Ea.	1,950	480	2,430
X	Valve, globe, bronze, rising stem, 3/4" diam, soldered	1.000	Ea.	139	33.50	172.50
	Copper tubing, type L, solder joints, hangers 10' OC 3/4" diam	20.000	L.F.	144	177	321
	Copper tubing, type M, solder joints, hangers 10' OC 3/4" diam	70.000	L.F.	381.50	602	983.50
	Sensor wire, #22-2 conductor, multistranded	.500	C.L.F.	7.83	31.25	39.08
	Wrought copper fittings & solder, 3/4" diam	76.000	Ea.	532	2,698	3,230
	TOTAL			8,640.40	5,721.15	14,361.55

D2020 275	Solar, Draindown, Hot Water Systems		MAT.	INST.	TOTAL
2550	Solar, draindown, hot water				
2580	1/2" tubing, 4 ea. 4' x 4'-4" vacuum tube collectors, 80 gal tank		8,250	5,575	13,825
2760	3/4" tubing, 3 ea. 3'x7' black chrome collectors, 120 gal tank	RD3010 -600	8,650	5,725	14,375
2780	3 ea. 3'x7' flat collectors, 120 gal tank		7,600	5,750	13,350
2800	2 ea. 4'x9' flat black & plastic glazing collectors, 120 gal tank		7,225	5,775	13,000
2840	1" tubing, 4 ea. 2'x9' plastic absorber & glazing collectors, 120 gal tank		12,000	6,450	18,450
2860	4 ea. 3'x7' black chrome absorber collectors, 120 gal tank		14,200	6,475	20,675
2880	4 ea. 3'x7' flat black absorber collectors, 120 gal tank		12,800	6,500	19,300

D20 Plumbing

D2020 Domestic Water Distribution

In this closed-loop indirect collection system, fluid with a low freezing temperature, such as propylene glycol, transports heat from the collectors to water storage. The transfer fluid is contained in a closed-loop consisting of collectors, supply and return piping, and a heat exchanger immersed in the storage tank. A typical two-or-three panel system contains 5 to 6 gallons of heat transfer fluid.

When the collectors become approximately 20°F warmer than the storage temperature, a controller activates the circulator. The circulator moves the fluid continuously through the collectors until the temperature difference between the collectors and storage is such that heat collection no longer occurs; at that point, the circulator shuts off. Since the heat transfer fluid has a very low freezing temperature, there is no need for it to be drained from the collectors between periods of collection.

D20 Plumbing

D2020 Domestic Water Distribution

System Components	QUANTITY	UNIT	COST EACH		
			MAT.	INST.	TOTAL
SYSTEM D2020 295 2760					
SOLAR, CLOSED LOOP, HOT WATER SYSTEM, IMMERSED HEAT EXCHANGER					
3/4" TUBING, THREE 3' X 7' BLACK CHROME COLLECTORS					
A, B Differential controller, 2 sensors, thermostat, solar energy system	1.000	Ea.	625	56	681
C Thermometer 2" dial	3.000	Ea.	69	126	195
D, T Fill & drain valves, brass, 3/4" connection	3.000	Ea.	30.15	84	114.15
E Air vent, manual, 1/8" fitting	1.000	Ea.	15.90	21	36.90
F Air purger	1.000	Ea.	51.50	56	107.50
G Expansion tank	1.000	Ea.	72.50	21	93.50
I Valve, gate, bronze, NRS, soldered 3/4" diam	3.000	Ea.	163.50	100.50	264
J Neoprene vent flashing	2.000	Ea.	27.40	67	94.40
K Circulator, solar heated liquid, 1/25 HP	1.000	Ea.	295	86.50	381.50
N-1, N Relief valve, temp & press 150 psi 210°F self-closing 3/4" IPS	2.000	Ea.	44	45	89
O Pipe covering, urethane ultraviolet cover, 1" wall, 3/4" diam	20.000	L.F.	60.20	121	181.20
P Pipe covering, fiberglass, all service jacket, 1" wall, 3/4" diam	50.000	L.F.	53.50	241.50	295
Roof clamps for solar energy collector panel	3.000	Set	8.37	51.90	60.27
Q Collector panel solar blk chrome on copper, 1/8" temp glass, 3'x7'	3.000	Ea.	3,375	381	3,756
R-1 Valve, swing check, bronze, regrinding disc, 3/4" diam, soldered	1.000	Ea.	98	33.50	131.50
S Pressure gauge, 60 psi, 2-1/2" dial	1.000	Ea.	25.50	21	46.50
U Valve, water tempering, bronze, sweat connections, 3/4" diam	1.000	Ea.	126	33.50	159.50
W-2, W Tank, water storage immersed heat exchr elec elem 2"x2# insul 120 Gal	1.000	Ea.	1,950	480	2,430
X Valve, globe, bronze, rising stem, 3/4" diam, soldered	1.000	Ea.	139	33.50	172.50
Copper tubing type L, solder joint, hanger 10' OC 3/4" diam	20.000	L.F.	144	177	321
Copper tubing, type M, solder joint, hanger 10' OC 3/4" diam	70.000	L.F.	381.50	602	983.50
Sensor wire, #22-2 conductor multistranded	.500	C.L.F.	7.83	31.25	39.08
Solar energy heat transfer fluid, propylene glycol, anti-freeze	6.000	Gal.	95.70	144	239.70
Wrought copper fittings & solder, 3/4" diam	76.000	Ea.	532	2,698	3,230
TOTAL			8,390.55	5,712.15	14,102.70

D2020 295	Solar, Closed Loop, Hot Water Systems		COST EACH		
			MAT.	INST.	TOTAL
2550	Solar, closed loop, hot water system, immersed heat exchanger				
2560	3/8" tubing, 3 ea. 4' x 4'-4" vacuum tube collectors, 80 gal. tank		7,550	5,175	12,725
2580	1/2" tubing, 4 ea. 4' x 4'-4" vacuum tube collectors, 80 gal. tank	RD3010 -600	8,025	5,575	13,600
2600	120 gal. tank		8,275	5,625	13,900
2640	2 ea. 3'x7' black chrome collectors, 80 gal. tank		6,475	5,325	11,800
2660	120 gal. tank		6,675	5,325	12,000
2760	3/4" tubing, 3 ea. 3'x7' black chrome collectors, 120 gal. tank		8,400	5,700	14,100
2780	3 ea. 3'x7' flat black collectors, 120 gal. tank		7,350	5,725	13,075
2840	1" tubing, 4 ea. 2'x9' plastic absorber & glazing collectors 120 gal. tank		8,550	6,425	14,975
2860	4 ea. 3'x7' black chrome collectors, 120 gal. tank		10,700	6,450	17,150

D20 Plumbing

D2040 Rain Water Drainage

Design Assumptions: Vertical conductor size is based on a maximum rate of rainfall of 4" per hour. To convert roof area to other rates multiply "Max. S.F. Roof Area" shown by four and divide the result by desired local rate. The answer is the local roof area that may be handled by the indicated pipe diameter.

Basic cost is for roof drain, 10' of vertical leader and 10' of horizontal, plus connection to the main.

Pipe Dia.	Max. S.F. Roof Area	Gallons per Min.
2"	544	23
3"	1610	67
4"	3460	144
5"	6280	261
6"	10,200	424
8"	22,000	913

System Components			COST EACH		
	QUANTITY	UNIT	MAT.	INST.	TOTAL
SYSTEM D2040 210 1880 **ROOF DRAIN, DWV PVC PIPE, 2" DIAM., 10' HIGH**					
Drain, roof, main, PVC, dome type 2" pipe size	1.000	Ea.	144	86.50	230.50
Clamp, roof drain, underdeck	1.000	Ea.	30	50.50	80.50
Pipe, Tee, PVC DWV, schedule 40, 2" pipe size	1.000	Ea.	8.10	60.50	68.60
Pipe, PVC, DWV, schedule 40, 2" diam.	20.000	L.F.	79	410	489
Pipe, elbow, PVC schedule 40, 2" diam.	2.000	Ea.	6.06	67	73.06
TOTAL			267.16	674.50	941.66

D2040 210	Roof Drain Systems	COST EACH		
		MAT.	INST.	TOTAL
1880	Roof drain, DWV PVC, 2" diam., piping, 10' high	267	675	942
1920	For each additional foot add	3.95	20.50	24.45
1960	3" diam., 10' high	360	790	1,150
2000	For each additional foot add	7.10	23	30.10
2040	4" diam., 10' high	445	880	1,325
2080	For each additional foot add	9.65	25	34.65
2120	5" diam., 10' high	1,325	1,025	2,350
2160	For each additional foot add	28.50	28	56.50
2200	6" diam., 10' high	1,350	1,125	2,475
2240	For each additional foot add	19.40	31	50.40
2280	8" diam., 10' high	2,950	1,925	4,875
2320	For each additional foot add	45	39	84
3940	C.I., soil, single hub, service wt., 2" diam. piping, 10' high	560	735	1,295
3980	For each additional foot add	10	19.20	29.20
4120	3" diam., 10' high	780	795	1,575
4160	For each additional foot add	14.05	20	34.05
4200	4" diam., 10' high	920	870	1,790
4240	For each additional foot add	18.60	22	40.60
4280	5" diam., 10' high	1,275	970	2,245
4320	For each additional foot add	25	25	50
4360	6" diam., 10' high	1,625	1,025	2,650
4400	For each additional foot add	31.50	26	57.50
4440	8" diam., 10' high	3,175	2,100	5,275
4480	For each additional foot add	48.50	43.50	92
6040	Steel galv. sch 40 threaded, 2" diam. piping, 10' high	730	715	1,445
6080	For each additional foot add	14.65	18.90	33.55
6120	3" diam., 10' high	1,425	1,025	2,450
6160	For each additional foot add	31	28	59

D20 Plumbing

D2040 Rain Water Drainage

D2040 210	Roof Drain Systems	COST EACH		
		MAT.	INST.	TOTAL
6200	4" diam., 10' high	2,050	1,325	3,375
6240	For each additional foot add	45.50	33.50	79
6280	5" diam., 10' high	2,325	1,100	3,425
6320	For each additional foot add	64.50	32.50	97
6360	6" diam, 10' high	2,775	1,425	4,200
6400	For each additional foot add	73	45	118
6440	8" diam., 10' high	4,375	2,050	6,425
6480	For each additional foot add	86	51	137

D20 Plumbing

D2090 Other Plumbing Systems

Pipe Material Considerations:

1. Malleable iron fittings should be used for gas service.
2. Malleable fittings are used where there are stresses/strains due to expansion and vibration.
3. Cast iron fittings may be broken as an aid to disassembling of piping joints frozen by long use, temperature and mineral deposits.
4. Cast iron pipe is extensively used for underground and submerged service.
5. Type M (light wall) copper tubing is available in hard temper only and is used for nonpressure and less severe applications than K and L.
6. Type L (medium wall) copper tubing, available hard or soft for interior service.
7. Type K (heavy wall) copper tubing, available in hard or soft temper for use where conditions are severe. For underground and interior service.
8. Hard drawn tubing requires fewer hangers or supports but should not be bent. Silver brazed fittings should be used.

Piping costs include hangers spaced for the material type and size, and the couplings required.

D2090 810	Piping - Installed - Unit Costs		COST PER L.F.		
			MAT.	INST.	TOTAL
0840	Cast iron, soil, B & S, service weight, 2" diameter	RD2010 -030	10	19.20	29.20
0860	3" diameter		14.05	20	34.05
0880	4" diameter		18.60	22	40.60
0900	5" diameter		25	25	50
0920	6" diameter		31.50	26	57.50
0940	8" diameter		48.50	43.50	92
0960	10" diameter		78	47.50	125.50
0980	12" diameter		111	53.50	164.50
1040	No hub, 1-1/2" diameter		9.90	17.05	26.95
1060	2" diameter		10.15	18.05	28.20
1080	3" diameter		14.05	18.90	32.95
1100	4" diameter		18.35	21	39.35
1120	5" diameter		24.50	22.50	47
1140	6" diameter		32	24	56
1160	8" diameter		55	37	92
1180	10" diameter		91.50	42	133.50
1220	Copper tubing, hard temper, solder, type K, 1/2" diameter		6.10	8.60	14.70
1260	3/4" diameter		10.75	9.10	19.85
1280	1" diameter		14.40	10.20	24.60
1300	1-1/4" diameter		17.90	12	29.90
1320	1-1/2" diameter		23.50	13.45	36.95
1340	2" diameter		36	16.80	52.80
1360	2-1/2" diameter		55	20	75
1380	3" diameter		77	22.50	99.50
1400	4" diameter		132	32	164
1480	5" diameter		292	38	330
1500	6" diameter		430	49.50	479.50
1520	8" diameter		720	55.50	775.50
1560	Type L, 1/2" diameter		4.66	8.30	12.96
1600	3/4" diameter		7.20	8.85	16.05
1620	1" diameter		10.55	9.90	20.45
1640	1-1/4" diameter		15.05	11.60	26.65
1660	1-1/2" diameter		19.30	12.95	32.25
1680	2" diameter		30.50	16	46.50
1700	2-1/2" diameter		47	19.50	66.50
1720	3" diameter		64	21.50	85.50
1740	4" diameter		111	31	142
1760	5" diameter		235	35.50	270.50
1780	6" diameter		335	47	382
1800	8" diameter		540	52.50	592.50
1840	Type M, 1/2" diameter		3.59	8	11.59
1880	3/4" diameter		5.45	8.60	14.05
1900	1" diameter		8.60	9.60	18.20
1920	1-1/4" diameter		12.60	11.20	23.80

D20 Plumbing

D2090 Other Plumbing Systems

D2090 810	Piping - Installed - Unit Costs	COST PER L.F.		
		MAT.	INST.	TOTAL
1940	1-1/2" diameter	17	12.45	29.45
1960	2" diameter	27.50	15.30	42.80
1980	2-1/2" diameter	41.50	18.90	60.40
2000	3" diameter	56	21	77
2020	4" diameter	102	30.50	132.50
2040	5" diameter	235	33.50	268.50
2060	6" diameter	335	45	380
2080	8" diameter	540	49.50	589.50
2120	Type DWV, 1-1/4" diameter	13.35	11.20	24.55
2160	1-1/2" diameter	16.75	12.45	29.20
2180	2" diameter	23	15.30	38.30
2200	3" diameter	43.50	21	64.50
2220	4" diameter	78	30.50	108.50
2240	5" diameter	213	33.50	246.50
2260	6" diameter	310	45	355
2280	8" diameter	675	49.50	724.50
2800	Plastic, PVC, DWV, schedule 40, 1-1/4" diameter	3.78	16	19.78
2820	1-1/2" diameter	3.43	18.70	22.13
2830	2" diameter	3.95	20.50	24.45
2840	3" diameter	7.10	23	30.10
2850	4" diameter	9.65	25	34.65
2890	6" diameter	19.40	31	50.40
3010	Pressure pipe 200 PSI, 1/2" diameter	2.97	12.45	15.42
3030	3/4" diameter	3.30	13.20	16.50
3040	1" diameter	3.56	14.60	18.16
3050	1-1/4" diameter	4.34	16	20.34
3060	1-1/2" diameter	4.67	18.70	23.37
3070	2" diameter	5.60	20.50	26.10
3080	2-1/2" diameter	10	21.50	31.50
3090	3" diameter	11.35	23	34.35
3100	4" diameter	18.25	25	43.25
3110	6" diameter	33	31	64
3120	8" diameter	50.50	39	89.50
4000	Steel, schedule 40, black, threaded, 1/2" diameter	3.25	10.65	13.90
4020	3/4" diameter	3.81	11	14.81
4030	1" diameter	5.60	12.70	18.30
4040	1-1/4" diameter	6.95	13.60	20.55
4050	1-1/2" diameter	10.45	15.15	25.60
4060	2" diameter	10.75	18.90	29.65
4070	2-1/2" diameter	16.80	24	40.80
4080	3" diameter	21.50	28	49.50
4090	4" diameter	32.50	33.50	66
4100	Grooved, 5" diameter	40.50	32.50	73
4110	6" diameter	44.50	45	89.50
4120	8" diameter	68	51	119
4130	10" diameter	85	60.50	145.50
4140	12" diameter	98	69.50	167.50
4200	Galvanized, threaded, 1/2" diameter	4.51	10.65	15.16
4220	3/4" diameter	5.35	11	16.35
4230	1" diameter	7.40	12.70	20.10
4240	1-1/4" diameter	9.35	13.60	22.95
4250	1-1/2" diameter	10.95	15.15	26.10
4260	2" diameter	14.65	18.90	33.55
4270	2-1/2" diameter	24.50	24	48.50
4280	3" diameter	31	28	59
4290	4" diameter	45.50	33.50	79
4300	Grooved, 5" diameter	64.50	32.50	97
4310	6" diameter	73	45	118

D20 Plumbing

D2090 Other Plumbing Systems

D2090 810	Piping - Installed - Unit Costs	COST PER L.F.		
		MAT.	INST.	TOTAL
4320	8" diameter	86	51	137
4330	10" diameter	137	60.50	197.50
4340	12" diameter	168	69.50	237.50
5010	Flanged, black, 1" diameter	11.15	18.15	29.30
5020	1-1/4" diameter	12.80	19.80	32.60
5030	1-1/2" diameter	13.50	22	35.50
5040	2" diameter	16	28.50	44.50
5050	2-1/2" diameter	23	35	58
5060	3" diameter	27	40	67
5070	4" diameter	35.50	48.50	84
5080	5" diameter	55	60.50	115.50
5090	6" diameter	59	78	137
5100	8" diameter	91	102	193
5110	10" diameter	132	122	254
5120	12" diameter	166	138	304
5720	Grooved joints, black, 3/4" diameter	5.15	9.45	14.60
5730	1" diameter	4.91	10.65	15.56
5740	1-1/4" diameter	6.35	11.60	17.95
5750	1-1/2" diameter	7.20	13.20	20.40
5760	2" diameter	8.85	16.80	25.65
5770	2-1/2" diameter	15.35	21	36.35
5900	3" diameter	19.05	24	43.05
5910	4" diameter	26	27	53
5920	5" diameter	40.50	32.50	73
5930	6" diameter	44.50	45	89.50
5940	8" diameter	68	51	119
5950	10" diameter	85	60.50	145.50
5960	12" diameter	98	69.50	167.50

D2090 820	Standard Pipe Fittings - Installed - Unit Costs	COST EACH		
		MAT.	INST.	TOTAL
0100	Cast iron, soil, B&S service weight, 1/8 Bend, 2" diameter	14.25	75.50	89.75
0110	3" diameter	22.50	86.50	109
0120	4" diameter	32.50	93	125.50
0130	5" diameter	45	105	150
0140	6" diameter	55.50	111	166.50
0150	8" diameter	165	233	398
0160	10" diameter	238	256	494
0170	12" diameter	450	285	735
0220	1/4 Bend, 2" diameter	20	75.50	95.50
0230	3" diameter	26.50	86.50	113
0240	4" diameter	42	93	135
0250	5" diameter	58.50	105	163.50
0260	6" diameter	73	111	184
0270	8" diameter	219	233	452
0280	10" diameter	320	256	576
0290	12" diameter	435	285	720
0360	Tee, 2" diameter	40.50	121	161.50
0370	3" diameter	60	134	194
0380	4" diameter	77.50	151	228.50
0390	5" diameter	164	157	321
0400	6" diameter	161	171	332
0410	8" diameter	291	365	656
0810	No hub, 1/8 Bend, 1-1/2" diameter	8.70		8.70
0820	2" diameter	9.65		9.65
0830	3" diameter	12.95		12.95
0840	4" diameter	16.95		16.95

D20 Plumbing

D2090 Other Plumbing Systems

D2090 820	Standard Pipe Fittings - Installed - Unit Costs	COST EACH		
		MAT.	INST.	TOTAL
0850	5" diameter	35.50		35.50
0860	6" diameter	37.50		37.50
0870	8" diameter	108		108
0880	10" diameter	206		206
0960	1/4 Bend, 1-1/2" diameter	10.30		10.30
0970	2" diameter	11.15		11.15
0980	3" diameter	15.65		15.65
0990	4" diameter	23		23
1000	5" diameter	56		56
1010	6" diameter	56		56
1020	8" diameter	157		157
1080	Sanitary tee, 1-1/2" diameter	14.50		14.50
1090	2" diameter	15.40		15.40
1110	3" diameter	19.10		19.10
1120	4" diameter	36		36
1130	5" diameter	84.50		84.50
1140	6" diameter	86		86
1150	8" diameter	350		350
1300	Coupling clamp & gasket, 1-1/2" diameter	8.15	25	33.15
1310	2" diameter	9.20	27.50	36.70
1320	3" diameter	10.95	32	42.95
1330	4" diameter	15.15	36.50	51.65
1340	5" diameter	26.50	43	69.50
1350	6" diameter	27.50	47	74.50
1360	8" diameter	105	77.50	182.50
1370	10" diameter	183	98.50	281.50
1510	Steel pipe fitting, CI, threaded, 125 Lb. black, 45° Elbow, 1/2" diameter	9.05	45	54.05
1520	3/4" diameter	9.10	48	57.10
1530	1" diameter	10.65	51.50	62.15
1540	1-1/4" diameter	14.35	55	69.35
1550	1-1/2" diameter	24	60.50	84.50
1560	2" diameter	27.50	67	94.50
1570	2-1/2" diameter	71.50	86.50	158
1580	3" diameter	113	121	234
1590	4" diameter	235	202	437
1600	5" diameter	248	122	370
1610	6" diameter	310	151	461
1620	8" diameter	565	180	745
1700	90° Elbow, 1/2" diameter	5.90	45	50.90
1710	3/4" diameter	6.15	48	54.15
1720	1" diameter	7.30	51.50	58.80
1730	1-1/4" diameter	10.35	55	65.35
1740	1-1/2" diameter	14.30	60.50	74.80
1750	2" diameter	22.50	67	89.50
1760	2-1/2" diameter	53.50	86.50	140
1770	3" diameter	88	121	209
1780	4" diameter	163	202	365
1790	5" diameter	248	122	370
1800	6" diameter	310	151	461
1810	8" diameter	565	180	745
1900	Tee, 1/2" diameter	9.20	74.50	83.70
1910	3/4" diameter	10.75	74.50	85.25
1920	1" diameter	9.55	84	93.55
1930	1-1/4" diameter	17.45	86.50	103.95
1940	1-1/2" diameter	22.50	93	115.50
1950	2" diameter	31.50	110	141.50
1960	2-1/2" diameter	82	134	216
1970	3" diameter	126	202	328

D20 Plumbing

D2090 Other Plumbing Systems

D2090 820	Standard Pipe Fittings - Installed - Unit Costs	COST EACH		
		MAT.	INST.	TOTAL
1980	4" diameter	246	305	551
1990	5" diameter	390	185	575
2000	6" diameter	480	224	704
2010	8" diameter	915	269	1,184
2300	Copper, wrought, solder joints, 45° Elbow, 1/2" diameter	5.75	33.50	39.25
2310	3/4" diameter	10	35.50	45.50
2320	1" diameter	25	42	67
2330	1-1/4" diameter	34	45	79
2340	1-1/2" diameter	41	51.50	92.50
2350	2" diameter	68.50	61	129.50
2360	2-1/2" diameter	146	93	239
2370	3" diameter	217	93	310
2380	4" diameter	460	134	594
2390	5" diameter	1,800	202	2,002
2400	6" diameter	2,825	209	3,034
2500	90° Elbow, 1/2" diameter	3.11	33.50	36.61
2510	3/4" diameter	7	35.50	42.50
2520	1" diameter	17.20	42	59.20
2530	1-1/4" diameter	26	45	71
2540	1-1/2" diameter	40.50	51.50	92
2550	2" diameter	74	61	135
2560	2-1/2" diameter	148	93	241
2610	3" diameter	197	110	307
2620	4" diameter	505	134	639
2630	5" diameter	2,125	202	2,327
2640	6" diameter	2,825	209	3,034
2700	Tee, 1/2" diameter	5.30	51.50	56.80
2710	3/4" diameter	12.85	56	68.85
2720	1" diameter	39.50	67	106.50
2730	1-1/4" diameter	54.50	74.50	129
2740	1-1/2" diameter	83.50	84	167.50
2750	2" diameter	130	96	226
2760	2-1/2" diameter	263	151	414
2770	3" diameter	400	173	573
2780	4" diameter	970	242	1,212
2790	5" diameter	3,175	305	3,480
2800	6" diameter	4,350	315	4,665
2880	Coupling, 1/2" diameter	2.35	30.50	32.85
2890	3/4" diameter	4.75	32	36.75
2900	1" diameter	9.45	37.50	46.95
2910	1-1/4" diameter	16.80	39.50	56.30
2920	1-1/2" diameter	22	45	67
2930	2" diameter	37	51.50	88.50
2940	2-1/2" diameter	79	80.50	159.50
2950	3" diameter	118	93	211
2960	4" diameter	249	173	422
2970	5" diameter	615	202	817
2980	6" diameter	1,025	235	1,260
3200	Malleable iron, 150 Lb. threaded, black, 45° elbow, 1/2" diameter	5.25	45	50.25
3210	3/4" diameter	6.50	48	54.50
3220	1" diameter	8.15	51.50	59.65
3230	1-1/4" diameter	14.65	55	69.65
3240	1-1/2" diameter	17.90	60.50	78.40
3250	2" diameter	27	67	94
3260	2-1/2" diameter	78.50	86.50	165
3270	3" diameter	102	121	223
3280	4" diameter	199	202	401
3290	5" diameter	248	122	370

D20 Plumbing

D2090 Other Plumbing Systems

D2090 820	Standard Pipe Fittings - Installed - Unit Costs	COST EACH		
		MAT.	INST.	TOTAL
3300	6" diameter	310	151	461
3400	90° Elbow, 1/2" diameter	3.18	45	48.18
3500	3/4" diameter	3.84	48	51.84
3510	1" diameter	6.65	51.50	58.15
3520	1-1/4" diameter	11	55	66
3530	1-1/2" diameter	14.45	60.50	74.95
3540	2" diameter	25	67	92
3550	2-1/2" diameter	55.50	86.50	142
3560	3" diameter	81	121	202
3570	4" diameter	174	202	376
3580	5" diameter	248	122	370
3590	6" diameter	310	151	461
3700	Tee, 1/2" diameter	4.25	74.50	78.75
3800	3/4" diameter	6.10	74.50	80.60
3810	1" diameter	10.45	84	94.45
3820	1-1/4" diameter	16.90	86.50	103.40
3830	1-1/2" diameter	21	93	114
3910	2" diameter	36	110	146
3920	2-1/2" diameter	77.50	134	211.50
3930	3" diameter	114	202	316
3940	4" diameter	275	305	580
3950	5" diameter	390	185	575
3960	6" diameter	233	113	346
4000	Coupling, 1/2" diameter	4.39	35.50	39.89
4010	3/4" diameter	5.15	37.50	42.65
4020	1" diameter	7.70	45	52.70
4030	1-1/4" diameter	10	46.50	56.50
4040	1-1/2" diameter	13.50	50.50	64
4050	2" diameter	20	57.50	77.50
4060	2-1/2" diameter	55	67	122
4070	3" diameter	74.50	86.50	161
4080	4" diameter	150	121	271
4090	5" diameter	58	30.50	88.50
4100	6" diameter	76.50	37.50	114
5100	Plastic, PVC, high impact/pressure sch 40, 45° Elbow, 1/2" diameter	.84	20	20.84
5110	3/4" diameter	1.29	23.50	24.79
5120	1" diameter	1.55	27	28.55
5130	1-1/4" diameter	2.17	30.50	32.67
5140	1-1/2" diameter	2.72	33.50	36.22
5150	2" diameter	3.53	33.50	37.03
5160	3" diameter	14.30	53	67.30
5170	4" diameter	25.50	66.50	92
5180	6" diameter	63.50	109	172.50
5190	8" diameter	152	183	335
5260	90° Elbow, 1/2" diameter	.51	20	20.51
5270	3/4" diameter	.56	23.50	24.06
5280	1" diameter	1.01	27	28.01
5290	1-1/4" diameter	1.79	30.50	32.29
5300	1-1/2" diameter	1.94	33.50	35.44
5400	2" diameter	3.03	33.50	36.53
5410	3" diameter	11	53	64
5420	4" diameter	19.65	66.50	86.15
5430	6" diameter	62.50	109	171.50
5440	8" diameter	161	183	344
5500	Tee, 1/2" diameter	.63	30.50	31.13
5510	3/4" diameter	.72	35.50	36.22
5520	1" diameter	1.34	40.50	41.84
5530	1-1/4" diameter	2.10	45.50	47.60

D20 Plumbing

D2090 Other Plumbing Systems

D2090 820	Standard Pipe Fittings - Installed - Unit Costs	COST EACH		
		MAT.	INST.	TOTAL
5540	1-1/2" diameter	2.56	50.50	53.06
5550	2" diameter	3.72	50	53.72
5560	3" diameter	16.15	79.50	95.65
5570	4" diameter	29	100	129
5580	6" diameter	98.50	164	262.50
5590	8" diameter	228	277	505
5680	Coupling, 1/2" diameter	.33	20	20.33
5690	3/4" diameter	.45	23.50	23.95
5700	1" diameter	.80	27	27.80
5710	1-1/4" diameter	1.11	30.50	31.61
5720	1-1/2" diameter	1.11	30.50	31.61
5730	2" diameter	1.80	33.50	35.30
5740	3" diameter	6.25	53	59.25
5750	4" diameter	9.05	66.50	75.55
6010	6" diameter	28.50	109	137.50
6020	8" diameter	53.50	183	236.50
7810	DWV, socket joints, sch. 40, 1/8 Bend, 1-1/4" diameter	7.25	33.50	40.75
7820	1-1/2" diameter	2.98	37	39.98
7830	2" diameter	4.30	36.50	40.80
7840	3" diameter	12.65	58	70.65
7850	4" diameter	22	73.50	95.50
7860	6" diameter	89.50	120	209.50
7960	1/4 bend, 1-1/4" diameter	11.10	33.50	44.60
8000	1-1/2" diameter	3.05	37	40.05
8010	2" diameter	4.81	36.50	41.31
8020	3" diameter	14.15	58	72.15
8030	4" diameter	28	73.50	101.50
8210	6" diameter	98	120	218
8300	Sanitary tee, 1-1/4" diameter	11.15	50	61.15
8310	1-1/2" diameter	5.50	55.50	61
8320	2" diameter	8.10	60.50	68.60
8330	3" diameter	21.50	87	108.50
8340	4" diameter	38	110	148
8400	Coupling, 1-1/4" diameter	26	33.50	59.50
8410	1-1/2" diameter	28	37	65
8420	2" diameter	33	36.50	69.50
8430	3" diameter	68.50	58	126.50
8440	4" diameter	140	73.50	213.50

D30 HVAC

D3010 Energy Supply

Basis for Heat Loss Estimate, Apartment Type Structures:

1. Masonry walls and flat roof are insulated. U factor is assumed at .08.
2. Window glass area taken as BOCA minimum, 1/10th of floor area. Double insulating glass with 1/4" air space, U = .65.
3. Infiltration = 0.3 C.F. per hour per S.F. of net wall.
4. Concrete floor loss is 2 BTUH per S.F.
5. Temperature difference taken as 70°F.
6. Ventilating or makeup air has not been included and must be added if desired. Air shafts are not used.

System Components	QUANTITY	UNIT	COST EACH		
			MAT.	INST.	TOTAL
SYSTEM D3010 510 1760					
HEATING SYSTEM, FIN TUBE RADIATION, FORCED HOT WATER					
1,000 S.F. AREA, 10,000 C.F. VOLUME					
Boiler, oil fired, CI, burner, ctrls/insul/breech/pipe/ftng/valves, 109 MBH	1.000	Ea.	4,200	3,806.25	8,006.25
Circulating pump, CI flange connection, 1/12 HP	1.000	Ea.	380	202	582
Expansion tank, painted steel, ASME 18 Gal capacity	1.000	Ea.	700	87.50	787.50
Storage tank, steel, above ground, 275 Gal capacity w/supports	1.000	Ea.	540	246	786
Copper tubing type L, solder joint, hanger 10' OC, 3/4" diam	100.000	L.F.	720	885	1,605
Radiation, 3/4" copper tube w/alum fin baseboard pkg, 7" high	30.000	L.F.	277.50	630	907.50
Pipe covering, calcium silicate w/cover, 1' wall, 3/4' diam	100.000	L.F.	363	655	1,018
TOTAL			7,180.50	6,511.75	13,692.25
COST PER S.F.			7.18	6.51	13.69

D3010 510	Apartment Building Heating - Fin Tube Radiation		COST PER S.F.		
			MAT.	INST.	TOTAL
1740	Heating systems, fin tube radiation, forced hot water	RD3020 -010			
1760	1,000 S.F. area, 10,000 C.F. volume		7.18	6.50	13.68
1800	10,000 S.F. area, 100,000 C.F. volume	RD3020 -020	3.48	3.93	7.41
1840	20,000 S.F. area, 200,000 C.F. volume		3.99	4.39	8.38
1880	30,000 S.F. area, 300,000 C.F. volume		3.82	4.25	8.07
1890					

D30 HVAC

D3010 Energy Supply

Fin Tube Radiator

Basis for Heat Loss Estimate, Factory or Commercial Type Structures:

1. Walls and flat roof are of lightly insulated concrete block or metal. U factor is assumed at .17.
2. Windows and doors are figured at 1-1/2 S.F. per linear foot of building perimeter. The U factor for flat single glass is 1.13.
3. Infiltration is approximately 5 C.F. per hour per S.F. of wall.
4. Concrete floor loss is 2 BTUH per S.F. of floor.
5. Temperature difference is assumed at 70°F.
6. Ventilation or makeup air has not been included and must be added if desired.

System Components	QUANTITY	UNIT	COST EACH MAT.	INST.	TOTAL
SYSTEM D3010 520 1960					
HEATING SYSTEM, FIN TUBE RADIATION, FORCED HOT WATER					
1,000 S.F. BLDG., ONE FLOOR					
Boiler, oil fired, CI, burner/ctrls/insul/breech/pipe/ftngs/valves, 109 MBH	1.000	Ea.	4,200	3,806.25	8,006.25
Expansion tank, painted steel, ASME 18 Gal capacity	1.000	Ea.	3,075	102	3,177
Storage tank, steel, above ground, 550 Gal capacity w/supports	1.000	Ea.	3,400	455	3,855
Circulating pump, CI flanged, 1/8 HP	1.000	Ea.	630	202	832
Pipe, steel, black, sch. 40, threaded, cplg & hngr 10'OC, 1-1/2" diam.	260.000	L.F.	2,717	3,939	6,656
Pipe covering, calcium silicate w/cover, 1" wall, 1-1/2" diam	260.000	L.F.	959.40	1,755	2,714.40
Radiation, steel 1-1/4" tube & 4-1/4" fin w/cover & damper, wall hung	42.000	L.F.	1,806	1,428	3,234
Rough in, steel fin tube radiation w/supply & balance valves	4.000	Set	1,192	3,000	4,192
TOTAL			17,979.40	14,687.25	32,666.65
COST PER S.F.			17.98	14.69	32.67

D3010 520	Commercial Building Heating - Fin Tube Radiation		COST PER S.F. MAT.	INST.	TOTAL
1940	Heating systems, fin tube radiation, forced hot water				
1960	1,000 S.F. bldg, one floor		18	14.70	32.70
2000	10,000 S.F., 100,000 C.F., total two floors	RD3020 -010	4.89	5.55	10.44
2040	100,000 S.F., 1,000,000 C.F., total three floors		1.89	2.48	4.37
2080	1,000,000 S.F., 10,000,000 C.F., total five floors	RD3020 -020	1.05	1.31	2.36
2090					

D30 HVAC

D3010 Energy Supply

Unit Heater

Basis for Heat Loss Estimate, Factory or Commercial Type Structures:

1. Walls and flat roof are of lightly insulated concrete block or metal. U factor is assumed at .17.
2. Windows and doors are figured at 1-1/2 S.F. per linear foot of building perimeter. The U factor for flat single glass is 1.13.
3. Infiltration is approximately 5 C.F. per hour per S.F. of wall.
4. Concrete floor loss is 2 BTUH per S.F. of floor.
5. Temperature difference is assumed at 70°F.
6. Ventilation or makeup air has not been included and must be added if desired.

System Components	QUANTITY	UNIT	COST EACH		
			MAT.	INST.	TOTAL
SYSTEM D3010 530 1880					
HEATING SYSTEM, TERMINAL UNIT HEATERS, FORCED HOT WATER					
1,000 S.F. BLDG., ONE FLOOR					
Boiler oil fired, CI, burner/ctrls/insul/breech/pipe/ftngs/valves, 109 MBH	1.000	Ea.	4,200	3,806.25	8,006.25
Expansion tank, painted steel, ASME 18 Gal capacity	1.000	Ea.	3,075	102	3,177
Storage tank, steel, above ground, 550 Gal capacity w/supports	1.000	Ea.	3,400	455	3,855
Circulating pump, CI, flanged, 1/8 HP	1.000	Ea.	630	202	832
Pipe, steel, black, schedule 40, threaded, cplg & hngr 10' OC 1-1/2" diam	260.000	L.F.	2,717	3,939	6,656
Pipe covering, calcium silicate w/cover, 1" wall, 1-1/2" diam	260.000	L.F.	959.40	1,755	2,714.40
Unit heater, 1 speed propeller, horizontal, 200° EWT, 26.9 MBH	2.000	Ea.	1,200	308	1,508
Unit heater piping hookup with controls	2.000	Set	1,480	2,900	4,380
TOTAL			17,661.40	13,467.25	31,128.65
COST PER S.F.			17.66	13.47	31.13

D3010 530	Commercial Bldg. Heating - Terminal Unit Heaters		COST PER S.F.		
			MAT.	INST.	TOTAL
1860	Heating systems, terminal unit heaters, forced hot water				
1880	1,000 S.F. bldg., one floor		17.70	13.50	31.20
1920	10,000 S.F. bldg., 100,000 C.F. total two floors	RD3020 -010	4.44	4.60	9.04
1960	100,000 S.F. bldg., 1,000,000 C.F. total three floors		1.93	2.27	4.20
2000	1,000,000 S.F. bldg., 10,000,000 C.F. total five floors	RD3020 -020	1.33	1.47	2.80
2010					

D30 HVAC

D3010 Energy Supply

This draindown pool system uses a differential thermostat similar to those used in solar domestic hot water and space heating applications. To heat the pool, the pool water passes through the conventional pump-filter loop and then flows through the collectors. When collection is not possible, or when the pool temperature is reached, all water drains from the solar loop back to the pool through the existing piping. The modes are controlled by solenoid valves or other automatic valves in conjunction with a vacuum breaker relief valve, which facilitates draindown.

D30 HVAC

D3010 Energy Supply

System Components	QUANTITY	UNIT	COST EACH		
			MAT.	INST.	TOTAL
SYSTEM D3010 660 2640 SOLAR SWIMMING POOL HEATER, ROOF MOUNTED COLLECTORS TEN 4' X 10' FULLY WETTED UNGLAZED PLASTIC ABSORBERS					
A Differential thermostat/controller, 110V, adj pool pump system	1.000	Ea.	330	335	665
A-1 Solenoid valve, PVC, normally 1 open 1 closed (included)	2.000	Ea.			
B Sensor, thermistor type (included)	2.000	Ea.			
E-1 Valve, vacuum relief	1.000	Ea.	31.50	21	52.50
Q Collector panel, solar energy, plastic, liquid full wetted, 4' x 10'	10.000	Ea.	2,650	2,420	5,070
R Valve, ball check, PVC, socket, 1-1/2" diam	1.000	Ea.	108	33.50	141.50
Z Valve, ball, PVC, socket, 1-1/2" diam	3.000	Ea.	202.50	100.50	303
Pipe, PVC, sch 40, 1-1/2" diam	80.000	L.F.	404	1,496	1,900
Pipe fittings, PVC sch 40, socket joint, 1-1/2" diam	10.000	Ea.	19.40	335	354.40
Sensor wire, #22-2 conductor, multistranded	.500	C.L.F.	7.83	31.25	39.08
Roof clamps for solar energy collector panels	10.000	Set	27.90	173	200.90
Roof strap, teflon for solar energy collector panels	26.000	L.F.	585	85.28	670.28
TOTAL			4,366.13	5,030.53	9,396.66

D3010 660	Solar Swimming Pool Heater Systems		COST EACH		
			MAT.	INST.	TOTAL
2530	Solar swimming pool heater systems, roof mounted collectors				
2540	10 ea. 3'x7' black chrome absorber, 1/8" temp. glass		13,000	3,875	16,875
2560	10 ea. 4'x8' black chrome absorber, 3/16" temp. glass	RD3010 -600	14,700	4,625	19,325
2580	10 ea. 3'8"x6' flat black absorber, 3/16" temp. glass		9,525	3,950	13,475
2600	10 ea. 4'x9' flat black absorber, plastic glazing		11,500	4,800	16,300
2620	10 ea. 2'x9' rubber absorber, plastic glazing		13,400	5,200	18,600
2640	10 ea. 4'x10' fully wetted unglazed plastic absorber		4,375	5,025	9,400
2660	Ground mounted collectors				
2680	10 ea. 3'x7' black chrome absorber, 1/8" temp. glass		13,100	4,325	17,425
2700	10 ea. 4'x8' black chrome absorber, 3/16" temp. glass		14,800	5,075	19,875
2720	10 ea. 3'8"x6' flat blk absorber, 3/16" temp. glass		9,625	4,400	14,025
2740	10 ea. 4'x9' flat blk absorber, plastic glazing		11,600	5,275	16,875
2760	10 ea. 2'x9' rubber absorber, plastic glazing		13,500	5,475	18,975
2780	10 ea. 4'x10' fully wetted unglazed plastic absorber		4,475	5,475	9,950

D30 HVAC

D3010 Energy Supply

Solar Direct Gain Glazing — Door Panels: In a Direct Gain System the collection, absorption, and storage of solar energy occur directly within the building's living space. In this system, a section of the building's south wall is removed and replaced with glazing. The glazing in the System Component example consists of three standard insulating glass door panels preassembled in wood frames. A direct gain system also includes a thermal storage mass and night insulation.

During the heating season, solar radiation enters the living space directly through the glazing and is converted to heat, which either warms the air or is stored in the mass. The storage mass is typically masonry or water. Masonry often serves as a basic structural component of the building (e.g., a wall or floor). In this system the cost of adding mass to the building is not included. To reduce heat loss through the south facing glass, thereby increasing overall thermal performance, night insulation is included in this system.

System Components	QUANTITY	UNIT	COST EACH		
			MAT.	INST.	TOTAL
SYSTEM D3010 682 2600					
SOLAR PASSIVE HEATING, DIRECT GAIN, 3'X6'8" DOUBLE GLAZED DOOR PANEL					
3 PANELS WIDE					
A Framing headers over openings, fir 2" x 12"	.025	M.B.F.	21.90	47.25	69.15
A-1 Treated wood sleeper, 1" x 2"	.002	M.B.F.	3.45	5.65	9.10
B Aluminum flashing, mill finish, .013 thick	5.000	S.F.	6.40	28.40	34.80
Studs, 8' high wall, 2" x 4"	.020	M.B.F.	11.90	24	35.90
L-2 Interior casing, stock pine, 11/16" x 2-1/2"	25.000	L.F.	42.25	55.25	97.50
L-4 Valance 1" x 8"	12.000	L.F.	15.36	33.24	48.60
M Vinyl clad sliding door, insulated glass, 6' wide	1.000	Ea.	2,336	354.56	2,690.56
N Caulking/sealants, silicone rubber, cartridges	.250	Gal.	11.25		11.25
O Drywall, gypsum plasterboard, nailed to studs, 5/8" standard	39.000	S.F.	11.70	21.45	33.15
O For taping and finishing, add	39.000	S.F.	1.95	21.45	23.40
P Mylar shade, dbl. layer heat reflective	49.640	S.F.	392.16	40.21	432.37
Q Stool cap, pine, 11/16" x 3-1/2"	12.000	L.F.	35.04	33.24	68.28
Paint wood trim	40.000	S.F.	5.60	23.20	28.80
Demolition, framing	84.000	S.F.		18.48	18.48
Demolition, shingles	84.000	S.F.		105	105
Demolition, drywall	84.000	S.F.		36.96	36.96
TOTAL			2,894.96	848.34	3,743.30

D3010 682	Solar Direct Gain Glazing, Door Panels	COST EACH		
		MAT.	INST.	TOTAL
2550	Solar passive heating, direct gain, 3'x6'8", double glazed door			
2560	One panel wide	1,450	460	1,910
2580	Two panels wide	2,225	660	2,885
2600	Three panels wide	2,900	850	3,750

D30 HVAC

D3010 Energy Supply

Solar Direct Gain Glazing — Window Units: In a Direct Gain System, collection, absorption, and storage of solar energy occur directly within the building's living space. In this system, a section of the building's south wall is removed and replaced with glazing. The glazing consists of standard insulating replacement glass units in wood site-built frames. A direct gain system also includes a thermal storage mass and night insulation.

During the heating season, solar radiation enters the living space directly through the glazing and is converted to heat, which either warms the air or is stored in the mass. The thermal storage mass is typically masonry or water. Masonry often serves as a basic structural component of the building (e.g., a wall or floor). In this system the cost of adding mass to the building is not included. To reduce heat loss through the south facing glass, thereby increasing overall thermal performance, night insulation is included in this system.

	System Components	QUANTITY	UNIT	COST EACH MAT.	COST EACH INST.	COST EACH TOTAL
	SYSTEM D3010 684 2600					
	SOLAR PASSIVE HEATING, DIRECT GAIN, 2'-6"X5' DBL. GLAZED WINDOW PANEL					
	3 PANELS WIDE					
A	Framing headers over openings, fir 2" x 8"	.022	M.B.F.	14.30	53.90	68.20
A-1	Wood exterior sill, oak, no horns, 8/4 x 8" deep	8.000	L.F.	42	88.40	130.40
B	Aluminum flashing, mill finish, .013 thick	4.160	S.F.	3.33	14.77	18.10
C	Glazing, insul glass 2 lite 3/16" float 5/8" thick unit, 15-30 S.F.	38.000	S.F.	564.30	442.70	1,007
F	2"x4" sill blocking	.013	M.B.F.	7.74	42.25	49.99
	Mullion	.027	M.B.F.	16.07	32.40	48.47
J-1	Exterior window stop 1" x 2"	45.000	L.F.	13.05	75.60	88.65
	Exterior sill stock water repellent 1" x 3"	8.000	L.F.	23.36	22.16	45.52
K	Interior window stop 1" x 3"	45.000	L.F.	52.65	103.50	156.15
L-1	Exterior trim and molding, corner board, pine, D & better, 1" x 6"	28.000	L.F.	28.80	66.48	95.28
L-2	Interior casing, 1" x 4"	28.000	L.F.	67.20	70.28	137.48
L-3	Apron	8.000	L.F.	21.44	20.08	41.52
L-4	Valance 1" x 8"	10.000	L.F.	12.80	27.70	40.50
M-1	Window frame	52.000	L.F.	270.40	119.60	390
O	Drywall, gypsum plasterboard, nailed/screw to studs, 5/8" standard	39.000	S.F.	11.70	21.45	33.15
O	For taping and finishing, add	39.000	S.F.	1.95	21.45	23.40
	Demolition, wall framing	60.000	S.F.		14.96	14.96
	Demolition, shingles	60.000	S.F.		75	75
P	Mylar triple layer heat reflective shade (see desc. sheet)	37.500	S.F.	391.88	30.38	422.26
	Demolition, drywall	60.000	S.F.		26.40	26.40
	Paint interior & exterior trim	200.000	L.F.	28	116	144
	TOTAL			1,570.97	1,485.46	3,056.43

D3010 684	Solar Direct Gain Glazing-Window Panels	COST EACH MAT.	COST EACH INST.	COST EACH TOTAL
2550	Solar passive heating, direct gain, 2'-6"x5', double glazed window			
2560	One panel wide	570	540	1,110
2580	Two panels wide	1,150	1,450	2,600
2600	Three panels wide	1,575	1,475	3,050

D30 HVAC

D3010 Energy Supply

Solar Indirect Gain — Thermal Wall: The thermal storage wall is an Indirect Gain System because collection, absorption, and storage of solar energy occur outside the living space. This system consists of an unvented existing masonry wall located between south facing glazing and the space to be heated. The outside surface of the masonry wall is painted black to enable it to absorb maximum thermal energy. Standard double glazed units in wooden site-built frames are added to the south side of the building.

Thermal energy conducts through the masonry wall and radiates into the living space. A masonry thermal storage wall can also contain vents that allow warm air to flow into the living space through convection.

Indirect gain solar systems can be retrofitted to masonry walls constructed from a wide variety of building materials, including concrete, concrete block (solid and filled), brick, stone and adobe.

System Components	QUANTITY	UNIT	COST EACH MAT.	COST EACH INST.	COST EACH TOTAL
SYSTEM D3010 686 2600					
SOLAR PASSIVE HEATING, INDIRECT GAIN, THERMAL STORAGE WALL					
3' X 6'-8" DOUBLE GLAZED PANEL, THREE PANELS WIDE					
B Aluminum flashing, mill finish, .013 thick	5.500	S.F.	4.40	19.53	23.93
C Glazing, insul glass 2 lite 3/16" float 5/8" thick unit, 15-30 S.F.	54.000	S.F.	801.90	629.10	1,431
F Framing 2" x 4" lumber	.030	M.B.F.	17.85	66.75	84.60
K Molding, miscellaneous, 1" x 3"	44.000	L.F.	51.48	101.20	152.68
L-2 Wood batten, 1" x 4", custom pine or cedar	44.000	L.F.	40.04	121.88	161.92
M Drill holes, 4 per panel 1/2" diameter	30.000	Inch		36.90	36.90
N Caulking/sealants, silicone rubber, cartridges	.250	Gal.	11.25		11.25
X Lag screws 3/8" x 3" long	20.000	Ea.	6.20	73.80	80
X-1 Reglet	8.000	L.F.	.48	33.84	34.32
Trim, primer & 1 coat	44.000	L.F.	9.24	32.56	41.80
Layout and drill holes for lag screws	20.000	Ea.	1.20	176	177.20
Expansion shields for lag screws	20.000	Ea.	8.20	69.20	77.40
Total			952.24	1,360.76	2,313

D3010 686	Passive Solar Indirect Gain Wall	MAT.	INST.	TOTAL
2550	Solar passive heating, indirect gain			
2560	3'x6'8", double glazed thermal storage wall, one panel wide	340	515	855
2580	Two panels wide	645	920	1,565
2600	Three panels wide	950	1,350	2,300

D30 HVAC

D3010 Energy Supply

MULLION DETAIL · SECTION · JAMB DETAIL

Solar Indirect Gain — Thermosyphon Panel: The Thermosyphon Air Panel (TAP) System is an indirect gain system used primarily for daytime heating because it does not include thermal mass. This site-built TAP System makes use of double glazing and a ribbed aluminum absorber plate. The panel is framed in wood and attached to the south side of a building. Solar radiation passes through the glass and is absorbed by the absorber plate. As the absorber plate heats, the air between it and the glass also heats. Vents cut through the building's south wall and the absorber plate near the top and bottom of the TAP allow hot air to circulate into the building by means of natural convection.

	System Components	QUANTITY	UNIT	MAT.	INST.	TOTAL
	SYSTEM D3010 688 2600					
	SOLAR PASSIVE HEATING, INDIRECT GAIN, THERMOSYPHON PANEL					
	3'X6'8" DOUBLE GLAZED PANEL, THREE PANELS WIDE					
B	Aluminum flashing and sleeves, mill finish, .013 thick	8.000	S.F.	18.40	81.65	100.05
B-1	Aluminum ribbed, 4" pitch, on steel frame .032" mill finish	45.000	S.F.	117.45	126.45	243.90
B-2	Angle brackets	9.000	Ea.	6.66	28.44	35.10
C	Glazing, insul glass 2 lite 3/16" float 5/8" thick unit, 15-30 S.F.	45.000	S.F.	668.25	524.25	1,192.50
D	Support brackets, fir 2"x6"	.003	M.B.F.	1.79	9.75	11.54
D-1	Framing fir 2" x 6"	.040	M.B.F.	28	111	139
E	Corrugated end closure	15.000	L.F.	14.10	40.80	54.90
E-1	Register, multilouvre operable 6" x 30"	3.000	Ea.	66	108	174
E-3	Grille 6" x 30"	3.000	Ea.	44.10	85.50	129.60
E-4	Back draft damper	.038	C.S.F.	.27	.57	.84
F	House wall blocking 2" x 4"	.012	M.B.F.	23.80	53	76.80
J	Absorber supports & glass stop	54.000	L.F.	21.06	110.70	131.76
K	Exterior & bottom trim & interior glazing stop	40.000	L.F.	45.12	207.74	252.86
L	Exterior trim and molding, cornice board, pine, #2, 1" x 6"	24.000	L.F.	22.56	53.04	75.60
L-2	Trim cap	8.000	L.F.	10.24	22.16	32.40
N	Caulking/sealants, silicone rubber, cartridges	.250	Gal.	11.25		11.25
O	Drywall, nailed & taped, 5/8" thick	60.000	S.F.	18	33	51
O	For taping and finishing, add	60.000	S.F.	3	33	36
O-1	Insulation foil faced exterior wall	45.000	S.F.	19.35	18.45	37.80
	Demolition, framing	60.000	S.F.		18.48	18.48
	Demolition, drywall	60.000	S.F.		26.40	26.40
	Paint absorber plate black	45.000	S.F.	6.30	52.20	58.50
	Finish paint	125.000	S.F.	17.50	72.50	90
	Paint walls & ceiling with roller, primer & 1 coat	60.000	S.F.	7.80	28.80	36.60
	TOTAL			**1,171**	**1,845.88**	**3,016.88**

D3010 688	Passive Solar Indirect Gain Panel	MAT.	INST.	TOTAL
2550	Solar passive heating, indirect gain			
2560	3'x6'8", double glazed thermosyphon panel, one panel wide	400	650	1,050
2580	Two panels wide	785	1,250	2,035
2600	Three panels wide	1,175	1,850	3,025

D30 HVAC

D3010 Energy Supply

Solar Attached Sunspace: The attached sunspace, a room adjacent to the south side of a building, is designed solely for heat collection. During the heating season, solar radiation enters the sunspace through south facing glass. Radiation is absorbed by elements in the space and converted to heat. This sunspace is thermally isolated from the building. A set of sliding glass doors connecting the sunspace to the building space is closed to prevent night heat losses from the living space. This attached sunspace uses standard size, double glazed units and has a concrete slab floor. Since the required depth of footings (if required) is dependent on local frost conditions, costs of footings at depths of 2′, 3′ and 4′ are provided with the base cost of system as shown.

	System Components	QUANTITY	UNIT	COST EACH MAT.	COST EACH INST.	COST EACH TOTAL
	SYSTEM D3010 690 2560					
	SOLAR PASSIVE HEATING, DIRECT GAIN, ATTACHED SUNSPACE					
	WOOD FRAMED, SLAB ON GRADE					
B	Aluminum, flashing, mill finish	21.000	S.F.	16.80	74.55	91.35
C	Glazing, double, tempered glass	72.000	S.F.	1,069.20	838.80	1,908
D-2-4	Knee & end wall framing	.188	M.B.F.	115.62	206.80	322.42
F	Ledgers, headers and rafters	.028	M.B.F.	107.70	239.83	347.53
	Building paper, asphalt felt sheathing paper 15 lb	240.000	S.F.	13.68	35.88	49.56
G	Sheathing plywood on roof CDX 1/2″	240.000	S.F.	153.60	189.60	343.20
H	Asphalt shingles inorganic class A 210-235 lb/sq	1.000	Sq.	89	93.50	182.50
I	Wood siding, match existing	164.000	S.F.	828.20	308.32	1,136.52
	Window stop, interior or external	96.000	L.F.	55.68	322.56	378.24
L-1	Exterior trim and molding, fascia, pine, D & better, 1″ x 6″	44.000	L.F.	52.80	97.24	150.04
L-2, L-3	Exterior trim 1″ x 4″	206.000	L.F.	187.46	570.62	758.08
O	Drywall, taping & finishing joints	200.000	S.F.	10	110	120
O	Drywall, 1/2″ on walls, standard, no finish	200.000	S.F.	52	110	162
S, S-1	Windows, wood, awning type, double glazed, with screens	4.000	Ea.	2,400	276	2,676
T	Insulation batts, fiberglass, faced	210.000	S.F.	100.80	86.10	186.90
T-1	Polyethylene vapor barrier, standard, .004″ thick	2.500	C.S.F.	6.42	29.90	36.32
Y	Button vents	6.000	Ea.	15.06	55.20	70.26
Z	Slab on grade, not including finish, 4″ thick	1.234	C.Y.	161.65	122.90	284.55
	Floor finishing, steel trowel	100.000	S.F.		88	88
Z-2	Mesh, welded wire 6 x 6, #10/10	1.000	C.S.F.	16.05	36	52.05
	Paint interior, exterior and trim	330.000	S.F.	46.20	254.10	300.30
	TOTAL			5,497.92	4,145.90	9,643.82

D3010 690	Passive Solar Sunspace	COST EACH MAT.	COST EACH INST.	COST EACH TOTAL
2550	Solar passive heating, direct gain, attached sunspace, 100 S.F.			
2560	Wood framed, slab on grade	5,500	4,150	9,650
2580	2′ deep frost wall	5,875	5,100	10,975
2600	3′ deep frost wall	6,000	5,450	11,450
2620	4′ deep frost wall	6,125	5,825	11,950

D30 HVAC

D3020 Heat Generating Systems

Boiler   **Baseboard Radiation**

Small Electric Boiler
System Considerations:
1. Terminal units are fin tube baseboard radiation rated at 720 BTU/hr with 200° water temperature or 820 BTU/hr steam.
2. Primary use being for residential or smaller supplementary areas, the floor levels are based on 7-1/2' ceiling heights.
3. All distribution piping is copper for boilers through 205 MBH. All piping for larger systems is steel pipe.

System Components	QUANTITY	UNIT	COST EACH		
			MAT.	INST.	TOTAL
SYSTEM D3020 102 1120					
SMALL HEATING SYSTEM, HYDRONIC, ELECTRIC BOILER					
1,480 S.F., 61 MBH, STEAM, 1 FLOOR					
Boiler, electric steam, std cntrls, trim, ftngs and valves, 18 KW, 61.4 MBH	1.000	Ea.	4,867.50	1,705	6,572.50
Copper tubing type L, solder joint, hanger 10'OC, 1-1/4" diam	160.000	L.F.	2,408	1,856	4,264
Radiation, 3/4" copper tube w/alum fin baseboard pkg 7" high	60.000	L.F.	555	1,260	1,815
Rough in baseboard panel or fin tube with valves & traps	10.000	Set	3,450	6,450	9,900
Pipe covering, calcium silicate w/cover, 1" wall 1-1/4" diam	160.000	L.F.	580.80	1,080	1,660.80
Low water cut-off, quick hookup, in gage glass tappings	1.000	Ea.	260	42.50	302.50
TOTAL			12,121.30	12,393.50	24,514.80
COST PER S.F.			8.19	8.37	16.56

D3020 102	Small Heating Systems, Hydronic, Electric Boilers		COST PER S.F.		
			MAT.	INST.	TOTAL
1100	Small heating systems, hydronic, electric boilers				
1120	Steam, 1 floor, 1480 S.F., 61 M.B.H.		8.18	8.38	16.56
1160	3,000 S.F., 123 M.B.H.	RD3020 -010	6.45	7.35	13.80
1200	5,000 S.F., 205 M.B.H.		5.75	6.80	12.55
1240	2 floors, 12,400 S.F., 512 M.B.H.	RD3020 -020	4.58	6.75	11.33
1280	3 floors, 24,800 S.F., 1023 M.B.H.		5	6.65	11.65
1320	34,750 S.F., 1,433 M.B.H.		4.62	6.45	11.07
1360	Hot water, 1 floor, 1,000 S.F., 41 M.B.H.		12.90	4.64	17.54
1400	2,500 S.F., 103 M.B.H.		9.85	8.35	18.20
1440	2 floors, 4,850 S.F., 205 M.B.H.		9.90	10.05	19.95
1480	3 floors, 9,700 S.F., 410 M.B.H.		10.95	10.35	21.30

D30 HVAC

D3020 Heat Generating Systems

Boiler

Unit Heater

Large Electric Boiler System Considerations:
1. Terminal units are all unit heaters of the same size. Quantities are varied to accommodate total requirements.
2. All air is circulated through the heaters a minimum of three times per hour.
3. As the capacities are adequate for commercial use, floor levels are based on 10′ ceiling heights.
4. All distribution piping is black steel pipe.

System Components	QUANTITY	UNIT	COST EACH MAT.	COST EACH INST.	COST EACH TOTAL
SYSTEM D3020 104 1240					
LARGE HEATING SYSTEM, HYDRONIC, ELECTRIC BOILER					
9,280 S.F., 135 KW, 461 MBH, 1 FLOOR					
Boiler, electric hot water, std ctrls, trim, ftngs, valves, 135 KW, 461 MBH	1.000	Ea.	11,850	3,712.50	15,562.50
Expansion tank, painted steel, 60 Gal capacity ASME	1.000	Ea.	4,025	205	4,230
Circulating pump, CI, close cpld, 50 GPM, 2 HP, 2″ pipe conn	1.000	Ea.	2,775	405	3,180
Unit heater, 1 speed propeller, horizontal, 200° EWT, 72.7 MBH	7.000	Ea.	6,125	1,561	7,686
Unit heater piping hookup with controls	7.000	Set	5,180	10,150	15,330
Pipe, steel, black, schedule 40, welded, 2-1/2″ diam	380.000	L.F.	5,947	10,351.20	16,298.20
Pipe covering, calcium silicate w/cover, 1″ wall, 2-1/2″ diam	380.000	L.F.	1,710	2,641	4,351
TOTAL			37,612	29,025.70	66,637.70
COST PER S.F.			4.05	3.13	7.18

D3020 104	Large Heating Systems, Hydronic, Electric Boilers		COST PER S.F. MAT.	COST PER S.F. INST.	COST PER S.F. TOTAL
1230	Large heating systems, hydronic, electric boilers				
1240	9,280 S.F., 135 K.W., 461 M.B.H., 1 floor		4.05	3.14	7.19
1280	14,900 S.F., 240 K.W., 820 M.B.H., 2 floors	RD3020 -010	5.40	5.05	10.45
1320	18,600 S.F., 296 K.W., 1,010 M.B.H., 3 floors		5.35	5.50	10.85
1360	26,100 S.F., 420 K.W., 1,432 M.B.H., 4 floors	RD3020 -020	5.25	5.40	10.65
1400	39,100 S.F., 666 K.W., 2,273 M.B.H., 4 floors		4.49	4.51	9
1440	57,700 S.F., 900 K.W., 3,071 M.B.H., 5 floors		4.34	4.42	8.76
1480	111,700 S.F., 1,800 K.W., 6,148 M.B.H., 6 floors		3.85	3.80	7.65
1520	149,000 S.F., 2,400 K.W., 8,191 M.B.H., 8 floors		3.80	3.80	7.60
1560	223,300 S.F., 3,600 K.W., 12,283 M.B.H., 14 floors		4.06	4.30	8.36

D30 HVAC

D3020 Heat Generating Systems

Boiler Selection: The maximum allowable working pressures are limited by ASME "Code for Heating Boilers" to 15 PSI for steam and 160 PSI for hot water heating boilers, with a maximum temperature limitation of 250°F. Hot water boilers are generally rated for a working pressure of 30 PSI. High pressure boilers are governed by the ASME "Code for Power Boilers" which is used almost universally for boilers operating over 15 PSIG. High pressure boilers used for a combination of heating/process loads are usually designed for 150 PSIG.

Boiler ratings are usually indicated as either Gross or Net Output. The Gross Load is equal to the Net Load plus a piping and pickup allowance. When this allowance cannot be determined, divide the gross output rating by 1.25 for a value equal to or greater than the next heat loss requirement of the building.

Table below lists installed cost per boiler and includes insulating jacket, standard controls, burner and safety controls. Costs do not include piping or boiler base pad. Outputs are Gross.

D3020 106	Boilers, Hot Water & Steam		COST EACH		
			MAT.	INST.	TOTAL
0600	Boiler, electric, steel, hot water, 12 K.W., 41 M.B.H.		5,275	1,425	6,700
0620	30 K.W., 103 M.B.H.		5,775	1,550	7,325
0640	60 K.W., 205 M.B.H.		6,425	1,675	8,100
0660	120 K.W., 410 M.B.H.	RD3020 -010	6,800	2,050	8,850
0680	210 K.W., 716 M.B.H.		8,125	3,100	11,225
0700	510 K.W., 1,739 M.B.H.	RD3020 -020	23,500	5,775	29,275
0720	720 K.W., 2,452 M.B.H.		29,000	6,500	35,500
0740	1,200 K.W., 4,095 M.B.H.		36,900	7,475	44,375
0760	2,100 K.W., 7,167 M.B.H.		62,500	9,400	71,900
0780	3,600 K.W., 12,283 M.B.H.		98,500	15,900	114,400
0820	Steam, 6 K.W., 20.5 M.B.H.		4,175	1,550	5,725
0840	24 K.W., 81.8 M.B.H.		5,125	1,675	6,800
0860	60 K.W., 205 M.B.H.		7,100	1,850	8,950
0880	150 K.W., 512 M.B.H.		10,200	2,850	13,050
0900	510 K.W., 1,740 M.B.H.		34,800	7,050	41,850
0920	1,080 K.W., 3,685 M.B.H.		43,400	10,200	53,600
0940	2,340 K.W., 7,984 M.B.H.		92,500	15,900	108,400
0980	Gas, cast iron, hot water, 80 M.B.H.		1,775	1,775	3,550
1000	100 M.B.H.		2,075	1,925	4,000
1020	163 M.B.H.		2,825	2,600	5,425
1040	280 M.B.H.		4,025	2,875	6,900
1060	544 M.B.H.		9,875	5,125	15,000
1080	1,088 M.B.H.		14,900	6,500	21,400
1100	2,000 M.B.H.		21,600	10,200	31,800
1120	2,856 M.B.H.		31,100	13,000	44,100
1140	4,720 M.B.H.		86,500	17,900	104,400
1160	6,970 M.B.H.		104,500	29,200	133,700
1180	For steam systems under 2,856 M.B.H., add 8%				
1520	Oil, cast iron, hot water, 109 M.B.H.		2,400	2,175	4,575
1540	173 M.B.H.		3,050	2,600	5,650
1560	236 M.B.H.		3,875	3,050	6,925
1580	1,084 M.B.H.		10,700	6,900	17,600
1600	1,600 M.B.H.		13,700	9,925	23,625
1620	2,480 M.B.H.		21,000	12,700	33,700
1640	3,550 M.B.H.		27,100	15,200	42,300
1660	Steam systems same price as hot water				

D30 HVAC

D3020 Heat Generating Systems

Unit Heater

Fossil Fuel Boiler System Considerations:

1. Terminal units are horizontal unit heaters. Quantities are varied to accommodate total heat loss per building.
2. Unit heater selection was determined by their capacity to circulate the building volume a minimum of three times per hour in addition to the BTU output.
3. Systems shown are forced hot water. Steam boilers cost slightly more than hot water boilers. However, this is compensated for by the smaller size or fewer terminal units required with steam.
4. Floor levels are based on 10' story heights.
5. MBH requirements are gross boiler output.

System Components	QUANTITY	UNIT	COST EACH		
			MAT.	INST.	TOTAL
SYSTEM D3020 108 1280					
HEATING SYSTEM, HYDRONIC, FOSSIL FUEL, TERMINAL UNIT HEATERS					
CAST IRON BOILER, GAS, 80 MBH, 1,070 S.F. BUILDING					
Boiler, gas, hot water, CI, burner, controls, insulation, breeching, 80 MBH	1.000	Ea.	1,863.75	1,863.75	3,727.50
Pipe, steel, black, schedule 40, threaded, cplg & hngr 10'OC, 2" diam	200.000	L.F.	2,150	3,780	5,930
Unit heater, 1 speed propeller, horizontal, 200° EWT, 72.7 MBH	2.000	Ea.	1,750	446	2,196
Unit heater piping hookup with controls	2.000	Set	1,480	2,900	4,380
Expansion tank, painted steel, ASME, 18 Gal capacity	1.000	Ea.	3,075	102	3,177
Circulating pump, CI, flange connection, 1/12 HP	1.000	Ea.	380	202	582
Pipe covering, calcium silicate w/cover, 1" wall, 2" diam	200.000	L.F.	844	1,390	2,234
TOTAL			11,542.75	10,683.75	22,226.50
COST PER S.F.			10.79	9.98	20.77

D3020 108	Heating Systems, Unit Heaters		COST PER S.F.		
			MAT.	INST.	TOTAL
1260	Heating systems, hydronic, fossil fuel, terminal unit heaters,				
1280	Cast iron boiler, gas, 80 M.B.H., 1,070 S.F. bldg.		10.75	10	20.75
1320	163 M.B.H., 2,140 S.F. bldg.	RD3020 -010	7.15	6.80	13.95
1360	544 M.B.H., 7,250 S.F. bldg.		5.30	4.81	10.11
1400	1,088 M.B.H., 14,500 S.F. bldg.	RD3020 -020	4.54	4.54	9.08
1440	3,264 M.B.H., 43,500 S.F. bldg.		3.62	3.42	7.04
1480	5,032 M.B.H., 67,100 S.F. bldg.		4.35	3.52	7.87
1520	Oil, 109 M.B.H., 1,420 S.F. bldg.		11.60	8.90	20.50
1560	235 M.B.H., 3,150 S.F. bldg.		7.60	6.35	13.95
1600	940 M.B.H., 12,500 S.F. bldg.		5.30	4.14	9.44
1640	1,600 M.B.H., 21,300 S.F. bldg.		4.89	3.96	8.85
1680	2,480 M.B.H., 33,100 S.F. bldg.		4.91	3.59	8.50
1720	3,350 M.B.H., 44,500 S.F. bldg.		4.56	3.65	8.21
1760	Coal, 148 M.B.H., 1,975 S.F. bldg.		10.25	5.80	16.05
1800	300 M.B.H., 4,000 S.F. bldg.		7.75	4.54	12.29
1840	2,360 M.B.H., 31,500 S.F. bldg.		5.35	3.65	9

D30 HVAC

D3020 Heat Generating Systems

Fin Tube Radiator

Fossil Fuel Boiler System Considerations:
1. Terminal units are commercial steel fin tube radiation. Quantities are varied to accommodate total heat loss per building.
2. Systems shown are forced hot water. Steam boilers cost slightly more than hot water boilers. However, this is compensated for by the smaller size or fewer terminal units required with steam.
3. Floor levels are based on 10' story heights.
4. MBH requirements are gross boiler output.

System Components	QUANTITY	UNIT	COST EACH		
			MAT.	INST.	TOTAL
SYSTEM D3020 110 3240					
HEATING SYSTEM, HYDRONIC, FOSSIL FUEL, FIN TUBE RADIATION					
CAST IRON BOILER, GAS, 80 MBH, 1,070 S.F. BLDG.					
Boiler, gas, hot water, CI, burner, controls, insulation, breeching, 80 MBH	1.000	Ea.	1,863.75	1,863.75	3,727.50
Pipe, steel, black, schedule 40, threaded, 2" diam	200.000	L.F.	2,150	3,780	5,930
Radiation, steel 1-1/4" tube & 4-1/4" fin w/cover & damper, wall hung	80.000	L.F.	3,440	2,720	6,160
Rough-in, wall hung steel fin radiation with supply & balance valves	8.000	Set	2,384	6,000	8,384
Expansion tank, painted steel, ASME, 18 Gal capacity	1.000	Ea.	3,075	102	3,177
Circulating pump, CI flanged, 1/12 HP	1.000	Ea.	380	202	582
Pipe covering, calcium silicate w/cover, 1" wall, 2" diam	200.000	L.F.	844	1,390	2,234
TOTAL			14,136.75	16,057.75	30,194.50
COST PER S.F.			13.21	15.01	28.22

D3020 110	Heating System, Fin Tube Radiation		COST PER S.F.		
			MAT.	INST.	TOTAL
3230	Heating systems, hydronic, fossil fuel, fin tube radiation				
3240	Cast iron boiler, gas, 80 MBH, 1,070 S.F. bldg.		13.18	15.05	28.23
3280	169 M.B.H., 2,140 S.F. bldg.	RD3020 -010	8.25	9.50	17.75
3320	544 M.B.H., 7,250 S.F. bldg.		7.10	8.15	15.25
3360	1,088 M.B.H., 14,500 S.F. bldg.	RD3020 -020	6.40	8	14.40
3400	3,264 M.B.H., 43,500 S.F. bldg.		5.65	7	12.65
3440	5,032 M.B.H., 67,100 S.F. bldg.		6.40	7.15	13.55
3480	Oil, 109 M.B.H., 1,420 S.F. bldg.		15.85	16.40	32.25
3520	235 M.B.H., 3,150 S.F. bldg.		9.35	9.60	18.95
3560	940 M.B.H., 12,500 S.F. bldg.		7.20	7.60	14.80
3600	1,600 M.B.H., 21,300 S.F. bldg.		6.90	7.55	14.45
3640	2,480 M.B.H., 33,100 S.F. bldg.		6.95	7.20	14.15
3680	3,350 M.B.H., 44,500 S.F. bldg.		6.55	7.25	13.80
3720	Coal, 148 M.B.H., 1,975 S.F. bldg.		11.95	9.05	21
3760	300 M.B.H., 4,000 S.F. bldg.		9.40	7.80	17.20
3800	2,360 M.B.H., 31,500 S.F. bldg.		7.35	7.20	14.55
4080	Steel boiler, oil, 97 M.B.H., 1,300 S.F. bldg.		14.05	13.45	27.50
4120	315 M.B.H., 4,550 S.F. bldg.		6.20	6.85	13.05
4160	525 M.B.H., 7,000 S.F. bldg.		8.70	7.90	16.60
4200	1,050 M.B.H., 14,000 S.F. bldg.		7.60	7.90	15.50
4240	2,310 M.B.H., 30,800 S.F. bldg.		7.35	7.30	14.65
4280	3,150 M.B.H., 42,000 S.F. bldg.		7.10	7.35	14.45

D30 HVAC

D3020 Heat Generating Systems

D3020 310	Gas Vent, Galvanized Steel, Double Wall	COST PER V.L.F.		
		MAT.	INST.	TOTAL
0030	Gas vent, galvanized steel, double wall, round pipe diameter, 3"	5.30	16.25	21.55
0040	4"	7.30	17.20	24.50
0050	5"	8.25	18.30	26.55
0060	6"	10.35	19.50	29.85
0070	7"	11.90	21	32.90
0080	8"	18.05	22.50	40.55
0090	10"	34.50	24.50	59
0100	12"	41	26.50	67.50
0110	14"	68.50	28	96.50
0120	16	98.50	29.50	128
0130	18"	122	31	153
0140	20"	143	50.50	193.50
0150	22"	182	53.50	235.50
0160	24"	225	57	282
0400				

D3020 310	Gas Vent Fittings, Galvanized Steel, Double Wall	COST EACH		
		MAT.	INST.	TOTAL
1020	Gas vent, fittings, 45° Elbow, adjustable thru 8" diameter			
1030	Pipe diameter, 3"	13.65	32.50	46.15
1040	4"	15.95	34.50	50.45
1050	5"	19.25	36.50	55.75
1060	6"	23.50	39	62.50
1070	7"	36	42	78
1080	8"	48	45	93
1090	10"	102	49	151
1100	12"	109	53	162
1110	14"	171	56	227
1120	16"	221	58.50	279.50
1130	18"	291	61.50	352.50
1140	20"	325	101	426
1150	22"	520	107	627
1170	24"	660	114	774
1410	90° Elbow, adjustable, pipe diameter, 3"	23	32.50	55.50
1420	4"	27	34.50	61.50
1430	5"	34	36.50	70.50
1440	6"	39.50	39	78.50
1450	7"	57	42	99
1460	8"	65	45	110
1510	Tees, standard, pipe diameter, 3"	35.50	43.50	79
1520	4"	38.50	45	83.50
1530	5"	42.50	47	89.50
1540	6"	46	49	95
1550	7"	60	51	111
1560	8"	68	53	121
1570	10"	182	56	238
1580	12"	187	58.50	245.50
1590	14"	475	65	540
1600	16"	495	73	568
1610	18"	595	83.50	678.50
1620	20"	1,150	107	1,257
1630	22"	1,175	140	1,315
1640	24"	1,200	152	1,352
1660	30"	2,500	182	2,682
1700	38"	3,525	228	3,753
1750	48"	5,325	365	5,690

D30 HVAC

D3020 Heat Generating Systems

D3020 310 Gas Vent Fittings, Galvanized Steel, Double Wall

		MAT.	INST.	TOTAL
1810	Tee, access caps, pipe diameter, 3"	3.56	26	29.56
1820	4"	4.02	28	32.02
1830	5"	5.30	29.50	34.80
1840	6"	6.20	31.50	37.70
1850	7"	10	33.50	43.50
1860	8"	11.80	34.50	46.30
1870	10"	24.50	36.50	61
1880	12"	32.50	39	71.50
1890	14"	67	42	109
1900	16"	77.50	47	124.50
1910	18"	87.50	49	136.50
1920	20"	93.50	67.50	161
1930	22"	108	83	191
1940	24"	121	86.50	207.50
2110	Tops, bird proof, pipe diameter, 3"	18.70	25.50	44.20
2120	4"	24.50	26.50	51
2130	5"	27.50	28	55.50
2140	6"	34.50	29.50	64
2150	7"	50.50	31	81.50
2160	8"	66	32.50	98.50
2170	10"	119	34.50	153.50
2180	12"	135	36.50	171.50
2190	14"	173	39	212
2200	16"	251	42	293
2210	18"	345	45	390
2220	20"	915	65	980
2230	22"	985	83	1,068
2240	24"	1,000	91	1,091
2410	Roof flashing, tall cone/adjustable, pipe diameter, 3"	8.55	32.50	41.05
2420	4"	9.95	34.50	44.45
2430	5"	18.40	36.50	54.90
2440	6"	24	39	63
2450	7"	30	42	72
2460	8"	36.50	45	81.50
2470	10"	43	49	92
2480	12"	60	53	113
2490	14"	171	58.50	229.50
2500	16"	211	65	276
2510	18"	225	73	298
2520	20"	370	101	471
2530	22"	470	130	600
2540	24"	515	152	667

Diagram labels: Insulated Pipe, Roof Thimble, Full Angle Ring, Wall Support, 15° Adjustable Elbow, Adjustable Length, Plate Support, Wall Guide, Standard Tee, Half Angle Ring, Drain

D3020 315 Chimney, Stainless Steel, Insulated, Wood & Oil

COST PER V.L.F.

		MAT.	INST.	TOTAL
0530	Chimney, stainless steel, insulated, diameter, 6"	68	19.50	87.50
0550	8"	78	22.50	100.50
0560	10"	87	24.50	111.50
0570	12"	99.50	26.50	126
0580	14"	112	28	140

D30 HVAC

D3030 Cooling Generating Systems

Design Assumptions: The chilled water, air cooled systems priced, utilize reciprocating hermetic compressors and propeller-type condenser fans. Piping with pumps and expansion tanks is included based on a two pipe system. No ducting is included and the fan-coil units are cooling only. Water treatment and balancing are not included. Chilled water piping is insulated. Area distribution is through the use of multiple fan coil units. Fewer but larger fan coil units with duct distribution would be approximately the same S.F. cost.

System Components	QUANTITY	UNIT	COST EACH		
			MAT.	INST.	TOTAL
SYSTEM D3030 110 1200 PACKAGED CHILLER, AIR COOLED, WITH FAN COIL UNIT APARTMENT CORRIDORS, 3,000 S.F., 5.50 TON					
Fan coil air conditioning unit, cabinet mounted & filters chilled water	1.000	Ea.	3,666	559.07	4,225.07
Water chiller, air conditioning unit, air cooled	1.000	Ea.	8,222.50	2,103.75	10,326.25
Chilled water unit coil connections	1.000	Ea.	1,300	1,525	2,825
Chilled water distribution piping	440.000	L.F.	11,440	20,680	32,120
TOTAL			24,628.50	24,867.82	49,496.32
COST PER S.F.			8.21	8.29	16.50

D3030 110	Chilled Water, Air Cooled Condenser Systems		COST PER S.F.		
			MAT.	INST.	TOTAL
1180	Packaged chiller, air cooled, with fan coil unit				
1200	Apartment corridors, 3,000 S.F., 5.50 ton		8.20	8.30	16.50
1360	40,000 S.F., 73.33 ton	RD3030 -010	5.10	3.68	8.78
1440	Banks and libraries, 3,000 S.F., 12.50 ton		12.65	9.75	22.40
1560	20,000 S.F., 83.33 ton		8.55	5	13.55
1680	Bars and taverns, 3,000 S.F., 33.25 ton		20.50	11.60	32.10
1760	10,000 S.F., 110.83 ton		13.95	2.88	16.83
1920	Bowling alleys, 3,000 S.F., 17.00 ton		15.85	10.95	26.80
2040	20,000 S.F., 113.33 ton		10	5.10	15.10
2160	Department stores, 3,000 S.F., 8.75 ton		11.20	9.25	20.45
2320	40,000 S.F., 116.66 ton		6.60	3.82	10.42
2400	Drug stores, 3,000 S.F., 20.00 ton		17.90	11.35	29.25
2520	20,000 S.F., 133.33 ton		12.10	5.40	17.50
2640	Factories, 2,000 S.F., 10.00 ton		11.10	9.35	20.45
2800	40,000 S.F., 133.33 ton		7.15	3.92	11.07
2880	Food supermarkets, 3,000 S.F., 8.50 ton		11.05	9.20	20.25
3040	40,000 S.F., 113.33 ton		6.30	3.82	10.12
3120	Medical centers, 3,000 S.F., 7.00 ton		9.95	8.95	18.90
3280	40,000 S.F., 93.33 ton		5.80	3.74	9.54
3360	Offices, 3,000 S.F., 9.50 ton		10.25	9.10	19.35
3520	40,000 S.F., 126.66 ton		7.05	3.96	11.01
3600	Restaurants, 3,000 S.F., 15.00 ton		14.15	10.15	24.30
3720	20,000 S.F., 100.00 ton		10.15	5.35	15.50
3840	Schools and colleges, 3,000 S.F., 11.50 ton		12	9.60	21.60
3960	20,000 S.F., 76.66 ton		8.35	5.05	13.40

D30 HVAC

D3030 Cooling Generating Systems

General: Water cooled chillers are available in the same sizes as air cooled units. They are also available in larger capacities.

Design Assumptions: The chilled water systems with water cooled condenser include reciprocating hermetic compressors, water cooling tower, pumps, piping and expansion tanks and are based on a two pipe system. Chilled water piping is insulated. No ducts are included and fan-coil units are cooling only. Area distribution is through use of multiple fan coil units. Fewer but larger fan coil units with duct distribution would be approximately the same S.F. cost. Water treatment and balancing are not included.

System Components	QUANTITY	UNIT	COST EACH		
			MAT.	INST.	TOTAL
SYSTEM D3030 115 1320					
PACKAGED CHILLER, WATER COOLED, WITH FAN COIL UNIT					
APARTMENT CORRIDORS, 4,000 S.F., 7.33 TON					
Fan coil air conditioner unit, cabinet mounted & filters, chilled water	2.000	Ea.	4,886	745.12	5,631.12
Water chiller, water cooled, 1 compressor, hermetic scroll,	1.000	Ea.	5,793.70	3,664	9,457.70
Cooling tower, draw thru single flow, belt drive	1.000	Ea.	1,663.91	153.93	1,817.84
Cooling tower pumps & piping	1.000	System	835.62	370.17	1,205.79
Chilled water unit coil connections	2.000	Ea.	2,600	3,050	5,650
Chilled water distribution piping	520.000	L.F.	13,520	24,440	37,960
TOTAL			29,299.23	32,423.22	61,722.45
COST PER S.F.			7.32	8.11	15.43

*Cooling requirements would lead to choosing a water cooled unit.

D3030 115	Chilled Water, Cooling Tower Systems		COST PER S.F.		
			MAT.	INST.	TOTAL
1300	Packaged chiller, water cooled, with fan coil unit				
1320	Apartment corridors, 4,000 S.F., 7.33 ton		7.33	8.10	15.43
1600	Banks and libraries, 4,000 S.F., 16.66 ton	RD3030 -010	12.85	8.90	21.75
1800	60,000 S.F., 250.00 ton		8.95	7.15	16.10
1880	Bars and taverns, 4,000 S.F., 44.33 ton		22	11.25	33.25
2000	20,000 S.F., 221.66 ton		21	9.35	30.35
2160	Bowling alleys, 4,000 S.F., 22.66 ton		14.65	9.80	24.45
2320	40,000 S.F., 226.66 ton		11.85	6.75	18.60
2440	Department stores, 4,000 S.F., 11.66 ton		8.35	8.80	17.15
2640	60,000 S.F., 175.00 ton		7.90	6.55	14.45
2720	Drug stores, 4,000 S.F., 26.66 ton		15.90	10.10	26
2880	40,000 S.F., 266.67 ton		12.10	7.65	19.75
3000	Factories, 4,000 S.F., 13.33 ton		11.10	8.50	19.60
3200	60,000 S.F., 200.00 ton		8.05	6.90	14.95
3280	Food supermarkets, 4,000 S.F., 11.33 ton		8.25	8.75	17
3480	60,000 S.F., 170.00 ton		7.85	6.55	14.40
3560	Medical centers, 4,000 S.F., 9.33 ton		7.20	8.05	15.25
3760	60,000 S.F., 140.00 ton		6.55	6.60	13.15
3840	Offices, 4,000 S.F., 12.66 ton		10.75	8.40	19.15
4040	60,000 S.F., 190.00 ton		7.75	6.80	14.55
4120	Restaurants, 4,000 S.F., 20.00 ton		13.15	9.10	22.25
4320	60,000 S.F., 300.00 ton		9.95	7.45	17.40
4400	Schools and colleges, 4,000 S.F., 15.33 ton		12.15	8.75	20.90
4600	60,000 S.F., 230.00 ton		8.25	6.90	15.15

D30 HVAC

D3030 Cooling Generating Systems

System Components	QUANTITY	UNIT	COST EACH		
			MAT.	INST.	TOTAL
SYSTEM D3030 214 1200					
HEATING/COOLING, GAS FIRED FORCED AIR,					
ONE ZONE, 1200 S.F. BLDG, SEER 14					
Thermostat manual	1.000	Ea.	50.50	85.50	136
Intermittent pilot	1.000	Ea.	165		165
Furnace, 3 Ton cooling, 115 MBH	1.000	Ea.	2,300	390	2,690
Cooling tubing 25 feet	1.000	Ea.	282		282
Ductwork	158.000	Lb.	121.66	1,224.50	1,346.16
Ductwork connection	12.000	Ea.	372	234	606
Supply ductwork	176.000	SF Surf	149.60	915.20	1,064.80
Supply grill	2.000	Ea.	58	59	117
Duct insulation	1.000	L.F.	396	648	1,044
Return register	1.000	Ea.	115.20	246	361.20
TOTAL			4,009.96	3,802.20	7,812.16

D3030 214		Heating/Cooling System	COST EACH		
			MAT.	INST.	TOTAL
1200		Heating/Cooling system, gas fired, SEER 14, 1200 S.F. Bldg	4,000	3,800	7,800
1300		2000 S.F. Bldg	5,075	6,000	11,075
1400		Heating/Cooling system, heat pump 3 ton, SEER 14, 1200 S.F. Bldg	6,500	4,925	11,425
1500		5 ton, SEER 14, 2000 S.F. Bldg	8,725	5,625	14,350

D30 HVAC

D3050 Terminal & Package Units

System Description: Rooftop single zone units are electric cooling and gas heat. Duct systems are low velocity, galvanized steel supply and return. Price variations between sizes are due to several factors. Jumps in the cost of the rooftop unit occur when the manufacturer shifts from the largest capacity unit on a small frame to the smallest capacity on the next larger frame, or changes from one compressor to two. As the unit capacity increases for larger areas the duct distribution grows in proportion. For most applications there is a tradeoff point where it is less expensive and more efficient to utilize smaller units with short simple distribution systems. Larger units also require larger initial supply and return ducts which can create a space problem. Supplemental heat may be desired in colder locations. The table below is based on one unit supplying the area listed. The 10,000 S.F. unit for bars and taverns is not listed because a nominal 110 ton unit would be required and this is above the normal single zone rooftop capacity.

System Components	QUANTITY	UNIT	COST EACH		
			MAT.	INST.	TOTAL
SYSTEM D3050 150 1280					
ROOFTOP, SINGLE ZONE, AIR CONDITIONER					
APARTMENT CORRIDORS, 500 S.F., .92 TON					
Rooftop air conditioner, 1 zone, electric cool, standard controls, curb	1.000	Ea.	1,886	805	2,691
Ductwork package for rooftop single zone units	1.000	System	265.88	1,127	1,392.88
TOTAL			2,151.88	1,932	4,083.88
COST PER S.F.			4.30	3.86	8.16

D3050 150	Rooftop Single Zone Unit Systems		COST PER S.F.		
			MAT.	INST.	TOTAL
1260	Rooftop, single zone, air conditioner				
1280	Apartment corridors, 500 S.F., .92 ton		4.30	3.85	8.15
1480	10,000 S.F., 18.33 ton	RD3030 -010	2.85	2.61	5.46
1560	Banks or libraries, 500 S.F., 2.08 ton		9.75	8.75	18.50
1760	10,000 S.F., 41.67 ton		5.30	5.85	11.15
1840	Bars and taverns, 500 S.F. 5.54 ton		14	11.65	25.65
2000	5,000 S.F., 55.42 ton		11.60	8.85	20.45
2080	Bowling alleys, 500 S.F., 2.83 ton		8.45	9.80	18.25
2280	10,000 S.F., 56.67 ton		6.75	8	14.75
2360	Department stores, 500 S.F., 1.46 ton		6.80	6.15	12.95
2560	10,000 S.F., 29.17 ton		3.79	4.11	7.90
2640	Drug stores, 500 S.F., 3.33 ton		9.90	11.55	21.45
2840	10,000 S.F., 66.67 ton		7.95	9.40	17.35
2920	Factories, 500 S.F., 1.67 ton		7.80	7	14.80
3120	10,000 S.F., 33.33 ton		4.34	4.70	9.04
3200	Food supermarkets, 500 S.F., 1.42 ton		6.65	5.95	12.60
3400	10,000 S.F., 28.33 ton		3.69	3.99	7.68
3480	Medical centers, 500 S.F., 1.17 ton		5.45	4.91	10.36
3680	10,000 S.F., 23.33 ton		3.62	3.32	6.94
3760	Offices, 500 S.F., 1.58 ton		7.40	6.65	14.05
3960	10,000 S.F., 31.67 ton		4.13	4.47	8.60
4000	Restaurants, 500 S.F., 2.50 ton		11.70	10.50	22.20
4200	10,000 S.F., 50.00 ton		5.95	7.05	13
4240	Schools and colleges, 500 S.F., 1.92 ton		8.95	8.05	17
4440	10,000 S.F., 38.33 ton		4.87	5.40	10.27

D30 HVAC

D3050 Terminal & Package Units

System Description: Rooftop units are multizone with up to 12 zones, and include electric cooling, gas heat, thermostats, filters, supply and return fans complete. Duct systems are low velocity, galvanized steel supply and return with insulated supplies.

Multizone units cost more per ton of cooling than single zone. However, they offer flexibility where load conditions are varied due to heat generating areas or exposure to radiational heating. For example, perimeter offices on the "sunny side" may require cooling at the same time "shady side" or central offices may require heating. It is possible to accomplish similar results using duct heaters in branches of the single zone unit. However, heater location could be a problem and total system operating energy efficiency could be lower.

System Components	QUANTITY	UNIT	COST EACH MAT.	COST EACH INST.	COST EACH TOTAL
SYSTEM D3050 155 1280					
ROOFTOP, MULTIZONE, AIR CONDITIONER					
APARTMENT CORRIDORS, 3,000 S.F., 5.50 TON					
Rooftop multizone unit, standard controls, curb	1.000	Ea.	29,920	1,881	31,801
Ductwork package for rooftop multizone units	1.000	System	2,172.50	12,925	15,097.50
TOTAL			32,092.50	14,806	46,898.50
COST PER S.F.			10.70	4.94	15.64

Note A: Small single zone unit recommended.

D3050 155	Rooftop Multizone Unit Systems		MAT.	INST.	TOTAL
1240	Rooftop, multizone, air conditioner				
1260	Apartment corridors, 1,500 S.F., 2.75 ton. See Note A.				
1280	3,000 S.F., 5.50 ton	RD3030 -010	10.70	4.93	15.63
1440	25,000 S.F., 45.80 ton		7.05	4.77	11.82
1520	Banks or libraries, 1,500 S.F., 6.25 ton		24.50	11.20	35.70
1640	15,000 S.F., 62.50 ton		11.90	10.75	22.65
1720	25,000 S.F., 104.00 ton		10.65	10.70	21.35
1800	Bars and taverns, 1,500 S.F., 16.62 ton		52.50	16.20	68.70
1840	3,000 S.F., 33.24 ton		43.50	15.60	59.10
1880	10,000 S.F., 110.83 ton		26	15.50	41.50
2080	Bowling alleys, 1,500 S.F., 8.50 ton		33	15.25	48.25
2160	10,000 S.F., 56.70 ton		22	14.75	36.75
2240	20,000 S.F., 113.00 ton		14.50	14.55	29.05
2640	Drug stores, 1,500 S.F., 10.00 ton		39	17.95	56.95
2680	3,000 S.F., 20.00 ton		27	17.30	44.30
2760	15,000 S.F., 100.00 ton		17.10	17.15	34.25
3760	Offices, 1,500 S.F., 4.75 ton, See Note A.				
3880	15,000 S.F., 47.50 ton		12.20	8.25	20.45
3960	25,000 S.F., 79.16 ton		9.05	8.20	17.25
4000	Restaurants, 1,500 S.F., 7.50 ton		29	13.45	42.45
4080	10,000 S.F., 50.00 ton		19.25	13	32.25
4160	20,000 S.F., 100.00 ton		12.85	12.85	25.70
4240	Schools and colleges, 1,500 S.F., 5.75 ton		22.50	10.30	32.80
4360	15,000 S.F., 57.50 ton		10.95	9.90	20.85
4440	25,000 S.F., 95.83 ton		10.30	9.90	20.20

D30 HVAC

D3050 Terminal & Package Units

System Description: Self-contained, single package water cooled units include cooling tower, pump, piping allowance. Systems for 1000 S.F. and up include duct and diffusers to provide for even distribution of air. Smaller units distribute air through a supply air plenum, which is integral with the unit.

Returns are not ducted and supplies are not insulated.

Hot water or steam heating coils are included but piping to boiler and the boiler itself is not included.

Where local codes or conditions permit single pass cooling for the smaller units, deduct 10%.

System Components	QUANTITY	UNIT	COST EACH MAT.	COST EACH INST.	COST EACH TOTAL
SYSTEM D3050 160 1300					
SELF-CONTAINED, WATER COOLED UNIT					
APARTMENT CORRIDORS, 500 S.F., .92 TON					
Self contained, water cooled, single package air conditioner unit	1.000	Ea.	1,327.20	638.40	1,965.60
Ductwork package for water or air cooled packaged units	1.000	System	75.44	855.60	931.04
Cooling tower, draw thru single flow, belt drive	1.000	Ea.	208.84	19.32	228.16
Cooling tower pumps & piping	1.000	System	104.88	46.46	151.34
TOTAL			1,716.36	1,559.78	3,276.14
COST PER S.F.			3.43	3.12	6.55

D3050 160	Self-contained, Water Cooled Unit Systems		COST PER S.F. MAT.	COST PER S.F. INST.	COST PER S.F. TOTAL
1280	Self-contained, water cooled unit		3.43	3.12	6.55
1300	Apartment corridors, 500 S.F., .92 ton		3.45	3.10	6.55
1440	10,000 S.F., 18.33 ton	RD3030 -010	3.84	2.16	6
1520	Banks or libraries, 500 S.F., 2.08 ton		7.45	3.20	10.65
1680	10,000 S.F., 41.66 ton		7.60	4.91	12.51
1760	Bars and taverns, 500 S.F., 5.54 ton		16.30	5.40	21.70
1920	10,000 S.F., 110.00 ton		18.70	7.65	26.35
2000	Bowling alleys, 500 S.F., 2.83 ton		10.15	4.35	14.50
2160	10,000 S.F., 56.66 ton		10.10	6.60	16.70
2200	Department stores, 500 S.F., 1.46 ton		5.20	2.24	7.44
2360	10,000 S.F., 29.17 ton		5.55	3.32	8.87
2440	Drug stores, 500 S.F., 3.33 ton		11.90	5.10	17
2600	10,000 S.F., 66.66 ton		11.85	7.75	19.60
2680	Factories, 500 S.F., 1.66 ton		5.95	2.56	8.51
2840	10,000 S.F., 33.33 ton		6.35	3.79	10.14
2920	Food supermarkets, 500 S.F., 1.42 ton		5.05	2.17	7.22
3080	10,000 S.F., 28.33 ton		5.40	3.21	8.61
3160	Medical centers, 500 S.F., 1.17 ton		4.17	1.79	5.96
3320	10,000 S.F., 23.33 ton		4.88	2.76	7.64
3400	Offices, 500 S.F., 1.58 ton		5.65	2.43	8.08
3560	10,000 S.F., 31.67 ton		6	3.60	9.60
3640	Restaurants, 500 S.F., 2.50 ton		8.95	3.84	12.79
3800	10,000 S.F., 50.00 ton		6.55	5.40	11.95
3880	Schools and colleges, 500 S.F., 1.92 ton		6.85	2.94	9.79
4040	10,000 S.F., 38.33 ton		7	4.52	11.52

D30 HVAC

D3050 Terminal & Package Units

System Description: Self-contained air cooled units with remote air cooled condenser and interconnecting tubing. Systems for 1000 S.F. and up include duct and diffusers. Smaller units distribute air directly.

Returns are not ducted and supplies are not insulated.

Potential savings may be realized by using a single zone rooftop system or through-the-wall unit, especially in the smaller capacities, if the application permits.

Hot water or steam heating coils are included but piping to boiler and the boiler itself is not included.

Condenserless models are available for 15% less where remote refrigerant source is available.

System Components	QUANTITY	UNIT	COST EACH		
			MAT.	INST.	TOTAL
SYSTEM D3050 165 1320					
SELF-CONTAINED, AIR COOLED UNIT					
APARTMENT CORRIDORS, 500 S.F., .92 TON					
Air cooled, package unit	1.000	Ea.	1,351.50	416.50	1,768
Ductwork package for water or air cooled packaged units	1.000	System	75.44	855.60	931.04
Refrigerant piping	1.000	System	556.60	565.80	1,122.40
Air cooled condenser, direct drive, propeller fan	1.000	Ea.	364.25	158.10	522.35
TOTAL			2,347.79	1,996	4,343.79
COST PER S.F.			4.70	3.99	8.69

D3050 165	Self-contained, Air Cooled Unit Systems	COST PER S.F.		
		MAT.	INST.	TOTAL
1300	Self-contained, air cooled unit			
1320	Apartment corridors, 500 S.F., .92 ton	4.70	4	8.70
1480	10,000 S.F., 18.33 ton	3.58	3.23	6.81
1560	Banks or libraries, 500 S.F., 2.08 ton	10.15	5.10	15.25
1720	10,000 S.F., 41.66 ton	8.40	7.20	15.60
1800	Bars and taverns, 500 S.F., 5.54 ton	21	11.15	32.15
1960	10,000 S.F., 110.00 ton	22.50	13.90	36.40
2040	Bowling alleys, 500 S.F., 2.83 ton	13.90	7	20.90
2200	10,000 S.F., 56.66 ton	11.65	9.75	21.40
2240	Department stores, 500 S.F., 1.46 ton	7.15	3.60	10.75
2400	10,000 S.F., 29.17 ton	5.85	5	10.85
2480	Drug stores, 500 S.F., 3.33 ton	16.35	8.20	24.55
2640	10,000 S.F., 66.66 ton	13.80	11.50	25.30
2720	Factories, 500 S.F., 1.66 ton	8.30	4.15	12.45
2880	10,000 S.F., 33.33 ton	6.70	5.75	12.45
3200	Medical centers, 500 S.F., 1.17 ton	5.75	2.88	8.63
3360	10,000 S.F., 23.33 ton	4.55	4.11	8.66
3440	Offices, 500 S.F., 1.58 ton	7.80	3.91	11.71
3600	10,000 S.F., 31.66 ton	6.35	5.45	11.80
3680	Restaurants, 500 S.F., 2.50 ton	12.20	6.15	18.35
3840	10,000 S.F., 50.00 ton	10.50	8.60	19.10
3920	Schools and colleges, 500 S.F., 1.92 ton	9.40	4.73	14.13
4080	10,000 S.F., 38.33 ton	7.75	6.60	14.35

D30 HVAC

D3050 Terminal & Package Units

General: Split systems offer several important advantages which should be evaluated when a selection is to be made. They provide a greater degree of flexibility in component selection which permits an accurate match-up of the proper equipment size and type with the particular needs of the building. This allows for maximum use of modern energy saving concepts in heating and cooling. Outdoor installation of the air cooled condensing unit allows space savings in the building and also isolates the equipment operating sounds from building occupants.

Design Assumptions: The systems below are comprised of a direct expansion air handling unit and air cooled condensing unit with interconnecting copper tubing. Ducts and diffusers are also included for distribution of air. Systems are priced for cooling only. Heat can be added as desired either by putting hot water/steam coils into the air unit or into the duct supplying the particular area of need. Gas fired duct furnaces are also available. Refrigerant liquid line is insulated.

System Components	QUANTITY	UNIT	COST EACH		
			MAT.	INST.	TOTAL
SYSTEM D3050 170 1280					
SPLIT SYSTEM, AIR COOLED CONDENSING UNIT					
APARTMENT CORRIDORS, 1,000 S.F., 1.80 TON					
Fan coil AC unit, cabinet mntd & filters direct expansion air cool	1.000	Ea.	567.30	150.06	717.36
Ductwork package, for split system, remote condensing unit	1.000	System	70.46	809.78	880.24
Refrigeration piping	1.000	System	448.35	933.30	1,381.65
Condensing unit, air cooled, incls compressor & standard controls	1.000	Ea.	1,677.50	597.80	2,275.30
TOTAL			2,763.61	2,490.94	5,254.55
COST PER S.F.			2.76	2.49	5.25

*Cooling requirements would lead to choosing a water cooled unit.

D3050 170	Split Systems With Air Cooled Condensing Units	COST PER S.F.		
		MAT.	INST.	TOTAL
1260	Split system, air cooled condensing unit			
1280	Apartment corridors, 1,000 S.F., 1.83 ton	2.78	2.50	5.28
1440	20,000 S.F., 36.66 ton	2.91	3.49	6.40
1520	Banks and libraries, 1,000 S.F., 4.17 ton	4.68	5.75	10.43
1680	20,000 S.F., 83.32 ton	7.75	8.25	16
1760	Bars and taverns, 1,000 S.F., 11.08 ton	11.80	11.25	23.05
1880	10,000 S.F., 110.84 ton	16.80	13.20	30
2000	Bowling alleys, 1,000 S.F., 5.66 ton	6.55	10.70	17.25
2160	20,000 S.F., 113.32 ton	11.65	11.70	23.35
2320	Department stores, 1,000 S.F., 2.92 ton	3.30	3.94	7.24
2480	20,000 S.F., 58.33 ton	4.62	5.55	10.17
2560	Drug stores, 1,000 S.F., 6.66 ton	7.70	12.55	20.25
2720	20,000 S.F., 133.32 ton*			
2800	Factories, 1,000 S.F., 3.33 ton	3.78	4.49	8.27
2960	20,000 S.F., 66.66 ton	5.75	6.65	12.40
3040	Food supermarkets, 1,000 S.F., 2.83 ton	3.21	3.81	7.02
3200	20,000 S.F., 56.66 ton	4.49	5.40	9.89
3280	Medical centers, 1,000 S.F., 2.33 ton	2.90	3.08	5.98
3440	20,000 S.F., 46.66 ton	3.71	4.44	8.15
3520	Offices, 1,000 S.F., 3.17 ton	3.59	4.27	7.86
3680	20,000 S.F., 63.32 ton	5.45	6.30	11.75
3760	Restaurants, 1,000 S.F., 5.00 ton	5.80	9.45	15.25
3920	20,000 S.F., 100.00 ton	10.25	10.35	20.60
4000	Schools and colleges, 1,000 S.F., 3.83 ton	4.32	5.30	9.62
4160	20,000 S.F., 76.66 ton	6.60	7.60	14.20

D30 HVAC

D3050 Terminal & Package Units

Computer rooms impose special requirements on air conditioning systems. A prime requirement is reliability, due to the potential monetary loss that could be incurred by a system failure. A second basic requirement is the tolerance of control with which temperature and humidity are regulated, and dust eliminated. As the air conditioning system reliability is so vital, the additional cost of reserve capacity and redundant components is often justified.

System Descriptions: Computer areas may be environmentally controlled by one of three methods as follows:

1. Self-contained Units
These are units built to higher standards of performance and reliability. They usually contain alarms and controls to indicate component operation failure, filter change, etc. It should be remembered that these units in the room will occupy space that is relatively expensive to build and that all alterations and service of the equipment will also have to be accomplished within the computer area.

2. Decentralized Air Handling Units In operation these are similar to the self-contained units except that their cooling capability comes from remotely located refrigeration equipment as refrigerant or chilled water. As no compressors or refrigerating equipment are required in the air units, they are smaller and require less service than self-contained units. An added plus for this type of system occurs if some of the computer components themselves also require chilled water for cooling.

3. Central System Supply Cooling is obtained from a central source which, since it is not located within the computer room, may have excess capacity and permit greater flexibility without interfering with the computer components. System performance criteria must still be met.

Note: The costs shown below do not include an allowance for ductwork or piping.

D3050 185	Computer Room Cooling Units		MAT.	INST.	TOTAL
		COST EACH			
0560	Computer room unit, air cooled, includes remote condenser				
0580	3 ton		19,500	2,450	21,950
0600	5 ton	RD3030	20,800	2,725	23,525
0620	8 ton	-010	38,800	4,550	43,350
0640	10 ton		40,600	4,900	45,500
0660	15 ton		44,700	5,575	50,275
0680	20 ton		54,000	7,950	61,950
0700	23 ton		66,500	9,100	75,600
0800	Chilled water, for connection to existing chiller system				
0820	5 ton		14,900	1,650	16,550
0840	8 ton		15,200	2,450	17,650
0860	10 ton		15,300	2,500	17,800
0880	15 ton		16,000	2,550	18,550
0900	20 ton		16,900	2,675	19,575
0920	23 ton		18,000	3,025	21,025
1000	Glycol system, complete except for interconnecting tubing				
1020	3 ton		24,400	3,075	27,475
1040	5 ton		26,700	3,225	29,925
1060	8 ton		42,900	5,350	48,250
1080	10 ton		45,700	5,850	51,550
1100	15 ton		56,500	7,350	63,850
1120	20 ton		62,000	7,950	69,950
1140	23 ton		65,500	8,300	73,800
1240	Water cooled, not including condenser water supply or cooling tower				
1260	3 ton		19,800	1,975	21,775
1280	5 ton		21,600	2,275	23,875
1300	8 ton		35,300	3,725	39,025
1320	15 ton		42,800	4,550	47,350
1340	20 ton		46,000	5,025	51,025
1360	23 ton		48,300	5,450	53,750

D40 Fire Protection

D4010 Sprinklers

Dry Pipe System: A system employing automatic sprinklers attached to a piping system containing air under pressure, the release of which from the opening of sprinklers permits the water pressure to open a valve known as a "dry pipe valve". The water then flows into the piping system and out the opened sprinklers.

All areas are assumed to be open.

System Components	QUANTITY	UNIT	COST EACH MAT.	COST EACH INST.	COST EACH TOTAL
SYSTEM D4010 310 0580					
DRY PIPE SPRINKLER, STEEL, BLACK, SCH. 40 PIPE					
LIGHT HAZARD, ONE FLOOR, 2000 S.F.					
Valve, gate, iron body 125 lb., OS&Y, flanged, 4" pipe size	1.000	Ea.	566.25	303.75	870
Valve, swing check, bronze, 125 lb, regrinding disc, 2-1/2" pipe size	1.000	Ea.	588.75	60.38	649.13
Valve, angle, bronze, 150 lb., rising stem, threaded, 2" pipe size	1.000	Ea.	555	45.75	600.75
*Alarm valve, 2-1/2" pipe size	1.000	Ea.	1,218.75	296.25	1,515
Alarm, water motor, complete with gong	1.000	Ea.	303.75	123.75	427.50
Fire alarm horn, electric	1.000	Ea.	49.13	70.13	119.26
Valve swing check w/balldrip Cl with brass trim, 4" pipe size	1.000	Ea.	243.75	296.25	540
Pipe, steel, black, schedule 40, 4" diam.	10.000	L.F.	195	255.53	450.53
Dry pipe valve, trim & gauges, 4" pipe size	1.000	Ea.	2,118.75	881.25	3,000
Pipe, steel, black, schedule 40, threaded, cplg & hngr 10'OC 2-1/2" diam.	20.000	L.F.	252	360	612
Pipe, steel, black, schedule 40, threaded, cplg & hngr 10'OC 2" diam.	12.500	L.F.	100.78	177.19	277.97
Pipe, steel, black, schedule 40, threaded, cplg & hngr 10'OC 1-1/4" diam.	37.500	L.F.	195.47	382.50	577.97
Pipe, steel, black, schedule 40, threaded, cplg & hngr 10'OC 1" diam.	112.000	L.F.	470.40	1,066.80	1,537.20
Pipe Tee, malleable iron black, 150 lb. threaded, 4" pipe size	2.000	Ea.	412.50	457.50	870
Pipe Tee, malleable iron black, 150 lb. threaded, 2-1/2" pipe size	2.000	Ea.	116.25	201	317.25
Pipe Tee, malleable iron black, 150 lb. threaded, 2" pipe size	1.000	Ea.	27	82.50	109.50
Pipe Tee, malleable iron black, 150 lb. threaded, 1-1/4" pipe size	5.000	Ea.	63.38	324.38	387.76
Pipe Tee, malleable iron black, 150 lb. threaded, 1" pipe size	4.000	Ea.	31.35	252	283.35
Pipe 90° elbow malleable iron black, 150 lb. threaded, 1" pipe size	6.000	Ea.	29.93	231.75	261.68
Sprinkler head dry 1/2" orifice 1" NPT, 3" to 4-3/4" length	12.000	Ea.	1,488	564	2,052
Air compressor, 200 Gal sprinkler system capacity, 1/3 HP	1.000	Ea.	607.50	378.75	986.25
*Standpipe connection, wall, flush, brs. w/plug & chain 2-1/2"x2-1/2"	1.000	Ea.	119.25	177.75	297
Valve gate bronze, 300 psi, NRS, class 150, threaded, 1" pipe size	1.000	Ea.	81.75	26.63	108.38
TOTAL			9,834.69	7,015.79	16,850.48
COST PER S.F.			4.92	3.51	8.43

*Not included in systems under 2000 S.F.

D4010 310	Dry Pipe Sprinkler Systems		COST PER S.F. MAT.	COST PER S.F. INST.	COST PER S.F. TOTAL
0520	Dry pipe sprinkler systems, steel, black, sch. 40 pipe				
0530	Light hazard, one floor, 500 S.F.		9	5.95	14.95
0560	1000 S.F.	RD4010 -100	5.20	3.50	8.70
0580	2000 S.F.		4.91	3.50	8.41
0600	5000 S.F.	RD4020 -300	2.64	2.39	5.03
0620	10,000 S.F.		1.92	1.98	3.90
0640	50,000 S.F.		1.53	1.75	3.28
0660	Each additional floor, 500 S.F.		2.08	2.90	4.98

D40 Fire Protection

D4010 Sprinklers

D4010 310	Dry Pipe Sprinkler Systems	COST PER S.F.		
		MAT.	INST.	TOTAL
0680	1000 S.F.	1.80	2.40	4.20
0700	2000 S.F.	1.74	2.21	3.95
0720	5000 S.F.	1.45	1.90	3.35
0740	10,000 S.F.	1.36	1.75	3.11
0760	50,000 S.F.	1.22	1.55	2.77
1000	Ordinary hazard, one floor, 500 S.F.	9.20	6	15.20
1020	1000 S.F.	5.30	3.52	8.82
1040	2000 S.F.	5.05	3.65	8.70
1060	5000 S.F.	3.04	2.54	5.58
1080	10,000 S.F.	2.51	2.60	5.11
1100	50,000 S.F.	2.28	2.45	4.73
1140	Each additional floor, 500 S.F.	2.26	2.97	5.23
1160	1000 S.F.	2.07	2.67	4.74
1180	2000 S.F.	2.06	2.42	4.48
1200	5000 S.F.	1.92	2.08	4
1220	10,000 S.F.	1.78	2.05	3.83
1240	50,000 S.F.	1.66	1.78	3.44
1500	Extra hazard, one floor, 500 S.F.	12.65	7.45	20.10
1520	1000 S.F.	7.80	5.40	13.20
1540	2000 S.F.	5.75	4.70	10.45
1560	5000 S.F.	3.63	3.57	7.20
1580	10,000 S.F.	3.69	3.40	7.09
1600	50,000 S.F.	4	3.28	7.28
1660	Each additional floor, 500 S.F.	3.12	3.68	6.80
1680	1000 S.F.	3.06	3.50	6.56
1700	2000 S.F.	2.93	3.50	6.43
1720	5000 S.F.	2.51	3.07	5.58
1740	10,000 S.F.	2.82	2.80	5.62
1760	50,000 S.F.	2.90	2.69	5.59
2020	Grooved steel, black, sch. 40 pipe, light hazard, one floor, 2000 S.F.	4.64	3	7.64
2060	10,000 S.F.	1.76	1.73	3.49
2100	Each additional floor, 2000 S.F.	1.65	1.77	3.42
2150	10,000 S.F.	1.20	1.50	2.70
2200	Ordinary hazard, one floor, 2000 S.F.	4.88	3.20	8.08
2250	10,000 S.F.	2.24	2.19	4.43
2300	Each additional floor, 2000 S.F.	1.89	1.97	3.86
2350	10,000 S.F.	1.69	1.96	3.65
2400	Extra hazard, one floor, 2000 S.F.	5.50	4	9.50
2450	10,000 S.F.	3.14	2.84	5.98
2500	Each additional floor, 2000 S.F.	2.69	2.86	5.55
2550	10,000 S.F.	2.44	2.46	4.90
3050	Grooved steel, black, sch. 10 pipe, light hazard, one floor, 2000 S.F.	4.52	2.97	7.49
3100	10,000 S.F.	1.69	1.70	3.39
3150	Each additional floor, 2000 S.F.	1.53	1.74	3.27
3200	10,000 S.F.	1.13	1.47	2.60
3250	Ordinary hazard, one floor, 2000 S.F.	4.78	3.18	7.96
3300	10,000 S.F.	2.12	2.15	4.27
3350	Each additional floor, 2000 S.F.	1.79	1.95	3.74
3400	10,000 S.F.	1.57	1.92	3.49
3450	Extra hazard, one floor, 2000 S.F.	5.45	3.98	9.43
3500	10,000 S.F.	2.93	2.79	5.72
3550	Each additional floor, 2000 S.F.	2.61	2.84	5.45
3600	10,000 S.F.	2.32	2.43	4.75
4050	Copper tubing, type M, light hazard, one floor, 2000 S.F.	6.05	2.97	9.02
4100	10,000 S.F.	2.99	1.72	4.71
4150	Each additional floor, 2000 S.F.	3.09	1.78	4.87
4200	10,000 S.F.	2.43	1.50	3.93
4250	Ordinary hazard, one floor, 2000 S.F.	6.50	3.33	9.83

D40 Fire Protection

D4010 Sprinklers

D4010 310	Dry Pipe Sprinkler Systems	COST PER S.F.		
		MAT.	INST.	TOTAL
4300	10,000 S.F.	3.69	2.04	5.73
4350	Each additional floor, 2000 S.F.	4.04	2.07	6.11
4400	10,000 S.F.	3	1.78	4.78
4450	Extra hazard, one floor, 2000 S.F.	7.75	4.05	11.80
4500	10,000 S.F.	8.10	3.08	11.18
4550	Each additional floor, 2000 S.F.	4.92	2.91	7.83
4600	10,000 S.F.	6.15	2.68	8.83
5050	Copper tubing, type M, T-drill system, light hazard, one floor			
5060	2000 S.F.	5.95	2.78	8.73
5100	10,000 S.F.	2.62	1.43	4.05
5150	Each additional floor, 2000 S.F.	2.97	1.59	4.56
5200	10,000 S.F.	2.06	1.21	3.27
5250	Ordinary hazard, one floor, 2000 S.F.	5.90	2.85	8.75
5300	10,000 S.F.	3.37	1.81	5.18
5350	Each additional floor, 2000 S.F.	2.92	1.62	4.54
5400	10,000 S.F.	2.68	1.50	4.18
5450	Extra hazard, one floor, 2000 S.F.	6.70	3.35	10.05
5500	10,000 S.F.	5.85	2.26	8.11
5550	Each additional floor, 2000 S.F.	3.86	2.21	6.07
5600	10,000 S.F.	3.93	1.86	5.79

D40 Fire Protection

D4010 Sprinklers

Pre-Action System: A system employing automatic sprinklers attached to a piping system containing air that may or may not be under pressure, with a supplemental heat responsive system of generally more sensitive characteristics than the automatic sprinklers themselves, installed in the same areas as the sprinklers. Actuation of the heat responsive system, as from a fire, opens a valve which permits water to flow into the sprinkler piping system and to be discharged from those sprinklers which were opened by heat from the fire.

All areas are assumed to be open.

System Components	QUANTITY	UNIT	COST EACH MAT.	COST EACH INST.	COST EACH TOTAL
SYSTEM D4010 350 0580					
PREACTION SPRINKLER SYSTEM, STEEL BLACK SCH. 40 PIPE					
LIGHT HAZARD, 1 FLOOR, 2000 S.F.					
Valve, gate, iron body 125 lb., OS&Y, flanged, 4" pipe size	1.000	Ea.	566.25	303.75	870
*Valve, swing check w/ball drip CI with brass trim 4" pipe size	1.000	Ea.	243.75	296.25	540
Valve, swing check, bronze, 125 lb, regrinding disc, 2-1/2" pipe size	1.000	Ea.	588.75	60.38	649.13
Valve, angle, bronze, 150 lb., rising stem, threaded, 2" pipe size	1.000	Ea.	555	45.75	600.75
*Alarm valve, 2-1/2" pipe size	1.000	Ea.	1,218.75	296.25	1,515
Alarm, water motor, complete with gong	1.000	Ea.	303.75	123.75	427.50
Fire alarm horn, electric	1.000	Ea.	49.13	70.13	119.26
Thermostatic release for release line	2.000	Ea.	990	49.50	1,039.50
Pipe, steel, black, schedule 40, 4" diam.	10.000	L.F.	195	255.53	450.53
Dry pipe valve, trim & gauges, 4" pipe size	1.000	Ea.	2,118.75	881.25	3,000
Pipe, steel, black, schedule 40, threaded, cplg. & hngr. 10'OC 2-1/2" diam.	20.000	L.F.	252	360	612
Pipe steel black, schedule 40, threaded, cplg. & hngr. 10'OC 2" diam.	12.500	L.F.	100.78	177.19	277.97
Pipe, steel, black, schedule 40, threaded, cplg. & hngr. 10'OC 1-1/4" diam.	37.500	L.F.	195.47	382.50	577.97
Pipe, steel, black, schedule 40, threaded, cplg. & hngr. 10'OC 1" diam.	112.000	L.F.	470.40	1,066.80	1,537.20
Pipe, Tee, malleable iron, black, 150 lb. threaded, 4" diam.	2.000	Ea.	412.50	457.50	870
Pipe, Tee, malleable iron, black, 150 lb. threaded, 2-1/2" pipe size	2.000	Ea.	116.25	201	317.25
Pipe, Tee, malleable iron, black, 150 lb. threaded, 2" pipe size	1.000	Ea.	27	82.50	109.50
Pipe, Tee, malleable iron, black, 150 lb. threaded, 1-1/4" pipe size	5.000	Ea.	63.38	324.38	387.76
Pipe, Tee, malleable iron, black, 150 lb. threaded, 1" pipe size	4.000	Ea.	31.35	252	283.35
Pipe, 90° elbow, malleable iron, blk., 150 lb. threaded, 1" pipe size	6.000	Ea.	29.93	231.75	261.68
Sprinkler head, std. spray, brass 135°-286°F 1/2" NPT, 3/8" orifice	12.000	Ea.	182.40	492	674.40
Air compressor auto complete 200 Gal sprinkler sys. cap., 1/3 HP	1.000	Ea.	607.50	378.75	986.25
*Standpipe conn.,wall, flush, brass w/plug & chain 2-1/2" x 2-1/2"	1.000	Ea.	119.25	177.75	297
Valve, gate, bronze, 300 psi, NRS, class 150, threaded, 1" pipe size	1.000	Ea.	81.75	26.63	108.38
TOTAL			9,519.09	6,993.29	16,512.38
COST PER S.F.			4.76	3.50	8.26

*Not included in systems under 2000 S.F.

D4010 350	Preaction Sprinkler Systems		COST PER S.F. MAT.	COST PER S.F. INST.	COST PER S.F. TOTAL
0520	Preaction sprinkler systems, steel, black, sch. 40 pipe				
0530	Light hazard, one floor, 500 S.F.	RD4010-100	8.95	4.76	13.71
0560	1000 S.F.		5.20	3.57	8.77
0580	2000 S.F.	RD4020-300	4.76	3.49	8.25

D40 Fire Protection

D4010 Sprinklers

D4010 350	Preaction Sprinkler Systems	COST PER S.F.		
		MAT.	INST.	TOTAL
0600	5000 S.F.	2.51	2.38	4.89
0620	10,000 S.F.	1.81	1.97	3.78
0640	50,000 S.F.	1.43	1.75	3.18
0660	Each additional floor, 500 S.F.	2.34	2.59	4.93
0680	1000 S.F.	1.81	2.39	4.20
0700	2000 S.F.	1.74	2.20	3.94
0720	5000 S.F.	1.32	1.89	3.21
0740	10,000 S.F.	1.25	1.74	2.99
0760	50,000 S.F.	1.23	1.61	2.84
1000	Ordinary hazard, one floor, 500 S.F.	9.20	5.15	14.35
1020	1000 S.F.	5.15	3.51	8.66
1040	2000 S.F.	5.15	3.66	8.81
1060	5000 S.F.	2.79	2.53	5.32
1080	10,000 S.F.	2.21	2.59	4.80
1100	50,000 S.F.	1.96	2.43	4.39
1140	Each additional floor, 500 S.F.	2.60	2.99	5.59
1160	1000 S.F.	1.78	2.41	4.19
1180	2000 S.F.	1.66	2.40	4.06
1200	5000 S.F.	1.80	2.25	4.05
1220	10,000 S.F.	1.66	2.36	4.02
1240	50,000 S.F.	1.55	2.10	3.65
1500	Extra hazard, one floor, 500 S.F.	12.35	6.60	18.95
1520	1000 S.F.	7.15	4.94	12.09
1540	2000 S.F.	5.25	4.67	9.92
1560	5000 S.F.	3.44	3.85	7.29
1580	10,000 S.F.	3.33	3.76	7.09
1600	50,000 S.F.	3.54	3.66	7.20
1660	Each additional floor, 500 S.F.	3.13	3.68	6.81
1680	1000 S.F.	2.58	3.47	6.05
1700	2000 S.F.	2.44	3.47	5.91
1720	5000 S.F.	2.07	3.08	5.15
1740	10,000 S.F.	2.22	2.82	5.04
1760	50,000 S.F.	2.19	2.66	4.85
2020	Grooved steel, black, sch. 40 pipe, light hazard, one floor, 2000 S.F.	4.64	2.99	7.63
2060	10,000 S.F.	1.65	1.72	3.37
2100	Each additional floor of 2000 S.F.	1.65	1.76	3.41
2150	10,000 S.F.	1.09	1.49	2.58
2200	Ordinary hazard, one floor, 2000 S.F.	4.72	3.19	7.91
2250	10,000 S.F.	1.94	2.18	4.12
2300	Each additional floor, 2000 S.F.	1.73	1.96	3.69
2350	10,000 S.F.	1.39	1.95	3.34
2400	Extra hazard, one floor, 2000 S.F.	5	3.97	8.97
2450	10,000 S.F.	2.54	2.80	5.34
2500	Each additional floor, 2000 S.F.	2.20	2.83	5.03
2550	10,000 S.F.	1.80	2.42	4.22
3050	Grooved steel, black, sch. 10 pipe light hazard, one floor, 2000 S.F.	4.52	2.96	7.48
3100	10,000 S.F.	1.58	1.69	3.27
3150	Each additional floor, 2000 S.F.	1.53	1.73	3.26
3200	10,000 S.F.	1.02	1.46	2.48
3250	Ordinary hazard, one floor, 2000 S.F.	4.55	2.97	7.52
3300	10,000 S.F.	1.57	2.12	3.69
3350	Each additional floor, 2000 S.F.	1.63	1.94	3.57
3400	10,000 S.F.	1.27	1.91	3.18
3450	Extra hazard, one floor, 2000 S.F.	4.94	3.95	8.89
3500	10,000 S.F.	2.29	2.75	5.04
3550	Each additional floor, 2000 S.F.	2.12	2.81	4.93
3600	10,000 S.F.	1.68	2.39	4.07
4050	Copper tubing, type M, light hazard, one floor, 2000 S.F.	6.05	2.96	9.01

D40 Fire Protection

D4010 Sprinklers

D4010 350	Preaction Sprinkler Systems	COST PER S.F.		
		MAT.	INST.	TOTAL
4100	10,000 S.F.	2.88	1.71	4.59
4150	Each additional floor, 2000 S.F.	3.10	1.77	4.87
4200	10,000 S.F.	2.07	1.47	3.54
4250	Ordinary hazard, one floor, 2000 S.F.	6.35	3.32	9.67
4300	10,000 S.F.	3.39	2.03	5.42
4350	Each additional floor, 2000 S.F.	3.11	1.83	4.94
4400	10,000 S.F.	2.45	1.63	4.08
4450	Extra hazard, one floor, 2000 S.F.	7.25	4.02	11.27
4500	10,000 S.F.	7.40	3.04	10.44
4550	Each additional floor, 2000 S.F.	4.43	2.88	7.31
4600	10,000 S.F.	5.50	2.64	8.14
5050	Copper tubing, type M, T-drill system, light hazard, one floor			
5060	2000 S.F.	5.95	2.77	8.72
5100	10,000 S.F.	2.51	1.42	3.93
5150	Each additional floor, 2000 S.F.	2.97	1.58	4.55
5200	10,000 S.F.	1.95	1.20	3.15
5250	Ordinary hazard, one floor, 2000 S.F.	5.75	2.84	8.59
5300	10,000 S.F.	3.07	1.80	4.87
5350	Each additional floor, 2000 S.F.	2.78	1.62	4.40
5400	10,000 S.F.	2.52	1.57	4.09
5450	Extra hazard, one floor, 2000 S.F.	6.20	3.32	9.52
5500	10,000 S.F.	5.15	2.22	7.37
5550	Each additional floor, 2000 S.F.	3.37	2.18	5.55
5600	10,000 S.F.	3.29	1.82	5.11

D40 Fire Protection

D4010 Sprinklers

Deluge System: A system employing open sprinklers attached to a piping system connected to a water supply through a valve which is opened by the operation of a heat responsive system installed in the same areas as the sprinklers. When this valve opens, water flows into the piping system and discharges from all sprinklers attached thereto.

All areas are assumed to be open.

System Components	QUANTITY	UNIT	COST EACH MAT.	COST EACH INST.	COST EACH TOTAL
SYSTEM D4010 370 0580					
DELUGE SPRINKLER SYSTEM, STEEL BLACK SCH. 40 PIPE					
LIGHT HAZARD, 1 FLOOR, 2000 S.F.					
Valve, gate, iron body 125 lb., OS&Y, flanged, 4" pipe size	1.000	Ea.	566.25	303.75	870
Valve, swing check w/ball drip, CI w/brass ftngs., 4" pipe size	1.000	Ea.	243.75	296.25	540
Valve, swing check, bronze, 125 lb, regrinding disc, 2-1/2" pipe size	1.000	Ea.	588.75	60.38	649.13
Valve, angle, bronze, 150 lb., rising stem, threaded, 2" pipe size	1.000	Ea.	555	45.75	600.75
*Alarm valve, 2-1/2" pipe size	1.000	Ea.	1,218.75	296.25	1,515
Alarm, water motor, complete with gong	1.000	Ea.	303.75	123.75	427.50
Fire alarm horn, electric	1.000	Ea.	49.13	70.13	119.26
Thermostatic release for release line	2.000	Ea.	990	49.50	1,039.50
Pipe, steel, black, schedule 40, 4" diam.	10.000	L.F.	195	255.53	450.53
Deluge valve trim, pressure relief, emergency release, gauge, 4" pipe size	1.000	Ea.	3,506.25	881.25	4,387.50
Deluge system, monitoring panel w/deluge valve & trim	1.000	Ea.	8,400	27.38	8,427.38
Pipe, steel, black, schedule 40, threaded, cplg & hngr 10' OC 2-1/2" diam.	20.000	L.F.	252	360	612
Pipe, steel, black, schedule 40, threaded, cplg & hngr 10' OC 2" diam.	12.500	L.F.	100.78	177.19	277.97
Pipe, steel, black, schedule 40, threaded, cplg & hngr 10' OC 1-1/4" diam.	37.500	L.F.	195.47	382.50	577.97
Pipe, steel, black, schedule 40, threaded, cplg & hngr 10' OC 1" diam.	112.000	L.F.	470.40	1,066.80	1,537.20
Pipe, Tee, malleable iron, black, 150 lb. threaded, 4" pipe size	2.000	Ea.	412.50	457.50	870
Pipe, Tee, malleable iron, black, 150 lb. threaded, 2-1/2" pipe size	2.000	Ea.	116.25	201	317.25
Pipe, Tee, malleable iron, black, 150 lb. threaded, 2" pipe size	1.000	Ea.	27	82.50	109.50
Pipe, Tee, malleable iron, black, 150 lb. threaded, 1-1/4" pipe size	5.000	Ea.	63.38	324.38	387.76
Pipe, Tee, malleable iron, black, 150 lb. threaded, 1" pipe size	4.000	Ea.	31.35	252	283.35
Pipe, 90° elbow, malleable iron, black, 150 lb. threaded 1" pipe size	6.000	Ea.	29.93	231.75	261.68
Sprinkler head, std spray, brass 135°-286°F 1/2" NPT, 3/8" orifice	9.720	Ea.	182.40	492	674.40
Air compressor, auto, complete, 200 Gal sprinkler sys. cap., 1/3 HP	1.000	Ea.	607.50	378.75	986.25
*Standpipe connection, wall, flush w/plug & chain 2-1/2" x 2-1/2"	1.000	Ea.	119.25	177.75	297
Valve, gate, bronze, 300 psi, NRS, class 150, threaded, 1" pipe size	1.000	Ea.	81.75	26.63	108.38
TOTAL			19,306.59	7,020.67	26,327.26
COST PER S.F.			9.65	3.51	13.16

*Not included in systems under 2000 S.F.

D4010 370	Deluge Sprinkler Systems		COST PER S.F. MAT.	COST PER S.F. INST.	COST PER S.F. TOTAL
0520	Deluge sprinkler systems, steel, black, sch. 40 pipe				
0530	Light hazard, one floor, 500 S.F.	RD4010-100	27.50	4.81	32.31
0560	1000 S.F.	RD4020-300	14.35	3.43	17.78
0580	2000 S.F.		9.65	3.50	13.15

D40 Fire Protection

D4010 Sprinklers

D4010 370	Deluge Sprinkler Systems	COST PER S.F.		
		MAT.	INST.	TOTAL
0600	5000 S.F.	4.47	2.39	6.86
0620	10,000 S.F.	2.79	1.97	4.76
0640	50,000 S.F.	1.63	1.75	3.38
0660	Each additional floor, 500 S.F.	2.34	2.59	4.93
0680	1000 S.F.	1.81	2.39	4.20
0700	2000 S.F.	1.74	2.20	3.94
0720	5000 S.F.	1.32	1.89	3.21
0740	10,000 S.F.	1.25	1.74	2.99
0760	50,000 S.F.	1.23	1.61	2.84
1000	Ordinary hazard, one floor, 500 S.F.	28	5.50	33.50
1020	1000 S.F.	14.35	3.54	17.89
1040	2000 S.F.	10.05	3.67	13.72
1060	5000 S.F.	4.75	2.54	7.29
1080	10,000 S.F.	3.19	2.59	5.78
1100	50,000 S.F.	2.21	2.46	4.67
1140	Each additional floor, 500 S.F.	2.60	2.99	5.59
1160	1000 S.F.	1.78	2.41	4.19
1180	2000 S.F.	1.66	2.40	4.06
1200	5000 S.F.	1.67	2.07	3.74
1220	10,000 S.F.	1.60	2.09	3.69
1240	50,000 S.F.	1.46	1.92	3.38
1500	Extra hazard, one floor, 500 S.F.	30.50	6.65	37.15
1520	1000 S.F.	16.75	5.15	21.90
1540	2000 S.F.	10.15	4.68	14.83
1560	5000 S.F.	5.10	3.55	8.65
1580	10,000 S.F.	4.10	3.45	7.55
1600	50,000 S.F.	4.13	3.36	7.49
1660	Each additional floor, 500 S.F.	3.13	3.68	6.81
1680	1000 S.F.	2.58	3.47	6.05
1700	2000 S.F.	2.44	3.47	5.91
1720	5000 S.F.	2.07	3.08	5.15
1740	10,000 S.F.	2.30	2.93	5.23
1760	50,000 S.F.	2.31	2.84	5.15
2000	Grooved steel, black, sch. 40 pipe, light hazard, one floor			
2020	2000 S.F.	9.55	3	12.55
2060	10,000 S.F.	2.65	1.74	4.39
2100	Each additional floor, 2,000 S.F.	1.65	1.76	3.41
2150	10,000 S.F.	1.09	1.49	2.58
2200	Ordinary hazard, one floor, 2000 S.F.	4.72	3.19	7.91
2250	10,000 S.F.	2.92	2.18	5.10
2300	Each additional floor, 2000 S.F.	1.73	1.96	3.69
2350	10,000 S.F.	1.39	1.95	3.34
2400	Extra hazard, one floor, 2000 S.F.	9.90	3.98	13.88
2450	10,000 S.F.	3.54	2.82	6.36
2500	Each additional floor, 2000 S.F.	2.20	2.83	5.03
2550	10,000 S.F.	1.80	2.42	4.22
3000	Grooved steel, black, sch. 10 pipe, light hazard, one floor			
3050	2000 S.F.	8.85	2.83	11.68
3100	10,000 S.F.	2.56	1.69	4.25
3150	Each additional floor, 2000 S.F.	1.53	1.73	3.26
3200	10,000 S.F.	1.02	1.46	2.48
3250	Ordinary hazard, one floor, 2000 S.F.	9.50	3.18	12.68
3300	10,000 S.F.	2.55	2.12	4.67
3350	Each additional floor, 2000 S.F.	1.63	1.94	3.57
3400	10,000 S.F.	1.27	1.91	3.18
3450	Extra hazard, one floor, 2000 S.F.	9.85	3.96	13.81
3500	10,000 S.F.	3.27	2.75	6.02
3550	Each additional floor, 2000 S.F.	2.12	2.81	4.93

D40 Fire Protection

D4010 Sprinklers

D4010 370	Deluge Sprinkler Systems	COST PER S.F.		
		MAT.	INST.	TOTAL
3600	10,000 S.F.	1.68	2.39	4.07
4000	Copper tubing, type M, light hazard, one floor			
4050	2000 S.F.	10.95	2.97	13.92
4100	10,000 S.F.	3.86	1.71	5.57
4150	Each additional floor, 2000 S.F.	3.09	1.77	4.86
4200	10,000 S.F.	2.07	1.47	3.54
4250	Ordinary hazard, one floor, 2000 S.F.	11.20	3.33	14.53
4300	10,000 S.F.	4.37	2.03	6.40
4350	Each additional floor, 2000 S.F.	3.11	1.83	4.94
4400	10,000 S.F.	2.45	1.63	4.08
4450	Extra hazard, one floor, 2000 S.F.	12.15	4.03	16.18
4500	10,000 S.F.	8.45	3.06	11.51
4550	Each additional floor, 2000 S.F.	4.43	2.88	7.31
4600	10,000 S.F.	5.50	2.64	8.14
5000	Copper tubing, type M, T-drill system, light hazard, one floor			
5050	2000 S.F.	10.85	2.78	13.63
5100	10,000 S.F.	3.49	1.42	4.91
5150	Each additional floor, 2000 S.F.	2.61	1.60	4.21
5200	10,000 S.F.	1.95	1.20	3.15
5250	Ordinary hazard, one floor, 2000 S.F.	10.65	2.85	13.50
5300	10,000 S.F.	4.05	1.80	5.85
5350	Each additional floor, 2000 S.F.	2.76	1.61	4.37
5400	10,000 S.F.	2.52	1.57	4.09
5450	Extra hazard, one floor, 2000 S.F.	11.10	3.33	14.43
5500	10,000 S.F.	6.15	2.22	8.37
5550	Each additional floor, 2000 S.F.	3.37	2.18	5.55
5600	10,000 S.F.	3.29	1.82	5.11

D40 Fire Protection

D4010 Sprinklers

Firecycle is a fixed fire protection sprinkler system utilizing water as its extinguishing agent. It is a time delayed, recycling, preaction type which automatically shuts the water off when heat is reduced below the detector operating temperature and turns the water back on when that temperature is exceeded.

The system senses a fire condition through a closed circuit electrical detector system which controls water flow to the fire automatically. Batteries supply up to 90 hour emergency power supply for system operation. The piping system is dry (until water is required) and is monitored with pressurized air. Should any leak in the system piping occur, an alarm will sound, but water will not enter the system until heat is sensed by a Firecycle detector.

All areas are assumed to be open.

System Components	QUANTITY	UNIT	COST EACH MAT.	COST EACH INST.	COST EACH TOTAL
SYSTEM D4010 390 0580					
FIRECYCLE SPRINKLER SYSTEM, STEEL BLACK SCH. 40 PIPE					
LIGHT HAZARD, ONE FLOOR, 2000 S.F.					
Valve, gate, iron body 125 lb., OS&Y, flanged, 4" pipe size	1.000	Ea.	566.25	303.75	870
Valve, angle, bronze, 150 lb., rising stem, threaded, 2" pipe size	1.000	Ea.	555	45.75	600.75
Valve, swing check, bronze, 125 lb, regrinding disc, 2-1/2" pipe size	1.000	Ea.	588.75	60.38	649.13
*Alarm valve, 2-1/2" pipe size	1.000	Ea.	1,218.75	296.25	1,515
Alarm, water motor, complete with gong	1.000	Ea.	303.75	123.75	427.50
Pipe, steel, black, schedule 40, 4" diam.	10.000	L.F.	195	255.53	450.53
Fire alarm, horn, electric	1.000	Ea.	49.13	70.13	119.26
Pipe, steel, black, schedule 40, threaded, cplg & hngr 10' OC 2-1/2" diam.	20.000	L.F.	252	360	612
Pipe, steel, black, schedule 40, threaded, cplg & hngr 10' OC 2" diam.	12.500	L.F.	100.78	177.19	277.97
Pipe, steel, black, schedule 40, threaded, cplg & hngr 10' OC 1-1/4" diam.	37.500	L.F.	195.47	382.50	577.97
Pipe, steel, black, schedule 40, threaded, cplg & hngr 10' OC 1" diam.	112.000	L.F.	470.40	1,066.80	1,537.20
Pipe, Tee, malleable iron, black, 150 lb. threaded, 4" pipe size	2.000	Ea.	412.50	457.50	870
Pipe, Tee, malleable iron, black, 150 lb. threaded, 2-1/2" pipe size	2.000	Ea.	116.25	201	317.25
Pipe, Tee, malleable iron, black, 150 lb. threaded, 2" pipe size	1.000	Ea.	27	82.50	109.50
Pipe, Tee, malleable iron, black, 150 lb. threaded, 1-1/4" pipe size	5.000	Ea.	63.38	324.38	387.76
Pipe, Tee, malleable iron, black, 150 lb. threaded, 1" pipe size	4.000	Ea.	31.35	252	283.35
Pipe, 90° elbow, malleable iron, black, 150 lb. threaded, 1" pipe size	6.000	Ea.	29.93	231.75	261.68
Sprinkler head std spray, brass 135°-286°F 1/2" NPT, 3/8" orifice	12.000	Ea.	182.40	492	674.40
Firecycle controls, incls panel, battery, solenoid valves, press switches	1.000	Ea.	15,225	1,875	17,100
Detector, firecycle system	2.000	Ea.	1,057.50	61.50	1,119
Firecycle pkg, swing check & flow control valves w/trim 4" pipe size	1.000	Ea.	4,200	881.25	5,081.25
Air compressor, auto, complete, 200 Gal sprinkler sys. cap., 1/3 HP	1.000	Ea.	607.50	378.75	986.25
*Standpipe connection, wall, flush, brass w/plug & chain 2-1/2"x2-1/2"	1.000	Ea.	119.25	177.75	297
Valve, gate, bronze 300 psi, NRS, class 150, threaded, 1" diam.	1.000	Ea.	81.75	26.63	108.38
TOTAL			26,649.09	8,584.04	35,233.13
COST PER S.F.			13.32	4.29	17.61

*Not included in systems under 2000 S.F.

D40 Fire Protection

D4010 Sprinklers

D4010 390	Firecycle Sprinkler Systems	COST PER S.F.		
		MAT.	INST.	TOTAL
0520	Firecycle sprinkler systems, steel black sch. 40 pipe			
0530	Light hazard, one floor, 500 S.F.	43.50	9.30	52.80
0560	1000 S.F.	22.50	5.85	28.35
0580	2000 S.F.	13.30	4.29	17.59
0600	5000 S.F.	5.95	2.69	8.64
0620	10,000 S.F.	3.59	2.13	5.72
0640	50,000 S.F.	1.82	1.78	3.60
0660	Each additional floor of 500 S.F.	2.41	2.60	5.01
0680	1000 S.F.	1.84	2.40	4.24
0700	2000 S.F.	1.51	2.20	3.71
0720	5000 S.F.	1.34	1.89	3.23
0740	10,000 S.F.	1.32	1.74	3.06
0760	50,000 S.F.	1.28	1.60	2.88
1000	Ordinary hazard, one floor, 500 S.F.	44	9.70	53.70
1020	1000 S.F.	22.50	5.80	28.30
1040	2000 S.F.	13.45	4.44	17.89
1060	5000 S.F.	6.20	2.84	9.04
1080	10,000 S.F.	3.99	2.75	6.74
1100	50,000 S.F.	2.62	2.75	5.37
1140	Each additional floor, 500 S.F.	2.67	3	5.67
1160	1000 S.F.	1.81	2.42	4.23
1180	2000 S.F.	1.84	2.21	4.05
1200	5000 S.F.	1.69	2.07	3.76
1220	10,000 S.F.	1.55	2.04	3.59
1240	50,000 S.F.	1.47	1.82	3.29
1500	Extra hazard, one floor, 500 S.F.	47	11.15	58.15
1520	1000 S.F.	24	7.20	31.20
1540	2000 S.F.	13.85	5.45	19.30
1560	5000 S.F.	6.55	3.85	10.40
1580	10,000 S.F.	5.05	3.90	8.95
1600	50,000 S.F.	4.26	4.34	8.60
1660	Each additional floor, 500 S.F.	3.20	3.69	6.89
1680	1000 S.F.	2.61	3.48	6.09
1700	2000 S.F.	2.48	3.48	5.96
1720	5000 S.F.	2.09	3.08	5.17
1740	10,000 S.F.	2.29	2.82	5.11
1760	50,000 S.F.	2.31	2.74	5.05
2020	Grooved steel, black, sch. 40 pipe, light hazard, one floor			
2030	2000 S.F.	13.20	3.79	16.99
2060	10,000 S.F.	3.71	2.60	6.31
2100	Each additional floor, 2000 S.F.	1.69	1.77	3.46
2150	10,000 S.F.	1.16	1.49	2.65
2200	Ordinary hazard, one floor, 2000 S.F.	13.30	3.99	17.29
2250	10,000 S.F.	3.96	2.47	6.43
2300	Each additional floor, 2000 S.F.	1.77	1.97	3.74
2350	10,000 S.F.	1.46	1.95	3.41
2400	Extra hazard, one floor, 2000 S.F.	13.60	4.77	18.37
2450	10,000 S.F.	4.25	2.94	7.19
2500	Each additional floor, 2000 S.F.	2.24	2.84	5.08
2550	10,000 S.F.	1.87	2.42	4.29
3050	Grooved steel, black, sch. 10 pipe light hazard, one floor,			
3060	2000 S.F.	13.10	3.76	16.86
3100	10,000 S.F.	3.36	1.85	5.21
3150	Each additional floor, 2000 S.F.	1.57	1.74	3.31
3200	10,000 S.F.	1.09	1.46	2.55
3250	Ordinary hazard, one floor, 2000 S.F.	13.20	3.97	17.17
3300	10,000 S.F.	3.60	2.30	5.90
3350	Each additional floor, 2000 S.F.	1.67	1.95	3.62

D40 Fire Protection

D4010 Sprinklers

D4010 390	Firecycle Sprinkler Systems	COST PER S.F.		
		MAT.	INST.	TOTAL
3400	10,000 S.F.	1.34	1.91	3.25
3450	Extra hazard, one floor, 2000 S.F.	13.50	4.75	18.25
3500	10,000 S.F.	4.04	2.89	6.93
3550	Each additional floor, 2000 S.F.	2.16	2.82	4.98
3600	10,000 S.F.	1.75	2.39	4.14
4060	Copper tubing, type M, light hazard, one floor, 2000 S.F.	14.65	3.76	18.41
4100	10,000 S.F.	4.66	1.87	6.53
4150	Each additional floor, 2000 S.F.	3.13	1.78	4.91
4200	10,000 S.F.	2.39	1.49	3.88
4250	Ordinary hazard, one floor, 2000 S.F.	14.90	4.12	19.02
4300	10,000 S.F.	5.15	2.19	7.34
4350	Each additional floor, 2000 S.F.	3.15	1.84	4.99
4400	10,000 S.F.	2.48	1.61	4.09
4450	Extra hazard, one floor, 2000 S.F.	15.80	4.82	20.62
4500	10,000 S.F.	9.20	3.23	12.43
4550	Each additional floor, 2000 S.F.	4.47	2.89	7.36
4600	10,000 S.F.	5.60	2.64	8.24
5060	Copper tubing, type M, T-drill system, light hazard, one floor 2000 S.F.	14.50	3.57	18.07
5100	10,000 S.F.	4.29	1.58	5.87
5150	Each additional floor, 2000 S.F.	3.18	1.68	4.86
5200	10,000 S.F.	2.02	1.20	3.22
5250	Ordinary hazard, one floor, 2000 S.F.	14.30	3.64	17.94
5300	10,000 S.F.	4.85	1.96	6.81
5350	Each additional floor, 2000 S.F.	2.80	1.62	4.42
5400	10,000 S.F.	2.59	1.57	4.16
5450	Extra hazard, one floor, 2000 S.F.	14.75	4.12	18.87
5500	10,000 S.F.	6.90	2.36	9.26
5550	Each additional floor, 2000 S.F.	3.41	2.19	5.60
5600	10,000 S.F.	3.36	1.82	5.18

D40 Fire Protection

D4010 Sprinklers

Wet Pipe System. A system employing automatic sprinklers attached to a piping system containing water and connected to a water supply so that water discharges immediately from sprinklers opened by heat from a fire.

All areas are assumed to be open.

System Components	QUANTITY	UNIT	COST EACH MAT.	COST EACH INST.	COST EACH TOTAL
SYSTEM D4010 410 0580					
WET PIPE SPRINKLER, STEEL, BLACK, SCH. 40 PIPE					
LIGHT HAZARD, ONE FLOOR, 2000 S.F.					
Valve, gate, iron body, 125 lb., OS&Y, flanged, 4" diam.	1.000	Ea.	566.25	303.75	870
Valve, swing check, bronze, 125 lb, regrinding disc, 2-1/2" pipe size	1.000	Ea.	588.75	60.38	649.13
Valve, angle, bronze, 150 lb., rising stem, threaded, 2" diam.	1.000	Ea.	555	45.75	600.75
*Alarm valve, 2-1/2" pipe size	1.000	Ea.	1,218.75	296.25	1,515
Alarm, water motor, complete with gong	1.000	Ea.	303.75	123.75	427.50
Valve, swing check, w/balldrip CI with brass trim 4" pipe size	1.000	Ea.	243.75	296.25	540
Pipe, steel, black, schedule 40, 4" diam.	10.000	L.F.	195	255.53	450.53
*Flow control valve, trim & gauges, 4" pipe size	1.000	Set	4,293.75	671.25	4,965
Fire alarm horn, electric	1.000	Ea.	49.13	70.13	119.26
Pipe, steel, black, schedule 40, threaded, cplg & hngr 10' OC, 2-1/2" diam.	20.000	L.F.	252	360	612
Pipe, steel, black, schedule 40, threaded, cplg & hngr 10' OC, 2" diam.	12.500	L.F.	100.78	177.19	277.97
Pipe, steel, black, schedule 40, threaded, cplg & hngr 10' OC, 1-1/4" diam.	37.500	L.F.	195.47	382.50	577.97
Pipe, steel, black, schedule 40, threaded cplg & hngr 10' OC, 1" diam.	112.000	L.F.	470.40	1,066.80	1,537.20
Pipe Tee, malleable iron black, 150 lb. threaded, 4" pipe size	2.000	Ea.	412.50	457.50	870
Pipe Tee, malleable iron black, 150 lb. threaded, 2-1/2" pipe size	2.000	Ea.	116.25	201	317.25
Pipe Tee, malleable iron black, 150 lb. threaded, 2" pipe size	1.000	Ea.	27	82.50	109.50
Pipe Tee, malleable iron black, 150 lb. threaded, 1-1/4" pipe size	5.000	Ea.	63.38	324.38	387.76
Pipe Tee, malleable iron black, 150 lb. threaded, 1" pipe size	4.000	Ea.	31.35	252	283.35
Pipe 90° elbow, malleable iron black, 150 lb. threaded, 1" pipe size	6.000	Ea.	29.93	231.75	261.68
Sprinkler head, standard spray, brass 135°-286°F 1/2" NPT, 3/8" orifice	12.000	Ea.	182.40	492	674.40
Valve, gate, bronze, NRS, class 150, threaded, 1" pipe size	1.000	Ea.	81.75	26.63	108.38
*Standpipe connection, wall, single, flush w/plug & chain 2-1/2"x2-1/2"	1.000	Ea.	119.25	177.75	297
TOTAL			10,096.59	6,355.04	16,451.63
COST PER S.F.			5.05	3.18	8.23

*Not included in systems under 2000 S.F.

D4010 410	Wet Pipe Sprinkler Systems		COST PER S.F. MAT.	COST PER S.F. INST.	COST PER S.F. TOTAL
0520	Wet pipe sprinkler systems, steel, black, sch. 40 pipe				
0530	Light hazard, one floor, 500 S.F.		2.83	3.06	5.89
0560	1000 S.F.	RD4010 -100	5.65	3.17	8.82
0580	2000 S.F.		5.05	3.18	8.23
0600	5000 S.F.	RD4020 -300	2.43	2.23	4.66
0620	10,000 S.F.		1.62	1.89	3.51
0640	50,000 S.F.		1.15	1.71	2.86
0660	Each additional floor, 500 S.F.		1.37	2.61	3.98

D40 Fire Protection

D4010 Sprinklers

D4010 410	Wet Pipe Sprinkler Systems	COST PER S.F.		
		MAT.	INST.	TOTAL
0680	1000 S.F.	1.33	2.44	3.77
0700	2000 S.F.	1.25	2.18	3.43
0720	5000 S.F.	.92	1.87	2.79
0740	10,000 S.F.	.90	1.72	2.62
0760	50,000 S.F.	.75	1.35	2.10
1000	Ordinary hazard, one floor, 500 S.F.	3.02	3.28	6.30
1020	1000 S.F.	5.60	3.11	8.71
1040	2000 S.F.	5.20	3.33	8.53
1060	5000 S.F.	2.71	2.38	5.09
1080	10,000 S.F.	2.02	2.51	4.53
1100	50,000 S.F.	1.65	2.36	4.01
1140	Each additional floor, 500 S.F.	1.61	2.94	4.55
1160	1000 S.F.	1.28	2.39	3.67
1180	2000 S.F.	1.41	2.39	3.80
1200	5000 S.F.	1.43	2.26	3.69
1220	10,000 S.F.	1.31	2.34	3.65
1240	50,000 S.F.	1.23	2.08	3.31
1500	Extra hazard, one floor, 500 S.F.	10.75	5.05	15.80
1520	1000 S.F.	6.70	4.38	11.08
1540	2000 S.F.	5.60	4.49	10.09
1560	5000 S.F.	3.57	3.92	7.49
1580	10,000 S.F.	3.07	3.73	6.80
1600	50,000 S.F.	3.38	3.60	6.98
1660	Each additional floor, 500 S.F.	2.14	3.63	5.77
1680	1000 S.F.	2.08	3.45	5.53
1700	2000 S.F.	1.95	3.45	5.40
1720	5000 S.F.	1.67	3.06	4.73
1740	10,000 S.F.	1.87	2.80	4.67
1760	50,000 S.F.	1.88	2.69	4.57
2020	Grooved steel, black sch. 40 pipe, light hazard, one floor, 2000 S.F.	4.94	2.68	7.62
2060	10,000 S.F.	1.89	1.70	3.59
2100	Each additional floor, 2000 S.F.	1.16	1.74	2.90
2150	10,000 S.F.	.74	1.47	2.21
2200	Ordinary hazard, one floor, 2000 S.F.	5	2.88	7.88
2250	10,000 S.F.	1.75	2.10	3.85
2300	Each additional floor, 2000 S.F.	1.24	1.94	3.18
2350	10,000 S.F.	1.04	1.93	2.97
2400	Extra hazard, one floor, 2000 S.F.	5.30	3.66	8.96
2450	10,000 S.F.	2.31	2.72	5.03
2500	Each additional floor, 2000 S.F.	1.71	2.81	4.52
2550	10,000 S.F.	1.45	2.40	3.85
3050	Grooved steel, black sch. 10 pipe, light hazard, one floor, 2000 S.F.	4.82	2.65	7.47
3100	10,000 S.F.	1.39	1.61	3
3150	Each additional floor, 2000 S.F.	1.04	1.71	2.75
3200	10,000 S.F.	.67	1.44	2.11
3250	Ordinary hazard, one floor, 2000 S.F.	4.92	2.86	7.78
3300	10,000 S.F.	1.63	2.06	3.69
3350	Each additional floor, 2000 S.F.	1.14	1.92	3.06
3400	10,000 S.F.	.92	1.89	2.81
3450	Extra hazard, one floor, 2000 S.F.	5.25	3.64	8.89
3500	10,000 S.F.	2.10	2.67	4.77
3550	Each additional floor, 2000 S.F.	1.63	2.79	4.42
3600	10,000 S.F.	1.33	2.37	3.70
4050	Copper tubing, type M, light hazard, one floor, 2000 S.F.	6.35	2.65	9
4100	10,000 S.F.	2.69	1.63	4.32
4150	Each additional floor, 2000 S.F.	2.60	1.75	4.35
4200	10,000 S.F.	1.97	1.47	3.44
4250	Ordinary hazard, one floor, 2000 S.F.	6.65	3.01	9.66

D40 Fire Protection

D4010 Sprinklers

D4010 410	Wet Pipe Sprinkler Systems	COST PER S.F.		
		MAT.	INST.	TOTAL
4300	10,000 S.F.	3.20	1.95	5.15
4350	Each additional floor, 2000 S.F.	2.96	1.96	4.92
4400	10,000 S.F.	2.35	1.75	4.10
4450	Extra hazard, one floor, 2000 S.F.	7.55	3.71	11.26
4500	10,000 S.F.	7.20	2.96	10.16
4550	Each additional floor, 2000 S.F.	3.94	2.86	6.80
4600	10,000 S.F.	5.15	2.62	7.77
5050	Copper tubing, type M, T-drill system, light hazard, one floor			
5060	2000 S.F.	6.25	2.46	8.71
5100	10,000 S.F.	2.32	1.34	3.66
5150	Each additional floor, 2000 S.F.	2.48	1.56	4.04
5200	10,000 S.F.	1.60	1.18	2.78
5250	Ordinary hazard, one floor, 2000 S.F.	6.05	2.53	8.58
5300	10,000 S.F.	2.88	1.72	4.60
5350	Each additional floor, 2000 S.F.	2.27	1.59	3.86
5400	10,000 S.F.	2.17	1.55	3.72
5450	Extra hazard, one floor, 2000 S.F.	6.50	3.01	9.51
5500	10,000 S.F.	4.98	2.14	7.12
5550	Each additional floor, 2000 S.F.	2.97	2.21	5.18
5600	10,000 S.F.	2.94	1.80	4.74

D40 Fire Protection

D4020 Standpipes

System Components	QUANTITY	UNIT	COST PER FLOOR		
			MAT.	INST.	TOTAL
SYSTEM D4020 310 0560					
WET STANDPIPE RISER, CLASS I, STEEL, BLACK, SCH. 40 PIPE, 10' HEIGHT					
4" DIAMETER PIPE, ONE FLOOR					
Pipe, steel, black, schedule 40, threaded, 4" diam.	20.000	L.F.	650	670	1,320
Pipe, Tee, malleable iron, black, 150 lb. threaded, 4" pipe size	2.000	Ea.	550	610	1,160
Pipe, 90° elbow, malleable iron, black, 150 lb threaded 4" pipe size	1.000	Ea.	174	202	376
Pipe, nipple, steel, black, schedule 40, 2-1/2" pipe size x 3" long	2.000	Ea.	24.50	151	175.50
Fire valve, gate, 300 lb., brass w/handwheel, 2-1/2" pipe size	1.000	Ea.	189	94	283
Fire valve, pressure restricting, adj, rgh brs, 2-1/2" pipe size	1.000	Ea.	352	188	540
Valve, swing check, w/ball drip, CI w/brs. ftngs., 4" pipe size	1.000	Ea.	325	395	720
Standpipe conn wall dble. flush brs. w/plugs & chains 2-1/2"x2-1/2"x4"	1.000	Ea.	515	237	752
Valve, swing check, bronze, 125 lb, regrinding disc, 2-1/2" pipe size	1.000	Ea.	785	80.50	865.50
Roof manifold, fire, w/valves & caps, horiz/vert brs 2-1/2"x2-1/2"x4"	1.000	Ea.	176	247	423
Fire, hydrolator, vent & drain, 2-1/2" pipe size	1.000	Ea.	74.50	55	129.50
Valve, gate, iron body 125 lb., OS&Y, threaded, 4" pipe size	1.000	Ea.	755	405	1,160
TOTAL			4,570	3,334.50	7,904.50

D4020 310	Wet Standpipe Risers, Class I	COST PER FLOOR		
		MAT.	INST.	TOTAL
0550	Wet standpipe risers, Class I, steel, black sch. 40, 10' height			
0560	4" diameter pipe, one floor	4,575	3,325	7,900
0580	Additional floors	1,075	1,025	2,100
0600	6" diameter pipe, one floor	7,600	5,850	13,450
0620	Additional floors	1,850	1,650	3,500
0640	8" diameter pipe, one floor	11,900	7,025	18,925
0660	Additional floors	2,700	1,975	4,675
0680				

D4020 310	Wet Standpipe Risers, Class II	COST PER FLOOR		
		MAT.	INST.	TOTAL
1030	Wet standpipe risers, Class II, steel black sch. 40, 10' height			
1040	2" diameter pipe, one floor	1,825	1,200	3,025
1060	Additional floors	545	465	1,010
1080	2-1/2" diameter pipe, one floor	2,675	1,750	4,425
1100	Additional floors	635	540	1,175
1120				

D40 Fire Protection

D4020 Standpipes

D4020 310	Wet Standpipe Risers, Class III	COST PER FLOOR		
		MAT.	INST.	TOTAL
1530	Wet standpipe risers, Class III, steel, black sch. 40, 10' height			
1540	4" diameter pipe, one floor	4,675	3,325	8,000
1560	Additional floors	920	865	1,785
1580	6" diameter pipe, one floor	7,725	5,850	13,575
1600	Additional floors	1,900	1,650	3,550
1620	8" diameter pipe, one floor	12,000	7,025	19,025
1640	Additional floors	2,750	1,975	4,725

D40 Fire Protection

D4020 Standpipes

- Roof — Roof connections with hose gate valves (for combustible roof)
- Hose connections on each floor (size based on class of service)
- Check Valve
- Siamese inlet connections (for fire department use)

System Components	QUANTITY	UNIT	COST PER FLOOR MAT.	INST.	TOTAL
SYSTEM D4020 330 0540					
DRY STANDPIPE RISER, CLASS I, PIPE, STEEL, BLACK, SCH 40, 10' HEIGHT					
4" DIAMETER PIPE, ONE FLOOR					
Pipe, steel, black, schedule 40, threaded, 4" diam.	20.000	L.F.	650	670	1,320
Pipe, Tee, malleable iron, black, 150 lb. threaded, 4" pipe size	2.000	Ea.	550	610	1,160
Pipe, 90° elbow, malleable iron, black, 150 lb threaded 4" pipe size	1.000	Ea.	174	202	376
Pipe, nipple, steel, black, schedule 40, 2-1/2" pipe size x 3" long	2.000	Ea.	24.50	151	175.50
Fire valve gate NRS 300 lb., brass w/handwheel, 2-1/2" pipe size	1.000	Ea.	189	94	283
Fire valve, pressure restricting, adj, rgh brs, 2-1/2" pipe size	1.000	Ea.	176	94	270
Standpipe conn wall dble. flush brs. w/plugs & chains 2-1/2"x2-1/2"x4"	1.000	Ea.	515	237	752
Valve swing check w/ball drip CI w/brs. ftngs., 4"pipe size	1.000	Ea.	325	395	720
Roof manifold, fire, w/valves & caps, horiz/vert brs 2-1/2"x2-1/2"x4"	1.000	Ea.	176	247	423
TOTAL			2,779.50	2,700	5,479.50

D4020 330	Dry Standpipe Risers, Class I	COST PER FLOOR MAT.	INST.	TOTAL
0530	Dry standpipe riser, Class I, steel, black sch. 40, 10' height			
0540	4" diameter pipe, one floor	2,775	2,700	5,475
0560	Additional floors	990	980	1,970
0580	6" diameter pipe, one floor	5,450	4,625	10,075
0600	Additional floors	1,775	1,575	3,350
0620	8" diameter pipe, one floor	8,750	5,600	14,350
0640	Additional floors	2,625	1,925	4,550
0660				

D4020 330	Dry Standpipe Risers, Class II	COST PER FLOOR MAT.	INST.	TOTAL
1030	Dry standpipe risers, Class II, steel, black sch. 40, 10' height			
1040	2" diameter pipe, one floor	1,500	1,250	2,750
1060	Additional floor	470	410	880
1080	2-1/2" diameter pipe, one floor	2,125	1,450	3,575
1100	Additional floors	560	485	1,045
1120				

D40 Fire Protection

D4020 Standpipes

D4020 330	Dry Standpipe Risers, Class III	COST PER FLOOR		
		MAT.	INST.	TOTAL
1530	Dry standpipe risers, Class III, steel, black sch. 40, 10' height			
1540	4" diameter pipe, one floor	2,825	2,650	5,475
1560	Additional floors	855	885	1,740
1580	6" diameter pipe, one floor	5,500	4,625	10,125
1600	Additional floors	1,825	1,575	3,400
1620	8" diameter pipe, one floor	8,800	5,600	14,400
1640	Additional floor	2,675	1,925	4,600

D40 Fire Protection

D4020 Standpipes

D4020 410	Fire Hose Equipment	COST EACH		
		MAT.	INST.	TOTAL
0100	Adapters, reducing, 1 piece, FxM, hexagon, cast brass, 2-1/2" x 1-1/2"	55.50		55.50
0200	Pin lug, 1-1/2" x 1"	47.50		47.50
0250	3" x 2-1/2"	125		125
0300	For polished chrome, add 75% mat.			
0400	Cabinets, D.S. glass in door, recessed, steel box, not equipped			
0500	Single extinguisher, steel door & frame	110	148	258
0550	Stainless steel door & frame	205	148	353
0600	Valve, 2-1/2" angle, steel door & frame	135	99	234
0650	Aluminum door & frame	164	99	263
0700	Stainless steel door & frame	220	99	319
0750	Hose rack assy, 2-1/2" x 1-1/2" valve & 100' hose, steel door & frame	252	198	450
0800	Aluminum door & frame	375	198	573
0850	Stainless steel door & frame	495	198	693
0900	Hose rack assy & extinguisher, 2-1/2"x1-1/2" valve & hose, steel door & frame	315	237	552
0950	Aluminum	475	237	712
1000	Stainless steel	575	237	812
1550	Compressor, air, dry pipe system, automatic, 200 gal., 3/4 H.P.	810	505	1,315
1600	520 gal., 1 H.P.	835	505	1,340
1650	Alarm, electric pressure switch (circuit closer)	86	25.50	111.50
2500	Couplings, hose, rocker lug, cast brass, 1-1/2"	53		53
2550	2-1/2"	69.50		69.50
3000	Escutcheon plate, for angle valves, polished brass, 1-1/2"	16.70		16.70
3050	2-1/2"	27		27
3500	Fire pump, electric, w/controller, fittings, relief valve			
3550	4" pump, 30 H.P., 500 G.P.M.	16,600	3,700	20,300
3600	5" pump, 40 H.P., 1000 G.P.M.	18,200	4,175	22,375
3650	5" pump, 100 H.P., 1000 G.P.M.	25,100	4,650	29,750
3700	For jockey pump system, add	2,675	595	3,270
5000	Hose, per linear foot, synthetic jacket, lined,			
5100	300 lb. test, 1-1/2" diameter	3.23	.46	3.69
5150	2-1/2" diameter	5.60	.54	6.14
5200	500 lb. test, 1-1/2" diameter	3.33	.46	3.79
5250	2-1/2" diameter	5.90	.54	6.44
5500	Nozzle, plain stream, polished brass, 1-1/2" x 10"	47.50		47.50
5550	2-1/2" x 15" x 13/16" or 1-1/2"	86.50		86.50
5600	Heavy duty combination adjustable fog and straight stream w/handle 1-1/2"	465		465
5650	2-1/2" direct connection	575		575
6000	Rack, for 1-1/2" diameter hose 100 ft. long, steel	60.50	59.50	120
6050	Brass	104	59.50	163.50
6500	Reel, steel, for 50 ft. long 1-1/2" diameter hose	131	84.50	215.50
6550	For 75 ft. long 2-1/2" diameter hose	212	84.50	296.50
7050	Siamese, w/plugs & chains, polished brass, sidewalk, 4" x 2-1/2" x 2-1/2"	630	475	1,105
7100	6" x 2-1/2" x 2-1/2"	810	595	1,405
7200	Wall type, flush, 4" x 2-1/2" x 2-1/2"	515	237	752
7250	6" x 2-1/2" x 2-1/2"	720	258	978
7300	Projecting, 4" x 2-1/2" x 2-1/2"	475	237	712
7350	6" x 2-1/2" x 2-1/2"	800	258	1,058
7400	For chrome plate, add 15% mat.			
8000	Valves, angle, wheel handle, 300 Lb., rough brass, 1-1/2"	94.50	55	149.50
8050	2-1/2"	174	94	268
8100	Combination pressure restricting, 1-1/2"	85	55	140
8150	2-1/2"	174	94	268
8200	Pressure restricting, adjustable, satin brass, 1-1/2"	125	55	180
8250	2-1/2"	176	94	270
8300	Hydrolator, vent and drain, rough brass, 1-1/2"	74.50	55	129.50
8350	2-1/2"	74.50	55	129.50
8400	Cabinet assy, incls. adapter, rack, hose, and nozzle	780	360	1,140

D40 Fire Protection

D4090 Other Fire Protection Systems

General: Automatic fire protection (suppression) systems other than water sprinklers may be desired for special environments, high risk areas, isolated locations or unusual hazards. Some typical applications would include:

Paint dip tanks
Securities vaults
Electronic data processing
Tape and data storage
Transformer rooms
Spray booths
Petroleum storage
High rack storage

Piping and wiring costs are dependent on the individual application and must be added to the component costs shown below.

All areas are assumed to be open.

D4090 910	Fire Suppression Unit Components	COST EACH		
		MAT.	INST.	TOTAL
0020	Detectors with brackets			
0040	Fixed temperature heat detector	55	78.50	133.50
0060	Rate of temperature rise detector	55	78.50	133.50
0080	Ion detector (smoke) detector	119	101	220
0200	Extinguisher agent			
0240	200 lb FM200, container	7,900	305	8,205
0280	75 lb carbon dioxide cylinder	1,300	202	1,502
0320	Dispersion nozzle			
0340	FM200 1-1/2" dispersion nozzle	68	48	116
0380	Carbon dioxide 3" x 5" dispersion nozzle	68	37.50	105.50
0420	Control station			
0440	Single zone control station with batteries	1,750	625	2,375
0470	Multizone (4) control station with batteries	3,350	1,250	4,600
0490				
0500	Electric mechanical release	170	325	495
0520				
0550	Manual pull station	61.50	112	173.50
0570				
0640	Battery standby power 10" x 10" x 17"	445	157	602
0700				
0740	Bell signalling device	85	78.50	163.50

D4090 920	FM200 Systems	COST PER C.F.		
		MAT.	INST.	TOTAL
0820	Average FM200 system, minimum			1.79
0840	Maximum			3.56

D50 Electrical

D5010 Electrical Service/Distribution

System Components	QUANTITY	UNIT	COST PER L.F. MAT.	COST PER L.F. INST.	COST PER L.F. TOTAL
SYSTEM D5010 110 0200					
HIGH VOLTAGE CABLE, NEUTRAL AND CONDUIT INCLUDED, COPPER #2, 5 kV					
Shielded cable, no splice/termn, copper, XLP shielding, 5 kV, #2	.030	C.L.F.	8.19	9.45	17.64
Wire, 600 volt, type THW, copper, stranded, #8	.010	C.L.F.	.41	.79	1.20
Rigid galv steel conduit to 15' H, 2" diam, w/term, ftng & support	1.000	L.F.	8.70	13.95	22.65
TOTAL			17.30	24.19	41.49

D5010 110	High Voltage Shielded Conductors	MAT.	INST.	TOTAL
0200	High voltage cable, neutral & conduit included, copper #2, 5 kV	17.30	24	41.30
0240	Copper #1, 5 kV	26.50	28.50	55
0280	15 kV	28.50	28.50	57
0320	Copper 1/0, 5 kV	31	36	67
0360	15 kV	34.50	36	70.50
0400	25 kV	39.50	36.50	76
0440	35 kV	40.50	37	77.50
0480	Copper 2/0, 5 kV	34	36.50	70.50
0520	15 kV	36.50	36.50	73
0560	25 kV	42	37.50	79.50
0600	35 kV	49.50	41.50	91
0640	Copper 4/0, 5 kV	43.50	41.50	85
0680	15 kV	47.50	41.50	89
0720	25 kV	53	42.50	95.50
0760	35 kV	56.50	43.50	100
0800	Copper 250 kcmil, 5 kV	47.50	42.50	90
0840	15 kV	49.50	42.50	92
0880	25 kV	59	43.50	102.50
0920	35 kV	65.50	47.50	113
0960	Copper 350 kcmil, 5 kV	58	47.50	105.50
1000	15 kV	60	47.50	107.50
1040	25 kV	69.50	48.50	118
1080	35 kV	72.50	50	122.50
1120	Copper 500 kcmil, 5 kV	93	59.50	152.50
1160	15 kV	95.50	59.50	155
1200	25 kV	104	61	165
1240	35 kV	109	62.50	171.50

D50 Electrical

D5010 Electrical Service/Distribution

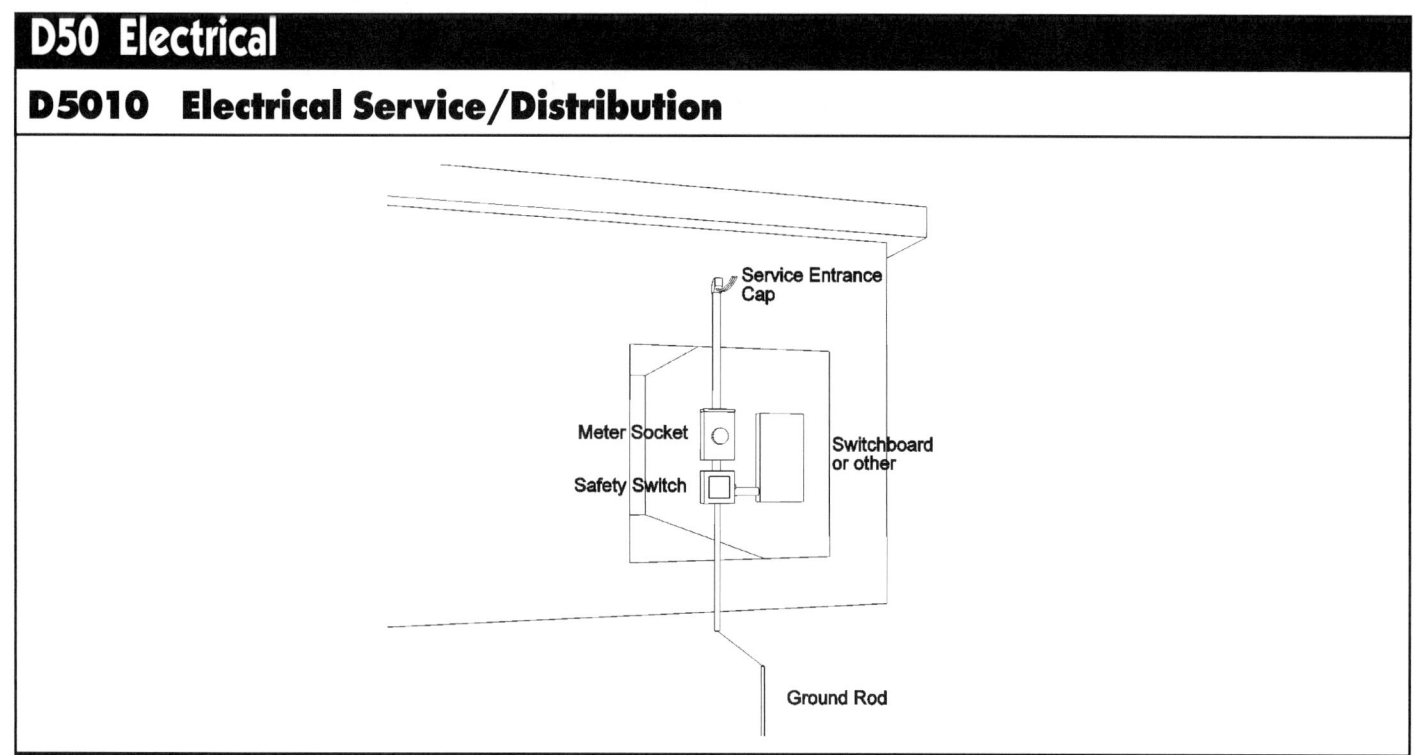

System Components	QUANTITY	UNIT	COST EACH		
			MAT.	INST.	TOTAL
SYSTEM D5010 120 0220					
SERVICE INSTALLATION, INCLUDES BREAKERS, METERING, 20' CONDUIT & WIRE					
3 PHASE, 4 WIRE, 60 A					
Wire, 600 volt, copper type XHHW, stranded #6	.600	C.L.F.	55.80	57.90	113.70
Rigid galvanized steel conduit, 3/4" including fittings	20.000	L.F.	59.20	157	216.20
Service entrance cap, 3/4" diameter	1.000	Ea.	9.90	48.50	58.40
Conduit LB fitting with cover, 3/4" diameter	1.000	Ea.	12.20	48.50	60.70
Meter socket, three phase, 100 A	1.000	Ea.	138	224	362
Safety switches, heavy duty, 240 volt, 3 pole NEMA 1 fusible, 60 amp	1.000	Ea.	305	273	578
Grounding, wire ground bare armored, #8-1 conductor	.200	C.L.F.	23.20	63	86.20
Grounding, clamp, bronze, 3/4" dia	1.000	Ea.	8.15	19.60	27.75
Grounding, rod, copper clad, 8' long, 3/4" diameter	1.000	Ea.	37	118	155
Wireway w/fittings, 2-1/2" x 2-1/2"	1.000	L.F.	11.45	13.95	25.40
TOTAL			659.90	1,023.45	1,683.35

D5010 120		Overhead Electric Service, 3 Phase - 4 Wire	COST EACH		
			MAT.	INST.	TOTAL
0200	Service installation, includes breakers, metering, 20' conduit & wire				
0220		3 phase, 4 wire, 120/208 volts, 60 A	660	1,025	1,685
0240		3 phase, 4 wire, 120/208 volts, 100 A	920	1,150	2,070
0245		100 A w/circuit breaker	1,650	1,425	3,075
0280		200 A	1,850	1,600	3,450
0285		200 A w/circuit breaker	3,400	2,000	5,400
0320		400 A	3,725	3,175	6,900
0325		400 A w/circuit breaker	6,625	3,950	10,575
0360		600 A	6,200	4,700	10,900
0365		600 A, w/switchboard	10,400	5,950	16,350
0400		800 A	8,925	5,625	14,550
0405		800 A, w/switchboard	13,100	7,050	20,150
0440		1000 A	10,800	6,925	17,725
0445		1000 A, w/switchboard	15,900	8,500	24,400
0480		1200 A	14,300	7,900	22,200
0485		1200 A, w/groundfault switchboard	36,100	9,800	45,900
0520		1600 A	17,900	10,300	28,200
0525		1600 A, w/groundfault switchboard	41,600	12,400	54,000

D50 Electrical

D5010 Electrical Service/Distribution

D5010 120	Overhead Electric Service, 3 Phase - 4 Wire	COST EACH		
		MAT.	INST.	TOTAL
0560	2000 A	23,700	12,500	36,200
0565	2000 A, w/groundfault switchboard	51,500	15,500	67,000
0610	1 phase, 3 wire, 120/240 volts, 100 A (no safety switch)	204	565	769
0615	100 A w/load center	550	1,100	1,650
0620	200 A	465	775	1,240
0625	200 A w/load center	1,225	1,575	2,800

D50 Electrical

D5010 Underground electric service

System Components	QUANTITY	UNIT	COST EACH		
			MAT.	INST.	TOTAL
SYSTEM D5010 130 1000					
2000 AMP UNDERGROUND ELECTRIC SERVICE WITH GROUNDFAULT SWITCHBOARD					
INCLUDING EXCAVATION, BACKFILL, AND COMPACTION					
Excavate Trench	44.440	B.C.Y.		294.64	294.64
4 inch conduit bank	108.000	L.F.	626.40	2,484	3,110.40
4 inch fitting	8.000	Ea.	212	420	632
4 inch bells	4.000	Ea.	16.08	210	226.08
Concrete material	16.580	C.Y.	1,774.06		1,774.06
Concrete placement	16.580	C.Y.		383	383
Backfill trench	33.710	L.C.Y.		94.06	94.06
Compact fill material in trench	25.930	E.C.Y.		121.36	121.36
Dispose of excess fill material on-site	24.070	L.C.Y.		268.62	268.62
500 kcmil power cable	18.000	C.L.F.	22,950	7,020	29,970
Wire, 600 volt, type THW, copper, stranded, 1/0	6.000	C.L.F.	1,638	1,140	2,778
Saw cutting, concrete walls, plain, per inch of depth	64.000	L.F.	2.56	501.12	503.68
Meter centers and sockets, single pos, 4 terminal, 400 amp	5.000	Ea.	4,600	1,850	6,450
Safety switch, 400 amp	5.000	Ea.	10,500	3,475	13,975
Ground rod clamp	1.000	Ea.	8.15	19.60	27.75
Wireway	1.000	L.F.	11.45	13.95	25.40
600 volt stranded copper wire, 1/0	1.200	C.L.F.	327.60	228	555.60
Flexible metallic conduit	120.000	L.F.	121.20	470.40	591.60
Ground rod	6.000	Ea.	468	942	1,410
TOTAL			43,255.50	19,935.75	63,191.25

D5010 130	Underground Electric Service	COST EACH		
		MAT.	INST.	TOTAL
0950	Underground electric service including excavation, backfill, and compaction			
1000	3 phase, 4 wire, 277/480 volts, 2000 A	43,300	19,900	63,200
1050	2000 A w/groundfault switch	69,000	22,300	91,300
1100	1600 A	34,400	16,100	50,500
1150	1600 A w/groundfault switchboard	58,000	18,200	76,200
1200	1200 A	26,200	12,800	39,000
1250	1200 A w/groundfault switchboard	48,000	14,700	62,700
1400	800 A	18,800	10,700	29,500
1450	800 A w/ switchboard	23,000	12,100	35,100
1500	600 A	11,900	9,800	21,700

D50 Electrical

D5010 Underground electric service

D5010 130	Underground Electric Service	COST EACH		
		MAT.	INST.	TOTAL
1550	600 A w/ switchboard	16,100	11,000	27,100
1600	1 phase, 3 wire, 120/240 volts, 200 A	3,375	3,400	6,775
1650	200 A w/load center	4,125	4,200	8,325
1700	100 A	2,250	2,775	5,025
1750	100 A w/load center	3,025	3,575	6,600

D50 Electrical

D5010 Electrical Service/Distribution

System Components	QUANTITY	UNIT	COST PER L.F. MAT.	COST PER L.F. INST.	COST PER L.F. TOTAL
SYSTEM D5010 230 0200					
FEEDERS, INCLUDING STEEL CONDUIT & WIRE, 60 A					
Rigid galvanized steel conduit, 3/4", including fittings	1.000	L.F.	2.96	7.85	10.81
Wire 600 volt, type XHHW copper stranded #6	.030	C.L.F.	2.79	2.90	5.69
Wire 600 volt, type XHHW copper stranded #8	.010	C.L.F.	.57	.79	1.36
TOTAL			6.32	11.54	17.86

D5010 230	Feeder Installation		COST PER L.F. MAT.	COST PER L.F. INST.	COST PER L.F. TOTAL
0200	Feeder installation 600 V, including RGS conduit and XHHW wire, 60 A		6.30	11.55	17.85
0240	100 A		11.75	15	26.75
0280	200 A	RD5010 -140	28	22.50	50.50
0320	400 A		56	45.50	101.50
0360	600 A		103	74	177
0400	800 A		135	87.50	222.50
0440	1000 A		168	115	283
0480	1200 A		207	148	355
0520	1600 A		271	175	446
0560	2000 A		335	230	565
1200	Branch installation 600 V, including EMT conduit and THW wire, 15 A		1.10	5.65	6.75
1240	20 A		1.32	5.95	7.27
1280	30 A		1.67	6.20	7.87
1320	50 A		3.50	8.60	12.10
1360	65 A		4.36	9.15	13.51
1400	85 A		6.80	10.60	17.40
1440	100 A		7.60	10.80	18.40
1480	130 A		11.30	12.65	23.95
1520	150 A		14.05	14.50	28.55
1560	200 A		19.15	16.55	35.70

D50 Electrical

D5010 Electrical Service/Distribution

System Components	QUANTITY	UNIT	COST EACH		
			MAT.	INST.	TOTAL
SYSTEM D5010 240 0240					
SWITCHGEAR INSTALLATION, INCL SWBD, PANELS & CIRC BREAKERS, 600 A					
Switchboards, 120/208 V, 600 amp	1.000	Ea.	4,200	1,250	5,450
Aluminum bus bars, 120/208 V, 600 amp	1.000	Ea.	2,075	1,250	3,325
Feeder section, circuit breakers, KA frame, 70 to 225 amp	1.000	Ea.	1,300	196	1,496
Feeder section, circuit breakers, LA frame, 125 to 400 amp	1.000	Ea.	2,950	273	3,223
TOTAL			10,525	2,969	13,494

D5010 240	Switchgear		COST EACH		
			MAT.	INST.	TOTAL
0190	Switchgear installation, including switchboard, panels, & circuit breaker				
0200	120/208 V, 1 phase, 400 A		8,050	2,675	10,725
0240	600 A	RD5010 -110	10,500	2,975	13,475
0280	800 A		13,500	3,425	16,925
0300	1000 A		16,500	3,825	20,325
0320	1200 A		17,500	4,325	21,825
0360	1600 A		26,600	4,975	31,575
0400	2000 A		31,800	5,425	37,225
0500	277/480 V, 3 phase, 400 A		11,400	4,975	16,375
0520	600 A		16,000	5,675	21,675
0540	800 A		18,900	6,275	25,175
0560	1000 A		23,500	6,975	30,475
0580	1200 A		25,700	7,825	33,525
0600	1600 A		34,800	8,575	43,375
0620	2000 A		42,500	9,475	51,975

D50 Electrical

D5010 Electrical Service/Distribution

System Components			COST EACH		
	QUANTITY	UNIT	MAT.	INST.	TOTAL
SYSTEM D5010 250 1020					
PANELBOARD INSTALLATION, INCLUDES PANELBOARD, CONDUCTOR, & CONDUIT					
Conduit, galvanized steel, 1-1/4"dia	37.000	L.F.	216.45	386.65	603.10
Panelboards, NQOD, 4 wire, 120/208 volts, 100 amp main, 24 circuits	1.000	Ea.	1,275	1,325	2,600
Wire, 600 volt, type THW, copper, stranded, #3	1.480	C.L.F.	202.76	185	387.76
TOTAL			1,694.21	1,896.65	3,590.86

D5010 250	Panelboard	COST EACH		
		MAT.	INST.	TOTAL
0900	Panelboards, NQOD, 4 wire, 120/208 volts w/conductor & conduit			
1000	100 A, 0 stories, 0' horizontal	1,275	1,325	2,600
1020	1 stories, 25' horizontal	1,700	1,900	3,600
1040	5 stories, 50' horizontal	2,525	3,025	5,550
1060	10 stories, 75' horizontal	3,475	4,350	7,825
1080	225A, 0 stories, 0' horizontal	2,375	1,750	4,125
2000	1 stories, 25' horizontal	3,650	2,775	6,425
2020	5 stories, 50' horizontal	6,175	4,800	10,975
2040	10 stories, 75' horizontal	9,100	7,175	16,275
2060	400A, 0 stories, 0' horizontal	3,500	2,625	6,125
2080	1 stories, 25' horizontal	5,175	4,300	9,475
3000	5 stories, 50' horizontal	8,500	7,600	16,100
3020	10 stories, 75' horizontal	17,800	14,900	32,700
3040	600 A, 0 stories, 0' horizontal	5,200	3,125	8,325
3060	1 stories, 25' horizontal	8,875	5,850	14,725
3080	5 stories, 50' horizontal	16,200	11,200	27,400
4000	10 stories, 75' horizontal	36,200	22,100	58,300
4010	Panelboards, NEHB, 4 wire, 277/480 volts w/conductor, conduit, & safety switch			
4020	100 A, 0 stories, 0' horizontal, includes safety switch	3,050	1,825	4,875
4040	1 stories, 25' horizontal	3,475	2,400	5,875
4060	5 stories, 50' horizontal	4,300	3,525	7,825
4080	10 stories, 75' horizontal	5,275	4,850	10,125
5000	225 A, 0 stories, 0' horizontal	4,600	2,225	6,825
5020	1 stories, 25' horizontal	5,875	3,250	9,125
5040	5 stories, 50' horizontal	8,400	5,300	13,700
5060	10 stories, 75' horizontal	11,300	7,650	18,950

D50 Electrical

D5010 Electrical Service/Distribution

D5010 250	Panelboard	COST EACH		
		MAT.	INST.	TOTAL
5080	400 A, 0 stories, 0' horizontal	7,675	3,425	11,100
6000	1 stories, 25' horizontal	9,350	5,100	14,450
6020	5 stories, 50' horizontal	12,700	8,400	21,100
6040	10 stories, 75' horizontal	22,000	15,700	37,700
6060	600 A, 0 stories, 0' horizontal	11,300	4,350	15,650
6080	1 stories, 25' horizontal	15,000	7,075	22,075
7000	5 stories, 50' horizontal	22,300	12,500	34,800
7020	10 stories, 75' horizontal	42,300	23,400	65,700

D50 Electrical

D5020 Lighting and Branch Wiring

Duplex Receptacle

System Components	QUANTITY	UNIT	COST PER S.F.		
			MAT.	INST.	TOTAL
SYSTEM D5020 110 0200 RECEPTACLES INCL. PLATE, BOX, CONDUIT, WIRE & TRANS. WHEN REQUIRED 2.5 PER 1000 S.F., .3 WATTS PER S.F.					
Steel intermediate conduit, (IMC) 1/2" diam	167.000	L.F.	.33	1.04	1.37
Wire 600V type THWN-THHN, copper solid #12	3.340	C.L.F.	.05	.19	.24
Wiring device, receptacle, duplex, 120V grounded, 15 amp	2.500	Ea.		.04	.04
Wall plate, 1 gang, brown plastic	2.500	Ea.		.02	.02
Steel outlet box 4" square	2.500	Ea.	.01	.08	.09
Steel outlet box 4" plaster rings	2.500	Ea.	.01	.02	.03
TOTAL			.40	1.39	1.79

D5020 110	Receptacle (by Wattage)	COST PER S.F.		
		MAT.	INST.	TOTAL
0190	Receptacles include plate, box, conduit, wire & transformer when required			
0200	2.5 per 1000 S.F., .3 watts per S.F.	.40	1.39	1.79
0240	With transformer	.47	1.46	1.93
0280	4 per 1000 S.F., .5 watts per S.F.	.45	1.63	2.08
0320	With transformer	.56	1.74	2.30
0360	5 per 1000 S.F., .6 watts per S.F.	.53	1.93	2.46
0400	With transformer	.67	2.07	2.74
0440	8 per 1000 S.F., .9 watts per S.F.	.56	2.13	2.69
0480	With transformer	.78	2.35	3.13
0520	10 per 1000 S.F., 1.2 watts per S.F.	.59	2.33	2.92
0560	With transformer	.87	2.61	3.48
0600	16.5 per 1000 S.F., 2.0 watts per S.F.	.69	2.89	3.58
0640	With transformer	1.14	3.35	4.49
0680	20 per 1000 S.F., 2.4 watts per S.F.	.73	3.15	3.88
0720	With transformer	1.28	3.71	4.99

(RD5010 -110)

D50 Electrical

D5020 Lighting and Branch Wiring

Underfloor Receptacle System

Description: Table D5020 115 includes installed costs of raceways and copper wire from panel to and including receptacle.

National Electrical Code prohibits use of undercarpet system in residential, school or hospital buildings. Can only be used with carpet squares.

Low density = (1) Outlet per 259 S.F. of floor area.

High density = (1) Outlet per 127 S.F. of floor area.

System Components	QUANTITY	UNIT	COST PER S.F.		
			MAT.	INST.	TOTAL
SYSTEM D5020 115 0200					
RECEPTACLE SYSTEMS, UNDERFLOOR DUCT, 5' ON CENTER, LOW DENSITY					
Underfloor duct 3-1/8" x 7/8" w/insert 24" on center	.190	L.F.	3.90	1.70	5.60
Vertical elbow for underfloor duct, 3-1/8", included					
Underfloor duct conduit adapter, 2" x 1-1/4", included					
Underfloor duct junction box, single duct, 3-1/8"	.003	Ea.	1.38	.47	1.85
Underfloor junction box carpet pan	.003	Ea.	1.14	.02	1.16
Underfloor duct outlet, high tension receptacle	.004	Ea.	.43	.31	.74
Wire 600V type THWN-THHN copper solid #12	.010	C.L.F.	.14	.57	.71
TOTAL			6.99	3.07	10.06

D5020 115	Receptacles, Floor		COST PER S.F.		
			MAT.	INST.	TOTAL
0200	Receptacle systems, underfloor duct, 5' on center, low density		7	3.07	10.07
0240	High density		7.55	3.96	11.51
0280	7' on center, low density	RD5010 -110	5.55	2.64	8.19
0320	High density		6.10	3.53	9.63
0400	Poke thru fittings, low density		1.20	1.50	2.70
0440	High density		2.37	3	5.37
0520	Telepoles, using Romex, low density		1.19	.93	2.12
0560	High density		2.38	1.86	4.24
0600	Using EMT, low density		1.27	1.24	2.51
0640	High density		2.51	2.46	4.97
0720	Conduit system with floor boxes, low density		1.29	1.04	2.33
0760	High density		2.58	2.10	4.68
0840	Undercarpet power system, 3 conductor with 5 conductor feeder, low density		1.49	.38	1.87
0880	High density		2.95	.79	3.74

D50 Electrical

D5020 Lighting and Branch Wiring

Duplex Receptacle

Wall Switch

System Components	QUANTITY	UNIT	COST PER EACH		
			MAT.	INST.	TOTAL
SYSTEM D5020 125 0520					
RECEPTACLES AND WALL SWITCHES, RECEPTICLE DUPLEX 120 V GROUNDED, 15 A					
Electric metallic tubing conduit, (EMT), 3/4" diam	22.000	L.F.	23.32	106.04	129.36
Wire, 600 volt, type THWN-THHN, copper, solid #12	.630	C.L.F.	8.63	35.91	44.54
Steel outlet box 4" square	1.000	Ea.	2.71	31.50	34.21
Steel outlet box, 4" square, plaster rings	1.000	Ea.	3.65	9.80	13.45
Receptacle, duplex, 120 volt grounded, 15 amp	1.000	Ea.	1.34	15.70	17.04
Wall plate, 1 gang, brown plastic	1.000	Ea.	.42	7.85	8.27
TOTAL			40.07	206.80	246.87

D5020 125		Receptacles & Switches by Each	COST PER EACH		
			MAT.	INST.	TOTAL
0460	Receptacles & Switches, with box, plate, 3/4" EMT conduit & wire				
0520		Receptacle duplex 120 V grounded, 15 A	40	207	247
0560		20 A	49.50	214	263.50
0600		Receptacle duplex ground fault interrupting, 15 A	79.50	214	293.50
0640		20 A	82	214	296
0680		Toggle switch single, 15 A	44.50	207	251.50
0720		20 A	48.50	214	262.50
0760		3 way switch, 15 A	46.50	219	265.50
0800		20 A	48	226	274
0840		4 way switch, 15 A	52.50	233	285.50
0880		20 A	84.50	248	332.50

D50 Electrical

D5020 Lighting and Branch Wiring

Description: Table D5020 130 includes the cost for switch, plate, box, conduit in slab or EMT exposed and copper wire. Add 20% for exposed conduit.

No power required for switches.

Federal energy guidelines recommend the maximum lighting area controlled per switch shall not exceed 1000 S.F. and that areas over 500 S.F. shall be so controlled that total illumination can be reduced by at least 50%.

System Components	QUANTITY	UNIT	COST PER S.F. MAT.	COST PER S.F. INST.	COST PER S.F. TOTAL
SYSTEM D5020 130 0360					
WALL SWITCHES, 5.0 PER 1000 S.F.					
Steel, intermediate conduit (IMC), 1/2" diameter	88.000	L.F.	.18	.55	.73
Wire, 600V type THWN-THHN, copper solid #12	1.710	C.L.F.	.02	.10	.12
Toggle switch, single pole, 15 amp	5.000	Ea.	.03	.08	.11
Wall plate, 1 gang, brown plastic	5.000	Ea.		.04	.04
Steel outlet box 4" plaster rings	5.000	Ea.	.01	.16	.17
Plaster rings	5.000	Ea.	.02	.05	.07
TOTAL			.26	.98	1.24

D5020 130	Wall Switch by Sq. Ft.	MAT.	INST.	TOTAL
0200	Wall switches, 1.0 per 1000 S.F.	.06	.23	.29
0240	1.2 per 1000 S.F.	.07	.25	.32
0280	2.0 per 1000 S.F.	.10	.36	.46
0320	2.5 per 1000 S.F.	.13	.45	.58
0360	5.0 per 1000 S.F.	.26	.98	1.24
0400	10.0 per 1000 S.F.	.54	1.98	2.52

D50 Electrical

D5020 Lighting and Branch Wiring

System D5020 135 includes all wiring and connections.

System Components	QUANTITY	UNIT	COST PER S.F.		
			MAT.	INST.	TOTAL
SYSTEM D5020 135 0200					
MISCELLANEOUS POWER, TO .5 WATTS					
Steel intermediate conduit, (IMC) 1/2" diam	15.000	L.F.	.03	.09	.12
Wire 600V type THWN-THHN, copper solid #12	.325	C.L.F.		.02	.02
TOTAL			.03	.11	.14

D5020 135	Miscellaneous Power	COST PER S.F.		
		MAT.	INST.	TOTAL
0200	Miscellaneous power, to .5 watts	.03	.11	.14
0240	.8 watts	.05	.16	.21
0280	1 watt	.07	.21	.28
0320	1.2 watts	.08	.25	.33
0360	1.5 watts	.09	.29	.38
0400	1.8 watts	.10	.34	.44
0440	2 watts	.12	.40	.52
0480	2.5 watts	.15	.50	.65
0520	3 watts	.18	.58	.76

D50 Electrical

D5020 Lighting and Branch Wiring

System D5020 140 includes all wiring and connections for central air conditioning units.

System Components	QUANTITY	UNIT	COST PER S.F.		
			MAT.	INST.	TOTAL
SYSTEM D5020 140 0200					
CENTRAL AIR CONDITIONING POWER, 1 WATT					
Steel intermediate conduit, 1/2" diam.	.030	L.F.	.06	.19	.25
Wire 600V type THWN-THHN, copper solid #12	.001	C.L.F.	.01	.06	.07
TOTAL			.07	.25	.32

D5020 140	Central A. C. Power (by Wattage)	COST PER S.F.		
		MAT.	INST.	TOTAL
0200	Central air conditioning power, 1 watt	.07	.25	.32
0220	2 watts	.08	.29	.37
0240	3 watts	.12	.32	.44
0280	4 watts	.18	.43	.61
0320	6 watts	.33	.56	.89
0360	8 watts	.43	.61	1.04
0400	10 watts	.61	.71	1.32

D50 Electrical

D5020 Lighting and Branch Wiring

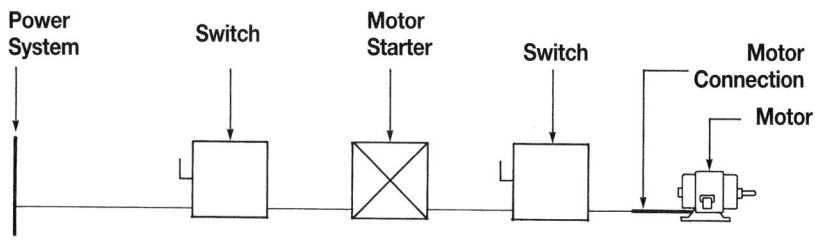

System D5020 145 installed cost of motor wiring using 50' of rigid conduit and copper wire. **Cost and setting of motor not included.**

System Components	QUANTITY	UNIT	COST EACH MAT.	COST EACH INST.	COST EACH TOTAL
SYSTEM D5020 145 0200					
MOTOR INST., SINGLE PHASE, 115V, TO AND INCLUDING 1/3 HP MOTOR SIZE					
Wire 600V type THWN-THHN, copper solid #12	1.250	C.L.F.	17.13	71.25	88.38
Steel intermediate conduit, (IMC) 1/2" diam.	50.000	L.F.	100	312.50	412.50
Magnetic FVNR, 115V, 1/3 HP, size 00 starter	1.000	Ea.	193	157	350
Safety switch, fused, heavy duty, 240V 2P 30 amp	1.000	Ea.	135	179	314
Safety switch, non fused, heavy duty, 600V, 3 phase, 30 A	1.000	Ea.	159	196	355
Flexible metallic conduit, Greenfield 1/2" diam.	1.500	L.F.	1.11	4.71	5.82
Connectors for flexible metallic conduit Greenfield 1/2" diam.	1.000	Ea.	3.45	7.85	11.30
Coupling for Greenfield to conduit 1/2" diam. flexible metalic conduit	1.000	Ea.	2.26	12.55	14.81
Fuse cartridge nonrenewable, 250V 30 amp	1.000	Ea.	2.21	12.55	14.76
TOTAL			613.16	953.41	1,566.57

D5020 145	Motor Installation		COST EACH MAT.	COST EACH INST.	COST EACH TOTAL
0200	Motor installation, single phase, 115V, 1/3 HP motor size		615	955	1,570
0240	1 HP motor size		635	955	1,590
0280	2 HP motor size	RD5010 -170	685	1,000	1,685
0320	3 HP motor size		765	1,025	1,790
0360	230V, 1 HP motor size		615	965	1,580
0400	2 HP motor size		650	965	1,615
0440	3 HP motor size		730	1,050	1,780
0520	Three phase, 200V, 1-1/2 HP motor size		730	1,050	1,780
0560	3 HP motor size		785	1,150	1,935
0600	5 HP motor size		850	1,275	2,125
0640	7-1/2 HP motor size		885	1,300	2,185
0680	10 HP motor size		1,400	1,650	3,050
0720	15 HP motor size		1,800	1,825	3,625
0760	20 HP motor size		2,325	2,100	4,425
0800	25 HP motor size		2,375	2,125	4,500
0840	30 HP motor size		3,675	2,500	6,175
0880	40 HP motor size		4,700	2,950	7,650
0920	50 HP motor size		7,875	3,450	11,325
0960	60 HP motor size		8,250	3,625	11,875
1000	75 HP motor size		10,800	4,150	14,950
1040	100 HP motor size		28,800	4,900	33,700
1080	125 HP motor size		29,700	5,375	35,075
1120	150 HP motor size		34,600	6,325	40,925
1160	200 HP motor size		35,300	7,500	42,800
1240	230V, 1-1/2 HP motor size		690	1,050	1,740
1280	3 HP motor size		745	1,150	1,895
1320	5 HP motor size		810	1,275	2,085
1360	7-1/2 HP motor size		810	1,275	2,085
1400	10 HP motor size		1,250	1,550	2,800
1440	15 HP motor size		1,450	1,700	3,150
1480	20 HP motor size		2,175	2,050	4,225
1520	25 HP motor size		2,325	2,100	4,425

D50 Electrical

D5020 Lighting and Branch Wiring

D5020 145	Motor Installation	COST EACH		
		MAT.	INST.	TOTAL
1560	30 HP motor size	2,375	2,125	4,500
1600	40 HP motor size	4,400	2,875	7,275
1640	50 HP motor size	4,700	3,050	7,750
1680	60 HP motor size	7,875	3,450	11,325
1720	75 HP motor size	9,750	3,925	13,675
1760	100 HP motor size	11,200	4,375	15,575
1800	125 HP motor size	29,300	5,050	34,350
1840	150 HP motor size	31,400	5,750	37,150
1880	200 HP motor size	33,400	6,400	39,800
1960	460V, 2 HP motor size	855	1,050	1,905
2000	5 HP motor size	910	1,150	2,060
2040	10 HP motor size	940	1,275	2,215
2080	15 HP motor size	1,225	1,450	2,675
2120	20 HP motor size	1,300	1,575	2,875
2160	25 HP motor size	1,425	1,650	3,075
2200	30 HP motor size	1,775	1,775	3,550
2240	40 HP motor size	2,300	1,900	4,200
2280	50 HP motor size	2,600	2,100	4,700
2320	60 HP motor size	3,775	2,475	6,250
2360	75 HP motor size	4,450	2,750	7,200
2400	100 HP motor size	4,975	3,050	8,025
2440	125 HP motor size	8,175	3,475	11,650
2480	150 HP motor size	10,100	3,875	13,975
2520	200 HP motor size	12,000	4,375	16,375
2600	575V, 2 HP motor size	855	1,050	1,905
2640	5 HP motor size	910	1,150	2,060
2680	10 HP motor size	940	1,275	2,215
2720	20 HP motor size	1,225	1,450	2,675
2760	25 HP motor size	1,300	1,575	2,875
2800	30 HP motor size	1,775	1,775	3,550
2840	50 HP motor size	1,925	1,825	3,750
2880	60 HP motor size	3,725	2,475	6,200
2920	75 HP motor size	3,775	2,475	6,250
2960	100 HP motor size	4,450	2,750	7,200
3000	125 HP motor size	7,700	3,400	11,100
3040	150 HP motor size	8,175	3,475	11,650
3080	200 HP motor size	10,500	3,925	14,425

D50 Electrical

D5020 Lighting and Branch Wiring

System Components	QUANTITY	UNIT	COST PER L.F.		
			MAT.	INST.	TOTAL
SYSTEM D5020 155 0200					
MOTOR FEEDER SYSTEMS, SINGLE PHASE, UP TO 115V, 1HP OR 230V, 2HP					
Steel intermediate conduit, (IMC) 1/2" diam	1.000	L.F.	2	6.25	8.25
Wire 600V type THWN-THHN, copper solid #12	.020	C.L.F.	.27	1.14	1.41
TOTAL			2.27	7.39	9.66

D5020 155	Motor Feeder	COST PER L.F.		
		MAT.	INST.	TOTAL
0200	Motor feeder systems, single phase, feed up to 115V 1HP or 230V 2 HP	2.27	7.40	9.67
0240	115V 2HP, 230V 3HP	2.43	7.50	9.93
0280	115V 3HP	2.81	7.80	10.61
0360	Three phase, feed to 200V 3HP, 230V 5HP, 460V 10HP, 575V 10HP	2.41	7.95	10.36
0440	200V 5HP, 230V 7.5HP, 460V 15HP, 575V 20HP	2.65	8.15	10.80
0520	200V 10HP, 230V 10HP, 460V 30HP, 575V 30HP	3.22	8.60	11.82
0600	200V 15HP, 230V 15HP, 460V 40HP, 575V 50HP	4.48	9.85	14.33
0680	200V 20HP, 230V 25HP, 460V 50HP, 575V 60HP	6.85	12.50	19.35
0760	200V 25HP, 230V 30HP, 460V 60HP, 575V 75HP	7.65	12.70	20.35
0840	200V 30HP	8.70	13.10	21.80
0920	230V 40HP, 460V 75HP, 575V 100HP	11.30	14.35	25.65
1000	200V 40HP	12.75	15.35	28.10
1080	230V 50HP, 460V 100HP, 575V 125HP	16	16.95	32.95
1160	200V 50HP, 230V 60HP, 460V 125HP, 575V 150HP	18.70	18	36.70
1240	200V 60HP, 460V 150HP	23	21	44
1320	230V 75HP, 575V 200HP	26.50	22	48.50
1400	200V 75HP	30	22.50	52.50
1480	230V 100HP, 460V 200HP	40.50	26	66.50
1560	200V 100HP	55	32.50	87.50
1640	230V 125HP	55	32.50	87.50
1720	200V 125HP, 230V 150HP	64	41	105
1800	200V 150HP	78	44	122
1880	200V 200HP	98	46.50	144.50
1960	230V 200HP	100	48.50	148.50

Note: RD5010-170 (referenced at row 0280/0360)

D50 Electrical

D5020 Lighting and Branch Wiring

Starters are full voltage, type NEMA 1 for general purpose indoor application with motor overload protection and include mounting and wire connections.

System Components	QUANTITY	UNIT	COST EACH		
			MAT.	INST.	TOTAL
SYSTEM D5020 160 0200					
MAGNETIC STARTER, SIZE 00 TO 1/3 HP, 1 PHASE 115V OR 1 HP 230V					
Magnetic starter, size 00, to 1/3 HP, 1 phase, 115V or 1 HP 230V	1.000	Ea.	193	157	350
TOTAL			193	157	350

D5020 160	Magnetic Starter	COST EACH		
		MAT.	INST.	TOTAL
0200	Magnetic starter, size 00, to 1/3 HP, 1 phase, 115V or 1 HP 230V	193	157	350
0280	Size 00, to 1-1/2 HP, 3 phase, 200-230V or 2 HP 460-575V	212	179	391
0360	Size 0, to 1 HP, 1 phase, 115V or 2 HP 230V	215	157	372
0440	Size 0, to 3 HP, 3 phase, 200-230V or 5 HP 460-575V	267	273	540
0520	Size 1, to 2 HP, 1 phase, 115V or 3 HP 230V	247	209	456
0600	Size 1, to 7-1/2 HP, 3 phase, 200-230V or 10 HP 460-575V	300	390	690
0680	Size 2, to 10 HP, 3 phase, 200V, 15 HP-230V or 25 HP 460-575V	565	570	1,135
0760	Size 3, to 25 HP, 3 phase, 200V, 30 HP-230V or 50 HP 460-575V	920	695	1,615
0840	Size 4, to 40 HP, 3 phase, 200V, 50 HP-230V or 100 HP 460-575V	2,050	1,050	3,100
0920	Size 5, to 75 HP, 3 phase, 200V, 100 HP-230V or 200 HP 460-575V	4,775	1,400	6,175
1000	Size 6, to 150 HP, 3 phase, 200V, 200 HP-230V or 400 HP 460-575V	21,400	1,575	22,975

D50 Electrical

D5020 Lighting and Branch Wiring

Safety switches are type NEMA 1 for general purpose indoor application, and include time delay fuses, insulation and wire terminations.

System Components	QUANTITY	UNIT	COST EACH		
			MAT.	INST.	TOTAL
SYSTEM D5020 165 0200					
SAFETY SWITCH, 30A FUSED, 1 PHASE, 115V OR 230V					
Safety switch fused, hvy duty, 240V 2p 30 amp	1.000	Ea.	135	179	314
Fuse, dual element time delay 250V, 30 amp	2.000	Ea.	15.80	25.10	40.90
TOTAL			150.80	204.10	354.90

D5020 165	Safety Switches	COST EACH		
		MAT.	INST.	TOTAL
0200	Safety switch, 30 A fused, 1 phase, 2 HP 115 V or 3 HP, 230 V	151	204	355
0280	3 phase, 5 HP, 200 V or 7 1/2 HP, 230 V	204	234	438
0360	15 HP, 460 V or 20 HP, 575 V	350	243	593
0440	60 A fused, 3 phase, 15 HP 200 V or 15 HP 230 V	345	310	655
0520	30 HP 460 V or 40 HP 575 V	450	320	770
0600	100 A fused, 3 phase, 20 HP 200 V or 25 HP 230 V	565	375	940
0680	50 HP 460 V or 60 HP 575 V	840	380	1,220
0760	200 A fused, 3 phase, 50 HP 200 V or 60 HP 230 V	1,025	530	1,555
0840	125 HP 460 V or 150 HP 575 V	1,300	545	1,845
0920	400 A fused, 3 phase, 100 HP 200 V or 125 HP 230 V	2,450	760	3,210
1000	250 HP 460 V or 350 HP 575 V	3,225	775	4,000
1020	600 A fused, 3 phase, 150 HP 200 V or 200 HP 230 V	4,225	1,125	5,350
1040	400 HP 460 V	5,225	1,150	6,375

D50 Electrical

D5020 Lighting and Branch Wiring

Straight Connector | Angle Connector | Flexible Conduit

Table below includes costs for the flexible conduit. Not included are wire terminations and testing motor for correct rotation.

System Components	QUANTITY	UNIT	COST EACH		
			MAT.	INST.	TOTAL
SYSTEM D5020 170 0200					
MOTOR CONNECTIONS, SINGLE PHASE, 115V/230V UP TO 1 HP					
Motor connection, flexible conduit & fittings, 1 HP motor 115V	1.000	Ea.	9.30	78.50	87.80
TOTAL			9.30	78.50	87.80

D5020 170		Motor Connections	COST EACH		
			MAT.	INST.	TOTAL
0200	Motor connections, single phase, 115/230V, up to 1 HP		9.30	78.50	87.80
0240	Up to 3 HP		17.55	96.50	114.05
0280	Three phase, 200/230/460/575V, up to 3 HP	RD5010 -170	11	92.50	103.50
0320	Up to 5 HP		9.45	115	124.45
0360	Up to 7-1/2 HP		13.40	136	149.40
0400	Up to 10 HP		30	149	179
0440	Up to 15 HP		30	190	220
0480	Up to 25 HP		42.50	232	274.50
0520	Up to 50 HP		85	285	370
0560	Up to 100 HP		200	420	620

D50 Electrical

D5020 Lighting and Branch Wiring

Manual Starter | **Magnetic Starter** | **Induction Motor**

For 230/460 Volt A.C., 3 phase, 60 cycle ball bearing squirrel cage induction motors, NEMA Class B standard line. Installation included.

No conduit, wire, or terminations included.

System Components	QUANTITY	UNIT	COST EACH		
			MAT.	INST.	TOTAL
SYSTEM D5020 175 0220					
MOTOR, DRIPPROOF CLASS B INSULATION, 1.15 SERVICE FACTOR, WITH STARTER					
1 H.P., 1200 RPM WITH MANUAL STARTER					
Motor, dripproof, class B insul, 1.15 serv fact, 1200 RPM, 1 HP	1.000	Ea.	405	139	544
Motor starter, manual, 3 phase, 1 HP motor	1.000	Ea.	223	179	402
TOTAL			628	318	946

D5020 175	Motor & Starter		COST EACH		
			MAT.	INST.	TOTAL
0190	Motor, dripproof, premium efficient, 1.15 service factor				
0200	1 HP, 1200 RPM, motor only		405	139	544
0220	With manual starter	RD5010 -170	630	320	950
0240	With magnetic starter		615	320	935
0260	1800 RPM, motor only		310	139	449
0280	With manual starter		535	320	855
0300	With magnetic starter		520	320	840
0320	2 HP, 1200 RPM, motor only		465	139	604
0340	With manual starter		690	320	1,010
0360	With magnetic starter		730	410	1,140
0380	1800 RPM, motor only		355	139	494
0400	With manual starter		580	320	900
0420	With magnetic starter		620	410	1,030
0440	3600 RPM, motor only		455	139	594
0460	With manual starter		680	320	1,000
0480	With magnetic starter		720	410	1,130
0500	3 HP, 1200 RPM, motor only		650	139	789
0520	With manual starter		875	320	1,195
0540	With magnetic starter		915	410	1,325
0560	1800 RPM, motor only		480	139	619
0580	With manual starter		705	320	1,025
0600	With magnetic starter		745	410	1,155
0620	3600 RPM, motor only		460	139	599
0640	With manual starter		685	320	1,005
0660	With magnetic starter		725	410	1,135
0680	5 HP, 1200 RPM, motor only		880	139	1,019
0700	With manual starter		1,150	455	1,605
0720	With magnetic starter		1,175	530	1,705
0740	1800 RPM, motor only		575	139	714
0760	With manual starter		840	455	1,295
0780	With magnetic starter		875	530	1,405
0800	3600 RPM, motor only		500	139	639

D50 Electrical

D5020 Lighting and Branch Wiring

D5020 175	Motor & Starter	COST EACH		
		MAT.	INST.	TOTAL
0820	With manual starter	765	455	1,220
0840	With magnetic starter	800	530	1,330
0860	7.5 HP, 1800 RPM, motor only	670	149	819
0880	With manual starter	935	465	1,400
0900	With magnetic starter	1,225	720	1,945
0920	10 HP, 1800 RPM, motor only	820	157	977
0940	With manual starter	1,075	470	1,545
0960	With magnetic starter	1,375	725	2,100
0980	15 HP, 1800 RPM, motor only	1,150	196	1,346
1000	With magnetic starter	1,725	765	2,490
1040	20 HP, 1800 RPM, motor only	1,475	241	1,716
1060	With magnetic starter	2,400	935	3,335
1100	25 HP, 1800 RPM, motor only	1,825	251	2,076
1120	With magnetic starter	2,750	945	3,695
1160	30 HP, 1800 RPM, motor only	2,125	261	2,386
1180	With magnetic starter	3,050	955	4,005
1220	40 HP, 1800 RPM, motor only	2,825	315	3,140
1240	With magnetic starter	4,875	1,375	6,250
1280	50 HP, 1800 RPM, motor only	2,675	390	3,065
1300	With magnetic starter	4,725	1,450	6,175
1340	60 HP, 1800 RPM, motor only	3,950	450	4,400
1360	With magnetic starter	8,725	1,850	10,575
1400	75 HP, 1800 RPM, motor only	3,450	525	3,975
1420	With magnetic starter	8,225	1,925	10,150
1460	100 HP, 1800 RPM, motor only	4,450	695	5,145
1480	With magnetic starter	9,225	2,100	11,325
1520	125 HP, 1800 RPM, motor only	5,125	895	6,020
1540	With magnetic starter	26,500	2,475	28,975
1580	150 HP, 1800 RPM, motor only	7,375	1,050	8,425
1600	With magnetic starter	28,800	2,625	31,425
1640	200 HP, 1800 RPM, motor only	9,000	1,250	10,250
1660	With magnetic starter	30,400	2,825	33,225
1680	Totally encl, premium efficient, 1.0 ser. fac., 1HP, 1200 RPM, motor only	385	139	524
1700	With manual starter	610	320	930
1720	With magnetic starter	595	320	915
1740	1800 RPM, motor only	400	139	539
1760	With manual starter	625	320	945
1780	With magnetic starter	610	320	930
1800	2 HP, 1200 RPM, motor only	445	139	584
1820	With manual starter	670	320	990
1840	With magnetic starter	710	410	1,120
1860	1800 RPM, motor only	490	139	629
1880	With manual starter	715	320	1,035
1900	With magnetic starter	755	410	1,165
1920	3600 RPM, motor only	350	139	489
1940	With manual starter	575	320	895
1960	With magnetic starter	615	410	1,025
1980	3 HP, 1200 RPM, motor only	605	139	744
2000	With manual starter	830	320	1,150
2020	With magnetic starter	870	410	1,280
2040	1800 RPM, motor only	580	139	719
2060	With manual starter	805	320	1,125
2080	With magnetic starter	845	410	1,255
2100	3600 RPM, motor only	430	139	569
2120	With manual starter	655	320	975
2140	With magnetic starter	695	410	1,105
2160	5 HP, 1200 RPM, motor only	825	139	964
2180	With manual starter	1,100	455	1,555

D50 Electrical

D5020 Lighting and Branch Wiring

D5020 175	Motor & Starter	COST EACH		
		MAT.	INST.	TOTAL
2200	With magnetic starter	1,125	530	1,655
2220	1800 RPM, motor only	625	139	764
2240	With manual starter	890	455	1,345
2260	With magnetic starter	925	530	1,455
2280	3600 RPM, motor only	540	139	679
2300	With manual starter	805	455	1,260
2320	With magnetic starter	840	530	1,370
2340	7.5 HP, 1800 RPM, motor only	900	149	1,049
2360	With manual starter	1,175	465	1,640
2380	With magnetic starter	1,475	720	2,195
2400	10 HP, 1800 RPM, motor only	955	157	1,112
2420	With manual starter	1,225	470	1,695
2440	With magnetic starter	1,525	725	2,250
2460	15 HP, 1800 RPM, motor only	1,350	196	1,546
2480	With magnetic starter	1,925	765	2,690
2500	20 HP, 1800 RPM, motor only	1,875	241	2,116
2520	With magnetic starter	2,800	935	3,735
2540	25 HP, 1800 RPM, motor only	2,150	251	2,401
2560	With magnetic starter	3,075	945	4,020
2580	30 HP, 1800 RPM, motor only	2,250	261	2,511
2600	With magnetic starter	3,175	955	4,130
2620	40 HP, 1800 RPM, motor only	2,750	315	3,065
2640	With magnetic starter	4,800	1,375	6,175
2660	50 HP, 1800 RPM, motor only	2,975	390	3,365
2680	With magnetic starter	5,025	1,450	6,475
2700	60 HP, 1800 RPM, motor only	4,975	450	5,425
2720	With magnetic starter	9,750	1,850	11,600
2740	75 HP, 1800 RPM, motor only	6,100	525	6,625
2760	With magnetic starter	10,900	1,925	12,825
2780	100 HP, 1800 RPM, motor only	7,750	695	8,445
2800	With magnetic starter	12,500	2,100	14,600
2820	125 HP, 1800 RPM, motor only	10,900	895	11,795
2840	With magnetic starter	32,300	2,475	34,775
2860	150 HP, 1800 RPM, motor only	11,900	1,050	12,950
2880	With magnetic starter	33,300	2,625	35,925
2900	200 HP, 1800 RPM, motor only	15,100	1,250	16,350
2920	With magnetic starter	36,500	2,825	39,325

D50 Electrical

D5020 Lighting and Branch Wiring

A. Strip Fixture B. Surface Mounted

C. Recessed D. Pendent Mounted

Design Assumptions:
1. A 100 footcandle average maintained level of illumination.
2. Ceiling heights range from 9' to 11'.
3. Average reflectance values are assumed for ceilings, walls and floors.
4. Cool white (CW) fluorescent lamps with 3150 lumens for 40 watt lamps and 6300 lumens for 8' slimline lamps.
5. Four 40 watt lamps per 4' fixture and two 8' lamps per 8' fixture.
6. Average fixture efficiency values and spacing to mounting height ratios.
7. Installation labor is average U.S. rate as of January 1.

System Components	QUANTITY	UNIT	COST PER S.F. MAT.	INST.	TOTAL
SYSTEM D5020 208 0520					
FLUORESCENT FIXTURES MOUNTED 9'-11" ABOVE FLOOR, 100 FC					
TYPE A, 8 FIXTURES PER 400 S.F.					
Conduit, steel intermediate, 1/2" diam.	.185	L.F.	.37	1.16	1.53
Wire, 600V, type THWN-THHN, copper, solid, #12	.004	C.L.F.	.05	.21	.26
Fluorescent strip fixture 8' long, surface mounted, two 75W SL	.020	Ea.	1.25	2.02	3.27
Steel outlet box 4" concrete	.020	Ea.	.37	.63	1
Steel outlet box plate with stud, 4" concrete	.020	Ea.	.25	.16	.41
Fixture hangers, flexible, 1/2" diameter, 4" long	.040	Ea.	.52	2.10	2.62
Fixture whip, THHN wire, three #12, 3/8" greenfield	.040	Ea.	.67	.78	1.45
TOTAL			3.48	7.06	10.54

D5020 208	Fluorescent Fixtures (by Type)		COST PER S.F. MAT.	INST.	TOTAL
0520	Fluorescent fixtures, type A, 8 fixtures per 400 S.F.		3.48	7.05	10.53
0560	11 fixtures per 600 S.F.		3.18	6.45	9.63
0600	17 fixtures per 1000 S.F.		2.95	6	8.95
0640	23 fixtures per 1600 S.F.	RD5020-200	2.48	5.05	7.53
0680	28 fixtures per 2000 S.F.		2.44	4.94	7.38
0720	41 fixtures per 3000 S.F.		2.39	4.84	7.23
0800	53 fixtures per 4000 S.F.		2.29	4.68	6.97
0840	64 fixtures per 5000 S.F.		2.26	4.58	6.84
0880	Type B, 11 fixtures per 400 S.F.		5.50	10.05	15.55
0920	15 fixtures per 600 S.F.		4.95	9.05	14
0960	24 fixtures per 1000 S.F.		4.76	8.75	13.51
1000	35 fixtures per 1600 S.F.		4.35	7.95	12.30
1040	42 fixtures per 2000 S.F.		4.15	7.60	11.75
1080	61 fixtures per 3000 S.F.		3.99	7.35	11.34
1160	80 fixtures per 4000 S.F.		2.89	4.66	7.55
1200	98 fixtures per 5000 S.F.		3.91	7.15	11.06
1240	Type C, 11 fixtures per 400 S.F.		4.78	10.45	15.23
1280	14 fixtures per 600 S.F.		3.99	8.75	12.74
1320	23 fixtures per 1000 S.F.		3.96	8.70	12.66
1360	34 fixtures per 1600 S.F.		3.64	8	11.64
1400	43 fixtures per 2000 S.F.		3.74	8.20	11.94
1440	63 fixtures per 3000 S.F.		3.61	7.90	11.51
1520	81 fixtures per 4000 S.F.		3.48	7.65	11.13
1560	101 fixtures per 5000 S.F.		3.45	7.55	11
1600	Type D, 8 fixtures per 400 S.F.		4.35	7.90	12.25
1640	12 fixtures per 600 S.F.		4.35	7.90	12.25
1680	19 fixtures per 1000 S.F.		4.13	7.50	11.63
1720	27 fixtures per 1600 S.F.		3.69	6.70	10.39
1760	34 fixtures per 2000 S.F.		3.69	6.70	10.39
1800	48 fixtures per 3000 S.F.		3.49	6.35	9.84
1880	64 fixtures per 4000 S.F.		3.49	6.35	9.84
1920	79 fixtures per 5000 S.F.		3.48	6.30	9.78

D50 Electrical

D5020 Lighting and Branch Wiring

Type C. Recessed, mounted on grid ceiling suspension system, 2' x 4', four 40 watt lamps, acrylic prismatic diffusers.

5.3 watts per S.F. for 100 footcandles.

3 watts per S.F. for 57 footcandles.

System Components	QUANTITY	UNIT	COST PER S.F.		
			MAT.	INST.	TOTAL
SYSTEM D5020 210 0200					
FLUORESCENT FIXTURES RECESS MOUNTED IN CEILING					
1 WATT PER S.F., 20 FC, 5 FIXTURES PER 1000 S.F.					
Steel intermediate conduit, (IMC) 1/2" diam.	.128	L.F.	.26	.80	1.06
Wire, 600 volt, type THW, copper, solid, #12	.003	C.L.F.	.04	.17	.21
Fluorescent fixture, recessed, 2'x 4', four 40W, w/lens, for grid ceiling	.005	Ea.	.32	.67	.99
Steel outlet box 4" square	.005	Ea.	.09	.16	.25
Fixture whip, Greenfield w/#12 THHN wire	.005	Ea.	.06	.04	.10
TOTAL			.77	1.84	2.61

D5020 210	Fluorescent Fixtures (by Wattage)		COST PER S.F.		
			MAT.	INST.	TOTAL
0190	Fluorescent fixtures recess mounted in ceiling				
0195	T12, standard 40 watt lamps				
0200	1 watt per S.F., 20 FC, 5 fixtures @40 watts per 1000 S.F.	RD5020 -200	.77	1.84	2.61
0240	2 watt per S.F., 40 FC, 10 fixtures @40 watt per 1000 S.F.		1.52	3.62	5.14
0280	3 watt per S.F., 60 FC, 15 fixtures @40 watt per 1000 S.F		2.30	5.45	7.75
0320	4 watt per S.F., 80 FC, 20 fixtures @40 watt per 1000 S.F.		3.04	7.20	10.24
0400	5 watt per S.F., 100 FC, 25 fixtures @40 watt per 1000 S.F.		3.82	9.05	12.87
0450	T8, energy saver 32 watt lamps				
0500	0.8 watt per S.F., 20 FC, 5 fixtures @32 watt per 1000 S.F.		.82	1.84	2.66
0520	1.6 watt per S.F., 40 FC, 10 fixtures @32 watt per 1000 S.F.		1.63	3.62	5.25
0540	2.4 watt per S.F., 60 FC, 15 fixtures @ 32 watt per 1000 S.F		2.45	5.45	7.90
0560	3.2 watt per S.F., 80 FC, 20 fixtures @32 watt per 1000 S.F.		3.25	7.20	10.45
0580	4 watt per S.F., 100 FC, 25 fixtures @32 watt per 1000 S.F.		4.08	9.05	13.13

D50 Electrical

D5020 Lighting and Branch Wiring

Type A. Recessed wide distribution reflector with flat glass lens 150 W.
  Maximum spacing = 1.2 x mounting height.
  13 watts per S.F. for 100 footcandles.

Type B. Recessed reflector down light with baffles 150 W.
  Maximum spacing = 0.8 x mounting height.
  18 watts per S.F. for 100 footcandles.

Type C. Recessed PAR-38 flood lamp with concentric louver 150 W.
  Maximum spacing = 0.5 x mounting height.
  19 watts per S.F. for 100 footcandles.

Type D. Recessed R-40 flood lamp with reflector skirt.
  Maximum spacing = 0.7 x mounting height.
  15 watts per S.F. for 100 footcandles.

System Components	QUANTITY	UNIT	COST PER S.F. MAT.	COST PER S.F. INST.	COST PER S.F. TOTAL
SYSTEM D5020 214 0400					
INCANDESCENT FIXTURE RECESS MOUNTED, 100 FC					
TYPE A, 34 FIXTURES PER 400 S.F.					
Steel intermediate conduit, (IMC) 1/2" diam	1.060	L.F.	2.12	6.63	8.75
Wire, 600V, type THWN THHN, copper, solid, #12	.033	C.L.F.	.45	1.88	2.33
Steel outlet box 4" square	.085	Ea.	1.59	2.68	4.27
Fixture whip, Greenfield w/#12 THHN wire	.085	Ea.	1.05	.67	1.72
Incandescent fixture, recessed, w/lens, prewired, square trim, 200W	.085	Ea.	8.93	7.95	16.88
TOTAL			14.14	19.81	33.95

D5020 214	Incandescent Fixture (by Type)	MAT.	INST.	TOTAL
0380	Incandescent fixture recess mounted, 100 FC			
0400	Type A, 34 fixtures per 400 S.F.	14.15	19.80	33.95
0440	49 fixtures per 600 S.F.	13.75	19.45	33.20
0480	63 fixtures per 800 S.F.	13.40	19.25	32.65
0520	90 fixtures per 1200 S.F.	12.90	18.85	31.75
0560	116 fixtures per 1600 S.F.	12.65	18.70	31.35
0600	143 fixtures per 2000 S.F.	12.55	18.65	31.20
0640	Type B, 47 fixtures per 400 S.F.	13.55	25	38.55
0680	66 fixtures per 600 S.F.	12.95	24.50	37.45
0720	88 fixtures per 800 S.F.	12.95	24.50	37.45
0760	127 fixtures per 1200 S.F.	12.65	24	36.65
0800	160 fixtures per 1600 S.F.	12.45	24	36.45
0840	206 fixtures per 2000 S.F.	12.45	24	36.45
0880	Type C, 51 fixtures per 400 S.F.	18.15	26	44.15
0920	74 fixtures per 600 S.F.	17.60	25.50	43.10
0960	97 fixtures per 800 S.F.	17.40	25.50	42.90
1000	142 fixtures per 1200 S.F.	17.10	25.50	42.60
1040	186 fixtures per 1600 S.F.	16.85	25	41.85
1080	230 fixtures per 2000 S.F.	16.80	25	41.80
1120	Type D, 39 fixtures per 400 S.F.	16.75	20.50	37.25
1160	57 fixtures per 600 S.F.	16.35	20.50	36.85
1200	75 fixtures per 800 S.F.	16.25	20.50	36.75
1240	109 fixtures per 1200 S.F.	15.85	20	35.85
1280	143 fixtures per 1600 S.F.	15.55	19.85	35.40
1320	176 fixtures per 2000 S.F.	15.45	19.80	35.25

RD5020-200

D50 Electrical

D5020 Lighting and Branch Wiring

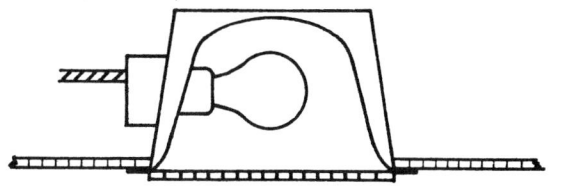

Type A. Recessed, wide distribution reflector with flat glass lens.

150 watt inside frost—2500 lumens per lamp.

PS-25 extended service lamp.

Maximum spacing = 1.2 x mounting height.

13 watts per S.F. for 100 footcandles.

System Components	QUANTITY	UNIT	COST PER S.F.		
			MAT.	INST.	TOTAL
SYSTEM D5020 216 0200					
INCANDESCENT FIXTURE RECESS MOUNTED, TYPE A					
1 WATT PER S.F., 8 FC, 6 FIXT PER 1000 S.F.					
Steel intermediate conduit, (IMC) 1/2" diam	.091	L.F.	.18	.57	.75
Wire, 600V, type THWN-THHN, copper, solid, #12	.002	C.L.F.	.03	.11	.14
Incandescent fixture, recessed, w/lens, prewired, square trim, 200W	.006	Ea.	.63	.56	1.19
Steel outlet box 4" square	.006	Ea.	.11	.19	.30
Fixture whip, Greenfield w/#12 THHN wire	.006	Ea.	.07	.05	.12
TOTAL			1.02	1.48	2.50

D5020 216	Incandescent Fixture (by Wattage)		COST PER S.F.		
			MAT.	INST.	TOTAL
0190	Incandescent fixture recess mounted, type A				
0200	1 watt per S.F., 8 FC, 6 fixtures per 1000 S.F.		1.02	1.48	2.50
0240	2 watt per S.F., 16 FC, 12 fixtures per 1000 S.F.	RD5020 -200	2.04	2.96	5
0280	3 watt per S.F., 24 FC, 18 fixtures, per 1000 S.F.		3.07	4.39	7.46
0320	4 watt per S.F., 32 FC, 24 fixtures per 1000 S.F.		4.10	5.85	9.95
0400	5 watt per S.F., 40 FC, 30 fixtures per 1000 S.F.		5.10	7.35	12.45

D50 Electrical

D5020 Lighting and Branch Wiring

High Bay Fluorescent Fixtures
Four T5 HO/54 watt lamps

System Components	QUANTITY	UNIT	COST PER S.F. MAT.	COST PER S.F. INST.	COST PER S.F. TOTAL
SYSTEM D5020 218 0200					
FLUORESCENT HIGH BAY FIXTURE, 4 LAMP, 8'-10' ABOVE WORK PLANE					
0.5 WATT/S.F., 29 FC, 2 FIXTURES/1000 S.F.					
Steel intermediate conduit, (IMC) 1/2" diam	.100	L.F.	.20	.63	.83
Wire, 600V, type THWN-THHN, copper, solid, #10	.002	C.L.F.	.04	.13	.17
Steel outlet box 4" concrete	.002	Ea.	.04	.06	.10
Steel outlet box plate with stud, 4" concrete	.002	Ea.	.02	.02	.04
Flourescent, hi bay, 4 lamp, T-5 fixture	.002	Ea.	.49	.28	.77
TOTAL			.79	1.12	1.91

D5020 218	Fluorescent Fixture, High Bay, 8'-10' (by Wattage)	COST PER S.F. MAT.	COST PER S.F. INST.	COST PER S.F. TOTAL
0190	Fluorescent high bay-4 lamp fixture, 8'-10' above work plane			
0200	.5 watt/SF, 29 FC, 2 fixtures per 1000 S.F.	.79	1.12	1.91
0400	1 watt/SF, 59 FC, 4 fixtures per 1000 S.F.	1.56	2.13	3.69
0600	1.5 watt/SF, 103 FC, 7 fixtures per 1000 S.F.	2.61	3.33	5.94
0800	2 watt/SF, 133 FC, 9 fixtures per 1000 S.F.	3.38	4.39	7.77
1000	2.5 watt/SF, 162 FC, 11 fixtures per 1000 S.F.	4.18	5.45	9.63

D50 Electrical

D5020 Lighting and Branch Wiring

HIGH BAY FIXTURES

B. Metal halide 400 watt

C. High pressure sodium 400 watt

E. Metal halide 1000 watt

F. High pressure sodium 1000 watt

G. Metal halide 1000 watt 125,000 lumen lamp

System Components	QUANTITY	UNIT	COST PER S.F.		
			MAT.	INST.	TOTAL
SYSTEM D5020 220 0880					
HIGH INTENSITY DISCHARGE FIXTURE, 8'-10' ABOVE WORK PLANE, 100 FC					
TYPE B, 8 FIXTURES PER 900 S.F.					
Steel intermediate conduit, (IMC) 1/2" diam	.460	L.F.	.92	2.88	3.80
Wire, 600V, type THWN-THHN, copper, solid, #10	.009	C.L.F.	.19	.56	.75
Steel outlet box 4" concrete	.009	Ea.	.17	.28	.45
Steel outlet box plate with stud 4" concrete	.009	Ea.	.11	.07	.18
Metal halide, hi bay, aluminum reflector, 400 W lamp	.009	Ea.	4.01	2.46	6.47
TOTAL			5.40	6.25	11.65

D5020 220	H.I.D. Fixture, High Bay, 8'-10' (by Type)		COST PER S.F.		
			MAT.	INST.	TOTAL
0500	High intensity discharge fixture, 8'-10' above work plane, 100 FC				
0880	Type B, 8 fixtures per 900 S.F.		5.40	6.25	11.65
0920	15 fixtures per 1800 S.F.	RD5020 -200	4.95	6	10.95
0960	24 fixtures per 3000 S.F.		4.95	6	10.95
1000	31 fixtures per 4000 S.F.	RD5020 -240	4.94	6	10.94
1040	38 fixtures per 5000 S.F.		4.94	6	10.94
1080	60 fixtures per 8000 S.F.		4.94	6	10.94
1120	72 fixtures per 10000 S.F.		4.43	5.55	9.98
1160	115 fixtures per 16000 S.F.		4.43	5.55	9.98
1200	230 fixtures per 32000 S.F.		4.43	5.55	9.98
1240	Type C, 4 fixtures per 900 S.F.		2.59	3.81	6.40
1280	8 fixtures per 1800 S.F.		2.59	3.81	6.40
1320	13 fixtures per 3000 S.F.		2.59	3.81	6.40
1360	17 fixtures per 4000 S.F.		2.59	3.81	6.40
1400	21 fixtures per 5000 S.F.		2.49	3.51	6
1440	33 fixtures per 8000 S.F.		2.47	3.44	5.91
1480	40 fixtures per 10000 S.F.		2.41	3.25	5.66
1520	63 fixtures per 16000 S.F.		2.41	3.25	5.66
1560	126 fixtures per 32000 S.F.		2.41	3.25	5.66

D50 Electrical

D5020 Lighting and Branch Wiring

HIGH BAY FIXTURES

B. Metal halide 400 watt

C. High pressure sodium 400 watt

E. Metal halide 1000 watt

F. High pressure sodium 1000 watt

G. Metal halide 1000 watt 125,000 lumen lamp

System Components	QUANTITY	UNIT	COST PER S.F.		
			MAT.	INST.	TOTAL
SYSTEM D5020 222 0240					
HIGH INTENSITY DISCHARGE FIXTURE, 8'-10' ABOVE WORK PLANE					
1 WATT/S.F., TYPE B, 29 FC, 2 FIXTURES/1000 S.F.					
Steel intermediate conduit, (IMC) 1/2" diam	.100	L.F.	.20	.63	.83
Wire, 600V, type THWN-THHN, copper, solid, #10	.002	C.L.F.	.04	.13	.17
Steel outlet box 4" concrete	.002	Ea.	.04	.06	.10
Steel outlet box plate with stud, 4" concrete	.002	Ea.	.02	.02	.04
Metal halide, hi bay, aluminum reflector, 400 W lamp	.002	Ea.	.89	.55	1.44
TOTAL			1.19	1.39	2.58

D5020 222	H.I.D. Fixture, High Bay, 8'-10' (by Wattage)		COST PER S.F.		
			MAT.	INST.	TOTAL
0190	High intensity discharge fixture, 8'-10' above work plane				
0240	1 watt/S.F., type B, 29 FC, 2 fixtures/1000 S.F.		1.19	1.39	2.58
0280	Type C, 54 FC, 2 fixtures/1000 S.F.		1.01	1.05	2.06
0400	2 watt/S.F., type B, 59 FC, 4 fixtures/1000 S.F.	RD5020 -200	2.36	2.66	5.02
0440	Type C, 108 FC, 4 fixtures/1000 S.F.	RD5020 -240	2.01	2.04	4.05
0560	3 watt/S.F., type B, 103 FC, 7 fixtures/1000 S.F.		4.01	4.25	8.26
0600	Type C, 189 FC, 6 fixtures/1000 S.F.		3.23	2.62	5.85
0720	4 watt/S.F., type B, 133 FC, 9 fixtures/1000 S.F.		5.20	5.60	10.80
0760	Type C, 243 FC, 9 fixtures/1000 S.F.		4.53	4.51	9.04
0880	5 watt/S.F., type B, 162 FC, 11 fixtures/1000 S.F.		6.35	6.90	13.25
0920	Type C, 297 FC, 11 fixtures/1000 S.F.		5.55	5.60	11.15

D50 Electrical

D5020 Lighting and Branch Wiring

HIGH BAY FIXTURES

B. Metal halide 400 watt

C. High pressure sodium 400 watt

E. Metal halide 1000 watt

F. High pressure sodium 1000 watt

G. Metal halide 1000 watt 125,000 lumen lamp

System Components	QUANTITY	UNIT	COST PER S.F.		
			MAT.	INST.	TOTAL
SYSTEM D5020 224 1240					
HIGH INTENSITY DISCHARGE FIXTURE, 16' ABOVE WORK PLANE, 100 FC					
TYPE C, 5 FIXTURES PER 900 S.F.					
Steel intermediate conduit, (IMC) 1/2" diam	.260	L.F.	.52	1.63	2.15
Wire, 600V, type THWN-THHN, copper, solid, #10	.007	C.L.F.	.15	.44	.59
Steel outlet box 4" concrete	.006	Ea.	.11	.19	.30
Steel outlet box plate with stud, 4" concrete	.006	Ea.	.07	.05	.12
High pressure sodium, hi bay, aluminum reflector, 400 W lamp	.006	Ea.	2.46	1.64	4.10
TOTAL			3.31	3.95	7.26

D5020 224	H.I.D. Fixture, High Bay, 16' (by Type)		COST PER S.F.		
			MAT.	INST.	TOTAL
0510	High intensity discharge fixture, 16' above work plane, 100 FC				
1240	Type C, 5 fixtures per 900 S.F.		3.31	3.95	7.26
1280	9 fixtures per 1800 S.F.	RD5020 -200	3.03	4.13	7.16
1320	15 fixtures per 3000 S.F.		3.03	4.13	7.16
1360	18 fixtures per 4000 S.F.	RD5020 -240	2.89	3.70	6.59
1400	22 fixtures per 5000 S.F.		2.89	3.70	6.59
1440	36 fixtures per 8000 S.F.		2.89	3.70	6.59
1480	42 fixtures per 10,000 S.F.		2.59	3.81	6.40
1520	65 fixtures per 16,000 S.F.		2.59	3.81	6.40
1600	Type G, 4 fixtures per 900 S.F.		4.24	6.25	10.49
1640	6 fixtures per 1800 S.F.		3.56	5.80	9.36
1720	9 fixtures per 4000 S.F.		2.94	5.65	8.59
1760	11 fixtures per 5000 S.F.		2.94	5.65	8.59
1840	21 fixtures per 10,000 S.F.		2.76	5.10	7.86
1880	33 fixtures per 16,000 S.F.		2.76	5.10	7.86

D50 Electrical

D5020 Lighting and Branch Wiring

HIGH BAY FIXTURES

B. Metal halide 400 watt
C. High pressure sodium 400 watt
E. Metal halide 1000 watt
F. High pressure sodium 1000 watt
G. Metal halide 1000 watt
125,000 lumen lamp

System Components	QUANTITY	UNIT	COST PER S.F.		
			MAT.	INST.	TOTAL
SYSTEM D5020 226 0240					
HIGH INTENSITY DISCHARGE FIXTURE, 16' ABOVE WORK PLANE					
1 WATT/S.F., TYPE E, 42 FC, 1 FIXTURE/1000 S.F.					
Steel intermediate conduit, (IMC) 1/2" diam	.160	L.F.	.32	1	1.32
Wire, 600V, type THWN-THHN, copper, solid, #10	.003	C.L.F.	.06	.19	.25
Steel outlet box 4" concrete	.001	Ea.	.02	.03	.05
Steel outlet box plate with stud, 4" concrete	.001	Ea.	.01	.01	.02
Metal halide, hi bay, aluminum reflector, 1000 W lamp	.001	Ea.	.64	.32	.96
TOTAL			1.05	1.55	2.60

D5020 226	H.I.D. Fixture, High Bay, 16' (by Wattage)		COST PER S.F.		
			MAT.	INST.	TOTAL
0190	High intensity discharge fixture, 16' above work plane				
0240	1 watt/S.F., type E, 42 FC, 1 fixture/1000 S.F.		1.05	1.55	2.60
0280	Type G, 52 FC, 1 fixture/1000 S.F.	RD5020 -200	1.05	1.55	2.60
0320	Type C, 54 FC, 2 fixture/1000 S.F.		1.23	1.73	2.96
0440	2 watt/S.F., type E, 84 FC, 2 fixture/1000 S.F.	RD5020 -240	2.13	3.15	5.28
0480	Type G, 105 FC, 2 fixture/1000 S.F.		2.13	3.15	5.28
0520	Type C, 108 FC, 4 fixture/1000 S.F.		2.47	3.46	5.93
0640	3 watt/S.F., type E, 126 FC, 3 fixture/1000 S.F.		3.20	4.69	7.89
0680	Type G, 157 FC, 3 fixture/1000 S.F.		3.20	4.69	7.89
0720	Type C, 162 FC, 6 fixture/1000 S.F.		3.71	5.20	8.91
0840	4 watt/S.F., type E, 168 FC, 4 fixture/1000 S.F.		4.26	6.30	10.56
0880	Type G, 210 FC, 4 fixture/1000 S.F.		4.26	6.30	10.56
0920	Type C, 243 FC, 9 fixture/1000 S.F.		5.40	7.20	12.60
1040	5 watt/S.F., type E, 210 FC, 5 fixture/1000 S.F.		5.30	7.85	13.15
1080	Type G, 262 FC, 5 fixture/1000 S.F.		5.30	7.85	13.15
1120	Type C, 297 FC, 11 fixture/1000 S.F.		6.65	8.95	15.60

D50 Electrical

D5020 Lighting and Branch Wiring

HIGH BAY FIXTURES

B. Metal halide 400 watt
C. High pressure sodium 400 watt
E. Metal halide 1000 watt
F. High pressure sodium 1000 watt
G. Metal halide 1000 watt
   125,000 lumen lamp

System Components	QUANTITY	UNIT	COST PER S.F. MAT.	INST.	TOTAL
SYSTEM D5020 228 1240					
HIGH INTENSITY DISCHARGE FIXTURE, 20' ABOVE WORK PLANE, 100 FC					
TYPE C, 6 FIXTURES PER 900 S.F.					
Steel intermediate conduit, (IMC) 1/2" diam	.350	L.F.	.70	2.19	2.89
Wire, 600V, type THWN-THHN, copper, solid, #10.	.011	C.L.F.	.24	.69	.93
Steel outlet box 4" concrete	.007	Ea.	.13	.22	.35
Steel outlet box plate with stud, 4" concrete	.007	Ea.	.09	.05	.14
High pressure sodium, hi bay, aluminum reflector, 400 W lamp	.007	Ea.	2.87	1.91	4.78
TOTAL			4.03	5.06	9.09

D5020 228	H.I.D. Fixture, High Bay, 20' (by Type)		COST PER S.F.		
			MAT.	INST.	TOTAL
0510	High intensity discharge fixture 20' above work plane, 100 FC				
1240	Type C, 6 fixtures per 900 S.F.		4.03	5.05	9.08
1280	10 fixtures per 1800 S.F.		3.58	4.76	8.34
1320	16 fixtures per 3000 S.F.	RD5020-200	3.18	4.57	7.75
1360	20 fixtures per 4000 S.F.	RD5020-210	3.18	4.57	7.75
1400	24 fixtures per 5000 S.F.		3.18	4.57	7.75
1440	38 fixtures per 8000 S.F.		3.12	4.38	7.50
1520	68 fixtures per 16000 S.F.		2.92	4.82	7.74
1560	132 fixtures per 32000 S.F.		2.92	4.82	7.74
1600	Type G, 4 fixtures per 900 S.F.		4.21	6.10	10.31
1640	6 fixtures per 1800 S.F.		3.57	5.80	9.37
1680	7 fixtures per 3000 S.F.		3.51	5.65	9.16
1720	10 fixtures per 4000 S.F.		3.49	5.55	9.04
1760	11 fixtures per 5000 S.F.		3.05	5.95	9
1800	18 fixtures per 8000 S.F.		3.05	5.95	9
1840	22 fixtures per 10000 S.F.		3.05	5.95	9
1880	34 fixtures per 16000 S.F.		2.99	5.80	8.79
1920	66 fixtures per 32000 S.F.		2.83	5.30	8.13

D50 Electrical

D5020 Lighting and Branch Wiring

HIGH BAY FIXTURES

B. Metal halide 400 watt
C. High pressure sodium 400 watt
E. Metal halide 1000 watt
F. High pressure sodium 1000 watt
G. Metal halide 1000 watt 125,000 lumen lamp

System Components	QUANTITY	UNIT	COST PER S.F.		
			MAT.	INST.	TOTAL
SYSTEM D5020 230 0240					
HIGH INTENSITY DISCHARGE FIXTURE, 20' ABOVE WORK PLANE					
1 WATT/S.F., TYPE E, 40 FC, 1 FIXTURE 1000 S.F.					
Steel intermediate conduit, (IMC) 1/2" diam	.160	L.F.	.32	1	1.32
Wire, 600V, type THWN-THHN, copper, solid, #10	.005	C.L.F.	.11	.31	.42
Steel outlet box 4" concrete	.001	Ea.	.02	.03	.05
Steel outlet box plate with stud, 4" concrete	.001	Ea.	.01	.01	.02
Metal halide, hi bay, aluminum reflector, 1000 W lamp	.001	Ea.	.64	.32	.96
TOTAL			1.10	1.67	2.77

D5020 230	H.I.D. Fixture, High Bay, 20' (by Wattage)	COST PER S.F.		
		MAT.	INST.	TOTAL
0190	High intensity discharge fixture, 20' above work plane			
0240	1 watt/S.F., type E, 40 FC, 1 fixture/1000 S.F.	1.10	1.67	2.77
0280	Type G, 50 FC, 1 fixture/1000 S.F.	1.10	1.67	2.77
0320	Type C, 52 FC, 2 fixtures/1000 S.F.	1.31	1.93	3.24
0440	2 watt/S.F., type E, 81 FC, 2 fixtures/1000 S.F.	2.20	3.34	5.54
0480	Type G, 101 FC, 2 fixtures/1000 S.F.	2.20	3.34	5.54
0520	Type C, 104 FC, 4 fixtures/1000 S.F.	2.58	3.79	6.37
0640	3 watt/S.F., type E, 121 FC, 3 fixtures/1000 S.F.	3.30	5	8.30
0680	Type G, 151 FC, 3 fixtures/1000 S.F.	3.30	5	8.30
0720	Type C, 155 FC, 6 fixtures/1000 S.F.	3.89	5.70	9.59
0840	4 watt/S.F., type E, 161 FC, 4 fixtures/1000 S.F.	4.39	6.65	11.04
0880	Type G, 202 FC, 4 fixtures/1000 S.F.	4.39	6.65	11.04
0920	Type C, 233 FC, 9 fixtures/1000 S.F.	5.60	7.90	13.50
1040	5 watt/S.F., type E, 202 FC, 5 fixtures/1000 S.F.	5.50	8.35	13.85
1080	Type G, 252 FC, 5 fixtures/1000 S.F.	5.50	8.35	13.85
1120	Type C, 285 FC, 11 fixtures/1000 S.F.	6.95	9.80	16.75

Note: RD5020-200 (rows 0280/0320), RD5020-240 (rows 0440/0480)

D50 Electrical

D5020 Lighting and Branch Wiring

HIGH BAY FIXTURES

B. Metal halide 400 watt

C. High pressure sodium 400 watt

E. Metal halide 1000 watt

F. High pressure sodium 1000 watt

G. Metal halide 1000 watt 125,000 lumen lamp

System Components	QUANTITY	UNIT	COST PER S.F.		
			MAT.	INST.	TOTAL
SYSTEM D5020 232 1240 HIGH INTENSITY DISCHARGE FIXTURE, 30' ABOVE WORK PLANE, 100 FC TYPE F, 4 FIXTURES PER 900 S.F.					
Steel intermediate conduit, (IMC) 1/2" diam	.580	L.F.	1.16	3.63	4.79
Wire, 600V, type THWN-THHN, copper, solid, #10	.018	C.L.F.	.39	1.13	1.52
Steel outlet box 4" concrete	.004	Ea.	.07	.13	.20
Steel outlet box plate with stud, 4" concrete	.004	Ea.	.05	.03	.08
High pressure sodium, hi bay, aluminum, 1000 W lamp	.004	Ea.	2.36	1.26	3.62
TOTAL			4.03	6.18	10.21

D5020 232	H.I.D. Fixture, High Bay, 30' (by Type)		COST PER S.F.		
			MAT.	INST.	TOTAL
0510	High intensity discharge fixture, 30' above work plane, 100 FC				
1240	Type F, 4 fixtures per 900 S.F.		4.03	6.20	10.23
1280	6 fixtures per 1800 S.F.		3.44	5.90	9.34
1320	8 fixtures per 3000 S.F.	RD5020 -200	3.44	5.90	9.34
1360	9 fixtures per 4000 S.F.		2.83	5.60	8.43
1400	10 fixtures per 5000 S.F.	RD5020 -240	2.83	5.60	8.43
1440	17 fixtures per 8000 S.F.		2.83	5.60	8.43
1480	18 fixtures per 10,000 S.F.		2.69	5.15	7.84
1520	27 fixtures per 16,000 S.F.		2.69	5.15	7.84
1560	52 fixtures per 32000 S.F.		2.67	5.10	7.77
1600	Type G, 4 fixtures per 900 S.F.		4.35	6.55	10.90
1640	6 fixtures per 1800 S.F.		3.61	5.95	9.56
1680	9 fixtures per 3000 S.F.		3.55	5.75	9.30
1720	11 fixtures per 4000 S.F.		3.49	5.55	9.04
1760	13 fixtures per 5000 S.F.		3.49	5.55	9.04
1800	21 fixtures per 8000 S.F.		3.49	5.55	9.04
1840	23 fixtures per 10,000 S.F.		3.04	5.90	8.94
1880	36 fixtures per 16,000 S.F.		3.04	5.90	8.94
1920	70 fixtures per 32,000 S.F.		3.04	5.90	8.94

D50 Electrical

D5020 Lighting and Branch Wiring

HIGH BAY FIXTURES

B. Metal halide 400 watt

C. High pressure sodium 400 watt

E. Metal halide 1000 watt

F. High pressure sodium 1000 watt

G. Metal halide 1000 watt 125,000 lumen lamp

System Components	QUANTITY	UNIT	COST PER S.F.		
			MAT.	INST.	TOTAL
SYSTEM D5020 234 0240					
HIGH INTENSITY DISCHARGE FIXTURE, 30' ABOVE WORK PLANE					
1 WATT/S.F., TYPE E, 37 FC, 1 FIXTURE/1000 S.F.					
Steel intermediate conduit, (IMC) 1/2" diam	.196	L.F.	.39	1.23	1.62
Wire, 600V type THWN-THHN, copper, solid, #10	.006	C.L.F.	.13	.38	.51
Steel outlet box 4" concrete	.001	Ea.	.02	.03	.05
Steel outlet box plate with stud, 4" concrete	.001	Ea.	.01	.01	.02
Metal halide, hi bay, aluminum reflector, 1000 W lamp	.001	Ea.	.64	.32	.96
TOTAL			1.19	1.97	3.16

D5020 234	H.I.D. Fixture, High Bay, 30' (by Wattage)		COST PER S.F.		
			MAT.	INST.	TOTAL
0190	High intensity discharge fixture, 30' above work plane				
0240	1 watt/S.F., type E, 37 FC, 1 fixture/1000 S.F.		1.19	1.97	3.16
0280	Type G, 45 FC, 1 fixture/1000 S.F.		1.19	1.97	3.16
0320	Type F, 50 FC, 1 fixture/1000 S.F.	RD5020 -200	1.01	1.54	2.55
0440	2 watt/S.F., type E, 74 FC, 2 fixtures/1000 S.F.		2.38	3.91	6.29
0480	Type G, 92 FC, 2 fixtures/1000 S.F.	RD5020 -240	2.38	3.91	6.29
0520	Type F, 100 FC, 2 fixtures/1000 S.F.		2.02	3.12	5.14
0640	3 watt/S.F., type E, 110 FC, 3 fixtures/1000 S.F.		3.61	5.95	9.56
0680	Type G, 138 FC, 3 fixtures/1000 S.F.		3.61	5.95	9.56
0720	Type F, 150 FC, 3 fixtures/1000 S.F.		3.04	4.65	7.69
0840	4 watt/S.F., type E, 148 FC, 4 fixtures/1000 S.F.		4.79	7.90	12.69
0880	Type G, 185 FC, 4 fixtures/1000 S.F.		4.79	7.90	12.69
0920	Type F, 200 FC, 4 fixtures/1000 S.F.		4.05	6.25	10.30
1040	5 watt/S.F., type E, 185 FC, 5 fixtures/1000 S.F.		6	9.85	15.85
1080	Type G, 230 FC, 5 fixtures/1000 S.F.		6	9.85	15.85
1120	Type F, 250 FC, 5 fixtures/1000 S.F.		5.05	7.80	12.85

D50 Electrical

D5020 Lighting and Branch Wiring

LOW BAY FIXTURES

J. Metal halide 250 watt

K. High pressure sodium 150 watt

System Components	QUANTITY	UNIT	COST PER S.F.		
			MAT.	INST.	TOTAL
SYSTEM D5020 236 0920					
HIGH INTENSITY DISCHARGE FIXTURE, 8'-10' ABOVE WORK PLANE, 50 FC					
TYPE J, 13 FIXTURES PER 1800 S.F.					
Steel intermediate conduit, (IMC) 1/2" diam	.550	L.F.	1.10	3.44	4.54
Wire, 600V, type THWN-THHN, copper, solid, #10	.012	C.L.F.	.26	.75	1.01
Steel outlet box 4" concrete	.007	Ea.	.13	.22	.35
Steel outlet box plate with stud, 4" concrete	.007	Ea.	.09	.05	.14
Metal halide, lo bay, aluminum reflector, 250 W DX lamp	.007	Ea.	2.73	1.37	4.10
TOTAL			4.31	5.83	10.14

D5020 236	H.I.D. Fixture, Low Bay, 8'-10' (by Type)		COST PER S.F.		
			MAT.	INST.	TOTAL
0510	High intensity discharge fixture, 8'-10' above work plane, 50 FC				
0880	Type J, 7 fixtures per 900 S.F.		4.67	5.90	10.57
0920	13 fixtures per 1800 S.F.	RD5020	4.31	5.85	10.16
0960	21 fixtures per 3000 S.F.	-200	4.31	5.85	10.16
1000	28 fixtures per 4000 S.F.		4.31	5.85	10.16
1040	35 fixtures per 5000 S.F.		4.31	5.85	10.16
1120	62 fixtures per 10,000 S.F.		3.88	5.60	9.48
1160	99 fixtures per 16,000 S.F.		3.88	5.60	9.48
1200	199 fixtures per 32,000 S.F.		3.88	5.60	9.48
1240	Type K, 9 fixtures per 900 S.F.		4.73	5.05	9.78
1280	16 fixtures per 1800 S.F.		4.37	4.86	9.23
1320	26 fixtures per 3000 S.F.		4.35	4.80	9.15
1360	31 fixtures per 4000 S.F.		4	4.69	8.69
1400	39 fixtures per 5000 S.F.		4	4.69	8.69
1440	62 fixtures per 8000 S.F.		4	4.69	8.69
1480	78 fixtures per 10,000 S.F.		4	4.69	8.69
1520	124 fixtures per 16,000 S.F.		3.96	4.57	8.53
1560	248 fixtures per 32,000 S.F.		3.82	5.10	8.92

D50 Electrical

D5020 Lighting and Branch Wiring

LOW BAY FIXTURES

J. Metal halide 250 watt

K. High pressure sodium 150 watt

System Components	QUANTITY	UNIT	COST PER S.F.		
			MAT.	INST.	TOTAL
SYSTEM D5020 238 0240					
HIGH INTENSITY DISCHARGE FIXTURE, 8'-10' ABOVE WORK PLANE					
1 WATT/S.F., TYPE J, 30 FC, 4 FIXTURES/1000 S.F.					
Steel intermediate conduit, (IMC) 1/2" diam	.280	L.F.	.56	1.75	2.31
Wire, 600V, type THWN-THHN, copper, solid, #10	.008	C.L.F.	.17	.50	.67
Steel outlet box 4" concrete	.004	Ea.	.07	.13	.20
Steel outlet box plate with stud, 4" concrete	.004	Ea.	.05	.03	.08
Metal halide, lo bay, aluminum reflector, 250 W DX lamp	.004	Ea.	1.56	.78	2.34
TOTAL			2.41	3.19	5.60

D5020 238	H.I.D. Fixture, Low Bay, 8'-10' (by Wattage)	COST PER S.F.		
		MAT.	INST.	TOTAL
0190	High intensity discharge fixture, 8'-10' above work plane			
0240	1 watt/S.F., type J, 30 FC, 4 fixtures/1000 S.F.	2.41	3.19	5.60
0280	Type K, 29 FC, 5 fixtures/1000 S.F.	2.48	2.87	5.35
0400	2 watt/S.F., type J, 52 FC, 7 fixtures/1000 S.F.	4.31	5.85	10.16
0440	Type K, 63 FC, 11 fixtures/1000 S.F.	5.40	6.05	11.45
0560	3 watt/S.F., type J, 81 FC, 11 fixtures/1000 S.F.	6.65	8.85	15.50
0600	Type K, 92 FC, 16 fixtures/1000 S.F.	7.85	8.90	16.75
0720	4 watt/S.F., type J, 103 FC, 14 fixtures/1000 S.F.	8.60	11.65	20.25
0760	Type K, 127 FC, 22 fixtures/1000 S.F.	10.75	12.05	22.80
0880	5 watt/S.F., type J, 133 FC, 18 fixtures/1000 S.F.	11	14.70	25.70
0920	Type K, 155 FC, 27 fixtures/1000 S.F.	13.20	14.90	28.10

RD5020-200

D50 Electrical

D5020 Lighting and Branch Wiring

LOW BAY FIXTURES

J. Metal halide 250 watt

K. High pressure sodium 150 watt

System Components	QUANTITY	UNIT	COST PER S.F.		
			MAT.	INST.	TOTAL
SYSTEM D5020 240 0880					
HIGH INTENSITY DISCHARGE FIXTURE, 16' ABOVE WORK PLANE, 50 FC					
TYPE J, 9 FIXTURES PER 900 S.F.					
Steel intermediate conduit, (IMC) 1/2" diam	.630	L.F.	1.26	3.94	5.20
Wire, 600V type, THWN-THHN, copper, solid, #10	.012	C.L.F.	.26	.75	1.01
Steel outlet box 4" concrete	.010	Ea.	.19	.32	.51
Steel outlet box plate with stud, 4" concrete	.010	Ea.	.12	.08	.20
Metal halide, lo bay, aluminum reflector, 250 W DX lamp	.010	Ea.	3.90	1.96	5.86
TOTAL			5.73	7.05	12.78

D5020 240	H.I.D. Fixture, Low Bay, 16' (by Type)	COST PER S.F.		
		MAT.	INST.	TOTAL
0510	High intensity discharge fixture, 16' above work plane, 50 FC			
0880	Type J, 9 fixtures per 900 S.F.	5.75	7.05	12.80
0920	14 fixtures per 1800 S.F.	4.91	6.65	11.56
0960	24 fixtures per 3000 S.F.	4.95	6.75	11.70
1000	32 fixtures per 4000 S.F.	4.95	6.75	11.70
1040	35 fixtures per 5000 S.F.	4.55	6.60	11.15
1080	56 fixtures per 8000 S.F.	4.55	6.60	11.15
1120	70 fixtures per 10,000 S.F.	4.55	6.60	11.15
1160	111 fixtures per 16,000 S.F.	4.55	6.60	11.15
1200	222 fixtures per 32,000 S.F.	4.55	6.60	11.15
1240	Type K, 11 fixtures per 900 S.F.	5.85	6.60	12.45
1280	20 fixtures per 1800 S.F.	5.40	6.10	11.50
1320	29 fixtures per 3000 S.F.	5.05	6	11.05
1360	39 fixtures per 4000 S.F.	5	5.95	10.95
1400	44 fixtures per 5000 S.F.	4.68	5.80	10.48
1440	62 fixtures per 8000 S.F.	4.68	5.80	10.48
1480	87 fixtures per 10,000 S.F.	4.68	5.80	10.48
1520	138 fixtures per 16,000 S.F.	4.68	5.80	10.48

D50 Electrical

D5020 Lighting and Branch Wiring

LOW BAY FIXTURES

J. Metal halide 250 watt

K. High pressure sodium 150 watt

System Components	QUANTITY	UNIT	COST PER S.F.		
			MAT.	INST.	TOTAL
SYSTEM D5020 242 0240 HIGH INTENSITY DISCHARGE FIXTURE, 16' ABOVE WORK PLANE 1 WATT/S.F., TYPE J, 28 FC, 4 FIXTURES/1000 S.F.					
Steel intermediate conduit, (IMC) 1/2" diam	.328	L.F.	.66	2.05	2.71
Wire, 600V, type THWN-THHN, copper, solid, #10	.010	C.L.F.	.22	.63	.85
Steel outlet box 4" concrete	.004	Ea.	.07	.13	.20
Steel outlet box plate with stud, 4" concrete	.004	Ea.	.05	.03	.08
Metal halide, lo bay, aluminum reflector, 250 W DX lamp	.004	Ea.	1.56	.78	2.34
TOTAL			2.56	3.62	6.18

D5020 242	H.I.D. Fixture, Low Bay, 16' (by Wattage)	COST PER S.F.		
		MAT.	INST.	TOTAL
0190	High intensity discharge fixture, mounted 16' above work plane			
0240	1 watt/S.F., type J, 28 FC, 4 fixt./1000 S.F.	2.56	3.62	6.18
0280	Type K, 27 FC, 5 fixt./1000 S.F.	2.87	4.06	6.93
0400	2 watt/S.F., type J, 48 FC, 7 fixt/1000 S.F.	4.69	7	11.69
0440	Type K, 58 FC, 11 fixt/1000 S.F.	6.10	8.30	14.40
0560	3 watt/S.F., type J, 75 FC, 11 fixt/1000 S.F.	7.25	10.65	17.90
0600	Type K, 85 FC, 16 fixt/1000 S.F.	8.95	12.35	21.30
0720	4 watt/S.F., type J, 95 FC, 14 fixt/1000 S.F.	9.35	14	23.35
0760	Type K, 117 FC, 22 fixt/1000 S.F.	12.20	16.55	28.75
0880	5 watt/S.F., type J, 122 FC, 18 fixt/1000 S.F.	11.95	17.60	29.55
0920	Type K, 143 FC, 27 fixt/1000 S.F.	15.05	20.50	35.55

RD5020-200

D50 Electrical

D5030 Communications and Security

Description: System below includes telephone fitting installed. Does not include cable.

When poke thru fittings and telepoles are used for power, they can also be used for telephones at a negligible additional cost.

System Components	QUANTITY	UNIT	COST PER S.F.		
			MAT.	INST.	TOTAL
SYSTEM D5030 310 0200					
TELEPHONE SYSTEMS, UNDERFLOOR DUCT, 5' ON CENTER, LOW DENSITY					
Underfloor duct 7-1/4" w/insert 2' O.C. 1-3/8" x 7-1/4" super duct	.190	L.F.	6.84	2.38	9.22
Vertical elbow for underfloor superduct, 7-1/4", included					
Underfloor duct conduit adapter, 2" x 1-1/4", included					
Underfloor duct junction box, single duct, 7-1/4" x 3 1/8"	.003	Ea.	1.61	.47	2.08
Underfloor junction box carpet pan	.003	Ea.	1.14	.02	1.16
Underfloor duct outlet, low tension	.004	Ea.	.43	.31	.74
TOTAL			10.02	3.18	13.20

D5030 310		Telephone Systems	COST PER S.F.		
			MAT.	INST.	TOTAL
0200	Telephone systems, underfloor duct, 5' on center, low density		10	3.18	13.18
0240		5' on center, high density	10.45	3.50	13.95
0280		7' on center, low density	8	2.65	10.65
0320		7' on center, high density	8.45	2.97	11.42
0400		Poke thru fittings, low density	1.12	1.13	2.25
0440		High density	2.24	2.25	4.49
0520	Telepoles, low density		1.15	.68	1.83
0560		High density	2.30	1.36	3.66
0640	Conduit system with floor boxes, low density		1.55	1.11	2.66
0680		High density	3.10	2.19	5.29
1020	Telephone wiring for offices & laboratories, 8 jacks/M.S.F.		.44	1.74	2.18

D50 Electrical

D5030 Communications and Security

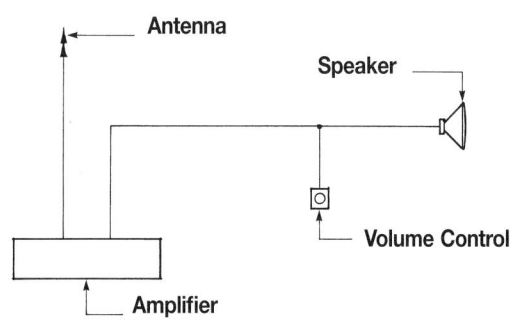

Sound System Includes AM–FM antenna, outlets, rigid conduit, and copper wire.
Fire Detection System Includes pull stations, signals, smoke and heat detectors, rigid conduit, and copper wire.
Intercom System Includes master and remote stations, rigid conduit, and copper wire.
Master Clock System Includes clocks, bells, rigid conduit, and copper wire.
Master TV Antenna Includes antenna, VHF–UHF reception and distribution, rigid conduit, and copper wire.

System Components	QUANTITY	UNIT	COST EACH MAT.	INST.	TOTAL
SYSTEM D5030 910 0220					
SOUND SYSTEM, INCLUDES OUTLETS, BOXES, CONDUIT & WIRE					
Steel intermediate conduit, (IMC) 1/2" diam	1200.000	L.F.	2,400	7,500	9,900
Wire sound shielded w/drain, #22-2 conductor	15.500	C.L.F.	306.90	1,216.75	1,523.65
Sound system speakers ceiling or wall	12.000	Ea.	1,524	942	2,466
Sound system volume control	12.000	Ea.	1,146	942	2,088
Sound system amplifier, 250 Watts	1.000	Ea.	1,350	625	1,975
Sound system antenna, AM FM	1.000	Ea.	147	157	304
Sound system monitor panel	1.000	Ea.	420	157	577
Sound system cabinet	1.000	Ea.	915	625	1,540
Steel outlet box 4" square	12.000	Ea.	32.52	378	410.52
Steel outlet box 4" plaster rings	12.000	Ea.	43.80	117.60	161.40
TOTAL			8,285.22	12,660.35	20,945.57

D5030 910	Communication & Alarm Systems		MAT.	INST.	TOTAL
0200	Communication & alarm systems, includes outlets, boxes, conduit & wire				
0210	Sound system, 6 outlets		5,750	7,925	13,675
0220	12 outlets	RD5010 -110	8,275	12,700	20,975
0240	30 outlets		14,800	23,900	38,700
0280	100 outlets		45,200	80,000	125,200
0320	Fire detection systems, non-addressable, 12 detectors		3,300	6,575	9,875
0360	25 detectors		5,775	11,100	16,875
0400	50 detectors		11,400	22,000	33,400
0440	100 detectors		21,200	40,000	61,200
0450	Addressable type, 12 detectors		4,950	6,625	11,575
0452	25 detectors		8,975	11,200	20,175
0454	50 detectors		17,300	22,000	39,300
0456	100 detectors		33,600	40,500	74,100
0458	Fire alarm control panel, 8 zone, excluding wire and conduit		845	1,250	2,095
0459	12 zone		2,575	1,875	4,450
0460	Fire alarm command center, addressable without voice, excl. wire & conduit		4,825	1,100	5,925
0462	Addressable with voice		10,200	1,725	11,925
0480	Intercom systems, 6 stations		3,950	5,400	9,350
0520	12 stations		7,600	10,800	18,400
0560	25 stations		12,700	20,700	33,400
0600	50 stations		24,700	39,000	63,700
0640	100 stations		48,700	76,000	124,700
0680	Master clock systems, 6 rooms		5,125	8,850	13,975
0720	12 rooms		8,050	15,100	23,150
0760	20 rooms		11,200	21,300	32,500
0800	30 rooms		18,300	39,300	57,600

D50 Electrical

D5030 Communications and Security

D5030 910	Communication & Alarm Systems	COST EACH		
		MAT.	INST.	TOTAL
0840	50 rooms	29,700	66,500	96,200
0880	100 rooms	57,500	131,500	189,000
0920	Master TV antenna systems, 6 outlets	2,625	5,600	8,225
0960	12 outlets	4,900	10,400	15,300
1000	30 outlets	11,400	24,100	35,500
1040	100 outlets	37,900	78,500	116,400

D5030 920	Data Communication	COST PER M.S.F.		
		MAT.	INST.	TOTAL
0100	Data communication system, incl data/voice outlets, boxes, conduit, & cable			
0102	Data and voice system, 2 data/voice outlets per 1000 S.F.	179	465	644
0104	4 data/voice outlets per 1000 S.F.	340	915	1,255
0106	6 data/voice outlets per 1000 S.F.	480	1,350	1,830
0110	8 data/voice outlets per 1000 S.F.	615	1,775	2,390

D50 Electrical

D5090 Other Electrical Systems

Description: System below tabulates the installed cost for generators by kW. Included in costs are battery, charger, muffler, and transfer switch.

No conduit, wire, or terminations included.

System Components	QUANTITY	UNIT	COST PER kW		
			MAT.	INST.	TOTAL
SYSTEM D5090 210 0200					
GENERATOR SET, INCL. BATTERY, CHARGER, MUFFLER & TRANSFER SWITCH					
GAS/GASOLINE OPER., 3 PHASE, 4 WIRE, 277/480V, 7.5 kW					
Generator set, gas or gasoline operated, 3 ph 4 W, 277/480 V, 7.5 kW	.133	Ea.	1,256.67	274.27	1,530.94
TOTAL			1,256.67	274.27	1,530.94

D5090 210	Generators (by kW)	COST PER kW		
		MAT.	INST.	TOTAL
0190	Generator sets, include battery, charger, muffler & transfer switch			
0200	Gas/gasoline operated, 3 phase, 4 wire, 277/480 volt, 7.5 kW	1,250	275	1,525
0240	11.5 kW	1,150	209	1,359
0280	20 kW	785	136	921
0320	35 kW	535	88.50	623.50
0360	80 kW	385	53.50	438.50
0400	100 kW	335	52	387
0440	125 kW	550	49	599
0480	185 kW	490	37	527
0560	Diesel engine with fuel tank, 30 kW	705	103	808
0600	50 kW	505	81	586
0720	125 kW	305	47	352
0760	150 kW	293	44	337
0800	175 kW	273	39	312
0840	200 kW	247	35.50	282.50
0880	250 kW	232	29.50	261.50
0920	300 kW	208	26	234
0960	350 kW	203	24	227
1000	400 kW	219	22.50	241.50
1040	500 kW	221	19	240
1200	750 kW	242	11.85	253.85
1400	1000 kW	224	12.55	236.55

Note on row 0280: RD5010-110

D50 Electrical

D5090 Special Construction

PV power system, stand alone, AC and DC loads

System Components	QUANTITY	UNIT	COST EACH		
			MAT.	INST.	TOTAL
SYSTEM D5090 420 0100					
PHOTOVOLTAIC POWER SYSTEM, STAND ALONE					
Alternative energy sources, photovoltaic module, 75 watt, 17 volts	12.000	Ea.	5,100	942	6,042
DC to AC inverter for, 48V, 5,500 watt	1.000	Ea.	4,075	315	4,390
PV components, combiner box, 10 lug, NEMA 3R enclosure	1.000	Ea.	206	157	363
Alternative energy sources, PV components, fuse, 15 A for combiner box	10.000	Ea.	156.50	157	313.50
Battery charger controller w/temperature sensor	1.000	Ea.	530	157	687
Digital readout panel, displays hours, volts, amps, etc.	1.000	Ea.	140	157	297
Deep cycle solar battery, 6 V, 180 Ah (C/20)	4.000	Ea.	1,012	314	1,326
Battery interconnection, 15" AWG #2/0, sealed w/copper ring lugs	3.000	Ea.	41.55	117	158.55
Battery interconn, 24" AWG #2/0, sealed w/copper ring lugs	2.000	Ea.	39.60	78	117.60
Battery interconn, 60" AWG #2/0, sealed w/copper ring lugs	2.000	Ea.	94	78	172
Batt temp computer probe, RJ11 jack, 15' cord	1.000	Ea.	21	39	60
System disconnect, DC 175 amp circuit breaker	1.000	Ea.	151	78.50	229.50
Conduit box for inverter	1.000	Ea.	47.50	78.50	126
Low voltage disconnect	1.000	Ea.	43	78.50	121.50
Vented battery enclosure, wood	1.000	Ea.	188	277	465
Grounding, rod, copper clad, 8' long, 5/8" diameter	1.000	Ea.	21	114	135
Grounding, clamp, bronze, 5/8" dia	1.000	Ea.	6.95	19.60	26.55
Bare copper wire, stranded, #8	1.000	C.L.F.	35.50	57	92.50
Wire, 600 volt, type THW, copper, stranded, #12	3.600	C.L.F.	56.52	205.20	261.72
Wire, 600 volt, type THW, copper, stranded, #8	1.050	C.L.F.	43.05	82.43	125.48
Rigid galvanized steel conduit, 3/4", including fittings	120.000	L.F.	355.20	942	1,297.20
Conduit,to 15' H,incl 2 termn,2 elb&11 bm clp per 100',(EMT), 1" dia	30.000	L.F.	55.80	163.50	219.30
Lightning surge suppressor	1.000	Ea.	55	19.60	74.60
General duty 240 volt, 2 pole, nonfusible, NEMA 3R, 60 amp	1.000	Ea.	298	273	571
Load centers, 3 wire, 120/240V, 100 amp main lugs, indoor, 8 circuits	2.000	Ea.	360	900	1,260
Plug-in panel or load center, 120/240 volt, to 60 amp, 1 pole	5.000	Ea.	59.75	262.50	322.25
Roof mounting frame, metal, for 6 modules	2.000	Ea.	1,930	258	2,188
Fuses, dual element, time delay, 250 volt, 50 amp	2.000	Ea.	22.20	25.10	47.30
TOTAL			15,144.12	6,345.43	21,489.55

D5090 420	Photovoltaic System, EACH	COST EACH		
		MAT.	INST.	TOTAL
0100	Photovoltaic power system, stand alone, AC and DC loads	15,100	6,350	21,450

D50 Electrical

D5090 Special Construction

Photovoltaic Modules, Photovoltaic power system, grid connected 10 kW, Combiner Box, Utility Connection, Isolation Transformer, AC Disconnect, Inverter, DC Disconnect

System Components	QUANTITY	UNIT	COST EACH MAT.	COST EACH INST.	COST EACH TOTAL
SYSTEM D5090 430 0100					
PHOTOVOLTAIC POWER SYSTEM, GRID CONNECTED 10 kW					
PV modules, 167 watt, 23.5 V	60.000	Ea.	42,300	4,710	47,010
Mounting frame 6 modules	10.000	Ea.	9,650	1,290	10,940
Steel angle support	400.000	L.F.	192	5,816	6,008
# 12 wire	4.000	C.L.F.	126	228	354
AC Disconnect switch, 60 A	1.000	Ea.	505	285	790
Fuse 75 A	3.000	Ea.	87	47.10	134.10
4/0 wire	2.000	C.L.F.	930	440	1,370
Module connection 4/0	4.000	Ea.	40.20	358	398.20
Combiner box 10 lug, NEMA 3R	1.000	Ea.	206	157	363
Utility connection, 3 pole breaker	1.000	Ea.	243	101	344
15 amp fuses	10.000	Ea.	156.50	157	313.50
Enclosure 24" x 24" x 10", NEMA 4	1.000	Ea.	5,150	1,250	6,400
Inverter 12 kW	1.000	Ea.	10,725	945	11,670
Conduit w/fittings & support	200.000	L.F.	372	1,090	1,462
# 6 ground wire	8.000	C.L.F.	556	772	1,328
Fuse, 60 A	3.000	Ea.	38.40	37.65	76.05
DC disconnect switch, 75 A	1.000	Ea.	940	350	1,290
10 kVA isolation tranformer	1.000	Ea.	1,850	785	2,635
# 6 ground wire	.400	C.L.F.	27.80	38.60	66.40
# 6 ground connection	2.000	Ea.	34.80	105	139.80
TOTAL			74,129.70	18,962.35	93,092.05

D5090 430	Photovoltaic System Grid Connected, AC and DC Loads	COST EACH MAT.	COST EACH INST.	COST EACH TOTAL
0100	Photovoltaic power system, grid connected 10 kW	74,000	19,000	93,000
1100	Photovoltaic power system, grid connected 20 kW	148,500	37,900	186,400

D50 Electrical

D5090 Other Electrical Systems

Low Density and Medium Density Baseboard Radiation

The costs shown in Table below are based on the following system considerations:

1. The heat loss per square foot is based on approximately 34 BTU/hr. per S.F. of floor or 10 watts per S.F. of floor.
2. Baseboard radiation is based on the low watt density type rated 187 watts per L.F. and the medium density type rated 250 watts per L.F.
3. Thermostat is not included.
4. Wiring costs include branch circuit wiring.

System Components	QUANTITY	UNIT	COST PER S.F.		
			MAT.	INST.	TOTAL
SYSTEM D5090 510 1000					
ELECTRIC BASEBOARD RADIATION, LOW DENSITY, 900 S.F., 31 MBH, 9 kW					
Electric baseboard radiator, 5' long 935 watt	.011	Ea.	.62	1.21	1.83
Steel intermediate conduit, (IMC) 1/2" diam	.170	L.F.	.34	1.06	1.40
Wire 600 volt, type THW, copper, solid, #12	.005	C.L.F.	.07	.29	.36
TOTAL			1.03	2.56	3.59

D5090 510	Electric Baseboard Radiation (Low Density)	COST PER S.F.		
		MAT.	INST.	TOTAL
1000	Electric baseboard radiation, low density, 900 S.F., 31 MBH, 9 kW	1.03	2.56	3.59
1200	1500 S.F., 51 MBH, 15 kW	1.01	2.50	3.51
1400	2100 S.F., 72 MBH, 21 kW	.90	2.24	3.14
1600	3000 S.F., 102 MBH, 30 kW	.81	2.04	2.85
2000	Medium density, 900 S.F., 31 MBH, 9 kW	.91	2.34	3.25
2200	1500 S.F., 51 MBH, 15 kW	.89	2.28	3.17
2400	2100 S.F., 72 MBH, 21 kW	.84	2.13	2.97
2600	3000 S.F., 102 MBH, 30 kW	.76	1.93	2.69

D50 Electrical

D5090 Other Electrical Systems

Commercial Duty Baseboard Radiation
The costs shown in Table below are based on the following system considerations:

1. The heat loss per square foot is based on approximately 41 BTU/hr. per S.F. of floor or 12 watts per S.F.
2. The baseboard radiation is of the commercial duty type rated 250 watts per L.F. served by 277 volt, single phase power.
3. Thermostat is not included.
4. Wiring costs include branch circuit wiring.

System Components	QUANTITY	UNIT	COST PER S.F.		
			MAT.	INST.	TOTAL
SYSTEM D5090 520 1000					
ELECTRIC BASEBOARD RADIATION, MEDIUM DENSITY, 1230 S.F., 51 MBH, 15 kW					
Electric baseboard radiator, 5' long	.013	Ea.	.73	1.43	2.16
Steel intermediate conduit, (IMC) 1/2" diam	.154	L.F.	.31	.96	1.27
Wire 600 volt, type THW, copper, solid, #12	.004	C.L.F.	.05	.23	.28
TOTAL			1.09	2.62	3.71

D5090 520	Electric Baseboard Radiation (Medium Density)	COST PER S.F.		
		MAT.	INST.	TOTAL
1000	Electric baseboard radiation, medium density, 1230 SF, 51 MBH, 15 kW	1.09	2.62	3.71
1200	2500 S.F. floor area, 106 MBH, 31 kW	1.01	2.46	3.47
1400	3700 S.F. floor area, 157 MBH, 46 kW	.99	2.39	3.38
1600	4800 S.F. floor area, 201 MBH, 59 kW	.92	2.21	3.13
1800	11,300 S.F. floor area, 464 MBH, 136 kW	.87	2.02	2.89
2000	30,000 S.F. floor area, 1229 MBH, 360 kW	.85	1.97	2.82

E Equipment & Furnishings

No part of this publication may be reproduced, stored in a retrieval system, or transmitted in any form or by any means without prior written permission of RSMeans, a division of Reed Construction Data.

E10 Equipment

E1010 Commercial Equipment

E1010 110	Security/Vault, EACH	COST EACH		
		MAT.	INST.	TOTAL
0100	Bank equipment, drive up window, drawer & mike, no glazing, economy	7,225	1,425	8,650
0110	Deluxe	9,775	2,850	12,625
0120	Night depository, economy	7,975	1,425	9,400
0130	Deluxe	11,300	2,850	14,150
0140	Pneumatic tube systems, 2 station, standard	27,100	4,775	31,875
0150	Teller, automated, 24 hour, single unit	47,500	4,775	52,275
0160	Teller window, bullet proof glazing, 44" x 60"	4,050	875	4,925
0170	Pass through, painted steel, 72" x 40"	4,425	1,775	6,200
0300	Safe, office type, 1 hr. rating, 34" x 20" x 20"	2,500		2,500
0310	4 hr. rating, 62" x 33" x 20"	9,475		9,475
0320	Data storage, 4 hr. rating, 63" x 44" x 16"	14,700		14,700
0330	Jewelers, 63" x 44" x 16"	15,700		15,700
0340	Money, "B" label, 9" x 14" x 14"	580		580
0350	Tool and torch resistive, 24" x 24" x 20"	8,375	218	8,593
0500	Security gates-scissors type, painted steel, single, 6' high, 5-1/2' wide	233	355	588
0510	Double gate, 7-1/2' high, 14' wide	710	715	1,425

E1010 510	Mercantile Equipment, EACH	COST EACH		
		MAT.	INST.	TOTAL
0015	Barber equipment, chair, hydraulic, economy	535	23	558
0020	Deluxe	3,425	34.50	3,459.50
0030	Console, including mirrors, deluxe	475	121	596
0040	Sink, hair washing basin	490	84	574
0100	Checkout counter, single belt	3,275	87.50	3,362.50
0110	Double belt, power take-away	4,725	97	4,822
0200	Display cases, freestanding, glass and aluminum, 3'-6" x 3' x 1'-0" deep	1,325	138	1,463
0210	5'-10" x 4' x 1'-6" deep	4,050	184	4,234
0220	Wall mounted, glass and aluminum, 3' x 4' x 1'-4" deep	2,350	221	2,571
0230	16' x 4' x 1'-4" deep	4,300	740	5,040
0240	Table exhibit, flat, 3' x 4' x 2'-0" wide	1,450	221	1,671
0250	Sloping, 3' x 4' x 3'-0" wide	830	370	1,200
0300	Refrigerated food cases, dairy, multi-deck, rear sliding doors, 6 ft. long	14,200	410	14,610
0310	Delicatessen case, service type, 12 ft. long	8,200	315	8,515
0320	Frozen food, chest type, 12 ft. long	8,775	370	9,145
0330	Glass door, reach-in, 5 door	15,300	410	15,710
0340	Meat case, 12 ft. long, single deck	7,925	370	8,295
0350	Multi-deck	11,300	395	11,695
0360	Produce case, 12 ft. long, single deck	7,450	370	7,820
0370	Multi-deck	9,475	395	9,870

E1010 610	Laundry/Dry Cleaning, EACH	COST EACH		
		MAT.	INST.	TOTAL
0100	Laundry equipment, dryers, gas fired, residential, 16 lb. capacity	720	224	944
0110	Commercial, 30 lb. capacity, single	3,525	224	3,749
0120	Dry cleaners, electric, 20 lb. capacity	36,600	6,500	43,100
0130	30 lb. capacity	55,000	8,675	63,675
0140	Ironers, commercial, 120" with canopy, 8 roll	192,000	18,600	210,600
0150	Institutional, 110", single roll	34,300	3,125	37,425
0160	Washers, residential, 4 cycle	945	224	1,169
0170	Commercial, coin operated, deluxe	3,650	224	3,874

E10 Equipment

E1020 Institutional Equipment

E1020 110 Ecclesiastical Equipment, EACH

		MAT.	INST.	TOTAL
0090	Church equipment, altar, wood, custom, plain	2,525	395	2,920
0100	Granite, custom, deluxe	41,300	5,025	46,325
0110	Baptistry, fiberglass, economy	5,650	1,450	7,100
0120	Bells & carillons, keyboard operation	21,200	8,500	29,700
0130	Confessional, wood, single, economy	3,450	920	4,370
0140	Double, deluxe	20,600	2,775	23,375
0150	Steeples, translucent fiberglas, 30" square, 15' high	8,275	1,750	10,025
0160	Porcelain enamel, custom, 60' high	27,000	11,700	38,700

E1020 130 Ecclesiastical Equipment, L.F.

		MAT.	INST.	TOTAL
0100	Arch. equip., church equip. pews, bench type, hardwood, economy	98	27.50	125.50
0110	Deluxe	194	37	231

E1020 210 Library Equipment, EACH

		MAT.	INST.	TOTAL
0110	Library equipment, carrels, metal, economy	280	111	391
0120	Hardwood, deluxe	1,475	138	1,613

E1020 230 Library Equipment, L.F.

		MAT.	INST.	TOTAL
0100	Library equipment, book shelf, metal, single face, 90" high x 10" shelf	127	46	173
0110	Double face, 90" high x 10" shelf	320	100	420
0120	Charging desk, built-in, with counter, plastic laminate	330	79	409

E1020 310 Theater and Stage Equipment, EACH

		MAT.	INST.	TOTAL
0200	Movie equipment, changeover, economy	510		510
0210	Film transport, incl. platters and autowind, economy	5,525		5,525
0220	Lamphouses, incl. rectifiers, xenon, 1000W	7,300	315	7,615
0230	4000W	10,400	420	10,820
0240	Projector mechanisms, 35 mm, economy	12,200		12,200
0250	Deluxe	16,700		16,700
0260	Sound systems, incl. amplifier, single, economy	3,625	695	4,320
0270	Dual, Dolby/super sound	18,200	1,575	19,775
0280	Projection screens, wall hung, manual operation, 50 S.F., economy	325	111	436
0290	Electric operation, 100 S.F., deluxe	2,175	555	2,730
0400	Stage equipment, control boards, incl. dimmers and breakers, economy	13,700	625	14,325
0410	Deluxe	140,500	3,125	143,625
0420	Spotlight, stationary, quartz, 6" lens	144	157	301
0430	Follow, incl. transformer, 2100W	3,375	157	3,532

E1020 320 Theater and Stage Equipment, S.F.

		MAT.	INST.	TOTAL
0090	Movie equipment, projection screens, rigid in wall, acrylic, 1/4" thick	46	5.40	51.40
0100	1/2" thick	53.50	8.05	61.55
0110	Stage equipment, curtains, velour, medium weight	8.70	1.84	10.54
0120	Silica based yarn, fireproof	16.70	22	38.70
0130	Stages, portable with steps, folding legs, 8" high	35.50		35.50
0140	Telescoping platforms, aluminum, deluxe	52	28.50	80.50

E1020 330 Theater and Stage Equipment, L.F.

		MAT.	INST.	TOTAL
0100	Stage equipment, curtain track, heavy duty	67.50	61.50	129
0110	Lights, border, quartz, colored	190	31.50	221.50

E10 Equipment

E1020 Institutional Equipment

E1020 610	Detention Equipment, EACH	COST EACH		
		MAT.	INST.	TOTAL
0110	Detention equipment, cell front rolling door, 7/8" bars, 5' x 7' high	5,800	1,525	7,325
0120	Cells, prefab., including front, 5' x 7' x 7' deep	10,600	2,025	12,625
0130	Doors and frames, 3' x 7', single plate	4,975	760	5,735
0140	Double plate	6,125	760	6,885
0150	Toilet apparatus, incl wash basin	3,700	960	4,660
0160	Visitor cubicle, vision panel, no intercom	3,475	1,525	5,000

E1020 710	Laboratory Equipment, EACH	COST EACH		
		MAT.	INST.	TOTAL
0110	Laboratory equipment, glassware washer, distilled water, economy	6,800	720	7,520
0120	Deluxe	14,000	1,300	15,300
0130	Glove box, fiberglass, bacteriological	18,600		18,600
0140	Radio isotope	18,600		18,600
0150	Titration unit, 4-2000 ml. reservoirs	6,050		6,050

E1020 720	Laboratory Equipment, S.F.	COST PER S.F.		
		MAT.	INST.	TOTAL
0100	Arch. equip., lab equip., counter tops, acid proof, economy	42	13.50	55.50
0110	Stainless steel	151	13.50	164.50

E1020 730	Laboratory Equipment, L.F.	COST PER L.F.		
		MAT.	INST.	TOTAL
0110	Laboratory equipment, cabinets, wall, open	172	55.50	227.50
0120	Base, drawer units	540	61.50	601.50
0130	Fume hoods, not incl. HVAC, economy	540	205	745
0140	Deluxe incl. fixtures	1,700	460	2,160

E1020 810	Medical Equipment, EACH	COST EACH		
		MAT.	INST.	TOTAL
0100	Dental equipment, central suction system, economy	1,700	560	2,260
0110	Compressor-air, deluxe	8,525	1,150	9,675
0120	Chair, hydraulic, economy	2,400	1,150	3,550
0130	Deluxe	9,050	2,275	11,325
0140	Drill console with accessories, economy	2,125	355	2,480
0150	Deluxe	5,625	355	5,980
0160	X-ray unit, portable	2,375	143	2,518
0170	Panoramic unit	17,100	955	18,055
0300	Medical equipment, autopsy table, standard	10,100	670	10,770
0310	Deluxe	17,100	1,125	18,225
0320	Incubators, economy	3,250		3,250
0330	Deluxe	15,800		15,800
0700	Station, scrub-surgical, single, economy	4,800	224	5,024
0710	Dietary, medium, with ice	18,200		18,200
0720	Sterilizers, general purpose, single door, 20" x 20" x 28"	13,400		13,400
0730	Floor loading, double door, 28" x 67" x 52"	224,000		224,000
0740	Surgery tables, standard	13,200	1,025	14,225
0750	Deluxe	25,300	1,425	26,725
0770	Tables, standard, with base cabinets, economy	1,075	370	1,445
0780	Deluxe	6,100	555	6,655
0790	X-ray, mobile, economy	18,300		18,300
0800	Stationary, deluxe	244,000		244,000

E10 Equipment

E1030 Vehicular Equipment

E1030 110	Vehicular Service Equipment, EACH	COST EACH		
		MAT.	INST.	TOTAL
0110	Automotive equipment, compressors, electric, 1-1/2 H.P., std. controls	485	1,050	1,535
0120	5 H.P., dual controls	2,950	1,550	4,500
0130	Hoists, single post, 4 ton capacity, swivel arms	6,050	3,900	9,950
0140	Dual post, 12 ton capacity, adjustable frame	10,400	820	11,220
0150	Lube equipment, 3 reel type, with pumps	9,550	3,125	12,675
0160	Product dispenser, 6 nozzles, w/vapor recovery, not incl. piping, installed	26,900		26,900
0800	Scales, dial type, built in floor, 5 ton capacity, 8' x 6' platform	8,250	3,325	11,575
0810	10 ton capacity, 9' x 7' platform	11,800	4,750	16,550
0820	Truck (including weigh bridge), 20 ton capacity, 24' x 10'	12,600	5,525	18,125
0830	Digital type, truck, 60 ton capacity, 75' x 10'	39,700	13,800	53,500
0840	Concrete foundations, 8' x 6'	1,075	4,200	5,275
0850	70' x 10'	4,675	14,000	18,675

E1030 210	Parking Control Equipment, EACH	COST EACH		
		MAT.	INST.	TOTAL
0110	Parking equipment, automatic gates, 8 ft. arm, one way	4,625	1,150	5,775
0120	Traffic detectors, single treadle	1,975	525	2,500
0130	Booth for attendant, economy	8,175		8,175
0140	Deluxe	30,300		30,300
0150	Ticket printer/dispenser, rate computing	9,750	895	10,645
0160	Key station on pedestal	875	305	1,180
0200	Concrete filled steel bollard, 8" diameter, 8' long	970	122	1,092

E1030 310	Loading Dock Equipment, EACH	COST EACH		
		MAT.	INST.	TOTAL
0110	Dock bumpers, rubber blocks, 4-1/2" thick, 10" high, 14" long	64	21.50	85.50
0120	6" thick, 20" high, 11" long	161	42.50	203.50
0130	Dock boards, H.D., 5' x 5', aluminum, 5000 lb. capacity	1,400		1,400
0140	16,000 lb. capacity	1,775		1,775
0150	Dock levelers, hydraulic, 7' x 8', 10 ton capacity	10,200	1,500	11,700
0160	Dock lifters, platform, 6' x 6', portable, 3000 lb. capacity	10,100		10,100
0170	Dock shelters, truck, scissor arms, economy	1,825	555	2,380
0180	Deluxe	2,625	1,100	3,725

E10 Equipment

E1090 Other Equipment

E1090 110	Maintenance Equipment, EACH	COST EACH		
		MAT.	INST.	TOTAL
0110	Vacuum cleaning, central, residential, 3 valves	1,175	635	1,810
0120	7 valves	1,825	1,425	3,250

E1090 210	Solid Waste Handling Equipment, EACH	COST EACH		
		MAT.	INST.	TOTAL
0110	Waste handling, compactors, single bag, 250 lbs./hr., hand fed	16,400	650	17,050
0120	Heavy duty industrial, 5 C.Y. capacity	35,200	3,125	38,325
0130	Incinerator, electric, 100 lbs./hr., economy	70,500	3,125	73,625
0140	Gas, 2000 lbs./hr., deluxe	440,000	25,600	465,600
0150	Shredder, no baling, 35 tons/hr.	331,000		331,000
0160	Incl. baling, 50 tons/day	661,000		661,000

E1090 350	Food Service Equipment, EACH	COST EACH		
		MAT.	INST.	TOTAL
0110	Kitchen equipment, bake oven, single deck	5,550	151	5,701
0120	Broiler, without oven	4,025	151	4,176
0130	Commercial dish washer, semiautomatic, 50 racks/hr.	8,450	930	9,380
0140	Automatic, 275 racks/hr.	37,900	3,950	41,850
0150	Cooler, beverage, reach-in, 6 ft. long	3,800	202	4,002
0160	Food warmer, counter, 1.65 kw	560		560
0170	Fryers, with submerger, single	1,325	173	1,498
0180	Double	2,600	242	2,842
0185	Ice maker, 1000 lb. per day, with bin	5,375	1,200	6,575
0190	Kettles, steam jacketed, 20 gallons	8,300	265	8,565
0200	Range, restaurant type, burners, 2 ovens and 24" griddle	4,700	202	4,902
0210	Range hood, incl. carbon dioxide system, economy	5,750	405	6,155
0220	Deluxe	42,000	1,200	43,200

E1090 360	Food Service Equipment, S.F.	COST PER S.F.		
		MAT.	INST.	TOTAL
0110	Refrigerators, prefab, walk-in, 7'-6" high, 6' x 6'	170	20	190
0120	12' x 20'	107	10.10	117.10

E1090 410	Residential Equipment, EACH	COST EACH		
		MAT.	INST.	TOTAL
0110	Arch. equip., appliances, range, cook top, 4 burner, economy	305	105	410
0120	Built in, single oven 30" wide, economy	815	105	920
0130	Standing, single oven-21" wide, economy	420	87.50	507.50
0140	Double oven-30" wide, deluxe	2,775	87.50	2,862.50
0150	Compactor, residential, economy	705	111	816
0160	Deluxe	1,025	184	1,209
0170	Dish washer, built-in, 2 cycles, economy	249	325	574
0180	4 or more cycles, deluxe	1,175	650	1,825
0190	Garbage disposer, sink type, economy	94.50	130	224.50
0200	Deluxe	239	130	369
0210	Refrigerator, no frost, 10 to 12 C.F., economy	430	87.50	517.50
0220	21 to 29 C.F., deluxe	3,400	291	3,691
0300	Washing machine, automatic	1,600	670	2,270

E1090 550	Darkroom Equipment, EACH	COST EACH		
		MAT.	INST.	TOTAL
0110	Darkroom equipment, developing tanks, 2' x 4', 5" deep	6,175	605	6,780
0120	2' x 9', 10" deep	11,300	805	12,105
0130	Combination, tray & tank sinks, washers & dry tables	11,700	2,700	14,400
0140	Dryers, dehumidified filtered air, 3' x 2' x 5'-8" high	5,575	310	5,885

E10 Equipment

E1090 Other Equipment

E1090 550 Darkroom Equipment, EACH

		COST EACH		
		MAT.	INST.	TOTAL
0150	Processors, manual, 16" x 20" print size	10,400	930	11,330
0160	Automatic, color print, deluxe	27,200	3,100	30,300
0170	Washers, round, 11" x 14" sheet size	3,750	605	4,355
0180	Square, 50" x 56" sheet size	5,325	1,525	6,850

E1090 610 School Equipment, EACH

		COST EACH		
		MAT.	INST.	TOTAL
0110	School equipment, basketball backstops, wall mounted, wood, fixed	1,425	970	2,395
0120	Suspended type, electrically operated	6,475	1,875	8,350
0130	Bleachers-telescoping, manual operation, 15 tier, economy (per seat)	99.50	34.50	134
0140	Power operation, 30 tier, deluxe (per seat)	380	61	441
0150	Weight lifting gym, universal, economy	550	875	1,425
0160	Deluxe	16,200	1,750	17,950
0170	Scoreboards, basketball, 1 side, economy	2,625	825	3,450
0180	4 sides, deluxe	6,250	11,400	17,650
0800	Vocational shop equipment, benches, metal	570	221	791
0810	Wood	630	221	851
0820	Dust collector, not incl. ductwork, 6' diam.	4,650	590	5,240
0830	Planer, 13" x 6"	1,100	277	1,377

E1090 620 School Equipment, S.F.

		COST PER S.F.		
		MAT.	INST.	TOTAL
0110	School equipment, gym mats, naugahyde cover, 2" thick	3.89		3.89
0120	Wrestling, 1" thick, heavy duty	5.90		5.90

E1090 810 Athletic, Recreational, and Therapeutic Equipment, EACH

		COST EACH		
		MAT.	INST.	TOTAL
0050	Bowling alley with gutters	49,900	11,600	61,500
0060	Combo. table and ball rack	1,250		1,250
0070	Bowling alley automatic scorer	10,900		10,900
0110	Sauna, prefabricated, incl. heater and controls, 7' high, 6' x 4'	5,375	845	6,220
0120	10' x 12'	13,200	1,850	15,050
0130	Heaters, wall mounted, to 200 C.F.	690		690
0140	Floor standing, to 1000 C.F., 12500 W	3,625	209	3,834
0610	Shooting range incl. bullet traps, controls, separators, ceilings, economy	38,500	7,325	45,825
0620	Deluxe	57,000	12,300	69,300
0650	Sport court, squash, regulation, in existing building, economy			40,400
0660	Deluxe			45,000
0670	Racketball, regulation, in existing building, economy	43,800	10,500	54,300
0680	Deluxe	47,400	21,000	68,400
0700	Swimming pool equipment, diving stand, stainless steel, 1 meter	7,475	410	7,885
0710	3 meter	12,800	2,775	15,575
0720	Diving boards, 16 ft. long, aluminum	4,300	410	4,710
0730	Fiberglass	3,125	410	3,535
0740	Filter system, sand, incl. pump, 6000 gal./hr.	2,075	745	2,820
0750	Lights, underwater, 12 volt with transformer, 300W	320	625	945
0760	Slides, fiberglass with aluminum handrails & ladder, 6' high, straight	3,700	690	4,390
0780	12' high, straight with platform	15,000	920	15,920

E1090 820 Athletic, Recreational, and Therapeutic Equipment, S.F.

		COST PER S.F.		
		MAT.	INST.	TOTAL
0110	Swimming pools, residential, vinyl liner, metal sides	17.80	7.75	25.55
0120	Concrete sides	21.50	14.05	35.55
0130	Gunite shell, plaster finish, 350 S.F.	39.50	29	68.50
0140	800 S.F.	32	16.80	48.80
0150	Motel, gunite shell, plaster finish	48.50	36.50	85
0160	Municipal, gunite shell, tile finish, formed gutters	187	63.50	250.50

E10 Equipment

E1090 Other Equipment

E1090 910	Other Equipment, EACH	COST EACH		
		MAT.	INST.	TOTAL
0200	Steam bath heater, incl. timer and head, single, to 140 C.F.	2,425	560	2,985
0210	Commercial size, to 2500 C.F.	12,500	840	13,340
0900	Wine Cellar, redwood, refrigerated, 6'-8" high, 6'W x 6'D	4,675	740	5,415
0910	6'W x 12'D	8,100	1,100	9,200

E20 Furnishings

E2010 Fixed Furnishings

E2010 310 Window Treatment, EACH

		COST EACH		
		MAT.	INST.	TOTAL
0110	Furnishings, blinds, exterior, aluminum, louvered, 1'-4" wide x 3'-0" long	75	55.50	130.50
0120	1'-4" wide x 6'-8" long	150	61.50	211.50
0130	Hemlock, solid raised, 1'-4" wide x 3'-0" long	79	55.50	134.50
0140	1'-4" wide x 6'-9" long	177	61.50	238.50
0150	Polystyrene, louvered, 1'-3" wide x 3'-3" long	71.50	55.50	127
0160	1'-3" wide x 6'-8" long	139	61.50	200.50
0200	Interior, wood folding panels, louvered, 7" x 20" (per pair)	86	32.50	118.50
0210	18" x 40" (per pair)	145	32.50	177.50

E2010 320 Window Treatment, S.F.

		COST PER S.F.		
		MAT.	INST.	TOTAL
0110	Furnishings, blinds-interior, venetian-aluminum, stock, 2" slats, economy	5.20	.94	6.14
0120	Custom, 1" slats, deluxe	7.20	1.26	8.46
0130	Vertical, PVC or cloth, T&B track, economy	9.05	1.20	10.25
0140	Deluxe	28	1.38	29.38
0150	Draperies, unlined, economy	3.38		3.38
0160	Lightproof, deluxe	9.70		9.70
0510	Shades, mylar, wood roller, single layer, non-reflective	7.55	.81	8.36
0520	Metal roller, triple layer, heat reflective	10.80	.81	11.61
0530	Vinyl, light weight, 4 ga.	.65	.81	1.46
0540	Heavyweight, 6 ga.	1.97	.81	2.78
0550	Vinyl coated cotton, lightproof decorator shades	3.23	.81	4.04
0560	Woven aluminum, 3/8" thick, light and fireproof	6.35	1.58	7.93

E2010 420 Fixed Floor Grilles and Mats, S.F.

		COST PER S.F.		
		MAT.	INST.	TOTAL
0110	Floor mats, recessed, inlaid black rubber, 3/8" thick, solid	23.50	2.82	26.32
0120	Colors, 1/2" thick, perforated	28.50	2.82	31.32
0130	Link-including nosings, steel-galvanized, 3/8" thick	22	2.82	24.82
0140	Vinyl, in colors	25.50	2.82	28.32

E2010 510 Fixed Multiple Seating, EACH

		COST EACH		
		MAT.	INST.	TOTAL
0110	Seating, painted steel, upholstered, economy	145	31.50	176.50
0120	Deluxe	465	39.50	504.50
0400	Seating, lecture hall, pedestal type, economy	206	50.50	256.50
0410	Deluxe	675	76.50	751.50
0500	Auditorium chair, veneer construction	214	50.50	264.50
0510	Fully upholstered, spring seat	268	50.50	318.50

E20 Furnishings

E2020 Moveable Furnishings

E2020 210	Furnishings/EACH	COST EACH		
		MAT.	INST.	TOTAL
0200	Hospital furniture, beds, manual, economy	865		865
0210	Deluxe	2,750		2,750
0220	All electric, economy	1,775		1,775
0230	Deluxe	4,850		4,850
0240	Patient wall systems, no utilities, economy, per room	1,550		1,550
0250	Deluxe, per room	2,100		2,100
0300	Hotel furnishings, standard room set, economy, per room	2,750		2,750
0310	Deluxe, per room	9,750		9,750
0500	Office furniture, standard employee set, economy, per person	635		635
0510	Deluxe, per person	2,525		2,525
0550	Posts, portable, pedestrian traffic control, economy	142		142
0560	Deluxe	460		460
0700	Restaurant furniture, booth, molded plastic, stub wall and 2 seats, economy	400	277	677
0710	Deluxe	1,525	370	1,895
0720	Upholstered seats, foursome, single-economy	815	111	926
0730	Foursome, double-deluxe	1,875	184	2,059

E2020 220	Furniture and Accessories, L.F.	COST PER L.F.		
		MAT.	INST.	TOTAL
0210	Dormitory furniture, desk top (built-in), laminated plastc, 24"deep, economy	47.50	22	69.50
0220	30" deep, deluxe	267	27.50	294.50
0230	Dressing unit, built-in, economy	183	92	275
0240	Deluxe	550	138	688
0310	Furnishings, cabinets, hospital, base, laminated plastic	288	111	399
0320	Stainless steel	530	111	641
0330	Countertop, laminated plastic, no backsplash	49.50	27.50	77
0340	Stainless steel	161	27.50	188.50
0350	Nurses station, door type, laminated plastic	335	111	446
0360	Stainless steel	635	111	746
0710	Restaurant furniture, bars, built-in, back bar	216	111	327
0720	Front bar	297	111	408
0910	Wardrobes & coatracks, standing, steel, single pedestal, 30" x 18" x 63"	116		116
0920	Double face rack, 39" x 26" x 70"	125		125
0930	Wall mounted rack, steel frame & shelves, 12" x 15" x 26"	77.50	7.85	85.35
0940	12" x 15" x 50"	47.50	4.12	51.62

F Special Construction

No part of this publication may be reproduced, stored in a retrieval system, or transmitted in any form or by any means without prior written permission of RSMeans, a division of Reed Construction Data.

F10 Special Construction

F1010 Special Structures

F1010 120	Air-Supported Structures, S.F.	COST PER S.F.		
		MAT.	INST.	TOTAL
0110	Air supported struc., polyester vinyl fabric, 24oz., warehouse, 5000 S.F.	27.50	.35	27.85
0120	50,000 S.F.	12.90	.28	13.18
0130	Tennis, 7,200 S.F.	26	.29	26.29
0140	24,000 S.F.	18.05	.29	18.34
0150	Woven polyethylene, 6 oz., shelter, 3,000 S.F.	7.25	.58	7.83
0160	24,000 S.F.	6.15	.29	6.44
0170	Teflon coated fiberglass, stadium cover, economy	63	.15	63.15
0180	Deluxe	75	.21	75.21
0190	Air supported storage tank covers, reinf. vinyl fabric, 12 oz., 400 S.F.	25.50	.49	25.99
0200	18,000 S.F.	8.10	.44	8.54

F1010 210	Pre-Engineered Structures, EACH	COST EACH		
		MAT.	INST.	TOTAL
0600	Radio towers, guyed, 40 lb. section, 50' high, 70 MPH basic wind speed	3,225	1,425	4,650
0610	90 lb. section, 400' high, wind load 70 MPH basic wind speed	40,000	16,100	56,100
0620	Self supporting, 60' high, 70 MPH basic wind speed	5,000	2,825	7,825
0630	190' high, wind load 90 MPH basic wind speed	32,800	11,300	44,100
0700	Shelters, aluminum frame, acrylic glazing, 8' high, 3' x 9'	4,175	1,250	5,425
0710	9' x 12'	6,575	1,950	8,525

F1010 320	Other Special Structures, S.F.	COST PER S.F.		
		MAT.	INST.	TOTAL
0110	Swimming pool enclosure, translucent, freestanding, economy	16.25	5.55	21.80
0120	Deluxe	98.50	15.80	114.30
0510	Tension structures, steel frame, polyester vinyl fabric, 12,000 S.F.	14.90	2.55	17.45
0520	20,800 S.F.	14.60	2.30	16.90

F1010 330	Special Structures, EACH	COST EACH		
		MAT.	INST.	TOTAL
0110	Kiosks, round, 5' diam., 1/4" fiberglass wall, 8' high	25,100		25,100
0120	Rectangular, 5' x 9', 1" insulated dbl. wall fiberglass, 7'-6" high	25,600		25,600
0220	Silos, steel prefab, 30,000 gal., painted, economy	24,300	5,700	30,000
0230	Epoxy-lined, deluxe	50,500	11,400	61,900

F1010 340	Special Structures, S.F.	COST PER S.F.		
		MAT.	INST.	TOTAL
0110	Comfort stations, prefab, mobile on steel frame, economy	207		207
0120	Permanent on concrete slab, deluxe	211	46.50	257.50
0210	Domes, bulk storage, wood framing, wood decking, 50' diam.	79.50	1.76	81.26
0220	116' diam.	38.50	2.03	40.53
0230	Steel framing, metal decking, 150' diam.	30.50	11.60	42.10
0240	400' diam.	25	8.90	33.90
0250	Geodesic, wood framing, wood panels, 30' diam.	32.50	2	34.50
0260	60' diam.	22.50	1.24	23.74
0270	Aluminum framing, acrylic panels, 40' diam.			103
0280	Aluminum panels, 400' diam.			23.50
0310	Garden house, prefab, wood, shell only, 48 S.F.	55.50	5.55	61.05
0320	200 S.F.	35.50	23	58.50
0410	Greenhouse, shell-stock, residential, lean-to, 8'-6" long x 3'-10" wide	37.50	32.50	70
0420	Freestanding, 8'-6" long x 13'-6" wide	27	10.25	37.25
0430	Commercial-truss frame, under 2000 S.F., deluxe			15.90
0440	Over 5,000 S.F., economy			16.05
0450	Institutional-rigid frame, under 500 S.F., deluxe			39
0460	Over 2,000 S.F., economy			12.90
0510	Hangar, prefab, galv. roof and walls, bottom rolling doors, economy	11.35	4.60	15.95
0520	Electric bifolding doors, deluxe	11.75	6	17.75

F10 Special Construction

F1020 Integrated Construction

F1020 110	Integrated Construction, EACH	COST EACH		
		MAT.	INST.	TOTAL
0110	Integrated ceilings, radiant electric, 2' x 4' panel, manila finish	99	25	124
0120	ABS plastic finish	112	48.50	160.50

F1020 120	Integrated Construction, S.F.	COST PER S.F.		
		MAT.	INST.	TOTAL
0110	Integrated ceilings, Luminaire, suspended, 5' x 5' modules, 50% lighted	4.32	13.25	17.57
0120	100% lighted	6.45	24	30.45
0130	Dimensionaire, 2' x 4' module tile system, no air bar	2.61	2.21	4.82
0140	With air bar, deluxe	3.83	2.21	6.04
0220	Pedestal access floors, stl pnls, no stringers, vinyl covering, >6000 S.F.	13.40	1.73	15.13
0230	With stringers, high pressure laminate covering, under 6000 S.F.	12.35	1.84	14.19
0240	Carpet covering, under 6000 S.F.	13.70	1.84	15.54
0250	Aluminum panels, no stringers, no covering	35.50	2.21	37.71

F1020 250	Special Purpose Room, EACH	COST EACH		
		MAT.	INST.	TOTAL
0110	Portable booth, acoustical, 27 db 1000 hz., 15 S.F. floor	3,925		3,925
0120	55 S.F. flr.	8,050		8,050

F1020 260	Special Purpose Room, S.F.	COST PER S.F.		
		MAT.	INST.	TOTAL
0110	Anechoic chambers, 7' high, 100 cps cutoff, 25 S.F.			2,600
0120	200 cps cutoff, 100 S.F.			1,350
0130	Audiometric rooms, under 500 S.F.	57.50	22.50	80
0140	Over 500 S.F.	55	18.45	73.45
0300	Darkrooms, shell, not including door, 240 S.F., 8' high	31	9.20	40.20
0310	64 S.F., 12' high	79	17.30	96.30
0510	Music practice room, modular, perforated steel, under 500 S.F.	35.50	15.80	51.30
0520	Over 500 S.F.	30	13.85	43.85

F1020 330	Special Construction, L.F.	COST PER L.F.		
		MAT.	INST.	TOTAL
0110	Spec. const., air curtains, shipping & receiving, 8'high x 5'wide, economy	740	130	870
0120	20' high x 8' wide, heated, deluxe	3,150	325	3,475
0130	Customer entrance, 10' high x 5' wide, economy	1,100	130	1,230
0140	12' high x 4' wide, heated, deluxe	2,000	163	2,163

F10 Special Construction

F1030 Special Construction Systems

F1030 120	Sound, Vibration, and Seismic Construction, S.F.	COST PER S.F.		
		MAT.	INST.	TOTAL
0020	Special construction, acoustical, enclosure, 4" thick, 8 psf panels	36	23	59
0030	Reverb chamber, 4" thick, parallel walls	51	27.50	78.50
0110	Sound absorbing panels, 2'-6" x 8', painted metal	13.55	7.70	21.25
0120	Vinyl faced	10.60	6.90	17.50
0130	Flexible transparent curtain, clear	8.20	9.10	17.30
0140	With absorbing foam, 75% coverage	11.40	9.10	20.50
0150	Strip entrance, 2/3 overlap	8.35	14.45	22.80
0160	Full overlap	10.50	16.95	27.45
0200	Audio masking system, plenum mounted, over 10,000 S.F.	.80	.29	1.09
0210	Ceiling mounted, under 5,000 S.F.	1.39	.52	1.91

F1030 210	Radiation Protection, EACH	COST EACH		
		MAT.	INST.	TOTAL
0110	Shielding, lead x-ray protection, radiography room, 1/16" lead, economy	10,100	3,725	13,825
0120	Deluxe	12,200	6,225	18,425

F1030 220	Radiation Protection, S.F.	COST PER S.F.		
		MAT.	INST.	TOTAL
0110	Shielding, lead, gypsum board, 5/8" thick, 1/16" lead	10.20	6.55	16.75
0120	1/8" lead	24	7.50	31.50
0130	Lath, 1/16" thick	11.05	6.90	17.95
0140	1/8" thick	22	7.75	29.75
0150	Radio frequency, galvanized steel, prefab type, economy	4.91	2.95	7.85
0160	Radio frequency, door, copper/wood laminate, 4" x 7'	6,350	740	7,100

F1030 910	Other Special Construction Systems, EACH	COST EACH		
		MAT.	INST.	TOTAL
0110	Disappearing stairways, folding, pine, 8'-6" ceiling	166	138	304
0120	9'-6" ceiling	185	138	323
0220	Automatic electric, wood, 8' to 9' ceiling	9,275	1,100	10,375
0230	Aluminum, 14' to 15' ceiling	10,600	1,575	12,175
0300	Fireplace prefabricated, freestanding or wall hung, economy	1,325	425	1,750
0310	Deluxe	3,000	615	3,615
0320	Woodburning stoves, cast iron, economy	1,050	850	1,900
0330	Deluxe	3,025	1,375	4,400

F10 Special Construction

F1040 Special Facilities

F1040 210	Ice Rinks, EACH	COST EACH		
		MAT.	INST.	TOTAL
0100	Ice skating rink, 85' x 200', 55° system, 5 mos., 100 ton			632,500
0110	90° system, 12 mos., 135 ton			715,000
0120	Dash boards, acrylic screens, polyethylene coated plywood	148,500	37,300	185,800
0130	Fiberglass and aluminum construction	170,500	37,300	207,800

F1040 510	Liquid & Gas Storage Tanks, EACH	COST EACH		
		MAT.	INST.	TOTAL
0100	Tanks, steel, ground level, 100,000 gal.			226,500
0110	10,000,000 gal.			5,151,500
0120	Elevated water, 50,000 gal.		185,000	445,500
0130	1,000,000 gal.		862,500	1,973,500
0150	Cypress wood, ground level, 3,000 gal.	10,200	11,000	21,200
0160	Redwood, ground level, 45,000 gal.	77,000	29,900	106,900

F1040 910	Special Construction, EACH	COST EACH		
		MAT.	INST.	TOTAL
0110	Special construction, bowling alley incl. pinsetter, scorer etc., economy	45,400	11,100	56,500
0120	Deluxe	57,500	12,300	69,800
0130	For automatic scorer, economy, add	9,050		9,050
0140	Deluxe, add	10,900		10,900
0300	Control tower, modular, 12' x 10', incl. instrumentation, economy			774,000
0310	Deluxe			1,080,000
0400	Garage costs, residential, prefab, wood, single car economy	5,625	1,100	6,725
0410	Two car deluxe	11,800	2,225	14,025
0500	Hangars, prefab, galv. steel, bottom rolling doors, economy (per plane)	15,000	5,100	20,100
0510	Electrical bi-folding doors, deluxe (per plane)	14,000	7,025	21,025

G Building Sitework

No part of this publication may be reproduced, stored in a retrieval system, or transmitted in any form or by any means without prior written permission of RSMeans, a division of Reed Construction Data.

G10 Site Preparation

G1030 Site Earthwork

Trenching Systems are shown on a cost per linear foot basis. The systems include: excavation; backfill and removal of spoil; and compaction for various depths and trench bottom widths. The backfill has been reduced to accommodate a pipe of suitable diameter and bedding.

The slope for trench sides varies from none to 1:1.

The Expanded System Listing shows Trenching Systems that range from 2' to 12' in width. Depths range from 2' to 25'.

System Components	QUANTITY	UNIT	COST PER L.F.		
			EQUIP.	LABOR	TOTAL
SYSTEM G1030 805 1310					
TRENCHING COMMON EARTH, NO SLOPE, 2' WIDE, 2' DP, 3/8 C.Y. BUCKET					
Excavation, trench, hyd. backhoe, track mtd., 3/8 C.Y. bucket	.148	B.C.Y.	.40	1	1.40
Backfill and load spoil, from stockpile	.153	L.C.Y.	.12	.30	.42
Compaction by vibrating plate, 6" lifts, 4 passes	.118	E.C.Y.	.03	.37	.40
Remove excess spoil, 8 C.Y. dump truck, 2 mile roundtrip	.040	L.C.Y.	.15	.15	.30
TOTAL			.70	1.82	2.52

G1030 805	Trenching Common Earth	COST PER L.F.		
		EQUIP.	LABOR	TOTAL
1310	Trenching, common earth, no slope, 2' wide, 2' deep, 3/8 C.Y. bucket	.70	1.82	2.52
1320	3' deep, 3/8 C.Y. bucket	1	2.75	3.75
1330	4' deep, 3/8 C.Y. bucket	1.30	3.67	4.97
1340	6' deep, 3/8 C.Y. bucket	1.68	4.75	6.43
1350	8' deep, 1/2 C.Y. bucket	2.21	6.30	8.51
1360	10' deep, 1 C.Y. bucket	3.59	7.50	11.09
1400	4' wide, 2' deep, 3/8 C.Y. bucket	1.62	3.58	5.20
1410	3' deep, 3/8 C.Y. bucket	2.21	5.40	7.61
1420	4' deep, 1/2 C.Y. bucket	2.57	6.10	8.67
1430	6' deep, 1/2 C.Y. bucket	4.12	9.80	13.92
1440	8' deep, 1/2 C.Y. bucket	6.80	12.65	19.45
1450	10' deep, 1 C.Y. bucket	8.20	15.65	23.85
1460	12' deep, 1 C.Y. bucket	10.55	20	30.55
1470	15' deep, 1-1/2 C.Y. bucket	9.30	17.85	27.15
1480	18' deep, 2-1/2 C.Y. bucket	12.90	25	37.90
1520	6' wide, 6' deep, 5/8 C.Y. bucket w/trench box	8.95	14.50	23.45
1530	8' deep, 3/4 C.Y. bucket	11.90	19.10	31
1540	10' deep, 1 C.Y. bucket	11.65	19.80	31.45
1550	12' deep, 1-1/2 C.Y. bucket	12.50	21.50	34
1560	16' deep, 2-1/2 C.Y. bucket	18	26.50	44.50
1570	20' deep, 3-1/2 C.Y. bucket	22	31.50	53.50
1580	24' deep, 3-1/2 C.Y. bucket	26	38	64
1640	8' wide, 12' deep, 1-1/2 C.Y. bucket w/trench box	17.50	27	44.50
1650	15' deep, 1-1/2 C.Y. bucket	23	35.50	58.50
1660	18' deep, 2-1/2 C.Y. bucket	26	35.50	61.50
1680	24' deep, 3-1/2 C.Y. bucket	35.50	49.50	85
1730	10' wide, 20' deep, 3-1/2 C.Y. bucket w/trench box	28.50	47	75.50
1740	24' deep, 3-1/2 C.Y. bucket	42.50	56.50	99
1780	12' wide, 20' deep, 3-1/2 C.Y. bucket w/trench box	45	59.50	104.50
1790	25' deep, bucket	56	75.50	131.50
1800	1/2 to 1 slope, 2' wide, 2' deep, 3/8 C.Y. bucket	1	2.75	3.75
1810	3' deep, 3/8 C.Y. bucket	1.68	4.82	6.50

G10 Site Preparation

G1030 Site Earthwork

G1030 805	Trenching Common Earth	COST PER L.F.		
		EQUIP.	LABOR	TOTAL
1820	4' deep, 3/8 C.Y. bucket	2.49	7.35	9.84
1840	6' deep, 3/8 C.Y. bucket	4.02	11.90	15.92
1860	8' deep, 1/2 C.Y. bucket	6.40	19.05	25.45
1880	10' deep, 1 C.Y. bucket	12.35	26.50	38.85
2300	4' wide, 2' deep, 3/8 C.Y. bucket	1.92	4.51	6.43
2310	3' deep, 3/8 C.Y. bucket	2.87	7.50	10.37
2320	4' deep, 1/2 C.Y. bucket	3.63	9.30	12.93
2340	6' deep, 1/2 C.Y. bucket	6.90	17.35	24.25
2360	8' deep, 1/2 C.Y. bucket	13.10	25.50	38.60
2380	10' deep, 1 C.Y. bucket	18.15	36	54.15
2400	12' deep, 1 C.Y. bucket	24	48.50	72.50
2430	15' deep, 1-1/2 C.Y. bucket	26.50	52	78.50
2460	18' deep, 2-1/2 C.Y. bucket	49	81.50	130.50
2840	6' wide, 6' deep, 5/8 C.Y. bucket w/trench box	13.25	21.50	34.75
2860	8' deep, 3/4 C.Y. bucket	19.15	32.50	51.65
2880	10' deep, 1 C.Y. bucket	18.75	33	51.75
2900	12' deep, 1-1/2 C.Y. bucket	23.50	43	66.50
2940	16' deep, 2-1/2 C.Y. bucket	41	63.50	104.50
2980	20' deep, 3-1/2 C.Y. bucket	55	85.50	140.50
3020	24' deep, 3-1/2 C.Y. bucket	77.50	117	194.50
3100	8' wide, 12' deep, 1-1/2 C.Y. bucket w/trench box	29.50	49	78.50
3120	15' deep, 1-1/2 C.Y. bucket	42.50	71.50	114
3140	18' deep, 2-1/2 C.Y. bucket	56.50	85	141.50
3180	24' deep, 3-1/2 C.Y. bucket	87	128	215
3270	10' wide, 20' deep, 3-1/2 C.Y. bucket w/trench box	55.50	98.50	154
3280	24' deep, 3-1/2 C.Y. bucket	96	140	236
3370	12' wide, 20' deep, 3-1/2 C.Y. bucket w/trench box	81	114	195
3380	25' deep, 3-1/2 C.Y. bucket	112	163	275
3500	1 to 1 slope, 2' wide, 2' deep, 3/8 C.Y. bucket	1.30	3.67	4.97
3520	3' deep, 3/8 C.Y. bucket	3.72	8.25	11.97
3540	4' deep, 3/8 C.Y. bucket	3.68	11.05	14.73
3560	6' deep, 3/8 C.Y. bucket	4.02	11.90	15.92
3580	8' deep, 1/2 C.Y. bucket	7.95	24	31.95
3600	10' deep, 1 C.Y. bucket	21	45.50	66.50
3800	4' wide, 2' deep, 3/8 C.Y. bucket	2.21	5.45	7.66
3820	3' deep, 3/8 C.Y. bucket	3.55	9.55	13.10
3840	4' deep, 1/2 C.Y. bucket	4.67	12.45	17.12
3860	6' deep, 1/2 C.Y. bucket	9.65	25	34.65
3880	8' deep, 1/2 C.Y. bucket	19.50	38.50	58
3900	10' deep, 1 C.Y. bucket	28	56	84
3920	12' deep, 1 C.Y. bucket	41.50	81.50	123
3940	15' deep, 1-1/2 C.Y. bucket	43.50	86.50	130
3960	18' deep, 2-1/2 C.Y. bucket	67	112	179
4030	6' wide, 6' deep, 5/8 C.Y. bucket w/trench box	17.35	29	46.35
4040	8' deep, 3/4 C.Y. bucket	25	40	65
4050	10' deep, 1 C.Y. bucket	27	48.50	75.50
4060	12' deep, 1-1/2 C.Y. bucket	36	66	102
4070	16' deep, 2-1/2 C.Y. bucket	64.50	100	164.50
4080	20' deep, 3-1/2 C.Y. bucket	88.50	139	227.50
4090	24' deep, 3-1/2 C.Y. bucket	129	196	325
4500	8' wide, 12' deep, 1-1/2 C.Y. bucket w/trench box	41.50	71.50	113
4550	15' deep, 1-1/2 C.Y. bucket	62.50	107	169.50
4600	18' deep, 2-1/2 C.Y. bucket	86	132	218
4650	24' deep, 3-1/2 C.Y. bucket	138	207	345
4800	10' wide, 20' deep, 3-1/2 C.Y. bucket w/trench box	82.50	150	232.50
4850	24' deep, 3-1/2 C.Y. bucket	147	219	366
4950	12' wide, 20' deep, 3-1/2 C.Y. bucket w/trench box	116	169	285
4980	25' deep, 3-1/2 C.Y. bucket	167	247	414

G10 Site Preparation

G1030 Site Earthwork

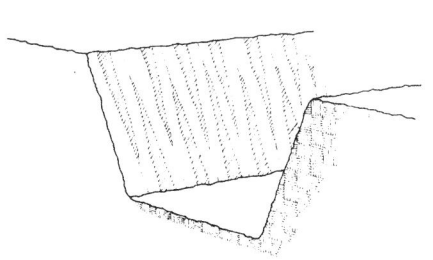

Trenching Systems are shown on a cost per linear foot basis. The systems include: excavation; backfill and removal of spoil; and compaction for various depths and trench bottom widths. The backfill has been reduced to accommodate a pipe of suitable diameter and bedding.

The slope for trench sides varies from none to 1:1.

The Expanded System Listing shows Trenching Systems that range from 2' to 12' in width. Depths range from 2' to 25'.

System Components	QUANTITY	UNIT	COST PER L.F.		
			EQUIP.	LABOR	TOTAL
SYSTEM G1030 806 1310					
TRENCHING LOAM & SANDY CLAY, NO SLOPE, 2' WIDE, 2' DP, 3/8 C.Y. BUCKET					
Excavation, trench, hyd. backhoe, track mtd., 3/8 C.Y. bucket	.148	B.C.Y.	.37	.93	1.30
Backfill and load spoil, from stockpile	.165	L.C.Y.	.13	.33	.46
Compaction by vibrating plate 18" wide, 6" lifts, 4 passes	.118	E.C.Y.	.03	.37	.40
Remove excess spoil, 8 C.Y. dump truck, 2 mile roundtrip	.042	L.C.Y.	.16	.16	.32
TOTAL			.70	1.82	2.52

G1030 806	Trenching Loam & Sandy Clay	COST PER L.F.		
		EQUIP.	LABOR	TOTAL
1310	Trenching, loam & sandy clay, no slope, 2' wide, 2' deep, 3/8 C.Y. bucket	.69	1.79	2.48
1320	3' deep, 3/8 C.Y. bucket	1.08	2.94	4.02
1330	4' deep, 3/8 C.Y. bucket	1.27	3.58	4.85
1340	6' deep, 3/8 C.Y. bucket	1.80	4.25	6.05
1350	8' deep, 1/2 C.Y. bucket	2.38	5.65	8.03
1360	10' deep, 1 C.Y. bucket	2.68	6.05	8.73
1400	4' wide, 2' deep, 3/8 C.Y. bucket	1.61	3.52	5.13
1410	3' deep, 3/8 C.Y. bucket	2.19	5.30	7.49
1420	4' deep, 1/2 C.Y. bucket	2.57	6	8.57
1430	6' deep, 1/2 C.Y. bucket	4.39	8.75	13.14
1440	8' deep, 1/2 C.Y. bucket	6.65	12.45	19.10
1450	10' deep, 1 C.Y. bucket	6.40	12.75	19.15
1460	12' deep, 1 C.Y. bucket	8.05	15.85	23.90
1470	15' deep, 1-1/2 C.Y. bucket	9.70	18.45	28.15
1480	18' deep, 2-1/2 C.Y. bucket	12.05	20.50	32.55
1520	6' wide, 6' deep, 5/8 C.Y. bucket w/trench box	8.65	14.25	22.90
1530	8' deep, 3/4 C.Y. bucket	11.45	18.75	30.20
1540	10' deep, 1 C.Y. bucket	10.70	19	29.70
1550	12' deep, 1-1/2 C.Y. bucket	12.20	21.50	33.70
1560	16' deep, 2-1/2 C.Y. bucket	17.50	27	44.50
1570	20' deep, 3-1/2 C.Y. bucket	20	32	52
1580	24' deep, 3-1/2 C.Y. bucket	25.50	39	64.50
1640	8' wide, 12' deep, 1-1/4 C.Y. bucket w/trench box	17.25	27	44.25
1650	15' deep, 1-1/2 C.Y. bucket	21	34.50	55.50
1660	18' deep, 2-1/2 C.Y. bucket	26.50	38.50	65
1680	24' deep, 3-1/2 C.Y. bucket	34.50	50	84.50
1730	10' wide, 20' deep, 3-1/2 C.Y. bucket w/trench box	35	50	85
1740	24' deep, 3-1/2 C.Y. bucket	43.50	62	105.50
1780	12' wide, 20' deep, 3-1/2 C.Y. bucket w/trench box	42	59.50	101.50
1790	25' deep, bucket	54.50	77	131.50
1800	1/2:1 slope, 2' wide, 2' deep, 3/8 C.Y. bucket	.98	2.68	3.66
1810	3' deep, 3/8 C.Y. bucket	1.64	4.70	6.34
1820	4' deep, 3/8 C.Y. bucket	2.43	7.20	9.63
1840	6' deep, 3/8 C.Y. bucket	4.32	10.65	14.97

G10 Site Preparation

G1030 Site Earthwork

G1030 806	Trenching Loam & Sandy Clay	COST PER L.F.		
		EQUIP.	LABOR	TOTAL
1860	8' deep, 1/2 C.Y. bucket	6.85	17	23.85
1880	10' deep, 1 C.Y. bucket	9.15	21.50	30.65
2300	4' wide, 2' deep, 3/8 C.Y. bucket	1.90	4.43	6.33
2310	3' deep, 3/8 C.Y. bucket	2.83	7.35	10.18
2320	4' deep, 1/2 C.Y. bucket	3.59	9.10	12.69
2340	6' deep, 1/2 C.Y. bucket	7.35	15.60	22.95
2360	8' deep, 1/2 C.Y. bucket	12.75	25	37.75
2380	10' deep, 1 C.Y. bucket	14.05	29.50	43.55
2400	12' deep, 1 C.Y. bucket	23.50	48.50	72
2430	15' deep, 1-1/2 C.Y. bucket	27.50	54	81.50
2460	18' deep, 2-1/2 C.Y. bucket	47.50	82.50	130
2840	6' wide, 6' deep, 5/8 C.Y. bucket w/trench box	12.60	21.50	34.10
2860	8' deep, 3/4 C.Y. bucket	18.45	32	50.45
2880	10' deep, 1 C.Y. bucket	19.10	36	55.10
2900	12' deep, 1-1/2 C.Y. bucket	23.50	43.50	67
2940	16' deep, 2-1/2 C.Y. bucket	40	64	104
2980	20' deep, 3-1/2 C.Y. bucket	53	86.50	139.50
3020	24' deep, 3-1/2 C.Y. bucket	75	119	194
3100	8' wide, 12' deep, 1-1/2 C.Y. bucket w/trench box	29	49.50	78.50
3120	15' deep, 1-1/2 C.Y. bucket	39	69.50	108.50
3140	18' deep, 2-1/2 C.Y. bucket	55	86	141
3180	24' deep, 3-1/2 C.Y. bucket	84.50	130	214.50
3270	10' wide, 20' deep, 3-1/2 C.Y. bucket w/trench box	67.50	105	172.50
3280	24' deep, 3-1/2 C.Y. bucket	93	142	235
3320	12' wide, 20' deep, 3-1/2 C.Y. bucket w/trench box	75	114	189
3380	25' deep, 3-1/2 C.Y. bucket w/trench box	102	153	255
3500	1:1 slope, 2' wide, 2' deep, 3/8 C.Y. bucket	1.27	3.58	4.85
3520	3' deep, 3/8 C.Y. bucket	2.28	6.75	9.03
3540	4' deep, 3/8 C.Y. bucket	3.57	10.80	14.37
3560	6' deep, 3/8 C.Y. bucket	4.32	10.65	14.97
3580	8' deep, 1/2 C.Y. bucket	11.35	28.50	39.85
3600	10' deep, 1 C.Y. bucket	15.60	36.50	52.10
3800	4' wide, 2' deep, 3/8 C.Y. bucket	2.19	5.30	7.49
3820	3' deep, 3/8 C.Y. bucket	3.49	9.35	12.84
3840	4' deep, 1/2 C.Y. bucket	4.61	12.25	16.86
3860	6' deep, 1/2 C.Y. bucket	10.30	22.50	32.80
3880	8' deep, 1/2 C.Y. bucket	18.95	38	56.95
3900	10' deep, 1 C.Y. bucket	21.50	46	67.50
3920	12' deep, 1 C.Y. bucket	31.50	65	96.50
3940	15' deep, 1-1/2 C.Y. bucket	45	89.50	134.50
3960	18' deep, 2-1/2 C.Y. bucket	65.50	114	179.50
4030	6' wide, 6' deep, 5/8 C.Y. bucket w/trench box	16.55	29	45.55
4040	8' deep, 3/4 C.Y. bucket	25.50	45	70.50
4050	10' deep, 1 C.Y. bucket	27.50	52.50	80
4060	12' deep, 1-1/2 C.Y. bucket	35	66	101
4070	16' deep, 2-1/2 C.Y. bucket	62.50	101	163.50
4080	20' deep, 3-1/2 C.Y. bucket	86	142	228
4090	24' deep, 3-1/2 C.Y. bucket	125	199	324
4500	8' wide, 12' deep, 1-1/4 C.Y. bucket w/trench box	40.50	71.50	112
4550	15' deep, 1-1/2 C.Y. bucket	57	104	161
4600	18' deep, 2-1/2 C.Y. bucket	83.50	133	216.50
4650	24' deep, 3-1/2 C.Y. bucket	134	210	344
4800	10' wide, 20' deep, 3-1/2 C.Y. bucket w/trench box	100	160	260
4850	24' deep, 3-1/2 C.Y. bucket	143	222	365
4950	12' wide, 20' deep, 3-1/2 C.Y. bucket w/trench box	108	169	277
4980	25' deep, 3-1/2 C.Y. bucket	162	250	412

G10 Site Preparation

G1030 Site Earthwork

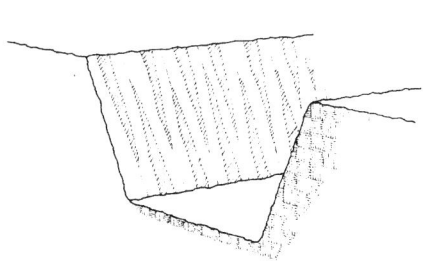

Trenching Systems are shown on a cost per linear foot basis. The systems include: excavation; backfill and removal of spoil; and compaction for various depths and trench bottom widths. The backfill has been reduced to accommodate a pipe of suitable diameter and bedding.

The slope for trench sides varies from none to 1:1.

The Expanded System Listing shows Trenching Systems that range from 2' to 12' in width. Depths range from 2' to 25'.

System Components	QUANTITY	UNIT	COST PER L.F.		
			EQUIP.	LABOR	TOTAL
SYSTEM G1030 807 1310					
TRENCHING SAND & GRAVEL, NO SLOPE, 2' WIDE, 2' DEEP, 3/8 C.Y. BUCKET					
Excavation, trench, hyd. backhoe, track mtd., 3/8 C.Y. bucket	.148	B.C.Y.	.36	.91	1.27
Backfill and load spoil, from stockpile	.140	L.C.Y.	.11	.28	.39
Compaction by vibrating plate 18" wide, 6" lifts, 4 passes	.118	E.C.Y.	.03	.37	.40
Remove excess spoil, 8 C.Y. dump truck, 2 mile roundtrip	.035	L.C.Y.	.14	.13	.27
TOTAL			.64	1.69	2.33

G1030 807	Trenching Sand & Gravel	COST PER L.F.		
		EQUIP.	LABOR	TOTAL
1310	Trenching, sand & gravel, no slope, 2' wide, 2' deep, 3/8 C.Y. bucket	.64	1.69	2.33
1320	3' deep, 3/8 C.Y. bucket	1.03	2.82	3.85
1330	4' deep, 3/8 C.Y. bucket	1.19	3.40	4.59
1340	6' deep, 3/8 C.Y. bucket	1.71	4.05	5.76
1350	8' deep, 1/2 C.Y. bucket	2.25	5.35	7.60
1360	10' deep, 1 C.Y. bucket	2.50	5.65	8.15
1400	4' wide, 2' deep, 3/8 C.Y. bucket	1.47	3.31	4.78
1410	3' deep, 3/8 C.Y. bucket	2.01	5.05	7.06
1420	4' deep, 1/2 C.Y. bucket	2.36	5.65	8.01
1430	6' deep, 1/2 C.Y. bucket	4.14	8.40	12.54
1440	8' deep, 1/2 C.Y. bucket	6.25	11.80	18.05
1450	10' deep, 1 C.Y. bucket	6	12.05	18.05
1460	12' deep, 1 C.Y. bucket	7.60	15	22.60
1470	15' deep, 1-1/2 C.Y. bucket	9.15	17.40	26.55
1480	18' deep, 2-1/2 C.Y. bucket	11.45	19.05	30.50
1520	6' wide, 6' deep, 5/8 C.Y. bucket w/trench box	8.20	13.50	21.70
1530	8' deep, 3/4 C.Y. bucket	10.85	17.75	28.60
1540	10' deep, 1 C.Y. bucket	10.10	18	28.10
1550	12' deep, 1-1/2 C.Y. bucket	11.50	20	31.50
1560	16' deep, 2 C.Y. bucket	16.75	25.50	42.25
1570	20' deep, 3-1/2 C.Y. bucket	19.10	30	49.10
1580	24' deep, 3-1/2 C.Y. bucket	24	36.50	60.50
1640	8' wide, 12' deep, 1-1/2 C.Y. bucket w/trench box	16.20	25.50	41.70
1650	15' deep, 1-1/2 C.Y. bucket	19.70	32.50	52.20
1660	18' deep, 2-1/2 C.Y. bucket	25	36.50	61.50
1680	24' deep, 3-1/2 C.Y. bucket	32.50	47	79.50
1730	10' wide, 20' deep, 3-1/2 C.Y. bucket w/trench box	32.50	47	79.50
1740	24' deep, 3-1/2 C.Y. bucket	41	58	99
1780	12' wide, 20' deep, 3-1/2 C.Y. bucket w/trench box	39.50	55.50	95
1790	25' deep, 3-1/2 C.Y. bucket	51.50	72	123.50
1800	1/2:1 slope, 2' wide, 2' deep, 3/8 C.Y. bucket	.91	2.55	3.46
1810	3' deep, 3/8 C.Y. bucket	1.53	4.48	6.01
1820	4' deep, 3/8 C.Y. bucket	2.28	6.85	9.13
1840	6' deep, 3/8 C.Y. bucket	4.10	10.15	14.25

G10 Site Preparation

G1030 Site Earthwork

G1030 807	Trenching Sand & Gravel	COST PER L.F.		
		EQUIP.	LABOR	TOTAL
1860	8' deep, 1/2 C.Y. bucket	6.50	16.25	22.75
1880	10' deep, 1 C.Y. bucket	8.55	20	28.55
2300	4' wide, 2' deep, 3/8 C.Y. bucket	1.74	4.17	5.91
2310	3' deep, 3/8 C.Y. bucket	2.62	6.95	9.57
2320	4' deep, 1/2 C.Y. bucket	3.31	8.65	11.96
2340	6' deep, 1/2 C.Y. bucket	6.95	14.95	21.90
2360	8' deep, 1/2 C.Y. bucket	12.10	24	36.10
2380	10' deep, 1 C.Y. bucket	13.25	27.50	40.75
2400	12' deep, 1 C.Y. bucket	22.50	46	68.50
2430	15' deep, 1-1/2 C.Y. bucket	26	51	77
2460	18' deep, 2-1/2 C.Y. bucket	45	77.50	122.50
2840	6' wide, 6' deep, 5/8 C.Y. bucket w/trench box	11.95	20.50	32.45
2860	8' deep, 3/4 C.Y. bucket	17.50	30	47.50
2880	10' deep, 1 C.Y. bucket	18.05	34	52.05
2900	12' deep, 1-1/2 C.Y. bucket	22.50	41	63.50
2940	16' deep, 2 C.Y. bucket	38.50	60.50	99
2980	20' deep, 3-1/2 C.Y. bucket	50.50	81.50	132
3020	24' deep, 3-1/2 C.Y. bucket	71.50	112	183.50
3100	8' wide, 12' deep, 1-1/4 C.Y. bucket w/trench box	27.50	46.50	74
3120	15' deep, 1-1/2 C.Y. bucket	37	65.50	102.50
3140	18' deep, 2-1/2 C.Y. bucket	52.50	81	133.50
3180	24' deep, 3-1/2 C.Y. bucket	80	122	202
3270	10' wide, 20' deep, 3-1/2 C.Y. bucket w/trench box	64	98.50	162.50
3280	24' deep, 3-1/2 C.Y. bucket	88.50	133	221.50
3370	12' wide, 20' deep, 3-1/2 C.Y. bucket w/trench box	71.50	109	180.50
3380	25' deep, 3-1/2 C.Y. bucket	103	154	257
3500	1:1 slope, 2' wide, 2' deep, 3/8 C.Y. bucket	2.04	4.18	6.22
3520	3' deep, 3/8 C.Y. bucket	2.14	6.40	8.54
3540	4' deep, 3/8 C.Y. bucket	3.37	10.30	13.67
3560	6' deep, 3/8 C.Y. bucket	4.10	10.15	14.25
3580	8' deep, 1/2 C.Y. bucket	10.80	27	37.80
3600	10' deep, 1 C.Y. bucket	14.60	34.50	49.10
3800	4' wide, 2' deep, 3/8 C.Y. bucket	2.01	5.05	7.06
3820	3' deep, 3/8 C.Y. bucket	3.24	8.90	12.14
3840	4' deep, 1/2 C.Y. bucket	4.26	11.65	15.91
3860	6' deep, 1/2 C.Y. bucket	9.80	21.50	31.30
3880	8' deep, 1/2 C.Y. bucket	17.95	36	53.95
3900	10' deep, 1 C.Y. bucket	20.50	43.50	64
3920	12' deep, 1 C.Y. bucket	29.50	61.50	91
3940	15' deep, 1-1/2 C.Y. bucket	42.50	84.50	127
3960	18' deep, 2-1/2 C.Y. bucket	62	107	169
4030	6' wide, 6' deep, 5/8 C.Y. bucket w/trench box	15.70	27.50	43.20
4040	8' deep, 3/4 C.Y. bucket	24	42.50	66.50
4050	10' deep, 1 C.Y. bucket	26	49.50	75.50
4060	12' deep, 1-1/2 C.Y. bucket	33.50	62.50	96
4070	16' deep, 2 C.Y. bucket	60	96	156
4080	20' deep, 3-1/2 C.Y. bucket	81.50	133	214.50
4090	24' deep, 3-1/2 C.Y. bucket	119	187	306
4500	8' wide, 12' deep, 1-1/2 C.Y. bucket w/trench box	38.50	68	106.50
4550	15' deep, 1-1/2 C.Y. bucket	54	98.50	152.50
4600	18' deep, 2-1/2 C.Y. bucket	79.50	125	204.50
4650	24' deep, 3-1/2 C.Y. bucket	127	197	324
4800	10' wide, 20' deep, 3-1/2 C.Y. bucket w/trench box	95	150	245
4850	24' deep, 3-1/2 C.Y. bucket	136	208	344
4950	12' wide, 20' deep, 3-1/2 C.Y. bucket w/trench box	102	159	261
4980	25' deep, 3-1/2 C.Y. bucket	154	235	389

G10 Site Preparation

G1030 Site Earthwork

The Pipe Bedding System is shown for various pipe diameters. Compacted bank sand is used for pipe bedding and to fill 12″ over the pipe. No backfill is included. Various side slopes are shown to accommodate different soil conditions. Pipe sizes vary from 6″ to 84″ diameter.

System Components	QUANTITY	UNIT	COST PER L.F.		
			MAT.	INST.	TOTAL
SYSTEM G1030 815 1440					
PIPE BEDDING, SIDE SLOPE 0 TO 1, 1′ WIDE, PIPE SIZE 6″ DIAMETER					
Borrow, bank sand, 2 mile haul, machine spread	.086	C.Y.	1.69	.67	2.36
Compaction, vibrating plate	.086	C.Y.		.21	.21
TOTAL			1.69	.88	2.57

G1030 815	Pipe Bedding	COST PER L.F.		
		MAT.	INST.	TOTAL
1440	Pipe bedding, side slope 0 to 1, 1′ wide, pipe size 6″ diameter	1.69	.88	2.57
1460	2′ wide, pipe size 8″ diameter	3.66	1.88	5.54
1480	Pipe size 10″ diameter	3.74	1.92	5.66
1500	Pipe size 12″ diameter	3.82	1.96	5.78
1520	3′ wide, pipe size 14″ diameter	6.20	3.18	9.38
1540	Pipe size 15″ diameter	6.25	3.21	9.46
1560	Pipe size 16″ diameter	6.30	3.25	9.55
1580	Pipe size 18″ diameter	6.45	3.31	9.76
1600	4′ wide, pipe size 20″ diameter	9.15	4.70	13.85
1620	Pipe size 21″ diameter	9.25	4.74	13.99
1640	Pipe size 24″ diameter	9.45	4.84	14.29
1660	Pipe size 30″ diameter	9.60	4.94	14.54
1680	6′ wide, pipe size 32″ diameter	16.45	8.45	24.90
1700	Pipe size 36″ diameter	16.80	8.65	25.45
1720	7′ wide, pipe size 48″ diameter	26.50	13.75	40.25
1740	8′ wide, pipe size 60″ diameter	32.50	16.70	49.20
1760	10′ wide, pipe size 72″ diameter	45.50	23.50	69
1780	12′ wide, pipe size 84″ diameter	60	30.50	90.50
2140	Side slope 1/2 to 1, 1′ wide, pipe size 6″ diameter	3.16	1.62	4.78
2160	2′ wide, pipe size 8″ diameter	5.35	2.75	8.10
2180	Pipe size 10″ diameter	5.75	2.96	8.71
2200	Pipe size 12″ diameter	6.10	3.13	9.23
2220	3′ wide, pipe size 14″ diameter	8.80	4.52	13.32
2240	Pipe size 15″ diameter	9	4.63	13.63
2260	Pipe size 16″ diameter	9.25	4.76	14.01
2280	Pipe size 18″ diameter	9.70	4.99	14.69
2300	4′ wide, pipe size 20″ diameter	12.85	6.60	19.45
2320	Pipe size 21″ diameter	13.10	6.75	19.85
2340	Pipe size 24″ diameter	13.95	7.15	21.10
2360	Pipe size 30″ diameter	15.45	7.95	23.40
2380	6′ wide, pipe size 32″ diameter	23	11.70	34.70
2400	Pipe size 36″ diameter	24.50	12.45	36.95
2420	7′ wide, pipe size 48″ diameter	38	19.45	57.45
2440	8′ wide, pipe size 60″ diameter	48	24.50	72.50
2460	10′ wide, pipe size 72″ diameter	66	34	100
2480	12′ wide, pipe size 84″ diameter	86.50	44.50	131
2620	Side slope 1 to 1, 1′ wide, pipe size 6″ diameter	4.62	2.37	6.99
2640	2′ wide, pipe size 8″ diameter	7.15	3.66	10.81

G10 Site Preparation

G1030 Site Earthwork

G1030 815	Pipe Bedding	COST PER L.F.		
		MAT.	INST.	TOTAL
2660	Pipe size 10" diameter	7.75	3.97	11.72
2680	Pipe size 12" diameter	8.45	4.33	12.78
2700	3' wide, pipe size 14" diameter	11.40	5.85	17.25
2720	Pipe size 15" diameter	11.80	6.05	17.85
2740	Pipe size 16" diameter	12.20	6.25	18.45
2760	Pipe size 18" diameter	13	6.65	19.65
2780	4' wide, pipe size 20" diameter	16.50	8.50	25
2800	Pipe size 21" diameter	16.95	8.70	25.65
2820	Pipe size 24" diameter	18.40	9.45	27.85
2840	Pipe size 30" diameter	21.50	10.90	32.40
2860	6' wide, pipe size 32" diameter	29	15.05	44.05
2880	Pipe size 36" diameter	31.50	16.25	47.75
2900	7' wide, pipe size 48" diameter	49	25	74
2920	8' wide, pipe size 60" diameter	63.50	32.50	96
2940	10' wide, pipe size 72" diameter	86.50	44.50	131
2960	12' wide, pipe size 84" diameter	113	58	171

G20 Site Improvements

G2010 Roadways

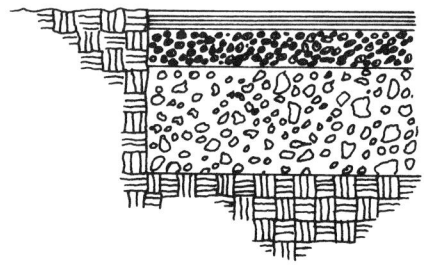

The Roadway System includes surveying; 8" of compacted bank-run gravel; fine grading with a grader and roller; and bituminous wearing course. Roadway Systems are shown on a cost per linear foot basis.

The Expanded System Listing shows Roadway Systems with widths that vary from 20' to 34'; for various pavement thicknesses.

System Components	QUANTITY	UNIT	COST PER L.F.		
			MAT.	INST.	TOTAL
SYSTEM G2010 210 1500					
ROADWAYS, BITUMINOUS CONC. PAVING, 2-1/2" THICK, 20' WIDE					
4 man surveying crew	.010	Ea.		24.11	24.11
Bank gravel, 2 mi. haul, place and spread by dozer	.494	C.Y.	13.09	3.85	16.94
Compaction, granular material to 98%	.494	C.Y.		3.12	3.12
Grading, fine grade, 3 passes with grader plus rolling	2.220	S.Y.		12.12	12.12
Bituminous, paving, wearing course 2-1/2" thick	2.220	S.Y.	24.20	3.64	27.84
Curbs, granite, split face, straight, 5" x 16"	2.000	L.F.	31.90	14.58	46.48
Painting lines, reflectorized, 4" wide	1.000	L.F.	.13	.17	.30
TOTAL			69.32	61.59	130.91

G2010 210	Roadway Pavement	COST PER L.F.		
		MAT.	INST.	TOTAL
1500	Roadways, bituminous conc. paving, 2-1/2" thick, 20' wide	69.50	62	131.50
1520	24' wide	77	63.50	140.50
1540	26' wide	80.50	66	146.50
1560	28' wide	84	67.50	151.50
1580	30' wide	88	69.50	157.50
1600	32' wide	92	72	164
1620	34' wide	95.50	73.50	169
1800	3" thick, 20' wide	74	60	134
1820	24' wide	82.50	64	146.50
1840	26' wide	86.50	66	152.50
1860	28' wide	91	68.50	159.50
1880	30' wide	95	70.50	165.50
1900	32' wide	99.50	72.50	172
1920	34' wide	103	74.50	177.50
2100	4" thick, 20' wide	81.50	60	141.50
2120	24' wide	91.50	64	155.50
2140	26' wide	96.50	66	162.50
2160	28' wide	101	68.50	169.50
2180	30' wide	106	70.50	176.50
2200	32' wide	111	72.50	183.50

G20 Site Improvements

G2010 Roadways

G2010 210	Roadway Pavement	COST PER L.F.		
		MAT.	INST.	TOTAL
2220	34' wide	116	74.50	190.50
2390				
2400	5" thick, 20' wide	98.50	61	159.50
2420	24' wide	112	65	177
2440	26' wide	118	67.50	185.50
2460	28' wide	125	69.50	194.50
2480	30' wide	132	72	204
2500	32' wide	138	74	212
2520	34' wide	145	76	221
3000	6" thick, 20' wide	95	62.50	157.50
3020	24' wide	108	67.50	175.50
3040	26' wide	114	70	184
3060	28' wide	120	72.50	192.50
3080	30' wide	127	74	201
3100	32' wide	133	76.50	209.50
3120	34' wide	139	79	218
3300	8" thick 20' wide	114	63	177
3320	24' wide	131	68	199
3340	26' wide	139	70.50	209.50
3360	28' wide	147	73	220
3380	30' wide	155	75	230
3400	32' wide	163	77.50	240.50
3420	34' wide	171	80	251
3600	12" thick 20' wide	144	67	211
3620	24' wide	166	72.50	238.50
3640	26' wide	178	75.50	253.50
3660	28' wide	189	78.50	267.50
3700	32' wide	211	84	295
3720	34' wide	222	87	309

G2010 305	Curbs & Berms	COST PER L.F.		
		MAT.	INST.	TOTAL
1000	Bituminous, curbs, 8" wide, 6" high	1.82	2.09	3.91
1100	8" high	2.43	2.33	4.76
1500	Berm, 12" wide, 3" to 6" high	.04	2.99	3.03
1600	1-1/2" to 4" high	.02	2.10	2.12
2000	Concrete, curb, 6" wide, 18" high, cast-in-place	2.99	6.45	9.44
2100	Precast	9.10	6.10	15.20
2500	Curb and gutter, monolithic, 6" high, 24" wide	16.60	8.50	25.10
2600	30" wide	18.25	9.35	27.60
3000	Granite, curb, 4-1/2" wide, 12" high precast	8	14.30	22.30
3100	5" wide, 16" high	15.95	7.30	23.25
3200	6" wide, 18" high	21	8.10	29.10
3290				

G20 Site Improvements

G2020 Parking Lots

The Parking Lot System includes: compacted bank-run gravel; fine grading with a grader and roller; and bituminous concrete wearing course. All Parking Lot systems are on a cost per car basis. There are three basic types of systems: 90° angle, 60° angle, and 45° angle. The gravel base is compacted to 98%. Final stall design and lay-out of the parking lot with precast bumpers, sealcoating and white paint is also included.

The Expanded System Listing shows the three basic parking lot types with various depths of both gravel base and wearing course. The gravel base depths range from 6″ to 10″. The bituminous paving wearing course varies from a depth of 3″ to 6″.

System Components	QUANTITY	UNIT	COST PER CAR		
			MAT.	INST.	TOTAL
SYSTEM G2020 210 1500					
PARKING LOT, 90° ANGLE PARKING, 3″ BITUMINOUS PAVING, 6″ GRAVEL BASE					
Surveying crew for layout, 4 man crew	.020	Day		48.21	48.21
Borrow, bank run gravel, haul 2 mi., spread w/dozer, no compaction	7.223	L.C.Y.	191.41	56.34	247.75
Grading, fine grade 3 passes with motor grader	43.333	S.Y.		98.78	98.78
Compact w/vibrating plate, 8″ lifts, granular mat'l. to 98%	7.223	E.C.Y.		45.57	45.57
Binder course paving 2″ thick	43.333	S.Y.	342.33	55.90	398.23
Wear course paving 1″ thick	43.333	S.Y.	169.87	36.83	206.70
Seal coating, petroleum resistant under 1,000 S.Y.	43.333	S.Y.	59.80	55.03	114.83
Lines on pvmt., parking stall, paint, white, 4″ wide	1.000	Ea.	4.94	3.60	8.54
Precast concrete parking bar, 6″ x 10″ x 6′-0″	1.000	Ea.	54	18.40	72.40
TOTAL			822.35	418.66	1,241.01

G2020 210	Parking Lots Gravel Base	COST PER CAR		
		MAT.	INST.	TOTAL
1500	Parking lot, 90° angle parking, 3″ bituminous paving, 6″ gravel base	820	420	1,240
1520	8″ gravel base	885	450	1,335
1540	10″ gravel base	950	485	1,435
1560	4″ bituminous paving, 6″ gravel base	1,025	445	1,470
1580	8″ gravel base	1,100	475	1,575
1600	10″ gravel base	1,175	515	1,690
1620	6″ bituminous paving, 6″ gravel base	1,375	475	1,850
1640	8″ gravel base	1,450	505	1,955
1660	10″ gravel base	1,500	540	2,040
1800	60° angle parking, 3″ bituminous paving, 6″ gravel base	820	420	1,240
1820	8″ gravel base	885	450	1,335
1840	10″ gravel base	950	485	1,435
1860	4″ bituminous paving, 6″ gravel base	1,025	445	1,470
1880	8″ gravel base	1,100	475	1,575
1900	10″ gravel base	1,175	515	1,690
1920	6″ bituminous paving, 6″ gravel base	1,375	475	1,850
1940	8″ gravel base	1,450	505	1,955
1960	10″ gravel base	1,500	540	2,040
2200	45° angle parking, 3″ bituminous paving, 6″ gravel base	840	425	1,265
2220	8″ gravel base	905	465	1,370
2240	10″ gravel base	975	500	1,475
2260	4″ bituminous paving, 6″ gravel base	1,050	455	1,505
2280	8″ gravel base	1,125	490	1,615
2300	10″ gravel base	1,200	525	1,725
2320	6″ bituminous paving, 6″ gravel base	1,400	490	1,890
2340	8″ gravel base	1,475	520	1,995
2360	10″ gravel base	1,550	555	2,105

Note: Row 1560 also shows reference "RG2020-500".

G20 Site Improvements

G2030 Pedestrian Paving

The Bituminous and Concrete Sidewalk Systems include excavation, hand grading and compacted gravel base. Pavements are shown for two conditions of each of the following variables; pavement thickness, gravel base thickness and pavement width. Costs are given on a linear foot basis.

The Plaza Systems listed include several brick and tile paving surfaces on two different bases: gravel and slab on grade. The type of bedding for the pavers depends on the base being used, and alternate bedding may be desirable. Also included in the paving costs are edging and precast grating costs and where concrete bases are involved, expansion joints. Costs are given on a square foot basis.

G2030 110	Bituminous Sidewalks	COST PER L.F.		
		MAT.	INST.	TOTAL
1580	Bituminous sidewalk, 1" thick paving, 4" gravel base, 3' width	2.27	4.43	6.70
1600	4' width	3.01	4.82	7.83
1640	6" gravel base, 3' width	2.77	4.67	7.44
1660	4' width	3.67	5.10	8.77
2120	2" thick paving, 4" gravel base, 3' width	3.54	5.30	8.84
2140	4' width	4.72	5.95	10.67
2180	6" gravel base, 3' width	4.04	5.55	9.59
2200	4' width	5.40	6.25	11.65

G2030 120	Concrete Sidewalks	COST PER L.F.		
		MAT.	INST.	TOTAL
1580	Concrete sidewalk, 4" thick, 4" gravel base, 3' wide	6.30	11.50	17.80
1600	4' wide	8.40	14.20	22.60
1640	6" gravel base, 3' wide	6.80	11.75	18.55
1660	4' wide	9.10	14.50	23.60
2120	6" thick concrete, 4" gravel base, 3' wide	9.30	13.20	22.50
2140	4' wide	12.40	16.25	28.65
2180	6" gravel base, 3' wide	9.80	13.40	23.20
2200	4' wide	13.05	16.55	29.60

G2030 150	Brick & Tile Plazas	COST PER S.F.		
		MAT.	INST.	TOTAL
2050	Brick pavers, 4" x 8" x 1-3/4", gravel base, stone dust bedding	6.05	5.70	11.75
2100	Slab on grade, asphalt bedding	8.15	8.35	16.50
3550	Concrete paving stone, 4" x 8" x 2-1/2", gravel base, sand bedding	4.33	4.02	8.35
3600	Slab on grade, asphalt bedding	6.05	6.15	12.20
4050	Concrete patio blocks, 8" x 16" x 2", gravel base, sand bedding	4.82	5.20	10.02
4100	Slab on grade, asphalt bedding	6.70	7.75	14.45
6050	Granite pavers, 3-1/2" x 3-1/2" x 3-1/2", gravel base, sand bedding	14	12.65	26.65
6100	Slab on grade, mortar bedding	16.80	20	36.80

G20 Site Improvements

G2040 Site Development

G2040 105	Fence & Guardrails	COST PER L.F.		
		MAT.	INST.	TOTAL
1000	Fence, chain link, 2" post, 1-5/8" rail, 9 ga. galv. wire, 5' high	21	5.35	26.35
1050	6' high w/barb. wire	21.50	6.70	28.20
1100	6 ga. galv. wire, 6' high	22	6.45	28.45
1150	6' high w/barb. wire	23.50	6.70	30.20
1200	1-3/8" rail, 11 ga., galv. wire, 10' high	25	14.35	39.35
1250	12' high	27	16	43
1300	9 ga. vinyl covered wire, 10' high	23	14.35	37.35
1350	12' high	27.50	16	43.50
1500	1-5/8" post, 1-3/8" rail, 11 ga. galv. wire, 3' high	2.31	3.22	5.53
1550	4' high	7.90	4.03	11.93
1600	9 ga. vinyl covered wire, 3' high	7.90	3.22	11.12
1650	4' high	9	4.03	13.03
2000	Guardrail, corrugated steel, galvanized steel posts	22.50	3.20	25.70
2100	Timber, 6" x 8" posts, 4" x 8" rails	11.60	2.84	14.44
2200	Steel box beam, 6" x 6" rails	35	22.50	57.50
2250	6" x 8" rails	46	12.70	58.70
3000	Barrier, median, precast concrete, single face	55.50	11.25	66.75
3050	Double face	64	12.60	76.60

G2040 810	Flagpoles	COST EACH		
		MAT.	INST.	TOTAL
0110	Flagpoles, on grade, aluminum, tapered, 20' high	1,075	675	1,750
0120	70' high	8,875	1,700	10,575
0130	Fiberglass, tapered, 23' high	650	675	1,325
0140	59' high	5,700	1,500	7,200
0150	Concrete, internal halyard, 20' high	1,300	540	1,840
0160	100' high	17,100	1,350	18,450

G2040 950	Other Site Development, EACH	COST EACH		
		MAT.	INST.	TOTAL
0110	Grandstands, permanent, closed deck, steel, economy (per seat)			60.50
0120	Deluxe (per seat)			121
0130	Composite design, economy (per seat)			143
0140	Deluxe (per seat)			286

G30 Site Mechanical Utilities

G3010 Water Supply

G3010 110	Water Distribution Piping	COST PER L.F.		
		MAT.	INST.	TOTAL
2000	Piping, excav. & backfill excl., ductile iron class 250, mech. joint			
2130	4" diameter	16.05	16.10	32.15
2150	6" diameter	18.20	20	38.20
2160	8" diameter	24	24	48
2170	10" diameter	31.50	28	59.50
2180	12" diameter	39.50	30.50	70
2210	16" diameter	67.50	44	111.50
2220	18" diameter	66	46.50	112.50
3000	Tyton joint			
3130	4" diameter	17.95	8.05	26
3150	6" diameter	18.55	9.65	28.20
3160	8" diameter	25	16.10	41.10
3170	10" diameter	36.50	17.70	54.20
3180	12" diameter	38.50	20	58.50
3210	16" diameter	57.50	28	85.50
3220	18" diameter	64	32	96
3230	20" diameter	66.50	36.50	103
3250	24" diameter	81	42	123
4000	Copper tubing, type K			
4050	3/4" diameter	9.55	3.03	12.58
4060	1" diameter	12.75	3.78	16.53
4080	1-1/2" diameter	20.50	4.57	25.07
4090	2" diameter	31	5.25	36.25
4110	3" diameter	64	9.05	73.05
4130	4" diameter	106	12.75	118.75
4150	6" diameter	262	23.50	285.50
5000	Polyvinyl chloride class 160, S.D.R. 26			
5130	1-1/2" diameter	.46	1.13	1.59
5150	2" diameter	.75	1.23	1.98
5160	4" diameter	2.89	5.55	8.44
5170	6" diameter	6.10	6.65	12.75
5180	8" diameter	9.05	8	17.05
6000	Polyethylene 160 psi, S.D.R. 7			
6050	3/4" diameter	.48	1.61	2.09
6060	1" diameter	.66	1.75	2.41
6080	1-1/2" diameter	1.36	1.88	3.24
6090	2" diameter	2.27	2.32	4.59

G30 Site Mechanical Utilities

G3020 Sanitary Sewer

G3020 110	Drainage & Sewage Piping	COST PER L.F.		
		MAT.	INST.	TOTAL
2000	Piping, excavation & backfill excluded, PVC, plain			
2130	4" diameter	1.60	3.92	5.52
2150	6" diameter	3.59	4.20	7.79
2160	8" diameter	7.70	4.39	12.09
2900	Box culvert, precast, 8' long			
3000	6' x 3'	275	33	308
3020	6' x 7'	345	37	382
3040	8' x 3'	340	41	381
3060	8' x 8'	415	46.50	461.50
3080	10' x 3'	475	42	517
3100	10' x 8'	550	58	608
3120	12' x 3'	715	46.50	761.50
3140	12' x 8'	915	69.50	984.50
4000	Concrete, nonreinforced			
4150	6" diameter	7.25	11.90	19.15
4160	8" diameter	8	14.15	22.15
4170	10" diameter	8.85	14.65	23.50
4180	12" diameter	10.55	15.80	26.35
4200	15" diameter	14.50	17.60	32.10
4220	18" diameter	17.10	22	39.10
4250	24" diameter	28.50	31.50	60
4400	Reinforced, no gasket			
4580	12" diameter	12.10	21	33.10
4600	15" diameter	15.35	21	36.35
4620	18" diameter	20.50	24	44.50
4650	24" diameter	29	31.50	60.50
4670	30" diameter	52.50	46.50	99
4680	36" diameter	68	57	125
4690	42" diameter	87.50	63	150.50
4700	48" diameter	107	70.50	177.50
4720	60" diameter	146	94.50	240.50
4730	72" diameter	225	113	338
4740	84" diameter	305	142	447
4800	With gasket			
4980	12" diameter	13.30	11.40	24.70
5000	15" diameter	16.90	12	28.90
5020	18" diameter	22.50	12.60	35.10
5050	24" diameter	36	14.05	50.05
5070	30" diameter	60.50	46.50	107
5080	36" diameter	78.50	57	135.50
5090	42" diameter	96	56.50	152.50
5100	48" diameter	120	70.50	190.50
5120	60" diameter	180	90.50	270.50
5130	72" diameter	245	113	358
5140	84" diameter	365	189	554
5700	Corrugated metal, alum. or galv. bit. coated			
5760	8" diameter	9.45	9.55	19
5770	10" diameter	9.85	12.15	22
5780	12" diameter	12.10	15.05	27.15
5800	15" diameter	16.55	15.80	32.35
5820	18" diameter	18.35	16.70	35.05
5850	24" diameter	22	19.80	41.80
5870	30" diameter	29.50	34	63.50
5880	36" diameter	36.50	34	70.50
5900	48" diameter	55.50	41	96.50
5920	60" diameter	83.50	60.50	144
5930	72" diameter	100	101	201
6000	Plain			

G30 Site Mechanical Utilities

G3020 Sanitary Sewer

G3020 110	Drainage & Sewage Piping	COST PER L.F.		
		MAT.	INST.	TOTAL
6060	8" diameter	8.55	8.95	17.50
6070	10" diameter	9.80	11.30	21.10
6080	12" diameter	10.90	14.40	25.30
6100	15" diameter	13.10	14.40	27.50
6120	18" diameter	16.55	15.40	31.95
6140	24" diameter	19.80	18.10	37.90
6170	30" diameter	26.50	31.50	58
6180	36" diameter	33	31.50	64.50
6200	48" diameter	50	37.50	87.50
6220	60" diameter	75.50	58	133.50
6230	72" diameter	195	148	343
6300	Steel or alum. oval arch, coated & paved invert			
6400	15" equivalent diameter	13.40	15.80	29.20
6420	18" equivalent diameter	17.05	21	38.05
6450	24" equivalent diameter	23	25	48
6470	30" equivalent diameter	32	31.50	63.50
6480	36" equivalent diameter	42	41	83
6490	42" equivalent diameter	49	46	95
6500	48" equivalent diameter	61.50	54.50	116
6600	Plain			
6700	15" equivalent diameter	12.10	14.10	26.20
6720	18" equivalent diameter	15.40	18.10	33.50
6750	24" equivalent diameter	21	21	42
6770	30" equivalent diameter	28.50	38	66.50
6780	36" equivalent diameter	37.50	38	75.50
6790	42" equivalent diameter	44	44.50	88.50
6800	48" equivalent diameter	55	54.50	109.50
8000	Polyvinyl chloride SDR 35			
8130	4" diameter	1.60	3.92	5.52
8150	6" diameter	3.59	4.20	7.79
8160	8" diameter	7.70	4.39	12.09
8170	10" diameter	12.55	5.80	18.35
8180	12" diameter	14.05	5.95	20
8200	15" diameter	15.75	8	23.75

G30 Site Mechanical Utilities

G3030 Storm Sewer

Manhole   Catch Basin

The Manhole and Catch Basin System includes: excavation with a backhoe; a formed concrete footing; frame and cover; cast iron steps and compacted backfill.

The Expanded System Listing shows manholes that have a 4', 5' and 6' inside diameter riser. Depths range from 4' to 14'. Construction material shown is either concrete, concrete block, precast concrete, or brick.

System Components	QUANTITY	UNIT	COST PER EACH		
			MAT.	INST.	TOTAL
SYSTEM G3030 210 1920					
MANHOLE/CATCH BASIN, BRICK, 4' I.D. RISER, 4' DEEP					
Excavation, hydraulic backhoe, 3/8 C.Y. bucket	14.815	B.C.Y.		117.93	117.93
Trim sides and bottom of excavation	64.000	S.F.		58.88	58.88
Forms in place, manhole base, 4 uses	20.000	SFCA	14.20	101	115.20
Reinforcing in place footings, #4 to #7	.019	Ton	20.90	22.80	43.70
Concrete, 3000 psi	.925	C.Y.	98.98		98.98
Place and vibrate concrete, footing, direct chute	.925	C.Y.		47.01	47.01
Catch basin or MH, brick, 4' ID, 4' deep	1.000	Ea.	430	980	1,410
Catch basin or MH steps; heavy galvanized cast iron	1.000	Ea.	19.80	13.55	33.35
Catch basin or MH frame and cover	1.000	Ea.	253	235	488
Fill, granular	12.954	L.C.Y.	297.94		297.94
Backfill, spread with wheeled front end loader	12.954	L.C.Y.		32.52	32.52
Backfill compaction, 12" lifts, air tamp	12.954	E.C.Y.		111.79	111.79
TOTAL			1,134.82	1,720.48	2,855.30

G3030 210		Manholes & Catch Basins	COST PER EACH		
			MAT.	INST.	TOTAL
1920	Manhole/catch basin, brick, 4' I.D. riser, 4' deep		1,125	1,725	2,850
1940		6' deep	1,600	2,375	3,975
1960		8' deep	2,150	3,250	5,400
1980		10' deep	2,600	4,050	6,650
3000		12' deep	3,300	4,375	7,675
3020		14' deep	4,150	6,150	10,300
3200	Block, 4' I.D. riser, 4' deep		1,100	1,400	2,500
3220		6' deep	1,525	1,950	3,475
3240		8' deep	2,050	2,700	4,750
3260		10' deep	2,425	3,350	5,775
3280		12' deep	3,100	4,275	7,375
3300		14' deep	3,900	5,200	9,100
4620	Concrete, cast-in-place, 4' I.D. riser, 4' deep		1,250	2,375	3,625
4640		6' deep	1,800	3,175	4,975
4660		8' deep	2,500	4,575	7,075
4680		10' deep	3,000	5,700	8,700
4700		12' deep	3,800	7,075	10,875
4720		14' deep	4,725	8,475	13,200
5820	Concrete, precast, 4' I.D. riser, 4' deep		1,500	1,275	2,775
5840		6' deep	2,025	1,700	3,725

G30 Site Mechanical Utilities

G3030 Storm Sewer

G3030 210	Manholes & Catch Basins	COST PER EACH		
		MAT.	INST.	TOTAL
5860	8' deep	2,550	2,375	4,925
5880	10' deep	3,075	2,925	6,000
5900	12' deep	3,875	3,600	7,475
5920	14' deep	4,800	4,625	9,425
6000	5' I.D. riser, 4' deep	2,450	1,425	3,875
6020	6' deep	3,100	2,025	5,125
6040	8' deep	3,975	2,675	6,650
6060	10' deep	5,075	3,425	8,500
6080	12' deep	6,300	4,325	10,625
6100	14' deep	7,650	5,300	12,950
6200	6' I.D. riser, 4' deep	3,300	1,900	5,200
6220	6' deep	4,125	2,500	6,625
6240	8' deep	5,225	3,500	8,725
6260	10' deep	6,600	4,450	11,050
6280	12' deep	8,100	5,600	13,700
6300	14' deep	9,725	6,800	16,525

G30 Site Mechanical Utilities

G3060 Fuel Distribution

G3060 110	Gas Service Piping	COST PER L.F.		
		MAT.	INST.	TOTAL
2000	Piping, excavation & backfill excluded, polyethylene			
2070	1-1/4" diam, SDR 10	1.91	3.88	5.79
2090	2" diam, SDR 11	3.12	4.32	7.44
2110	3" diam, SDR 11.5	6.10	5.15	11.25
2130	4" diam, SDR 11	13.30	9.10	22.40
2150	6" diam, SDR 21	36.50	9.85	46.35
2160	8" diam, SDR 21	56	11.80	67.80
3000	Steel, schedule 40, plain end, tarred & wrapped			
3060	1" diameter	6.40	8.75	15.15
3090	2" diameter	10.05	9.35	19.40
3110	3" diameter	16.65	10.05	26.70
3130	4" diameter	21.50	15.90	37.40
3140	5" diameter	31.50	18.50	50
3150	6" diameter	38.50	22.50	61
3160	8" diameter	61	29	90
3170	10" diameter	86.50	40.50	127
3180	12" diameter	100	51	151
3190	14" diameter	106	54	160
3200	16" diameter	116	58.50	174.50
3210	18" diameter	150	62.50	212.50
3220	20" diameter	233	67.50	300.50
3230	24" diameter	266	81.50	347.50

G40 Site Electrical Utilities

G4020 Site Lighting

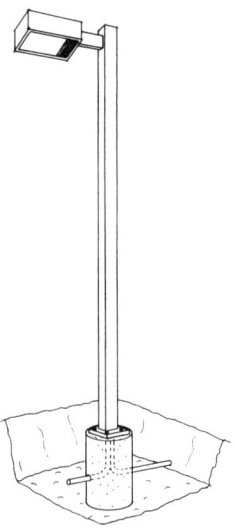

The Site Lighting System includes the complete unit from foundation to electrical fixtures. Each system includes: excavation; concrete base; backfill by hand; compaction with a plate compacter; pole of specified material; all fixtures; and lamps.

The Expanded System Listing shows Site Lighting Systems that use one of two types of lamps: high pressure sodium; and metal halide. Systems are listed for 400-watt and 1000-watt lamps. Pole height varies from 20' to 40'. There are four types of poles possibly listed: aluminum, fiberglass, steel and wood.

G4020 110	Site Lighting	COST EACH		
		MAT.	INST.	TOTAL
2320	Site lighting, high pressure sodium, 400 watt, aluminum pole, 20' high	1,725	1,350	3,075
2360	40' high	3,150	2,150	5,300
2920	Wood pole, 20' high	1,025	1,325	2,350
2960	40' high	1,600	2,050	3,650
3120	1000 watt, aluminum pole, 20' high	1,800	1,400	3,200
3160	40' high	3,225	2,175	5,400
3520	Wood pole, 20' high	1,100	1,350	2,450
3560	40' high	1,700	2,075	3,775
5820	Metal halide, 400 watt, aluminum pole, 20' high	1,700	1,350	3,050
5860	40' high	3,125	2,150	5,275
7620	1000 watt, aluminum pole, 20' high	1,775	1,400	3,175
7660	40' high	3,200	2,175	5,375

G40 Site Electrical Utilities

G4020 Site Lighting

Table G4020 210 Procedure for Calculating Floodlights Required for Various Footcandles

Poles should not be spaced more than 4 times the fixture mounting height for good light distribution.

Estimating Chart

Select Lamp type.

Determine total square feet.

Chart will show quantity of fixtures to provide 1 footcandle initial, at intersection of lines. Multiply fixture quantity by desired footcandle level.

Chart based on use of wide beam luminaires in an area whose dimensions are large compared to mounting height and is approximate only.

To maintain 1 footcandle over a large area use these watts per square foot:
- Incandescent 0.15
- Metal Halide 0.032
- Mercury Vapor 0.05
- High Pressure Sodium 0.024

System Components	QUANTITY	UNIT	COST EACH		
			MAT.	INST.	TOTAL
SYSTEM G4020 210 0200					
LIGHT POLES, ALUMINUM, 20' HIGH, 1 ARM BRACKET					
Aluminum light pole, 20', no concrete base	1.000	Ea.	1,000	587	1,587
Bracket arm for Aluminum light pole	1.000	Ea.	130	78.50	208.50
Excavation by hand, pits to 6' deep, heavy soil or clay	2.368	C.Y.		258.11	258.11
Footing, concrete incl forms, reinforcing, spread, under 1 C.Y.	.465	C.Y.	82.77	123.83	206.60
Backfill by hand	1.903	C.Y.		75.17	75.17
Compaction vibrating plate	1.903	C.Y.		10.26	10.26
TOTAL			1,212.77	1,132.87	2,345.64

G4020 210	Light Pole (Installed)	COST EACH		
		MAT.	INST.	TOTAL
0200	Light pole, aluminum, 20' high, 1 arm bracket	1,225	1,125	2,350
0240	2 arm brackets	1,350	1,125	2,475
0280	3 arm brackets	1,475	1,175	2,650
0320	4 arm brackets	1,600	1,175	2,775
0360	30' high, 1 arm bracket	2,125	1,425	3,550
0400	2 arm brackets	2,275	1,425	3,700
0440	3 arm brackets	2,400	1,475	3,875
0480	4 arm brackets	2,525	1,475	4,000
0680	40' high, 1 arm bracket	2,625	1,900	4,525
0720	2 arm brackets	2,750	1,900	4,650
0760	3 arm brackets	2,900	1,950	4,850
0800	4 arm brackets	3,025	1,950	4,975
0840	Steel, 20' high, 1 arm bracket	1,450	1,200	2,650
0880	2 arm brackets	1,550	1,200	2,750
0920	3 arm brackets	1,575	1,225	2,800
0960	4 arm brackets	1,725	1,225	2,950
1000	30' high, 1 arm bracket	1,700	1,525	3,225
1040	2 arm brackets	1,800	1,525	3,325
1080	3 arm brackets	1,825	1,575	3,400
1120	4 arm brackets	1,950	1,575	3,525
1320	40' high, 1 arm bracket	2,225	2,075	4,300
1360	2 arm brackets	2,325	2,075	4,400
1400	3 arm brackets	2,350	2,100	4,450
1440	4 arm brackets	2,475	2,100	4,575

Note: RD5020-250 (reference at row 0320)

Reference Section

All the reference information is in one section, making it easy to find what you need to know ... and easy to use the book on a daily basis. This section is visually identified by a vertical gray bar on the page edges.

In this section, you'll see the background that relates to the reference numbers that appeared in the Assembly Cost Sections. You'll find reference tables, explanations, and estimating information that support how we derive the systems data. Also included are alternate pricing methods, technical data, and estimating procedures, along with information on design and economy in construction.

Also in this Reference Section, we've included Historical Cost Indexes for cost comparison over time; City Cost Indexes and Location Factors for adjusting costs to the region you are in; Square Foot Costs; Reference Aids; Estimating Forms; and an explanation of all Abbreviations used in the book.

Table of Contents

Reference Tables

A10 Foundations
RA1010 Standard Foundations 442
RA1020 Special Foundations 449
RA1030 Slab on Grade 452

A20 Basement Construction
RA2020 Basement Walls 453

B10 Superstructure
RB1010 Floor Construction 455
RB1020 Roof Construction 462

B20 Exterior Enclosure
RB2010 Exterior Walls 464

B30 Roofing
RB3010 Roof Coverings 472

C20 Stairs
RC2010 Stair Construction 474

D10 Conveying
RD1010 Elevators & Lifts 475

D20 Plumbing
RD2010 Plumbing Fixtures 477
RD2020 Domestic Water Distribution 481

D30 HVAC
RD3010 Energy Supply 482
RD3020 Heat Generating Systems 483
RD3030 Cooling Generating Systems 486

D40 Fire Protection
RD4010 Sprinklers 489
RD4020 Standpipes 490

D50 Electrical
RD5010 Electrical Service & Distribution. 491
RD5020 Lighting & Branch Wiring 509

G10 Site Preparation
RG1010 Site Clearing 514
RG1030 Site Earthwork 514

G20 Site Improvements
RG2020 Parking Lots 515

Reference Tables (cont.)

H General Conditions
RH1010 Overhead, Design & Engineering, Worker's Compensation 517
RH1020 Overhead & Miscellaneous Data & Overtime 523
RH1030 Sales Tax 523
RH1040 Temporary Supports 524

City Adjustments
RJ1010 Component Indexes 525
RJ1020 Historical Cost Indexes 526
RJ1030 City Cost Indexes 527
RJ1040 Location Factors 538

Square Foot Costs
RK1010 Square Foot Costs & Area Requirements 544

Reference Aids
RL1010 Live Loads, Dead Loads, Wind Loads, & Snow Loads 554
RL1020 Area Requirements & Building Code Requirements 561
RL1030 "U" & "R" Values & Thermal Properties 564
RL1040 Construction Classification 572
RL1090 Conversion Factors 574

Estimating Forms 577

Abbreviations 583

A10 Foundations — RA1010-120 Spread Footings

General: A spread footing is used to convert a concentrated load (from one superstructure column, or substructure grade beams) into an allowable area load on supporting soil.

Because of punching action from the column load, a spread footing is usually thicker than strip footings which support wall loads. One or two story commercial or residential buildings should have no less than 1' thick spread footings. Heavier loads require no less than 2' thick. Spread footings may be square, rectangular or octagonal in plan.

Spread footings tend to minimize excavation and foundation materials, as well as labor and equipment. Another advantage is that footings and soil conditions can be readily examined. They are the most widely used type of footing, especially in mild climates and for buildings of four stories or under. This is because they are usually more economical than other types, if suitable soil and site conditions exist.

They are used when suitable supporting soil is located within several feet of the surface or line of subsurface excavation. Suitable soil types include sands and gravels, gravels with a small amount of clay or silt, hardpan, chalk, and rock. Pedestals may be used to bring the column base load down to the top of footing. Alternately, undesirable soil between underside of footing and top of bearing level can be removed and replaced with lean concrete mix or compacted granular material.

Depth of footing should be below topsoil, uncompacted fill, muck, etc. It must be lower than frost penetration (see local code or Table L1030-502 in Section L) but should be above the water table. It must not be at the ground surface because of potential surface erosion. If the ground slopes, approximately three horizontal feet of edge protection must remain. Differential footing elevations may overlap soil stresses or cause excavation problems if clear spacing between footings is less than the difference in depth.

Other footing types are usually used for the following reasons:
   A. Bearing capacity of soil is low.
   B. Very large footings are required, at a cost disadvantage.
   C. Soil under footing (shallow or deep) is very compressible, with probability of causing excessive or differential settlement.
   D. Good bearing soil is deep.
   E. Potential for scour action exists.
   F. Varying subsoil conditions within building perimeter.

Cost of spread footings for a building is determined by:
1. The soil bearing capacity.
2. Typical bay size.
3. Total load (live plus dead) per S.F. for roof and elevated floor levels.
4. The size and shape of the building.
5. Footing configuration. Does the building utilize outer spread footings or are there continuous perimeter footings only or a combination of spread footings plus continuous footings?

COST DETERMINATION

1. Determine Soil Bearing Capacity by a known value or using Table A1010-121 as a guide.

Table A1010-121 Soil Bearing Capacity in Kips per S.F.

Bearing Material	Typical Allowable Bearing Capacity
Hard sound rock	120 KSF
Medium hard rock	80
Hardpan overlaying rock	24
Compact gravel and boulder-gravel; very compact sandy gravel	20
Soft rock	16
Loose gravel; sandy gravel; compact sand; very compact sand-inorganic silt	12
Hard dry consolidated clay	10
Loose coarse to medium sand; medium compact fine sand	8
Compact sand-clay	6
Loose fine sand; medium compact sand-inorganic silts	4
Firm or stiff clay	3
Loose saturated sand-clay; medium soft clay	2

A10 Foundations — RA1010-120 Spread Footings

Table A1010-122 Working Load Determination

2. **Determine Bay Size in S.F.**
3. **Determine Load to the Footing**

Type Load	Reference	Working Loads			
		Roof	Floor	Column	Total
1. Total Load	Floor System	PSF	PSF		
2. Whole Bay Load/Level	(Line 1) x Bay Area ÷ 1000	Kips	Kips		
3. Load to Column	[(Roof + (Floor x No. Floors)]	Kips	Kips		Kips
4. Column Weight	Systems B1010 201 to B1010 208			Kips	Kips
5. Fireproofing	System B1010 720			Kips	Kips
6. Total Load to Footing	Roof + Floor + Column				Kips

Working loads represent the highest actual load a structural member is designed to support.

4. Determine Size and Shape of Building

Figure A1010-123

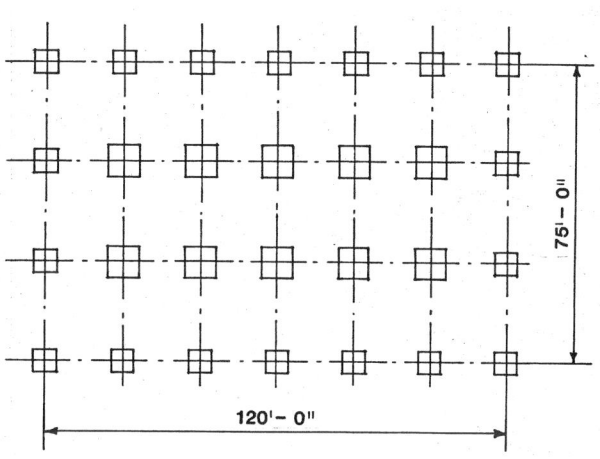

Figure A1010-124

5. Footing Configuration Possibilities

a. Building has spread footings throughout (including exterior and corners).

b. Building has continuous perimeter footings to support exterior columns, corners and exterior walls (as well as interior spread footings).

c. Building has both perimeter continuous footings for walls plus exterior spread footings for column loads. If there are interior spread footings also, use "b".
  1. Figure continuous footings from tables in Section A1010-110.
  2. Add total dollars for spread footings and continuous footings to determine total foundation dollars.

A10 Foundations — RA1010-120 Spread Footings

Table A1010-126 Working Load Determination

Tabulate interior footing costs and add continuous footing costs from Systems A1010 210 and A1010 110.

Example: Spread Footing Cost Determination

Office Building	5 Story
Story Height	12'
Dimensions	75' x 120'
Bay Size	20' x 25'
Soil Bearing Capacity	6 KSF
Foundation	Spread Footings
Slab on Grade	5" Reinforced
Code	BOCA
Construction	From Example RB1010-112
	Using concrete flat plate

Figure A1010-125

Determine load to footings Table A1010-122 or Example RB1010-112.

Type Load	Reference	Working Loads			
		Roof	Floor	Column	Total
Determine floor system	System in Div. B1010 (Example: Flat Plate)				
1. Total Load	System # B1010 223	146PSF	188PSF		
2. Whole Bay Load/Level	Bay Area x Line 1/1000	73K	94K		
3. Load To Column	Roof + (Floor x No. Floors)	73K	276K		449K
4. Column Weight	Systems B1010 201 to B1010 208			25K	25K
5. Fireproofing	N/A			—	
6. Load to Footing	Roof + Floor + Column				474K

1. Total load working from System B1010 223: 40# superimposed load for roof and 75# superimposed load for floors.
2. Bay Area 20' x 25' = 500 S.F. x Line 1
   500 x 146/1000 = 73K roof
   500 x 188/1000 = 94K floor
3. Roof + (floor x no. floors)
   73K + (4 x 94K) = 449K

4. Enter System B1010 203 with a total working load of 474K and 12' story height. A 16" square tied column with 6 KSI concrete is most economical. Minimum column size from System B1010 223 is 20". Use the larger. Column weight = 394 #/V.L.F. x 12' = approx. 5K.
5. No fireproofing is required for concrete column.
6. Add roof, floors and column loads for a total load to each interior footing.

Footing Cost: Enter System A1010 210 with total working load and allowable soil pressure. Determine the cost per footing using the closest higher load on the table. Add the total costs for interior, exterior, and corner footings.

A10 Foundations — RA1010-120 Spread Footings

Table A1010-127 Typical Spread Footing, Square

EXTERIOR FOOTINGS
Exterior vs. Interior Footings

System A1010 210 contains the cost for individual spread interior footings for various loading and soil bearing conditions. To determine the loads for exterior footings:

1. Multiply whole bay working load by 60% for half bay footings and by 45% for quarter bay (or corner) footings. This will give the approximate column load for half and quarter bay footings.
2. Enter System A1010 210 and find cost of footing with loads determined in step 1.
3. Determine the number of each type of footing on the job and multiply the cost per footing by the appropriate quantity. This will give the total dollar cost for exterior and interior spread footings.

Example: Using soil capacity, 6KSF
Total working load from Table A1010-126 .... = 474K
Half Bay Footings .... 474K x 60% = 284K
Corner Footings .... 474K x 45% = 213K

Table A1010-128 Spread Footing Cost

Footing Type	Working Load (Kips)	Total Quantity	System A1010 210 Unit Cost	Total
Interior	474	10	$2,500	$25,000
Exterior	284	14	1,385	19,390
Corner	213	4	1,385	5,540
Total				$49,930

A10 Foundations — RA1010-140 Strip Footings

Table A1010-141 Section Thru Typical Strip Footing

General: Strip (continuous) footings are a type of spread footing, but are used to convert the lineal type wall into an allowable area load on supporting soil.

Many of the general comments on isolated spread footings (Section A1010 210) apply here as well.

Strip footings may be used under walls of concrete, brick, block or stone. They are constructed of continuously placed concrete, 2000 psi or greater. Normally they are not narrower than twice the wall thickness, nor are they thinner than the wall thickness. Plain concrete footings should be at least 8" thick, and reinforced footings require at least 6" of thickness above the bottom reinforcement. Where there is no basement, the bottom of footing is usually 3' to 5' below finished exterior grade, and 12" below the average frost penetration (see Figure A1010-141), resting on undisturbed soil. In some cases strip footings serve as leveling pads (stone walls, etc.).

Method of construction is affected by reinforcement and soil type. If steel mats and/or dowels are present and require accurate placement (frequently for reinforced concrete or reinforced masonry walls), then side forms are required, as they are in wet or sandy soil. If there is no reinforcement or only longitudinal reinforcement, and the soil will "stand" forms might be eliminated and economically should be.

Preferably, strip footings are used on sand and gravel. The presence of a small amount of clay or dense silty sand in gravel is acceptable, as is bearing on rock or chalk. They are sometimes used in uniform, firm or stiff clays with little nearby ground vegetation, and placed at least 3-1/2' into the clay. If the clay is sloped, there is potential for downhill creep. Footings up to 3-1/2' wide are sometimes used in soft or silty clays, but settlement must be expected and provision made.

Sloping terrain requires stepping the footing to maintain depth, since a sloped footing will create horizontal thrust which may distress the structure. Steps should be at least 2' long horizontally and each vertical step no greater than three-fourths the horizontal distance between steps. Vertical risers must be at least 6" thick and of footing width. Potential for erosion of surrounding soil must be considered.

Strip footings should be used under walls constructed of unreinforced concrete, or of units bonded with mortar. They are also used when bearing capacity of soil is inadequate to support wall thickness alone.

Footings require horizontal reinforcement at right angles to the wall line when side projection of footings beyond face of wall is greater than 1/2 to 2/3 footing thickness. Horizontal reinforcement parallel to wall line should be used if soil bearing capacity is low, or soil compressibility is variable, or the footing spans pipe trenches, etc. Vertical reinforcing dowels depend upon wall design.

Alternate foundation types include bored piles, grade beams, and mat. Short, bored compacted concrete piles are sometimes competitive with strip footings, especially in shrinkable clay soils where they may provide an added factor of safety. When soil bearing capacity is inadequate, grade beams can be used to transfer wall loads across the inadequate ground to column support foundations. Mat footings replace both strip and isolated spread footings. They may eliminate differential settlement problems.

Figure A1010-141
Section Thru Typical Strip Footing

A10 Foundations — RA1010-140 Strip Footings

Example: Strip Footing Cost Determination for Bearing Wall

Office Building .... 5 Story
Story Height .... 12 VLF
Dimensions .... 50' x 80' = 4000 S.F.G.
Exterior Wall .... 12" Brick and Block
Interior Wall .... 8" Block
Floor .... Prestressed Concrete Slabs + 2" Topping
Roof .... Prestressed Concrete Slabs, No Topping
Soil Bearing Capacity .... 6 KSF
Superimposed Load:
    floor .... 100 PSF
    roof .... 40 PSF

Figure A1010-142 Footing Plan

A. Total load to each footing

Table A1010-143 Strip Footing, Load Determination

Type Load	Reference	Working Loads			
		Roof	Floor	Wall	Total
1. Total Load	System B1010 229	90 PSF	175 PSF	95 PSF	
2. Load Span or Height	Building Design	10 LF	10 LF	12 VLF	
3. Whole Bay Load/Level	(Line 1 x Line 2) ÷ 1000	.9 KIPS	1.75 KIPS	1.1 KIPS	
4. Load to Footing/L.F.	Roof + (Floor + Wall)	.9 KLF	7.0 KLF	5.9 KLF	13.8 KLF

1. Total load from System B1010-229.
2. Roof and floor load spans for exterior strip footings are 1/2 total span of 20' or 10' (wall height for our example is 12').
3. Multiply Line 1 by Line 2 in each column and divide by 1,000 Lb/Kip.
4. Add 1 roof + 4 elevated floors + 5.33 stories of wall (includes .33 added for a frost wall) for the total load to the footing.

B. Enter System A1010-110 with a load of 13.8 KLF and interpolate between line items 2700 and 3100.

Interpolating: $\$38.50 + \left[\dfrac{13.8 - 11.1}{14.8 - 11.1} \right] \times (\$13.80 - \$38.50) = \$20.40$

C. If the building design incorporates other strip footing loads, repeat steps A and B for each type of footing load.

Table A1010-144 Cost Determination, Strip Footings

Description	Strip Footing Type or Size	Total Length Each Type (Ft.)	Cost From System A1010 110 ($/L.F.)	Cost x Length ($/Each Type)	Total Cost All Types
End Bearing Wall	1	100	$42.40	$ 4,240	
Non Bearing Wall	2	160	38.50	$ 6,160	
Interior Bearing Wall	3	150	87.00	$13,050	$23,450

Note: Type 2 wall: use "wall" loads only in Step A above.
Type 3 wall: same as Step A above with load span increased to 20'.
Interpolated values from System A1010 110 are rounded to $.05.

A10 Foundations — RA1010-330 Pile Caps

General: The function of a reinforced concrete pile cap is to transfer superstructure load from isolated column or pier to each pile in its supporting cluster. To do this, the cap must be thick and rigid, with all piles securely embedded into and bonded to it.

Figure A1010-331 Section Through Pile Cap

Table A1010-332 Concrete Quantities for Pile Caps

Load Working (K)	Number of Piles @ 3'-0" O.C. Per Footing Cluster									
	2 (CY)	4 (CY)	6 (CY)	8 (CY)	10 (CY)	12 (CY)	14 (CY)	16 (CY)	18 (CY)	20 (CY)
50	(.9)	(1.9)	(3.3)	(4.9)	(5.7)	(7.8)	(9.9)	(11.1)	(14.4)	(16.5)
100	(1.0)	(2.2)	(3.3)	(4.9)	(5.7)	(7.8)	(9.9)	(11.1)	(14.4)	(16.5)
200	(1.0)	(2.2)	(4.0)	(4.9)	(5.7)	(7.8)	(9.9)	(11.1)	(14.4)	(16.5)
400	(1.1)	(2.6)	(5.2)	(6.3)	(7.4)	(8.2)	(13.7)	(11.1)	(14.4)	(16.5)
800		(2.9)	(5.8)	(7.5)	(9.2)	(13.6)	(17.6)	(15.9)	(19.7)	(22.1)
1200			(5.8)	(8.3)	(9.7)	(14.2)	(18.3)	(20.4)	(21.2)	(22.7)
1600				(9.8)	(11.4)	(14.5)	(19.5)	(20.4)	(24.6)	(27.2)
2000				(9.8)	(11.4)	(16.6)	(24.1)	(21.7)	(26.0)	(28.8)
3000						(17.5)		(26.5)	(30.3)	(32.9)
4000								(30.2)	(30.7)	(36.5)

Table A1010-333 Concrete Quantities for Pile Caps

Load Working (K)	Number of Piles @ 4'-6" O.C. Per Footing					
	2 (CY)	3 (CY)	4 (CY)	5 (CY)	6 (CY)	7 (CY)
50	(2.3)	(3.6)	(5.6)	(11.0)	(13.7)	(12.9)
100	(2.3)	(3.6)	(5.6)	(11.0)	(13.7)	(12.9)
200	(2.3)	(3.6)	(5.6)	(11.0)	(13.7)	(12.9)
400	(3.0)	(3.6)	(5.6)	(11.0)	(13.7)	(12.9)
800			(6.2)	(11.5)	(14.0)	(12.9)
1200				(13.0)	(13.7)	(13.4)
1600						(14.0)

A10 Foundations — RA1020-100 Piles, General

General: Piles are column-like shafts which receive superstructure loads, overturning forces, or uplift forces. They receive these loads from isolated column or pier foundations (pile caps), foundation walls, grade beams, or foundation mats. The piles then transfer these loads through shallower poor soil strata to deeper soil of adequate support strength and acceptable settlement with load.

Be sure that other foundation types aren't better suited to the job. Consider ground and settlement, as well as loading, when reviewing. Piles usually are associated with difficult foundation problems and substructure condition. Ground conditions determine type of pile (different pile types have been developed to suit ground conditions). Decide each case by technical study, experience, and sound engineering judgment—not rules of thumb. A full investigation of ground conditions, early, is essential to provide maximum information for professional foundation engineering and an acceptable structure.

Piles support loads by end bearing and friction. Both are generally present; however, piles are designated by their principal method of load transfer to soil.

Boring should be taken at expected pile locations. Ground strata (to bedrock or depth of 1-1/2 building width) must be located and identified with appropriate strengths and compressibilities. The sequence of strata determines if end-bearing or friction piles are best suited. See **Table A1020-105** for site investigation costs.

End-bearing piles have shafts which pass through soft strata or thin hard strata and tip bear on bedrock or penetrate some distance into a dense, adequate soil (sand or gravel).

Friction piles have shafts which may be entirely embedded in cohesive soil (moist clay), and develop required support mainly by adhesion or "skin-friction" between soil and shaft area.

Piles pass through soil by either one of two ways:
1. Displacement piles force soil out of the way. This may cause compaction, ground heaving, remolding of sensitive soils, damage to adjacent structures, or hard driving.
2. Non-displacement piles have either a hole bored and the pile cast or placed in the hole, or open-ended pipe (casing) driven and the soil core removed. They tend to eliminate heaving or lateral pressure damage to adjacent structures of piles. Steel "HP" piles are considered of small displacement.

Placement of piles (attitude) is most often vertical; however, they are sometimes battered (placed at a small angle from vertical) to advantageously resist lateral loads. Seldom are piles installed singly but rather in clusters (or groups). Codes require a minimum of three piles per major column load or two per foundation wall or grade beam. Single pile capacity is limited by pile structural strength or support strength of soil. Support capacity of a pile cluster is almost always less than the sum of its individual pile capacities due to overlapping of bearing the friction stresses. See **Table A1020-101** for minimum pile spacing requirements.

Large rigs for heavy, long piles create large soil surface loads and additional expense on weak ground. See **Table A1020-103** for percent of cost increases.

Fewer piles create higher costs per pile. See **Table A1020-102** for effect of mobilization.

Pile load tests are frequently required by code, ground situation, or pile type. See **Table A1020-104** for costs. Test load is twice the design load.

Table A1020-101 Min. Pile Spacing by Pile Type

Type Support		Min. Pile Spacing	
		X's Butt Diameter	Foot
End Bearing	Bedrock	2	2'-0"
	Hard Strata	2.5	2'-6"
Friction		3 to 5	3'-6"

Table A1020-102 Add cost for Mobilization (Setup and Removal)

Job Size	Cost	
	$/Job	$/L.F. Pile
Small (12,000 L.F.)	$14,600	$1.22
Large (25,000 L.F.)	24,200	.97

Table A1020-103 Add Costs for Special Soil Conditions

Special Conditions	Add % of Total
Soft, damp ground	40%
Swampy, wet ground	40%
Barge-mounted drilling rig	30%

Table A1020-104 Testing Costs, if Required, Any Pile Type

Test Weight (Tons)	$/Test
50-100T	$15,500
100-200T	22,000
150-300T	28,500
200-400T	31,000

Table A1020-105 Cost/L.F. of Boring Types (4") and Per Job Cost of Other Items

Type Soil	Type Boring	Sample	$/L.F. Boring	Total $
Earth	Auger	None	$ 31.50	
	Cased	Yes	68.00	
Rock, "BX"	Core	None	77.50	
	Cased	Yes	109.00	
Filed survey, mobilization, demobilization & engineering report				$3,165

A10 Foundations — RA1020-200 Caissons

General: Caissons, as covered in this section, are drilled cylindrical foundation shafts which function primarily as short column-like compression members. They transfer superstructure loads through inadequate soils to bedrock or hard stratum. They may be either reinforced or unreinforced and either straight or belled out at the bearing level.

Shaft diameters range in size from 20" to 84" with the most usual sizes beginning at 34". If inspection of bottom is required, the minimum diameter practical is 30". If handwork is required (in addition to mechanical belling, etc.) the minimum diameter is 32". The most frequently used shaft diameter is probably 36" with a 5' or 6' bell diameter. The maximum bell diameter practical is three times the shaft diameter.

Plain concrete is commonly used, poured directly against the excavated face of soil. Permanent casings add to cost and economically should be avoided. Wet or loose strata are undesirable. The associated installation sometimes involves a mudding operation with bentonite clay slurry to keep walls of excavation stable (costs not included here).

Reinforcement is sometimes used, especially for heavy loads. It is required if uplift, bending moment, or lateral loads exist. A small amount of reinforcement is desirable at the top portion of each caisson, even if the above conditions theoretically are not present. This will provide for construction eccentricities and other possibilities. Reinforcement, if present, should extend below the soft strata. Horizontal reinforcement is not required for belled bottoms.

There are three basic types of caisson bearing details:

1. Belled, which are generally recommended to provide reduced bearing pressure on soil. These are not for shallow depths or poor soils. Good soils for belling include most clays, hardpan, soft shale, and decomposed rock.

   Soils requiring handwork include hard shale, limestone, and sandstone.

   Soils not recommended include sand, gravel, silt, and igneous rock. Compact sand and gravel above water table may stand. Water in the bearing strata is undesirable.

2. Straight shafted, which have no bell but the entire length is enlarged to permit safe bearing pressures. They are most economical for light loads on high bearing capacity soil.

3. Socketed (or keyed), which are used for extremely heavy loads. They involve sinking the shaft into rock for combined friction and bearing support action. Reinforcement of shaft is usually necessary. Wide flange cores are frequently used here.

Advantages include:
  A. Shafts can pass through soils that piles cannot
  B. No soil heaving or displacement during installation
  C. No vibration during installation
  D. Less noise than pile driving
  E. Bearing strata can be visually inspected & tested

Uses include:
  A. Situations where unsuitable soil exists to moderate depth
  B. Tall structures
  C. Heavy structures
  D. Underpinning (extensive use)

See **Table A1010-121** for Soil Bearing Capacities.

Figure A1020-201 Design Assumptions

Figure A1020-202 Size Range

A10 Foundations — RA1020-230 Grade Beams

Table A1020-231 Grade Beam Detail

General: Grade beams are stiff self-supporting structural members, partly exposed above grade. They support wall loads and carry them across unacceptable soil to column footings, support piles or caissons.

They should be deep enough to be below frost depth. Therefore, they are more frequently used in mild climates.

They must be stiff enough to prevent cracking of the supported wall. Usually, they are designed as simply supported beams, so as to minimize effects of unequal footing settlement.

Heavy column loads and light wall loads tend to make grade beams an economical consideration. Light column loads but heavy wall loads tend to make strip footings more economical.

Conditions which favor other foundation types include:
- A. Deep frost penetration requiring deep members
- B. Good soil bearing near surface
- C. Large live loads to grade beam
- D. Ground floor 2' or 3' above finished grade requiring wall depth beams (not included here).
- E. Varying ground elevations
- F. Basement, requiring walls

Design Assumption: See Figure A1020-231
  Concrete placed by chute
  Forms, four uses
  Simply supported
  Max. deflection = L/480
  Design span = Bay width less 2'

Figure A1020-231 Grade Beam Detail

A10 Foundations — RA1030-200 Slab on Grade

Table A1030-202 Thickness and Loading Assumptions by Type of Use

General: Grade slabs are classified on the basis of use. Thickness is generally controlled by the heaviest concentrated load supported. If load area is greater than 80 sq. in., soil bearing may be important. The base granular fill must be a uniformly compacted material of limited capillarity, such as gravel or crushed rock. Concrete is placed on this surface or the vapor barrier on top of base.

Grade slabs are either single or two course floors. Single course are widely used. Two course floors have a subsequent wear resistant topping.

Reinforcement is provided to maintain tightly closed cracks.

Control joints limit crack locations and provide for differential horizontal movement only. Isolation joints allow both horizontal and vertical differential movement.

Use of Table: Determine appropriate type of slab (A, B, C, or D) by considering type of use or amount of abrasive wear of traffic type.

Determine thickness by maximum allowable wheel load or uniform load, opposite 1st column thickness. Increase the controlling thickness if details require, and select either plain or reinforced slab thickness and type.

Figure A1030-201 Section, Slab-on-Grade

Table A1030-202 Thickness and Loading Assumptions by Type of Use

SLAB THICKNESS (IN.)	TYPE	A Non Little Foot Only Load* (K)	B Light Light Pneumatic Wheels Load* (K)	C Normal Moderate Solid Rubber Wheels Load* (K)	D Heavy Severe Steel Tires Load* (K)	◀ Slab I.D. ◀ Industrial ◀ Abrasion ◀ Type of Traffic Max. Uniform Load to Slab ▼ (PSF)
4"	Reinf. Plain	4K				100
5"	Reinf. Plain	6K	4K			200
6"	Reinf. Plain		8K	6K	6K	500 to 800
7"	Reinf. Plain			9K	8K	1,500
8"	Reinf. Plain				11K	* Max. Wheel Load in Kips (incl. impact)
10"	Reinf. Plain				14K	
12"	Reinf. Plain					
DESIGN ASSUMPTIONS	Concrete, Chuted	f'c = 3.5 KSI	4 KSI	4.5 KSI	Slab @ 3.5 KSI	ASSUMPTIONS BY SLAB TYPE
	Topping			1" Integral	1" Bonded	
	Finish	Steel Trowel	Steel Trowel	Steel Trowel	Screed & Steel Trowel	
	Compacted Granular Base	4" deep for 4" slab thickness 6" deep for 5" slab thickness & greater				ASSUMPTIONS FOR ALL SLAB TYPES
	Vapor Barrier	6 mil polyethylene				
	Forms & Joints	Allowances included				
	Reinforcement	WWF As required ≥ 60,000 psi				

A20 Basement Const. — RA2020-210 Foundation Walls

Table A2020-212 Minimum Wall Reinforcement Weight (PSF)

General: Foundation wall heights depend upon depth of frost penetration, basement configuration or footing requirements. For certain light load conditions, good soil bearing and adequate wall thickness, strip footings may not be required.

Thickness requirements are based upon code requirements (usually 8″ min. for foundation, exposed basement, fire and party walls), unsupported height or length, configuration of wall above, and structural load considerations.

Shrinkage and temperature forces make longitudinal reinforcing desirable, while high vertical and lateral loads make vertical reinforcing necessary. Earthquake design also necessitates reinforcing.

Figure A2020-211 shows the range in wall sizes for which costs are provided in System A2020 110. Other pertinent information is also given.

Design Assumptions: Reinforced concrete wall designs are based upon ACI 318-71-14, Empirical Design of Walls. Plain concrete wall designs are based upon nonreinforced load bearing masonry criteria (fm = 1700 psi).

Earthquake, lateral or shear load requirements may add to reinforcement costs, and are not included.

Minimum Wall Reinforcement: Table below lists the weight of wall for minimum quantities of reinforcing steel for various wall thicknesses.

Design Assumptions:
Reinforcing steel is A 615, grade 60 (fy = 60 k).
Total percent of steel area is 0.20% for horizontal steel, 0.12% for vertical steel.
For steel at both faces, mats are identical.
For single layer, mat is in center of wall.
For grade 40 steel, add 25% to weights and costs.
If other than the minimum amount of steel area is required, factor weights directly with increased percentage.

**Figure A2020-211
Concrete Foundation Walls
Size Range for System A2020 110**

Note: Excavation and fill costs are not included in System A2020 110, but are provided in STRIP FOOTING COST, System A1010 110

Table A2020-212 Minimum Wall Reinforcement Weight (PSF)

Location	Wall Thickness	Horizontal Steel				Vertical Steel				Horizontal & Vertical Steel
		Bar Size	Spacing C/C	Sq. In. per L.F.	Total Wt. per S.F.	Bar Size	Spacing C/C	Sq. In. per L.F.	Total Wt. per S.F.	Total Wt. per S.F.
Both Faces	10″	#4	18″	.13	.891#	#3	18″	.07	.501#	1.392#
	12″	#4	16″	.15	1.002	#3	15″	.09	.602	1.604
	14″	#4	14″	.17	1.145	#3	13″	.10	.694	1.839
	16″	#4	12″	.20	1.336	#3	11″	.12	.820	2.156
	18″	#5	17″	.22	1.472	#4	18″	.13	.891	2.363
One Face	6″	#3	9″	.15	.501	#3	18″	.07	.251	.752
	8″	#4	12″	.20	.668	#3	11″	.12	.410	1.078
	10″	#5	15″	.25	.834	#4	16″	.15	.501	1.335

A20 Basement Const. — RA2020-210 Foundation Walls

Table A2020-213 Weight of Steel Reinforcing per S.F. in Walls (PSF)

Reinforced Weights: Table below lists the weight per S.F. for reinforcing steel in walls.

Design Assumptions: Reinforcing Steel is A615 grade 60 (for costs).

Weights will be correct for any grade steel reinforcing. For bars in two directions, add weights for each size and spacing.

C/C Spacing in Inches	#3 Wt. (PSF)	#4 Wt. (PSF)	#5 Wt. (PSF)	#6 Wt. (PSF)	#7 Wt. (PSF)	#8 Wt. (PSF)	#9 Wt. (PSF)	#10 Wt. (PSF)	#11 Wt. (PSF)
2"	2.26	4.01	6.26	9.01	12.27				
3"	1.50	2.67	4.17	6.01	8.18	0.68	13.60	17.21	21.25
4"	1.13	2.01	3.13	4.51	6.13	8.10	10.20	12.91	15.94
5"	.90	1.60	2.50	3.60	4.91	6.41	8.16	10.33	12.75
6"	.752	1.34	2.09	3.00	4.09	5.34	6.80	8.61	10.63
8"	.564	1.00	1.57	2.25	3.07	4.01	5.10	6.46	7.97
10"	.451	.802	1.25	1.80	2.45	3.20	4.08	5.16	6.38
12"	.376	.668	1.04	1.50	2.04	2.67	3.40	4.30	5.31
18"	.251	.445	.695	1.00	1.32	1.78	2.27	2.86	3.54
24"	.188	.334	.522	.751	1.02	1.34	1.70	2.15	2.66
30"	.150	.267	.417	.600	.817	1.07	1.36	1.72	2.13
36"	.125	.223	.348	.501	.681	.890	1.13	1.43	1.77
42"	.107	.191	.298	.429	.584	.763	.97	1.17	1.52
48"	.094	.167	.261	.376	.511	.668	.85	1.08	1.33

B10 Superstructure — RB1010-010 Floor Systems

Selecting a Floor System:
1. Determine size of building — total S.F. and occupancy (i.e., office building, dormitory, etc.).
2. Determine number of floors and S.F./floor.
3. Select a possible bay size and layout.
4. From Table L1010-101 determine the minimum design live load. Add partition load, ceiling load and miscellaneous loads such as mechanical, light fixtures and flooring.

Example:

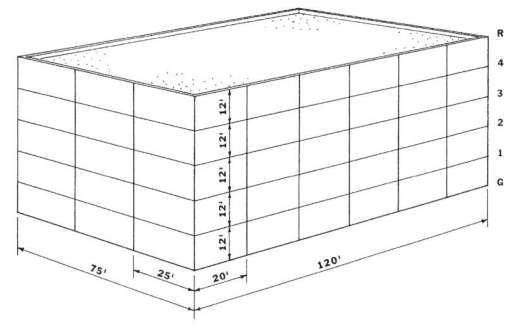

Live Load Office Building	50	PSF	BOCA Code Table L1010-101
Partitions	20		Assume using Tables L1010-201
Ceiling	5		through L1010-226
	75	PSF	Total Superimposed Load

Suspended Floors					
Building Dimension (Ft.)	Suspended Floor (S.F.)	Number of Suspended Floors	Total Suspended Floors (S.F.)	Bay Size (Ft.)	Superimposed Load PSF
75 x 120	9,000	4	36,000	20 x 25	75

5. Enter Tables with 75 PSF superimposed load and 20 x 25 bay size.

Table B1010-223	C.I.P. Flat Plate	$14.80 /S.F.
Table B1010-256	Composite Beam, Deck & Slab	18.45 /S.F.
Table B1010-250	Steel Joists & Beams on Col. (col. not included)	16.15 /S.F.

It appears, from the costs listed above, that the most economical system for framing the suspended floors with 20' x 25' bays, 75 PSF superimposed load, is the open web joists, beams, slab form and 2-1/2" concrete slabs. However, the system will require a fireproof ceiling for fireproofing the structure, which will ultimately raise the cost.

The concrete flat plate may be more economical if no hung ceiling is required and the soil capacity is sufficient to justify the additional load to the foundations.

Some of the factors affecting the selection of a floor system are:
1. Location
2. Owner's preference
3. Fire code
4. Availability of materials
5. Subsoil conditions
6. Economy
7. Clear span
8. Acoustical characteristics
9. Contractor capability

B10 Superstructure — RB1010-100 Floor Systems Cost

Table B1010-101 Comparative Costs ($/S.F.) of Floor Systems/Type (Table Number), Bay Size, & Load

Bay Size	Cast-In-Place Concrete						Precast Concrete			Structural Steel					Wood	
	1 Way BM & Slab B1010-219	2 Way BM & Slab B1010-220	Flat Slab B1010-222	Flat Plate B1010-223	Joist Slab B1010-226	Waffle Slab B1010-227	Beams & Hollow Core Slabs B1010-236	Beams & Hollow Core Slabs Topped B1010-238	Beams & Double Tees Topped B1010-239	Bar Joists on Cols. & Brg. Walls B1010-248	Bar Joists & Beams on cols. B1010-250	Composite Beams & C.I.P. Slab B1010-252	Composite Deck & Slab, W Shapes B1010-254	Composite Beam & DK., Lt. Wt. Slab B1010-256	Wood Beams & Joists B1010-264	Laminated Wood Beams & Joists B1010-265
Superimposed Load = 40 PSF																
15 x 15	15.54	14.69	13.06	12.50	16.55	—	—	—	—	—	—	—	—	—	13.36	14.21
15 x 20	15.56	15.24	13.51	12.31	16.65	—	—	—	—	10.62	12.51	—	17.95	—	16.05	14.21
20 x 20	15.74	16.00	13.75	13.33	16.70	18.10	25.62	28.00	—	10.90	13.46	—	19.30	—	15.44	14.47
20 x 25	15.97	16.85	14.90	14.55	16.80	18.35	23.10	26.90	—	11.40	14.34	20.50	19.45	17.95	—	—
25 x 25	16.05	17.25	15.30	14.80	16.65	18.70	23.91	27.10	—	12.17	15.35	21.10	21.70	17.45	—	—
25 x 30	16.35	17.95	16.20	—	17.40	18.95	22.01	24.55	25.24	13.51	16.95	22.80	21.80	17.65	—	—
30 x 30	18.05	19.35	17.00	—	17.85	19.70	22.15	25.00	25.60	13.38	17.00	22.90	23.90	17.85	—	—
30 x 35	18.80	20.80	18.05	—	18.50	20.00	21.51	24.61	—	14.87	18.90	23.80	25.85	18.65	—	—
35 x 35	19.95	21.65	18.40	—	18.60	20.85	22.40	25.05	—	15.32	19.45	24.70	26.30	19.55	—	—
35 x 40	20.40	22.75	—	—	19.25	21.30	22.40	25.79	23.27	—	—	26.70	27.55	21.45	—	—
40 x 40	—	—	—	—	19.75	21.95	23.02	25.26	25.13	—	—	—	—	—	—	—
40 x 45	—	—	—	—	20.50	22.50	—	—	—	—	—	—	—	—	—	—
40 x 50	—	—	—	—	—	—	—	—	22.42	—	—	—	—	—	—	—
Superimposed Load = 75 PSF																
15 x 15	15.73	14.97	13.26	12.53	16.60	—	—	—	—	—	—	—	—	—	16.67	17.70
15 x 20	16.12	16.45	13.95	13.80	17.40	—	—	—	—	11.49	14.18	—	19.55	—	20.15	19.50
20 x 20	17.00	17.35	14.50	13.85	17.55	18.40	20.36	29.05	—	12.31	15.45	—	21.55	—	20.72	19.30
20 x 25	17.40	18.75	15.70	14.80	17.65	18.70	23.10	26.45	—	12.85	16.15	22.90	23.75	18.45	—	—
25 x 25	17.30	18.60	15.90	15.40	17.60	19.15	23.91	22.20	—	14.01	17.75	24.20	24.70	18.90	—	—
25 x 30	17.50	19.45	17.05	—	18.00	19.40	22.62	24.77	25.24	14.23	18.10	25.70	25.85	18.70	—	—
30 x 30	19.40	20.95	17.90	—	18.40	20.15	22.06	26.00	25.60	15.32	19.50	25.40	27.55	19.10	—	—
30 x 35	19.65	21.75	19.05	—	18.75	20.00	22.47	24.86	—	17.40	21.85	27.35	29.75	20.00	—	—
35 x 35	21.70	22.55	19.50	—	19.45	21.25	23.05	26.85	—	18.70	23.40	28.20	30.30	21.95	—	—
35 x 40	22.00	23.60	—	—	20.10	21.95	—	—	23.29	—	—	30.05	31.80	23.00	—	—
40 x 40	—	—	—	—	20.30	22.70	—	—	24.17	—	—	—	—	—	—	—
40 x 45	—	—	—	—	20.65	23.25	—	—	—	—	—	—	—	—	—	—
40 x 50	—	—	—	—	—	—	—	—	21.09	—	—	—	—	—	—	—
Superimposed Load = 125 PSF																
15 x 15	15.96	15.57	13.66	12.83	16.85	—	—	—	—	—	—	—	—	—	24.55	25.80
15 x 20	16.78	17.80	14.50	14.60	17.80	—	—	—	—	12.35	15.65	—	22.40	—	28.90	29.60
20 x 20	17.80	17.95	15.45	14.55	17.80	18.65	—	—	—	13.94	17.60	—	24.10	—	38.80	28.55
20 x 25	18.50	19.20	16.65	15.60	18.60	19.05	—	—	—	15.10	18.85	25.60	27.60	22.30	—	—
25 x 25	20.00	20.20	16.85	16.05	19.40	19.60	—	—	—	16.35	20.60	27.45	29.40	19.95	—	—
25 x 30	20.10	21.10	17.60	—	19.20	19.85	—	—	—	17.25	21.70	30.00	29.20	21.60	—	—
30 x 30	20.65	22.05	18.55	—	19.45	20.40	—	—	—	19.15	24.25	29.95	32.40	22.75	—	—
30 x 35	21.90	23.75	19.70	—	19.50	20.85	—	—	—	19.65	24.75	33.55	34.35	23.40	—	—
35 x 35	23.05	24.65	20.00	—	19.50	21.65	—	—	—	21.80	25.90	32.55	35.65	25.70	—	—
35 x 40	23.35	24.75	—	—	20.00	22.75	—	—	—	—	—	35.35	37.10	26.25	—	—
40 x 40	—	—	—	—	21.10	23.05	—	—	—	—	—	—	—	—	—	—
40 x 45	—	—	—	—	21.45	23.60	—	—	—	—	—	—	—	—	—	—
40 x 50	—	—	—	—	—	—	—	—	—	—	—	—	—	—	—	—
Superimposed Load = 200 PSF																
15 x 15	16.61	16.70	14.15	—	17.40	—	—	—	—	—	—	—	—	—	46.35	45.00
15 x 20	18.15	19.15	14.80	—	18.30	—	—	—	—	—	—	—	26.95	—	40.50	40.00
20 x 20	19.45	19.45	15.70	—	18.65	19.55	—	—	—	—	—	—	28.50	—	—	41.60
20 x 25	19.95	20.90	17.25	—	19.45	19.75	—	—	—	—	—	31.35	31.10	24.80	—	—
25 x 25	21.70	22.70	17.45	—	20.05	20.10	—	—	—	—	—	32.55	33.80	25.90	—	—
25 x 30	21.85	22.95	18.40	—	20.30	21.20	—	—	—	—	—	34.45	35.60	26.00	—	—
30 x 30	22.90	23.60	19.55	—	20.45	21.85	—	—	—	—	—	36.35	42.90	27.00	—	—
30 x 35	23.15	25.05	—	—	20.95	22.70	—	—	—	—	—	39.60	41.45	27.20	—	—
35 x 35	25.00	25.90	—	—	20.95	23.05	—	—	—	—	—	40.25	45.10	29.35	—	—
35 x 40	25.15	26.55	—	—	21.35	24.45	—	—	—	—	—	43.56	46.30	31.20	—	—
40 x 40	—	—	—	—	—	—	—	—	—	—	—	—	—	—	—	—
40 x 45	—	—	—	—	—	—	—	—	—	—	—	—	—	—	—	—
40 x 50	—	—	—	—	—	—	—	—	—	—	—	—	—	—	—	—

B10 Superstructure — RB1010-112 Concrete Column Example

Design Assumptions:
Bay Size 20′ x 25′ = 500 S.F.
Use Concrete Flat Plate

Roof Load:	Superimposed Load	40 PSF
	Total Load	146 PSF
Floor Load:	Superimposed Load	75 PSF
	Total Load	188 PSF

Minimum Column size 20″

ROOF Total Roof Load .146 KSF x 500 S.F. = 73 K

73 K

Column Load 1.67′ x 1.67′ x .15 KCF x 12′ = 5 K

78 K

FOURTH Total Floor Load .188 K x 500 S.F. = 94 K

172 K

Column 5 K

177 K

THIRD Total Floor Load 94 K

271 K

Column 5 K

276 K

SECOND Total Floor Load 94 K

370 K

Column 5 K

375 K

FIRST Total Floor Load 94 K

469 K

Column 5 K

474 K

GROUND

474 K

Total Load →

Description:
1. Multiply roof load x bay area.
2. Show total load at top of column.
3. Multiply est. column weight x floor to floor height.
4. Add to roof load.
5. Multiply floor load x bay area.
6. Add to roof and column load.
7. Multiply est. column weight x floor to floor height.
8. Add to total loads above.
9. Repeat steps above for remainder of floors & columns.
10. Enter the minimum reinforced portion of the table with total load on the column and the minimum allowable column size for the selected cast-in-place floor system.

    If the total load on the column does not exceed the allowable load shown, use the cost per L.F. multiplied by the length of columns required to obtain the column cost.
11. If the total load on the column exceeds the allowable working load shown in the minimum reinforced portion of the table enter the first portion of the table with the total load on the column and the minimum allowable column size from the selected cast-in-place floor system.

    Select a cost per L.F. for bottom level columns by total load or minimum allowable column size.

    Select a cost per L.F. for top level columns using the column size required for bottom level columns from the minimum reinforced portion of the table.

$$\frac{\text{Bottom \& Top Col. Costs/L.F.}}{2} = \text{Average Column Cost/L.F.}$$

$$\text{Column Cost} = \text{Average Col. Cost/L.F.} \times \text{Length of Cols. Req'd.}$$

B10 Superstructure — RB1010-130 Steel Columns

Description: Below is an example of steel column determination when roof and floor loads are known.

Design Assumptions:

Bay Size	35′ x 35′ = 1,225 S.F.	
Roof Load:	Superimposed Load	40 PSF
	Dead Load	44 PSF
	Total Load	84 PSF
Floor Load:	Superimposed Load	125 PSF
	Dead Load	45 PSF
	Total Load	170 PSF

Calculation of Total Loads

ROOF — Total Roof Load .084 KSF x 1225 S.F. = 102.9 K
→ 102.9 K

Column Estimated .087 K x 12′ = + 1.0 K
→ 103.9 K (W12 x 79)

FOURTH — Total Floor Load = .17 KSF x 1225 S.F. = 208.3 K
→ 312.2 K

Column Estimated .087 K x 12′ = + 1.0 K
→ 313.2 K   — Assumed Splice (±3′)

THIRD — Total Floor Load = .17 KSF x 1225 S.F. = 208.3 K
→ 521.5 K

Column Estimated .145 K x 12′ = + 1.7 K
→ 523.2 K   (W14 x 145)

SECOND — Total Floor Load = .17 KSF x 1225 S.F. = 208.3 K
→ 731.5 K

Column Estimated .145 K x 12′ = + 1.7 K
→ 733.2 K   — Assumed Splice (±3′)

FIRST — Total Floor Load = .17 KSF x 1225 S.F. = 208.3 K
→ 941.5 K

Column Estimated .176 K x 12′ = + 2.1 K
→ 943.6 K   (W14 x 176)

Slab on grade
GROUND — Total Load to Foundation → 944 K

(5 Floors at 12′-0″ = 60′-0″)

Description:

1. Multiply roof load x bay area.
2. Show total load at top of column.
3. Multiply estimated column weight* x floor to floor height.
4. Add to roof load.
5. Multiply floor load x bay area.
6. Add to roof and column load.
7. Multiply estimated column weight x floor to floor height.
8. Add to total loads above.
9. Choose column from **System B1010 208** using unsupported height.
10. Interpolate or use higher loading to obtain cost/L.F.
11. Repeat steps above for remainder of floors and columns.
12. Multiply number of columns by the height of the column times the cost per foot to obtain the cost of each type of column.

* To Estimate Column Weight

Roof Load	.084 KSF
Floor Load x No Floors above Splice	
170 x 1	.170 KSF
Total	.254 KSF

Total Load (KSF) x Bay Area (S.F.) = Load to Col.
.254 KSF x 1,225 KSF = 311 K

From **System B1010 208**, choose a column by:

Load,	Height,	Weight
400 K	10′	79 lb.

B10 Superstructure — RB1010-240 Wood Beams

Table B1010-241 One, Two and Three Member Beams, Maximum Load for Various Spans

Description: Table below lists the maximum uniform load (W) or concentrated load (P) allowable for various size beams at various spans.

Design Assumptions: Fiber strength (f) is 1,000 psi.

Maximum deflection does not exceed 1/360 the span of the beam.

Modulus of elasticity (E) is 1,100,000 psi. Anything less will result in excessive deflection in the longer members so that spans must be reduced or member size increased.

The uniform loads (W) are in pounds per foot; concentrated loads (P), at the midpoint, are in pounds.

The span is in feet and is the unsupported clear span.

The member sizes are from 2″ x 6″ to 4″ x 12″ and include one, two and three pieces of each member size to arrive at spans and loading.

Individual Member Size	No.	6′ W #/L.F.	6′ P #	8′ W #/L.F.	8′ P #	10′ W #/L.F.	10′ P #	12′ W #/L.F.	12′ P #	14′ W #/L.F.	14′ P #	16′ W #/L.F.	16′ P #	18′ W #/L.F.	18′ P #
2″ x 6″	2	238	840	—	—										
	3	357	1260	—	—										
2″ x 8″	2	314	1460	235	1095										
	3	471	2190	353	1642	232	1164								
2″ x 10″	2	400	2376	300	1782	240	1426	—	—						
	3	600	3565	450	2673	360	2139	279	1679						
2″ x 12″	2	487	2925	365	2636	292	2109	243	1757	208	1479	—	—		
	3	731	4387	548	3955	548	3164	365	2636	313	2219	212	1699		
2″ x 14″	2	574	3445	430	3445	344	2926	287	2438	246	2090	215	1828	—	—
	3	861	5167	645	5167	516	4389	430	3657	369	3135	322	2743	243	2193
3″ x 6″	1	—	—	—	—										
	2	397	1400	200	882										
	3	595	2100	300	1323										
3″ x 8″	1	261	1216	—	—	—	—	—	—						
	2	523	2433	392	1825	258	1293	—	—						
	3	785	3650	589	2737	388	1940	224	1347						
3″ x 10″	1	334	1980	250	1485	200	1188	—	—	—	—	—	—		
	2	668	3961	500	2970	400	2376	310	1865	—	—				
	3	1002	5941	750	4456	600	3565	466	2798	293	2056				
3″ x 12″	1	406	2437	304	2197	243	1757	203	1464	—	—	—	—	—	—
	2	812	4875	608	4394	487	3515	406	2929	348	2466	236	1888	—	—
	3	1218	7312	912	6591	731	5273	609	4394	522	3699	354	2832	248	2237
4″ x 6″	1	278	980	—	—	—	—								
	2	556	1960	308	1235										
4″ x 8″	1	366	1703	274	1277	—	—	—	—						
	2	733	3406	549	2555	362	1811	209	1257						
4″ x 10″	1	467	2772	350	2079	280	1663	217	1306	—	—				
	2	935	5545	700	4159	560	3327	435	2612	274	1919				
4″ x 12″	1	568	3412	426	3076	341	2460	284	2050	243	1726	—	—		
	2	1137	6825	853	6152	682	4921	568	4100	487	3452	330	2643	232	2088

B10 Superstructure — RB1010-710 Wood Floor Joists

Table B1010-711 Maximum Floor Joist Spans

Description: Table below lists the maximum clear spans and the framing lumber quantity required per S.F. of floor.

Design Assumptions: Dead load = joist weight plus floor weight of 5 psf plus ceiling or partition load of 10 psf.

Maximum L.L. deflection is 1/360 of the clear span.

Modulus of elasticity is 1,100,000 psi.

Fiber strength (f_b) is 1,000 psi.

10% allowance has been added to framing quantities for overlaps, waste, double joists at openings, etc. 5% added to subfloor for waste.

Maximum span is in feet and is the unsupported clear span.

Floor Joist Size in Inches	Spacing C/C	Framing Lumber B.F./S.F.	Live Load in Pounds per Square Foot							
			30 Span	40 Span	50 Span	60 Span	70 Span	80 Span	90 Span	100 Span
2" x 6"	12"	1.10	9'-0"	8'-6"	8'-0"	7'-8"	7'-4"	7'-0"	6'-10"	6'-7"
	16"	.83	8'-4"	7'-9"	7'-4"	7'-0"	6'-8"	6'-4"	6'-0"	—
	24"	.55	7'-2"	6'-9"	6'-2"	5'-9"	—	—	—	—
2" x 8"	12"	1.46	12'-0"	11'-2"	10'-7"	10'-0"	9'-8"	9'-4"	9'-0"	8'-8"
	16"	1.10	10'-10"	10'-2"	9'-7"	9'-2"	8'-9"	8'-4"	7'-10"	7'-6"
	24"	.74	9'-6"	8'-10"	8'-2"	7'-6"	7'-0"	6'-9"	6'-5"	6'-0"
2" x 10"	12"	1.84	15'-4"	14'-4"	13'-6"	12'-10"	12'-4"	11'-10"	11'-6"	11'-2"
	16"	1.38	13'-10"	13'-0"	12'-4"	11'-8"	11'-2"	10'-7"	10'-0"	9'-7"
	24"	.91	12'-0"	11'-4"	10'-5"	9'-9"	9'-0"	8'-7"	8'-3"	7'-10"
2" x 12"	12"	2.20	18'-7"	17'-4"	16'-5"	15'-8"	15'-0"	14'-6"	14'-0"	13'-6"
	16"	1.65	16'-10"	15'-9"	14'-10"	14'-3"	13'-7"	12'-10"	12'-3"	11'-8"
	24"	1.10	14'-9"	13'-9"	12'-8"	11'-9"	11'-2"	10'-6"	10'-0"	9'-6"
2" x 14"	12"	2.56	21'-10"	20'-6"	19'-4"	18'-5"	17'-9"	17'-0"	16'-6"	15'-10"
	16"	1.93	19'-10"	18'-7"	17'-7"	16'-9"	16'-0"	15'-2"	14'-5"	13'-9"
	24"	1.29	17'-4"	16'-3"	15'-0"	13'-10"	13'-2"	12'-4"	11'-9"	11'-3"
3" x 6"	12"	1.65	10'-9"	10'-0"	9'-6"	9'-0"	8'-8"	8'-4"	8'-0"	7'-10"
	16"	1.24	9'-9"	9'-0"	8'-8"	8'-3"	7'-10"	7'-7"	7'-4"	7'-0"
	24"	.83	8'-6"	8'-0"	7'-6"	7'-2"	6'-10"	6'-8"	6'-4"	6'-0"
3" x 8"	12"	2.20	14'-2"	13'-3"	12'-6"	11'-10"	11'-6"	11'-0"	10'-8"	10'-4"
	16"	1.65	12'-10"	12'-0"	11'-4"	10'-10"	10'-5"	10'-0"	9'-9"	9'-5"
	24"	1.10	11'-3"	10'-6"	9'-10"	9'-6"	9'-0"	8'-9"	8'-3"	7'-10"
3" x 10"	12"	2.75	18'-0"	16'-10"	16'-0"	15'-4"	14'-8"	14'-0"	13'-8"	13'-3"
	16"	2.07	16'-5"	15'-4"	14'-6"	13'-10"	13'-3"	12'-10"	12'-4"	12'-0"
	24"	1.38	14'-4"	13'-5"	12'-8"	12'-0"	11'-7"	11'-0"	10'-7"	10'-0"
3" x 12"	12"	3.30	22'-0"	20'-7"	19'-6"	18'-7"	17'-10"	17'-0"	16'-7"	16'-0"
	16"	2.48	20'-0"	18'-9"	17'-8"	16'-10"	16'-2"	15'-7"	15'-0"	14'-7"
	24"	1.65	17'-6"	16'-4"	15'-6"	14'-9"	14'-0"	13'-7"	12'-10"	12'-4"
4" x 6"	12"	2.20	—	—	—	10'-0"	9'-9"	9'-4"	9'-0"	8'-9"
	16"	1.65	—	—	—	9'-2"	8'-10"	8'-6"	8'-3"	8'-0"
	24"	1.10	—	—	—	8'-0"	7'-9"	7'-4"	7'-2"	7'-0"
4" x 8"	12"	2.93	—	—	—	13'-4"	12'-10"	12'-4"	12'-0"	11'-7"
	16"	2.20	—	—	—	12'-0"	11'-8"	11'-3"	10'-10"	10'-6"
	24"	1.47	—	—	—	10'-7"	9'-2"	9'-9"	9'-6"	9'-2"
4" x 10"	12"	3.67	—	—	—	17'-0"	16'-4"	15'-9"	15'-3"	14'-9"
	16"	2.75	—	—	—	15'-6"	14'-7"	14'-4"	13'-10"	13'-5"
	24"	1.83	—	—	—	13'-6"	13'-0"	12'-6"	12'-0"	11'-9"

B10 Superstructure — RB1010-780 Wood Decking

Table B1010-781 Decking Material Characteristics

Description: The table below lists the maximum spans for commonly used wood decking materials for various loading conditions.

Design Assumptions: Applied load is the total load, live plus dead.

Maximum deflection is 1/180 of the clear span which is not suitable if plaster ceilings will be supported by the roof or floor.

Modulus of elasticity (E) and fiber strength (F) are as shown in table to the right.

No allowance for waste has been included.

Deck Material	Modulus of Elasticity	Fiber Strength
Cedar	1,100,000 psi	1,000 psi
Douglas Fir	1,760,000	1,200
Hemlock	1,600,000	1,200
White Spruce	1,320,000	1,200

Assume Dead Load of wood plank is 33.3 lbs. per C.F.

Table B1010-782 Maximum Spans for Wood Decking

Nominal Thickness	Type Wood	B.F. per S.F.	40	50	70	80	100	150	200	250
2"	Cedar	2.40	6.5'	6'	5.7'	5.3'	5'	4.5'	4'	3.5'
	Douglas fir	2.40	8'	7'	6.7'	6.3'	6'	5'	4.5'	4'
	Hemlock	2.40	7.5'	7'	6.5'	6'	5.5'	5'	4.5'	4'
	White spruce	2.40	7'	6.5'	6'	5.5'	5'	4.5'	4'	3.5'
3"	Cedar	3.65	11'	10'	9'	8.5'	8'	7'	6.5'	6'
	Douglas fir	3.65	13'	12'	11'	10.5'	10'	8'	7.5'	7'
	Hemlock	3.65	12'	11'	10.5'	10'	9'	8'	7.5'	7'
	White spruce	3.65	11'	10'	9.5'	9'	8'	7.5'	7'	6'
4"	Cedar	4.65	15'	14'	13'	12'	11'	10'	9'	8'
	Douglas fir	4.65	18'	17'	15'	14'	13'	11'	10.5'	10'
	Hemlock	4.65	17'	16'	14'	13.5'	13'	11'	10'	9'
	White spruce	4.65	16'	15'	13'	12.5'	12'	10'	9'	8'
6"	Cedar	6.65	23'	21'	19'	18'	17'	15'	14'	13'
	Douglas fir	6.65	24'+	24'	23'	22'	21'	18'	16'	15'
	Hemlock	6.65	24'+	24'	22'	21'	20'	17'	16'	15'
	White spruce	6.65	24'	23'	21'	20'	18'	16'	14'	13'

Total Uniform Load per S.F. (column headings 40–250)

B10 Superstructure — RB1020-510 Wood Joists & Rafters

Table B1020-511 Maximum Roof Joist or Rafter Spans

Description: Table below lists the maximum clear spans and the framing lumber quantity required per S.F. of roof. Spans and loads are based on flat roofs.

Design Assumptions: Dead load = Joist weight plus weight of roof sheathing at 2.5 psf plus either of two alternate roofing weights.

Maximum deflection is 1/360 of the clear span.

Modulus of elasticity is 1,100,000 psi.

Fiber strength (f) is 1,000 psi.

Allowance: 10% has been added to framing quantities for overlaps, waste, double joists at openings, etc.

Maximum span is measured horizontally in feet and is the clear span.

Note: To convert to inclined measurements or to add for overhangs, multiply quantities in Table B1020-511 by factors in Table B1020-512.

Joist Size in Inches	Center to Center Spacing	Framing B.F. per S.F.	Group I Covering (2.5 psf)				Group II Covering (8.0 psf)			
			15 Span	20 Span	30 Span	40 Span	20 Span	30 Span	40 Span	50 Span
2" x 4"	12"	0.73	7'-11"	7'-3"	6'-5"	5'-10"	6'-9"	6'-1"	5'-7"	5'-3"
	16"	0.55	7'-2"	6'-7"	5'-10"	5'-4"	6'-1"	5'-6"	5'-1"	4'-10"
	24"	0.37	6'-3"	5'-9"	5'-0"	4'-8"	5'-4"	4'-10"	4'-6"	4'-2"
2" x 6"	12"	1.10	12'-5"	11'-5"	10'-0"	9'-3"	10'-7"	9'-7"	8'-10"	8'-4"
	16"	0.83	11'-3"	10'-4"	9'-2"	8'-5"	9'-8"	8'-9"	8'-0"	7'-6"
	24"	0.55	9'-10"	9'-0"	8'-0"	7'-4"	8'-5"	7'-7"	7'-0"	6'-6"
2" x 8"	12"	1.46	16'-5"	15'-0"	13'-4"	12'-2"	14'-0"	12'-8"	11'-8"	11'-0"
	16"	1.10	14'-11"	13'-8"	12'-0"	11'-0"	12'-9"	11'-6"	10'-7"	11'-0"
	24"	0.74	13'-0"	12'-0"	10'-7"	9'-8"	11'-1"	10'-0"	9'-3"	10'-0"
2" x 10"	12"	1.84	20'-11"	19'-3"	17'-0"	19'-7"	17'-11"	16'-2"	14'-11"	14'-0"
	16"	1.38	19'-0"	17'-6"	15'-6"	14'-2"	16'-3"	14'-8"	13'-7"	12'-9"
	24"	0.91	16'-7"	15'-3"	13'-6"	12'-4"	14'-2"	12'-10"	11'-10"	11'-0"
2" x 12"	12"	2.20	25'-6"	23'-5"	20'-9"	18'-11"	21'-9"	19'-8"	18'-2"	17'-1"
	16"	1.65	23'-1"	21'-3"	18'-10"	17'-2"	19'-9"	17'-10"	16'-6"	15'-6"
	24"	1.10	20'-2"	18'-7"	16'-5"	15'-0"	17'-3"	15'-7"	14'-5"	13'-5"
2" x 14"	12"	2.56	30'-0"	27'-7"	24'-5"	22'-4"	25'-8"	23'-2"	21'-5"	20'-1"
	16"	1.93	27'-3"	25'-0"	22'-2"	20'-3"	23'-3"	21'-0"	19'-6"	18'-3"
	24"	1.29	23'-10"	21'-11"	19'-4"	17'-8"	20'-4"	18'-4"	17'-0"	15'-10"
3" x 6"	12"	1.65	14'-9"	13'-7"	12'-0"	11'-0"	12'-7"	11'-4"	10'-6"	9'-11"
	16"	1.24	13'-4"	12'-4"	10'-11"	9'-11"	11'-5"	10'-4"	9'-7"	9'-0"
	24"	.83	11'-9"	10'-9"	9'-6"	8'-8"	10'-0"	9'-0"	8'-4"	7'-10"
3" x 8"	12"	2.20	19'-5"	17'-11"	15'-10"	14'-6"	16'-7"	15'-0"	13'-11"	13'-0"
	16"	1.65	17'-8"	16'-9"	14'-4"	13'-2"	15'-1"	13'-7"	12'-7"	11'-10"
	24"	1.10	15'-5"	14'-2"	12'-6"	11'-6"	13'-2"	11'-11"	11'-0"	10'-4"
3" x 10"	12"	2.75	24'-10"	22'-10"	22'-2"	18'-6"	21'-9"	19'-2"	17'-9"	16'-8"
	16"	2.07	22'-6"	20'-9"	18'-4"	16'-9"	19'-9"	17'-4"	16'-1"	15'-1"
	24"	1.38	19'-8"	18'-1"	16'-0"	14'-8"	16'-10"	15'-2"	14'-0"	13'-2"
3" x 12"	12"	3.30	30'-2"	27'-9"	24'-6"	22'-5"	25'-10"	23'-4"	21'-6"	20'-3"
	16"	2.48	27'-5"	25'-3"	22'-3"	20'-4"	23'-5"	21'-2"	19'-7"	18'-4"
	24"	1.65	24'-0"	22'-0"	19'-6"	17'-10"	20'-6"	18'-6"	17'-1"	16'-0"

B10 Superstructure — RB1020-510 Wood Joists & Rafters

Table B1020-512 Factors for Converting Inclined to Horizontal

Rafters: The quantities shown in Table B1020-511 are for flat roofs. For inclined roofs using rafters, the quantities should be multiplied by the factors in Table B1020-512 to allow for the increased area of the inclined roofs.

Roof Slope	Approx. Angle	Factor	Roof Slope	Approx. Angle	Factor
Flat	0°	1.000	12 in 12	45.0	1.414
1 in 12	4.8°	1.003	13 in 12	47.3	1.474
2 in 12	9.5°	1.014	14 in 12	49.4	1.537
3 in 12	14.0°	1.031	15 in 12	51.3	1.601
4 in 12	18.4°	1.054	16 in 12	53.1	1.667
5 in 12	22.6°	1.083	17 in 12	54.8	1.734
6 in 12	26.6°	1.118	18 in 12	56.3	1.803
7 in 12	30.3°	1.158	19 in 12	57.7	1.873
8 in 12	33.7°	1.202	20 in 12	59.0	1.943
9 in 12	36.9°	1.250	21 in 12	60.3	2.015
10 in 12	39.8°	1.302	22 in 12	61.4	2.088
11 in 12	42.5°	1.357	23 in 12	62.4	2.162

INCLINED TO HORIZONTAL SECTIONS

Table B1020-513 below can be used two ways:
1. Use 1/2 span with overhang on one side only.
2. Use whole span with total overhang from both sides.

Roof or Deck Overhangs: The quantities shown in table do not include cantilever overhangs. The S.F. quantities should be multiplied by the factors in Table B1020-513 to allow for the increased area of the overhang. All dimensions are horizontal. For inclined overhangs also multiply by factors in Table B1020-512 above.

Table B1020-513 Allowance Factors for Including Roof or Deck Overhangs

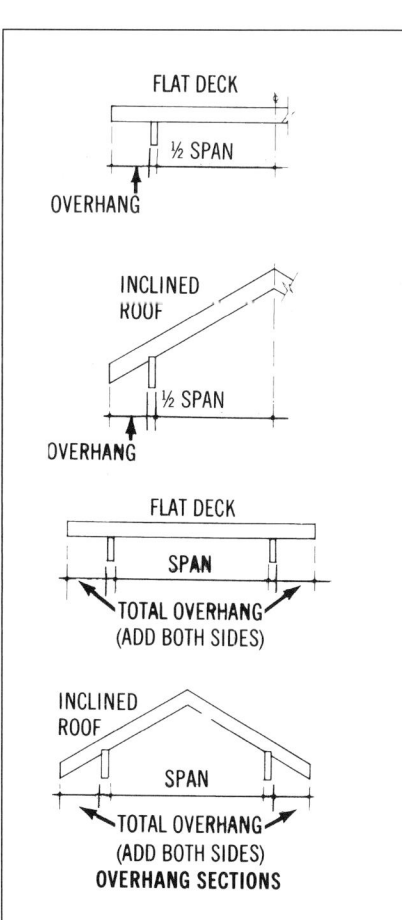

Horizontal Span	Roof overhang measured horizontally							
	0'-6"	1'-0"	1'-6"	2'-0"	2'-6"	3'-0"	3'-6"	4'-0"
6'	1.083	1.167	1.250	1.333	1.417	1.500	1.583	1.667
7'	1.071	1.143	1.214	1.286	1.357	1.429	1.500	1.571
8'	1.063	1.125	1.188	1.250	1.313	1.375	1.438	1.500
9'	1.056	1.111	1.167	1.222	1.278	1.333	1.389	1.444
10'	1.050	1.100	1.150	1.200	1.250	1.300	1.350	1.400
11'	1.045	1.091	1.136	1.182	1.227	1.273	1.318	1.364
12'	1.042	1.083	1.125	1.167	1.208	1.250	1.292	1.333
13'	1.038	1.077	1.115	1.154	1.192	1.231	1.269	1.308
14'	1.036	1.071	1.107	1.143	1.179	1.214	1.250	1.286
15'	1.033	1.067	1.100	1.133	1.167	1.200	1.233	1.267
16'	1.031	1.063	1.094	1.125	1.156	1.188	1.219	1.250
17'	1.029	1.059	1.088	1.118	1.147	1.176	1.206	1.235
18'	1.028	1.056	1.083	1.111	1.139	1.167	1.194	1.222
19'	1.026	1.053	1.079	1.105	1.132	1.158	1.184	1.211
20'	1.025	1.050	1.075	1.100	1.125	1.150	1.175	1.200
21'	1.024	1.048	1.071	1.095	1.119	1.143	1.167	1.190
22'	1.023	1.045	1.068	1.091	1.114	1.136	1.159	1.182
23'	1.022	1.043	1.065	1.087	1.109	1.130	1.152	1.174
24'	1.021	1.042	1.063	1.083	1.104	1.125	1.146	1.167
25'	1.020	1.040	1.060	1.080	1.100	1.120	1.140	1.160
26'	1.019	1.038	1.058	1.077	1.096	1.115	1.135	1.154
27'	1.019	1.037	1.056	1.074	1.093	1.111	1.130	1.148
28'	1.018	1.036	1.054	1.071	1.089	1.107	1.125	1.143
29'	1.017	1.034	1.052	1.069	1.086	1.103	1.121	1.138
30'	1.017	1.033	1.050	1.067	1.083	1.100	1.117	1.133
32'	1.016	1.031	1.047	1.063	1.078	1.094	1.109	1.125

B20 Exterior Enclosure — RB2010-010 Exterior Walls

The "Exterior Closure" Division has many types of wall systems which are most commonly used in current United States and Canadian building construction.

Systems referenced have an illustration as well as description and cost breakdown. Total costs per square foot of each system are provided. Each system is itemized with quantities, extended and summed for material, labor, or total.

Exterior walls usually consist of several component systems. The weighted square foot cost of the enclosure is made up of individual component systems costs, prorated for their percent of wall area. For instance, a given wall system may take up 70% of the enclosure wall's composite area. The fenestration may make up the balance of 30%. The components should be factored by the proportion of area each contributes to the total.

Example: An enclosure wall is 100' long and 15' high.
The total wall area is therefore 1,500 S.F.
70% of the wall area is a brick & block system.
@ $31.50/S.F. (B2010-132-1240)

30% is a window system:

Framing	$27.85	(B2020-210-1700)
Glazing	31.80	(B2020-220-1400)
	$59.65	

Solution: Brick & block = 1,500 S.F. x .70 x $31.50 = $33,075
Windows = 1,500 S.F. x .30 x $59.65 = $26,843
Total enclosure wall cost = $59,918
or $39.95 per S.F. of wall

The square foot unit costs include the installing contractors markup for overhead and profit. However, fees for professional services must be added along with other items. See H1010, General Conditions, for these costs. Note that interior finishes (drywall, etc.) are included elsewhere.

There may be times when the wall systems costs will need to be expressed in units of square feet of floor area.

Example: A building 25' x 25' requires a wall 100' long and 15' high.

Cost of the wall is $39.95 per S.F. of wall.
The floor area is 625 S.F.

Solution: Wall cost per S.F of floor area

$$\frac{100 \text{ L.F.} \times 15' \times \$39.95}{625 \text{ S.F.}} = \$95.87 \text{ per S.F. of floor}$$

B20 Exterior Enclosure — RB2010-020 Masonry Block Lintels

Concrete Block Lintels

Design Assumptions:
Bearing length each end = 8 inches
f'c = 2,000 p.s.i.
Wall load = 300 lbs. per linear foot (p.l.f.)
Floor & roof load (includes wall load) = 100 p.l.f.
Weight of 8" deep lintel (included in wall load) = 50 p.l.f.
Weight of 16" deep lintel (included in wall load) = 100 p.l.f.

Load to be supported by the lintel is shown below. The weight of the masonry above the opening can be safely assumed as the weight of a triangular section whose height is one-half the clear span of the opening. Corbelling action of the masonry above the top of the opening may be counted upon to support the weight of the masonry outside the triangle. To the dead load of a wall must be added the uniform dead & live loads of floors & roof that frame into the wall above the opening and below the apex of the 45° triangle. Any load above the apex may be neglected.

B20 Exterior Enclosure — RB2010-030 "U" & "R" Factors

Table B2010-031 Example Using 14" Masonry Cavity Wall with 1" Plaster for the Exterior Closure

Total Heat Transfer is found using the equation
$Q = AU(T_2 - T_1)$ where:
- Q = Heat flow, BTU per hour
- A = Area, square feet
- U = Overall heat transfer coefficient
- $(T_2 - T_1)$ = Difference in temperature of the air on each side of the construction component in degrees Fahrenheit

Coefficients of Transmission ("U") are expressed in BTU per (hour)(square foot) (Fahrenheit degree difference in temperature between the air on two sides) and are based on 15 mph outside wind velocity.

The lower the U-value the higher the insulating value.

$U = 1/R$ where "R" is the summation of the resistances of air films, materials and air spaces that make up the assembly.

Figure B2010-031 Example Using 14" Masonry Cavity Wall with 1" Plaster for the Exterior Closure

Construction	Resistance R.
1. Outside surface (15 mph wind)	0.17
2. Face Brick (4")	0.44
3. Air space (2", 50° mean temp, 10° diff)	1.02
4. Concrete block (8", lightweight)	2.12
5. Plaster (1", sand aggregate)	0.18
6. Inside Surface (still air)	0.68
Total resistance	**4.61**

$U = 1/R = 1/4.61 = 0.22$
Weight of system = 77 psf
I.S. = Initial System

Assume 20,000 S.F. of wall in Boston, MA (5600 degree days) Table L1030-501
Initial system effective $U = U_w \times M = 0.22 \times 0.905 = 0.20$
  (M = weight correction factor from Table L1030-203)
Modified system effective $U = U_w \times M = 0.067 \times 0.915 = 0.06$
Initial systems: $Q = 20,000 \times 0.20 \times (72° - 10°) = 248,000$ BTU/hr.
Modified system: $Q = 20,000 \times 0.06 \times (72° - 10°) = 74,400$ BTU/hr.

Reduction in BTU/hr. heat loss; $\dfrac{\text{I.S.} - \text{M.S.}}{\text{I.S.}} \times 100 = \dfrac{248,000 - 74,400}{248,000} \times 100 = 70\%$

Replace item 3 with 2" smooth rigid polystyrene insulation and item 5 with 3/4" furring and 1/2" drywall.

Total resistance	4.61
Deduct 3. Airspace	1.02
5. Plaster	0.18
	1.20

Difference $4.61 - 1.20 = 3.41$
Add rigid polystyrene insulation 10.00
3/4" air space 1.01
1/2" gypsum board 0.45
Total resistance 14.87

$U = 1/R = 1/14.87 = 0.067$
Weight of system = 70 psf
M.S. = Modified System

B20 Exterior Enclosure   RB2010-030 "U" & "R" Factors

Table B2010-032 Thermal Coefficients of Exterior Closures

EXAMPLE:
Construction	Resistance R
1. Outside surface (15 MPH wind)	0.17
2. Common brick, 4"	0.80
3. Nonreflective air space, 0.75"	1.01
4. Concrete block, S&G aggregate, 8"	1.46
5. Nonreflective air space, 0.75"	1.01
6. Gypsum board, 0.5"	0.45
7. Inside surface (still air)	0.68
Total resistance	**5.58**

U = 1/R = 1/5.58 = 0.18

System weight = 97 psf

Substitution
Replace item 3 with perlite loose fill insulation and fill block cavities with the same.

Total resistance (Example)	R =	5.58
Deduct 3. Air space	1.01	
4. Concrete block	1.46	
	2.47	– 2.47
Difference		3.11
Add 0.75" perlite cavity fill		2.03
Perlite filled 8" block		2.94
Net total resistance		**8.08**

U = 1/R = 1/8.08 = 0.12

System weight = 97 psf

EXAMPLE:
Construction	Resistance R
1. Outside surface (15 MPH wind)	0.17
2. Common brick, 8"	1.60
3. Nonreflective air space, 0.75"	1.01
4. Gypsum board, 0.625"	0.56
5. Inside surface (still air)	0.68
Total resistance	**4.02**

U = 1/R = 1/4.02 = 0.25

System weight = 82 psf

Substitution
Replace item 3 with 4" blanket insulation and item 4 with 0.75" Gypsum plaster (sand agg.).

Total resistance (Example)	R =	4.02
Deduct 3. Air space	1.01	
4. Gypsum board	0.56	
	1.57	– 1.57
Difference		2.45
ADD Blanket insulation		13.00
Gypsum plaster		0.14
Net total resistance		**15.59**

U = 1/R = 1/15.59 = 0.06

System weight = 87 psf

EXAMPLE:
Construction	Resistance R
1. Outside surface (15 MPH wind)	0.17
2. Cement stucco, 0.75"	0.15
3. Concrete block, 8" light weight	2.12
4. Reflective air space, 0.75"	2.77
5. Gypsum board, 0.5"	0.45
6. Inside surface (still air)	0.68
Total resistance	**6.34**

U = 1/R = 1/6.34 = 0.16

System weight = 47 psf

Substitution
Replace item 4 with 2" insulation board and item 5 with 0.75" Perlite plaster.

Total resistance (Example)	R =	6.34
Deduct 4. Air space	2.77	
5. Gypsum board	0.45	
	3.22	– 3.22
Difference		3.12
Add Insulation board 2" (polyurethane)		12.50
Perlite plaster		1.34
Net total resistance		**16.96**

U = 1/R = 1/16.96 = 0.06

System weight = 48 psf

B20 Exterior Enclosure — RB2010-100 Loading Tables

Table B2010-112 Partially and Fully Grouted Reinforced Concrete Masonry Wall Capacities Per L.F. (Kips & In-Kips)

Thk. T (Nom.) (in)	Earthquake Zones 1, 2 & 3 — Length Or Height h' (Ft.)	h'/T (in/in)	Grouted Core & Rebar (spacing) (in O.C.)	Rebar Size (@ d.)	Allowable Vertical Wall Loads — Eccentric Loads 7.0 in-K/Ft. Inspection No (K/Ft.)	Eccentric 7.0 Yes (K/Ft.)	Eccentric 3.5 in-K/Ft. Inspection No (K/Ft.)	Eccentric 3.5 Yes (K/Ft.)	Without Wind or Eccentric Loads Inspection No (K/Ft.)	Without Wind Yes (K/Ft.)	With Wind Inspection No 15 psf (K/Ft.)	With Wind No 30 psf (K/Ft.)	With Wind Yes 15 & 30 (K/Ft.)	Allowable Wall Moments (Without Vertical Wall Loads) Not Wind or Earthquake Inspection No (in.-K/Ft.)	Yes (in.-K/Ft.)
8" Conc. Block	8'	12	48"	#8	5.10	12.55	6.25	13.70	7.75	14.90	7.45	7.45	14.90	7.55	12.20
			32"	#5	5.45	13.35	6.65	14.60	7.90	15.80	7.90	7.90	15.80	6.45	9.60
			16"	↓	6.50	15.80	7.90	17.15	9.25	18.50	9.25	9.25	18.50	7.95	12.85
			8"	↓	10.10	23.40	11.70	25.00	13.30	26.55	13.30	13.30	26.55	10.20	17.15
	12'	18	48"	#8	4.70	11.60	5.80	12.65	6.85	13.75	6.85	6.85	13.75	7.55	12.20
			32"	#5	5.05	12.35	6.15	13.45	7.30	14.60	7.30	7.30	14.60	6.45	9.60
			16"	↓	6.00	14.55	7.30	15.85	8.55	17.10	8.55	8.55	17.10	7.95	12.85
			8"	↓	9.30	21.60	10.80	23.05	12.25	24.55	12.25	12.25	24.55	10.20	17.15
	16'	24	48"	#8	3.95	9.70	4.85	10.60	5.75	11.55	5.75	—	11.55	7.55	12.20
			32"	#5	4.20	12.25	5.15	11.30	6.10	12.25	6.10	—	12.25	6.45	9.60
			16"	↓	5.05	12.20	6.10	13.30	7.15	14.35	7.15	—	14.35	7.95	12.85
			8"	↓	7.80	18.10	9.05	19.35	10.30	20.60	10.30	9.65	20.60	10.20	17.15
10" Conc. Block	8'	9.6	48"	#8	7.25	16.45	8.20	17.40	9.15	—	9.15	9.15	18.35	13.10	21.05
			32"	↓	7.80	17.60	8.80	18.60	9.80	19.55	9.80	9.80	19.55	14.55	24.15
			16"	#5	9.45	21.15	10.55	22.25	11.65	23.35	11.65	11.65	23.35	13.90	22.35
			8"	↓	14.50	31.65	15.80	32.95	17.10	34.25	17.10	17.10	34.25	18.45	30.15
	12	14.4	48"	#8	7.05	16.00	8.00	16.90	8.90	17.85	8.90	8.90	17.85	13.10	21.05
			32"	↓	7.60	17.10	8.55	18.05	9.50	19.05	9.50	9.50	19.05	14.55	24.15
			16"	#5	9.20	20.55	10.25	21.60	11.35	22.70	11.35	11.35	22.70	13.90	22.35
			8"	↓	14.10	30.75	15.35	32.05	16.65	33.00	16.65	16.65	33.30	18.35	30.15
	16'	19.2	48"	#8	6.70	15.10	7.55	16.00	8.40	16.85	8.40	8.40	16.85	13.10	21.05
			32"	↓	7.15	16.15	8.05	17.05	9.00	17.95	9.00	9.00	17.95	14.55	24.15
			16"	#5	8.70	19.40	9.70	20.40	10.70	21.45	10.70	10.70	21.45	13.90	22.35
			8"	↓	13.35	29.05	14.50	30.25	15.70	31.45	15.70	15.70	31.45	18.35	30.15
	20'	24	48"	#8	6.05	13.65	6.80	14.45	7.60	15.20	7.60	—	15.20	13.10	21.05
			32"	↓	6.45	14.60	7.30	15.40	8.10	16.25	8.10	0.70	16.25	14.55	24.15
			16"	#5	7.85	17.50	8.75	18.45	9.65	19.35	9.65	0.25	19.35	13.90	22.35
			8"	↓	12.05	26.25	13.10	27.30	14.20	28.40	14.20	13.40	28.40	18.35	30.15
12" Conc. Block	8'	8	48"	#8	9.20	20.00	10.00	20.75	10.75	21.55	10.75	10.75	21.55	15.30	24.40
			32"	↓	9.90	21.50	10.75	22.30	11.55	23.15	11.55	11.55	23.15	17.10	28.10
			16"	#5	12.10	26.05	13.00	26.95	13.90	27.85	13.90	13.90	27.85	16.20	25.90
			8"	↓	18.60	39.35	19.65	40.40	20.75	41.50	20.75	20.75	41.50	21.45	35.10
	12'	12	48"	#8	9.00	19.55	9.75	20.35	10.55	21.10	10.55	10.55	21.10	15.30	24.40
			32"	↓	9.70	21.05	10.50	21.85	11.30	22.65	11.30	11.30	22.65	17.10	28.10
			16"	#5	11.85	25.50	12.75	26.40	13.65	27.30	13.65	13.65	27.30	16.20	25.90
			8"	↓	18.20	38.50	19.25	39.55	20.30	40.60	20.30	20.30	40.60	21.45	35.10
	16'	16	48"	#8	8.65	18.75	9.35	19.50	10.10	20.20	10.10	10.10	20.20	15.30	24.40
			32"	↓	9.30	20.15	10.05	20.90	10.85	21.70	10.85	10.85	21.70	17.10	28.10
			16"	#5	11.35	24.40	12.20	25.25	13.05	26.15	13.05	13.05	26.15	16.20	25.90
			8"	↓	17.45	36.90	18.45	37.90	19.45	38.90	19.45	19.45	38.90	21.45	35.10
	24'	24	48"	#8	7.05	15.35	7.65	15.95	8.30	16.55	8.30	—	16.55	15.30	24.40
			32"	↓	7.60	16.55	8.25	17.15	8.90	17.80	8.90	—	17.80	17.10	28.10
			16"	#5	9.30	20.00	10.00	20.70	10.70	21.45	10.40*	—	21.45	16.20	25.90
			8"	↓	14.30	30.25	15.10	31.10	15.95	31.90	15.55*	15.15	31.90	21.45	35.10

*Zone 3 only

B20 Exterior Enclosure — RB2010-100 Loading Tables

Table B2010-114 Fully Grouted Reinforced Masonry Wall Capacities Per L.F. (Kips & In-Kips)

Earthquake Zones 1, 2 & 3				Allowable Vertical Wall Loads									Allowable Wall Moments Without Vertical Wall Loads	
Thk.	Length Or Height		Type Wall	Rebar	Eccentric Loads				Without Wind or Eccentric Loads		With Wind			Not Wind or Earthquake	
					7.0 in-K/Ft.		3.5 in-K/Ft.				Inspection				
			Brick Grout Conc. M.U.	Size and Spacing (in. O.C.)	Inspection		Inspection		Inspection		No		Yes	Inspection	
T (Nom.) (in)	h′ (Ft.)	h′/T (in/in)			No (K/Ft.)	Yes (K/Ft.)	No (K/Ft.)	Yes (K/Ft.)	No (K/Ft.)	Yes (K/Ft.)	15 psf (K/Ft.)	30 psf (K/Ft.)	15 & 30 (K/Ft.)	No (in.-K/Ft.)	Yes (in.-K/Ft.)
8″	8′	12	4″0″4″ Solid CMU	#5@32	10.10 12.75	23.35 28.70	11.70 14.35	24.50 30.30	13.30 15.95	26.55 31.90	13.30 15.95	13.30 15.95	26.55 31.90	6.90 7.80	9.65 9.75
	12′	18	4″0″4″ Solid CMU		9.30 16.65	21.60 26.50	10.80 18.15	23.05 28.00	12.25 14.70	24.55 29.45	12.25 14.70	12.25 14.70	24.55 29.45	6.90 7.80	9.65 9.75
	16′	24	4″0″4″ Solid CMU		7.80 9.85	18.10 22.20	9.05 11.10	19.35 23.45	10.30 12.35	20.60 24.70	10.30 12.35	9.65 12.35	20.60 12.70	6.90 7.80	9.65 9.75
10″	8′	9.6	4″0″6″ Solid CMU 4″2″4″	#5@24	14.45 17.85 14.45	31.50 38.35 31.50	15.75 19.15 15.75	32.80 39.65 32.80	17.05 20.45 17.05	34.10 40.90 34.10	17.05 20.45 17.05	17.05 20.45 17.05	34.10 40.90 34.10	11.20 12.65 11.20	16.20 16.35 16.20
	12′	14.4	4″0″6″ Solid CMU 4″2″4″		13.90 17.20 13.90	30.35 36.90 30.35	15.15 18.45 15.15	31.60 38.15 31.60	16.40 19.70 16.40	32.80 39.40 32.80	16.40 19.70 16.40	16.40 19.70 16.40	32.80 39.40 32.80	11.20 12.65 11.20	16.20 16.35 16.20
	16′	19.2	4″0″6″ Solid CMU 4″2″4″		12.85 15.90 12.85	28.05 34.10 23.05	14.00 17.05 14.00	29.20 35.25 29.20	15.15 18.20 15.15	30.35 36.40 30.35	15.15 13.20 15.15	15.15 13.20 15.15	30.35 36.40 30.35	11.20 12.65 11.20	16.20 16.35 16.20
	20′	24	4″0″6″ Solid CMU 4″2″4″		11.15 13.75 11.15	24.25 29.50 24.25	12.10 14.75 12.10	25.25 30.50 25.25	13.10 15.75 13.10	26.20 31.50 26.20	13.10 15.75 13.10	12.40 15.75 12.40	26.20 31.50 26.20	11.20 12.65 11.20	16.20 16.35 16.20
12″	8′	8	4″0″8″ Solid CMU 4″2″6″	#8@48 ↓ #5@20	18.60 22.75 18.60	39.35 47.65 39.35	19.65 23.80 19.65	40.40 48.70 40.40	20.75 24.90 20.75	41.50 49.80 41.50	20.75 24.90 20.75	20.75 24.90 20.75	41.50 49.80 41.50	16.65 18.80 11.90	24.90 25.10 18.75
	12′	12	4″0″8″ Solid CMU 4″2″6″	#8@48 ↓ #5@20	18.20 22.25 18.20	38.50 46.65 38.50	19.25 23.30 19.25	39.55 47.70 39.55	20.30 24.35 20.30	40.60 48.75 40.60	20.30 24.35 20.30	20.30 24.35 20.30	40.60 48.75 40.60	16.65 18.80 11.90	24.90 25.10 18.75
	16′	16	4″0″8″ Solid CMU 4″2″6″	#8@48 ↓ #5@20	17.45 21.30 17.45	36.90 44.70 36.90	18.45 22.35 18.45	37.90 45.70 37.90	19.45 23.35 19.45	38.90 46.70 38.90	19.45 23.25 19.45	19.45 23.35 19.45	38.90 46.70 38.90	16.65 18.80 11.90	24.90 25.10 18.75
	24′	24	4″0″8″ Solid CMU 4″2″6″	#8@48 ↓ #5@20	14.30 17.15 14.30	30.25 36.65 30.25	15.10 18.30 15.10	31.10 37.45 31.10	15.95 19.15 15.95	31.90 38.30 31.90	15.95 19.15 15.95	15.15 19.15 15.15	31.90 38.30 31.90	16.65 18.80 11.90	24.90 25.10 18.75
16″	8′	6	4″0″12″ Solid CMU 4″2″10″	#8@32 ↓ #5@15	26.40 32.00 26.40	54.40 65.65 54.40	27.20 32.80 27.20	55.20 66.45 55.20	28.00 33.60 28.00	56.05 67.25 56.05	28.00 33.60 28.00	28.00 33.60 28.00	56.05 67.25 56.05	31.20 35.25 14.75	49.10 50.40 24.10
	12′	9	4″0″12″ Solid CMU 4″2″10″	#8@32 ↓ #5@15	26.15 31.70 26.15	53.95 66.05 53.95	26.95 32.50 26.95	54.75 65.85 54.75	27.75 33.30 27.75	55.55 66.65 55.55	27.75 33.00 27.75	27.75 33.30 27.75	55.55 66.65 55.55	31.20 35.25 14.75	49.10 50.40 24.10
	16′	12	4″0″12″ Solid CMU 4″2″10″	#8@32 ↓ #5@15	25.70 31.20 25.70	53.05 63.95 53.05	26.50 31.95 26.50	53.80 64.75 53.80	28.30 32.75 27.30	54.60 65.50 54.60	27.30 32.75 27.30	27.30 32.75 27.30	54.60 65.50 59.60	31.20 35.25 14.75	49.10 50.40 24.10
	32′	24	4″0″12″ Solid CMU 4″2″10″	#8@32 ↓ #5@15	20.35 24.65 20.35	41.95 50.60 41.95	20.95 25.30 20.95	42.55 51.20 42.55	21.60 25.90 21.60	43.20 51.85 43.20	18.75* 24.55* 18.75*	18.75* 24.55* 18.75*	43.20 51.85 43.20	31.20 35.25 14.75	49.10 50.40 24.10

*Zone 3 Only

B20 Exterior Enclosure — RB2010-100 Loading Tables

Table B2010-116 Unreinforced Masonry Wall Capacities Per L.F. (Kips & In-Kips)

Thk. T (Nom.) (in)	Length Or Height h' (Ft.)	h'/t (in/in)	Type of Wall	Eccentric Loads 7.0 (K/Ft.)	Eccentric Loads 3.5 (K/Ft.)	Without Wind or Eccentric Loads (K/Ft.)	With Wind 15 psf (K/Ft.)	With Wind 30 psf (K/Ft.)	Not Wind or Earthquake Inspection No (in-K/Ft.)	Not Wind or Earthquake Inspection Yes (in-K/Ft.)	Wind or Earthquake Inspection No (in-K/Ft.)	Wind or Earthquake Inspection Yes (in-K/Ft.)
8"	8'	12	Solid Brick	4.85	7.50	10.15	10.15	10.15	1.15	2.30	1.55	3.10
			Solid CM Units	8.85	11.50	14.15	14.15	14.15	.70	1.40	.95	1.85
			Hollow CM Units	7.95	10.60	13.30	13.30	13.30	.70	1.40	.60	1.25
	10'	15	Solid Brick	4.70	7.30	9.85	9.85	9.85	1.15	2.30	1.55	3.10
			Solid CM Units	8.55	11.15	13.75	13.75	13.75	.70	1.40	.90	1.85
			Hollow CM Units	—	4.45	6.30	6.30	6.05	.45	.95	.60	1.25
	12'	18	Solid Brick	—	6.95	9.40	9.40	7.95	1.15	2.30	1.55	3.10
			Solid CM Units	8.15	10.60	13.10	13.10	12.90	.70	1.40	.90	1.85
			Hollow CM Units	—	4.25	6.00	6.00	4.75	.45	.95	.60	1.25
12"	8'	8	Solid Brick	12.30	14.10	15.90	15.90	15.90	2.70	5.40	3.60	7.20
			Solid CM Units	18.50	20.30	22.10	22.10	22.10	1.60	3.20	2.15	4.30
			Hollow CM Units	7.85	9.10	10.40	10.40	10.40	1.10	2.25	1.50	3.00
			Brick & Hollow CMU	5.50	7.80	10.05	10.05	10.05	1.35	2.70	1.80	3.60
	12'	12	Solid Brick	12.05	13.80	15.55	15.55	15.55	2.70	5.40	3.60	7.20
			Solid CM Units	13.45	14.75	16.05	16.05	16.05	1.60	3.20	2.15	4.30
			Hollow CM Units	7.65	8.90	10.15	10.15	10.15	1.10	2.25	1.50	3.00
			Brick & Hollow CMU	5.40	7.65	9.85	9.85	9.05	1.35	2.70	1.80	3.60
	16'	16	Solid Brick	11.55	13.20	14.90	14.90	14.35	2.70	5.40	3.60	7.20
			Solid CM Units	17.35	19.05	20.70	20.70	20.70	1.60	3.20	2.15	4.30
			Hollow CM Units	7.35	8.55	9.75	9.75	9.05	1.10	2.25	1.50	3.00
			Brick & Hollow CMU	5.20	7.30	9.45	9.10	—	1.35	2.70	1.80	3.60
16"	12'	9	Solid Brick	18.60	19.95	21.30	21.30	21.30	4.85	9.75	6.50	13.00
			Solid CM Units	26.95	28.30	29.60	29.60	29.60	2.90	5.85	3.90	7.80
			Hollow CM Units	10.85	12.10	13.30	13.30	13.30	1.50	3.05	2.00	4.05
			Brick & Hollow CMU	9.30	11.10	12.85	12.85	12.85	2.20	4.40	2.90	5.85
	16'	12	Solid Brick	18.30	19.60	20.90	20.90	20.90	4.85	9.75	6.50	13.00
			Solid CM Units	26.50	27.80	29.10	29.10	29.10	2.90	5.85	3.90	7.80
			Hollow CM Units	10.70	11.90	13.10	13.10	13.10	1.50	3.05	2.00	4.05
			Brick & Hollow CMU	9.15	10.90	12.65	12.65	11.10	2.20	4.40	2.90	5.85
	20'	15	Solid Brick	17.80	19.05	20.30	20.30	20.30	4.85	9.75	6.50	13.00
			Solid CM Units	25.75	27.00	28.25	28.25	28.25	2.90	5.85	3.90	7.80
			Hollow CM Units	10.40	11.55	12.70	12.70	10.95	1.50	3.05	2.00	4.05
			Brick & Hollow CMU	8.90	10.60	12.30	12.00	—	2.20	4.40	2.90	5.85
20"	12'	7.2	Solid Brick	24.80	25.85	26.90	26.90	26.90	7.70	15.40	10.25	20.50
			Solid CM Units	35.30	36.40	37.45	37.45	37.45	4.60	9.20	6.15	12.30
			Hollow CM Units	16.15	17.15	18.15	18.15	18.15	2.50	5.05	3.35	6.75
			Brick & Hollow CMU	16.10	17.25	18.45	18.45	18.45	4.05	8.10	5.40	10.80
	16'	9.6	Solid Brick	24.55	25.60	26.70	26.70	26.70	7.70	15.40	10.25	20.50
			Solid CM Units	35.00	36.10	37.15	37.15	37.15	4.60	9.20	6.15	12.30
			Hollow CM Units	16.00	17.00	18.00	18.00	18.00	2.50	5.05	3.35	6.75
			Brick & Hollow CMU	15.95	17.10	18.30	18.30	18.30	4.05	8.10	5.40	10.80
	24'	14.4	Solid Brick	23.70	24.70	25.75	25.75	25.75	7.70	15.40	10.25	20.50
			Solid CM Units	33.80	34.80	35.80	35.80	35.80	4.60	9.20	6.15	12.30
			Hollow CM Units	15.45	16.40	17.40	17.40	16.05	2.50	5.05	3.35	6.75
			Brick & Hollow CMU	15.40	16.50	17.65	17.65	15.25	4.05	8.10	5.40	10.80

B20 Exterior Enclosure — RB2010-200 Specialties - Concrete Block

Table B2010-202 Mix Proportions by Volume, Compressive Strength of Mortar

Where Used	Mortar Type	Allowable Proportions by Volume				Compressive Strength @ 28 days
		Portland Cement	Masonry Cement	Hydrated Lime	Masonry Sand	
Plain Masonry	M	1	1	—	6	2500 psi
		1	—	1/4	3	
	S	1/2	1	—	4	1800 psi
		1	—	1/4 to 1/2	4	
	N	—	1	—	3	750 psi
		1	—	1/2 to 1-1/4	6	
	O	—	1	—	3	350 psi
		1	—	1-1/4 to 2-1/2	9	
	K	1	—	2-1/2 to 4	12	75 psi
Reinforced Masonry	PM	1	1	—	6	2500 psi
	PL	1	—	1/4 to 1/2	4	2500 psi

Note: The total aggregate should be between 2.25 to 3 times the sum of the cement and lime used.

Table B2010-203 Volume and Cost Per S.F. of Grout Fill for Concrete Block Walls

Center to Center Spacing Grouted Cores	6" C.M.U. Per S.F.				8" C.M.U. Per S.F.				12" C.M.U. Per S.F.			
	Volume in C.F.		Cost		Volume in C.F.		Cost		Volume in C.F.		Cost	
	40% Solid	75% Solid	40% Solid	75% Solid	40% Solid	75% Solid	40% Solid	75% Solid	40% Solid	75% Solid	40% Solid	75% Solid
All cores grouted solid	0.27	0.11	$2.92	$1.19	0.36	0.15	$3.89	$1.62	0.55	0.23	$5.94	$2.48
cores grouted 16" O.C.	0.14	0.06	1.51	0.65	0.18	0.08	1.94	0.86	0.28	0.12	3.02	1.30
cores grouted 24" O.C.	0.09	0.04	0.97	0.43	0.12	0.05	1.30	0.54	0.18	0.08	1.94	0.86
cores grouted 32" O.C.	0.07	0.03	0.76	0.32	0.09	0.04	0.97	0.43	0.14	0.06	1.51	0.65
cores grouted 40" O.C.	0.05	0.02	0.54	0.22	0.07	0.03	0.76	0.32	0.11	0.05	1.19	0.54
cores grouted 48" O.C.	0.04	0.02	0.43	0.22	0.06	0.03	0.65	0.32	0.09	0.04	0.97	0.43

Installed costs, including O&P, based on High-Lift Grouting method

Low-Lift Grouting is used when the wall is built to a maximum height of 5'. The grout is pumped or poured into the cores of the concrete block. The operation is repeated after each five additional feet of wall height has been completed.

High-Lift Grouting is used when the wall has been built to the full-story height. Some of the advantages are: the vertical reinforcing steel can be placed after the wall is completed; and the grout can be supplied by a ready-mix concrete supplier so that it may be pumped in a continuous operation.

B20 Exterior Enclosure — RB2010-430 Pre-engin. Steel Building

Pre-engineered Steel Buildings

These buildings are manufactured by many companies and normally erected by franchised dealers throughout the U.S. The four basic types are: Rigid Frames, Truss type, Post and Beam and the Sloped Beam type. Most popular roof slope is low pitch of 1" in 12". The minimum economical area of these buildings is about 3000 S.F. of floor area. Bay sizes are usually 20' to 24' but can go as high as 30' with heavier girts and purlins. Eave heights are usually 12' to 24' with 18' to 20' most typical.

Material prices generally do not include floors, foundations, interior finishes or utilities. Costs assume at least three bays of 24' each, a 1" in 12" roof slope, and they are based on 30 psf roof load and 20 psf wind load (wind load is a function of wind speed, building height, and terrain characteristics; this should be determined by a Registered Structural Engineer) and no unusual requirements. Costs include the structural frame, 26 ga. non-insulated colored corrugated or ribbed roofing or siding panels, fasteners, closures, trim and flashing but no allowance for insulation, doors, windows, skylights, gutters or downspouts. Very large projects would generally cost less for materials than the prices shown.

Conditions at the site, weather, shape and size of the building, and labor availability will affect the erection cost of the building.

B30 Roofing — RB3010-010 "U" & "R" Factors

Table B3010-011 Thermal Coefficients for Roof Systems

EXAMPLE:

Construction	Resistance R
1. Outside surface (15 MPH wind)	0.17
2. Built up roofing .375" thick	0.33
3. Rigid roof insulation, 2"	5.26
4. Metal decking	0.00
5. Air space (non reflective)	1.14
6. Structural members	0.00
7. Gypsum board ceiling 5/8" thick	0.56
8. Inside surface (still air)	0.61
Total resistance	**8.07**

$U = 1/R = 1/8.07 = 0.12$

Substitution
Replace item 3 with Perlite lightweight concrete and item 7 with acoustical tile.

Total resistance (Example)		R =	8.07
Deduct 3. Rigid insulation	5.26		
7. Gypsum board ceiling	0.05		
	5.31		-5.31
Difference			2.76
Add Perlite concrete, 4"			4.32
Acoustical tile, 3/4"			1.56
		Total resistance	**8.64**

$U = 1/R = 1/8.64 = 0.12$

EXAMPLE

Construction	Resistance R
1. Outside surface (15 MPH wind)	0.17
2. Built up roofing .375" thick	0.33
3. Rigid roof insulation, 2" thick	1.20
4. Form bd., 1/2" thick	0.45
5. Air space (non reflective)	1.14
6. Structural members	0.00
7. Gypsum board ceiling 3/4" thick	0.14
8. Inside surface (still air)	0.61
Total resistance	**4.04**

$U = 1/R = 1/4.04 = 0.25$

(for modified system see next page)

B30 Roofing — RB3010-010 "U" & "R" Factors

Table B3010-011 Thermal Coefficients for Roof Systems

Substitution
Replace item 3 with Tectum board and item 7 with acoustical tile.

Total resistance (Example)		R = 4.04
Deduct 3. Gypsum concrete	1.20	
4. Form bd.	0.45	
7. Plaster	0.14	
	1.79	<-1.79>
Difference		2.25
Add Tectum bd. 3" thick		5.25
Acoustical tile, 3/4"		1.56
	Total resistance	9.06

$U = 1/R = 1/9.06 = 0.11$

(for limited system see previous page)

EXAMPLE:

Construction	Resistance R
1. Outside surface (15 MPH wind)	0.17
2. Asphalt shingle roof	0.44
3. Felt building paper	0.06
4. Plywood sheathing, 5/8" thick	0.78
5. Reflective air space, 3-1/2"	2.17
6. Gypsum board, 1/2" foil backed	0.45
7. Inside surface (still air)	0.61
Total resistance	4.68

$U = 1/R = 1/4.68 = 0.21$

Substitution
Replace item 2 with slate, item 5 with insulation and item 6 with gypsum plaster

Total resistance (Example)		R = 4.68
Deduct 2. Asphalt shingles	0.44	
5. Air space	2.17	
6. Gypsum board	0.45	
	3.06	<-3.06>
Difference		1.62
Add Slate, 1/4" thick		0.03
Insulation, 6" blanket		19.00
Gypsum sand plaster 3/4"		0.14
	Total resistance	20.79

$U = 1/R = 1/20.79 = 0.05$

C20 Stairs — RC2010-100 Concrete, Metal & Wood Stairs

Table C2010-101 Typical Range of Risers for Various Story Heights

General Design: See Table L1010-101 for code requirements.
Maximum height between landings is 12'; usual stair angle is 20°
to 50° with 30° to 35° best. Usual relation of riser to treads is:
  Riser + tread = 17.5
  2x (Riser) + tread = 25
  Riser x tread = 70 or 75
Maximum riser height is 7" for commercial, 8-1/4" for residential.
Usual riser height is 6-1/2" to 7-1/4".
Minimum tread width is 11" for commercial, 9" for residential.

Story Height	Minimum Risers	Maximum Riser Height	Tread Width	Maximum Risers	Minimum Riser Height	Tread Width	Average Risers	Average Riser Height	Tread Width
7'-6"	12	7.50"	10.00"	14	6.43"	11.07"	13	6.92"	10.58"
8'-0"	13	7.38	10.12	15	6.40	11.10	14	6.86	10.64
8'-6"	14	7.29	10.21	16	6.38	11.12	15	6.80	10.70
9'-0"	15	7.20	10.30	17	6.35	11.15	16	6.75	10.75
9'-6"	16	7.13	10.37	18	6.33	11.17	17	6.71	10.79
10'-0"	16	7.50	10.00	19	6.32	11.18	18	6.67	10.83
10'-6"	17	7.41	10.09	20	6.30	11.20	18	7.00	10.50
11'-0"	18	7.33	10.17	21	6.29	11.21	19	6.95	10.55
11'-6"	19	7.26	10.24	22	6.27	11.23	20	6.90	10.60
12'-0"	20	7.20	10.30	23	6.26	11.24	21	6.86	10.64
12'-6"	20	7.50	10.00	24	6.25	11.25	22	6.82	10.68
13'-0"	21	7.43	10.07	25	6.24	11.26	22	7.09	10.41
13'-6"	22	7.36	10.14	25	6.48	11.02	23	7.04	10.46
14'-0"	23	7.30	10.20	26	6.46	11.04	24	7.00	10.50

D10 Conveying — RD1010-010 Passenger Elevators

Table D1010-011 Elevator Hoistway Sizes

Elevator Type	Floors	Building Type	Capacity Lbs.	Capacity Passengers	Entry	Hoistway Width	Hoistway Depth	S.F. Area per Floor
Hydraulic	5	Apt./Small office	1500	10	S	6'-7"	4'-6"	29.6
			2000	13	S	7'-8"	4'-10"	37.4
	7	Average office/Hotel	2500	16	S	8'-4"	5'-5"	45.1
			3000	20	S	8'-4"	5'-11"	49.3
		Large office/Store	3500	23	S	8'-4"	6'-11"	57.6
		Freight light duty	2500		D	7'-2"	7'-10"	56.1
		Heavy duty	5000		D	10'-2"	10'-10"	110.1
			7500		D	10'-2"	12'-10"	131
			5000		D	10'-2"	10'-10"	110.1
			7500		D	10'-2"	12'-10"	131
			10000		D	10'-4"	14'-10"	153.3
		Hospital	3500		D	6'-10"	9'-2"	62.7
			4000		D	7'-4"	9'-6"	69.6
Electric Traction, high speed	High Rise	Apt./Small office	2000	13	S	7'-8"	5'-10"	44.8
			2500	16	S	8'-4"	6'-5"	54.5
			3000	20	S	8'-4"	6'-11"	57.6
			3500	23	S	8'-4"	7'-7"	63.1
		Store	3500	23	S	9'-5"	6'-10"	64.4
		Large office	4000	26	S	9'-4"	7'-6"	70
		Hospital	3500		D	7'-6"	9'-2"	69.4
			4000		D	7'-10"	9'-6"	58.8
Geared, low speed		Apartment	1200	8	S	6'-4"	5'-3"	33.2
			2000	13	S	7'-8"	5'-8"	43.5
			2500	16	S	8'-4"	6'-3"	52
		Office	3000	20	S	8'-4"	6'-9"	56
		Store	3500	23	S	9'-5"	6'-10"	64.4
						Add 4" width for multiple units		

Figure D1010-012 Elevator Passenger Capacity During Peak Periods

Figure D1010-013 Speed & Travel Chart

D10 Conveying — RD1010-010 Passenger Elevators

Table D1010-014 Elevator Speed vs. Height Requirements

Building Type and Elevator Capacities	Travel Speeds in Feet per Minute								
	100 fpm	200 fpm	250 fpm	350 fpm	400 fpm	500 fpm	700 fpm	800 fpm	1000 fpm
Apartments 1200 lb. to 2500 lb.	to 70	to 100	to 125	to 150	to 175	to 250	to 350		
Department Stores 1200 lb. to 2500 lb.		100		125	175	250	350		
Hospitals 3500 lb. to 4000 lb.	70	100	125	150	175	250	350		
Office Buildings 2000 lb. to 4000 lb.		100	125	150		175	250	to 350	over 350

Note: Vertical transportation capacity may be determined by code occupancy requirements of the number of square feet per person divided into the total square feet of building type. If we are contemplating an office building, we find that the Occupancy Code Requirement is 100 S.F. per person. For a 20,000 S.F. building, we would have a legal capacity of two hundred people. Elevator handling capacity is subject to the five minute evacuation recommendation, but it may vary from 11 to 18%. Speed required is a function of the travel height, number of stops and capacity of the elevator.

D20 Plumbing — RD2010-030 Piping

Table D2010-031 Plumbing Approximations for Quick Estimating

Water Control
Water Meter; Backflow Preventer,
Shock Absorbers; Vacuum
Breakers; Mixer .... 10 to 15% of Fixtures
Pipe And Fittings .... 30 to 60% of Fixtures

    Note: Lower percentage for compact buildings or larger buildings with plumbing in one area.
    Larger percentage for large buildings with plumbing spread out.
    In extreme cases pipe may be more than 100% of fixtures.
    Percentages **do not** include special purpose or process piping.

Plumbing Labor
1 & 2 Story Residential .... Rough-in Labor = 80% of Materials
Apartment Buildings .... Rough-in Labor = 90 to 100% of Materials
Labor for handling and placing fixtures is approximately 25 to 30% of fixtures
Quality/Complexity Multiplier (for all installations)
Economy installation, add .... 0 to 5%
Good quality, medium complexity, add .... .5 to 15%
Above average quality and complexity, add .... 15 to 25%

Table D2010-032 Pipe Material Considerations

1. Malleable fittings should be used for gas service.
2. Malleable fittings are used where there are stresses/strains due to expansion and vibration.
3. Cast fittings may be broken as an aid to disassembling of heating lines frozen by long use, temperature and minerals.
4. Cast iron pipe is extensively used for underground and submerged service.
5. Type M (light wall) copper tubing is available in hard temper only and is used for nonpressure and less severe applications than K and L.
6. Type L (medium wall) copper tubing, available hard or soft for interior service.
7. Type K (heavy wall) copper tubing, available in hard or soft temper for use where conditions are severe. For underground and interior service.
8. Hard drawn tubing requires fewer hangers or supports but should not be bent. Silver brazed fittings are recommended, however soft solder is normally used.
9. Type DMV (very light wall) copper tubing designed for drainage, waste and vent plus other non-critical pressure services.

Table D2010-033 Domestic/Imported Pipe and Fittings Cost

The prices shown in this publication for steel/cast iron pipe and steel, cast iron, malleable iron fittings are based on domestic production sold at the normal trade discounts. The above listed items of foreign manufacture may be available at prices of 1/3 to 1/2 those shown. Some imported items after minor machining or finishing operations are being sold as domestic to further complicate the system.

Caution: Most pipe prices in this book also include a coupling and pipe hangers which for the larger sizes can add significantly to the per foot cost and should be taken into account when comparing "book cost" with quoted supplier's cost.

D20 Plumbing — RD2010-400 Individual Fixture Systems

Table D2010-401 Minimum Plumbing Fixture Requirements

Minimum Plumbing Fixture Requirements

Type of Building or Occupancy (2)	Water Closets (14) (Fixtures per Person) Male	Water Closets (14) (Fixtures per Person) Female	Urinals (5,10) (Fixtures per Person) Male	Urinals (5,10) (Fixtures per Person) Female	Lavatories (Fixtures per Person) Male	Lavatories (Fixtures per Person) Female	Bathtubs or Showers (Fixtures per Person)	Drinking Fountains (Fixtures per Person) (3, 13)
Assembly Places- Theatres, Auditoriums, Convention Halls, etc.-for permanent employee use	1: 1 - 15 2: 16 - 35 3: 36 - 55 Over 55, add 1 fixture for each additional 40 persons	1: 1 - 15 2: 16 - 35 3: 36 - 55	0: 1 - 9 1: 10 - 50 Add one fixture for each additional 50 males		1 per 40	1 per 40		
Assembly Places- Theatres, Auditoriums, Convention Halls, etc. - for public use	1: 1 - 100 2: 101 - 200 3: 201 - 400 Over 400, add 1 fixture for each additional 500 males and 1 for each additional 125 females	3: 1 - 50 4: 51 - 100 8: 101 - 200 11: 201 - 400	1: 1 - 100 2: 101 - 200 3: 201 - 400 4: 401 - 600 Over 600, add 1 fixture for each additional 300 males		1: 1 - 200 2: 201 - 400 3: 401 - 750 Over 750, add 1 fixture for each additional 500 persons	1: 1 - 200 2: 201 - 400 3: 401 - 750		1: 1 - 150 2: 151 - 400 3: 401 - 750 Over 750, add one fixture for each additional 500 persons
Dormitories (9) School or Labor	1 per 10 Add 1 fixture for each additional 25 males (over 10) and 1 for each additional 20 females (over 8)	1 per 8	1 per 25 Over 150, add 1 fixture for each additional 50 males		1 per 12 Over 12 add 1 fixture for each additional 20 males and 1 for each 15 additional females	1 per 12	1 per 8 For females add 1 bathtub per 30. Over 150, add 1 per 20	1 per 150 (12)
Dormitories- for Staff Use	1: 1 - 15 2: 16 - 35 3: 36 - 55 Over 55, add 1 fixture for each additional 40 persons	1: 1 - 15 3: 16 - 35 4: 36 - 55	1 per 50		1 per 40	1 per 40	1 per 8	
Dwellings: Single Dwelling Multiple Dwelling or Apartment House	1 per dwelling	1 per dwelling or apartment unit			1 per dwelling	1 per dwelling or apartment unit	1 per dwelling 1 per dwelling or apartment unit	
Hospital Waiting rooms	1 per room				1 per room			1 per 150 (12)
Hospitals- for employee use	1: 1 - 15 2: 16 - 35 3: 36 - 55 Over 55, add 1 fixture for each additional 40 persons	1: 1 - 15 3: 16 - 35 4: 36 - 55	0: 1 - 9 1: 10 - 50 Add 1 fixture for each additional 50 males		1 per 40	1 per 40		
Hospitals: Individual Room Ward Room	1 per room 1 per 8 patients				1 per room 1 per 10 patients		1 per room 1 per 20 patients	1 per 150 (12)
Industrial (6) Warehouses Workshops, Foundries and similar establishments- for employee use	1: 1 -10 2: 11 - 25 3: 26 - 50 4: 51 - 75 5: 76 - 100 Over 100, add 1 fixture for each additional 30 persons	1: 1 -10 2: 11 - 25 3: 26 - 50 4: 51 - 75 5: 76 - 100			Up to 100, per 10 persons Over 100, 1 per 15 persons (7, 8)		1 shower for each 15 persons exposed to excessive heat or to skin contamination with poisonous, infectious or irritating material	1 per 150 (12)
Institutional - Other than Hospitals or Penal Institutions (on each occupied floor)	1 per 25	1 per 20	0: 1 - 9 1: 10 - 50 Add 1 fixture for each additional 50 males		1 per 10	1 per 10	1 per 8	1 per 150 (12)
Institutional - Other than Hospitals or Penal Institutions (on each occupied floor)- for employee use	1: 1 - 15 2: 16 - 35 3: 36 - 55 Over 55, add 1 fixture for each additional 40 persons	1: 1 - 15 3: 16 - 35 4: 36 - 55	0: 1 - 9 1: 10 - 50 Add 1 fixture for each additional 50 males		1 per 40	1 per 40	1 per 8	1 per 150 (12)
Office or Public Buildings	1: 1 - 100 2: 101 - 200 3: 201 - 400 Over 400, add 1 fixture for each additional 500 males and 1 for each additional 150 females	3: 1 - 50 4: 51 - 100 8: 101 - 200 11: 201 - 400	1: 1 - 100 2: 101 - 200 3: 201 - 400 4: 401 - 600 Over 600, add 1 fixture for each additional 300 males		1: 1 - 200 2: 201 - 400 3: 401 - 750 Over 750, add 1 fixture for each additional 500 persons	1: 1 - 200 2: 201 - 400 3: 401 - 750		1 per 150 (12)
Office or Public Buildings - for employee use	1: 1 - 15 2: 16 - 35 3: 36 - 55 Over 55, add 1 fixture for each additional 40 persons	1: 1 - 15 3: 16 - 35 4: 36 - 55	0: 1 - 9 1: 10 - 50 Add 1 fixture for each additional 50 males		1 per 40	1 per 40		

D20 Plumbing — RD2010-400 Individual Fixture Systems

Table D2010-401 Minimum Plumbing Fixture Requirements (cont.)

Minimum Plumbing Fixture Requirements

Type of Building or Occupancy	Water Closets (14) (Fixtures per Person) Male	Water Closets (14) (Fixtures per Person) Female	Urinals (5, 10) (Fixtures per Person) Male	Urinals (5, 10) (Fixtures per Person) Female	Lavatories (Fixtures per Person) Male	Lavatories (Fixtures per Person) Female	Bathtubs or Showers (Fixtures per Person)	Drinking Fountains (Fixtures per Person) (3, 13)
Penal Institutions - for employee use	1: 1 - 15 2: 16 - 35 3: 36 - 55 Over 55, add 1 fixture for each additional 40 persons	1: 1 - 15 3: 16 - 35 4: 36 - 55	0: 1 - 9 1: 10 - 50 Add 1 fixture for each additional 50 males		1 per 40	1 per 40		1 per 150 (12)
Penal Institutions - for prison use Cell	1 per cell	1 per cell			1 per cell	1 per cell		1 per cellblock floor
Exercise room	1 per exercise room	1 per exercise room	1 per exercise room	1 per exercise room	1 per exercise room	1 per exercise room		1 per exercise room
Restaurants, Pubs and Lounges (11)	1: 1 - 50 2: 51 - 150 3: 151 - 300 Over 300, add 1 fixture for each additional 200 persons	1: 1 - 50 2: 51 - 150 4: 151 - 300	1: 1 - 150 Over 150, add 1 fixture for each additional 150 males		1: 1 - 150 2: 151 - 200 3: 201 - 400 Over 400, add 1 fixture for each additional 400 persons	1: 1 - 150 2: 151 - 200 3: 201 - 400		
Schools - for staff use All Schools	1: 1 - 15 2: 16 - 35 3: 36 - 55 Over 55, add 1 fixture for each additional 40 persons	1: 1 - 15 3: 16 - 35 4: 36 - 55	1 per 50		1 per 40	1 per 40		
Schools - for student use: Nursery	1: 1 - 20 2: 21 - 50 Over 50, add 1 fixture for each additional 50 persons	1: 1 - 20 2: 21 - 50			1: 1 - 25 2: 26 - 50 Over 50, add 1 fixture for each additional 50 persons	1: 1 - 25 2: 26 - 50		1 per 150 (12)
Elementary	1 per 30	1 per 25	1 per 75		1 per 35	1 per 35		1 per 150 (12)
Secondary	1 per 40	1 per 30	1 per 35		1 per 40	1 per 40		1 per 150 (12)
Others (Colleges, Universities, Adult Centers, etc.)	1 per 40	1 per 30	1 per 35		1 per 40	1 per 40		1 per 150 (12)
Worship Places Educational and Activities Unit	1 per 150	1 per 75	1 per 150		1 per 2 water closets			1 per 150 (12)
Worship Places Principal Assembly Place	1 per 150	1 per 75	1 per 150		1 per 2 water closets			1 per 150 (12)

Notes:
1. The figures shown are based upon one (1) fixture being the minimum required for the number of persons indicated or any fraction thereof.
2. Building categories not shown on this table shall be considered separately by the Administrative Authority.
3. Drinking fountains shall not be installed in toilet rooms.
4. Laundry trays. One (1) laundry tray or one (1) automatic washer standpipe for each dwelling unit or one (1) laundry trays or one (1) automatic washer standpipes, or combination thereof, for each twelve (12) apartments. Kitchen sinks, one (1) for each dwelling or apartment unit.
5. For each urinal added in excess of the minimum required, one water closet may be deducted. The number of water closets shall not be reduced to less than two-thirds (2/3) of the minimum requirement.
6. As required by ANSI Z4.1-1968, Sanitation in Places of Employment.
7. Where there is exposure to skin contamination with poisonous, infectious, or irritating materials, provide one (1) lavatory for each five (5) persons.
8. Twenty-four (24) lineal inches of wash sink or eighteen (18) inches of a circular basin, when provided with water outlets for such space shall be considered equivalent to one (1) lavatory.
9. Laundry trays, one (1) for each fifty (50) persons. Service sinks, one (1) for each hundred (100) persons.
10. General. In applying this schedule of facilities, consideration shall be given to the accessibility of the fixtures. Conformity purely on a numerical basis may not result in an installation suited to the need of the individual establishment. For example, schools should be provided with toilet facilities on each floor having classrooms.
    a. Surrounding materials, wall and floor space to a point two (2) feet in front of urinal lip and four (4) feet above the floor, and at least two (2) feet to each side of the urinal shall be lined with non-absorbent materials.
    b. Trough urinals shall be prohibited.
11. A restaurant is defined as a business which sells food to be consumed on the premises.
    a. The number of occupants for a drive-in restaurant shall be considered as equal to the number of parking stalls.
    b. Employee toilet facilities shall not to be included in the above restaurant requirements. Hand washing facilities shall be available in the kitchen for employees.
12. Where food is consumed indoors, water stations may be substituted for drinking fountains. Offices, or public buildings for use by more than six (6) persons shall have one (1) drinking fountain for the first one hundred fifty (150) persons and one additional fountain for each three hundred (300) persons thereafter.
13. There shall be a minimum of one (1) drinking fountain per occupied floor in schools, theaters, auditoriums, dormitories, offices of public building.
14. The total number of water closets for females shall be at least equal to the total number of water closets and urinals required for males.

D20 Plumbing — RD2010-400 Individual Fixture Systems

Table D2010-403 Water Cooler Application

Type of Service	Requirement
Office, School or Hospital	12 persons per gallon per hour
Office, Lobby or Department Store	4 or 5 gallons per hour per fountain
Light manufacturing	7 persons per gallon per hour
Heavy manufacturing	5 persons per gallon per hour
Hot heavy manufacturing	4 persons per gallon per hour
Hotel	.08 gallons per hour per room
Theater	1 gallon per hour per 100 seats

D20 Plumbing — RD2020-100 Hot Water

Table D2020-101 Hot Water Consumption Rates

Type of Building	Size Factor	Maximum Hourly Demand	Average Day Demand
Apartment Dwellings	No. of Apartments:		
	Up to 20	12.0 Gal. per apt.	42.0 Gal. per apt.
	21 to 50	10.0 Gal. per apt.	40.0 Gal. per apt.
	51 to 75	8.5 Gal. per apt.	38.0 Gal. per apt.
	76 to 100	7.0 Gal. per apt.	37.0 Gal. per apt.
	101 to 200	6.0 Gal. per apt.	36.0 Gal. per apt.
	201 up	5.0 Gal. per apt.	35.0 Gal. per apt.
Dormitories	Men	3.8 Gal. per man	13.1 Gal. per man
	Women	5.0 Gal. per woman	12.3 Gal. per woman
Hospitals	Per bed	23.0 Gal. per patient	90.0 Gal. per patient
Hotels	Single room with bath	17.0 Gal. per unit	50.0 Gal. per unit
	Double room with bath	27.0 Gal. per unit	80.0 Gal. per unit
Motels	No. of units:		
	Up to 20	6.0 Gal. per unit	20.0 Gal. per unit
	21 to 100	5.0 Gal. per unit	14.0 Gal. per unit
	101 Up	4.0 Gal. per unit	10.0 Gal. per unit
Nursing Homes		4.5 Gal. per bed	18.4 Gal. per bed
Office buildings		0.4 Gal. per person	1.0 Gal. per person
Restaurants	Full meal type	1.5 Gal./max. meals/hr.	2.4 Gal. per meal
	Drive-in snack type	0.7 Gal./max. meals/hr.	0.7 Gal. per meal
Schools	Elementary	0.6 Gal. per student	0.6 Gal. per student
	Secondary & High	1.0 Gal. per student	1.8 Gal. per student

For evaluation purposes, recovery rate and storage capacity are inversely proportional. Water heaters should be sized so that the maximum hourly demand anticipated can be met in addition to allowance for the heat loss from the pipes and storage tank.

Table D2020-102 Fixture Demands in Gallons Per Fixture Per Hour

Table below is based on 140°F final temperature except for dishwashers in public places (*) where 180°F water is mandatory.

Fixture	Apartment House	Club	Gym	Hospital	Hotel	Indust. Plant	Office	Private Home	School
Bathtubs	20	20	30	20	20			20	
Dishwashers, automatic	15	50-150*		50-150*	50-200*	20-100*		15	20-100*
Kitchen sink	10	20		20	30	20	20	10	20
Laundry, stationary tubs	20	28		28	28			20	
Laundry, automatic wash	75	75		100	150			75	
Private lavatory	2	2	2	2	2	2	2	2	2
Public lavatory	4	6	8	6	8	12	6		15
Showers	30	150	225	75	75	225	30	30	225
Service sink	20	20		20	30	20	20	15	20
Demand factor	0.30	0.30	0.40	0.25	0.25	0.40	0.30	0.30	0.40
Storage capacity factor	1.25	0.90	1.00	0.60	0.80	1.00	2.00	0.70	1.00

To obtain the probable maximum demand multiply the total demands for the fixtures (gal./fixture/hour) by the demand factor. The heater should have a heating capacity in gallons per hour equal to this maximum. The storage tank should have a capacity in gallons equal to the probable maximum demand multiplied by the storage capacity factor.

D30 HVAC — RD3010-600 Solar Collectors

Table D3010-601 Collector Tilt for Domestic Hot Water

Optimum collector tilt is usually equal to the site latitude. Variations of plus or minus 10 degrees are acceptable, and orientation of 20 degrees on either side of true south is acceptable. However, local climate and collector type may influence the choice between east or west deviations.

Flat plate collectors consist of a number of components as follows: Insulation to reduce heat loss through the bottom and sides of the collector. The enclosure which contains all the components in this assembly is usually weatherproof and prevents dust, wind and water from coming in contact with the absorber plate. The cover plate usually consists of one or more layers of a variety of glass or plastic and reduces the reradiation. It creates an air space which traps the heat by reducing radiation losses between the cover plate and the absorber plate.

The absorber plate must have a good thermal bond with the fluid passages.

The absorber plate is usually metallic and treated with a surface coating which improves absorptivity. Black or dark paints or selective coatings are used for this purpose, and the design of this passage and plate combination helps determine a solar system's effectiveness.

Heat transfer fluid passage tubes are attached above and below or integral with an absorber plate for the purpose of transferring thermal energy from the absorber plate to a heat transfer medium. The heat exchanger is a device for transferring thermal energy from one fluid to another. The rule of thumb of space heating sizing is one S.F. of collector per 2.5 S.F. of floor space.

For domestic hot water the rule of thumb is 3/4 S.F. of collector for one gallon of water used per day, on an average use of twenty-five gallons per day per person, plus ten gallons per dishwasher or washing machine.

D30 HVAC — RD3020-010 Heating

Table D3020-011

Heating Systems

The basic function of a heating system is to bring an enclosed volume up to a desired temperature and then maintain that temperature within a reasonable range. To accomplish this, the selected system must have sufficient capacity to offset transmission losses resulting from the temperature difference on the interior and exterior of the enclosing walls in addition to losses due to cold air infiltration through cracks, crevices and around doors and windows. The amount of heat to be furnished is dependent upon the building size, construction, temperature difference, air leakage, use, shape, orientation and exposure. Air circulation is also an important consideration. Circulation will prevent stratification which could result in heat losses through uneven temperatures at various levels. For example, the most efficient use of unit heaters can usually be achieved by circulating the space volume through the total number of units once every 20 minutes or 3 times an hour. This general rule must, of course, be adapted for special cases such as large buildings with low ratios of heat transmitting surface to cubical volume. The type of occupancy of a building will have considerable bearing on the number of heat transmitting units and the location selected. It is axiomatic, however, that the basis of any successful heating system is to provide the maximum amount of heat at the points of maximum heat loss such as exposed walls, windows, and doors. Large roof areas, wind direction, and wide doorways create problems of excessive heat loss and require special consideration and treatment.

Heat Transmission

Heat transfer is an important parameter to be considered during selection of the exterior wall style, material and window area. A high rate of transfer will permit greater heat loss during the wintertime with the resultant increase in heating energy costs and a greater rate of heat gain in the summer with proportionally greater cooling cost. Several terms are used to describe various aspects of heat transfer. However, for general estimating purposes this book lists U valves for systems of construction materials. U is the "overall heat transfer coefficient." It is defined as the heat flow per hour through one square foot when the temperature difference in the air on either side of the structure wall, roof, ceiling or floor is one degree Fahrenheit. The structural segment may be a single homogeneous material or a composite.

Total heat transfer is found using the following equation:

$Q = AU(T_2 - T_1)$ where
Q = Heat flow, BTU per hour
A = Area, square feet
U = Overall heat transfer coefficient
$(T_2 - T_1)$ = Difference in temperature of air on each side of the construction component. (Also abbreviated TD)

Note that heat can flow through all surfaces of any building and this flow is in addition to heat gain or loss due to ventilation, infiltration and generation (appliances, machinery, people).

Table D3020-012 Heating Approximations for Quick Estimating

Oil Piping & Boiler Room Piping:
- Small System .... 20 to 30% of Boiler
- Complex System with Pumps, Headers, Etc. .... 80 to 110% of Boiler

Breeching With Insulation:
- Small .... 10 to 15% of Boiler
- Large .... 15 to 25% of Boiler

Coils: .... 15 to 30% of Containing Unit
Balancing (Independent) .... 1/2% of H.V.A.C. Estimating

Quality/Complexity Adjustment: For all heating installations add these adjustments to the estimate to more closely allow for the equipment and conditions of the particular job under consideration.
- Economy installation, add .... 0 to 5% of System
- Good quality, medium complexity, add .... 5 to 15% of System
- Above average quality and complexity, add .... 15 to 25% of System

D30 HVAC

RD3020-020 Heat Transfer

Table D3020-021 Factor for Determining Heat Loss for Various Types of Buildings

General: While the most accurate estimates of heating requirements would naturally be based on detailed information about the building being considered, it is possible to arrive at a reasonable approximation using the following procedure:
1. Calculate the cubic volume of the room or building.
2. Select the appropriate factor from Table D3020-021 below. Note that the factors apply only to inside temperatures listed in the first column and to 0° F outside temperature.
3. If the building has bad north and west exposures, multiply the heat loss factor by 1.1.
4. If the outside design temperature is other than 0°F, multiply the factor from Table D3020-021 by the factor from Table D3020-022.
5. Multiply the cubic volume by the factor selected from Table D3020-021. This will give the estimated BTUH heat loss which must be made up to maintain inside temperature.

Building Type	Conditions	Qualifications	Loss Factor*
Factories & Industrial Plants General Office Areas at 70° F	One Story	Skylight in Roof	6.2
		No Skylight in Roof	5.7
	Multiple Story	Two Story	4.6
		Three Story	4.3
		Four Story	4.1
		Five Story	3.9
		Six Story	3.6
	All Walls Exposed	Flat Roof	6.9
		Heated Space Above	5.2
	One Long Warm Common Wall	Flat Roof	6.3
		Heated Space Above	4.7
	Warm Common Walls on Both Long Sides	Flat Roof	5.8
		Heated Space Above	4.1
Warehouses at 60° F	All Walls Exposed	Skylights in Roof	5.5
		No Skylight in Roof	5.1
		Heated Space Above	4.0
	One Long Warm Common Wall	Skylight in Roof	5.0
		No Skylight in Roof	4.9
		Heated Space Above	3.4
	Warm Common Walls on Both Long Sides	Skylight in Roof	4.7
		No Skylight in Roof	4.4
		Heated Space Above	3.0

*Note: This table tends to be conservative particularly for new buildings designed for minimum energy consumption.

Table D3020-022 Outside Design Temperature Correction Factor (for Degrees Fahrenheit)

Outside Design Temperature	50	40	30	20	10	0	–10	–20	–30
Correction Factor	0.29	0.43	0.57	0.72	0.86	1.00	1.14	1.28	1.43

Tables D3020-023 and D3020-024 provide a way to calculate heat transmission of various construction materials from their U values and the TD (Temperature Difference).
1. From the exterior Enclosure Division or elsewhere, determine U values for the construction desired.
2. Determine the coldest design temperature. The difference between this temperature and the desired interior temperature is the TD (temperature difference).
3. Enter Figure D3020-023 or D3020-024 at correct U value. Cross horizontally to the intersection with appropriate TD. Read transmission per square foot from bottom of figure.
4. Multiply this value of BTU per hour transmission per square foot of area by the total surface area of that type of construction.

D30 HVAC — RD3020-020 Heat Transfer

Table D3020-023 Transmission of Heat

Table D3020-024 Transmission of Heat (Low Rate)

D30 HVAC — RD3030-010 Air Conditioning & Ventilation

Table D3030-012 Air Conditioning Requirements

BTU's per hour per S.F. of floor area and S.F. per ton of air conditioning.

Type of Building	BTU/Hr per S.F.	S.F. per Ton	Type of Building	BTU/Hr per S.F.	S.F. per Ton	Type of Building	BTU/Hr per S.F.	S.F. per Ton
Apartments, Individual	26	450	Dormitory, Rooms	40	300	Libraries	50	240
Corridors	22	550	Corridors	30	400	Low Rise Office, Exterior	38	320
Auditoriums & Theaters	40	300/18*	Dress Shops	43	280	Interior	33	360
Banks	50	240	Drug Stores	80	150	Medical Centers	28	425
Barber Shops	48	250	Factories	40	300	Motels	28	425
Bars & Taverns	133	90	High Rise Office—Ext. Rms.	46	263	Office (small suite)	43	280
Beauty Parlors	66	180	Interior Rooms	37	325	Post Office, Individual Office	42	285
Bowling Alleys	68	175	Hospitals, Core	43	280	Central Area	46	260
Churches	36	330/20*	Perimeter	46	260	Residences	20	600
Cocktail Lounges	68	175	Hotel, Guest Rooms	44	275	Restaurants	60	200
Computer Rooms	141	85	Corridors	30	400	Schools & Colleges	46	260
Dental Offices	52	230	Public Spaces	55	220	Shoe Stores	55	220
Dept. Stores, Basement	34	350	Industrial Plants, Offices	38	320	Shop'g. Ctrs., Supermarkets	34	350
Main Floor	40	300	General Offices	34	350	Retail Stores	48	250
Upper Floor	30	400	Plant Areas	40	300	Specialty	60	200

*Persons per ton
12,000 BTU = 1 ton of air conditioning

Table D3030-013 Psychrometric Table

Dewpoint or Saturation Temperature (F)

	32	35	40	45	50	55	60	65	70	75	80	85	90	95	100
100	32	35	40	45	50	55	60	65	70	75	80	85	90	95	100
90	30	33	37	42	47	52	57	62	67	72	77	82	87	92	97
80	27	30	34	39	44	49	54	58	64	68	73	78	83	88	93
70	24	27	31	36	40	45	50	55	60	64	69	74	79	84	88
60	20	24	28	32	36	41	46	51	55	60	65	69	74	79	83
50	16	20	24	28	33	36	41	46	50	55	60	64	69	73	78
40	12	15	18	23	27	31	35	40	45	49	53	58	62	67	71
30	8	10	14	18	21	25	29	33	37	42	46	50	54	59	62
20	6	7	8	9	13	16	20	24	28	31	35	40	43	48	52
10	4	4	5	5	6	8	9	10	13	17	20	24	27	30	34
	32	35	40	45	50	55	60	65	70	75	80	85	90	95	100

Dry bulb temperature (F)

This table shows the relationship between RELATIVE HUMIDITY, DRY BULB TEMPERATURE AND DEWPOINT. As an example, assume that the thermometer in a room reads 75°F, and we know that the relative humidity is 50%. The chart shows the dewpoint temperature to be 55°. That is, any surface colder than 55°F will "sweat" or collect condensing moisture. This surface could be the outside of an uninsulated chilled water pipe in the summertime, or the inside surface of a wall or deck in the wintertime. After determining the extreme ambient parameters, the table at the left is useful in determining which surfaces need insulation or vapor barrier protection.

D30 HVAC — RD3030-010 Air Conditioning & Ventilation

Table D3030-014 Recommended Ventilation Air Changes

Table below lists range of time in minutes per change for various types of facilities.

Facility	Min/Change	Facility	Min/Change	Facility	Min/Change
Assembly Halls	2-10	Dance Halls	2-10	Laundries	1-3
Auditoriums	2-10	Dining Rooms	3-10	Markets	2-10
Bakeries	2-3	Dry Cleaners	1-5	Offices	2-10
Banks	3-10	Factories	2-5	Pool Rooms	2-5
Bars	2-5	Garages	2-10	Recreation Rooms	2-10
Beauty Parlors	2-5	Generator Rooms	2-5	Sales Rooms	2-10
Boiler Rooms	1-5	Gymnasiums	2-10	Theaters	2-8
Bowling Alleys	2-10	Kitchens-Hospitals	2-5	Toilets	2-5
Churches	5-10	Kitchens-Restaurant	1-3	Transformer Rooms	1-5

CFM air required for changes = Volume of room in cubic feet ÷ Minutes per change.

Table D3030-015 Ductwork

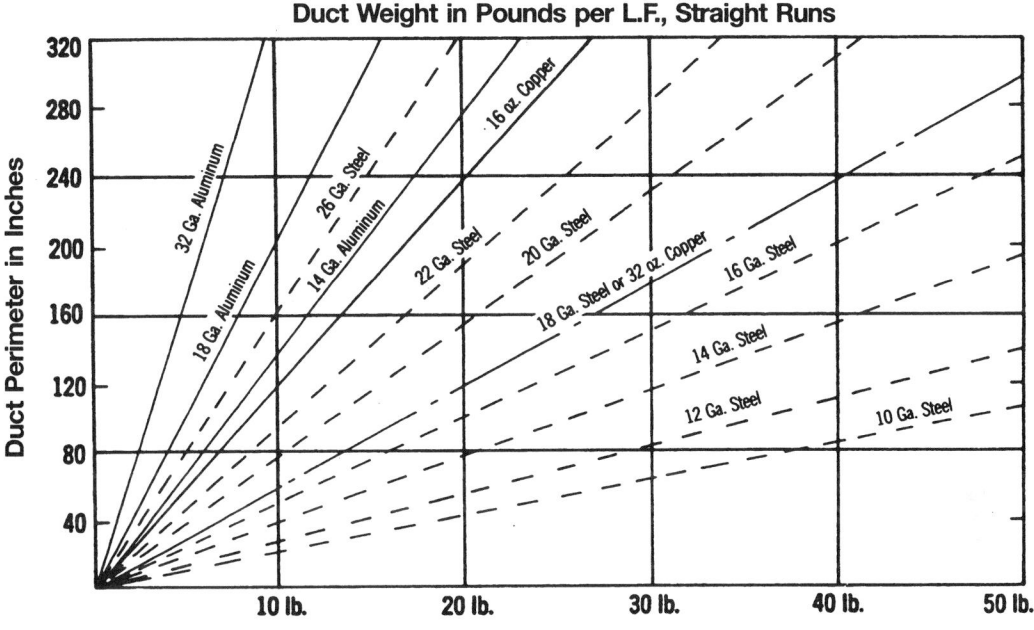

Duct Weight in Pounds per L.F., Straight Runs

Add to the above for fittings; 90° elbow is 3 L.F.; 45° elbow is 2.5 L.F.; offset is 4 L.F.; transition offset is 6 L.F.; square-to-round transition is 4 L.F.; 90° reducing elbow is 5 L.F. For bracing and waste, add 20% to aluminum and copper, 15% to steel.

Table D3030-017 Ductwork Packages (per Ton of Cooling)

System	Sheet Metal	Insulation	Diffusers	Return Register
Roof Top Unit Single Zone	120 Lbs.	52 S.F.	1	1
Roof Top Unit Multizone	240 Lbs.	104 S.F.	2	1
Self-contained Air or Water Cooled	108 Lbs.	—	2	—
Split System Air Cooled	102 Lbs.	—	2	—

Systems reflect most common usage. Refer to system graphics for duct layout.

Table D3030-018 Quality/Complexity Adjustment for Air Conditioning Systems

Economy installation, add	0 to 5%
Good quality, medium complexity, add	5 to 15%
Above average quality and complexity, add	15 to 25%

Add the above adjustments to the estimate to more closely allow for the equipment and conditions of the particular job under consideration.

D30 HVAC — RD3030-010 Air Conditioning & Ventilation

Table D3030-019 Sheet Metal Calculator (Weight in Lb./Ft. of Length)

Gauge	26	24	22	20	18	16	Gauge	26	24	22	20	18	16
Wt.-Lb./S.F.	.906	1.156	1.406	1.656	2.156	2.656	Wt.-Lb./S.F.	.906	1.156	1.406	1.656	2.156	2.656
SMACNA Max. Dimension – Long Side		30"	54"	84"	85" Up		SMACNA Max. Dimension – Long Side		30"	54"	84"	85" Up	
Sum-2 sides							Sum-2 Sides						
2	.3	.40	.50	.60	.80	.90	56	9.3	12.0	14.0	16.2	21.3	25.2
3	.5	.65	.80	.90	1.1	1.4	57	9.5	12.3	14.3	16.5	21.7	25.7
4	.7	.85	1.0	1.2	1.5	1.8	58	9.7	12.5	14.5	16.8	22.0	26.1
5	.8	1.1	1.3	1.5	1.9	2.3	59	9.8	12.7	14.8	17.1	22.4	26.6
6	1.0	1.3	1.5	1.7	2.3	2.7	60	10.0	12.9	15.0	17.4	22.8	27.0
7	1.2	1.5	1.8	2.0	2.7	3.2	61	10.2	13.1	15.3	17.7	23.2	27.5
8	1.3	1.7	2.0	2.3	3.0	3.6	62	10.3	13.3	15.5	18.0	23.6	27.9
9	1.5	1.9	2.3	2.6	3.4	4.1	63	10.5	13.5	15.8	18.3	24.0	28.4
10	1.7	2.2	2.5	2.9	3.8	4.5	64	10.7	13.7	16.0	18.6	24.3	28.8
11	1.8	2.4	2.8	3.2	4.2	5.0	65	10.8	13.9	16.3	18.9	24.7	29.3
12	2.0	2.6	3.0	3.5	4.6	5.4	66	11.0	14.1	16.5	19.1	25.1	29.7
13	2.2	2.8	3.3	3.8	4.9	5.9	67	11.2	14.3	16.8	19.4	25.5	30.2
14	2.3	3.0	3.5	4.1	5.3	6.3	68	11.3	14.6	17.0	19.7	25.8	30.6
15	2.5	3.2	3.8	4.4	5.7	6.8	69	11.5	14.8	17.3	20.0	26.2	31.1
16	2.7	3.4	4.0	4.6	6.1	7.2	70	11.7	15.0	17.5	20.3	26.6	31.5
17	2.8	3.7	4.3	4.9	6.5	7.7	71	11.8	15.2	17.8	20.6	27.0	32.0
18	3.0	3.9	4.5	5.2	6.8	8.1	72	12.0	15.4	18.0	20.9	27.4	32.4
19	3.2	4.1	4.8	5.5	7.2	8.6	73	12.2	15.6	18.3	21.2	27.7	32.9
20	3.3	4.3	5.0	5.8	7.6	9.0	74	12.3	15.8	18.5	21.5	28.1	33.3
21	3.5	4.5	5.3	6.1	8.0	9.5	75	12.5	16.1	18.8	21.8	28.5	33.8
22	3.7	4.7	5.5	6.4	8.4	9.9	76	12.7	16.3	19.0	22.0	28.9	34.2
23	3.8	5.0	5.8	6.7	8.7	10.4	77	12.8	16.5	19.3	22.3	29.3	34.7
24	4.0	5.2	6.0	7.0	9.1	10.8	78	13.0	16.7	19.5	22.6	29.6	35.1
25	4.2	5.4	6.3	7.3	9.5	11.3	79	13.2	16.9	19.8	22.9	30.0	35.6
26	4.3	5.6	6.5	7.5	9.9	11.7	80	13.3	17.1	20.0	23.2	30.4	36.0
27	4.5	5.8	6.8	7.8	10.3	12.2	81	13.5	17.3	20.3	23.5	30.8	36.5
28	4.7	6.0	7.0	8.1	10.6	12.6	82	13.7	17.5	20.5	23.8	31.2	36.9
29	4.8	6.2	7.3	8.4	11.0	13.1	83	13.8	17.8	20.8	24.1	31.5	37.4
30	5.0	6.5	7.5	8.7	11.4	13.5	84	14.0	18.0	21.0	24.4	31.9	37.8
31	5.2	6.7	7.8	9.0	11.8	14.0	85	14.2	18.2	21.3	24.7	32.3	38.3
32	5.3	6.9	8.0	9.3	12.2	14.4	86	14.3	18.4	21.5	24.9	32.7	38.7
33	5.5	7.1	8.3	9.6	12.5	14.9	87	14.5	18.6	21.8	25.2	33.1	39.2
34	5.7	7.3	8.5	9.9	12.9	15.3	88	14.7	18.8	22.0	25.5	33.4	39.6
35	5.8	7.5	8.8	10.2	13.3	15.8	89	14.8	19.0	22.3	25.8	33.8	40.1
36	6.0	7.8	9.0	10.4	13.7	16.2	90	15.0	19.3	22.5	26.1	34.2	40.5
37	6.2	8.0	9.3	10.7	14.1	16.7	91	15.2	19.5	22.8	26.4	34.6	41.0
38	6.3	8.2	9.5	11.0	14.4	17.1	92	15.3	19.7	23.0	26.7	35.0	41.4
39	6.5	8.4	9.8	11.3	14.8	17.6	93	15.5	19.9	23.3	27.0	35.3	41.9
40	6.7	8.6	10.0	11.6	15.2	18.0	94	15.7	20.1	23.5	27.3	35.7	42.3
41	6.8	8.8	10.3	11.9	15.6	18.5	95	15.8	20.3	23.8	27.6	36.1	42.8
42	7.0	9.0	10.5	12.2	16.0	18.9	96	16.0	20.5	24.0	27.8	36.5	43.2
43	7.2	9.2	10.8	12.5	16.3	19.4	97	16.2	20.8	24.3	28.1	36.9	43.7
44	7.3	9.5	11.0	12.8	16.7	19.8	98	16.3	21.0	24.5	28.4	37.2	44.1
45	7.5	9.7	11.3	13.1	17.1	20.3	99	16.5	21.2	24.8	28.7	37.6	44.6
46	7.7	9.9	11.5	13.3	17.5	20.7	100	16.7	21.4	25.0	29.0	38.0	45.0
47	7.8	10.1	11.8	13.6	17.9	21.2	101	16.8	21.6	25.3	29.3	38.4	45.5
48	8.0	10.3	12.0	13.9	18.2	21.6	102	17.0	21.8	25.5	29.6	38.8	45.9
49	8.2	10.5	12.3	14.2	18.6	22.1	103	17.2	22.0	25.8	29.9	39.1	46.4
50	8.3	10.7	12.5	14.5	19.0	22.5	104	17.3	22.3	26.0	30.2	39.5	46.8
51	8.5	11.0	12.8	14.8	19.4	23.0	105	17.5	22.5	26.3	30.5	39.9	47.3
52	8.7	11.2	13.0	15.1	19.8	23.4	106	17.7	22.7	26.5	30.7	40.3	47.7
53	8.8	11.4	13.3	15.4	20.1	23.9	107	17.8	22.9	26.8	31.0	40.7	48.2
54	9.0	11.6	13.5	15.7	20.5	24.3	108	18.0	23.1	27.0	31.3	41.0	48.6
55	9.2	11.8	13.8	16.0	20.9	24.8	109	18.2	23.3	27.3	31.6	41.4	49.1
							110	18.3	23.5	27.5	31.9	41.8	49.5

Example: If duct is 34" x 20" x 15' long, 34" is greater than 30" maximum, for 24 ga. so must be 22 ga. 34" + 20" = 54" going across from 54" find 13.5 lb. per foot. 13.5 x 15' = 202.5 lbs.

For S.F. of surface area 202.5 ÷ 1.406 = 144 S.F.
Note: Figures include an allowance for scrap.

D40 Fire Protection

RD4010-100 Fire Protection Classification

Table D4010-102 System Classification

System Classification
Rules for installation of sprinkler systems vary depending on the classification of occupancy falling into one of three categories as follows:

Light Hazard Occupancy
The protection area allotted per sprinkler should not exceed 225 S.F., with the maximum distance between lines and sprinklers on lines being 15'. The sprinklers do not need to be staggered. Branch lines should not exceed eight sprinklers on either side of a cross main. Each large area requiring more than 100 sprinklers and without a sub-dividing partition should be supplied by feed mains or risers sized for ordinary hazard occupancy.
Maximum system area = 52,000 S.F.

Included in this group are:

Churches	Nursing Homes
Clubs	Offices
Educational	Residential
Hospitals	Restaurants
Institutional	Theaters and Auditoriums
Libraries	(except stages and prosceniums)
(except large stack rooms)	Unused Attics
Museums	

Ordinary Hazard Occupancy
The protection area allotted per sprinkler shall not exceed 130 S.F. of noncombustible ceiling and 130 S.F. of combustible ceiling. The maximum allowable distance between sprinkler lines and sprinklers on line is 15'. Sprinklers shall be staggered if the distance between heads exceeds 12'. Branch lines should not exceed eight sprinklers on either side of a cross main.
Maximum system area = 52,000 S.F.

Included in this group are:

Group 1	Group 2
Automotive Parking and Showrooms	Cereal Mills
Bakeries	Chemical Plants—Ordinary
Beverage manufacturing	Confectionery Products
Canneries	Distilleries
Dairy Products Manufacturing/Processing	Dry Cleaners
Electronic Plans	Feed Mills
Glass and Glass Products Manufacturing	Horse Stables
Laundries	Leather Goods Manufacturing
Restaurant Service Areas	Libraries—Large Stack Room Areas
	Machine Shops
	Metal Working
	Mercantile
	Paper and Pulp Mills
	Paper Process Plants
	Piers and Wharves
	Post Offices
	Printing and Publishing
	Repair Garages
	Stages
	Textile Manufacturing
	Tire Manufacturing
	Tobacco Products Manufacturing
	Wood Machining
	Wood Product Assembly

Extra Hazard Occupancy
The protection area allotted per sprinkler shall not exceed 100 S.F. of noncombustible ceiling and 100 S.F. of combustible ceiling. The maximum allowable distance between lines and between sprinklers on lines is 12'. Sprinklers on alternate lines shall be staggered if the distance between sprinklers on lines exceeds 8'. Branch lines should not exceed six sprinklers on either side of a cross main.
Maximum system area:
　Design by pipe schedule = 25,000 S.F.
　Design by hydraulic calculation = 40,000 S.F.

Included in this group are:

Group 1	Group 2
Aircraft hangars	Asphalt Saturating
Combustible Hydraulic Fluid Use Area	Flammable Liquids Spraying
Die Casting	Flow Coating
Metal Extruding	Manufactured/Modular Home Building Assemblies (where finished enclosure is present and has combustible interiors)
Plywood/Particle Board Manufacturing	
Printing (inks with flash points < 100 degrees F)	
Rubber Reclaiming, Compounding, Drying, Milling, Vulcanizing	Open Oil Quenching
Saw Mills	Plastics Processing
Textile Picking, Opening, Blending, Garnetting, Carding, Combing of Cotton, Synthetics, Wood Shoddy, or Burlap	Solvent Cleaning
Upholstering with Plastic Foams	Varnish and Paint Dipping

D40 Fire Protection — RD4020-300 Standpipe

Table D4020-301 Standpipe Systems

The basis for standpipe system design is National Fire Protection Association NFPA 14. However, the authority having jurisdiction should be consulted for special conditions, local requirements and approval.

Standpipe systems, properly designed and maintained, are an effective and valuable time saving aid for extinguishing fires, especially in the upper stories of tall buildings, the interior of large commercial or industrial malls, or other areas where construction features or access make the laying of temporary hose lines time consuming and/or hazardous. Standpipes are frequently installed with automatic sprinkler systems for maximum protection.

There are three general classes of service for standpipe systems:
Class I for use by fire departments and personnel with special training for heavy streams (2-1/2" hose connections).
Class II for use by building occupants until the arrival of the fire department (1-1/2" hose connector with hose).
Class III for use by either fire departments and trained personnel or by the building occupants (both 2-1/2" and 1-1/2" hose connections or one 2-1/2" hose valve with an easily removable 2-1/2" by 1-1/2" adapter).

Standpipe systems are also classified by the way water is supplied to the system. The four basic types are:
Type 1: Wet standpipe system having supply valve open and water pressure maintained at all times.
Type 2: Standpipe system so arranged through the use of approved devices as to admit water to the system automatically by opening a hose valve.
Type 3: Standpipe system arranged to admit water to the system through manual operation of approved remote control devices located at each hose station.
Type 4: Dry standpipe having no permanent water supply.

Table D4020-302 NFPA 14 Basic Standpipe Design

Class	Design-Use	Pipe Size Minimums	Water Supply Minimums
Class I	2 1/2" hose connection on each floor. All areas within 150' of an exit in every exit stairway. Fire Department Trained Personnel	Height to 100', 4" dia. Heights above 100', 6" dia. (275' max. except with pressure regulators 400' max.)	For each standpipe riser 500 GPM flow. For common supply pipe allow 500 GPM for first standpipe plus 250 GPM for each additional standpipe (2500 GPM max. total). 30 min. duration 65 PSI at 500 GPM
Class II	1 1/2" hose connection with hose on each floor. All areas within 130' of hose connection measured along path of hose travel. Occupant personnel	Height to 50', 2" dia. Height above 50', 2 1/2" dia.	For each standpipe riser 100 GPM flow. For multiple riser common supply pipe 100 GPM. 300 min. duration, 65 PSI at 100 GPM
Class III	Both of above. Class I valved connections will meet Class III with additional 2 1/2" by 1 1/2" adapter and 1 1/2" hose.	Same as Class I	Same as Class I

*Note: Where 2 or more standpipes are installed in the same building or section of building they shall be interconnected at the bottom.

Combined Systems

Combined systems are systems where the risers supply both automatic sprinklers and 2-1/2" hose connection outlets for fire department use. In such a system the sprinkler spacing pattern shall be in accordance with NFPA 13 while the risers and supply piping will be sized in accordance with NFPA 14. When the building is completely sprinklered the risers may be sized by hydraulic calculation. The minimum size riser for buildings not completely sprinklered is 6".

The minimum water supply of a completely sprinklered, light hazard, high-rise occupancy building will be 500 GPM while the supply required for other types of completely sprinklered high-rise buildings is 1000 GPM.

General System Requirements

1. Approved valves will be provided at the riser for controlling branch lines to hose outlets.
2. A hose valve will be provided at each outlet for attachment of hose.
3. Where pressure at any standpipe outlet exceeds 100 PSI a pressure reducer must be installed to limit the pressure to 100 PSI. Note that the pressure head due to gravity in 100' of riser is 43.4 PSI. This must be overcome by city pressure, fire pumps, or gravity tanks to provide adequate pressure at the top of the riser.
4. Each hose valve on a wet system having linen hose shall have an automatic drip connection to prevent valve leakage from entering the hose.
5. Each riser will have a valve to isolate it from the rest of the system.
6. One or more fire department connections as an auxiliary supply shall be provided for each Class I or Class III standpipe system. In buildings having two or more zones, a connection will be provided for each zone.
7. There will be no shutoff valve in the fire department connection, but a check valve will be located in the line before it joins the system.
8. All hose connections street side will be identified on a cast plate or fitting as to purpose.

Table D4020-303 Quality/Complexity Adjustment for Sprinkler/Standpipe Systems

Economy installation, add	0 to 5%
Good quality, medium complexity, add	5 to 15%
Above average quality and complexity, add	15 to 25%

D50 Electrical — RD5010-110 Electrical Systems Estimating

Figure D5010-111 Typical Overhead Service Entrance

Figure D5010-112 Typical Commercial Electric System

A = 1" Conduit w/1-#1/0 Wire XHHW
B = 2-3" Conduits w/4-350kcmil XHHW in Each
C = 1¼" Conduit w/4 #3 THHN
D = 1¼" Conduit w/3 #1 XHHW
E = 1" Conduit w/4 #6 THHN
F = ½" Conduit w/2 #12 THHN

D50 Electrical
RD5010-110 Electrical Systems Estimating

Figure D5010-113 Preliminary Procedure

1. Determine building size and use.
2. Develop total load in watts.
   a. Lighting
   b. Receptacle
   c. Air Conditioning
   d. Elevator
   e. Other power requirements
3. Determine best voltage available from utility company.
4. Determine cost from tables for loads (a) thru (e) above.
5. Determine size of service from formulas (D5010-116).
6. Determine costs for service, panels, and feeders from tables.

Figure D5010-114 Office Building 90' x 210', 3 story, w/garage

Garage Area = 18,900 S.F.
Office Area = 56,700 S.F.
Elevator = 2 @ 125 FPM

Tables	Power Required	Watts
D5010-1151	Garage Lighting .5 Watt/S.F.	9,450
	Office Lighting 3 Watts/S.F.	170,100
D5010-1151	Office Receptacles 2 Watts/S.F.	113,400
D5010-1151, R262213-27	Low Rise Office A.C. 4.3 Watts/S.F.	243,810
D5010-1152, 1153	Elevators - 2 @ 20 HP = 2 @ 17,404 Watts/Ea.	34,808
D5010-1151	Misc. Motors + Power 1.2 Watts/S.F.	68,040
	Total	639,608 Watts

Voltage Available

277/480V, 3 Phase, 4 Wire

Formula

D5010-116

$$\text{Amperes} = \frac{\text{Watts}}{\text{Volts} \times \text{Power Factor} \times 1.73} = \frac{639,608}{480V \times .8 \times 1.73} = 963 \text{ Amps}$$

Use 1200 Amp Service

System	Description	Unit Cost	Unit	Total
D5020 210 0200	Garage Lighting (Interpolated)	$1.31	S.F.	$ 24,759
D5020 210 0540	Office Lighting (Using 32 W lamps)	7.90	S.F.	447,930
D5020 115 0880	Receptacle-Undercarpet	3.74	S.F.	212,058
D5020 140 0280	Air Conditioning	0.61	S.F.	34,587
D5020 135 0320	Misc. Pwr.	0.33	S.F.	18,711
D5020 145 2120	Elevators - 2 @ 20HP	2,875.00	Ea.	5,750
D5010 120 0480	Service-1200 Amp (add 25% for 277/480V)	22,200.00	Ea.	27,750
D5010 230 0480	Feeder - Assume 200 feet	355.00	L.F.	71,000
D5010 240 0320	Panels - 1200 Amp (add 20% for 277/480V)	21,825.00	Ea.	26,190
D5030 910 0400	Fire Detection	33,400.00	Ea.	33,400
	Total			$902,135
	or			$902,100

D50 Electrical — RD5010-110 Electrical Systems Estimating

Table D5010-1151 Nominal Watts Per S.F. for Electric Systems for Various Building Types

Type Construction	1. Lighting	2. Devices	3. HVAC	4. Misc.	5. Elevator	Total Watts
Apartment, luxury high rise	2	2.2	3	1		
Apartment, low rise	2	2	3	1		
Auditorium	2.5	1	3.3	.8		
Bank, branch office	3	2.1	5.7	1.4		
Bank, main office	2.5	1.5	5.7	1.4		
Church	1.8	.8	3.3	.8		
College, science building	3	3	5.3	1.3		
College, library	2.5	.8	5.7	1.4		
College, physical education center	2	1	4.5	1.1		
Department store	2.5	.9	4	1		
Dormitory, college	1.5	1.2	4	1		
Drive-in donut shop	3	4	6.8	1.7		
Garage, commercial	.5	.5	0	.5		
Hospital, general	2	4.5	5	1.3		
Hospital, pediatric	3	3.8	5	1.3		
Hotel, airport	2	1	5	1.3		
Housing for the elderly	2	1.2	4	1		
Manufacturing, food processing	3	1	4.5	1.1		
Manufacturing, apparel	2	1	4.5	1.1		
Manufacturing, tools	4	1	4.5	1.1		
Medical clinic	2.5	1.5	3.2	1		
Nursing home	2	1.6	4	1		
Office building, hi rise	3	2	4.7	1.2		
Office building, low rise	3	2	4.3	1.2		
Radio-TV studio	3.8	2.2	7.6	1.9		
Restaurant	2.5	2	6.8	1.7		
Retail store	2.5	.9	5.5	1.4		
School, elementary	3	1.9	5.3	1.3		
School, junior high	3	1.5	5.3	1.3		
School, senior high	2.3	1.7	5.3	1.3		
Supermarket	3	1	4	1		
Telephone exchange	1	.6	4.5	1.1		
Theater	2.5	1	3.3	.8		
Town Hall	2	1.9	5.3	1.3		
U.S. Post Office	3	2	5	1.3		
Warehouse, grocery	1	.6	0	.5		

Rule of Thumb: 1 KVA = 1 HP (Single Phase)

Three Phase:

Watts = 1.73 x Volts x Current x Power Factor x Efficiency

$$\text{Horsepower} = \frac{\text{Volts} \times \text{Current} \times 1.73 \times \text{Power Factor}}{746 \text{ Watts}}$$

Table D5010-1152 Horsepower Requirements for Elevators with 3 Phase Motors

Type	Maximum Travel Height in Ft.	Travel Speeds in FPM	Capacity of Cars in Lbs.		
			1200	1500	1800
Hydraulic	70	70	10	15	15
		85	15	15	15
		100	15	15	20
		110	20	20	20
		125	20	20	20
		150	25	25	25
		175	25	30	30
		200	30	30	40
Geared Traction	300	200			
		350			
			2000	2500	3000
Hydraulic	70	70	15	20	20
		85	20	20	25
		100	20	25	30
		110	20	25	30
		125	25	30	40
		150	30	40	50
		175	40	50	50
		200	40	50	60
Geared Traction	300	200	10	10	15
		350	15	15	23
			3500	4000	4500
Hydraulic	70	70	20	25	30
		85	25	30	30
		100	30	40	40
		110	40	40	50
		125	40	50	50
		150	50	50	60
		175	60		
		200	60		
Geared Traction	300	200	15		23
		350	23		35

The power factor of electric motors varies from 80% to 90% in larger size motors. The efficiency likewise varies from 80% on a small motor to 90% on a large motor.

Table D5010-1153 Watts per Motor

90% Power Factor & Efficiency @ 200 or 460V			
HP	Watts	HP	Watts
10	9024	30	25784
15	13537	40	33519
20	17404	50	41899
25	21916	60	49634

D50 Electrical

RD5010-110 Electrical Systems Estimating

Table D5010-116 Electrical Formulas

Ohm's Law

Ohm's Law is a method of explaining the relation existing between voltage, current, and resistance in an electrical circuit. It is practically the basis of all electrical calculations. The term "electromotive force" is often used to designate pressure in volts. This formula can be expressed in various forms.

To find the current in amperes:

$$\text{Current} = \frac{\text{Voltage}}{\text{Resistance}} \quad \text{or} \quad \text{Amperes} = \frac{\text{Volts}}{\text{Ohms}} \quad \text{or} \quad I = \frac{E}{R}$$

The flow of current in amperes through any circuit is equal to the voltage or electromotive force divided by the resistance of that circuit.

To find the pressure or voltage:

Voltage = Current x Resistance   or   Volts = Amperes x Ohms   or $E = I \times R$

The voltage required to force a current through a circuit is equal to the resistance of the circuit multiplied by the current.

To find the resistance:

$$\text{Resistance} = \frac{\text{Voltage}}{\text{Current}} \quad \text{or} \quad \text{Ohms} = \frac{\text{Volts}}{\text{Amperes}} \quad \text{or} \quad R = \frac{E}{I}$$

The resistance of a circuit is equal to the voltage divided by the current flowing through that circuit.

Power Formulas

One horsepower = 746 watts   One kilowatt = 1000 watts

The power factor of electric motors varies from 80% to 90% in the larger size motors.

Single-Phase Alternating Current Circuits

Power in Watts = Volts x Amperes x Power Factor

To find current in amperes:

$$\text{Current} = \frac{\text{Watts}}{\text{Volts} \times \text{Power Factor}} \quad \text{or}$$

$$\text{Amperes} = \frac{\text{Watts}}{\text{Volts} \times \text{Power Factor}} \quad \text{or} \quad I = \frac{W}{E \times PF}$$

To find current of a motor, single phase:

$$\text{Current} = \frac{\text{Horsepower} \times 746}{\text{Volts} \times \text{Power Factor} \times \text{Efficiency}} \quad \text{or}$$

$$I = \frac{HP \times 746}{E \times PF \times \text{Eff.}}$$

To find horsepower of a motor, single phase:

$$\text{Horsepower} = \frac{\text{Volts} \times \text{Current} \times \text{Power Factor} \times \text{Efficiency}}{746 \text{ Watts}}$$

$$HP = \frac{E \times I \times PF \times \text{Eff.}}{746}$$

To find power in watts of a motor, single phase:

Watts = Volts x Current x Power Factor x Efficiency   or
Watts = $E \times I \times PF \times \text{Eff.}$

To find single phase kVA:

$$1 \text{ Phase kVA} = \frac{\text{Volts} \times \text{Amps}}{1000}$$

Three-Phase Alternating Current Circuits

Power in Watts = Volts x Amperes x Power Factor x 1.73

To find current in amperes in each wire:

$$\text{Current} = \frac{\text{Watts}}{\text{Voltage} \times \text{Power Factor} \times 1.73} \quad \text{or}$$

$$\text{Amperes} = \frac{\text{Watts}}{\text{Volts} \times \text{Power Factor} \times 1.73} \quad \text{or} \quad I = \frac{W}{E \times PF \times 1.73}$$

To find current of a motor, 3 phase:

$$\text{Current} = \frac{\text{Horsepower} \times 746}{\text{Volts} \times \text{Power Factor} \times \text{Efficiency} \times 1.73} \quad \text{or}$$

$$I = \frac{HP \times 746}{E \times PF \times \text{Eff.} \times 1.73}$$

To find horsepower of a motor, 3 phase:

$$\text{Horsepower} = \frac{\text{Volts} \times \text{Current} \times 1.73 \times \text{Power Factor}}{746 \text{ Watts}}$$

$$HP = \frac{E \times I \times 1.73 \times PF}{746}$$

To find power in watts of a motor, 3 phase:

Watts = Volts x Current x 1.73 x Power Factor x Efficiency   or
Watts = $E \times I \times 1.73 \times PF \times \text{Eff.}$

To find 3 phase kVA:

$$3 \text{ phase kVA} = \frac{\text{Volts} \times \text{Amps} \times 1.73}{1000} \quad \text{or}$$

$$kVA = \frac{V \times A \times 1.73}{1000}$$

Power Factor (PF) is the percentage ratio of the measured watts (effective power) to the volt-amperes (apparent watts)

$$\text{Power Factor} = \frac{\text{Watts}}{\text{Volts} \times \text{Amperes}} \times 100\%$$

D50 Electrical — RD5010-110 Electrical Systems Estimating

A **Conceptual Estimate** of the costs for a building when final drawings are not available can be quickly figured by using **Table D5010-117 Cost Per S.F. for Electrical Systems for Various Building Types.** The following definitions apply to this table.

1. **Service and Distribution:** This system includes the incoming primary feeder from the power company, main building transformer, metering arrangement, switchboards, distribution panel boards, stepdown transformers, and power and lighting panels. Items marked (*) include the cost of the primary feeder and transformer. In all other projects the cost of the primary feeder and transformer is paid for by the local power company.
2. **Lighting:** Includes all interior fixtures for decor, illumination, exit and emergency lighting. Fixtures for exterior building lighting are included, but parking area lighting is not included unless mentioned. See also Section D5020 for detailed analysis of lighting requirements and costs.
3. **Devices:** Includes all outlet boxes, receptacles, switches for lighting control, dimmers and cover plates.
4. **Equipment Connections:** Includes all materials and equipment for making connections for heating, ventilating and air conditioning, food service and other motorized items requiring connections.
5. **Basic Materials:** This category includes all disconnect power switches not part of service equipment, raceways for wires, pull boxes, junction boxes, supports, fittings, grounding materials, wireways, busways, and wire and cable systems.
6. **Special Systems:** Includes installed equipment only for the particular system such as fire detection and alarm, sound, emergency generator and others as listed in the table.

Table D5010-117 Cost per S.F. for Electric Systems for Various Building Types

Building Type	1. Service & Distrib.	2. Lighting	3. Devices	4. Equipment Connections	5. Basic Materials	6. Special Systems — Fire Alarm & Detection	Telephone	Television Distribution
Apartment, Low Rise	$0.77	$0.84	$0.21	$0.28	$ 2.11	$ 0.49	0.77	$0.21
Apartment, High Rise	2.35	1.30	0.39	0.52	4.04	0.78	1.3	0.26
Assisted Living	4.90	1.52	0.34	0.68	3.21	1.35	0.84	0.84
Auditorium	2.34	4.04	0.43	0.64	6.81	1.28	1.06	
Bank	3.95	2.82	0.28	0.85	6.21	3.39	2.26	
Church	1.03	3.59	0.34	0.86	6.16	1.71	0.68	
College, Classroom	3.49	3.08	0.41	0.62	5.54	0.82	0.82	0.82
College, Laboratory	6.37	6.79	0.85	0.85	12.74	2.55	2.12	
Community Center	1.69	2.63	0.38	0.75	5.64	2.45	0.75	0.38
Department Store	1.04	2.89	0.12	0.46	3.93	0.69	0.46	
Dormitory, Low Rise	0.89	1.77	0.59	0.59	4.88	1.18	1.33	0.30
Factory	1.29	5.33	0.16	0.48	5.49	1.29	0.48	0.16
Funeral Home	1.52	1.72	0.20	0.51	3.23	1.21	1.31	
Garage, Commercial	3.25	0.89	0.15	0.30	2.66	5.62	$0.74	
Garage, Parking	0.20	0.55	0.12	0.04	1.70	0.08	0.0	
Gymnasium	0.99	2.69	0.28	0.57	5.66	1.56	0.42	
Hospital, Low Rise	2.40	3.61	1.60	1.60	13.62	2.80	2.00	0.80
Hospital, High Rise	4.25	3.48	1.55	1.16	15.46	1.93	1.55	0.39
Housing, Elderly	4.00	1.24	0.28	0.55	2.62	1.10	0.69	0.69
Library	1.59	4.99	0.45	1.36	9.30	2.04	0.68	
Medical Office	2.28	1.33	0.57	0.38	4.36	1.52	1.52	0.38
Motel	1.42	0.98	0.55	0.11	3.61	1.20	1.09	0.22
Nursing Home	2.21	2.41	0.40	0.60	5.21	1.20	1.40	1.4
Office, Low Rise	2.94	2.13	0.33	0.65	5.07	1.80	1.14	0.16
Office, High Rise	0.69	2.22	0.28	0.42	5.13	1.39	1.11	0.14
Post Office	2.85	1.51	0.17	0.50	5.70	3.52	1.51	
Restaurant	8.36	3.04	0.25	1.01	6.33	4.05	0.51	
Restaurant, Fast Food	5.97	3.53	0.27	1.08	8.41	5.42	0.54	
Retail Store	1.75	2.22	0.12	0.47	3.97	1.05	0.35	
School, Elementary	1.57	2.80	0.35	0.52	5.77	2.10	0.87	0.35
School, Junior High	1.40	2.80	0.52	0.70	5.59	0.87	1.05	0.35
School, High	1.32	3.31	0.44	0.66	6.84	1.10	1.32	1.32
Supermarket	2.99	2.59	0.14	0.41	4.36	1.36	0.27	
Telephone Exchange	6.44	5.06	0.92	2.30	13.81	10.59	5.52	
Theater (Movie)	2.63	4.78	0.24	0.72	5.49	3.34	0.96	
Town Hall	1.55	3.29	0.58	1.16	7.36	1.55	1.55	
Warehouse	0.33	1.22	0.24	0.16	2.93	1.95	0.41	

*Includes cost of primary feeder and transformer. Cont'd on next page.

D50 Electrical
RD5010-110 Electrical Systems Estimating

COST ASSUMPTIONS:

Each of the projects analyzed in Table D5010-117 were bid within the last 10 years in the northeastern part of the United States. Bid prices have been adjusted to Jan. 1 levels. The list of projects is by no means all-inclusive, yet by carefully examining the various systems for a particular building type, certain cost relationships will emerge. The use of the electrical portion of the Square Foot Costs, in the Reference Section, along with the Project Size Modifier, should produce a budget S.F. cost for the electrical portion of a job that is consistent with the amount of design information normally available at the conceptual estimate stage.

Table D5010-117 Cost per S.F. for Electric Systems for Various Building Types (cont.)

Building Type	6. Special Systems, (cont.)						
	Intercom Systems	Sound Systems	Video Monitoring	Internet	Emergency Generator	Access Control	Master Clock Sys.
Apartment, Low Rise	0.63			0.28	$0.14	0.28	
Apartment, High Rise	0.91			0.52	0.13	$0.52	
Assisted Living	2.03	0.17	0.51		0.17	0.34	
Auditorium		2.34	0.85		1.28	0.21	
Bank			2.82	2.54	0.56	2.54	
Church		2.22			0.17	0.34	
College, Classroom	0.82	1.03		1.64	0.41	0.41	0.62
College, Laboratory	2.12			4.25	0.42	1.27	2.12
Community Center		1.88		0.94	0.38	0.94	
Department Store			0.81	0.35		0.81	
Dormitory, Low Rise	1.33			1.48		0.44	
Factory			0.48		0.16	0.48	0.32
Funeral Home				0.1	0.10	0.2	
Garage, Commercial					0.44	0.74	
Garage, Parking			0.4		0.04	0.79	
Gymnasium		1.27			0.14	0.57	
Hospital, Low Rise	2.4	0.40	0.8	0.4	6.01	1.2	0.4
Hospital, High Rise	1.16	0.39	0.39	$0.39	5.02	1.16	0.39
Housing, Elderly	1.66	0.14	0.41		0.14	0.28	
Library		0.45		0.91	0.45	0.45	
Medical Office	2.28		0.95	1.71	0.76	0.95	
Motel				1.2	0.11	0.44	
Nursing Home	3.41	0.20	0.8		0.4	0.4	
Office, Low Rise			0.33	1.31	0.16	0.33	
Office, High Rise			0.42	1.25	0.42	0.42	
Post Office				0.5	0.17	0.34	
Restaurant				0.51	0.51	0.76	
Restaurant, Fast Food				0.54	0.54	0.81	
Retail Store		0.12	0.7	0.12	0.12	0.70	
School, Elementary	1.05	0.87		0.35		0.52	0.35
School, Junior High	1.22	0.87		0.52	0.35	0.52	0.70
School, High	1.32	1.1	0.88	0.44	0.66	0.66	0.66
Supermarket		0.14	0.54	0.14		0.68	
Telephone Exchange					0.46	0.92	
Theater (Movie)		4.3		0.24	0.24	0.96	
Town Hall				1.74	0.19	0.39	
Warehouse			0.41	0.08		0.41	

*Includes cost of primary feeder and transformer. Cont'd on next page.

D50 Electrical

RD5010-110 Electrical Systems Estimating

General: Variations in the following square foot costs are due to the type of structural systems of the buildings, geographical location, local electrical codes, designer's preference for specific materials and equipment, and the owner's particular requirements.

Table D5010-117 Cost per S.F. for Electric Systems for Various Building Types (cont.)

Building Type	Additional Information	Total Floor Area in Square Feet	Total Cost per Square Foot for Total Electric Systems
Apartment, Low Rise	Stories = 3, Height = 10', Perimeter = 715'	40,200	7.04
Apartment, High Rise	Stories = 15, Height = 10', Perimeter = 350'	115,000	13.04
Assisted Living	Stories = 1, Height = 10', Perimeter = 400'	10,000	16.89
Auditorium	Stories = 1, Height = 24', Perimeter = 747'	28,000	21.29
Bank	Stories = 1, Height = 14', Perimeter = 168'	2,700	28.22
Church	Stories = 1, Height = 24', Perimeter = 562'	17,700	17.10
College, Classroom	Stories = 2, Height = 12', Perimeter = 630'	50,000	20.50
College, Laboratory	Stories = 1, Height = 12', Perimeter = 550'	27,500	42.46
Community Center	Stories = 1, Height = 12', Perimeter = 453'	10,000	18.81
Department Store	Stories = 1, Height = 14', Perimeter = 1039'	85,800	11.56
Dormitory, Low Rise	Stories = 3, Height = 12', Perimeter = 1008'	63,000	14.78
Factory	Stories = 1, Height = 20', Perimeter = 730'	30,000	16.14
Funeral Home	Stories = 1, Height = 12', Perimeter = 424'	10,000	10.10
Garage, Commercial	Stories = 1, Height = 12', Perimeter = 153'	1,400	14.80
Garage, Parking	Stories = 5, Height = 10', Perimeter = 723'	145,000	3.96
Gymnasium	Stories = 1, Height = 25', Perimeter = 660'	22,000	14.16
Hospital, Low Rise	Stories = 3, Height = 12', Perimeter = 566'	55,000	40.07
Hospital, High Rise	Stories = 6, Height = 12', Perimeter = 866'	200,000	38.64
Housing, Elderly	Stories = 1, Height = 10', Perimeter = 400'	10,000	13.81
Library	Stories = 2, Height = 14', Perimeter = 435'	22,000	22.68
Medical Office	Stories = 2, Height = 10', Perimeter = 240'	7,000	18.96
Motel	Stories = 3, Height = 9', Perimeter = 606'	49,000	10.93
Nursing Home	Stories = 2, Height = 10', Perimeter = 384'	21,000	20.05
Office, Low Rise	Stories = 3, Height = 12', Perimeter = 360'	20,000	16.35
Office, High Rise	Stories = 15, Height = 10', Perimeter = 634'	311,200	13.87
Post Office	Stories = 1, Height = 14', Perimeter = 486'	13,000	16.75
Restaurant	Stories = 1, Height = 12', Perimeter = 174'	2,900	25.32
Restaurant, Fast Food	Stories = 1, Height = 10', Perimeter = 260'	4,000	27.12
Retail Store	Stories = 1, Height = 14', Perimeter = 360'	8,000	11.68
School, Elementary	Stories = 1, Height = 15', Perimeter = 1325'	39,500	17.49
School, Junior High	Stories = 2, Height = 15', Perimeter = 1700'	110,000	17.48
School, High	Stories = 2, Height = 15', Perimeter = 2411'	158300	22.08
Supermarket	Stories = 1, Height = 18', Perimeter = 584'	30600	13.61
Telephone Exchange	Stories = 1, Height = 12', Perimeter = 286'	5000	46.04
Theater (Movie)	Stories = 1, Height = 20', Perimeter = 513'	14000	23.89
Town Hall	Stories = 3, Height = 12', Perimeter = 356'	20000	19.37
Warehouse	Stories = 1, Height = 24', Perimeter = 700'	30000	8.13

*Includes cost of primary feeder and transformer.

D50 Electrical

RD5010-120 Electrical Circuit Voltages

D5010-120 Electric Circuit Voltages

General: The following method provides the user with a simple non-technical means of obtaining comparative costs of wiring circuits. The circuits considered serve the electrical loads of motors, electric heating, lighting and transformers, for example, that require low voltage 60 Hertz alternating current.

The method used here is suitable only for obtaining estimated costs. It is **not** intended to be used as a substitute for electrical engineering design applications.

Conduit and wire circuits can represent from twenty to thirty percent of the total building electrical cost. By following the described steps and using the tables the user can translate the various types of electric circuits into estimated costs.

Wire Size: Wire size is a function of the electric load which is usually listed in one of the following units:

1. Amperes (A)
2. Watts (W)
3. Kilowatts (kW)
4. Volt amperes (VA)
5. Kilovolt amperes (kVA)
6. Horsepower (HP)

These units of electric load must be converted to amperes in order to obtain the size of wire necessary to carry the load. To convert electric load units to amperes one must have an understanding of the voltage classification of the power source and the voltage characteristics of the electrical equipment or load to be energized. The seven A.C. circuits commonly used are illustrated in Figures D5010-121 thru D5010-127 showing the tranformer load voltage and the point of use voltage at the point on the circuit where the load is connected. The difference between the source and point of use voltages is attributed to the circuit voltage drop and is considered to be approximately 4%.

Motor Voltages: Motor voltages are listed by their point of use voltage and not the power source voltage.

For example: 460 volts instead of 480 volts
200 instead of 208 volts
115 volts instead of 120 volts

Lighting and Heating Voltages: Lighting and heating equipment voltages are listed by the power source voltage and not the point of wire voltage.

For example: 480, 277, 120 volt lighting
480 volt heating or air conditioning unit
208 volt heating unit

Transformer Voltages: Transformer primary (input) and secondary (output) voltages are listed by the power source voltage.

For example: Single phase 10 kVA
Primary 240/480 volts
Secondary 120/240 volts

In this case, the primary voltage may be 240 volts with a 120 volts secondary or may be 480 volts with either a 120V or a 240V secondary.

For example: Three phase 10 kVA
Primary 480 volts
Secondary 208Y/120 volts

In this case the transformer is suitable for connection to a circuit with a 3 phase 3 wire or 3 phase 4 wire circuit with a 480 voltage. This application will provide a secondary circuit of 3 phase 4 wire with 208 volts between phase wires and 120 volts between any phase wire and the neutral (white) wire.

Figure D5010-121

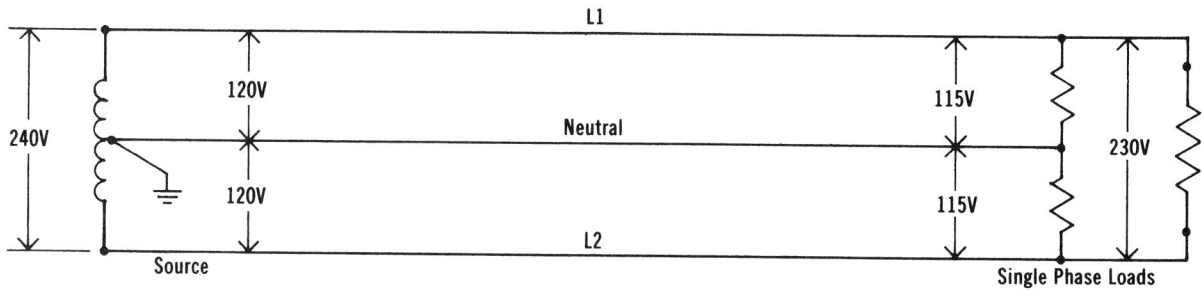

3 Wire, 1 Phase, 120/240 Volt System

Figure D5010-122

4 wire, 3 Phase, 208Y/120 Volt System

D50 Electrical — RD5010-120 Electrical Circuit Voltages

Figure D5010-123 Electric Circuit Voltages (cont.)

Figure D5010-124

Figure D5010-125

Figure D5010-126

D50 Electrical RD5010-120 Electrical Circuit Voltages

Figure D5010-127 Electric Circuit Voltages (cont.)

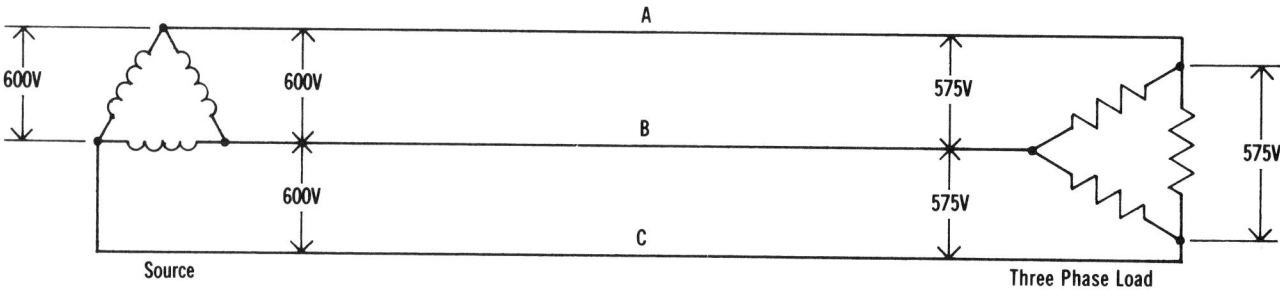

Maximum Circuit Length: Table D5010-130 indicates typical maximum installed length a circuit can have and still maintain an adequate voltage level at the point of use. The circuit length is similar to the conduit length.

If the circuit length for an ampere load and a copper wire size exceeds the length obtained from Table D5010-130, use the next largest wire size to compensate voltage drop.

Example: A 130 ampere load at 480 volts, 3 phase, 3 wire with No. 1 wire can be run a maximum of 555 L.F. and provide satisfactory operation. If the same load is to be wired at the end of a 625 L.F. circuit, then a larger wire must be used.

Table D5010-130 Maximum Circuit Length (approximate) for Various Power Requirements Assuming THW, Copper Wire @ 75°C, Based Upon a 4% Voltage Drop

Amperes	Wire Size	Maximum Circuit Length in Feet				
		2 Wire, 1 Phase		3 Wire, 3 Phase		
		120V	240V	240V	480V	600V
15	14*	50	105	120	240	300
	14	50	100	120	235	295
20	12*	60	125	145	290	360
	12	60	120	140	280	350
30	10*	65	130	155	305	380
	10	65	130	150	300	375
50	8	60	125	145	285	355
65	6	75	150	175	345	435
85	4	90	185	210	425	530
115	2	110	215	250	500	620
130	1	120	240	275	555	690
150	1/0	130	260	305	605	760
175	2/0	140	285	330	655	820
200	3/0	155	315	360	725	904
230	4/0	170	345	395	795	990
255	250	185	365	420	845	1055
285	300	195	395	455	910	1140
310	350	210	420	485	975	1220
380	500	245	490	565	1130	1415

*Solid Conductor

Note: The circuit length is the one-way distance between the origin and the load.

D50 Electrical — RD5010-140 Design & Cost Tables

RD5010-141 Minimum Copper and Aluminum Wire Size Allowed for Various Types of Insulation

Minimum Wire Sizes

Amperes	Copper THW THWN or XHHW	Copper THHN XHHW *	Aluminum THW XHHW	Aluminum THHN XHHW *	Amperes	Copper THW THWN or XHHW	Copper THHN XHHW *	Aluminum THW XHHW	Aluminum THHN XHHW *
15A	#14	#14	#12	#12	195	3/0	2/0	250kcmil	4/0
20	#12	#12	#10	#10	200	3/0	3/0	250kcmil	4/0
25	#10	#10	#10	#10	205	4/0	3/0	250kcmil	4/0
30	#10	#10	#8	#8	225	4/0	3/0	300kcmil	250kcmil
40	#8	#8	#8	#8	230	4/0	4/0	300kcmil	250kcmil
45	#8	#8	#6	#8	250	250kcmil	4/0	350kcmil	300kcmil
50	#8	#8	#6	#6	255	250kcmil	4/0	400kcmil	300kcmil
55	#6	#8	#4	#6	260	300kcmil	4/0	400kcmil	350kcmil
60	#6	#6	#4	#6	270	300kcmil	250kcmil	400kcmil	350kcmil
65	#6	#6	#4	#4	280	300kcmil	250kcmil	500kcmil	350kcmil
75	#4	#6	#3	#4	285	300kcmil	250kcmil	500kcmil	400kcmil
85	#4	#4	#2	#3	290	350kcmil	250kcmil	500kcmil	400kcmil
90	#3	#4	#2	#2	305	350kcmil	300kcmil	500kcmil	400kcmil
95	#3	#4	#1	#2	310	350kcmil	300kcmil	500kcmil	500kcmil
100	#3	#3	#1	#2	320	400kcmil	300kcmil	600kcmil	500kcmil
110	#2	#3	1/0	#1	335	400kcmil	350kcmil	600kcmil	500kcmil
115	#2	#2	1/0	#1	340	500kcmil	350kcmil	600kcmil	500kcmil
120	#1	#2	1/0	1/0	350	500kcmil	350kcmil	700kcmil	500kcmil
130	#1	#2	2/0	1/0	375	500kcmil	400kcmil	700kcmil	600kcmil
135	1/0	#1	2/0	1/0	380	500kcmil	400kcmil	750kcmil	600kcmil
150	1/0	#1	3/0	2/0	385	600kcmil	500kcmil	750kcmil	600kcmil
155	2/0	1/0	3/0	3/0	420	600kcmil	500kcmil		700kcmil
170	2/0	1/0	4/0	3/0	430		500kcmil		750kcmil
175	2/0	2/0	4/0	3/0	435		600kcmil		750kcmil
180	3/0	2/0	4/0	4/0	475		600kcmil		

*Dry Locations Only

Notes:
1. Size #14 to 4/0 is in AWG units (American Wire Gauge).
2. Size 250 to 750 is in kcmil units (Thousand Circular Mils).
3. Use next higher ampere value if exact value is not listed in table.
4. For loads that operate continuously increase ampere value by 25% to obtain proper wire size.
5. Refer to Table D5010-130 for the maximum circuit length for the various size wires.
6. Table D5010-141 has been written for estimating purpose only, based on ambient temperature of 30°C (86° F); for ambient temperature other than 30°C (86° F), ampacity correction factors will be applied.

D50 Electrical — RD5010-140 Design & Cost Tables

Table D5010-142 Conductors in Conduit

Table below lists maximum number of conductors for various sized conduit using THW, TW or THWN insulations.

Copper Wire Size	1/2"			3/4"			1"			1-1/4"			1-1/2"			2"			2-1/2"			3"			3-1/2"			4"	
	TW	THW	THWN	TW	THW	THWN	TW	THW	THWN	TW	THW	THWN	TW	THW	THWN	TW	THW	THWN	TW	THW	THWN	THW	THWN	THW	THWN	THW	THWN
#14	9	6	13	15	10	24	25	16	39	44	29	69	60	40	94	99	65	154	142	93		143		192			
#12	7	4	10	12	8	18	19	13	29	35	24	51	47	32	70	78	53	114	111	76	164	117		157			
#10	5	4	6	9	6	11	15	11	18	26	19	32	36	26	44	60	43	73	85	61	104	95	160	127		163	
#8	2	1	3	4	3	5	7	5	9	12	10	16	17	13	22	28	22	36	40	32	51	49	79	66	106	85	136
#6		1	1		2	4		4	6		7	11		10	15		16	26		23	37	36	57	48	76	62	98
#4		1	1		1	2		3	4		5	7		7	9		12	16		17	22	27	35	36	47	47	60
#3		1	1		1	1		2	3		4	6		6	8		10	13		15	19	23	29	31	39	40	51
#2		1	1		1	1		2	3		4	5		5	7		9	11		13	16	20	25	27	33	34	43
#1					1	1		1	1		3	3		4	5		6	8		9	12	14	18	19	25	25	32
1/0					1	1		1	1		2	3		3	4		5	7		8	10	12	15	16	21	21	27
2/0					1	1		1	1		1	2		3	3		5	6		7	8	10	13	14	17	18	22
3/0					1	1		1	1		1	1		2	3		4	5		6	7	9	11	12	14	15	18
4/0						1		1	1		1	1		1	2		3	4		5	6	7	9	10	12	13	15
250 MCM								1	1		1	1		1	1		2	3		4	4	6	7	8	10	10	12
300								1	1		1	1		1	1		2	3		3	4	5	6	7	8	9	11
350									1		1	1		1	1		1	2		3	3	4	5	6	7	8	9
400											1	1		1	1		1	1		2	3	4	5	5	6	7	8
500											1	1		1	1		1	1		1	2	3	4	4	5	6	7
600														1	1		1	1		1	1	3	3	4	4	5	5
700																	1	1		1	1	2	3	3	4	4	5
750																	1	1		1	1	2	2	3	3	4	4

Table D5010-146A Metric Equivalent, Conduit

U.S. vs. European Conduit – Approximate Equivalents			
United States		European	
Trade Size	Inside Diameter Inch/MM	Trade Size	Inside Diameter MM
1/2	.622/15.8	11	16.4
3/4	.824/20.9	16	19.9
1	1.049/26.6	21	25.5
1 1/4	1.380/35.0	29	34.2
1 1/2	1.610/40.9	36	44.0
2	2.067/52.5	42	51.0
2 1/2	2.469/62.7		
3	3.068/77.9		
3 1/2	3.548/90.12		
4	4.026/102.3		
5	5.047/128.2		
6	6.065/154.1		

Table D5010-146B Metric Equivalent, Wire

U.S. vs. European Wire – Approximate Equivalents			
United States		European	
Size AWG or MCM	Area Cir. Mils.(CM) MM^2	Size MM^2	Area Cir. Mils.
18	1620/.82	.75	1480
16	2580/1.30	1.0	1974
14	4110/2.08	1.5	2961
12	6530/3.30	2.5	4935
10	10,380/5.25	4	7896
8	16,510/8.36	6	11,844
6	26,240/13.29	10	19,740
4	41,740/21.14	16	31,584
3	52,620/26.65	25	49,350
2	66,360/33.61	–	–
1	83,690/42.39	35	69,090
1/0	105,600/53.49	50	98,700
2/0	133,100/67.42	–	–
3/0	167,800/85.00	70	138,180
4/0	211,600/107.19	95	187,530
250	250,000/126.64	120	236,880
300	300,000/151.97	150	296,100
350	350,000/177.30	–	–
400	400,000/202.63	185	365,190
500	500,000/253.29	240	473,760
600	600,000/303.95	300	592,200
700	700,000/354.60	–	–
750	750,000/379.93	–	–

D50 Electrical — RD5010-140 Design & Cost Tables

D5010-147 Concrete for Conduit Encasement

Table below lists C.Y. of concrete for 100 L.F. of trench. Conduits separation center to center should meet 7.5" (N.E.C.).

Number of Conduits	1	2	3	4	6	8	9	Number of Conduits
Trench Dimension	11.5" x 11.5"	11.5" x 19"	11.5" x 27"	19" x 19"	19" x 27"	19" x 38"	27" x 27"	Trench Dimension
Conduit Diameter 2.0"	3.29	5.39	7.64	8.83	12.51	17.66	17.72	Conduit Diameter 2.0"
2.5"	3.23	5.29	7.49	8.62	12.19	17.23	17.25	2.5"
3.0"	3.15	5.13	7.24	8.29	11.71	16.59	16.52	3.0"
3.5"	3.08	4.97	7.02	7.99	11.26	15.98	15.84	3.5"
4.0"	2.99	4.80	6.76	7.65	10.74	15.30	15.07	4.0"
5.0"	2.78	4.37	6.11	6.78	9.44	13.57	13.12	5.0"
6.0"	2.52	3.84	5.33	5.74	7.87	11.48	10.77	6.0"

Table D5010-148 Size Required and Weight (Lbs./1000 L.F.) of Aluminum and Copper THW Wire by Ampere Load

Amperes	Copper Size	Aluminum Size	Copper Weight	Aluminum Weight
15	14	12	24	11
20	12	10	33	17
30	10	8	48	39
45	8	6	77	52
65	6	4	112	72
85	4	2	167	101
100	3	1	205	136
115	2	1/0	252	162
130	1	2/0	324	194
150	1/0	3/0	397	233
175	2/0	4/0	491	282
200	3/0	250	608	347
230	4/0	300	753	403
255	250	400	899	512
285	300	500	1068	620
310	350	500	1233	620
335	400	600	1396	772
380	500	750	1732	951

Table D5010-149 Weight (Lbs./L.F.) of 4 Pole Aluminum and Copper Bus Duct by Ampere Load

Amperes	Aluminum Feeder	Copper Feeder	Aluminum Plug-In	Copper Plug-In
225			7	7
400			8	13
600	10	10	11	14
800	10	19	13	18
1000	11	19	16	22
1350	14	24	20	30
1600	17	26	25	39
2000	19	30	29	46
2500	27	43	36	56
3000	30	48	42	73
4000	39	67		
5000		78		

Table D5010-150 Transformer Weight (Lbs.) by kVA

Oil Filled 3 Phase 5/15 kV To 480/277			
kVA	Lbs.	kVA	Lbs.
150	1800	1000	6200
300	2900	1500	8400
500	4700	2000	9700
750	5300	3000	15000

Dry 240/480 To 120/240 Volt			
1 Phase		3 Phase	
kVA	Lbs.	kVA	Lbs.
1	23	3	90
2	36	6	135
3	59	9	170
5	73	15	220
7.5	131	30	310
10	149	45	400
15	205	75	600
25	255	112.5	950
37.5	295	150	1140
50	340	225	1575
75	550	300	1870
100	670	500	2850
167	900	750	4300

Table D5010-151 Generator Weight (Lbs.) by kW

3 Phase 4 Wire 277/480 Volt			
Gas		Diesel	
kW	Lbs.	kW	Lbs.
7.5	600	30	1800
10	630	50	2230
15	960	75	2250
30	1500	100	3840
65	2350	125	4030
85	2570	150	5500
115	4310	175	5650
170	6530	200	5930
		250	6320
		300	7840
		350	8220
		400	10750
		500	11900

D50 Electrical — RD5010-140 Design & Cost Tables

Table D5010-152A Conduit Weight Comparisons (Lbs. per 100 ft.) Empty

Type	1/2"	3/4"	1"	1-1/4"	1-1/2"	2"	2-1/2"	3"	3-1/2"	4"	5"	6"
Rigid Aluminum	28	37	55	72	89	119	188	246	296	350	479	630
Rigid Steel	79	105	153	201	249	332	527	683	831	972	1314	1745
Intermediate Steel (IMC)	60	82	116	150	182	242	401	493	573	638		
Electrical Metallic Tubing (EMT)	29	45	65	96	111	141	215	260	365	390		
Polyvinyl Chloride, Schedule 40	16	22	32	43	52	69	109	142	170	202	271	350
Polyvinyl Chloride Encased Burial						38		67	88	105	149	202
Fibre Duct Encased Burial						127		164	180	206	400	511
Fibre Duct Direct Burial						150		251	300	354		
Transite Encased Burial						160		240	290	330	450	550
Transite Direct Burial						220		310		400	540	640

Table D5010-152B Conduit Weight Comparisons (Lbs. per 100 ft.) with Maximum Cable Fill*

Type	1/2"	3/4"	1"	1-1/4"	1-1/2"	2"	2-1/2"	3"	3-1/2"	4"	5"	6"
Rigid Galvanized Steel (RGS)	104	140	235	358	455	721	1022	1451	1749	2148	3083	4343
Intermediate Steel (IMC)	84	113	186	293	379	611	883	1263	1501	1830		
Electrical Metallic Tubing (EMT)	54	116	183	296	368	445	641	930	1215	1540		

*Conduit & Heaviest Conductor Combination

D5010-153 Weight Comparisons of Common Size Cast Boxes in Lbs.

Size NEMA 4 or 9	Cast Iron	Cast Aluminum	Size NEMA 7	Cast Iron	Cast Aluminum
6" x 6" x 6"	17	7	6" x 6" x 6"	40	15
8" x 6" x 6"	21	8	8" x 6" x 6"	50	19
10" x 6" x 6"	23	9	10" x 6" x 6"	55	21
12" x 12" x 6"	52	20	12" x 6" x 6"	100	37
16" x 16" x 6"	97	36	16" x 16" x 6"	140	52
20" x 20" x 6"	133	50	20" x 20" x 6"	180	67
24" x 18" x 8"	149	56	24" x 18" x 8"	250	93
24" x 24" x 10"	238	88	24" x 24" x 10"	358	133
30" x 24" x 12"	324	120	30" x 24" x 10"	475	176
36" x 36" x 12"	500	185	30" x 24" x 12"	510	189

D50 Electrical

RD5010-155 kW Value/Cost Determination

General: Lighting and electric heating loads are expressed in watts and kilowatts.

Cost Determination:

The proper ampere values can be obtained as follows:
1. Convert watts to kilowatts (watts 1000 ÷ kilowatts)
2. Determine voltage rating of equipment.
3. Determine whether equipment is single phase or three phase.
4. Refer to Table D5010-158 to find ampere value from kW, Ton and Btu/hr. values.
5. Determine type of wire insulation – TW, THW, THWN.
6. Determine if wire is copper or aluminum.
7. Refer to Table D5010-141 to obtain copper or aluminum wire size from ampere values.
8. Next refer to Table D5010-142 for the proper conduit size to accommodate the number and size of wires in each particular case.
9. Next refer to unit price data for the per linear foot cost of the conduit.
10. Next refer to unit price data for the per linear foot cost of the wire. Multiply cost of wire per L.F. x number of wires in the circuits to obtain total wire cost per L.F..
11. Add values obtained in Step 9 and 10 for total cost per linear foot for conduit and wire x length of circuit = Total Cost.

Notes:
1. Phase refers to single phase, 2 wire circuits.
2. Phase refers to three phase, 3 wire circuits.
3. For circuits which operate continuously for 3 hours or more, multiply the ampere values by 1.25 for a given kW requirement.
4. For kW ratings not listed, add ampere values.

For example: Find the ampere value of 9 kW at 208 volt, single phase.

$$\begin{array}{r} 4 \text{ kW} = 19.2A \\ 5 \text{ kW} = 24.0A \\ \hline 9 \text{ kW} = 43.2A \end{array}$$

5. "Length of Circuit" refers to the one way distance of the run, not to the total sum of wire lengths.

Table D5010-158 Ampere Values as Determined by kW Requirements, BTU/HR or Ton, Voltage and Phase Values

			Ampere Values						
			120V	208V		240V		277V	480V
kW	Ton	BTU/HR	1 Phase	1 Phase	3 Phase	1 Phase	3 Phase	1 Phase	3 Phase
0.5	.1422	1,707	4.2A	2.4A	1.4A	2.1A	1.2A	1.8A	0.6A
0.75	.2133	2,560	6.2	3.6	2.1	3.1	1.9	2.7	.9
1.0	.2844	3,413	8.3	4.9	2.8	4.2	2.4	3.6	1.2
1.25	.3555	4,266	10.4	6.0	3.5	5.2	3.0	4.5	1.5
1.5	.4266	5,120	12.5	7.2	4.2	6.3	3.1	5.4	1.8
2.0	.5688	6,826	16.6	9.7	5.6	8.3	4.8	7.2	2.4
2.5	.7110	8,533	20.8	12.0	7.0	10.4	6.1	9.1	3.1
3.0	.8532	10,239	25.0	14.4	8.4	12.5	7.2	10.8	3.6
4.0	1.1376	13,652	33.4	19.2	11.1	16.7	9.6	14.4	4.8
5.0	1.4220	17,065	41.6	24.0	13.9	20.8	12.1	18.1	6.1
7.5	2.1331	25,598	62.4	36.0	20.8	31.2	18.8	27.0	9.0
10.0	2.8441	34,130	83.2	48.0	27.7	41.6	24.0	36.5	12.0
12.5	3.5552	42,663	104.2	60.1	35.0	52.1	30.0	45.1	15.0
15.0	4.2662	51,195	124.8	72.0	41.6	62.4	37.6	54.0	18.0
20.0	5.6883	68,260	166.4	96.0	55.4	83.2	48.0	73.0	24.0
25.0	7.1104	85,325	208.4	120.2	70.0	104.2	60.0	90.2	30.0
30.0	8.5325	102,390		144.0	83.2	124.8	75.2	108.0	36.0
35.0	9.9545	119,455		168.0	97.1	145.6	87.3	126.0	42.1
40.0	11.3766	136,520		192.0	110.8	166.4	96.0	146.0	48.0
45.0	12.7987	153,585			124.8	187.5	112.8	162.0	54.0
50.0	14.2208	170,650			140.0	208.4	120.0	180.4	60.0
60.0	17.0650	204,780			166.4		150.4	216.0	72.0
70.0	19.9091	238,910			194.2		174.6		84.2
80.0	22.7533	273,040			221.6		192.0		96.0
90.0	25.5975	307,170					225.6		108.0
100.0	28.4416	341,300							120.0

D50 Electrical — RD5010-160 kVA Value/Cost Determination

General: Control transformers are listed in VA. Step-down and power transformers are listed in kVA.

Cost Determination:

1. Convert VA to kVA. Volt amperes (VA) ÷ 1000 = Kilovolt amperes (kVA).
2. Determine voltage rating of equipment.
3. Determine whether equipment is single phase or three phase.
4. Refer to Table D5010-161 to find ampere value from kVA value.
5. Determine type of wire insulation – TW, THW, THWN.
6. Determine if wire is copper or aluminum.
7. Refer to Table D5010-161 to obtain copper or aluminum wire size from ampere values.
8. Next refer to Table D5010-142 for the proper conduit size to accommodate the number and size of wires in each particular case.
9. Next refer to unit price data for the per linear foot cost of the conduit.
10. Next refer to unit price data for the per linear foot cost of the wire. Multiply cost of wire per L.F. x number of wires in the circuits to obtain total wire cost.
11. Add values obtained in Step 9 and 10 for total cost per linear foot for conduit and wire x length of circuit = Total Cost.

Example: A transformer rated 10 kVA 480 volts primary, 240 volts secondary, 3 phase has the capacity to furnish the following:

1. Primary amperes = 10 kVA x 1.20 = 12 amperes (from Table D5010-161)
2. Secondary amperes = 10 kVA x 2.40 = 24 amperes (from Table D5010-161)

Note: Transformers can deliver generally 125% of their rated kVA. For instance, a 10 kVA rated transformer can safely deliver 12.5 kVA.

Table D5010-161 Multiplier Values for kVA to Amperes Determined by Voltage and Phase Values

Volts	Multiplier for Circuits	
	2 Wire, 1 Phase	3 Wire, 3 Phase
115	8.70	
120	8.30	
230	4.30	2.51
240	4.16	2.40
200	5.00	2.89
208	4.80	2.77
265	3.77	2.18
277	3.60	2.08
460	2.17	1.26
480	2.08	1.20
575	1.74	1.00
600	1.66	0.96

D50 Electrical — RD5010-170 HP Value/Cost Determination

General: Motors can be powered by any of the seven systems shown in Figure D5010-121 thru Figure D5010-127 provided the motor voltage characteristics are compatible with the power system characteristics.

Cost Determination:

Motor Amperes for the various size H.P. and voltage are listed in Table D5010-171. To find the amperes, locate the required H.P. rating and locate the amperes under the appropriate circuit characteristics.

For example:

A. 100 H.P., 3 phase, 460 volt motor = 124 amperes (Table D5010-171)

B. 10 H.P., 3 phase, 200 volt motor = 32.2 amperes (Table D5010-171)

Motor Wire Size: After the amperes are found in Table D5010-171 the amperes must be increased 25% to compensate for power losses. Next refer to Table D5010-141. Find the appropriate insulation column for copper or aluminum wire to determine the proper wire size.

For example:

A. 100 H.P., 3 phase, 460 volt motor has an ampere value of 124 amperes from Table D5010-171

B. 124A x 1.25 = 155 amperes

C. Refer to Table D5010-141 for THW or THWN wire insulations to find the proper wire size. For a 155 ampere load using copper wire a size 2/0 wire is needed.

D. For the 3 phase motor three wires of 2/0 size are required.

Conduit Size: To obtain the proper conduit size for the wires and type of insulation used, refer to Table D5010-142.

For example: For the 100 H.P., 460V, 3 phase motor, it was determined that three 2/0 wires are required. Assuming THWN insulated copper wire, use Table D5010-142 to determine that three 2/0 wires require 1-1/2" conduit.

Material Cost of the conduit and wire system depends on:

1. Wire size required
2. Copper or aluminum wire
3. Wire insulation type selected
4. Steel or plastic conduit
5. Type of conduit raceway selected.

Labor Cost of the conduit and wire system depends on:

1. Type and size of conduit
2. Type and size of wires installed
3. Location and height of installation in building or depth of trench
4. Support system for conduit.

Table D5010-171 Ampere Values Determined by Horsepower, Voltage and Phase Values

H.P.	Single Phase		Three Phase			
	115V	230V	200 V	230V	460V	575V
1/6	4.4A	2.2A				
1/4	5.8	2.9				
1/3	7.2	3.6				
1/2	9.8	4.9	2.3A	2.0A	1.0A	0.8A
3/4	13.8	6.9	3.2	2.8	1.4	1.1
1	16	8	4.1	3.6	1.8	1.4
1-1/2	20	10	6.0	5.2	2.6	2.1
2	24	12	7.8	6.8	3.4	2.7
3	34	17	11.0	9.6	4.8	3.9
5			17.5	15.2	7.6	6.1
7-1/2			25.3	22	11	9
10			32.2	28	14	11
15			48.3	42	21	17
20			62.1	54	27	22
25			78.2	68	34	27
30			92.0	80	40	32
40			119.6	104	52	41
50			149.5	130	65	52
60			177	154	77	62
75			221	192	96	77
100			285	248	124	99
125			359	312	156	125
150			414	360	180	144
200			552	480	240	192

D50 Electrical — RD5010-170 HP Value/Cost Determination

Cost Determination (cont.)

Magnetic starters, switches, and motor connection:

To complete the cost picture from H.P. to Costs additional items must be added to the cost of the conduit and wire system to arrive at a total cost.

1. Assembly D5020 160 Magnetic Starters Installed Cost lists the various size starters for single phase and three phase motors.
2. Assembly D5020 165 Heavy Duty Safety Switches Installed Cost lists safety switches required at the beginning of a motor circuit and also one required in the vicinity of the motor location.
3. Assembly D5020 170 Motor Connection lists the various costs for single and three phase motors.

Worksheet to obtain total motor wiring costs:

It is assumed that the motors or motor driven equipment are furnished and installed under other sections for this estimate and the following work is done under this section:

1. Conduit
2. Wire (add 10% for additional wire beyond conduit ends for connections to switches, boxes, starters, etc.)
3. Starters
4. Safety switches
5. Motor connections

Table D5010-172

				Cost	
Item	Type	Size	Quantity	Unit	Total
Wire					
Conduit					
Switch					
Starter					
Switch					
Motor Connection					
Other					
Total Cost					

Table D5010-173 Maximum Horsepower for Starter Size by Voltage

Starter Size	Maximum HP (3ϕ)			
	208V	240V	480V	600V
00	1½	1½	2	2
0	3	3	5	5
1	7½	7½	10	10
2	10	15	25	25
3	25	30	50	50
4	40	50	100	100
5		100	200	200
6		200	300	300
7		300	600	600
8		450	900	900
8L		700	1500	1500

D50 Electrical — RD5020-200 Illumination

General: The cost of the lighting portion of the electrical costs is dependent upon:
1. The footcandle requirement of the proposed building.
2. The type of fixtures required.
3. The ceiling heights of the building.
4. Reflectance value of ceilings, walls and floors.
5. Fixture efficiencies and spacing vs. mounting height ratios.

Footcandle Requirements: See Table D5020-204 for Footcandle and Watts per S.F. determination.

Table D5020-201 IESNA* Recommended Illumination Levels in Footcandles

Commercial Buildings			Industrial Buildings		
Type	Description	Footcandles	Type	Description	Footcandles
Bank	Lobby	50	Assembly Areas	Rough bench & machine work	50
	Customer Areas	70		Medium bench & machine work	100
	Teller Stations	150		Fine bench & machine work	500
	Accounting Areas	150	Inspection Areas	Ordinary	50
Offices	Routine Work	100		Difficult	100
	Accounting	150		Highly Difficult	200
	Drafting	200	Material Handling	Loading	20
	Corridors, Halls, Washrooms	30		Stock Picking	30
Schools	Reading or Writing	70		Packing, Wrapping	50
	Drafting, Labs, Shops	100	Stairways	Service Areas	20
	Libraries	70	Washrooms	Service Areas	20
	Auditoriums, Assembly	15	Storage Areas	Inactive	5
	Auditoriums, Exhibition	30		Active, Rough, Bulky	10
Stores	Circulation Areas	30		Active, Medium	20
	Stock Rooms	30		Active, Fine	50
	Merchandise Areas, Service	100	Garages	Active Traffic Areas	20
	Self-Service Areas	200		Service & Repair	100

*IESNA - Illuminating Engineering Society of North America

D50 Electrical — RD5020-200 Illumination

Table D5020-202 General Lighting Loads by Occupancies

Type of Occupancy	Unit Load per S.F. (Watts)
Armories and Auditoriums	1
Banks	5
Barber Shops and Beauty Parlors	3
Churches	1
Clubs	2
Court Rooms	2
*Dwelling Units	3
Garages — Commercial (storage)	½
Hospitals	2
*Hotels and Motels, including apartment houses without provisions for cooking by tenants	2
Industrial Commercial (Loft) Buildings	2
Lodge Rooms	1½
Office Buildings	5
Restaurants	2
Schools	3
Stores	3
Warehouses (storage)	¼
*In any of the above occupancies except one-family dwellings and individual dwelling units of multi-family dwellings:	
Assembly Halls and Auditoriums	1
Halls, Corridors, Closets	½
Storage Spaces	¼

Table D5020-203 Lighting Limit (Connected Load) for Listed Occupancies: New Building Proposed Energy Conservation Guideline

Type of Use	Maximum Watts per S.F.
Interior	3.00
Category A: Classrooms, office areas, automotive mechanical areas, museums, conference rooms, drafting rooms, clerical areas, laboratories, merchandising areas, kitchens, examining rooms, book stacks, athletic facilities.	
Category B: Auditoriums, waiting areas, spectator areas, restrooms, dining areas, transportation terminals, working corridors in prisons and hospitals, book storage areas, active inventory storage, hospital bedrooms, hotel and motel bedrooms, enclosed shopping mall concourse areas, stairways.	1.00
Category C: Corridors, lobbies, elevators, inactive storage areas.	0.50
Category D: Indoor parking.	0.25
Exterior	
Category E: Building perimeter: wall-wash, facade, canopy.	5.00 (per linear foot)
Category F: Outdoor parking.	0.10

D50 Electrical — RD5020-200 Illumination

Table D5020-204 Procedure for Calculating Footcandles and Watts Per Square Foot

1. Initial footcandles = No. of fixtures × lamps per fixture × lumens per lamp × coefficient of utilization ÷ square feet
2. Maintained footcandles = initial footcandles × maintenance factor
3. Watts per square foot = No. of fixtures × lamps × (lamp watts + ballast watts) ÷ square feet

Example: To find footcandles and watts per S.F. for an office 20′ x 20′ with 11 fluorescent fixtures each having 4–40 watt C.W. lamps.

Based on good reflectance and clean conditions:

Lumens per lamp = 40 watt cool white at 3150 lumens per lamp RD5020-250, Table RD5020-251
Coefficient of utilization = .42 (varies from .62 for light colored areas to .27 for dark)
Maintenance factor = .75 (varies from .80 for clean areas with good maintenance to .50 for poor)
Ballast loss = 8 watts per lamp. (Varies with manufacturer. See manufacturers' catalog.)

1. Initial footcandles:

$$\frac{11 \times 4 \times 3150 \times .42}{400} = \frac{58,212}{400} = 145 \text{ footcandles}$$

2. Maintained footcandles:

$$145 \times .75 = 109 \text{ footcandles}$$

3. Watts per S.F.

$$\frac{11 \times 4 (40 + 8)}{400} = \frac{2,112}{400} = 5.3 \text{ watts per S.F.}$$

Table D5020-205 Approximate Watts Per Square Foot for Popular Fixture Types

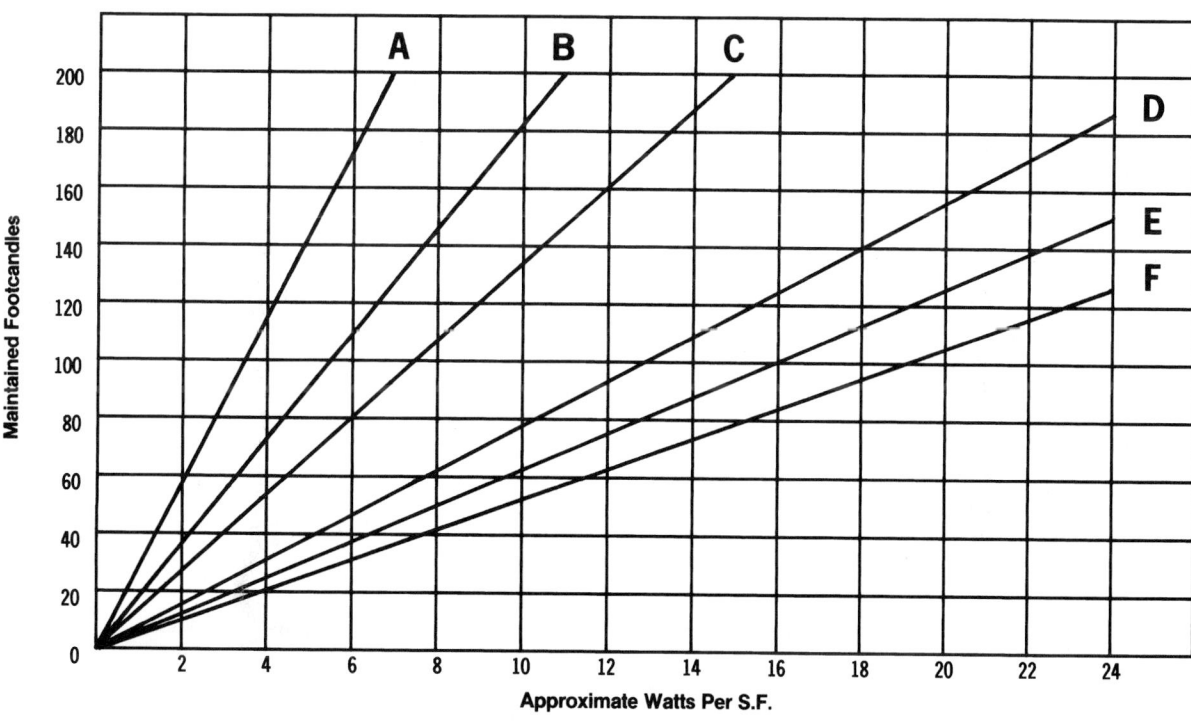

Due to the many variables involved, use for preliminary estimating only:
 a. Fluorescent – industrial System D5020 208
 b. Fluorescent – lens unit System D5020 208 Fixture types B & C
 c. Fluorescent – louvered unit
 d. Incandescent – open reflector System D5020 214, Type D
 e. Incandescent – lens unit System D5020 214, Type A
 f. Incandescent – down light System D5020 214, Type B

D50 Electrical — RD5020-240 High Intensity Discharge

Table D5020-241 Comparison - Cost of Operation of High Intensity Discharge Lamps

Lamp Type	Wattage	Life (Hours)	1 Circuit Wattage	Average Initial Lumens	2 L.L.D.	3 Mean Lumens
M.V.	100 DX	24,000	125	4,000	61%	2,440
L.P.S.	SOX-35	18,000	65	4,800	100%	4,800
H.P.S.	LU-70	12,000	84	5,800	90%	5,220
M.H.	No Equivalent					
M.V.	175 DX	24,000	210	8,500	66%	5,676
M.V.	250 DX	24,000	295	13,000	66%	7,986
L.P.S.	SOX-55	18,000	82	8,000	100%	8,000
H.P.S.	LU-100	12,000	120	9,500	90%	8,550
M.H.	No Equivalent					
M.V.	400 DX	24,000	465	24,000	64%	14,400
L.P.S.	SOX-90	18,000	141	13,500	100%	13,500
H.P.S.	LU-150	16,000	188	16,000	90%	14,400
M.H.	MH-175	7,500	210	14,000	73%	10,200
M.V.	No Equivalent					
L.P.S.	SOX-135	18,000	147	22,500	100%	22,500
H.P.S.	LU-250	20,000	310	25,500	92%	23,205
M.H.	MH-250	7,500	295	20,500	78%	16,000
M.V.	1000 DX	24,000	1,085	63,000	61%	37,820
L.P.S.	SOX-180	18,000	248	33,000	100%	33,000
H.P.S.	LU-400	20,000	480	50,000	90%	45,000
M.H.	MH-400	15,000	465	34,000	72%	24,600
H.P.S.	LU-1000	15,000	1,100	140,000	91%	127,400

1. Includes ballast losses and average lamp watts
2. Lamp lumen depreciation (% of initial light output at 70% rated life)
3. Lamp lumen output at 70% rated life (L.L.D. x initial)

M.V. = Mercury Vapor
L.P.S. = Low Pressure Sodium
H.P.S. = High Pressure Sodium
L.H. = Metal Halide

Table D5020-242 For Other than Regular Cool White (CW) Lamps

	Multiply Material Costs as Follows:				
Regular Lamps	Cool white deluxe (CWX)	x 1.35	Energy Saving Lamps	Cool white (CW/ES)	x 1.35
	Warm white deluxe (WWX)	x 1.35		Cool white deluxe (CWX/ES)	x 1.65
	Warm white (WW)	x 1.30		Warm white (WW/ES)	x 1.55
	Natural (N)	x 2.05		Warm white deluxe (WWX/ES)	x 1.65

D50 Electrical — RD5020-250 Lighting

Table D5020-251 Lamp Comparison Chart with Enclosed Floodlight, Ballast, & Lamp for Pole

Type	Watts	Initial Lumens	Lumens per Watt	Lumens @ 40% Life	Life (Hours)
Incandescent	150	2,880	19	85	750
	300	6,360	21	84	750
	500	10,850	22	80	1,000
	1,000	23,740	24	80	1,000
	1,500	34,400	23	80	1,000
Tungsten	500	10,950	22	97	2,000
Halogen	1,500	35,800	24	97	2,000
Fluorescent	40	3,150	79	88	20,000
Cool	110	9,200	84	87	12,000
White	215	16,000	74	81	12,000
Deluxe	250	12,100	48	86	24,000
Mercury	400	22,500	56	85	24,000
	1,000	63,000	63	75	24,000
Metal	175	14,000	80	77	7,500
Halide	400	34,000	85	75	15,000
	1,000	100,000	100	83	10,000
	1,500	155,000	103	92	1,500
High	70	5,800	83	90	20,000
Pressure	100	9,500	95	90	20,000
Sodium	150	16,000	107	90	24,000
	400	50,000	125	90	24,000
	1,000	140,000	140	90	24,000
Low	55	4,600	131	98	18,000
Pressure	90	12,750	142	98	18,000
Sodium	180	33,000	183	98	18,000

Color: High Pressure Sodium — Slightly Yellow
Low Pressure Sodium — Yellow
Mercury Vapor — Green-Blue
Metal Halide — Blue White

Note: Pole not included.

D50 Electrical — RD5020-600 Air Conditioning

Table D5020-602 Central Air Conditioning Watts per S.F., BTU's per Hour per S.F. of Floor Area and S.F. per Ton of Air Conditioning

Type Building	Watts per S.F.	BTUH per S.F.	S.F. per Ton	Type Building	Watts per S.F.	BTUH per S.F.	S.F. per Ton	Type Building	Watts per S.F.	BTUH per S.F.	S.F. per Ton
Apartments, Individual	3	26	450	Dormitory, Rooms	4.5	40	300	Libraries	5.7	50	240
Corridors	2.5	22	550	Corridors	3.4	30	400	Low Rise Office, Ext.	4.3	38	320
Auditoriums & Theaters	3.3	40	300/18*	Dress Shops	4.9	43	280	Interior	3.8	33	360
Banks	5.7	50	240	Drug Stores	9	80	150	Medical Centers	3.2	28	425
Barber Shops	5.5	48	250	Factories	4.5	40	300	Motels	3.2	28	425
Bars & Taverns	15	133	90	High Rise Off.-Ext. Rms.	5.2	46	263	Office (small suite)	4.9	43	280
Beauty Parlors	7.6	66	180	Interior Rooms	4.2	37	325	Post Office, Int. Office	4.9	42	285
Bowling Alleys	7.8	68	175	Hospitals, Core	4.9	43	280	Central Area	5.3	46	260
Churches	3.3	36	330/20*	Perimeter	5.3	46	260	Residences	2.3	20	600
Cocktail Lounges	7.8	68	175	Hotels, Guest Rooms	5	44	275	Restaurants	6.8	60	200
Computer Rooms	16	141	85	Public Spaces	6.2	55	220	Schools & Colleges	5.3	46	260
Dental Offices	6	52	230	Corridors	3.4	30	400	Shoe Stores	6.2	55	220
Dept. Stores, Basement	4	34	350	Industrial Plants, Offices	4.3	38	320	Shop'g. Ctrs., Sup. Mkts.	4	34	350
Main Floor	4.5	40	300	General Offices	4	34	350	Retail Stores	5.5	48	250
Upper Floor	3.4	30	400	Plant Areas	4.5	40	300	Specialty Shops	6.8	60	200

*Persons per ton

12,000 BTUH = 1 ton of air conditioning

G10 Site Preparation — RG1010-010 Information

The Building Site Work Division is divided into five sections:

G10 Site Preparation
G20 Site Improvements
G30 Site Mechanical Utilities
G40 Site Electrical Utilities
G90 Other Site Construction

Clear and Grub, Demolition and Landscaping must be added to these costs as they apply to individual sites.

The Site Preparation section includes typical bulk excavation, trench excavation, and backfill. To this must be added sheeting, dewatering, drilling and blasting, traffic control, and any unusual items relating to a particular site.

The Site Improvements section, including Parking Lot and Roadway construction, has quantities and prices for typical designs with gravel base and asphaltic concrete pavement. Different climate, soil conditions, material availability, and owner's requirements can easily be adapted to these parameters.

The Site Utilities sections tabulate the installed price for pipe. Excavation and backfill costs for the pipe are derived from the Site Preparation section.

By following the examples and illustrations of typical site work conditions in this division and adapting them to your particular site, an accurate price, including O&P, can be determined.

Table G1010-011 Site Preparation

Description	Unit	Total Costs
Clear and Grub Average	Acre	$10,750
Bulk Excavation with Front End Loader	C.Y.	2.47
Spread and Compact Dumped Material	C.Y.	2.99
Fine Grade and Seed	S.Y.	3.62

G10 Site Preparation — RG1030-400 Excavation, General

Bulk excavation on building jobs today is done entirely by machine. Hand work is used only to square out corners, fine grading, cutting out trenches and footings below the general excavating grade, and for minor miscellaneous cuts. Hand trimming of material other than sand and gravel is generally done with picks and shovels.

The following figures on machine excavating apply to basement pits, large footings and other excavation for buildings. On grading excavation this performance can be bettered by 33% to 50%. The figures presume loading directly into trucks. Allowance has been made for idle time due to truck delay and moving about on the job. Figures include no water problems or slow down due to sheeting and bracing operations. They also do not include truck spotters or labor for cleanup.

Table G1030-401 Bulk Excavation Cost, 15' Deep

1½ C.Y. Hydraulic Excavator Backhoe, Four Tandem Trucks with a Two Mile round Trip Haul Distance

Item	Daily Cost	Labor & Equipment	Unit Price (C.Y.)
Equipment operator @ $ 73.75 per hr. x 8 hrs.	$ 590.00		
Laborer @ $ 54.60 per hr. x 8 hrs.	436.80	$2,820.40 Labor	$3.92 Labor
Four truck drivers @ $ 56.05 per hr. x 8 hrs.	1,793.60		
1½ C.Y. backhoe @ $1,036.00 per day	$1,036.00	$3,805.60 Equipment	$5.29 Equipment
Four 16 ton trucks @ $ 692.40 per day	2,769.60		
Production = 720 C.Y. per day	$6,626.00	$6,626.00 Labor & Equip.	$9.21 Labor & Equip.

Table G1030-402 Total Cost Per C.Y. Excavation for Various Size Jobs

No. of C.Y. to be Excavated	Cost Per Cubic Yard	Mobilizing & Demobilizing Cost*	Total Cost Per Cubic Yard
500 C.Y.	$9.21	$2.02	$11.23
1,000		1.01	$10.22
1,500		.67	$ 9.88
2,000		.51	$ 9.72
3,000		.34	$ 9.55
5,000		.20	$ 9.41
10,000		.10	$ 9.31
15,000		.07	$ 9.28

*$505.00 for Mobilizing or Demobilizing 1½ C.Y. Hydraulic Excavator

Table G1030-403 Trench Bottom Widths for Various Outside Diameters of Buried Pipe

Outside Diameter	Trench Bottom Width
24 inches	4.1 feet
30	4.9
36	5.6
42	6.3
48	6.3
60	8.5
72	10.0
84	11.4

G20 Site Improvements — RG2020-500 Parking Lots

SINGLE UNIT

OVERLAPPING UNITS

Table G2020-502 Layout Data Based on 9' x 19' Parking Stall Size

φ Angle of Stall	P Parking Depth	A Aisle Width	C Curb Length	W Width Overall	N Net Car Area	G Gross Car Area	D Distance Last Car	L Lost Area	P' Parking Depth	W' Width Overall
90°	19'.	24'	9'.	62'.	171 S.F.	171 S.F.	9'.	0	19'.	62'.
60°	21'.8	18'	10.4'	60'.	171 S.F.	217 S.F.	7.8'	205 S.F.	18.8'	55.5'
45°	19.8'	13'	12.8'	52.7'	171 S.F.	252 S.F.	6.4'	286 S.F.	16.6'	46.2'

Note: Square foot per car areas do not include the area of the travel lane.

90° Stall Angle: The main reason for use of this stall angle is to achieve the highest car capacity. This may be sound reasoning for employee lots with all day parking, but in most (in & out) lots there is difficulty in entering the stalls and no traffic lane direction. This may outweigh the advantage of high capacity.

60° Stall Angle: This layout is used most often due to the ease of entering and backing out, also the traffic aisle may be smaller.

45° Stall Angle: Requires a small change of direction from the traffic aisle to the stall, so the aisle may be reduced in width.

G20 Site Improvements — RG2020-500 Parking Lots

PARKING LOT PLAN (50 CAR)

Preliminary Design Data: The space required for parking and maneuvering is between 300 and 400 S.F. per car, depending upon engineering layout and design.

Ninety degree (90°) parking, with a central driveway and two rows of parked cars, will provide the best economy.

Diagonal parking is easier than 90° for the driver and reduces the necessary driveway width, but requires more total space.

General Conditions — RH1010-100 Overhead

Table H1010-101 General Contractor's Overhead

There are two distinct types of overhead on a construction project: Project Overhead and Main Office Overhead. Project Overhead includes those costs at a construction site not directly associated with the installation of construction materials. Examples of Project Overhead costs include the following:

1. Superintendent
2. Construction office and storage trailers
3. Temporary sanitary facilities
4. Temporary utilities
5. Security fencing
6. Photographs
7. Clean up
8. Performance and payment bonds

The above Project Overhead items are also referred to as General Requirements and therefore are estimated in Division 1. Division 1 is the first division listed in the CSI MasterFormat but it is usually the last division estimated. The sum of the costs in Divisions 1 through 49 is referred to as the sum of the direct costs.

All construction projects also include indirect costs. The primary components of indirect costs are the contractor's Main Office Overhead and profit. The amount of the Main Office Overhead expense varies depending on the the following:

1. Owner's compensation
2. Project managers and estimator's wages
3. Clerical support wages
4. Office rent and utilities
5. Corporate legal and accounting costs
6. Advertising
7. Automobile expenses
8. Association dues
9. Travel and entertainment expenses

These costs are usually calculated as a percentage of annual sales volume. This percentage can range from 35% for a small contractor doing less than $500,000 to 5% for a large contractor with sales in excess of $100 million.

Table H1010-102 Main Office Expense

A General Contractor's main office expense consists of many items not detailed in the front portion of the book. The percentage of main office expense declines with increased annual volume of the contractor. Typical main office expense ranges from 2% to 20% with the median about 7.2% of total volume. This equals about 7.7% of direct costs. The following are approximate percentages of total overhead for different items usually included in a General Contractor's main office overhead. With different accounting procedures, these percentages may vary.

Item	Typical Range	Average
Managers', clerical and estimators' salaries	40 % to 55 %	48%
Profit sharing, pension and bonus plans	2 to 20	12
Insurance	5 to 8	6
Estimating and project management (not including salaries)	5 to 9	7
Legal, accounting and data processing	0.5 to 5	3
Automobile and light truck expense	2 to 8	5
Depreciation of overhead capital expenditures	2 to 6	4
Maintenance of office equipment	0.1 to 1.5	1
Office rental	3 to 5	4
Utilities including phone and light	1 to 3	2
Miscellaneous	5 to 15	8
Total		100%

General Conditions — RH1010-200 Design & Engineering

Table H1010-201 Architectural Fees

Tabulated below are typical percentage fees by project size, for good professional architectural service. Fees may vary from those listed depending upon degree of design difficulty and economic conditions in any particular area.

Rates can be interpolated horizontally and vertically. Various portions of the same project requiring different rates should be adjusted proportionately. For alterations, add 50% to the fee for the first $500,000 of project cost and add 25% to the fee for project cost over $500,000.

Architectural fees tabulated below include Structural, Mechanical and Electrical Engineering Fees. They do not include the fees for special consultants such as kitchen planning, security, acoustical, interior design, etc.

Building Types	Total Project Size in Thousands of Dollars						
	100	250	500	1,000	5,000	10,000	50,000
Factories, garages, warehouses, repetitive housing	9.0%	8.0%	7.0%	6.2%	5.3%	4.9%	4.5%
Apartments, banks, schools, libraries, offices, municipal buildings	12.2	12.3	9.2	8.0	7.0	6.6	6.2
Churches, hospitals, homes, laboratories, museums, research	15.0	13.6	12.7	11.9	9.5	8.8	8.0
Memorials, monumental work, decorative furnishings	—	16.0	14.5	13.1	10.0	9.0	8.3

Table H1010-202 Engineering Fees

Typical **Structural Engineering Fees** based on type of construction and total project size. These fees are included in Architectural Fees.

Type of Construction	Total Project Size (in thousands of dollars)			
	$500	$500-$1,000	$1,000-$5,000	Over $5000
Industrial buildings, factories & warehouses	Technical payroll times 2.0 to 2.5	1.60%	1.25%	1.00%
Hotels, apartments, offices, dormitories, hospitals, public buildings, food stores		2.00%	1.70%	1.20%
Museums, banks, churches and cathedrals		2.00%	1.75%	1.25%
Thin shells, prestressed concrete, earthquake resistive		2.00%	1.75%	1.50%
Parking ramps, auditoriums, stadiums, convention halls, hangars & boiler houses		2.50%	2.00%	1.75%
Special buildings, major alterations, underpinning & future expansion		Add to above 0.5%	Add to above 0.5%	Add to above 0.5%

For complex reinforced concrete or unusually complicated structures, add 20% to 50%.

Table H1010-203 Mechanical and Electrical Fees

Typical **Mechanical and Electrical Engineering Fees** based on the size of the subcontract. The fee structure for both are shown below. These fees are included in Architectural Fees.

Type of Construction	Subcontract Size							
	$25,000	$50,000	$100,000	$225,000	$350,000	$500,000	$750,000	$1,000,000
Simple structures	6.4%	5.7%	4.8%	4.5%	4.4%	4.3%	4.2%	4.1%
Intermediate structures	8.0	7.3	6.5	5.6	5.1	5.0	4.9	4.8
Complex structures	10.1	9.0	9.0	8.0	7.5	7.5	7.0	7.0

For renovations, add 15% to 25% to applicable fee.

Table H1010-204 Construction Time Requirements

Table below is average construction time in months for different types of building projects. Table at right is the construction time in months for different size projects. Design time runs 25% to 40% of construction time.

Type Building	Construction Time	Project Value	Construction Time
Industrial Buildings	12 Months	Under $1,400,000	10 Months
Commercial Buildings	15 Months	Up to $3,800,000	15 Months
Research & Development	18 Months	Up to $19,000,000	21 Months
Institutional Buildings	20 Months	over $19,000,000	28 Months

General Conditions — RH1010-300 Individual Insurance Bonds

Table H1010-301 Builder's Risk Insurance

Builder's Risk Insurance is insurance on a building during construction. Premiums are paid by the owner or the contractor. Blasting, collapse and underground insurance would raise total insurance costs above those listed. Floater policy for materials delivered to the job runs $.75 to $1.25 per $100 value. Contractor equipment insurance runs $.50 to $1.50 per $100 value. Insurance for miscellaneous tools to $1,500 value runs from $3.00 to $7.50 per $100 value.

Tabulated below are New England Builder's Risk insurance rates in dollars per $100 value for $1,000 deductible. For $25,000 deductible, rates can be reduced 13% to 34%. On contracts over $1,000,000, rates may be lower than those tabulated. Policies are written annually for the total completed value in place. For "all risk" insurance (excluding flood, earthquake and certain other perils) add $.025 to total rates below.

Coverage	Frame Construction (Class 1) Range	Frame Construction (Class 1) Average	Brick Construction (Class 4) Range	Brick Construction (Class 4) Average	Fire Resistive (Class 6) Range	Fire Resistive (Class 6) Average
Fire Insurance	$.350 to $.850	$.600	$.158 to $.189	$.174	$.052 to $.080	$.070
Extended Coverage	.115 to .200	.158	.080 to .105	.101	.081 to .105	.100
Vandalism	.012 to .016	.014	.008 to .011	.011	.008 to .011	.010
Total Annual Rate	$.477 to $1.066	$.772	$.246 to $.305	$.286	$.141 to $.196	$.180

Table H1010-302 Performance Bond

This table shows the cost of a Performance Bond for a construction job scheduled to be completed in 12 months. Add 1% of the premium cost per month for jobs requiring more than 12 months to complete. The rates are "standard" rates offered to contractors that the bonding company considers financially sound and capable of doing the work. Preferred rates are offered by some bonding companies based upon financial strength of the contractor. Actual rates vary from contractor to contractor and from bonding company to bonding company. Contractors should prequalify through a bonding agency before submitting a bid on a contract that requires a bond.

Contract Amount	Building Construction Class B Projects	Highways & Bridges Class A New Construction	Highways & Bridges Class A-1 Highway Resurfacing
First $100,000 bid	$25.00 per M	$15.00 per M	$9.40 per M
Next 400,000 bid	$2,500 plus $15.00 per M	$1,500 plus $10.00 per M	$940 plus $7.20 per M
Next 2,000,000 bid	8,500 plus 10.00 per M	5,500 plus 7.00 per M	3,820 plus 5.00 per M
Next 2,500,000 bid	28,500 plus 7.50 per M	19,500 plus 5.50 per M	15,820 plus 4.50 per M
Next 2,500,000 bid	47,250 plus 7.00 per M	33,250 plus 5.00 per M	28,320 plus 4.50 per M
Over 7,500,000 bid	64,750 plus 6.00 per M	45,750 plus 4.50 per M	39,570 plus 4.00 per M

General Conditions — RH1010-400 Workers' Compensation

Table H1010-401 Workers' Compensation Insurance Rates by Trade

The table below tabulates the national averages for workers' compensation insurance rates by trade and type of building. The average "Insurance Rate" is multiplied by the "% of Building Cost" for each trade. This produces the "Workers' Compensation" cost by % of total labor cost, to be added for each trade by building type to determine the weighted average workers' compensation rate for the building types analyzed.

Trade	Insurance Rate (% Labor Cost) Range	Insurance Rate (% Labor Cost) Average	% of Building Cost Office Bldgs.	% of Building Cost Schools & Apts.	% of Building Cost Mfg.	Workers' Compensation Office Bldgs.	Workers' Compensation Schools & Apts.	Workers' Compensation Mfg.
Excavation, Grading, etc.	3.9% to 18.0%	9.2%	4.8%	4.9%	4.5%	0.44%	0.45%	0.41%
Piles & Foundations	5.5 to 50.2	15.3	7.1	5.2	8.7	1.09	0.80	1.33
Concrete	4.7 to 28.1	12.6	5.0	14.8	3.7	0.63	1.86	0.47
Masonry	5.1 to 33.0	12.5	6.9	7.5	1.9	0.86	0.94	0.24
Structural Steel	5.5 to 125.8	36.5	10.7	3.9	17.6	3.91	1.42	6.42
Miscellaneous & Ornamental Metals	4.5 to 22.6	10.6	2.8	4.0	3.6	0.30	0.42	0.38
Carpentry & Millwork	5.5 to 36.6	15.1	3.7	4.0	0.5	0.56	0.60	0.08
Metal or Composition Siding	4.3 to 61.9	16.4	2.3	0.3	4.3	0.38	0.05	0.71
Roofing	5.5 to 90.6	29.6	2.3	2.6	3.1	0.68	0.77	0.92
Doors & Hardware	3.5 to 36.6	10.8	0.9	1.4	0.4	0.10	0.15	0.04
Sash & Glazing	5.4 to 35.7	12.5	3.5	4.0	1.0	0.44	0.50	0.13
Lath & Plaster	3.3 to 28.8	10.9	3.3	6.9	0.8	0.36	0.75	0.09
Tile, Marble & Floors	3.0 to 17.6	8.0	2.6	3.0	0.5	0.21	0.24	0.04
Acoustical Ceilings	3.3 to 36.7	7.9	2.4	0.2	0.3	0.19	0.02	0.02
Painting	4.6 to 31.5	11.1	1.5	1.6	1.6	0.17	0.18	0.18
Interior Partitions	5.5 to 36.6	15.1	3.9	4.3	4.4	0.59	0.65	0.66
Miscellaneous Items	2.5 to 125.0	14.4	5.2	3.7	9.7	0.75	0.53	1.40
Elevators	1.8 to 13.1	5.6	2.1	1.1	2.2	0.12	0.06	0.12
Sprinklers	2.3 to 17.2	6.8	0.5	—	2.0	0.03	—	0.14
Plumbing	2.6 to 12.9	6.7	4.9	7.2	5.2	0.33	0.48	0.35
Heat., Vent., Air Conditioning	3.7 to 20.5	8.6	13.5	11.0	12.9	1.16	0.95	1.11
Electrical	2.8 to 13.1	5.7	10.1	8.4	11.1	0.58	0.48	0.63
Total	1.8% to 125.8%	—	100.0%	100.0%	100.0%	13.88%	12.30%	15.87%
Overall Weighted Average		14.02%						

Workers' Compensation Insurance Rates by States

The table below lists the weighted average workers' compensation base rate for each state with a factor comparing this with the national average of 13.7%.

State	Weighted Average	Factor	State	Weighted Average	Factor	State	Weighted Average	Factor
Alabama	19.5%	142	Kentucky	16.0%	117	North Dakota	9.4%	69
Alaska	16.0	117	Louisiana	16.9	123	Ohio	8.8	64
Arizona	9.5	69	Maine	13.2	96	Oklahoma	14.5	106
Arkansas	9.6	70	Maryland	15.6	114	Oregon	12.2	89
California	26.5	193	Massachusetts	12.3	90	Pennsylvania	17.0	124
Colorado	7.4	54	Michigan	18.9	138	Rhode Island	11.9	87
Connecticut	24.0	175	Minnesota	24.7	180	South Carolina	20.1	147
Delaware	10.4	76	Mississippi	13.8	101	South Dakota	16.1	118
District of Columbia	11.6	85	Missouri	16.9	123	Tennessee	14.6	107
Florida	10.7	78	Montana	11.5	84	Texas	10.5	77
Georgia	30.3	221	Nebraska	18.9	138	Utah	8.4	61
Hawaii	8.8	64	Nevada	9.0	66	Vermont	14.3	104
Idaho	10.7	78	New Hampshire	22.9	167	Virginia	9.4	69
Illinois	22.5	164	New Jersey	14.1	103	Washington	11.2	82
Indiana	5.7	42	New Mexico	16.1	118	West Virginia	10.7	78
Iowa	10.7	78	New York	6.2	45	Wisconsin	14.1	103
Kansas	9.1	66	North Carolina	17.5	128	Wyoming	6.1	45
Weighted Average for U.S. is			14.1% of payroll = 100%					

Rates in the following table are the base or manual costs per $100 of payroll for workers' compensation in each state. Rates are usually applied to straight time wages only and not to premium time wages and bonuses.

The weighted average skilled worker rate for 35 trades is 13.7%. For bidding purposes, apply the full value of workers' compensation directly to total labor costs, or if labor is 38%, materials 42% and overhead and profit 20% of total cost, carry 38/80 × 13.7% = 6.5% of cost (before overhead and profit) into overhead. Rates vary not only from state to state but also with the experience rating of the contractor.

Rates are the most current available at the time of publication.

General Conditions — RH1010-400 Workers' Compensation

Table H1010-403 Workers' Compensation Insurance Rates by Trade and State

State	Carpentry — 3 stories or less	Carpentry — interior cab. work	Carpentry — general	Concrete Work — NOC	Concrete Work — flat (flr, sdwk.)	Electrical Wiring — inside	Excavation — earth NOC	Excavation — rock	Glaziers	Insulation Work	Lathing	Masonry	Painting & Decorating	Pile Driving	Plastering	Plumbing	Roofing	Sheet Metal Work (HVAC)	Steel Erection — door & sash	Steel Erection — inter., ornam.	Steel Erection — structure	Steel Erection — NOC	Tile Work — (interior ceramic)	Waterproofing	Wrecking
	5651	5437	5403	5213	5221	5190	6217	6217	5462	5479	5443	5022	5474	6003	5480	5183	5551	5538	5102	5102	5040	5057	5348	9014	5701
AL	33.16	14.38	24.57	12.56	9.62	9.05	13.51	13.51	17.48	16.31	8.34	22.18	17.10	17.88	15.08	7.53	46.99	9.68	11.01	11.01	42.67	23.81	12.67	9.00	42.67
AK	15.60	10.19	12.62	10.76	10.50	6.72	9.35	9.35	31.18	16.18	10.41	14.44	12.20	30.90	11.89	6.88	28.31	7.92	7.95	7.95	29.64	33.41	5.25	5.38	29.64
AZ	13.84	5.84	16.22	7.78	4.67	4.54	5.48	5.48	6.75	10.41	3.94	8.00	7.92	8.19	5.86	4.82	18.65	6.85	11.02	11.02	20.46	9.79	3.05	3.72	20.46
AR	12.64	5.52	10.90	6.45	5.52	4.56	6.98	6.98	7.80	8.42	4.23	8.42	7.44	9.52	13.32	4.58	18.55	5.18	8.04	8.04	19.86	14.63	6.30	2.90	19.86
CA	36.56	36.56	36.56	18.18	18.18	13.06	15.93	15.93	27.92	15.97	17.81	25.50	23.38	22.56	28.82	6.44	67.92	20.51	16.42	16.42	32.53	20.51	12.92	23.38	20.51
CO	9.14	4.76	5.76	6.29	4.46	2.90	5.10	5.10	5.36	7.05	3.69	7.41	5.85	6.81	5.42	3.94	14.82	3.74	4.51	4.51	26.55	9.93	3.98	2.88	19.25
CT	24.20	17.13	32.36	23.82	14.23	7.81	14.76	14.76	20.15	20.20	10.64	29.90	19.14	20.98	18.42	11.02	51.86	13.90	18.99	18.99	65.52	27.25	14.41	6.82	65.52
DE	10.15	10.15	7.92	8.38	7.06	3.56	6.26	6.26	8.82	7.92	8.82	9.44	10.62	12.26	8.82	5.43	21.11	6.82	10.50	10.50	19.57	10.50	4.61	9.44	19.57
DC	9.43	7.29	8.15	13.63	7.29	4.11	9.26	9.26	12.04	6.67	8.28	10.03	6.01	15.97	8.79	7.69	16.04	6.43	19.51	19.51	22.71	9.28	7.60	4.71	22.71
FL	10.88	8.00	11.48	11.88	5.72	5.65	6.59	6.59	9.56	10.12	4.66	9.77	8.86	17.52	10.48	5.31	17.10	7.37	9.25	9.25	22.79	11.00	5.25	4.78	22.79
GA	61.87	20.95	29.49	21.35	16.06	11.54	17.98	17.98	19.86	18.38	12.52	33.03	31.51	24.06	20.88	12.92	90.59	17.73	22.61	22.61	73.34	31.15	13.64	11.47	73.34
HI	8.48	6.62	14.65	6.90	3.88	4.10	5.15	5.15	8.04	5.53	7.12	7.86	6.88	7.68	7.24	4.84	14.83	4.22	6.73	6.73	27.79	7.53	5.28	5.39	27.79
ID	13.87	7.07	11.31	12.31	5.33	3.88	6.92	6.92	9.01	8.49	4.97	9.39	9.72	8.39	10.73	5.22	22.39	7.91	9.62	9.62	28.90	8.36	6.12	4.95	28.90
IL	23.38	15.06	20.46	28.05	11.77	7.55	10.37	10.37	17.20	16.69	10.75	19.99	12.36	17.62	22.12	9.56	33.65	11.09	20.83	20.83	62.17	19.02	17.57	5.48	62.17
IN	7.76	4.17	6.49	4.81	3.67	2.95	3.93	3.93	5.43	6.15	3.51	5.11	4.63	6.30	3.60	2.64	10.53	4.15	5.10	5.10	11.78	4.77	2.98	2.81	11.78
IA	10.34	8.42	12.90	12.72	8.24	4.47	6.88	6.88	9.20	5.82	4.64	8.40	6.76	8.73	10.98	6.40	19.92	4.80	5.70	5.70	36.26	14.65	6.70	4.60	24.10
KS	14.38	6.41	9.59	8.14	6.41	4.94	4.91	4.91	8.62	7.32	4.14	7.47	8.33	8.31	6.21	5.27	16.38	5.64	6.57	6.57	21.39	14.64	5.42	4.41	14.64
KY	18.75	12.00	24.00	11.00	7.25	5.25	12.50	12.50	19.84	11.27	7.23	7.50	10.00	18.59	12.49	5.50	34.00	12.00	10.67	10.67	48.00	13.78	15.43	4.00	48.00
LA	23.72	13.00	21.51	13.00	10.19	7.26	12.45	12.45	15.57	10.97	6.96	14.17	15.60	20.40	13.16	6.05	35.41	13.54	15.88	15.88	43.26	9.29	8.36	6.48	43.26
ME	12.86	9.34	18.66	15.82	8.40	4.84	10.34	10.34	14.07	13.62	5.16	10.73	13.40	9.74	10.18	6.24	23.06	6.89	7.53	7.53	42.08	13.38	5.77	4.87	42.08
MD	12.84	14.69	13.94	19.92	5.91	6.13	7.34	7.34	13.74	13.95	6.39	10.39	7.28	18.75	9.37	6.93	30.19	11.62	12.36	12.36	62.69	20.35	6.49	5.91	34.01
MA	8.68	5.23	9.61	18.85	6.24	2.84	4.35	4.35	9.58	7.78	5.27	10.55	5.09	12.92	4.68	3.50	30.99	5.72	6.89	6.89	54.08	33.00	5.81	2.48	23.75
MI	18.76	12.20	18.76	17.13	11.09	6.05	14.34	14.34	12.16	12.23	12.20	17.36	15.21	50.15	10.05	7.06	41.77	10.97	12.36	12.36	50.15	11.72	12.17	6.73	50.15
MN	19.08	19.63	32.94	10.69	13.05	5.59	10.82	10.82	35.70	23.79	9.93	14.01	15.96	34.92	9.93	7.70	68.07	18.46	9.02	9.02	125.84	6.99	13.08	7.38	6.99
MS	18.84	9.11	13.10	9.88	6.18	5.44	10.83	10.83	13.36	9.02	5.50	11.97	10.74	14.86	10.19	8.67	34.07	7.00	13.91	13.91	38.39	8.30	8.30	5.37	38.39
MO	20.18	12.02	13.09	14.94	10.39	6.92	11.53	11.53	10.35	12.21	8.01	15.54	11.65	16.04	13.35	9.62	38.70	8.65	12.57	12.57	44.61	30.73	11.93	6.78	44.61
MT	10.19	10.19	16.22	8.28	7.10	4.79	10.08	10.08	10.07	12.78	5.36	9.49	8.64	11.25	8.99	5.69	34.41	5.97	8.60	8.60	20.45	7.77	7.08	6.02	7.77
NE	21.40	15.45	20.10	19.40	10.83	8.43	17.60	17.60	11.63	21.53	8.53	19.73	12.58	15.20	14.93	9.58	36.13	12.95	13.88	13.88	52.18	18.15	8.40	7.08	67.55
NV	9.92	6.70	12.24	9.79	5.97	4.28	8.50	8.50	6.28	7.41	3.32	6.54	5.68	8.18	5.81	5.97	17.37	7.36	7.25	7.25	19.00	11.91	4.01	4.36	11.91
NH	22.39	11.73	18.76	25.40	13.80	5.76	12.76	12.76	13.34	13.60	7.44	18.16	25.71	16.95	11.36	10.77	39.32	7.96	11.58	11.58	125.75	36.22	13.21	7.76	125.75
NJ	16.94	12.03	16.94	13.84	11.18	4.86	9.96	9.96	9.40	11.59	14.39	17.16	11.07	13.95	14.39	6.66	36.02	6.81	11.88	11.88	14.78	12.26	9.40	6.16	16.81
NM	24.37	9.11	19.35	12.09	10.28	7.25	7.11	7.11	11.76	10.64	6.21	14.95	11.19	13.88	14.09	6.83	37.02	9.49	11.71	11.71	53.67	22.25	7.23	6.93	53.67
NY	4.33	3.54	6.47	7.73	5.97	2.75	3.94	3.94	6.58	3.77	4.72	6.95	4.90	7.10	3.33	3.81	11.62	4.70	6.32	6.32	11.22	6.07	3.67	3.28	6.06
NC	15.50	11.55	18.63	17.12	8.14	9.96	13.11	13.11	11.87	12.57	9.34	11.05	11.85	16.82	13.63	8.92	33.88	11.87	11.97	11.97	70.40	20.61	8.12	5.41	70.40
ND	9.54	9.54	9.54	5.43	5.43	3.75	4.13	4.13	9.54	9.54	8.78	6.85	5.64	9.33	8.78	5.76	16.48	5.76	9.33	9.33	9.33	9.33	9.54	16.48	9.04
OH	7.73	4.31	5.52	4.67	4.92	3.43	4.79	4.79	8.83	11.38	36.65	6.93	8.15	6.99	7.80	3.52	14.89	4.00	6.58	6.58	11.33	5.84	4.85	4.58	5.84
OK	18.56	9.92	13.23	13.71	8.14	7.02	11.13	11.13	12.71	12.47	5.73	15.58	11.38	12.79	17.44	7.23	24.19	10.52	11.32	11.32	33.69	18.85	7.23	6.79	33.69
OR	22.03	8.09	13.21	9.91	9.68	5.04	9.25	9.25	11.50	11.30	6.46	13.57	12.29	11.17	10.00	6.02	25.38	6.95	7.37	7.37	21.11	12.91	7.22	6.11	21.11
PA	17.84	17.84	13.93	17.58	12.59	6.86	10.29	10.29	13.07	13.93	13.07	14.91	15.32	17.84	13.07	8.68	34.96	8.55	17.39	17.39	27.42	17.39	10.29	14.94	27.42
RI	9.69	10.13	12.42	13.18	10.51	4.12	8.56	8.56	10.26	13.56	4.93	8.95	9.48	20.95	8.02	5.77	19.37	6.98	6.06	6.06	28.84	20.64	4.60	4.61	24.05
SC	24.05	18.13	22.49	16.89	9.59	11.45	11.78	11.78	16.99	18.58	11.06	13.93	15.50	20.61	15.75	12.36	50.08	13.06	13.91	13.91	41.22	32.53	10.50	6.33	41.22
SD	20.93	15.47	19.14	18.15	9.09	5.14	9.75	9.75	12.75	14.72	6.60	13.21	10.64	14.24	10.47	11.11	35.39	8.50	13.06	13.06	39.43	23.87	5.91	5.78	39.43
TN	23.85	11.59	12.99	12.86	6.71	6.45	12.53	12.53	12.51	11.17	7.83	13.19	9.84	18.04	12.47	6.13	34.24	8.48	12.29	12.29	21.99	22.19	8.37	4.68	21.99
TX	9.68	7.76	9.68	8.61	6.25	5.65	7.33	7.33	7.82	10.32	4.91	8.32	8.04	12.94	8.04	5.47	17.96	13.94	7.16	7.16	29.32	11.95	5.11	5.97	8.11
UT	12.88	5.69	8.30	6.49	5.64	3.19	7.02	7.02	6.89	5.87	5.20	7.33	6.66	6.78	5.02	3.49	20.72	5.58	5.07	5.07	22.57	10.74	4.22	4.58	18.77
VT	15.08	11.19	16.10	12.21	10.23	4.29	10.77	10.77	14.25	15.36	5.94	12.50	8.94	11.44	9.33	7.98	27.37	7.17	10.63	10.63	40.22	21.76	6.92	7.60	40.22
VA	10.37	7.72	8.56	9.91	5.11	4.15	7.23	7.23	7.71	6.23	6.03	7.08	8.23	7.98	6.24	4.64	21.77	5.23	6.70	6.70	28.82	11.93	4.63	2.80	28.82
WA	9.36	9.36	9.36	8.89	8.84	3.25	6.31	6.31	13.96	7.06	9.36	12.14	14.60	21.48	11.32	4.26	20.35	4.48	8.15	8.15	8.15	8.15	7.73	20.35	8.15
WV	13.06	8.27	9.67	9.38	5.25	5.27	7.90	7.90	9.46	7.92	5.44	8.80	8.73	12.01	8.28	5.23	24.07	6.86	7.49	7.49	28.94	13.43	6.48	3.54	29.69
WI	11.59	12.19	16.32	11.25	8.29	5.40	7.88	7.88	12.04	12.28	4.23	15.07	12.86	27.64	10.08	6.29	26.71	9.04	11.42	11.42	20.66	22.20	15.04	5.37	20.66
WY	5.50	5.50	5.50	5.50	5.50	5.50	5.50	5.50	5.50	5.50	5.50	5.50	5.50	5.50	5.50	5.50	5.50	5.50	5.50	5.50	5.50	5.50	5.50	5.50	5.50
AVG.	16.40	10.76	15.13	12.62	8.36	5.68	9.20	9.20	12.46	11.44	7.89	12.48	11.12	15.32	10.91	6.66	29.63	8.56	10.55	10.55	36.45	16.12	7.97	6.57	31.38

General Conditions — RH1010-400 Workers' Compensation

Table H1010-404 Workers' Compensation (Canada in Canadian dollars)

Province		Alberta	British Columbia	Manitoba	Ontario	New Brunswick	Newfndld. & Labrador	Northwest Territories	Nova Scotia	Prince Edward Island	Quebec	Saskatchewan	Yukon
Carpentry—3 stories or less	Rate	5.89	3.35	4.04	4.44	3.80	8.38	4.16	5.71	6.55	12.17	3.54	6.63
	Code	42143	721028	40102	723	238130	4226	4-41	4226	403	80110	B1226	202
Carpentry—interior cab. work	Rate	2.21	2.75	4.04	4.44	3.80	4.10	4.16	5.39	3.89	12.17	2.03	6.63
	Code	42133	721021	40102	723	238350	4279	4-41	4274	402	80110	B1127	202
CARPENTRY—general	Rate	5.89	3.35	4.04	4.44	3.80	4.10	4.16	5.71	6.55	12.17	3.54	6.63
	Code	42143	721028	40102	723	238130	4299	4-41	4226	403	80110	B1202	202
CONCRETE WORK—NOC	Rate	3.86	3.24	6.35	17.86	3.80	8.38	4.16	5.04	3.89	12.32	3.38	6.63
	Code	42104	721010	40110	748	238110	4224	4-41	4224	402	80100	B1314	203
CONCRETE WORK—flat (flr. sidewalk)	Rate	3.86	3.24	6.35	17.86	3.80	8.38	4.16	5.04	3.89	12.32	3.38	6.63
	Code	42104	721010	40110	748	238110	4224	4-41	4224	402	80100	B1314	203
ELECTRICAL Wiring—inside	Rate	1.66	1.81	1.70	3.60	1.73	2.42	2.45	2.14	3.89	4.94	2.03	3.89
	Code	42124	721019	40203	704	238210	4261	4-46	4261	402	80170	B1105	206
EXCAVATION—earth NOC	Rate	2.13	3.55	4.09	5.16	3.80	3.22	3.68	3.38	3.89	7.71	2.17	3.89
	Code	40604	721031	40706	711	238190	4214	4-43	4214	402	80030	R1106	207
EXCAVATION—rock	Rate	2.13	3.55	4.09	5.16	3.80	3.22	3.68	3.38	3.89	7.71	2.17	3.89
	Code	40604	721031	40706	711	238190	4214	4-43	4214	402	80030	R1106	207
GLAZIERS	Rate	2.38	3.15	4.04	10.00	3.80	5.14	4.16	5.71	3.89	13.74	3.38	3.89
	Code	42121	715020	40109	751	238150	4233	4-41	4233	402	80150	B1304	212
INSULATION WORK	Rate	1.81	8.22	4.04	10.00	3.80	5.14	4.16	5.71	3.89	12.17	2.54	6.63
	Code	42184	721029	40102	751	238310	4234	4-41	4234	402	80110	B1207	202
LATHING	Rate	4.16	6.65	4.04	4.44	3.80	4.10	4.16	5.39	3.89	12.17	3.38	6.63
	Code	42135	721042	40102	723	238390	4279	4-41	4271	402	80110	B1316	202
MASONRY	Rate	3.86	2.90	4.04	12.39	3.80	5.14	4.16	5.71	3.89	12.32	3.38	6.63
	Code	42102	721037	40102	741	238140	4231	4-41	4231	402	80100	B1318	202
PAINTING & DECORATING	Rate	3.51	3.30	3.39	7.33	3.80	4.10	4.16	5.39	3.89	12.17	3.54	6.63
	Code	42111	721041	40105	719	238320	4275	4-41	4275	402	80110	B1201	202
PILE DRIVING	Rate	3.82	3.47	4.09	6.86	3.80	3.22	3.68	5.04	6.55	7.71	3.38	6.63
	Code	42159	722004	40706	732	238190	4129	4-43	4221	403	80030	B1310	202
PLASTERING	Rate	4.16	6.65	4.64	7.33	3.80	4.10	4.16	5.39	3.89	12.17	3.54	6.63
	Code	42135	721042	40108	719	238390	4271	4-41	4271	402	80110	B1221	202
PLUMBING	Rate	1.66	2.39	2.93	4.06	1.73	2.50	2.45	3.38	2.20	6.38	2.03	3.89
	Code	42122	721043	40204	707	238220	4241	4-46	4241	401	80160	B1101	214
ROOFING	Rate	5.53	5.19	6.85	14.44	3.80	8.38	4.16	7.06	3.89	19.12	3.38	6.63
	Code	42118	721036	40403	728	238160	4236	4-41	4236	402	80130	B1320	202
SHEET METAL WORK (HVAC)	Rate	1.66	2.39	6.85	4.06	1.73	2.50	2.45	2.78	2.20	6.38	2.03	3.89
	Code	42117	721043	40402	707	238220	4244	4-46	4244	401	80160	B1107	208
STEEL ERECTION—door & sash	Rate	1.81	11.51	7.42	17.86	3.80	8.38	4.16	5.71	3.89	18.60	3.38	6.63
	Code	42106	722005	40502	748	238120	4227	4-41	4227	402	80080	B1322	202
STEEL ERECTION—inter., ornam.	Rate	1.81	11.51	7.42	17.86	3.80	8.38	4.16	5.71	3.89	18.68	3.38	6.63
	Code	42106	722005	40502	748	238120	4227	4-41	4227	402	80080	B1322	202
STEEL ERECTION—structure	Rate	1.81	11.51	7.42	17.86	3.80	8.38	4.16	5.71	3.89	18.68	3.38	6.63
	Code	42106	722005	40502	748	238120	4227	4-41	4227	402	80080	B1322	202
STEEL ERECTION—NOC	Rate	1.81	11.51	7.42	17.86	3.80	8.38	4.16	5.71	3.89	18.68	3.38	6.63
	Code	42106	722005	40502	748	238120	4227	4-41	4227	402	80080	B1322	202
TILE WORK—inter. (ceramic)	Rate	2.84	3.44	1.87	7.33	3.80	4.10	4.16	4.83	3.89	12.17	3.38	6.63
	Code	42113	721054	40103	719	238340	4276	4-41	4276	402	80110	B1301	202
WATERPROOFING	Rate	3.51	3.29	4.04	4.44	3.80	4.10	4.16	5.71	3.89	19.12	3.54	6.63
	Code	42139	721016	40102	723	238190	4299	4-41	4239	402	80130	B1217	202
WRECKING	Rate	2.13	5.32	6.86	17.86	3.80	3.22	3.68	3.38	3.89	12.17	2.17	6.63
	Code	40604	721005	40106	748	238190	4211	4-43	4211	402	80110	R1125	202

General Conditions — RH1020-300 Overhead & Miscellaneous Data

Unemployment Taxes and Social Security Taxes

State unemployment tax rates vary not only from state to state, but also with the experience rating of the contractor. The federal unemployment tax rate is 6.2% of the first $7,000 of wages. This is reduced by a credit of up to 5.4% for timely payment to the state. The minimum federal unemployment tax is 0.8% after all credits.

Social security (FICA) for 2013 is estimated at time of publication to be 7.65% of wages up to $114,000.

General Conditions — RH1020-400 Overtime

Overtime

One way to improve the completion date of a project or eliminate negative float from a schedule is to compress activity duration times. This can be achieved by increasing the crew size or working overtime with the proposed crew.

To determine the costs of working overtime to compress activity duration times, consider the following examples. Below is an overtime efficiency and cost chart based on a five, six, or seven day week with an eight through twelve hour day. Payroll percentage increases for time and one half and double time are shown for the various working days.

Days per Week	Hours per Day	Production Efficiency					Payroll Cost Factors	
		1 Week	2 Weeks	3 Weeks	4 Weeks	Average 4 Weeks	@ 1-1/2 Times	@ 2 Times
5	8	100%	100%	100%	100%	100 %	100 %	100 %
	9	100	100	95	90	96.25	105.6	111.1
	10	100	95	90	85	91.25	110.0	120.0
	11	95	90	75	65	81.25	113.6	127.3
	12	90	85	70	60	76.25	116.7	133.3
6	8	100	100	95	90	96.25	108.3	116.7
	9	100	95	90	85	92.50	113.0	125.9
	10	95	90	85	80	87.50	116.7	133.3
	11	95	85	70	65	78.75	119.7	139.4
	12	90	80	65	60	73.75	122.2	144.4
7	8	100	95	85	75	88.75	114.3	128.6
	9	95	90	80	70	83.75	118.3	136.5
	10	90	85	75	65	78.75	121.4	142.9
	11	85	80	65	60	72.50	124.0	148.1
	12	85	75	60	55	68.75	126.2	152.4

General Conditions — RH1030-100 General

Sales Tax by State

State sales tax on materials is tabulated below (5 states have no sales tax). Many states allow local jurisdictions, such as a county or city, to levy additional sales tax.

Some projects may be sales tax exempt, particularly those constructed with public funds.

State	Tax (%)	State	Tax (%)	State	Tax (%)	State	Tax (%)
Alabama	4	Illinois	6.25	Montana	0	Rhode Island	7
Alaska	0	Indiana	7	Nebraska	5.5	South Carolina	6
Arizona	6.6	Iowa	6	Nevada	6.85	South Dakota	4
Arkansas	6	Kansas	5.3	New Hampshire	0	Tennessee	7
California	7.25	Kentucky	6	New Jersey	7	Texas	6.25
Colorado	2.9	Louisiana	4	New Mexico	5.13	Utah	4.75
Connecticut	6	Maine	5	New York	4	Vermont	6
Delaware	0	Maryland	6	North Carolina	5.75	Virginia	4
District of Columbia	6	Massachusetts	6.25	North Dakota	5	Washington	6.5
Florida	6	Michigan	6	Ohio	5.5	West Virginia	6
Georgia	4	Minnesota	6.88	Oklahoma	4.5	Wisconsin	5
Hawaii	4	Mississippi	7	Oregon	0	Wyoming	4
Idaho	6	Missouri	4.23	Pennsylvania	6	Average	5.03 %

General Conditions — RH1040-100 Scaffolding

Table H1040-101 Steel Tubular Scaffolding

On new construction, tubular scaffolding is efficient up to 60' high or five stories. Above this it is usually better to use a hung scaffolding if construction permits. Swing scaffolding operations may interfere with tenants. In this case, the tubular is more practical at all heights.

In repairing or cleaning the front of an existing building, the cost of tubular scaffolding per S.F. of building front increases as the height increases above the first tier. The first tier cost is relatively high due to leveling and alignment.

The minimum efficient crew for erection is three workers. For heights over 50', a crew of four is more efficient. Use two or more on top and two at the bottom for handing up or hoisting. Four workers can erect and dismantle about nine frames per hour up to five stories. From five to eight stories, they will average six frames per hour. With 7' horizontal spacing, this will run about 400 S.F. and 265 S.F. of wall surface, respectively. Time for placing planks must be added to the above. On heights above 50', five planks can be placed per labor-hour.

The cost per 1,000 S.F. of building front in the table below was developed by pricing the materials required for a typical tubular scaffolding system eleven frames long and two frames high. Planks were figured five wide for standing plus two wide for materials.

Frames are 5' wide and usually spaced 7' O.C. horizontally. Sidewalk frames are 6' wide. Rental rates will be lower for jobs over three months' duration.

For jobs under twenty-five frames, add 50% to rental cost. These figures do not include accessories which are listed separately below. Large quantities for long periods can reduce rental rates by 20%.

Item	Unit	Monthly Rent	Per 1,000 S.F. of Building Front — No. of Pieces	Per 1,000 S.F. of Building Front — Rental per Month
5' Wide Standard Frame, 6'-4" High	Ea.	$ 5.15	24	$123.60
Leveling Jack & Plate		2.00	24	48.00
Cross Brace		1.00	44	44.00
Side Arm Bracket, 21"		2.00	12	24.00
Guardrail Post		1.00	12	12.00
Guardrail, 7' section		1.00	22	22.00
Stairway Section		40.00	2	80.00
Walk-Thru Frame Guardrail		2.50	2	5.00
			Total	$358.60
			Per C.S.F., 1 Use/Mo.	$ 35.86

Scaffolding is often used as falsework over 15' high during construction of cast-in-place concrete beams and slabs. Two-foot wide scaffolding is generally used for heavy beam construction. The span between frames depends upon the load to be carried, with a maximum span of 5'.

Heavy duty shoring frames with a capacity of 10,000#/leg can be spaced up to 10' O. C., depending upon form support design and loading.

Scaffolding used as horizontal shoring requires less than half the material required with conventional shoring.

On new construction, erection is done by carpenters.

Rolling towers supporting horizontal shores can reduce labor and speed the job. For maintenance work, catwalks with spans up to 70' can be supported by the rolling towers.

City Adjustments RJ1010-010

General: The following information on current city cost indexes is calculated for over 930 zip code locations in the United States and Canada. Index figures for both material and installation are based upon the 30 major city average of 100 and represent the cost relationship on July 1, 2013.

In addition to index adjustment, the user should consider:
1. productivity
2. management efficiency
3. competitive conditions
4. automation
5. restrictive union practices
6. unique local requirements
7. regional variations due to specific building codes

The weighted-average index is calculated from about 66 materials and equipment types and 21 building trades. The component contribution of these in a model building is tabulated in Table J1010-011 below.

If the systems component distribution of a building is unknown, the weighted-average index can be used to adjust the cost for any city.

Table J1010-011 Labor, Material, and Equipment Cost Distribution for Weighted Average Listed by System Division

Division No.	Building System	Percentage	Division No.	Building System	Percentage
A	Substructure	6.2%	D10	Services: Conveying	4.3%
B10	Shell: Superstructure	18.8	D20-40	Mechanical	22.7
B20	Exterior Closure	11.6	D50	Electrical	11.5
B30	Roofing	2.9	E	Equipment & Furnishings	2.3
C	Interior Construction	15.5	G	Site Work	4.2
				Total weighted average (Div. A-G)	100.0%

How to Use the Component Indexes: Table J1010-012 below shows how the average costs obtained from this book for each division should be adjusted for your particular building in your city. The example is a building adjusted for New York, NY. Indexes for other cities are tabulated in Division RJ1030. These indexes should also be used when compiling data on the form in Division K1010.

Table J1010-012 Adjustment of "Book" Costs to a Particular City

Systems Division Number	System Description	New York, NY							
		City Cost Index	Cost Factor	Book Cost	Adjusted Cost	City Cost Index	Cost Factor	Book Cost	Adjusted Cost
A	Substructure	142.8	1.428	$ 97,900	$ 139,800				
B10	Shell: Superstructure	133.1	1.331	270,400	359,900				
B20	Exterior Closure	136.9	1.369	183,300	250,900				
B30	Roofing	126.4	1.264	40,400	51,100				
C	Interior Construction	134.2	1.342	260,900	350,100				
D10	Services: Conveying	108.2	1.082	52,200	56,500				
D20-40	Mechanical	127.7	1.277	323,800	413,500				
D50	Electrical	143.4	1.434	181,200	259,800				
E	Equipment	104.4	1.044	27,400	28,600				
G	Site Work	122.4	1.224	77,000	94,200				
Total Cost, (Div. A-G)				$1,514,500	$2,004,400				

Alternate Method: Use the weighted-average index for your city rather than the component indexes. **For example:**

(Total weighted-average index A-G)/100 x (total cost) = Adjusted city cost
New York, NY 1.319 x $1,514,500 = $1,997,600

City Adjustments RJ1020-010

City to City: To convert known or estimated costs from one city to another, cost indexes can be used as follows:

$$\text{Unknown City Cost} = \text{Known Cost} \times \frac{\text{Unknown City Index}}{\text{Known City Index}}$$

For example: If the building cost in Boston, MA, is $2,000,000, how much would a duplicated building cost in Los Angeles, CA?

$$\text{L.A. Cost} = \$2,000,000 \times \frac{\text{(Los Angeles) } 106.9}{\text{(Boston) } 117.5} = \$1,819,500$$

The table below lists both the Means City Cost Index based on January 1, 1993 = 100 as well as the computed value of an index based on January 1, 2013 costs. Since the January 1, 2013 figure is estimated, space is left to write in the actual index figures as they become available through either the quarterly *Means Construction Cost Indexes* or as printed in the *Engineering News-Record*. To compute the actual index based on January 1, 2013 = 100, divide the Quarterly City Cost Index for a particular year by the actual January 1, 2013 Quarterly City Cost Index. Space has been left to advance the index figures as the year progresses.

Table J1020-011 Historical Cost Indexes

Year	Historical Cost Index Jan. 1, 1993 = 100		Current Index Based on Jan. 1, 2013 = 100		Year	Historical Cost Index Jan. 1, 1993 = 100	Current Index Based on Jan. 1, 2013 = 100		Year	Historical Cost Index Jan. 1, 1993 = 100	Current Index Based on Jan. 1, 2013 = 100	
	Est.	Actual	Est.	Actual		Actual	Est.	Actual		Actual	Est.	Actual
Oct 2013*					July 1998	115.1	59.7		July 1980	62.9	31.8	
July 2013*					1997	112.8	57.1		1979	57.8	29.3	
April 2013*					1996	110.2	55.8		1978	53.5	27.1	
Jan 2013*	197.6		100.0	100.0	1995	107.6	54.5		1977	49.5	25.1	
July 2012		194.6	98.5		1994	104.4	52.8		1976	46.9	23.7	
2011		191.2	96.8		1993	101.7	51.5		1975	44.8	22.7	
2010		183.5	92.9		1992	99.4	50.3		1974	41.4	21.0	
2009		180.1	91.1		1991	96.8	49.0		1973	37.7	19.1	
2008		180.4	91.3		1990	94.3	47.7		1972	34.8	17.6	
2007		169.4	85.7		1989	92.1	46.6		1971	32.1	16.2	
2006		162.0	82.0		1988	89.9	45.5		1970	28.7	14.5	
2005		151.6	76.7		1987	87.7	44.4		1969	26.9	13.6	
2004		143.7	72.7		1986	84.2	42.6		1968	24.9	12.6	
2003		132.0	66.8		1985	82.6	41.8		1967	23.5	11.9	
2002		128.7	65.1		1984	82.0	41.5		1966	22.7	11.5	
2001		125.1	63.3		1983	80.2	40.6		1965	21.7	11.0	
2000		120.9	61.2		1982	76.1	38.5		1964	21.2	10.7	
▼ 1999		117.6	59.5		▼ 1981	70.0	35.4		▼ 1963	20.7	10.5	

To find the **current cost** from a project built previously in either the same city or a different city, the following formula is used:

$$\text{Present Cost (City X)} = \frac{\text{Current HCI} \times \text{CCI (City X)}}{\text{Previous HCI} \times \text{CCI (City Y)}} \times \text{Former Cost (City Y)}$$

For example: Find the construction cost of a building to be built in San Francisco, CA, as of January 1, 2013 when the identical building cost $500,000 in Boston, MA, on July 1, 1968.

$$\text{Jan. 1, 2013 (San Francisco)} = \frac{\text{(San Francisco) } 197.6 \times 122.5}{\text{(Boston) } 24.9 \times 118.1} \times \$500,000 = \$4,115,500$$

Note: The City Cost Indexes for Canada can be used to convert U.S. national averages to local costs in Canadian dollars.

To Project Future Construction Costs: Using the results of the last five years average percentage increase as a basis, an average increase of 4.4% could be used.

The historical index figures above are compiled from the Means Construction Index Service.

City Cost Indexes

RJ1030-010 Building Systems

DIV. NO.	BUILDING SYSTEMS	ALABAMA														
		BIRMINGHAM			HUNTSVILLE			MOBILE			MONTGOMERY			TUSCALOOSA		
		MAT.	INST.	TOTAL	MAT.	INST.	TOTAL	MAT.	INST.	TOTAL	MAT.	INST.	TOTAL	MAT.	INST.	TOTAL
A	Substructure	96.8	77.1	84.8	90.7	72.6	79.7	96.6	71.0	81.0	97.5	61.7	75.7	93.0	64.6	75.7
B10	Shell: Superstructure	95.3	82.8	90.3	95.2	78.3	88.5	95.2	78.5	88.5	95.1	72.3	86.0	95.1	73.6	86.5
B20	Exterior Closure	92.8	78.5	86.0	92.7	60.8	77.6	93.0	57.6	76.2	92.2	45.5	70.0	92.2	52.8	73.4
B30	Roofing	95.5	81.4	90.2	94.9	75.5	87.7	94.9	74.5	87.3	94.7	67.2	84.5	95.0	70.6	85.9
C	Interior Construction	96.7	68.4	84.4	96.5	62.2	81.6	98.8	56.3	80.3	98.5	44.1	74.9	96.5	47.3	75.1
D10	Services: Conveying	100.0	91.4	97.6	100.0	88.2	96.7	100.0	85.5	96.0	100.0	85.5	96.0	100.0	87.3	96.5
D20-40	Mechanical	99.9	64.7	85.5	99.9	54.5	81.3	99.9	63.5	85.0	100.0	35.8	73.7	99.9	35.6	73.6
D50	Electrical	103.3	61.3	81.2	98.8	64.4	80.7	98.8	59.6	78.2	99.4	63.4	80.5	98.3	61.3	78.8
E	Equipment & Furnishings	100.0	69.2	98.4	100.0	63.6	98.1	100.0	54.9	97.7	100.0	44.6	97.1	100.0	46.7	97.2
G	Site Work	87.2	94.9	92.4	82.7	94.0	90.3	94.7	89.1	90.9	93.6	89.0	90.5	83.2	94.0	90.5
A-G	WEIGHTED AVERAGE	97.4	73.8	87.0	96.6	67.4	83.7	97.5	66.8	84.0	97.5	56.1	79.2	96.5	58.1	79.6

DIV. NO.	BUILDING SYSTEMS	ALASKA									ARIZONA					
		ANCHORAGE			FAIRBANKS			JUNEAU			FLAGSTAFF			MESA/TEMPE		
		MAT.	INST.	TOTAL	MAT.	INST.	TOTAL	MAT.	INST.	TOTAL	MAT.	INST.	TOTAL	MAT.	INST.	TOTAL
A	Substructure	140.8	119.5	127.9	134.3	119.6	125.3	132.9	119.5	124.8	94.6	78.3	84.7	95.6	77.3	84.5
B10	Shell: Superstructure	125.8	109.2	119.2	122.3	109.3	117.1	121.6	109.2	116.7	95.5	74.0	86.9	93.0	73.4	85.2
B20	Exterior Closure	146.7	121.5	134.7	138.9	121.5	130.6	135.5	121.5	128.9	117.0	63.9	91.8	107.6	62.3	86.1
B30	Roofing	186.7	115.8	160.3	180.1	117.2	156.7	177.3	115.8	154.4	96.0	67.9	85.6	100.2	64.4	86.9
C	Interior Construction	131.7	119.7	126.5	126.9	120.2	124.0	124.7	119.7	122.6	102.8	62.5	85.3	98.8	63.5	83.5
D10	Services: Conveying	100.0	104.2	101.2	100.0	104.2	101.2	100.0	104.2	101.2	100.0	86.2	96.2	100.0	85.9	96.1
D20-40	Mechanical	100.3	104.3	101.9	100.3	107.6	103.3	100.3	103.2	101.5	100.2	81.6	92.6	100.2	75.5	90.1
D50	Electrical	133.9	116.5	124.8	145.4	116.5	130.2	130.4	116.5	123.1	100.3	61.1	79.6	92.7	61.1	76.0
E	Equipment & Furnishings	100.0	117.7	100.9	100.0	117.2	100.9	100.0	117.7	100.9	100.0	66.5	98.3	100.0	66.3	98.3
G	Site Work	124.7	129.7	128.1	117.4	129.7	125.8	124.3	129.7	128.0	83.1	95.9	91.7	90.5	97.3	95.1
A-G	WEIGHTED AVERAGE	123.8	114.6	119.8	122.0	115.4	119.1	119.7	114.3	117.3	100.7	72.7	88.4	98.1	71.3	86.3

DIV. NO.	BUILDING SYSTEMS	ARIZONA									ARKANSAS					
		PHOENIX			PRESCOTT			TUCSON			FORT SMITH			JONESBORO		
		MAT.	INST.	TOTAL	MAT.	INST.	TOTAL	MAT.	INST.	TOTAL	MAT.	INST.	TOTAL	MAT.	INST.	TOTAL
A	Substructure	95.5	79.6	85.8	92.6	63.2	74.7	92.9	79.3	84.6	90.2	68.2	76.8	87.9	66.4	74.8
B10	Shell: Superstructure	94.1	75.4	86.7	95.3	60.5	81.4	93.1	74.6	85.8	98.5	66.4	85.7	92.6	68.2	82.9
B20	Exterior Closure	101.7	66.7	85.1	103.4	46.9	76.5	100.3	64.0	83.1	100.3	52.1	77.4	98.8	47.5	74.4
B30	Roofing	99.9	67.9	88.0	94.8	49.9	78.1	101.2	63.8	87.3	98.3	54.9	82.2	102.2	52.4	83.7
C	Interior Construction	100.8	66.6	86.0	101.6	43.9	76.6	95.7	64.2	82.1	94.1	58.8	78.8	92.4	51.1	74.5
D10	Services: Conveying	100.0	86.5	96.2	100.0	83.7	95.5	100.0	86.5	96.2	100.0	80.2	94.5	100.0	46.6	85.2
D20-40	Mechanical	100.1	81.6	92.5	100.2	69.9	87.8	100.2	72.2	88.7	100.3	50.7	80.0	100.5	53.6	81.3
D50	Electrical	99.8	69.5	83.9	99.9	61.0	79.4	94.9	58.7	75.8	94.9	67.3	80.3	102.3	62.2	81.2
E	Equipment & Furnishings	100.0	69.4	98.4	100.0	43.1	97.1	100.0	68.9	98.4	100.0	59.4	97.9	100.0	50.3	97.4
G	Site Work	90.9	97.8	95.5	72.3	94.0	86.9	86.5	97.6	94.0	78.4	86.2	83.7	96.0	100.2	98.8
A-G	WEIGHTED AVERAGE	98.7	75.4	88.4	98.5	61.1	82.0	96.9	70.9	85.5	97.4	61.8	81.7	97.0	60.0	80.7

DIV. NO.	BUILDING SYSTEMS	ARKANSAS									CALIFORNIA					
		LITTLE ROCK			PINE BLUFF			TEXARKANA			ANAHEIM			BAKERSFIELD		
		MAT.	INST.	TOTAL	MAT.	INST.	TOTAL	MAT.	INST.	TOTAL	MAT.	INST.	TOTAL	MAT.	INST.	TOTAL
A	Substructure	91.1	69.1	77.7	84.2	69.1	75.0	91.2	56.5	70.0	93.6	117.9	108.4	97.6	116.8	109.3
B10	Shell: Superstructure	95.1	66.4	83.7	95.3	66.3	83.7	91.1	57.2	77.6	100.4	110.4	104.4	97.1	109.7	102.1
B20	Exterior Closure	98.9	52.5	76.8	112.3	52.5	83.9	97.3	37.8	69.0	95.7	116.0	105.4	101.4	114.0	107.4
B30	Roofing	95.9	55.6	80.9	97.3	55.6	81.8	98.0	46.8	79.0	100.0	124.1	109.0	101.2	112.2	105.3
C	Interior Construction	95.2	62.2	80.9	90.1	62.2	78.0	93.8	43.2	71.9	101.5	118.1	108.7	102.3	116.9	108.6
D10	Services: Conveying	100.0	80.2	94.5	100.0	80.2	94.5	100.0	32.8	81.3	100.0	109.9	102.8	100.0	101.7	100.5
D20-40	Mechanical	100.1	58.2	82.9	100.3	53.0	80.9	100.3	56.3	82.3	100.0	116.6	106.8	100.0	107.7	103.2
D50	Electrical	102.6	68.4	84.6	94.2	68.4	80.6	96.2	37.5	65.3	92.4	105.2	99.1	102.5	100.1	101.3
E	Equipment & Furnishings	100.0	64.8	98.2	100.0	64.8	98.2	100.0	42.8	97.0	100.0	120.4	101.1	100.0	119.9	101.0
G	Site Work	84.0	86.2	85.5	80.0	86.2	84.2	90.3	86.8	88.0	97.3	108.1	104.6	98.6	105.7	103.4
A-G	WEIGHTED AVERAGE	97.5	64.2	82.8	97.1	63.3	82.1	96.0	50.6	76.0	98.5	113.7	105.4	100.1	109.8	104.4

DIV. NO.	BUILDING SYSTEMS	CALIFORNIA														
		FRESNO			LOS ANGELES			OAKLAND			OXNARD			REDDING		
		MAT.	INST.	TOTAL	MAT.	INST.	TOTAL	MAT.	INST.	TOTAL	MAT.	INST.	TOTAL	MAT.	INST.	TOTAL
A	Substructure	96.3	119.1	110.2	98.4	117.9	110.2	111.7	126.5	120.7	100.5	116.6	110.3	108.2	119.0	114.8
B10	Shell: Superstructure	100.8	112.1	105.3	98.3	112.2	103.8	102.5	116.9	108.2	97.7	110.0	102.6	102.8	112.2	106.7
B20	Exterior Closure	105.5	121.3	113.0	99.6	119.6	109.1	116.4	133.8	124.7	103.1	113.6	108.0	106.6	116.6	111.4
B30	Roofing	93.8	113.5	101.1	102.2	123.0	109.9	107.8	135.8	118.2	103.2	121.0	109.8	103.6	116.0	108.2
C	Interior Construction	104.4	131.1	116.0	101.0	118.2	108.5	102.4	144.3	120.6	101.7	117.7	108.6	103.8	127.6	114.1
D10	Services: Conveying	100.0	120.8	105.8	100.0	110.2	102.8	100.0	122.8	106.3	100.0	110.2	102.8	100.0	120.8	105.8
D20-40	Mechanical	100.1	111.3	104.7	100.1	116.7	106.9	100.1	145.6	118.7	100.0	116.7	106.8	100.0	107.0	102.9
D50	Electrical	91.8	101.5	96.9	99.8	120.8	110.8	106.1	140.8	124.4	96.4	105.4	101.2	99.7	101.8	100.8
E	Equipment & Furnishings	100.0	133.4	101.7	100.0	121.0	101.1	100.0	152.6	102.7	100.0	120.2	101.0	100.0	133.3	101.7
G	Site Work	101.1	105.0	103.8	99.7	107.7	105.1	123.8	105.0	111.1	101.4	103.8	103.0	104.4	104.6	104.6
A-G	WEIGHTED AVERAGE	100.3	115.0	106.8	99.8	116.6	107.2	104.6	133.3	117.3	99.9	112.9	105.6	102.4	113.1	107.1

DIV. NO.	BUILDING SYSTEMS	CALIFORNIA														
		RIVERSIDE			SACRAMENTO			SAN DIEGO			SAN FRANCISCO			SAN JOSE		
		MAT.	INST.	TOTAL	MAT.	INST.	TOTAL	MAT.	INST.	TOTAL	MAT.	INST.	TOTAL	MAT.	INST.	TOTAL
A	Substructure	97.7	117.4	109.7	96.1	121.8	111.8	100.3	107.5	104.7	116.9	128.9	124.2	113.3	125.5	120.7
B10	Shell: Superstructure	101.5	110.4	105.0	96.4	110.9	102.1	98.9	104.9	101.3	108.4	122.8	114.1	99.3	121.3	108.0
B20	Exterior Closure	92.4	115.7	103.5	106.1	122.2	113.8	100.7	112.6	106.3	117.5	137.2	126.9	115.5	134.5	124.5
B30	Roofing	100.2	124.1	109.1	116.6	122.2	118.6	101.2	105.1	102.6	109.6	137.6	120.0	99.6	139.2	114.3
C	Interior Construction	101.5	118.1	108.7	108.2	131.6	118.3	98.2	108.4	102.6	105.8	145.6	123.1	96.1	144.2	117.0
D10	Services: Conveying	100.0	109.9	102.8	100.0	121.1	105.9	100.0	109.1	102.5	100.0	123.3	106.5	100.0	122.4	106.2
D20-40	Mechanical	100.0	116.6	108.8	100.1	119.3	108.1	100.0	115.2	106.2	100.1	167.3	127.6	100.0	144.8	118.3
D50	Electrical	92.5	106.5	99.9	101.2	113.3	107.6	99.2	99.1	99.1	106.1	157.1	133.0	102.5	153.2	129.2
E	Equipment & Furnishings	100.0	120.5	101.1	100.0	136.9	101.9	100.0	105.6	100.3	100.0	153.5	102.8	100.0	153.3	102.8
G	Site Work	96.0	106.1	102.8	101.2	111.4	108.1	100.6	102.8	102.1	125.2	111.4	115.8	129.5	99.9	109.4
A-G	WEIGHTED AVERAGE	98.8	113.7	105.4	101.7	119.2	109.5	99.6	108.0	103.3	106.8	142.4	122.5	102.5	135.3	116.9

City Cost Indexes — RJ1030-010 Building Systems

DIV. NO.	BUILDING SYSTEMS	CALIFORNIA									COLORADO					
		SANTA BARBARA			STOCKTON			VALLEJO			COLORADO SPRINGS			DENVER		
		MAT.	INST.	TOTAL	MAT.	INST.	TOTAL	MAT.	INST.	TOTAL	MAT.	INST.	TOTAL	MAT.	INST.	TOTAL
A	Substructure	100.4	117.0	110.5	100.3	119.3	111.9	102.2	126.5	117.0	102.9	84.6	91.7	99.8	86.5	91.7
B10	Shell: Superstructure	98.0	110.1	102.8	99.0	113.3	104.7	100.3	115.3	106.3	102.6	84.9	95.6	104.3	84.9	96.6
B20	Exterior Closure	102.1	116.2	108.8	102.9	122.6	112.3	92.3	133.6	111.9	100.3	76.8	89.1	98.2	76.8	88.1
B30	Roofing	100.2	119.0	107.2	103.3	119.4	109.3	109.8	132.0	118.0	103.6	82.0	95.6	102.0	75.2	92.0
C	Interior Construction	102.4	117.4	108.9	102.1	128.7	113.6	109.1	143.5	124.0	102.1	75.5	90.6	104.1	81.2	94.2
D10	Services: Conveying	100.0	110.2	102.8	100.0	115.2	104.2	100.0	120.9	105.8	100.0	91.8	97.7	100.0	91.4	97.6
D20 - 40	Mechanical	100.0	116.7	106.8	100.0	111.3	104.6	100.1	132.3	113.3	100.2	74.2	89.5	99.9	79.5	91.6
D50	Electrical	89.4	109.7	100.1	99.3	112.1	106.0	98.3	127.4	113.6	103.2	83.6	92.8	104.9	85.8	94.8
E	Equipment & Furnishings	100.0	120.3	101.1	100.0	133.9	101.8	100.0	151.7	102.7	100.0	80.4	99.0	100.0	80.6	99.0
G	Site Work	101.3	105.7	104.3	99.1	104.9	103.1	102.1	111.3	108.3	94.4	93.2	93.6	93.7	101.4	98.9
A-G	WEIGHTED AVERAGE	99.2	113.9	105.7	100.5	116.6	107.6	101.0	128.5	113.1	101.4	80.6	92.2	101.7	83.4	93.6

DIV. NO.	BUILDING SYSTEMS	COLORADO									CONNECTICUT					
		FORT COLLINS			GRAND JUNCTION			GREELEY			PUEBLO			BRIDGEPORT		
		MAT.	INST.	TOTAL	MAT.	INST.	TOTAL	MAT.	INST.	TOTAL	MAT.	INST.	TOTAL	MAT.	INST.	TOTAL
A	Substructure	110.5	82.9	93.7	112.5	82.8	94.4	97.6	78.3	85.8	103.8	84.4	92.0	108.9	120.1	115.7
B10	Shell: Superstructure	103.0	80.9	94.2	100.9	80.2	92.7	104.0	77.5	91.3	100.1	85.5	94.3	100.2	124.5	109.9
B20	Exterior Closure	108.8	79.6	94.9	122.9	75.9	100.6	100.3	56.1	79.3	100.8	76.8	89.4	103.8	131.3	116.9
B30	Roofing	103.2	71.1	91.3	103.5	68.2	90.4	102.6	69.8	90.4	103.1	81.5	95.1	97.0	128.5	108.7
C	Interior Construction	100.7	72.6	88.5	106.5	75.1	92.9	100.0	74.9	89.1	102.1	77.9	91.6	102.7	125.9	112.8
D10	Services: Conveying	100.0	91.4	97.6	100.0	91.6	97.7	100.0	91.4	97.6	100.0	92.5	97.9	100.0	106.8	101.9
D20 - 40	Mechanical	100.0	79.3	91.5	99.9	87.0	94.6	100.0	79.2	91.5	99.9	74.2	89.4	100.0	117.6	107.2
D50	Electrical	99.3	85.1	91.8	94.8	56.1	74.4	99.3	85.7	92.1	95.8	75.1	84.9	101.9	107.8	105.0
E	Equipment & Furnishings	100.0	80.2	99.0	100.0	78.7	98.9	100.0	85.8	99.3	100.0	80.0	99.0	100.0	121.6	101.1
G	Site Work	105.2	94.7	98.1	125.8	95.6	105.3	92.4	94.1	93.6	117.4	91.5	99.9	113.5	104.1	107.1
A-G	WEIGHTED AVERAGE	102.3	80.8	92.8	104.4	78.2	92.9	99.8	77.3	89.9	100.7	79.8	91.5	101.7	119.7	109.6

DIV. NO.	BUILDING SYSTEMS	CONNECTICUT														
		BRISTOL			HARTFORD			NEW BRITAIN			NEW HAVEN			NORWALK		
		MAT.	INST.	TOTAL	MAT.	INST.	TOTAL	MAT.	INST.	TOTAL	MAT.	INST.	TOTAL	MAT.	INST.	TOTAL
A	Substructure	104.8	120.1	114.1	107.1	120.1	115.0	105.8	120.0	114.5	106.8	120.2	115.0	107.8	120.7	115.7
B10	Shell: Superstructure	99.3	124.4	109.3	103.9	124.4	112.0	97.0	124.4	107.9	97.3	124.4	108.1	100.0	125.1	110.0
B20	Exterior Closure	100.5	131.3	115.1	100.8	131.3	115.3	100.9	131.3	115.4	114.0	131.3	122.2	100.3	132.6	115.7
B30	Roofing	97.1	124.5	107.3	103.0	124.5	111.0	97.1	123.4	106.9	97.2	124.5	107.3	97.2	128.8	108.9
C	Interior Construction	102.8	125.9	112.8	102.6	125.9	112.7	102.8	125.9	112.8	102.8	125.9	112.8	102.8	125.9	112.8
D10	Services: Conveying	100.0	106.8	101.9	100.0	106.8	101.9	100.0	106.8	101.9	100.0	106.8	101.9	100.0	106.8	101.9
D20 - 40	Mechanical	100.0	117.6	107.2	100.0	117.6	107.2	100.0	117.6	107.2	100.0	117.6	107.2	100.0	117.7	107.2
D50	Electrical	101.9	110.4	106.4	101.3	111.4	106.6	102.0	110.4	106.4	101.8	110.4	106.4	101.9	159.2	132.1
E	Equipment & Furnishings	100.0	121.7	101.1	100.0	121.7	101.1	100.0	121.7	101.1	100.0	121.7	101.1	100.0	124.0	101.2
G	Site Work	112.7	104.1	106.8	106.4	104.1	104.8	112.8	104.1	106.9	112.6	104.8	107.3	113.3	104.1	107.1
A-G	WEIGHTED AVERAGE	100.9	120.0	109.3	102.0	120.1	109.9	100.6	119.9	109.1	102.1	120.0	110.0	101.2	127.1	112.6

DIV. NO.	BUILDING SYSTEMS	CONNECTICUT						D.C.			DELAWARE			FLORIDA		
		STAMFORD			WATERBURY			WASHINGTON			WILMINGTON			DAYTONA BEACH		
		MAT.	INST.	TOTAL	MAT.	INST.	TOTAL	MAT.	INST.	TOTAL	MAT.	INST.	TOTAL	MAT.	INST.	TOTAL
A	Substructure	108.9	120.7	116.1	108.8	120.0	115.6	117.2	87.3	99.0	92.1	104.8	99.8	98.2	77.2	85.4
B10	Shell: Superstructure	100.2	125.1	110.1	100.2	124.4	109.8	102.0	96.4	99.8	96.2	110.2	101.7	96.0	83.2	90.9
B20	Exterior Closure	100.7	132.6	115.8	100.7	131.3	115.2	96.6	84.0	90.6	104.8	101.0	103.0	94.0	69.6	82.4
B30	Roofing	97.1	128.8	108.9	97.1	124.5	107.3	100.4	86.3	95.1	99.1	113.7	104.5	97.0	76.8	89.5
C	Interior Construction	102.8	125.9	112.8	102.7	125.9	112.7	100.9	83.3	93.3	97.6	103.8	100.3	101.2	72.4	88.7
D10	Services: Conveying	100.0	106.8	101.9	100.0	106.8	101.9	100.0	101.9	100.5	100.0	89.1	97.0	100.0	76.0	93.4
D20 - 40	Mechanical	100.0	117.7	107.2	100.0	117.6	107.2	100.1	94.6	97.9	100.1	116.9	107.0	99.9	75.6	90.0
D50	Electrical	101.9	159.2	132.1	101.3	107.8	104.7	99.5	103.9	101.8	97.3	111.0	104.5	97.4	57.3	76.3
E	Equipment & Furnishings	100.0	124.0	101.2	100.0	121.6	101.1	100.0	79.2	98.9	100.0	103.1	100.2	100.0	71.2	98.5
G	Site Work	114.0	104.1	107.3	113.2	104.1	107.0	108.4	93.4	98.3	93.9	113.9	107.4	111.6	90.3	97.2
A-G	WEIGHTED AVERAGE	101.4	127.1	112.7	101.3	119.6	109.3	101.1	92.4	97.3	98.6	108.9	103.1	98.6	74.3	87.9

DIV. NO.	BUILDING SYSTEMS	FLORIDA														
		FORT LAUDERDALE			JACKSONVILLE			MELBOURNE			MIAMI			ORLANDO		
		MAT.	INST.	TOTAL	MAT.	INST.	TOTAL	MAT.	INST.	TOTAL	MAT.	INST.	TOTAL	MAT.	INST.	TOTAL
A	Substructure	98.2	76.4	84.9	98.7	70.2	81.4	110.5	78.3	90.9	100.1	76.8	85.9	112.4	78.4	91.7
B10	Shell: Superstructure	96.4	84.5	91.6	95.0	75.9	87.4	104.7	83.9	96.4	100.8	84.0	94.1	103.4	84.0	95.7
B20	Exterior Closure	92.5	71.7	82.6	94.0	65.0	80.2	95.6	74.2	85.4	92.9	74.8	84.3	97.9	69.5	84.4
B30	Roofing	97.2	86.2	93.1	97.3	64.2	85.0	97.4	79.7	90.8	102.8	78.9	93.9	96.9	77.6	89.7
C	Interior Construction	99.4	71.0	87.1	101.2	58.2	82.6	100.1	74.9	89.2	100.7	72.0	88.3	102.4	75.8	90.9
D10	Services: Conveying	100.0	86.8	96.3	100.0	74.3	92.9	100.0	77.6	93.8	100.0	87.4	96.5	100.0	76.0	93.4
D20 - 40	Mechanical	99.9	70.2	87.8	99.9	67.4	86.6	99.9	77.8	90.9	99.9	70.1	87.7	99.9	59.6	83.4
D50	Electrical	97.4	72.9	84.5	97.1	64.9	80.1	98.8	66.7	81.9	102.0	76.9	88.8	101.9	62.5	81.2
E	Equipment & Furnishings	100.0	70.0	98.4	100.0	54.4	97.6	100.0	71.6	98.5	100.0	70.3	98.5	100.0	77.1	98.8
G	Site Work	98.2	78.0	84.5	111.6	89.7	96.8	119.5	89.6	99.3	100.2	77.8	85.1	112.7	89.5	97.0
A-G	WEIGHTED AVERAGE	97.9	75.2	87.9	98.4	68.6	85.2	101.2	77.3	90.7	99.8	76.0	89.3	101.8	72.3	88.8

DIV. NO.	BUILDING SYSTEMS	FLORIDA														
		PANAMA CITY			PENSACOLA			ST. PETERSBURG			TALLAHASSEE			TAMPA		
		MAT.	INST.	TOTAL	MAT.	INST.	TOTAL	MAT.	INST.	TOTAL	MAT.	INST.	TOTAL	MAT.	INST.	TOTAL
A	Substructure	104.3	64.0	79.7	118.4	69.2	88.7	106.0	70.2	84.2	102.9	62.8	78.4	104.3	82.5	91.0
B10	Shell: Superstructure	96.7	73.3	87.4	100.6	77.5	91.4	99.9	79.3	91.7	92.1	71.2	83.8	98.9	91.5	96.0
B20	Exterior Closure	100.9	50.9	77.2	108.2	59.4	85.0	118.4	55.4	88.5	95.6	57.4	77.5	94.6	80.0	87.7
B30	Roofing	97.5	59.9	83.5	97.4	64.9	85.3	97.0	60.9	83.6	102.7	76.1	92.8	97.2	88.2	93.9
C	Interior Construction	100.3	47.7	77.5	99.4	57.3	81.1	99.8	55.2	80.5	102.7	50.0	79.9	101.3	75.4	90.1
D10	Services: Conveying	100.0	48.7	85.8	100.0	47.6	85.5	100.0	59.0	88.6	100.0	72.0	92.2	100.0	75.9	93.3
D20 - 40	Mechanical	99.9	54.8	81.5	99.9	55.2	81.6	99.9	61.5	84.2	100.0	41.1	75.9	99.9	85.2	93.9
D50	Electrical	96.0	61.8	78.0	100.3	56.1	77.0	97.8	64.2	80.1	103.1	61.8	81.4	97.5	64.2	80.0
E	Equipment & Furnishings	100.0	43.9	97.1	100.0	53.0	97.6	100.0	52.4	97.5	100.0	43.5	97.1	100.0	79.6	98.9
G	Site Work	124.7	88.6	100.3	124.8	89.0	100.6	112.7	89.2	96.9	108.7	88.8	95.2	112.9	89.8	97.3
A-G	WEIGHTED AVERAGE	99.8	60.2	82.4	102.2	63.4	85.1	102.2	65.6	86.1	99.1	59.1	81.5	99.5	81.1	91.4

City Cost Indexes — RJ1030-010 Building Systems

DIV. NO.	BUILDING SYSTEMS	GEORGIA ALBANY			GEORGIA ATLANTA			GEORGIA AUGUSTA			GEORGIA COLUMBUS			GEORGIA MACON		
		MAT.	INST.	TOTAL	MAT.	INST.	TOTAL	MAT.	INST.	TOTAL	MAT.	INST.	TOTAL	MAT.	INST.	TOTAL
A	Substructure	99.3	59.4	75.0	107.9	79.7	90.8	103.7	69.7	83.0	99.1	65.7	78.7	98.9	66.5	79.2
B10	Shell: Superstructure	96.2	70.2	85.8	95.4	77.1	88.1	93.6	68.7	83.7	95.8	75.3	87.6	92.5	73.4	84.9
B20	Exterior Closure	93.4	46.2	71.0	93.7	69.1	82.0	91.2	44.3	68.9	93.4	58.8	77.0	97.5	42.9	71.5
B30	Roofing	94.9	64.8	83.7	93.2	74.3	86.2	93.0	61.2	81.2	94.8	65.8	84.0	93.3	66.6	83.3
C	Interior Construction	99.3	43.5	75.1	94.4	74.2	85.7	91.2	62.4	78.7	99.3	56.2	80.6	94.2	51.0	75.5
D10	Services: Conveying	100.0	86.8	96.3	100.0	87.7	96.6	100.0	49.3	85.9	100.0	86.8	96.3	100.0	84.8	96.2
D20 - 40	Mechanical	99.9	67.7	86.7	99.9	72.7	88.8	100.0	64.5	85.5	99.9	49.1	79.1	99.9	65.6	85.9
D50	Electrical	92.8	59.5	75.3	100.2	74.1	86.4	101.6	64.4	82.0	92.8	72.5	82.1	91.9	62.1	76.2
E	Equipment & Furnishings	100.0	40.4	96.9	100.0	75.9	98.8	100.0	72.5	98.6	100.0	54.8	97.7	100.0	53.3	97.6
G	Site Work	98.5	79.5	85.6	94.7	94.8	94.8	91.7	92.6	92.3	98.4	79.6	85.7	98.8	94.8	96.1
A-G	WEIGHTED AVERAGE	97.4	61.1	81.4	97.5	75.9	88.0	96.3	64.2	82.1	97.4	63.9	82.6	96.2	63.9	82.0

DIV. NO.	BUILDING SYSTEMS	GEORGIA SAVANNAH			GEORGIA VALDOSTA			HAWAII HONOLULU			HAWAII STATES & POSS., GUAM			IDAHO BOISE		
		MAT.	INST.	TOTAL	MAT.	INST.	TOTAL	MAT.	INST.	TOTAL	MAT.	INST.	TOTAL	MAT.	INST.	TOTAL
A	Substructure	105.5	61.0	78.4	98.8	58.8	74.4	140.0	124.9	130.8	166.4	82.9	115.5	93.7	80.6	85.7
B10	Shell: Superstructure	94.4	70.1	84.7	94.7	67.7	84.0	132.5	119.5	127.3	144.6	75.9	117.3	99.6	76.4	90.4
B20	Exterior Closure	93.3	52.7	74.0	100.4	51.8	77.3	108.2	128.2	117.7	135.1	46.9	93.1	113.6	62.7	89.4
B30	Roofing	95.9	60.1	82.6	95.0	67.3	84.7	115.6	126.1	119.5	118.9	72.5	101.7	94.4	67.5	84.4
C	Interior Construction	101.5	50.2	79.2	95.9	44.5	73.6	115.2	137.5	124.9	143.0	60.8	107.4	97.1	73.0	86.6
D10	Services: Conveying	100.0	51.7	86.6	100.0	51.7	86.6	100.0	112.5	103.5	100.0	79.4	94.3	100.0	58.0	88.3
D20 - 40	Mechanical	100.0	63.2	85.0	99.9	67.5	86.7	100.1	112.9	105.3	102.0	39.3	76.3	100.0	68.3	87.0
D50	Electrical	97.7	59.8	77.7	90.8	58.6	73.9	108.8	124.7	117.2	156.9	42.9	96.9	95.9	75.5	85.4
E	Equipment & Furnishings	100.0	46.9	97.2	100.0	39.5	96.9	100.0	138.4	102.0	100.0	60.2	97.9	100.0	78.6	98.9
G	Site Work	99.6	80.4	86.6	107.2	79.5	88.5	124.3	106.9	112.5	156.4	102.7	120.1	82.4	95.5	91.3
A-G	WEIGHTED AVERAGE	98.2	61.1	81.9	97.4	60.4	81.1	113.6	122.3	117.4	130.4	60.0	99.4	99.6	73.3	88.0

DIV. NO.	BUILDING SYSTEMS	IDAHO LEWISTON			IDAHO POCATELLO			IDAHO TWIN FALLS			ILLINOIS CHICAGO			ILLINOIS DECATUR		
		MAT.	INST.	TOTAL	MAT.	INST.	TOTAL	MAT.	INST.	TOTAL	MAT.	INST.	TOTAL	MAT.	INST.	TOTAL
A	Substructure	103.7	79.7	89.1	95.9	80.5	86.5	98.6	54.3	71.6	102.1	139.6	124.9	96.7	109.2	104.3
B10	Shell: Superstructure	96.5	79.4	89.7	106.1	76.2	94.2	106.6	57.9	87.2	96.5	142.8	115.0	98.1	109.8	102.8
B20	Exterior Closure	123.2	80.1	102.7	110.8	62.5	87.8	118.6	34.6	78.7	99.2	150.1	123.4	87.3	116.3	101.1
B30	Roofing	147.6	73.6	120.1	94.5	61.0	82.0	95.3	37.9	73.9	103.4	143.7	118.4	100.0	109.0	103.3
C	Interior Construction	139.2	63.2	106.2	98.4	73.1	87.4	99.9	32.4	70.7	99.6	157.3	124.6	98.9	116.9	106.7
D10	Services: Conveying	100.0	64.5	90.1	100.0	58.0	88.3	100.0	34.9	81.9	100.0	119.8	105.5	100.0	102.8	100.3
D20 - 40	Mechanical	100.7	89.4	96.1	99.9	68.3	87.0	99.9	58.4	83.0	99.9	133.3	113.6	99.9	100.8	100.3
D50	Electrical	85.3	80.8	83.0	93.3	72.1	82.1	88.9	62.6	75.0	97.8	130.8	115.2	98.0	88.5	93.0
E	Equipment & Furnishings	100.0	51.5	97.5	100.0	78.4	98.9	100.0	28.7	96.3	100.0	157.2	103.0	100.0	115.3	100.8
G	Site Work	87.1	92.0	90.4	82.4	95.5	91.3	88.8	93.8	92.2	105.9	94.3	98.0	93.7	97.7	96.4
A-G	WEIGHTED AVERAGE	108.1	79.5	95.5	100.7	72.5	88.3	101.8	52.7	80.1	99.3	138.3	116.5	97.6	105.8	101.2

DIV. NO.	BUILDING SYSTEMS	ILLINOIS EAST ST. LOUIS			ILLINOIS JOLIET			ILLINOIS PEORIA			ILLINOIS ROCKFORD			ILLINOIS SPRINGFIELD		
		MAT.	INST.	TOTAL	MAT.	INST.	TOTAL	MAT.	INST.	TOTAL	MAT.	INST.	TOTAL	MAT.	INST.	TOTAL
A	Substructure	94.6	108.6	103.1	102.3	133.5	121.3	96.5	109.1	104.1	97.0	121.8	112.1	92.0	108.0	101.7
B10	Shell: Superstructure	95.2	115.1	103.1	94.9	135.0	110.9	97.8	111.2	103.1	97.9	128.2	110.0	98.7	109.1	102.9
B20	Exterior Closure	81.5	116.1	98.0	100.5	143.6	121.0	106.7	112.0	109.2	94.9	133.4	113.2	89.7	117.2	102.8
B30	Roofing	94.2	111.3	100.6	103.2	139.7	116.8	100.1	111.6	104.4	102.5	127.4	111.8	100.7	112.3	105.0
C	Interior Construction	92.7	110.9	100.6	98.1	150.7	120.9	99.0	114.7	105.8	99.0	128.7	111.9	100.0	114.8	106.5
D10	Services: Conveying	100.0	98.9	99.7	100.0	118.2	105.0	100.0	102.7	100.8	100.0	85.8	96.1	100.0	103.2	100.9
D20 - 40	Mechanical	99.9	95.3	98.0	100.0	130.9	112.6	99.9	103.8	101.5	99.9	113.6	105.5	99.8	103.9	101.5
D50	Electrical	94.7	103.3	99.2	97.0	135.3	117.2	97.1	95.3	96.2	97.5	129.3	114.2	101.1	90.7	95.6
E	Equipment & Furnishings	100.0	108.2	100.4	100.0	153.2	102.8	100.0	112.7	100.7	100.0	129.4	101.5	100.0	115.5	100.8
G	Site Work	104.9	98.3	100.4	105.8	93.2	97.3	100.3	96.6	97.8	99.7	97.9	98.5	98.6	97.8	98.0
A-G	WEIGHTED AVERAGE	95.0	106.6	100.1	98.8	134.5	114.5	99.7	106.7	102.8	98.6	122.3	109.0	98.3	106.5	101.9

DIV. NO.	BUILDING SYSTEMS	INDIANA ANDERSON			INDIANA BLOOMINGTON			INDIANA EVANSVILLE			INDIANA FORT WAYNE			INDIANA GARY		
		MAT.	INST.	TOTAL	MAT.	INST.	TOTAL	MAT.	INST.	TOTAL	MAT.	INST.	TOTAL	MAT.	INST.	TOTAL
A	Substructure	102.2	83.8	91.0	96.7	83.9	88.9	95.7	91.3	93.0	105.8	79.5	89.8	105.2	104.8	104.9
B10	Shell: Superstructure	93.4	84.8	90.0	96.3	78.9	89.4	90.4	82.3	87.2	94.2	80.7	88.8	93.9	105.1	98.4
B20	Exterior Closure	90.4	81.0	85.9	100.6	78.1	89.9	97.9	81.4	90.0	92.1	77.8	85.3	91.0	111.7	100.9
B30	Roofing	99.9	78.7	92.0	94.1	82.6	89.8	98.1	83.5	92.7	99.7	81.4	92.9	98.6	108.0	102.1
C	Interior Construction	96.6	79.5	89.2	102.4	80.7	93.0	97.7	79.8	90.0	96.5	74.4	86.9	96.2	110.8	102.5
D10	Services: Conveying	100.0	93.6	98.2	100.0	84.9	95.8	100.0	90.1	97.2	100.0	85.8	96.1	100.0	98.7	99.6
D20 - 40	Mechanical	99.9	76.0	92.2	99.8	79.4	91.4	99.9	79.7	91.6	99.9	73.2	89.0	99.9	103.9	101.6
D50	Electrical	86.6	84.7	85.6	99.4	84.0	91.3	95.4	84.0	89.4	87.4	72.7	79.7	99.1	112.1	106.0
E	Equipment & Furnishings	100.0	80.8	99.0	100.0	81.3	99.0	100.0	77.4	98.8	100.0	73.0	98.6	100.0	107.6	100.4
G	Site Work	94.6	95.3	95.0	83.2	94.8	91.1	88.4	124.6	112.9	95.9	95.0	95.3	95.2	98.6	97.5
A-G	WEIGHTED AVERAGE	95.7	82.3	89.8	98.8	81.6	91.2	96.5	85.2	91.5	96.3	77.7	88.1	97.2	107.0	101.5

DIV. NO.	BUILDING SYSTEMS	INDIANA INDIANAPOLIS			INDIANA MUNCIE			INDIANA SOUTH BEND			INDIANA TERRE HAUTE			IOWA CEDAR RAPIDS		
		MAT.	INST.	TOTAL	MAT.	INST.	TOTAL	MAT.	INST.	TOTAL	MAT.	INST.	TOTAL	MAT.	INST.	TOTAL
A	Substructure	100.1	87.6	92.5	100.9	83.6	90.3	101.5	82.4	89.9	93.9	92.6	93.1	106.1	84.4	92.9
B10	Shell: Superstructure	94.4	81.8	89.4	98.2	84.8	92.9	93.1	89.1	91.5	90.6	84.7	88.3	94.2	87.2	91.4
B20	Exterior Closure	97.3	81.7	89.9	97.0	81.0	89.4	86.0	79.9	83.1	105.4	81.1	93.9	100.2	79.5	90.3
B30	Roofing	97.8	84.0	92.6	96.8	79.0	90.2	91.9	83.0	88.6	98.2	81.5	92.0	105.5	79.6	95.9
C	Interior Construction	102.7	84.5	94.8	97.6	79.3	89.7	91.5	79.0	86.1	98.0	80.6	90.4	106.6	80.1	95.1
D10	Services: Conveying	100.0	94.8	98.5	100.0	93.0	98.1	100.0	67.4	90.9	100.0	96.3	99.0	100.0	86.0	96.1
D20 - 40	Mechanical	99.9	82.0	92.6	99.8	76.0	90.0	99.9	76.0	91.1	99.9	78.5	91.2	100.3	80.5	92.2
D50	Electrical	101.0	84.6	92.4	90.9	78.8	84.5	100.6	82.6	91.1	93.5	84.0	88.5	98.8	79.4	88.6
E	Equipment & Furnishings	100.0	84.2	99.2	100.0	80.4	99.0	100.0	77.7	98.8	100.0	77.7	98.8	100.0	79.0	98.9
G	Site Work	94.1	98.8	97.3	83.4	94.7	91.0	97.0	95.2	95.8	90.4	124.9	113.7	101.0	96.0	97.6
A-G	WEIGHTED AVERAGE	98.9	84.7	92.6	97.5	81.4	90.4	95.5	81.8	89.5	97.2	85.7	92.1	100.3	82.8	92.6

529

City Cost Indexes — RJ1030-010 Building Systems

DIV. NO.	BUILDING SYSTEMS	IOWA														
		COUNCIL BLUFFS			DAVENPORT			DES MOINES			DUBUQUE			SIOUX CITY		
		MAT.	INST.	TOTAL	MAT.	INST.	TOTAL	MAT.	INST.	TOTAL	MAT.	INST.	TOTAL	MAT.	INST.	TOTAL
A	Substructure	109.3	78.6	90.6	103.4	90.5	95.6	93.8	86.6	89.4	102.6	81.9	90.0	107.4	70.0	84.6
B10	Shell: Superstructure	98.1	82.8	92.0	93.7	94.7	94.1	92.1	87.6	90.3	91.9	85.1	89.2	94.1	73.0	85.7
B20	Exterior Closure	101.8	76.2	89.6	98.8	85.1	92.3	94.8	82.9	89.1	99.5	76.7	88.7	97.2	56.5	77.8
B30	Roofing	105.0	67.3	91.0	105.1	85.2	97.7	105.1	78.6	95.2	105.3	70.9	92.5	105.1	66.4	90.7
C	Interior Construction	101.5	72.9	89.1	104.5	88.1	97.4	102.9	78.4	92.3	103.7	70.8	89.4	105.4	63.3	87.2
D10	Services: Conveying	100.0	87.1	96.4	100.0	87.3	96.5	100.0	86.5	96.2	100.0	85.5	96.0	100.0	83.5	95.4
D20 - 40	Mechanical	100.3	76.9	90.7	100.3	88.0	95.3	100.1	79.9	91.8	100.3	75.1	90.0	100.3	68.1	87.1
D50	Electrical	104.5	82.1	92.7	95.9	85.0	90.2	109.5	80.9	94.4	102.9	77.6	89.6	98.8	72.9	85.2
E	Equipment & Furnishings	100.0	70.4	98.5	100.0	85.3	99.2	100.0	75.0	98.7	100.0	66.4	98.3	100.0	64.3	98.1
G	Site Work	105.1	92.3	96.4	99.5	99.0	99.1	89.8	99.8	96.5	99.3	93.2	95.2	110.4	95.7	100.5
A-G	WEIGHTED AVERAGE	101.2	79.1	91.5	99.3	89.2	94.8	98.9	83.5	92.1	99.5	78.6	90.3	100.0	69.7	86.7

DIV. NO.	BUILDING SYSTEMS	IOWA			KANSAS											
		WATERLOO			DODGE CITY			KANSAS CITY			SALINA			TOPEKA		
		MAT.	INST.	TOTAL	MAT.	INST.	TOTAL	MAT.	INST.	TOTAL	MAT.	INST.	TOTAL	MAT.	INST.	TOTAL
A	Substructure	106.0	66.9	82.2	110.4	66.7	83.8	92.1	94.1	93.3	99.6	64.8	78.4	92.2	62.9	74.3
B10	Shell: Superstructure	94.3	74.9	86.6	98.2	68.3	86.3	99.0	96.4	98.0	95.9	69.4	85.4	98.7	75.9	89.6
B20	Exterior Closure	97.2	72.9	85.7	111.6	59.3	86.8	105.3	94.5	100.2	113.3	57.0	86.5	100.4	59.0	80.7
B30	Roofing	104.8	71.1	92.3	100.7	60.1	85.6	99.6	96.1	98.3	100.2	59.0	84.9	100.5	61.3	85.9
C	Interior Construction	102.9	55.1	82.2	98.6	55.2	79.8	96.6	92.6	94.8	97.8	55.7	79.6	99.0	49.1	77.4
D10	Services: Conveying	100.0	86.1	96.1	100.0	40.3	83.4	100.0	59.7	88.8	100.0	51.8	86.6	100.0	50.3	86.2
D20 - 40	Mechanical	100.3	79.8	91.9	99.9	66.2	86.1	99.8	97.3	98.8	99.9	66.3	86.2	99.9	56.4	82.1
D50	Electrical	98.8	58.0	77.3	97.4	72.3	84.2	103.3	94.1	98.5	97.2	72.3	84.1	101.9	72.2	86.3
E	Equipment & Furnishings	100.0	39.6	96.9	100.0	60.2	97.9	100.0	95.8	99.8	100.0	54.6	97.6	100.0	39.6	96.9
G	Site Work	100.0	96.1	97.4	114.1	93.4	100.1	96.8	91.7	93.4	104.6	93.8	97.3	96.8	90.6	92.6
A-G	WEIGHTED AVERAGE	99.4	71.2	86.9	101.2	65.8	85.6	99.7	93.9	97.2	100.1	65.9	85.0	99.4	63.8	83.7

DIV. NO.	BUILDING SYSTEMS	KANSAS			KENTUCKY											
		WICHITA			BOWLING GREEN			LEXINGTON			LOUISVILLE			OWENSBORO		
		MAT.	INST.	TOTAL	MAT.	INST.	TOTAL	MAT.	INST.	TOTAL	MAT.	INST.	TOTAL	MAT.	INST.	TOTAL
A	Substructure	90.9	62.2	73.4	87.0	89.9	88.8	94.5	86.2	89.4	95.2	83.5	88.1	89.9	96.8	94.1
B10	Shell: Superstructure	98.5	68.6	86.6	92.1	86.1	89.7	95.2	84.7	91.0	101.0	83.6	94.1	86.4	87.5	86.9
B20	Exterior Closure	97.5	54.4	77.0	99.2	80.1	90.1	94.4	74.1	84.7	94.9	79.3	87.5	104.1	82.9	94.0
B30	Roofing	97.0	53.0	80.6	85.5	82.8	84.5	98.1	89.0	94.9	95.8	77.5	89.0	98.1	91.6	95.7
C	Interior Construction	98.7	49.6	77.4	96.5	79.6	89.2	98.8	75.0	88.5	98.4	79.9	90.3	95.3	83.4	90.1
D10	Services: Conveying	100.0	48.9	85.8	100.0	56.7	88.0	100.0	91.5	97.6	100.0	88.2	96.7	100.0	97.1	99.2
D20 - 40	Mechanical	99.7	62.4	84.5	99.9	82.4	92.8	99.9	75.4	89.9	100.0	80.8	92.1	99.9	75.8	90.1
D50	Electrical	102.1	72.3	86.4	94.3	82.1	87.9	94.8	82.1	88.1	101.7	79.2	89.9	93.7	82.3	87.7
E	Equipment & Furnishings	100.0	44.8	97.1	100.0	77.7	98.8	100.0	71.8	98.5	100.0	77.4	98.8	100.0	78.9	98.9
G	Site Work	96.8	92.5	93.9	75.5	98.3	90.9	84.2	101.1	95.6	73.3	98.1	90.1	88.7	125.0	113.3
A-G	WEIGHTED AVERAGE	98.8	63.2	83.1	95.6	83.3	90.1	97.0	81.0	90.0	98.6	82.2	91.3	95.6	86.7	91.7

DIV. NO.	BUILDING SYSTEMS	LOUISIANA														
		ALEXANDRIA			BATON ROUGE			LAKE CHARLES			MONROE			NEW ORLEANS		
		MAT.	INST.	TOTAL	MAT.	INST.	TOTAL	MAT.	INST.	TOTAL	MAT.	INST.	TOTAL	MAT.	INST.	TOTAL
A	Substructure	94.7	51.7	68.5	97.0	65.9	78.1	101.9	66.3	80.2	94.4	56.0	71.0	101.9	72.3	83.9
B10	Shell: Superstructure	90.4	52.7	75.4	100.8	66.1	87.0	97.6	65.5	84.8	107.8	57.9	77.9	108.7	70.9	93.1
B20	Exterior Closure	109.7	49.1	80.9	102.5	51.4	78.2	101.8	56.2	80.1	107.4	47.3	78.8	102.2	62.6	83.3
B30	Roofing	98.4	48.3	79.8	98.4	56.3	82.7	96.3	56.0	81.3	98.4	48.3	79.8	96.5	71.1	87.1
C	Interior Construction	94.1	40.4	70.8	101.5	61.3	84.1	100.7	58.3	82.4	94.0	40.2	70.7	101.6	67.8	86.9
D10	Services: Conveying	100.0	51.1	86.4	100.0	82.1	95.0	100.0	82.1	95.0	100.0	46.5	85.1	100.0	86.0	96.1
D20 - 40	Mechanical	100.3	50.3	79.8	100.0	58.1	82.8	100.1	58.7	83.2	100.3	49.2	79.4	100.0	67.1	86.5
D50	Electrical	93.0	55.4	73.2	110.6	61.1	84.6	102.1	65.7	82.9	95.1	56.6	74.8	102.7	72.9	87.0
E	Equipment & Furnishings	100.0	34.7	96.6	100.0	65.2	98.2	100.0	59.1	97.9	100.0	35.1	96.6	100.0	68.8	98.4
G	Site Work	95.8	86.2	89.3	100.0	85.9	90.5	102.3	85.7	91.1	95.7	85.6	88.9	104.4	88.6	93.7
A-G	WEIGHTED AVERAGE	97.2	52.8	77.6	101.5	62.6	84.4	100.1	63.4	83.9	97.1	52.9	77.6	102.4	70.5	88.4

DIV. NO.	BUILDING SYSTEMS	LOUISIANA			MAINE											
		SHREVEPORT			AUGUSTA			BANGOR			LEWISTON			PORTLAND		
		MAT.	INST.	TOTAL	MAT.	INST.	TOTAL	MAT.	INST.	TOTAL	MAT.	INST.	TOTAL	MAT.	INST.	TOTAL
A	Substructure	95.8	53.8	70.2	93.6	90.8	91.9	86.6	106.4	98.6	92.2	106.4	100.8	103.8	106.4	105.4
B10	Shell: Superstructure	88.5	57.0	76.0	87.9	89.9	88.7	85.8	98.4	90.8	89.4	98.4	93.0	93.1	98.4	95.2
B20	Exterior Closure	101.9	47.5	76.0	103.3	68.2	86.6	109.0	99.3	104.4	99.3	99.3	99.3	99.4	99.3	99.4
B30	Roofing	100.0	46.6	80.2	102.1	67.5	89.3	97.3	85.2	92.8	97.1	85.2	92.7	104.6	85.2	97.4
C	Interior Construction	95.5	42.0	72.3	99.6	92.2	96.4	99.0	97.5	98.4	101.7	97.5	99.9	101.6	97.5	99.8
D10	Services: Conveying	100.0	65.7	90.5	100.0	101.3	100.4	100.0	113.4	103.7	100.0	113.4	103.7	100.0	113.4	103.7
D20 - 40	Mechanical	100.1	54.1	81.3	99.8	66.7	86.3	100.0	77.2	90.7	100.0	77.2	90.7	99.8	77.2	90.9
D50	Electrical	100.8	65.3	82.1	100.9	78.6	89.2	99.7	78.6	88.6	102.0	78.6	89.7	104.5	78.6	90.9
E	Equipment & Furnishings	100.0	38.3	96.8	100.0	106.6	100.3	100.0	98.8	99.9	100.0	98.8	99.9	100.0	98.8	99.9
G	Site Work	97.7	85.5	89.4	91.5	100.9	97.9	93.7	101.2	98.8	90.0	101.2	97.6	88.5	101.2	97.0
A-G	WEIGHTED AVERAGE	97.0	55.6	78.8	97.5	81.7	90.5	97.1	92.1	94.9	97.6	92.1	95.2	99.2	92.1	96.1

DIV. NO.	BUILDING SYSTEMS	MARYLAND						MASSACHUSETTS								
		BALTIMORE			HAGERSTOWN			BOSTON			BROCKTON			FALL RIVER		
		MAT.	INST.	TOTAL	MAT.	INST.	TOTAL	MAT.	INST.	TOTAL	MAT.	INST.	TOTAL	MAT.	INST.	TOTAL
A	Substructure	111.5	82.2	93.7	96.2	82.7	88.0	106.0	139.6	126.4	102.9	135.2	122.6	100.9	137.2	123.0
B10	Shell: Superstructure	101.4	87.3	95.8	95.2	86.1	91.6	101.8	139.1	116.7	98.9	135.2	113.3	98.4	139.4	114.7
B20	Exterior Closure	99.4	78.2	89.3	96.3	82.9	90.0	109.9	161.4	134.4	107.5	158.7	131.8	107.9	157.3	131.4
B30	Roofing	97.7	79.7	90.8	96.4	76.1	88.9	97.4	149.2	116.6	97.3	145.0	115.1	97.4	138.0	112.5
C	Interior Construction	95.5	75.7	86.9	92.5	78.3	86.3	98.7	151.5	121.6	97.2	142.2	116.7	97.2	143.2	117.1
D10	Services: Conveying	100.0	89.0	97.0	100.0	92.3	97.9	100.0	116.6	104.6	100.0	115.8	104.4	100.0	116.5	104.6
D20 - 40	Mechanical	100.0	82.4	92.8	99.9	83.4	93.2	100.0	128.2	111.6	100.0	112.3	105.0	100.0	112.5	105.1
D50	Electrical	101.1	93.3	97.0	98.5	83.1	90.4	101.1	139.7	121.4	100.2	99.1	99.6	100.2	99.1	99.6
E	Equipment & Furnishings	100.0	76.0	98.8	100.0	76.1	98.8	100.0	143.2	102.2	100.0	129.5	101.5	100.0	129.4	101.5
G	Site Work	101.4	94.7	96.9	91.1	91.3	91.3	99.3	108.3	105.4	96.5	104.9	102.2	95.4	105.0	101.9
A-G	WEIGHTED AVERAGE	100.1	84.1	93.0	96.8	83.5	90.9	101.5	139.2	118.1	100.1	127.1	112.0	99.9	127.9	112.3

City Cost Indexes

RJ1030-010 Building Systems

DIV. NO.	BUILDING SYSTEMS	MASSACHUSETTS															
		HYANNIS			LAWRENCE			LOWELL			NEW BEDFORD			PITTSFIELD		
		MAT.	INST.	TOTAL	MAT.	INST.	TOTAL	MAT.	INST.	TOTAL	MAT.	INST.	TOTAL	MAT.	INST.	TOTAL
A	Substructure	91.3	136.9	119.1	103.2	133.1	121.4	97.7	132.9	119.1	93.7	137.2	120.2	98.9	112.8	107.4
B10	Shell: Superstructure	93.6	139.0	111.7	97.0	132.1	111.0	95.7	130.2	109.5	96.9	139.4	113.8	95.9	112.0	102.3
B20	Exterior Closure	104.4	157.4	129.6	107.1	158.3	131.4	97.5	148.5	121.7	107.0	157.3	130.9	97.8	119.7	108.2
B30	Roofing	97.0	138.9	112.6	97.3	144.5	114.9	97.1	142.8	114.1	97.3	138.0	112.4	97.2	114.5	103.6
C	Interior Construction	93.6	143.0	115.0	97.0	141.7	116.4	101.5	141.7	118.9	97.1	143.2	117.1	101.5	113.7	106.8
D10	Services: Conveying	100.0	115.8	104.4	100.0	115.8	104.4	100.0	115.8	104.4	100.0	116.5	104.6	100.0	103.4	101.0
D20 - 40	Mechanical	100.0	112.4	105.1	100.0	124.2	109.9	100.0	124.1	109.9	100.0	112.5	105.1	100.0	98.0	99.2
D50	Electrical	97.6	99.1	98.4	101.4	126.0	114.4	102.0	126.0	114.6	101.1	99.1	100.0	102.0	94.5	98.0
E	Equipment & Furnishings	100.0	129.5	101.5	100.0	130.8	101.6	100.0	130.8	101.6	100.0	129.4	101.5	100.0	106.4	100.3
G	Site Work	93.9	104.7	101.2	97.3	104.9	102.4	96.2	104.9	102.1	93.9	105.0	101.4	97.3	103.0	101.1
A-G	WEIGHTED AVERAGE	97.3	127.7	110.7	99.8	132.5	114.2	99.0	130.9	113.1	99.3	127.9	111.9	99.1	107.1	102.7

DIV. NO.	BUILDING SYSTEMS	MASSACHUSETTS						MICHIGAN								
		SPRINGFIELD			WORCESTER			ANN ARBOR			DEARBORN			DETROIT		
		MAT.	INST.	TOTAL	MAT.	INST.	TOTAL	MAT.	INST.	TOTAL	MAT.	INST.	TOTAL	MAT.	INST.	TOTAL
A	Substructure	100.0	114.6	108.9	99.7	130.3	118.3	89.9	108.7	101.4	88.7	110.8	102.2	96.0	110.8	105.0
B10	Shell: Superstructure	98.1	112.9	104.0	98.1	130.0	110.8	93.2	115.4	102.0	92.9	117.0	102.5	97.2	107.5	101.3
B20	Exterior Closure	97.6	124.5	110.4	97.4	148.7	121.8	98.1	107.4	102.5	98.0	112.7	105.0	97.8	111.5	104.3
B30	Roofing	97.1	117.5	104.7	97.1	135.6	111.4	101.1	108.7	103.9	99.6	114.9	105.3	98.1	114.9	104.3
C	Interior Construction	101.5	114.9	107.3	101.5	136.7	116.7	97.3	109.6	102.6	97.3	113.3	104.2	99.0	113.3	105.2
D10	Services: Conveying	100.0	105.7	101.6	100.0	108.4	102.3	100.0	106.0	101.7	100.0	107.4	102.1	100.0	107.4	102.1
D20 - 40	Mechanical	100.0	103.1	101.3	100.0	115.9	106.5	100.0	98.4	99.4	100.0	110.5	104.3	100.0	111.7	104.8
D50	Electrical	102.0	94.5	98.0	102.0	100.6	101.3	96.3	108.3	102.6	96.3	107.1	102.0	97.3	107.0	102.4
E	Equipment & Furnishings	100.0	106.7	100.3	100.0	124.2	101.3	100.0	108.3	100.4	100.0	112.7	100.7	100.0	112.7	100.7
G	Site Work	96.7	103.3	101.2	96.6	104.8	102.2	80.7	96.8	91.6	80.4	97.1	91.7	94.1	99.2	97.5
A-G	WEIGHTED AVERAGE	99.6	109.5	104.0	99.6	124.3	110.4	96.8	106.7	101.2	96.6	111.0	102.9	98.4	109.6	103.4

DIV. NO.	BUILDING SYSTEMS	MICHIGAN															
		FLINT			GRAND RAPIDS			KALAMAZOO			LANSING			MUSKEGON		
		MAT.	INST.	TOTAL	MAT.	INST.	TOTAL	MAT.	INST.	TOTAL	MAT.	INST.	TOTAL	MAT.	INST.	TOTAL
A	Substructure	88.8	97.2	93.9	99.4	86.8	91.7	96.5	91.6	93.5	94.9	95.6	95.3	95.2	89.5	91.7
B10	Shell: Superstructure	93.4	106.7	98.7	96.1	85.2	91.8	94.5	88.8	92.2	92.9	105.3	97.8	92.5	87.0	90.3
B20	Exterior Closure	98.1	96.1	97.1	94.8	81.1	88.3	97.5	84.1	91.1	95.1	94.0	94.6	92.7	76.8	85.1
B30	Roofing	98.8	97.9	98.5	97.3	75.0	89.0	94.2	85.8	91.1	98.6	90.1	95.4	93.2	77.7	87.5
C	Interior Construction	96.9	89.0	93.5	98.4	77.5	89.3	96.1	85.1	91.3	99.8	86.3	94.0	93.8	82.4	88.9
D10	Services: Conveying	100.0	95.3	98.7	100.0	101.1	100.3	100.0	101.3	100.3	100.0	95.1	98.6	100.0	101.2	100.3
D20 - 40	Mechanical	100.0	92.5	96.9	99.9	81.7	92.5	99.9	81.6	92.4	99.9	90.5	96.0	99.8	83.7	93.2
D50	Electrical	96.3	93.2	94.6	99.7	73.6	86.0	94.2	85.2	89.5	99.9	91.2	95.3	94.7	78.5	86.2
E	Equipment & Furnishings	100.0	87.6	99.4	100.0	73.9	98.6	100.0	84.1	99.2	100.0	84.5	99.2	100.0	82.7	99.1
G	Site Work	71.2	96.0	88.0	90.1	87.6	88.4	88.9	87.7	88.1	88.5	95.7	93.4	86.4	87.7	87.2
A-G	WEIGHTED AVERAGE	96.4	95.7	96.1	98.0	81.6	90.8	96.8	86.2	92.1	97.4	93.8	95.8	95.4	83.6	90.2

DIV. NO.	BUILDING SYSTEMS	MICHIGAN			MINNESOTA											
		SAGINAW			DULUTH			MINNEAPOLIS			ROCHESTER			SAINT PAUL		
		MAT.	INST.	TOTAL	MAT.	INST.	TOTAL	MAT.	INST.	TOTAL	MAT.	INST.	TOTAL	MAT.	INST.	TOTAL
A	Substructure	88.1	95.6	92.7	101.7	108.9	106.0	109.4	120.0	115.8	107.8	106.5	107.1	108.1	117.8	114.0
B10	Shell: Superstructure	93.1	105.4	98.0	98.4	118.5	106.4	102.3	128.8	112.8	99.7	120.2	107.9	99.2	127.9	110.6
B20	Exterior Closure	98.2	89.7	94.2	100.1	118.5	108.8	101.6	127.3	113.8	99.4	117.0	107.8	107.4	127.1	116.8
B30	Roofing	99.6	95.4	98.1	106.0	118.5	110.7	104.9	133.0	115.4	105.4	101.2	103.8	106.4	132.4	116.1
C	Interior Construction	95.8	86.6	91.8	97.1	118.2	106.2	99.7	131.6	113.5	97.3	113.9	104.5	95.3	129.1	109.9
D10	Services: Conveying	100.0	94.3	98.4	100.0	100.0	100.0	100.0	106.7	101.9	100.0	102.5	100.7	100.0	106.4	101.8
D20 - 40	Mechanical	100.0	86.0	94.2	99.9	99.2	99.6	99.9	120.1	108.2	100.0	97.7	99.1	99.9	115.1	106.1
D50	Electrical	93.5	89.5	91.4	102.1	106.8	104.6	101.7	111.0	106.6	100.8	87.3	93.7	99.2	111.0	105.4
E	Equipment & Furnishings	100.0	86.0	99.3	100.0	114.6	100.8	100.0	120.0	101.5	100.0	107.2	100.4	100.0	125.7	101.3
G	Site Work	74.2	95.6	88.7	100.3	100.5	100.4	102.8	105.5	104.8	101.9	100.1	100.7	102.9	101.1	101.7
A-G	WEIGHTED AVERAGE	96.0	92.2	94.3	99.7	110.3	104.3	101.4	122.0	110.5	100.1	106.1	102.7	100.4	119.9	109.0

DIV. NO.	BUILDING SYSTEMS	MINNESOTA			MISSISSIPPI												
		ST. CLOUD			BILOXI			GREENVILLE			JACKSON			MERIDIAN		
		MAT.	INST.	TOTAL	MAT.	INST.	TOTAL	MAT.	INST.	TOTAL	MAT.	INST.	TOTAL	MAT.	INST.	TOTAL
A	Substructure	96.9	117.0	109.1	104.6	64.2	80.0	105.8	69.6	83.7	94.5	66.8	77.6	100.6	69.7	81.7
B10	Shell: Superstructure	94.4	125.5	106.8	95.0	71.0	85.5	93.2	75.6	86.2	92.8	72.5	84.7	92.2	74.4	85.1
B20	Exterior Closure	96.7	124.7	110.1	93.2	49.4	72.4	116.6	53.0	86.4	93.8	52.5	74.2	91.4	57.9	75.5
B30	Roofing	105.3	122.1	111.5	94.9	56.9	80.8	95.0	63.7	83.4	96.8	62.6	84.1	94.6	65.5	83.8
C	Interior Construction	91.6	126.1	106.6	99.7	55.3	80.4	97.3	66.5	83.9	99.9	60.7	82.9	96.8	66.1	83.5
D10	Services: Conveying	100.0	98.2	99.5	100.0	57.8	88.3	100.0	59.8	88.8	100.0	59.8	88.8	100.0	61.4	89.3
D20 - 40	Mechanical	99.6	119.7	107.9	99.9	49.5	79.3	99.9	50.2	79.6	100.0	55.5	81.8	99.9	57.7	82.7
D50	Electrical	102.7	111.0	107.0	98.2	57.9	77.0	97.4	65.3	80.5	100.0	65.3	81.7	97.1	58.1	76.6
E	Equipment & Furnishings	100.0	120.9	101.1	100.0	57.8	97.8	100.0	70.1	98.4	100.0	61.8	98.0	100.0	67.1	98.3
G	Site Work	93.5	105.2	101.4	100.8	89.5	93.2	104.3	89.6	94.3	95.9	89.6	91.6	96.9	89.8	92.1
A-G	WEIGHTED AVERAGE	97.2	119.4	107.0	98.1	59.5	81.0	99.9	64.3	84.2	97.4	63.6	82.5	96.5	65.3	82.7

DIV. NO.	BUILDING SYSTEMS	MISSOURI															
		CAPE GIRARDEAU			COLUMBIA			JOPLIN			KANSAS CITY			SPRINGFIELD		
		MAT.	INST.	TOTAL	MAT.	INST.	TOTAL	MAT.	INST.	TOTAL	MAT.	INST.	TOTAL	MAT.	INST.	TOTAL
A	Substructure	93.6	84.1	87.8	91.7	90.8	91.2	102.8	86.0	92.5	96.0	107.9	103.3	96.3	86.5	90.4
B10	Shell: Superstructure	95.0	93.4	94.4	92.4	105.3	97.5	99.1	92.9	96.6	104.0	113.5	107.8	96.9	96.6	96.8
B20	Exterior Closure	109.6	79.0	95.0	108.6	93.4	101.4	97.4	87.7	92.8	100.6	109.6	104.9	95.3	87.8	91.7
B30	Roofing	101.6	82.3	94.4	94.3	88.1	92.0	96.8	77.5	89.6	96.7	109.4	101.4	98.7	78.9	91.3
C	Interior Construction	98.7	76.5	89.1	94.5	82.9	89.5	97.5	77.0	88.6	98.3	110.4	103.5	99.7	78.4	90.5
D10	Services: Conveying	100.0	90.4	97.3	100.0	98.4	99.5	100.0	93.2	98.1	100.0	100.8	100.2	100.0	95.9	98.8
D20 - 40	Mechanical	99.9	93.2	97.2	99.9	95.1	97.9	100.2	73.5	89.3	100.0	109.6	103.9	99.9	73.5	89.1
D50	Electrical	99.9	105.8	103.0	96.4	87.3	91.6	93.2	67.7	79.8	104.5	101.9	103.1	100.7	71.1	85.1
E	Equipment & Furnishings	100.0	74.4	98.7	100.0	76.3	98.8	100.0	72.8	98.6	100.0	108.8	100.5	100.0	73.8	98.6
G	Site Work	95.3	93.8	94.3	103.7	95.9	98.4	102.6	99.4	100.4	96.5	97.8	97.4	99.3	93.8	95.6
A-G	WEIGHTED AVERAGE	99.5	89.5	95.1	97.7	93.2	95.7	98.6	81.7	91.2	100.7	108.2	104.0	98.6	82.8	91.7

City Cost Indexes — RJ1030-010 Building Systems

DIV. NO.	BUILDING SYSTEMS	MISSOURI ST. JOSEPH			MISSOURI ST. LOUIS			MONTANA BILLINGS			MONTANA BUTTE			MONTANA GREAT FALLS		
		MAT.	INST.	TOTAL	MAT.	INST.	TOTAL	MAT.	INST.	TOTAL	MAT.	INST.	TOTAL	MAT.	INST.	TOTAL
A	Substructure	95.8	98.1	97.2	93.1	104.2	99.9	107.1	78.3	89.5	116.1	79.9	94.1	120.1	77.1	93.9
B10	Shell: Superstructure	100.8	104.6	102.3	99.0	113.0	104.6	109.1	82.5	98.5	104.6	83.9	96.4	108.9	82.0	98.2
B20	Exterior Closure	100.4	98.1	99.3	100.1	112.6	106.1	108.9	74.0	92.3	103.9	77.1	91.1	108.7	75.3	92.8
B30	Roofing	97.2	96.0	96.7	101.6	106.4	103.4	105.2	70.2	92.2	105.1	71.4	92.5	105.6	68.1	91.6
C	Interior Construction	97.9	93.6	96.0	101.0	103.5	102.1	101.2	69.5	87.5	100.6	73.5	88.9	104.1	70.0	89.3
D10	Services: Conveying	100.0	98.2	99.5	100.0	102.0	100.6	100.0	98.2	99.5	100.0	98.2	99.5	100.0	98.2	99.5
D20 - 40	Mechanical	100.2	90.4	96.2	99.9	103.3	101.3	100.2	74.9	89.9	100.3	70.5	88.1	100.3	73.6	89.4
D50	Electrical	102.8	81.6	91.7	102.5	106.2	104.5	97.6	75.2	85.8	105.7	72.2	88.1	97.7	72.2	84.3
E	Equipment & Furnishings	100.0	88.7	99.4	100.0	103.3	100.2	100.0	68.6	98.4	100.0	73.5	98.6	100.0	68.5	98.4
G	Site Work	98.1	93.5	95.0	95.0	97.4	96.6	98.8	97.0	97.6	106.2	96.4	99.6	109.8	96.9	101.0
A-G	WEIGHTED AVERAGE	99.9	94.3	97.4	99.9	106.3	102.7	103.3	77.5	91.9	103.1	77.5	91.8	104.5	76.8	92.3

DIV. NO.	BUILDING SYSTEMS	MONTANA HELENA			MONTANA MISSOULA			NEBRASKA GRAND ISLAND			NEBRASKA LINCOLN			NEBRASKA NORTH PLATTE		
		MAT.	INST.	TOTAL	MAT.	INST.	TOTAL	MAT.	INST.	TOTAL	MAT.	INST.	TOTAL	MAT.	INST.	TOTAL
A	Substructure	102.0	75.6	85.9	94.8	78.1	84.6	109.5	77.8	90.2	89.5	73.0	79.4	109.8	75.5	88.9
B10	Shell: Superstructure	104.7	80.4	95.0	98.2	82.8	92.1	96.6	80.5	90.2	96.6	76.1	88.4	96.9	78.8	89.7
B20	Exterior Closure	104.9	74.7	90.5	108.4	76.3	93.1	104.4	77.7	91.7	98.8	68.0	84.1	97.0	85.5	91.5
B30	Roofing	104.8	71.1	92.3	104.4	80.8	95.6	99.7	74.0	90.2	99.0	69.8	88.2	99.7	75.4	90.7
C	Interior Construction	102.8	65.0	86.4	100.3	72.7	88.3	95.1	78.9	88.1	98.7	64.3	83.8	94.8	76.5	86.9
D10	Services: Conveying	100.0	98.2	99.5	100.0	98.2	99.5	100.0	87.1	96.4	100.0	87.1	96.4	100.0	67.2	90.9
D20 - 40	Mechanical	100.2	70.5	88.0	100.3	70.8	88.2	99.9	76.9	90.5	99.8	76.9	90.4	99.9	72.3	88.6
D50	Electrical	104.9	72.4	87.8	103.4	70.6	86.1	91.4	66.6	78.4	103.1	66.6	83.9	93.5	80.3	86.5
E	Equipment & Furnishings	100.0	65.1	98.2	100.0	70.6	98.5	100.0	74.5	98.7	100.0	55.2	97.7	100.0	74.8	98.7
G	Site Work	99.7	96.0	97.2	88.1	96.5	93.8	104.5	92.0	96.1	93.0	92.0	92.3	105.7	91.2	95.9
A-G	WEIGHTED AVERAGE	102.7	75.0	90.5	100.6	77.0	90.2	98.7	77.8	89.5	98.6	73.0	87.3	98.1	78.3	89.4

DIV. NO.	BUILDING SYSTEMS	NEBRASKA OMAHA			NEVADA CARSON CITY			NEVADA LAS VEGAS			NEVADA RENO			NEW HAMPSHIRE MANCHESTER		
		MAT.	INST.	TOTAL	MAT.	INST.	TOTAL	MAT.	INST.	TOTAL	MAT.	INST.	TOTAL	MAT.	INST.	TOTAL
A	Substructure	93.6	79.1	84.8	97.5	96.6	97.0	97.7	110.4	105.4	103.2	97.5	99.7	107.8	101.3	103.8
B10	Shell: Superstructure	97.0	76.9	89.0	95.7	99.4	97.2	100.8	111.0	104.9	97.5	101.3	99.0	99.8	97.0	98.7
B20	Exterior Closure	101.0	80.0	91.0	109.8	82.2	96.7	107.7	107.6	107.6	110.5	80.4	96.2	99.9	101.1	100.5
B30	Roofing	94.1	79.9	88.8	99.7	88.1	95.4	112.7	100.5	108.2	99.0	88.1	95.0	103.5	94.3	100.1
C	Interior Construction	111.5	73.5	95.1	99.2	94.3	97.1	98.9	114.2	105.5	98.2	94.7	96.7	101.4	95.5	98.9
D10	Services: Conveying	100.0	85.8	96.1	100.0	92.8	98.0	100.0	86.7	96.3	100.0	92.8	98.0	100.0	104.9	101.4
D20 - 40	Mechanical	99.7	79.0	91.2	99.9	82.7	92.9	100.1	106.6	102.7	99.9	82.7	92.9	99.8	88.8	95.3
D50	Electrical	100.5	82.9	91.2	96.2	96.6	96.4	98.3	114.3	106.7	94.3	96.6	95.5	102.0	79.9	90.4
E	Equipment & Furnishings	100.0	72.6	98.6	100.0	98.3	99.9	100.0	113.3	100.7	100.0	98.2	99.9	100.0	88.4	99.4
G	Site Work	81.5	91.3	88.1	72.4	98.5	90.1	67.1	101.0	90.1	64.9	98.5	87.7	94.1	101.6	99.2
A-G	WEIGHTED AVERAGE	100.4	79.5	91.2	98.9	91.8	95.8	100.2	109.0	104.1	99.0	92.0	95.9	100.6	94.0	97.7

DIV. NO.	BUILDING SYSTEMS	NEW HAMPSHIRE NASHUA			NEW HAMPSHIRE PORTSMOUTH			NEW JERSEY CAMDEN			NEW JERSEY ELIZABETH			NEW JERSEY JERSEY CITY		
		MAT.	INST.	TOTAL	MAT.	INST.	TOTAL	MAT.	INST.	TOTAL	MAT.	INST.	TOTAL	MAT.	INST.	TOTAL
A	Substructure	98.7	101.3	100.3	89.2	101.5	96.7	92.3	120.2	109.3	94.1	121.3	110.7	92.3	121.3	110.0
B10	Shell: Superstructure	97.7	97.0	97.4	92.9	97.4	94.7	96.0	113.8	103.1	93.8	118.1	103.5	96.1	116.9	104.4
B20	Exterior Closure	99.5	101.1	100.3	95.7	100.0	97.7	97.0	125.8	110.7	105.9	126.5	115.7	95.6	126.4	110.3
B30	Roofing	97.3	94.3	96.2	97.0	117.1	104.5	96.9	122.8	106.5	97.5	125.6	107.9	96.9	125.6	107.5
C	Interior Construction	101.8	95.5	99.1	99.9	95.5	98.0	102.0	128.8	113.6	104.7	131.9	116.5	102.0	131.9	115.0
D10	Services: Conveying	100.0	104.9	101.4	100.0	104.9	101.4	100.0	108.8	102.5	100.0	113.1	103.8	100.0	113.1	103.6
D20 - 40	Mechanical	100.0	88.8	95.4	100.0	88.9	95.4	100.0	123.4	109.6	100.0	126.6	110.9	100.0	127.2	111.1
D50	Electrical	101.9	79.9	90.3	99.6	79.9	89.3	102.3	136.9	120.5	98.8	134.4	117.6	103.5	134.3	119.7
E	Equipment & Furnishings	100.0	88.4	99.4	100.0	88.4	99.4	100.0	126.7	101.4	100.0	126.7	101.4	100.0	126.8	101.4
G	Site Work	98.4	101.6	100.6	92.8	101.7	98.8	101.9	104.2	103.5	118.7	105.0	109.4	101.9	105.2	104.1
A-G	WEIGHTED AVERAGE	99.8	94.0	97.2	97.3	94.5	96.1	99.0	122.8	109.5	100.1	124.8	111.0	99.0	124.7	110.3

DIV. NO.	BUILDING SYSTEMS	NEW JERSEY NEWARK			NEW JERSEY PATERSON			NEW JERSEY TRENTON			NEW MEXICO ALBUQUERQUE			NEW MEXICO FARMINGTON		
		MAT.	INST.	TOTAL	MAT.	INST.	TOTAL	MAT.	INST.	TOTAL	MAT.	INST.	TOTAL	MAT.	INST.	TOTAL
A	Substructure	110.0	121.3	116.9	106.9	121.4	115.7	102.8	119.8	113.2	101.1	77.7	86.8	103.8	77.7	87.9
B10	Shell: Superstructure	99.5	118.1	106.9	95.1	118.2	104.3	94.5	114.5	102.4	101.5	79.2	92.6	100.3	79.2	91.9
B20	Exterior Closure	101.2	126.5	113.2	99.1	126.5	112.1	95.5	125.6	109.8	105.5	64.0	85.7	111.1	64.0	88.8
B30	Roofing	102.9	125.6	111.3	97.4	124.3	107.4	101.1	123.2	109.3	98.7	74.6	89.7	98.9	74.6	89.9
C	Interior Construction	106.3	131.9	117.4	106.6	131.9	117.5	102.1	129.1	113.8	98.1	65.6	84.0	98.8	65.6	84.4
D10	Services: Conveying	100.0	113.1	103.6	100.0	113.1	103.6	100.0	108.7	102.4	100.0	78.8	94.1	100.0	78.8	94.1
D20 - 40	Mechanical	100.0	126.6	110.9	100.0	127.2	111.1	100.0	122.5	109.2	100.1	69.1	87.4	99.9	69.1	87.3
D50	Electrical	107.7	134.3	121.7	103.5	134.4	119.8	106.1	134.9	121.3	89.4	72.9	80.7	87.6	72.9	79.8
E	Equipment & Furnishings	100.0	126.7	101.4	100.0	126.8	101.4	100.0	122.2	101.1	100.0	65.9	98.2	100.0	65.9	98.2
G	Site Work	117.9	105.0	109.2	116.2	105.2	108.7	99.4	105.1	103.2	82.0	103.9	96.8	89.2	103.9	99.1
A-G	WEIGHTED AVERAGE	102.7	124.8	112.4	100.9	124.9	111.5	99.5	122.5	109.6	99.1	73.5	87.9	99.7	73.5	88.2

DIV. NO.	BUILDING SYSTEMS	NEW MEXICO LAS CRUCES			NEW MEXICO ROSWELL			NEW MEXICO SANTA FE			NEW YORK ALBANY			NEW YORK BINGHAMTON		
		MAT.	INST.	TOTAL	MAT.	INST.	TOTAL	MAT.	INST.	TOTAL	MAT.	INST.	TOTAL	MAT.	INST.	TOTAL
A	Substructure	95.0	70.1	79.8	99.2	77.7	86.1	104.3	77.7	88.1	81.9	101.8	94.0	96.8	101.0	99.4
B10	Shell: Superstructure	98.1	74.3	88.6	99.1	79.2	91.2	100.3	79.2	91.9	95.5	105.4	99.3	95.8	109.6	101.3
B20	Exterior Closure	88.8	63.3	76.7	114.5	64.0	90.5	105.3	64.0	85.7	100.3	100.3	100.3	99.0	124.7	111.3
B30	Roofing	85.8	69.2	79.6	99.1	74.6	90.0	98.7	74.6	89.7	92.3	96.9	94.0	104.7	89.9	99.2
C	Interior Construction	98.8	64.7	84.0	97.1	65.6	83.4	98.2	65.6	84.1	96.7	95.7	96.3	94.4	84.4	90.0
D10	Services: Conveying	100.0	75.8	93.3	100.0	78.8	94.1	100.0	78.8	94.1	100.0	93.9	98.3	100.0	96.8	99.1
D20 - 40	Mechanical	100.3	68.8	87.4	99.9	69.1	87.3	99.9	69.1	87.3	100.0	97.8	99.1	100.4	86.9	94.9
D50	Electrical	88.1	66.6	76.8	88.3	72.9	80.2	102.3	72.9	86.8	101.5	97.3	99.3	100.7	109.8	105.5
E	Equipment & Furnishings	100.0	64.9	98.2	100.0	65.9	98.2	100.0	65.9	98.2	100.0	92.8	99.6	100.0	80.6	99.0
G	Site Work	91.8	83.5	86.2	93.7	103.9	100.6	86.2	103.9	98.2	70.0	107.1	95.1	92.4	92.1	92.2
A-G	WEIGHTED AVERAGE	96.2	69.4	84.4	99.6	73.5	88.1	100.4	73.5	88.5	97.0	99.8	98.2	98.2	100.1	99.0

City Cost Indexes — RJ1030-010 Building Systems

DIV. NO.	BUILDING SYSTEMS	NEW YORK BUFFALO MAT.	INST.	TOTAL	HICKSVILLE MAT.	INST.	TOTAL	NEW YORK MAT.	INST.	TOTAL	RIVERHEAD MAT.	INST.	TOTAL	ROCHESTER MAT.	INST.	TOTAL
A	Substructure	101.1	109.2	106.0	93.5	150.7	128.4	103.3	168.2	142.8	95.0	150.3	128.7	95.9	98.2	97.3
B10	Shell: Superstructure	100.7	103.0	101.7	102.2	145.3	119.4	113.7	162.5	133.1	103.1	144.6	119.6	97.3	99.9	98.3
B20	Exterior Closure	105.9	111.7	108.7	109.9	159.1	133.3	104.1	173.1	136.9	111.8	159.0	134.2	114.2	98.8	106.9
B30	Roofing	100.8	108.1	103.6	105.8	149.7	122.1	103.5	165.0	126.4	106.4	149.7	122.5	99.7	97.8	99.0
C	Interior Construction	95.8	113.4	103.4	95.8	145.3	117.3	98.6	180.6	134.2	96.1	145.1	117.4	98.9	93.6	96.6
D10	Services: Conveying	100.0	104.7	101.3	100.0	119.6	105.4	100.0	129.7	108.2	100.0	119.6	105.4	100.0	95.7	98.8
D20-40	Mechanical	100.0	96.8	98.7	99.8	147.8	119.4	100.1	167.5	127.7	99.9	147.7	119.5	99.9	89.1	95.5
D50	Electrical	98.6	97.8	98.2	99.8	138.3	120.1	104.8	178.2	143.4	101.6	138.3	120.9	101.5	94.0	97.5
E	Equipment & Furnishings	100.0	113.6	100.7	100.0	144.0	102.3	100.0	185.1	104.4	100.0	144.0	102.3	100.0	90.1	99.5
G	Site Work	97.9	98.3	98.2	111.9	128.5	123.1	118.5	124.3	122.4	113.5	128.5	123.6	74.7	109.1	98.0
A-G	WEIGHTED AVERAGE	100.0	104.1	101.8	101.0	145.4	120.5	104.2	167.0	131.9	101.7	145.1	120.9	100.1	96.0	98.3

DIV. NO.	BUILDING SYSTEMS	NEW YORK SCHENECTADY MAT.	INST.	TOTAL	SYRACUSE MAT.	INST.	TOTAL	UTICA MAT.	INST.	TOTAL	WATERTOWN MAT.	INST.	TOTAL	WHITE PLAINS MAT.	INST.	TOTAL
A	Substructure	90.0	101.0	96.7	92.7	98.5	96.2	86.3	92.2	89.9	94.2	95.3	94.9	92.5	132.1	116.6
B10	Shell: Superstructure	97.6	105.0	100.5	97.6	100.2	98.6	95.1	96.0	95.5	96.1	98.6	97.1	100.9	134.2	114.1
B20	Exterior Closure	100.0	98.2	99.2	102.8	100.5	101.7	99.4	97.9	98.7	106.7	100.7	103.8	97.2	131.9	113.7
B30	Roofing	89.2	95.6	91.6	100.1	95.3	98.3	89.0	92.9	90.4	89.2	94.2	91.1	104.2	135.5	115.8
C	Interior Construction	96.6	95.2	96.0	95.2	89.9	92.9	96.6	85.7	91.9	95.5	89.4	92.9	91.9	138.5	112.1
D10	Services: Conveying	100.0	93.0	98.0	100.0	97.2	99.2	100.0	91.8	97.7	100.0	96.3	99.0	100.0	114.1	103.9
D20-40	Mechanical	100.2	96.4	98.6	100.2	90.1	96.1	100.2	88.5	95.4	100.2	83.4	93.3	100.3	122.3	109.3
D50	Electrical	100.9	97.3	99.0	100.9	97.3	99.0	99.0	97.3	98.1	100.9	90.0	95.1	94.6	154.5	126.2
E	Equipment & Furnishings	100.0	92.7	99.6	100.0	87.1	99.3	100.0	81.2	99.0	100.0	87.6	99.4	100.0	130.5	101.6
G	Site Work	71.7	107.0	95.6	91.3	106.6	101.6	69.8	104.7	93.4	77.1	106.3	96.9	111.9	117.5	115.7
A-G	WEIGHTED AVERAGE	97.6	99.0	98.2	98.7	96.2	97.6	96.7	93.3	95.2	98.2	93.1	95.9	98.3	133.0	113.6

DIV. NO.	BUILDING SYSTEMS	NEW YORK YONKERS MAT.	INST.	TOTAL	NORTH CAROLINA ASHEVILLE MAT.	INST.	TOTAL	CHARLOTTE MAT.	INST.	TOTAL	DURHAM MAT.	INST.	TOTAL	FAYETTEVILLE MAT.	INST.	TOTAL
A	Substructure	101.8	132.0	120.2	107.9	54.0	75.1	112.4	54.6	77.2	109.7	56.9	77.5	111.8	63.4	82.3
B10	Shell: Superstructure	110.0	134.2	119.7	92.8	65.3	81.8	97.1	64.9	84.3	103.9	65.1	88.5	107.6	70.0	92.6
B20	Exterior Closure	103.8	131.9	117.2	90.8	45.9	69.4	93.6	53.8	74.7	93.4	39.8	67.9	91.3	43.6	68.6
B30	Roofing	104.4	135.5	116.0	100.0	42.5	78.6	100.5	44.7	79.7	100.7	45.7	80.3	99.9	46.2	79.9
C	Interior Construction	96.4	138.5	114.6	94.6	39.9	70.9	96.9	41.3	72.8	97.6	44.6	74.6	95.0	58.1	79.0
D10	Services: Conveying	100.0	117.8	104.9	100.0	83.8	95.5	100.0	84.6	95.7	100.0	77.6	93.8	100.0	78.5	94.0
D20-40	Mechanical	100.3	122.3	109.3	100.3	53.7	81.2	99.9	54.6	81.4	100.4	52.3	80.7	100.2	53.5	81.1
D50	Electrical	103.0	154.5	130.1	97.9	58.1	76.9	96.8	57.6	76.2	98.7	46.4	71.2	96.6	52.8	73.6
E	Equipment & Furnishings	100.0	130.5	101.6	100.0	36.9	96.7	100.0	38.9	96.8	100.0	45.3	97.2	100.0	65.7	98.2
G	Site Work	117.9	117.3	117.5	107.7	75.6	85.9	108.3	75.7	86.3	107.6	84.4	91.9	106.6	84.6	91.7
A-G	WEIGHTED AVERAGE	102.9	133.1	116.2	97.1	55.1	78.6	98.7	56.5	80.1	100.3	53.9	79.9	100.2	59.1	82.1

DIV. NO.	BUILDING SYSTEMS	NORTH CAROLINA GREENSBORO MAT.	INST.	TOTAL	RALEIGH MAT.	INST.	TOTAL	WILMINGTON MAT.	INST.	TOTAL	WINSTON-SALEM MAT.	INST.	TOTAL	NORTH DAKOTA BISMARCK MAT.	INST.	TOTAL
A	Substructure	108.9	57.7	77.7	113.9	59.7	80.9	107.9	57.7	77.3	110.7	59.3	79.4	101.3	63.0	78.0
B10	Shell: Superstructure	98.8	65.5	85.6	98.0	66.7	85.6	92.6	66.8	82.4	97.3	66.7	85.1	96.9	69.5	86.0
B20	Exterior Closure	92.5	41.7	68.3	92.1	43.7	69.1	86.5	44.2	66.4	92.5	42.8	68.9	99.4	59.4	80.4
B30	Roofing	100.5	43.6	79.3	100.7	46.2	80.4	100.0	49.1	81.1	100.5	44.8	79.8	110.3	49.9	87.8
C	Interior Construction	97.7	44.5	74.7	95.7	47.3	74.7	95.0	49.2	75.1	97.7	48.3	76.3	106.7	45.2	80.0
D10	Services: Conveying	100.0	82.9	95.3	100.0	78.8	94.1	100.0	79.5	94.3	100.0	84.3	95.6	100.0	90.3	97.3
D20-40	Mechanical	100.3	53.5	81.2	100.0	54.0	81.2	100.3	54.9	81.7	100.3	54.2	81.4	100.2	61.2	84.3
D50	Electrical	97.6	58.1	76.8	101.1	41.9	70.0	98.1	52.8	74.2	97.6	58.1	76.8	102.5	72.4	86.7
E	Equipment & Furnishings	100.0	46.2	97.2	100.0	48.1	97.3	100.0	50.1	97.4	100.0	50.6	97.4	100.0	37.9	96.8
G	Site Work	107.4	84.6	92.0	108.0	84.6	92.2	108.7	75.8	86.4	107.8	84.7	92.2	98.2	97.3	97.6
A-G	WEIGHTED AVERAGE	99.0	56.2	80.2	99.0	55.1	79.7	96.7	56.5	79.0	98.8	57.5	80.6	101.0	64.4	84.9

DIV. NO.	BUILDING SYSTEMS	NORTH DAKOTA FARGO MAT.	INST.	TOTAL	GRAND FORKS MAT.	INST.	TOTAL	MINOT MAT.	INST.	TOTAL	OHIO AKRON MAT.	INST.	TOTAL	CANTON MAT.	INST.	TOTAL
A	Substructure	99.2	63.8	77.6	108.2	59.1	78.3	108.2	71.6	85.9	93.2	95.9	94.9	93.8	89.4	91.1
B10	Shell: Superstructure	98.2	70.1	87.0	97.9	64.5	84.6	98.0	76.6	89.5	95.3	86.8	91.9	95.4	78.9	88.9
B20	Exterior Closure	110.1	59.5	86.0	106.4	63.1	85.8	104.3	65.8	86.0	92.8	93.8	93.3	92.4	83.6	88.2
B30	Roofing	107.1	51.1	86.3	106.4	45.3	83.7	106.2	54.2	86.9	108.0	96.2	103.6	109.1	92.9	103.1
C	Interior Construction	106.5	49.4	81.8	107.6	35.8	76.5	107.1	67.2	89.8	106.7	91.2	100.0	103.2	79.9	93.1
D10	Services: Conveying	100.0	90.3	97.3	100.0	33.9	81.7	100.0	90.3	97.3	100.0	90.4	97.3	100.0	89.1	97.0
D20-40	Mechanical	100.3	66.9	86.6	100.5	35.0	73.7	100.5	61.3	84.4	100.0	96.9	98.7	100.0	82.4	92.8
D50	Electrical	99.7	67.8	82.9	100.5	55.8	77.0	104.0	78.9	90.8	98.9	92.7	95.6	98.1	89.1	93.4
E	Equipment & Furnishings	100.0	38.2	96.8	100.0	33.7	96.6	100.0	69.3	98.4	100.0	89.5	99.5	100.0	79.4	98.9
G	Site Work	98.8	97.3	97.8	110.6	93.7	99.2	107.9	97.3	100.7	94.5	105.0	101.6	94.6	104.5	101.3
A-G	WEIGHTED AVERAGE	101.9	65.8	86.0	102.4	52.6	80.5	102.4	71.6	88.9	99.1	93.6	96.6	98.5	85.0	92.5

DIV. NO.	BUILDING SYSTEMS	OHIO CINCINNATI MAT.	INST.	TOTAL	CLEVELAND MAT.	INST.	TOTAL	COLUMBUS MAT.	INST.	TOTAL	DAYTON MAT.	INST.	TOTAL	LORAIN MAT.	INST.	TOTAL
A	Substructure	93.3	88.0	90.1	92.3	101.8	98.1	95.0	91.4	92.8	89.2	87.3	88.0	90.6	96.2	94.0
B10	Shell: Superstructure	95.5	83.6	90.8	96.4	91.7	94.5	97.4	84.2	92.2	94.1	79.3	88.2	95.2	87.1	92.0
B20	Exterior Closure	92.5	85.7	89.2	93.4	103.3	98.1	97.2	92.8	95.1	92.3	82.1	87.4	90.6	98.1	94.2
B30	Roofing	100.4	91.4	97.1	106.7	110.1	108.0	101.0	96.5	99.3	105.8	88.5	99.4	109.1	101.8	106.4
C	Interior Construction	101.8	82.6	93.4	100.7	99.7	100.2	100.5	87.3	94.7	102.2	80.0	92.6	103.1	90.4	97.6
D10	Services: Conveying	100.0	93.9	98.3	100.0	95.0	98.6	100.0	95.7	98.8	100.0	93.2	98.1	100.0	92.8	98.0
D20-40	Mechanical	100.0	82.7	92.9	100.0	100.2	100.1	100.0	92.4	96.9	100.9	83.3	93.7	100.0	83.7	93.3
D50	Electrical	97.4	79.7	88.1	98.4	108.1	103.5	96.6	86.4	91.2	95.9	82.2	88.7	98.3	88.2	93.0
E	Equipment & Furnishings	100.0	79.8	99.0	100.0	97.4	99.9	100.0	85.8	99.3	100.0	79.1	98.9	100.0	83.4	99.1
G	Site Work	93.6	103.7	100.4	94.4	105.3	101.8	94.6	99.3	97.7	92.6	103.8	100.1	93.9	103.9	100.7
A-G	WEIGHTED AVERAGE	97.9	85.1	92.3	98.2	100.7	99.3	98.6	90.0	94.8	97.7	83.9	91.6	98.1	90.8	94.9

City Cost Indexes — RJ1030-010 Building Systems

DIV. NO.	BUILDING SYSTEMS	OHIO SPRINGFIELD MAT.	INST.	TOTAL	OHIO TOLEDO MAT.	INST.	TOTAL	OHIO YOUNGSTOWN MAT.	INST.	TOTAL	OKLAHOMA ENID MAT.	INST.	TOTAL	OKLAHOMA LAWTON MAT.	INST.	TOTAL
A	Substructure	90.6	88.6	89.4	94.9	97.0	96.2	92.7	93.6	93.2	95.0	55.5	70.9	93.3	59.8	72.9
B10	Shell: Superstructure	94.4	80.4	88.8	97.2	90.9	94.7	95.2	83.7	90.7	93.4	56.8	78.8	96.0	60.0	81.6
B20	Exterior Closure	91.8	81.9	87.1	100.9	97.1	99.1	92.2	90.8	91.5	105.1	54.6	81.1	100.3	55.1	78.8
B30	Roofing	105.7	89.5	99.6	102.6	102.3	102.4	109.3	93.8	103.5	98.6	56.2	82.8	98.4	58.0	83.4
C	Interior Construction	101.2	84.0	93.8	98.8	95.3	97.3	103.1	85.6	95.5	93.3	36.3	68.6	94.9	45.8	73.6
D10	Services: Conveying	100.0	93.4	98.2	100.0	98.4	99.6	100.0	89.8	97.2	100.0	74.2	92.8	100.0	74.2	92.8
D20-40	Mechanical	100.9	83.5	93.8	100.1	100.3	100.2	100.0	90.8	96.3	100.3	60.2	83.9	100.3	60.2	83.9
D50	Electrical	95.9	82.4	88.8	96.7	102.1	99.5	98.3	84.4	91.0	94.6	67.2	80.2	96.4	67.2	81.0
E	Equipment & Furnishings	100.0	84.7	99.2	100.0	93.9	99.7	100.0	82.2	99.1	100.0	30.8	96.4	100.0	45.2	97.2
G	Site Work	92.9	102.4	99.4	93.9	99.9	98.0	94.4	105.5	101.9	101.1	90.0	93.6	96.9	91.5	93.2
A-G	WEIGHTED AVERAGE	97.6	84.8	91.9	98.8	97.5	98.2	98.4	89.1	94.3	97.5	58.0	80.1	97.7	60.6	81.3

DIV. NO.	BUILDING SYSTEMS	OKLAHOMA MUSKOGEE MAT.	INST.	TOTAL	OKLAHOMA CITY MAT.	INST.	TOTAL	TULSA MAT.	INST.	TOTAL	OREGON EUGENE MAT.	INST.	TOTAL	OREGON MEDFORD MAT.	INST.	TOTAL
A	Substructure	88.0	45.0	61.8	93.7	58.6	72.3	93.6	56.7	71.1	98.5	102.8	101.1	100.5	102.7	101.9
B10	Shell: Superstructure	92.3	44.7	73.4	97.6	58.6	82.1	95.8	64.2	83.2	90.2	99.7	94.0	89.9	99.6	93.8
B20	Exterior Closure	106.1	43.7	76.4	100.1	56.2	79.2	99.9	56.9	79.5	102.6	106.5	104.5	107.3	106.5	106.9
B30	Roofing	98.2	37.9	75.8	93.4	57.8	80.2	98.2	51.5	80.8	103.0	90.2	98.2	103.6	91.8	99.2
C	Interior Construction	93.9	31.2	66.7	95.7	40.7	71.9	94.6	40.7	71.2	99.3	99.8	99.5	101.7	98.7	100.4
D10	Services: Conveying	100.0	74.0	92.8	100.0	74.8	93.0	100.0	75.3	93.1	100.0	103.8	101.0	100.0	103.8	101.0
D20-40	Mechanical	100.3	23.0	68.7	100.0	61.0	84.1	100.3	46.5	78.3	100.0	99.2	99.7	100.0	99.2	99.7
D50	Electrical	94.2	27.5	59.0	102.1	67.2	83.8	96.4	35.8	64.5	99.5	95.9	97.6	103.2	82.8	92.4
E	Equipment & Furnishings	100.0	28.4	96.3	100.0	37.1	96.7	100.0	36.6	96.7	100.0	100.0	100.0	100.0	100.0	100.0
G	Site Work	88.9	85.5	86.6	95.7	92.0	93.2	94.7	86.9	89.4	100.4	103.1	102.2	107.9	103.1	104.7
A-G	WEIGHTED AVERAGE	96.8	38.8	71.3	98.5	59.8	81.4	97.5	52.8	77.8	98.2	100.3	99.1	99.7	98.3	99.1

DIV. NO.	BUILDING SYSTEMS	OREGON PORTLAND MAT.	INST.	TOTAL	OREGON SALEM MAT.	INST.	TOTAL	PENNSYLVANIA ALLENTOWN MAT.	INST.	TOTAL	PENNSYLVANIA ALTOONA MAT.	INST.	TOTAL	PENNSYLVANIA ERIE MAT.	INST.	TOTAL
A	Substructure	100.3	102.9	101.9	93.3	102.8	99.1	87.5	109.7	101.0	92.4	92.8	92.7	92.7	93.7	93.3
B10	Shell: Superstructure	91.4	99.9	94.8	89.5	99.8	93.6	96.8	118.4	105.4	92.2	105.2	97.4	93.1	105.2	97.9
B20	Exterior Closure	102.4	106.5	104.4	104.9	106.5	105.6	100.2	105.4	102.7	92.7	72.5	83.1	86.5	91.8	89.0
B30	Roofing	102.9	96.6	100.6	102.1	93.6	98.9	100.1	118.5	107.0	99.6	91.9	96.8	100.1	93.1	97.5
C	Interior Construction	98.1	99.4	98.7	99.8	99.8	99.8	94.1	111.7	101.7	89.9	87.8	89.0	90.8	90.1	90.5
D10	Services: Conveying	100.0	103.8	101.0	100.0	103.8	101.0	100.0	104.7	101.3	100.0	97.6	99.3	100.0	102.8	100.8
D20-40	Mechanical	100.0	99.2	99.7	100.0	99.2	99.7	100.2	110.5	104.4	99.8	85.6	94.0	99.8	93.3	97.1
D50	Electrical	99.7	101.3	100.5	108.2	95.9	101.7	99.9	98.8	99.3	89.4	108.4	99.4	91.4	89.0	90.1
E	Equipment & Furnishings	100.0	100.9	100.0	100.0	100.6	100.0	100.0	118.1	100.9	100.0	82.8	99.1	100.0	85.7	99.3
G	Site Work	103.4	103.1	103.2	96.2	103.1	100.9	90.2	105.9	100.8	92.3	105.7	101.4	89.2	106.1	100.7
A-G	WEIGHTED AVERAGE	98.4	101.2	99.6	98.9	100.4	99.5	97.7	109.5	102.9	94.4	93.2	93.9	94.2	95.2	94.6

DIV. NO.	BUILDING SYSTEMS	PENNSYLVANIA HARRISBURG MAT.	INST.	TOTAL	PHILADELPHIA MAT.	INST.	TOTAL	PITTSBURGH MAT.	INST.	TOTAL	READING MAT.	INST.	TOTAL	SCRANTON MAT.	INST.	TOTAL
A	Substructure	88.1	97.9	94.1	97.8	126.4	115.2	100.1	101.8	101.1	81.9	98.9	92.3	89.7	97.5	94.5
B10	Shell: Superstructure	98.6	108.8	102.6	99.2	134.2	113.1	98.7	111.6	103.8	93.0	107.6	98.8	98.9	109.6	103.1
B20	Exterior Closure	98.4	93.4	96.0	99.4	130.4	114.1	98.8	106.2	102.3	100.0	96.7	98.4	100.3	97.7	99.0
B30	Roofing	103.6	108.9	105.6	101.3	132.4	112.9	105.3	101.4	104.1	100.9	109.9	104.2	100.0	97.0	98.9
C	Interior Construction	94.5	91.6	93.2	98.1	139.9	116.2	99.0	101.4	100.1	95.2	92.3	93.9	95.2	94.3	94.8
D10	Services: Conveying	100.0	99.1	99.8	100.0	117.5	104.8	100.0	103.8	101.0	100.0	101.9	100.5	100.0	98.4	99.6
D20-40	Mechanical	100.0	93.6	97.4	100.0	129.0	111.9	99.9	99.3	99.7	100.1	106.6	102.8	100.2	101.5	100.7
D50	Electrical	99.4	87.0	92.9	97.5	149.1	124.7	97.1	108.4	103.1	99.8	92.8	96.1	99.9	93.3	96.4
E	Equipment & Furnishings	100.0	88.3	99.4	100.0	138.4	102.0	100.0	97.5	99.9	100.0	88.5	99.4	100.0	87.0	99.3
G	Site Work	81.7	104.9	97.4	101.5	99.9	100.4	103.1	106.9	105.6	98.9	112.8	108.3	90.6	106.3	101.2
A-G	WEIGHTED AVERAGE	97.8	96.5	97.2	99.2	132.2	113.7	99.4	104.7	101.8	97.1	101.1	98.8	98.4	99.9	99.1

DIV. NO.	BUILDING SYSTEMS	PENNSYLVANIA YORK MAT.	INST.	TOTAL	PUERTO RICO SAN JUAN MAT.	INST.	TOTAL	RHODE ISLAND PROVIDENCE MAT.	INST.	TOTAL	SOUTH CAROLINA CHARLESTON MAT.	INST.	TOTAL	SOUTH CAROLINA COLUMBIA MAT.	INST.	TOTAL
A	Substructure	83.4	97.7	92.1	127.7	37.5	72.8	102.7	121.1	113.9	104.1	56.1	74.8	105.6	55.9	75.3
B10	Shell: Superstructure	95.4	108.7	100.7	113.4	28.0	79.4	98.8	126.8	110.0	92.3	62.5	80.4	92.8	61.9	80.5
B20	Exterior Closure	101.8	93.0	97.6	97.5	16.8	59.1	106.3	127.6	116.4	96.1	41.3	70.0	93.5	39.5	67.9
B30	Roofing	99.1	108.6	102.7	124.5	22.0	86.4	102.1	117.7	107.9	100.1	42.0	78.5	98.5	41.0	77.1
C	Interior Construction	91.3	91.5	91.4	176.7	15.9	107.0	97.1	127.4	110.2	98.1	43.5	74.4	96.9	40.8	72.6
D10	Services: Conveying	100.0	98.9	99.7	100.0	16.4	76.8	100.0	124.4	100.7	100.0	71.3	92.0	100.0	71.3	92.0
D20-40	Mechanical	100.2	93.3	97.3	102.8	13.1	66.1	99.8	111.1	104.4	100.3	39.2	75.3	99.9	34.7	73.2
D50	Electrical	93.1	87.0	89.9	131.0	12.4	68.5	101.2	97.1	99.1	97.4	92.5	94.8	97.8	47.3	71.2
E	Equipment & Furnishings	100.0	88.3	99.4	100.0	15.2	95.6	100.0	119.7	101.0	100.0	41.1	96.9	100.0	38.5	96.8
G	Site Work	80.7	104.9	97.0	135.0	91.0	105.2	92.5	103.9	100.3	99.8	84.3	89.3	97.8	84.3	88.7
A-G	WEIGHTED AVERAGE	96.0	96.4	96.2	121.0	23.8	78.2	100.1	116.7	107.4	97.7	56.7	79.6	97.3	48.8	75.9

DIV. NO.	BUILDING SYSTEMS	SOUTH CAROLINA FLORENCE MAT.	INST.	TOTAL	GREENVILLE MAT.	INST.	TOTAL	SPARTANBURG MAT.	INST.	TOTAL	SOUTH DAKOTA ABERDEEN MAT.	INST.	TOTAL	SOUTH DAKOTA PIERRE MAT.	INST.	TOTAL
A	Substructure	95.2	55.8	71.2	95.5	54.0	70.2	95.7	54.0	70.3	102.8	54.2	73.2	92.3	56.6	70.6
B10	Shell: Superstructure	88.7	62.0	78.1	89.4	58.9	77.3	89.6	58.9	77.4	102.6	54.4	83.4	100.3	60.3	84.4
B20	Exterior Closure	94.9	41.3	69.4	91.9	40.8	67.6	92.8	40.8	68.1	101.0	56.0	79.6	99.0	55.8	78.4
B30	Roofing	100.4	41.9	78.7	100.4	41.9	78.6	100.4	41.9	78.7	105.0	46.4	83.2	107.1	44.8	84.0
C	Interior Construction	94.8	41.2	71.5	95.6	42.5	72.6	96.0	42.5	72.8	102.1	40.0	75.2	102.2	40.0	75.3
D10	Services: Conveying	100.0	71.3	92.0	100.0	71.3	92.0	100.0	71.3	92.0	100.0	42.9	84.2	100.0	82.4	95.1
D20-40	Mechanical	100.3	34.7	73.5	100.3	33.3	72.9	100.3	33.3	72.9	100.2	41.9	76.4	100.0	41.7	76.2
D50	Electrical	95.2	45.0	68.8	97.4	48.7	71.7	97.4	48.7	71.7	101.1	53.3	75.9	104.2	52.1	76.8
E	Equipment & Furnishings	100.0	38.4	96.8	100.0	38.5	96.8	100.0	38.5	96.8	100.0	37.6	96.8	100.0	38.4	96.8
G	Site Work	109.5	84.3	92.5	104.5	83.9	90.6	104.2	83.9	90.5	97.8	94.5	95.6	93.2	94.5	94.1
A-G	WEIGHTED AVERAGE	96.0	48.7	75.2	96.0	48.4	75.0	96.3	48.4	75.2	101.3	51.7	79.4	100.5	53.7	79.8

City Cost Indexes

RJ1030-010 Building Systems

DIV. NO.	BUILDING SYSTEMS	SOUTH DAKOTA						TENNESSEE								
		RAPID CITY			SIOUX FALLS			CHATTANOOGA			JACKSON			JOHNSON CITY		
		MAT.	INST.	TOTAL	MAT.	INST.	TOTAL	MAT.	INST.	TOTAL	MAT.	INST.	TOTAL	MAT.	INST.	TOTAL
A	Substructure	99.8	56.1	73.2	94.6	58.1	72.4	98.2	69.1	80.5	100.0	62.6	77.2	86.1	60.7	70.6
B10	Shell: Superstructure	103.4	60.1	86.2	102.3	61.4	86.0	96.5	74.3	87.6	94.1	70.6	84.8	91.0	69.3	82.4
B20	Exterior Closure	100.0	56.8	79.5	97.0	55.9	77.5	97.1	52.8	76.0	106.0	52.0	80.3	118.0	48.9	85.2
B30	Roofing	105.3	46.5	83.5	111.7	47.5	87.8	99.4	59.3	84.5	94.8	52.5	79.1	94.7	54.8	79.9
C	Interior Construction	104.0	44.6	78.3	103.3	48.1	79.4	98.1	57.5	80.5	96.5	47.2	75.1	97.8	41.1	73.2
D10	Services: Conveying	100.0	82.7	95.2	100.0	82.4	95.1	100.0	41.0	83.6	100.0	44.6	84.6	100.0	45.9	84.9
D20 - 40	Mechanical	100.2	42.2	76.5	100.1	40.5	75.7	100.2	62.2	84.6	100.0	70.3	87.9	99.9	55.3	81.7
D50	Electrical	97.4	52.1	73.5	102.0	66.3	83.2	102.3	67.9	84.2	101.6	61.0	80.2	91.9	48.3	68.9
E	Equipment & Furnishings	100.0	35.1	96.6	100.0	41.5	97.0	100.0	57.7	97.8	100.0	47.2	97.3	100.0	38.3	96.8
G	Site Work	95.8	94.6	95.0	93.7	96.2	95.4	102.3	99.8	100.6	98.1	98.1	98.1	109.1	88.4	95.1
A-G	**WEIGHTED AVERAGE**	101.2	54.6	80.6	100.9	57.1	81.6	98.9	65.5	84.2	98.9	63.2	83.2	98.4	56.0	79.8

DIV. NO.	BUILDING SYSTEMS	TENNESSEE									TEXAS					
		KNOXVILLE			MEMPHIS			NASHVILLE			ABILENE			AMARILLO		
		MAT.	INST.	TOTAL	MAT.	INST.	TOTAL	MAT.	INST.	TOTAL	MAT.	INST.	TOTAL	MAT.	INST.	TOTAL
A	Substructure	92.2	58.5	71.7	92.8	70.6	79.2	92.9	73.6	81.2	94.9	54.8	70.5	94.6	61.5	74.4
B10	Shell: Superstructure	96.1	68.3	85.0	95.9	77.7	88.7	97.7	78.6	90.1	98.8	57.1	82.2	98.6	61.4	83.8
B20	Exterior Closure	85.2	55.1	70.9	92.5	64.9	79.4	89.7	60.4	75.8	100.4	56.1	79.3	100.4	57.9	80.2
B30	Roofing	92.8	60.4	80.8	96.0	63.5	83.9	95.6	61.4	82.9	98.4	46.9	79.2	98.0	50.5	80.3
C	Interior Construction	93.0	45.8	72.6	96.9	57.9	80.0	99.9	65.2	84.9	93.3	45.8	72.7	94.2	56.6	77.9
D10	Services: Conveying	100.0	48.7	85.8	100.0	84.3	95.6	100.0	84.5	95.7	100.0	81.6	94.9	100.0	70.4	91.8
D20 - 40	Mechanical	99.9	65.5	85.9	100.0	69.5	87.5	100.0	78.8	91.3	100.3	44.4	77.4	100.0	50.2	79.7
D50	Electrical	97.9	56.9	76.3	102.1	66.0	83.1	96.9	64.3	79.7	96.5	49.5	71.8	102.0	65.8	82.9
E	Equipment & Furnishings	100.0	39.2	96.8	100.0	62.2	98.0	100.0	64.4	98.2	100.0	37.9	96.8	100.0	55.3	97.7
G	Site Work	89.7	88.9	89.1	91.6	94.4	93.4	98.2	100.9	100.1	95.1	86.6	89.3	95.1	87.6	90.1
A-G	**WEIGHTED AVERAGE**	95.5	60.8	80.2	97.4	70.0	85.3	97.6	73.1	86.8	98.1	53.6	78.5	98.6	60.1	81.7

DIV. NO.	BUILDING SYSTEMS	TEXAS														
		AUSTIN			BEAUMONT			CORPUS CHRISTI			DALLAS			EL PASO		
		MAT.	INST.	TOTAL	MAT.	INST.	TOTAL	MAT.	INST.	TOTAL	MAT.	INST.	TOTAL	MAT.	INST.	TOTAL
A	Substructure	94.9	59.3	73.2	85.7	58.4	69.1	108.7	52.2	74.3	97.5	66.3	78.5	90.7	53.9	68.3
B10	Shell: Superstructure	94.9	58.4	80.4	95.2	57.6	80.2	96.1	59.7	81.6	100.0	70.9	88.9	97.8	54.4	80.5
B20	Exterior Closure	96.1	49.1	73.7	105.5	57.0	82.4	95.8	49.1	73.6	102.8	58.6	81.7	99.3	47.3	74.6
B30	Roofing	91.6	49.4	75.9	103.0	53.9	84.8	98.7	42.9	78.0	94.0	63.5	82.6	95.4	51.8	79.2
C	Interior Construction	95.2	49.9	75.6	93.8	52.2	75.8	102.4	43.3	76.8	99.0	61.3	82.7	94.2	47.1	73.7
D10	Services: Conveying	100.0	81.3	94.8	100.0	83.7	95.5	100.0	82.5	95.2	100.0	84.5	95.7	100.0	82.3	95.1
D20 - 40	Mechanical	100.0	54.3	81.3	100.2	58.2	83.0	100.1	39.4	75.3	99.9	63.3	84.9	100.0	34.8	73.3
D50	Electrical	97.1	63.2	79.3	94.3	71.2	82.2	92.6	53.7	72.1	93.5	66.3	79.2	93.1	52.2	71.6
E	Equipment & Furnishings	100.0	54.4	97.7	100.0	49.9	97.4	100.0	37.2	96.7	100.0	66.7	98.3	100.0	46.0	97.2
G	Site Work	101.1	86.8	91.4	88.1	87.0	87.3	137.6	82.2	100.1	105.0	88.0	93.5	92.7	85.9	88.1
A-G	**WEIGHTED AVERAGE**	97.1	58.1	79.9	97.3	61.3	81.4	99.7	51.7	78.6	99.5	66.5	85.0	97.2	50.6	76.7

DIV. NO.	BUILDING SYSTEMS	TEXAS														
		FORT WORTH			HOUSTON			LAREDO			LUBBOCK			ODESSA		
		MAT.	INST.	TOTAL	MAT.	INST.	TOTAL	MAT.	INST.	TOTAL	MAT.	INST.	TOTAL	MAT.	INST.	TOTAL
A	Substructure	96.7	64.4	77.0	90.1	70.5	78.2	86.3	56.6	68.2	99.3	59.1	74.8	95.0	53.6	69.8
B10	Shell: Superstructure	96.5	63.9	83.5	98.3	77.3	89.9	93.7	55.6	78.5	101.7	67.0	87.9	98.3	55.7	81.3
B20	Exterior Closure	100.8	57.7	80.3	102.2	67.3	85.6	100.7	55.2	79.0	100.6	55.7	79.3	100.4	44.0	73.5
B30	Roofing	95.3	52.2	79.3	93.6	67.0	83.7	94.4	47.6	77.0	88.0	49.5	73.7	98.4	42.2	77.5
C	Interior Construction	91.8	59.9	78.0	102.3	62.8	85.2	94.8	42.9	72.3	99.4	50.4	78.2	93.3	41.3	70.8
D10	Services: Conveying	100.0	84.1	95.6	100.0	86.6	96.3	100.0	80.2	94.5	100.0	81.3	94.8	100.0	82.8	95.2
D20 - 40	Mechanical	100.1	58.8	83.2	100.0	66.4	86.3	100.0	38.1	74.7	99.8	43.5	76.8	100.3	45.7	77.9
D50	Electrical	93.7	61.2	76.6	96.0	68.7	81.6	94.4	61.0	76.8	95.1	50.5	71.6	96.7	41.0	67.4
E	Equipment & Furnishings	100.0	66.4	98.3	100.0	65.4	98.2	100.0	37.4	96.8	100.0	54.4	97.6	100.0	37.9	96.8
G	Site Work	96.7	87.0	90.1	103.0	85.7	91.3	100.5	86.6	91.1	115.4	84.5	94.5	95.1	86.9	89.5
A-G	**WEIGHTED AVERAGE**	97.1	62.9	82.0	99.3	70.3	86.5	96.7	53.2	77.5	99.7	56.2	80.6	98.0	50.1	76.9

DIV. NO.	BUILDING SYSTEMS	TEXAS									UTAH					
		SAN ANTONIO			WACO			WICHITA FALLS			LOGAN			OGDEN		
		MAT.	INST.	TOTAL	MAT.	INST.	TOTAL	MAT.	INST.	TOTAL	MAT.	INST.	TOTAL	MAT.	INST.	TOTAL
A	Substructure	86.6	66.4	74.3	89.1	55.8	68.9	92.8	53.3	68.7	93.7	74.2	81.8	92.8	74.2	81.5
B10	Shell: Superstructure	94.3	63.7	82.1	97.5	56.3	81.1	98.3	56.1	81.5	99.1	71.5	88.1	99.6	71.5	88.4
B20	Exterior Closure	100.6	66.6	84.4	102.0	53.2	78.8	102.2	56.1	80.3	114.4	60.6	88.9	101.0	60.6	81.8
B30	Roofing	94.1	64.1	82.9	98.9	46.4	79.4	98.9	48.0	80.0	97.3	63.3	84.7	96.2	63.3	84.0
C	Interior Construction	96.1	55.7	78.6	90.7	39.2	68.4	90.9	44.2	70.7	94.8	57.7	78.7	93.8	57.7	78.1
D10	Services: Conveying	100.0	84.7	95.7	100.0	84.1	95.6	100.0	70.1	91.7	100.0	54.5	87.4	100.0	54.5	87.4
D20 - 40	Mechanical	99.9	66.2	86.1	100.3	50.4	79.9	100.3	45.8	78.0	99.9	70.7	88.0	99.9	70.7	88.0
D50	Electrical	94.8	61.1	77.0	96.7	66.3	80.7	98.4	56.8	76.5	94.3	72.3	82.7	94.7	72.3	82.9
E	Equipment & Furnishings	100.0	54.3	97.6	100.0	36.9	96.7	100.0	36.0	96.7	100.0	57.9	97.8	100.0	57.9	97.8
G	Site Work	100.2	91.4	94.3	93.8	86.9	89.2	94.6	86.6	89.2	92.1	95.8	94.6	81.5	95.8	91.2
A-G	**WEIGHTED AVERAGE**	97.0	65.5	83.2	97.3	55.8	79.0	97.9	54.1	78.6	99.4	69.1	86.1	97.6	69.1	85.1

DIV. NO.	BUILDING SYSTEMS	UTAH						VERMONT						VIRGINIA		
		PROVO			SALT LAKE CITY			BURLINGTON			RUTLAND			ALEXANDRIA		
		MAT.	INST.	TOTAL	MAT.	INST.	TOTAL	MAT.	INST.	TOTAL	MAT.	INST.	TOTAL	MAT.	INST.	TOTAL
A	Substructure	95.0	73.8	82.1	98.4	74.2	83.7	103.1	94.7	98.0	97.1	94.6	95.6	108.0	82.1	92.2
B10	Shell: Superstructure	98.2	71.4	87.5	104.4	71.4	91.3	100.2	86.1	94.6	99.7	86.1	93.1	99.7	88.3	95.2
B20	Exterior Closure	116.5	60.6	89.9	122.1	60.6	92.9	104.2	68.8	87.4	95.4	68.8	82.8	96.9	74.2	86.1
B30	Roofing	99.1	63.3	85.8	103.2	63.3	88.4	103.7	78.1	94.2	96.7	78.6	89.9	98.8	83.3	93.0
C	Interior Construction	97.1	56.5	79.5	95.2	56.5	78.4	100.6	79.3	91.4	100.4	79.3	91.3	98.2	76.2	88.7
D10	Services: Conveying	100.0	54.5	87.4	100.0	54.5	87.4	100.0	77.5	93.8	100.0	77.5	93.8	100.0	91.3	97.6
D20 - 40	Mechanical	99.9	70.7	88.0	100.1	70.7	88.1	99.8	71.5	88.2	100.0	71.5	88.3	100.3	89.3	95.8
D50	Electrical	95.2	62.3	77.9	97.7	62.3	79.1	100.2	63.7	81.0	101.7	63.7	81.7	97.5	100.6	99.1
E	Equipment & Furnishings	100.0	57.5	97.8	100.0	57.5	97.8	100.0	84.0	99.2	100.0	84.0	99.2	100.0	73.8	98.6
G	Site Work	89.0	94.1	92.5	81.2	95.7	91.0	88.9	98.7	95.5	89.4	98.5	95.5	119.3	88.9	98.7
A-G	**WEIGHTED AVERAGE**	100.0	67.4	85.6	101.9	67.5	86.7	100.5	77.8	90.5	98.8	77.8	89.6	99.9	86.0	93.8

City Cost Indexes — RJ1030-010 Building Systems

DIV. NO.	BUILDING SYSTEMS	VIRGINIA ARLINGTON			NEWPORT NEWS			NORFOLK			PORTSMOUTH			RICHMOND		
		MAT.	INST.	TOTAL	MAT.	INST.	TOTAL	MAT.	INST.	TOTAL	MAT.	INST.	TOTAL	MAT.	INST.	TOTAL
A	Substructure	110.0	81.4	92.6	109.9	73.1	87.5	116.4	73.4	90.2	108.2	69.0	84.3	107.2	67.5	83.0
B10	Shell: Superstructure	98.6	88.1	94.5	100.0	80.6	92.3	100.8	80.7	92.8	98.4	77.1	89.9	101.2	77.6	91.8
B20	Exterior Closure	106.9	73.6	91.1	96.4	58.3	78.3	99.1	58.3	79.7	98.9	57.8	79.3	95.4	57.5	77.4
B30	Roofing	100.7	81.8	93.7	100.1	71.4	89.5	101.6	71.4	90.4	100.2	69.6	88.8	103.3	71.7	91.6
C	Interior Construction	97.2	75.7	87.9	97.8	65.6	83.8	97.3	68.8	85.0	96.4	58.9	80.1	98.0	63.7	83.2
D10	Services: Conveying	100.0	85.9	96.1	100.0	77.7	93.8	100.0	77.7	93.8	100.0	77.7	93.8	100.0	78.8	94.1
D20-40	Mechanical	100.3	88.3	95.4	100.3	66.0	86.3	100.0	65.5	85.9	100.3	65.4	86.0	100.0	69.0	87.3
D50	Electrical	95.0	100.6	97.9	97.2	61.4	78.4	101.9	58.7	79.2	95.4	58.7	76.1	100.2	72.1	85.4
E	Equipment & Furnishings	100.0	73.7	98.6	100.0	66.9	98.3	100.0	66.8	98.3	100.0	52.4	97.5	100.0	59.8	97.9
G	Site Work	129.1	86.7	100.5	111.0	88.0	95.4	109.3	89.0	95.5	109.2	87.7	94.7	108.4	88.1	94.7
A-G	WEIGHTED AVERAGE	100.8	85.3	93.9	99.8	69.3	86.3	100.8	69.4	87.0	99.2	66.7	84.9	100.1	70.1	86.9

DIV. NO.	BUILDING SYSTEMS	VIRGINIA ROANOKE			WASHINGTON EVERETT			RICHLAND			SEATTLE			SPOKANE		
		MAT.	INST.	TOTAL	MAT.	INST.	TOTAL	MAT.	INST.	TOTAL	MAT.	INST.	TOTAL	MAT.	INST.	TOTAL
A	Substructure	117.8	68.6	87.9	102.7	103.1	103.0	101.2	84.9	91.3	105.2	103.0	103.9	104.2	84.7	92.3
B10	Shell: Superstructure	101.7	76.4	91.6	103.7	96.0	100.6	91.1	83.2	88.0	105.1	97.2	101.9	93.9	82.8	89.5
B20	Exterior Closure	96.9	57.2	78.0	104.8	99.3	102.2	100.2	80.1	90.6	103.9	99.5	101.8	100.7	80.2	90.9
B30	Roofing	100.1	69.8	88.8	102.4	101.2	102.0	153.8	79.8	126.3	102.3	101.2	101.9	150.0	79.4	123.8
C	Interior Construction	97.6	63.9	83.0	105.0	95.2	100.8	111.7	81.1	97.1	105.6	96.9	101.8	111.2	78.3	97.0
D10	Services: Conveying	100.0	75.1	93.1	100.0	92.5	97.9	100.0	63.1	89.8	100.0	100.5	100.1	100.0	63.1	89.8
D20-40	Mechanical	100.3	62.7	84.9	100.1	102.6	101.1	100.4	108.1	103.6	100.0	120.0	108.2	100.3	88.1	95.3
D50	Electrical	97.2	56.7	75.9	103.6	99.0	101.2	97.5	93.9	95.6	103.4	107.2	105.4	95.2	79.1	86.7
E	Equipment & Furnishings	100.0	63.7	98.1	100.0	97.3	99.9	100.0	77.8	98.8	100.0	97.8	99.9	100.0	77.1	98.8
G	Site Work	109.4	85.5	93.2	91.2	110.9	104.5	101.1	90.6	94.0	94.6	109.1	104.4	100.2	90.6	93.7
A-G	WEIGHTED AVERAGE	100.4	66.2	85.3	102.4	99.7	101.2	101.8	88.7	96.0	102.8	105.1	103.8	102.0	82.4	93.4

DIV. NO.	BUILDING SYSTEMS	WASHINGTON TACOMA			VANCOUVER			YAKIMA			WEST VIRGINIA CHARLESTON			HUNTINGTON		
		MAT.	INST.	TOTAL	MAT.	INST.	TOTAL	MAT.	INST.	TOTAL	MAT.	INST.	TOTAL	MAT.	INST.	TOTAL
A	Substructure	103.8	103.2	103.4	112.7	97.4	103.4	108.3	94.4	99.8	101.8	91.3	95.4	113.1	96.0	102.7
B10	Shell: Superstructure	104.8	96.1	101.3	104.4	94.5	100.5	104.3	88.3	98.0	98.8	93.8	96.8	100.9	97.7	99.6
B20	Exterior Closure	103.7	99.4	101.6	108.8	98.1	103.7	101.8	78.3	90.6	95.1	91.2	93.3	96.6	90.8	93.8
B30	Roofing	102.2	98.8	100.9	102.6	92.8	98.9	102.3	81.5	94.6	99.9	87.5	95.3	100.3	89.1	96.2
C	Interior Construction	104.6	95.6	100.7	100.6	92.7	97.1	104.0	85.8	96.1	98.1	90.6	94.8	97.2	94.5	96.0
D10	Services: Conveying	100.0	100.5	100.1	100.0	58.9	88.6	100.0	97.5	99.3	100.0	97.1	99.2	100.0	98.2	99.5
D20-40	Mechanical	100.1	102.7	101.2	100.2	91.4	96.6	100.1	107.4	103.1	100.0	83.9	93.4	100.3	86.0	94.4
D50	Electrical	103.4	101.4	102.4	108.5	99.1	103.5	106.5	93.9	99.9	98.7	92.6	95.5	97.2	93.7	95.4
E	Equipment & Furnishings	100.0	97.5	99.9	100.0	93.1	99.6	100.0	96.6	99.8	100.0	86.4	99.3	100.0	94.4	99.7
G	Site Work	93.9	110.9	105.4	103.2	98.1	99.8	96.6	110.3	105.9	101.7	88.9	93.0	108.3	89.8	95.8
A-G	WEIGHTED AVERAGE	102.5	100.3	101.6	103.5	94.1	99.4	102.7	93.6	98.7	98.9	90.1	95.0	99.9	92.5	96.6

DIV. NO.	BUILDING SYSTEMS	WEST VIRGINIA PARKERSBURG			WHEELING			WISCONSIN EAU CLAIRE			GREEN BAY			KENOSHA		
		MAT.	INST.	TOTAL	MAT.	INST.	TOTAL	MAT.	INST.	TOTAL	MAT.	INST.	TOTAL	MAT.	INST.	TOTAL
A	Substructure	108.0	88.8	96.3	108.0	89.5	96.8	100.0	97.9	98.7	103.2	97.8	99.9	107.8	100.7	103.4
B10	Shell: Superstructure	98.1	92.4	95.8	98.3	92.4	96.0	96.2	99.7	97.6	99.0	98.6	98.8	100.2	100.8	100.4
B20	Exterior Closure	93.9	87.9	91.0	104.7	87.2	96.4	94.3	96.0	95.1	110.8	98.1	104.8	99.5	105.9	102.5
B30	Roofing	100.2	86.6	95.1	100.5	86.0	95.1	104.1	81.6	95.8	105.8	83.2	97.4	104.7	93.3	100.5
C	Interior Construction	96.7	87.1	92.5	97.2	86.2	92.4	102.1	99.9	101.2	101.2	99.5	100.5	100.2	103.5	101.6
D10	Services: Conveying	100.0	96.8	99.1	100.0	98.0	99.4	100.0	87.7	96.6	100.0	91.6	97.7	100.0	98.5	99.6
D20-40	Mechanical	100.2	88.1	95.2	100.3	89.7	96.0	100.2	83.3	93.3	100.5	84.3	93.8	100.1	93.2	97.3
D50	Electrical	97.7	92.6	95.0	94.7	96.9	95.8	104.7	88.5	96.2	98.8	81.2	89.5	102.6	97.5	99.9
E	Equipment & Furnishings	100.0	83.6	99.2	100.0	82.1	99.1	100.0	98.0	99.9	100.0	97.7	99.9	100.0	101.0	100.1
G	Site Work	112.5	88.8	96.5	113.3	87.9	96.1	97.8	103.1	101.4	100.6	99.3	99.7	102.2	105.1	104.2
A-G	WEIGHTED AVERAGE	98.9	89.6	94.8	100.0	90.3	95.7	99.5	93.5	96.9	101.5	92.6	97.6	100.8	99.8	100.4

DIV. NO.	BUILDING SYSTEMS	WISCONSIN LA CROSSE			MADISON			MILWAUKEE			RACINE			WYOMING CASPER		
		MAT.	INST.	TOTAL	MAT.	INST.	TOTAL	MAT.	INST.	TOTAL	MAT.	INST.	TOTAL	MAT.	INST.	TOTAL
A	Substructure	91.9	95.8	94.3	95.3	98.0	96.9	99.8	110.8	106.5	99.9	101.5	100.9	100.2	68.2	80.7
B10	Shell: Superstructure	94.1	95.8	94.8	98.6	95.5	97.4	100.2	105.1	102.1	98.9	100.7	99.6	98.4	64.5	84.9
B20	Exterior Closure	91.0	94.0	92.5	99.6	100.2	99.9	100.9	118.9	109.5	100.7	105.9	103.1	102.9	39.1	72.6
B30	Roofing	103.6	81.2	95.3	107.9	87.4	100.3	102.9	113.7	106.9	105.0	92.8	100.5	100.4	53.4	83.0
C	Interior Construction	100.4	98.0	99.4	101.8	97.2	99.8	104.2	122.4	112.1	101.6	103.9	102.6	100.0	42.2	75.0
D10	Services: Conveying	100.0	88.4	96.8	100.0	88.1	96.7	100.0	101.6	100.5	100.0	98.5	99.6	100.0	94.3	98.4
D20-40	Mechanical	100.2	84.1	93.6	99.9	93.0	97.0	99.9	102.7	101.0	99.9	93.2	97.2	100.0	66.7	86.4
D50	Electrical	105.0	88.7	96.4	101.3	92.3	96.6	103.6	100.9	102.2	101.9	95.5	98.5	94.5	66.8	79.9
E	Equipment & Furnishings	100.0	98.1	99.9	100.0	95.3	99.8	100.0	123.2	101.2	100.0	100.8	100.0	100.0	42.5	97.0
G	Site Work	91.5	103.2	99.4	92.3	108.3	103.1	97.2	98.7	98.2	96.2	108.9	104.8	96.3	95.0	95.4
A-G	WEIGHTED AVERAGE	98.0	92.4	95.5	99.9	96.0	98.2	101.1	108.6	104.4	100.3	99.9	100.1	99.4	61.5	82.7

DIV. NO.	BUILDING SYSTEMS	WYOMING CHEYENNE			ROCK SPRINGS			CANADA CALGARY, ALBERTA			EDMONTON, ALBERTA			HALIFAX, NOVA SCOTIA		
		MAT.	INST.	TOTAL	MAT.	INST.	TOTAL	MAT.	INST.	TOTAL	MAT.	INST.	TOTAL	MAT.	INST.	TOTAL
A	Substructure	97.1	72.4	82.0	97.4	68.7	79.9	174.7	100.2	129.3	185.7	100.2	133.6	154.0	83.2	110.9
B10	Shell: Superstructure	99.0	68.5	86.8	96.2	67.2	84.7	148.3	94.0	126.7	150.6	94.0	128.1	135.8	80.5	113.8
B20	Exterior Closure	102.3	50.9	77.9	127.1	50.1	90.5	158.3	90.9	126.3	152.8	90.9	123.4	136.4	83.5	111.3
B30	Roofing	97.3	55.9	81.9	98.9	59.8	84.4	123.0	97.9	113.7	126.7	97.9	116.0	115.1	78.9	101.6
C	Interior Construction	100.1	54.8	80.5	101.0	61.6	83.9	113.1	99.2	107.0	114.0	99.2	107.6	99.3	81.3	91.5
D10	Services: Conveying	100.0	94.3	98.4	100.0	83.4	95.4	131.1	100.2	122.5	131.1	100.2	122.5	131.1	68.7	113.8
D20-40	Mechanical	100.0	66.7	86.4	99.9	64.6	85.5	100.5	95.2	98.3	100.6	95.2	98.4	99.8	80.2	91.8
D50	Electrical	97.5	71.5	83.8	93.3	65.6	78.7	120.9	102.7	111.3	115.3	102.7	108.7	126.0	81.9	102.8
E	Equipment & Furnishings	100.0	56.7	97.8	100.0	67.0	98.3	131.1	99.0	129.5	131.1	99.0	129.5	131.1	79.1	128.4
G	Site Work	90.1	95.1	93.5	87.0	94.5	92.1	129.1	105.9	113.5	150.5	105.9	120.4	105.9	98.8	101.1
A-G	WEIGHTED AVERAGE	99.4	66.7	85.0	101.2	65.6	85.5	127.9	97.4	114.5	128.5	97.4	114.8	119.5	82.2	103.0

City Cost Indexes

RJ1030-010 Building Systems

DIV. NO.	BUILDING SYSTEMS	CANADA															
		HAMILTON, ONTARIO			KITCHENER, ONTARIO			LAVAL, QUEBEC			LONDON, ONTARIO			MONTREAL, QUEBEC		
		MAT.	INST.	TOTAL	MAT.	INST.	TOTAL	MAT.	INST.	TOTAL	MAT.	INST.	TOTAL	MAT.	INST.	TOTAL
A	Substructure	147.9	101.5	119.6	127.0	94.3	107.1	132.8	94.3	109.4	150.0	96.9	117.6	161.7	98.8	123.4
B10	Shell: Superstructure	130.2	95.4	116.3	118.6	92.0	108.0	112.6	90.3	103.7	131.2	93.1	116.0	139.2	95.6	121.8
B20	Exterior Closure	140.8	100.5	121.6	122.9	96.5	110.3	131.7	86.2	110.1	143.6	97.6	121.7	134.8	92.7	114.8
B30	Roofing	124.1	95.7	113.5	110.8	92.1	103.9	105.5	93.1	100.9	124.9	93.2	113.1	121.4	98.9	113.0
C	Interior Construction	111.4	97.5	105.4	101.0	90.7	96.6	104.0	92.0	98.8	112.0	92.3	103.4	107.8	97.5	103.4
D10	Services: Conveying	131.1	101.3	122.9	131.1	100.2	122.5	131.1	87.0	118.9	131.1	100.6	122.7	131.1	88.3	119.2
D20 - 40	Mechanical	100.8	88.0	95.5	99.7	89.2	95.4	99.7	94.4	97.5	100.8	85.2	94.4	100.7	95.6	98.6
D50	Electrical	117.0	97.8	106.9	117.7	95.3	105.9	118.9	74.4	95.5	113.5	95.3	103.9	122.8	88.1	104.5
E	Equipment & Furnishings	131.1	95.9	129.3	131.1	87.7	128.9	131.1	89.6	129.0	131.1	88.8	128.9	131.1	94.8	129.3
G	Site Work	120.4	113.2	115.5	105.0	105.4	105.2	99.4	100.4	100.1	119.7	105.5	110.1	109.1	99.9	102.9
A-G	WEIGHTED AVERAGE	120.4	97.0	110.1	112.6	93.5	104.2	112.9	89.7	102.7	120.8	93.5	108.8	121.8	94.9	110.0

DIV. NO.	BUILDING SYSTEMS	CANADA															
		OSHAWA, ONTARIO			OTTAWA, ONTARIO			QUEBEC, QUEBEC			REGINA, SASKATCHEWAN			SASKATOON, SASKATCHEWAN		
		MAT.	INST.	TOTAL	MAT.	INST.	TOTAL	MAT.	INST.	TOTAL	MAT.	INST.	TOTAL	MAT.	INST.	TOTAL
A	Substructure	155.3	93.7	117.8	155.4	98.1	120.5	140.7	99.0	115.3	151.7	94.2	116.7	129.5	94.2	108.0
B10	Shell: Superstructure	117.8	91.7	107.4	132.8	94.0	117.4	133.8	95.9	118.7	128.6	89.4	113.0	110.2	89.4	102.0
B20	Exterior Closure	132.3	96.2	115.1	141.4	97.7	120.6	138.5	95.4	118.0	128.0	92.8	111.3	126.3	92.8	110.4
B30	Roofing	111.8	88.5	103.1	127.2	93.2	114.6	120.4	99.1	112.5	114.1	85.8	103.6	105.0	85.8	97.9
C	Interior Construction	105.2	94.9	100.8	114.1	93.7	105.2	109.8	97.5	104.5	105.7	88.6	98.3	102.2	88.6	96.3
D10	Services: Conveying	131.1	100.9	122.7	131.1	98.0	121.9	131.1	88.4	119.3	131.1	71.3	114.5	131.1	72.2	114.8
D20 - 40	Mechanical	99.7	104.5	101.7	100.8	85.6	94.6	100.5	95.6	98.5	99.9	86.9	94.6	99.8	86.9	94.5
D50	Electrical	119.0	92.7	105.2	109.7	96.1	102.6	113.7	88.1	100.2	118.0	93.1	104.9	122.0	93.1	106.8
E	Equipment & Furnishings	131.1	90.7	129.0	131.1	92.7	129.1	131.1	94.8	129.3	131.1	93.4	129.2	131.1	93.4	129.2
G	Site Work	117.5	104.7	108.8	117.5	105.2	109.2	106.4	100.0	102.1	120.6	98.5	105.6	110.5	98.5	102.4
A-G	WEIGHTED AVERAGE	115.8	96.8	107.4	121.1	94.1	109.2	119.5	95.3	108.9	117.6	90.1	105.5	112.0	90.1	102.4

DIV. NO.	BUILDING SYSTEMS	CANADA															
		ST CATHARINES, ONTARIO			ST JOHNS, NEWFOUNDLAND			THUNDER BAY, ONTARIO			TORONTO, ONTARIO			VANCOUVER, BRITISH COLUMBIA		
		MAT.	INST.	TOTAL	MAT.	INST.	TOTAL	MAT.	INST.	TOTAL	MAT.	INST.	TOTAL	MAT.	INST.	TOTAL
A	Substructure	122.9	98.7	108.2	158.3	88.1	115.6	130.0	97.0	109.9	152.3	104.0	122.8	153.0	98.0	119.5
B10	Shell: Superstructure	110.1	95.2	104.2	131.6	84.1	112.7	111.7	93.5	104.4	131.2	98.1	118.1	153.0	93.4	129.2
B20	Exterior Closure	122.5	100.0	111.8	148.7	88.6	120.1	126.0	99.0	113.2	146.3	104.9	126.6	146.5	94.9	122.0
B30	Roofing	110.8	96.8	105.6	118.2	86.8	106.5	111.0	93.7	104.6	122.7	100.7	114.5	132.5	89.9	116.6
C	Interior Construction	99.2	97.0	98.2	108.9	79.3	96.1	101.4	94.5	98.4	111.9	102.7	107.9	114.0	94.2	105.4
D10	Services: Conveying	131.1	74.0	115.3	131.1	72.5	114.9	131.1	74.2	115.3	131.1	102.5	123.2	131.1	99.2	122.3
D20 - 40	Mechanical	99.7	86.8	94.4	100.1	76.2	90.3	99.7	86.7	94.4	100.6	96.0	98.8	100.6	84.9	94.2
D50	Electrical	119.4	96.2	107.2	120.8	73.7	96.0	117.7	94.6	105.6	113.7	98.3	105.6	119.2	82.8	100.0
E	Equipment & Furnishings	131.1	95.8	129.3	131.1	82.2	128.6	131.1	91.8	129.1	131.1	100.9	129.6	131.1	92.9	129.2
G	Site Work	106.4	101.9	103.4	119.7	99.9	106.3	111.2	101.8	104.8	138.8	106.0	116.6	123.6	108.3	113.3
A-G	WEIGHTED AVERAGE	110.6	94.6	103.6	121.7	81.9	104.2	111.9	93.3	103.7	121.6	100.5	112.3	126.8	91.9	111.4

DIV. NO.	BUILDING SYSTEMS	CANADA															
		WINDSOR, ONTARIO			WINNIPEG, MANITOBA											
		MAT.	INST.	TOTAL	MAT.	INST.	TOTAL	MAT.	INST.	TOTAL	MAT.	INST.	TOTAL	MAT.	INST.	TOTAL
A	Substructure	125.1	97.4	108.2	168.8	77.5	113.2									
B10	Shell: Superstructure	110.8	93.9	104.1	156.4	72.8	123.2									
B20	Exterior Closure	122.3	98.2	110.8	147.0	69.9	110.4									
B30	Roofing	110.8	92.7	104.1	118.1	72.0	101.0									
C	Interior Construction	100.1	93.9	97.4	110.5	68.9	92.5									
D10	Services: Conveying	131.1	73.6	115.2	131.1	68.8	113.8									
D20 - 40	Mechanical	99.7	86.3	94.2	100.5	67.8	87.1									
D50	Electrical	123.0	96.2	108.9	119.7	68.1	92.6									
E	Equipment & Furnishings	131.1	91.4	129.1	131.1	69.7	128.0									
G	Site Work	100.4	101.8	101.3	126.4	101.4	109.5									
A-G	WEIGHTED AVERAGE	111.2	93.4	103.3	127.3	72.3	103.1									

Location Factors — RJ1040-010 Building Systems

Costs shown in RSMeans cost data publications are based on national averages for materials and installation. To adjust these costs to a specific location, simply multiply the base cost by the factor and divide by 100 for that city. The data is arranged alphabetically by state and postal zip code numbers. For a city not listed, use the factor for a nearby city with similar economic characteristics.

STATE/ZIP	CITY	MAT.	INST.	TOTAL
ALABAMA				
350-352	Birmingham	97.4	73.8	87.0
354	Tuscaloosa	96.5	58.1	79.6
355	Jasper	96.8	56.9	79.2
356	Decatur	96.5	59.4	80.1
357-358	Huntsville	96.6	67.4	83.7
359	Gadsden	96.5	57.1	79.1
360-361	Montgomery	97.5	56.1	79.2
362	Anniston	95.9	63.8	81.7
363	Dothan	96.4	51.4	76.6
364	Evergreen	95.9	52.8	77.0
365-366	Mobile	97.5	66.8	84.0
367	Selma	96.1	51.3	76.4
368	Phenix City	96.8	55.4	78.5
369	Butler	96.3	51.4	76.5
ALASKA				
995-996	Anchorage	123.8	114.6	119.8
997	Fairbanks	122.0	115.4	119.1
998	Juneau	119.7	114.3	117.3
999	Ketchikan	132.8	114.3	124.7
ARIZONA				
850,853	Phoenix	98.7	75.4	88.4
851,852	Mesa/Tempe	98.1	71.3	86.3
855	Globe	98.2	61.4	82.0
856-857	Tucson	96.9	70.9	85.5
859	Show Low	98.3	61.8	82.2
860	Flagstaff	100.7	72.7	88.4
863	Prescott	98.5	61.1	82.0
864	Kingman	96.8	72.3	86.0
865	Chambers	96.8	62.0	81.5
ARKANSAS				
716	Pine Bluff	97.1	63.1	82.1
717	Camden	94.4	49.5	74.6
718	Texarkana	96.0	50.6	76.0
719	Hot Springs	93.7	52.2	75.4
720-722	Little Rock	97.5	64.2	82.8
723	West Memphis	96.0	62.0	81.0
724	Jonesboro	97.0	60.0	80.7
725	Batesville	94.6	55.3	77.3
726	Harrison	95.8	51.6	76.4
727	Fayetteville	93.3	53.4	75.7
728	Russellville	94.5	54.6	76.9
729	Fort Smith	97.4	61.8	81.7
CALIFORNIA				
900-902	Los Angeles	99.8	116.6	107.2
903-905	Inglewood	95.0	112.1	102.5
906-908	Long Beach	96.5	112.2	103.4
910-912	Pasadena	95.4	112.1	102.8
913-916	Van Nuys	98.3	112.1	104.4
917-918	Alhambra	97.3	112.1	103.8
919-921	San Diego	99.6	108.0	103.3
922	Palm Springs	96.3	109.5	102.2
923-924	San Bernardino	94.3	109.0	100.8
925	Riverside	98.8	113.7	105.4
926-927	Santa Ana	96.1	109.5	102.0
928	Anaheim	98.8	113.7	105.4
930	Oxnard	99.9	112.9	105.6
931	Santa Barbara	99.2	113.9	105.7
932-933	Bakersfield	100.1	109.8	104.4
934	San Luis Obispo	99.4	107.9	103.2
935	Mojave	96.8	105.4	100.6
936-938	Fresno	100.3	115.0	106.8
939	Salinas	100.3	121.8	109.8
940-941	San Francisco	106.8	142.4	122.5
942,956-958	Sacramento	101.7	119.2	109.5
943	Palo Alto	99.8	128.5	112.5
944	San Mateo	102.1	130.3	114.5
945	Vallejo	101.0	128.5	113.1
946	Oakland	104.6	133.3	117.3
947	Berkeley	103.8	132.5	116.5
948	Richmond	102.9	128.7	114.3
949	San Rafael	104.4	128.7	115.1
950	Santa Cruz	104.4	121.9	112.1

STATE/ZIP	CITY	MAT.	INST.	TOTAL
CALIFORNIA (CONT'D)				
951	San Jose	102.5	135.3	116.9
952	Stockton	100.5	116.6	107.6
953	Modesto	100.4	116.0	107.3
954	Santa Rosa	100.0	134.2	115.1
955	Eureka	101.3	112.7	106.3
959	Marysville	100.5	113.1	106.1
960	Redding	102.4	113.1	107.1
961	Susanville	101.1	113.1	106.4
COLORADO				
800-802	Denver	101.7	83.4	93.6
803	Boulder	98.4	83.3	91.7
804	Golden	100.4	80.9	91.8
805	Fort Collins	102.3	80.8	92.8
806	Greeley	99.8	77.3	89.9
807	Fort Morgan	99.0	80.9	91.0
808-809	Colorado Springs	101.4	80.6	92.2
810	Pueblo	100.7	79.8	91.5
811	Alamosa	101.8	75.7	90.3
812	Salida	101.5	77.4	90.9
813	Durango	102.1	78.3	91.6
814	Montrose	101.0	78.0	90.8
815	Grand Junction	104.4	78.2	92.9
816	Glenwood Springs	101.8	80.9	92.6
CONNECTICUT				
060	New Britain	100.6	119.9	109.1
061	Hartford	102.0	120.1	109.9
062	Willimantic	101.2	120.2	109.6
063	New London	97.2	120.0	107.2
064	Meriden	99.2	120.0	108.4
065	New Haven	102.1	120.0	110.0
066	Bridgeport	101.7	119.7	109.6
067	Waterbury	101.3	119.6	109.3
068	Norwalk	101.2	127.1	112.6
069	Stamford	101.4	127.1	112.7
D.C.				
200-205	Washington	101.1	92.4	97.3
DELAWARE				
197	Newark	98.8	108.8	103.2
198	Wilmington	98.6	108.9	103.1
199	Dover	99.3	108.8	103.5
FLORIDA				
320,322	Jacksonville	98.4	68.6	85.2
321	Daytona Beach	98.6	74.3	87.9
323	Tallahassee	99.1	59.1	81.5
324	Panama City	99.8	60.2	82.4
325	Pensacola	102.2	63.4	85.1
326,344	Gainesville	100.0	68.9	86.3
327-328,347	Orlando	101.8	72.3	88.8
329	Melbourne	101.2	77.3	90.7
330-332,340	Miami	99.8	76.0	89.3
333	Fort Lauderdale	97.9	75.2	87.9
334,349	West Palm Beach	96.6	73.5	86.4
335-336,346	Tampa	99.5	81.1	91.4
337	St. Petersburg	102.2	65.6	86.1
338	Lakeland	98.7	81.0	90.9
339,341	Fort Myers	97.9	75.1	87.8
342	Sarasota	99.7	76.9	89.7
GEORGIA				
300-303,399	Atlanta	97.5	75.9	88.0
304	Statesboro	96.9	53.3	77.7
305	Gainesville	95.6	63.1	81.3
306	Athens	95.0	63.3	81.0
307	Dalton	96.9	59.3	80.3
308-309	Augusta	96.3	64.2	82.1
310-312	Macon	96.2	63.9	82.0
313-314	Savannah	98.2	61.1	81.9
315	Waycross	97.1	60.6	81.0
316	Valdosta	97.4	60.4	81.1
317,398	Albany	97.4	61.1	81.4
318-319	Columbus	97.4	63.9	82.6

Location Factors — RJ1040-010 Building Systems

STATE/ZIP	CITY	MAT.	INST.	TOTAL
HAWAII				
967	Hilo	109.5	122.3	115.2
968	Honolulu	113.6	122.3	117.4
STATES & POSS.				
969	Guam	130.4	60.0	99.4
IDAHO				
832	Pocatello	100.7	72.5	88.3
833	Twin Falls	101.8	52.7	80.1
834	Idaho Falls	99.4	58.3	81.3
835	Lewiston	108.1	79.5	95.5
836-837	Boise	99.6	73.3	88.0
838	Coeur d'Alene	107.4	77.1	94.0
ILLINOIS				
600-603	North Suburban	98.8	130.3	112.7
604	Joliet	98.8	134.5	114.5
605	South Suburban	98.8	130.3	112.7
606-608	Chicago	99.3	138.3	116.5
609	Kankakee	94.8	125.4	108.3
610-611	Rockford	98.6	122.3	109.0
612	Rock Island	96.0	98.3	97.0
613	La Salle	97.1	119.0	106.8
614	Galesburg	97.0	103.2	99.7
615-616	Peoria	99.7	106.7	102.8
617	Bloomington	96.4	107.8	101.4
618-619	Champaign	99.8	106.8	102.9
620-622	East St. Louis	95.0	106.6	100.1
623	Quincy	96.1	98.6	97.2
624	Effingham	95.4	102.3	98.4
625	Decatur	97.6	105.8	101.2
626-627	Springfield	98.3	106.5	101.9
628	Centralia	93.4	105.4	98.7
629	Carbondale	93.1	100.5	96.4
INDIANA				
460	Anderson	95.7	82.3	89.8
461-462	Indianapolis	98.9	84.7	92.6
463-464	Gary	97.2	107.0	101.5
465-466	South Bend	95.5	81.8	89.5
467-468	Fort Wayne	96.3	77.7	88.1
469	Kokomo	93.3	81.1	87.9
470	Lawrenceburg	93.0	78.4	86.6
471	New Albany	94.1	75.4	85.9
472	Columbus	96.4	80.1	89.3
473	Muncie	97.5	81.4	90.4
474	Bloomington	98.8	81.6	91.2
475	Washington	94.7	82.4	89.3
476-477	Evansville	96.5	85.2	91.5
478	Terre Haute	97.2	85.7	92.1
479	Lafayette	96.1	82.0	89.9
IOWA				
500-503,509	Des Moines	98.9	83.5	92.1
504	Mason City	97.3	66.9	83.9
505	Fort Dodge	97.5	58.9	80.5
506-507	Waterloo	99.4	71.2	86.9
508	Creston	97.6	77.7	88.8
510-511	Sioux City	100.0	69.7	86.7
512	Sibley	98.5	55.8	79.7
513	Spencer	100.1	55.8	80.5
514	Carroll	97.0	71.4	85.7
515	Council Bluffs	101.2	79.1	91.5
516	Shenandoah	97.8	71.3	86.1
520	Dubuque	99.5	78.6	90.3
521	Decorah	98.1	64.8	83.4
522-524	Cedar Rapids	100.3	82.8	92.6
525	Ottumwa	98.0	69.7	85.5
526	Burlington	97.2	76.2	88.0
527-528	Davenport	99.3	89.2	94.8
KANSAS				
660-662	Kansas City	99.7	93.9	97.2
664-666	Topeka	99.4	63.8	83.7
667	Fort Scott	97.8	72.9	86.8
668	Emporia	97.9	68.7	85.0
669	Belleville	99.5	64.6	84.1
670-672	Wichita	98.8	63.2	83.1
673	Independence	99.2	68.4	85.6
674	Salina	100.1	65.9	85.0
675	Hutchinson	95.0	64.5	81.5
676	Hays	99.0	64.8	83.9
677	Colby	99.5	64.8	84.2

STATE/ZIP	CITY	MAT.	INST.	TOTAL
KANSAS (CONT'D)				
678	Dodge City	101.2	65.8	85.6
679	Liberal	98.8	64.8	83.8
KENTUCKY				
400-402	Louisville	98.6	82.2	91.3
403-405	Lexington	97.0	81.0	90.0
406	Frankfort	98.0	82.3	91.1
407-409	Corbin	94.2	71.4	84.1
410	Covington	95.0	94.1	94.6
411-412	Ashland	93.7	98.0	95.6
413-414	Campton	94.9	80.0	88.3
415-416	Pikeville	96.0	90.2	93.4
417-418	Hazard	94.4	80.4	88.2
420	Paducah	93.0	84.5	89.2
421-422	Bowling Green	95.6	83.3	90.1
423	Owensboro	95.6	86.7	91.7
424	Henderson	92.8	86.6	90.1
425-426	Somerset	92.5	79.3	86.7
427	Elizabethtown	91.9	82.6	87.8
LOUISIANA				
700-701	New Orleans	102.4	70.5	88.4
703	Thibodaux	100.1	62.7	83.6
704	Hammond	97.5	55.4	78.9
705	Lafayette	99.9	61.0	82.7
706	Lake Charles	100.1	63.4	83.9
707-708	Baton Rouge	101.5	62.6	84.4
710-711	Shreveport	97.0	55.6	78.8
712	Monroe	97.1	52.9	77.6
713-714	Alexandria	97.2	52.8	77.6
MAINE				
039	Kittery	94.2	84.0	89.7
040-041	Portland	99.2	92.1	96.1
042	Lewiston	97.6	92.1	95.2
043	Augusta	97.5	81.7	90.5
044	Bangor	97.1	92.1	94.9
045	Bath	95.3	83.6	90.1
046	Machias	94.9	87.5	91.6
047	Houlton	95.0	87.6	91.7
048	Rockland	94.2	87.5	91.2
049	Waterville	95.4	81.7	89.4
MARYLAND				
206	Waldorf	97.0	82.7	90.7
207-208	College Park	97.0	84.5	91.5
209	Silver Spring	96.3	82.6	90.3
210-212	Baltimore	100.1	84.1	93.0
214	Annapolis	100.4	82.7	92.6
215	Cumberland	95.8	81.9	89.7
216	Easton	97.3	69.7	85.1
217	Hagerstown	96.8	83.5	90.9
218	Salisbury	97.7	64.6	83.1
219	Elkton	94.8	83.2	89.7
MASSACHUSETTS				
010-011	Springfield	99.6	109.5	104.0
012	Pittsfield	99.1	107.1	102.7
013	Greenfield	96.9	109.9	102.6
014	Fitchburg	95.7	124.1	108.2
015-016	Worcester	99.6	124.3	110.4
017	Framingham	94.9	131.2	110.9
018	Lowell	99.0	130.9	113.1
019	Lawrence	99.8	132.5	114.2
020-022, 024	Boston	101.5	139.2	118.1
023	Brockton	100.1	127.1	112.0
025	Buzzards Bay	94.3	127.7	109.0
026	Hyannis	97.3	127.7	110.7
027	New Bedford	99.3	127.9	111.9
MICHIGAN				
480,483	Royal Oak	94.4	104.5	98.9
481	Ann Arbor	96.8	106.7	101.2
482	Detroit	98.4	109.6	103.4
484-485	Flint	96.4	95.7	96.1
486	Saginaw	96.0	92.2	94.3
487	Bay City	96.2	92.3	94.5
488-489	Lansing	97.4	93.8	95.8
490	Battle Creek	96.5	87.7	92.6
491	Kalamazoo	96.8	86.0	92.1
492	Jackson	94.7	94.5	94.6
493,495	Grand Rapids	98.0	81.6	90.8
494	Muskegon	95.4	83.6	90.2

Location Factors — RJ1040-010 Building Systems

STATE/ZIP	CITY	MAT.	INST.	TOTAL
MICHIGAN (CONT'D)				
496	Traverse City	94.2	77.4	86.8
497	Gaylord	95.2	71.7	84.8
498-499	Iron mountain	97.2	86.4	92.4
MINNESOTA				
550-551	Saint Paul	100.4	119.9	109.0
553-555	Minneapolis	101.4	122.0	110.5
556-558	Duluth	99.7	110.3	104.3
559	Rochester	100.1	106.1	102.7
560	Mankato	97.3	103.2	99.9
561	Windom	96.0	92.8	94.6
562	Willmar	95.6	103.9	99.2
563	St. Cloud	97.2	119.4	107.0
564	Brainerd	97.1	103.7	100.0
565	Detroit Lakes	99.1	97.9	98.6
566	Bemidji	98.4	99.8	99.0
567	Thief River Falls	98.0	94.5	96.4
MISSISSIPPI				
386	Clarksdale	96.1	54.9	77.9
387	Greenville	99.9	64.3	84.2
388	Tupelo	97.7	57.3	79.9
389	Greenwood	97.3	54.4	78.4
390-392	Jackson	97.4	63.6	82.5
393	Meridian	96.5	65.3	82.7
394	Laurel	97.5	58.1	80.1
395	Biloxi	98.1	59.5	81.0
396	Mccomb	96.1	54.0	77.5
397	Columbus	97.4	56.2	79.3
MISSOURI				
630-631	St. Louis	99.9	106.3	102.7
633	Bowling Green	98.2	94.5	96.6
634	Hannibal	97.1	91.8	94.7
635	Kirksville	100.3	89.4	95.5
636	Flat River	99.1	95.2	97.4
637	Cape Girardeau	99.5	89.5	95.1
638	Sikeston	97.5	89.9	94.2
639	Poplar Bluff	97.0	89.7	93.8
640-641	Kansas City	100.7	108.2	104.0
644-645	St. Joseph	99.9	94.3	97.4
646	Chillicothe	96.6	95.2	96.0
647	Harrisonville	96.2	102.9	99.1
648	Joplin	98.6	81.7	91.2
650-651	Jefferson City	96.6	91.7	94.4
652	Columbia	97.7	93.2	95.7
653	Sedalia	96.7	92.4	94.8
654-655	Rolla	95.1	95.7	95.3
656-658	Springfield	98.6	82.8	91.7
MONTANA				
590-591	Billings	103.3	77.5	91.9
592	Wolf Point	102.5	75.4	90.6
593	Miles City	100.3	75.1	89.2
594	Great Falls	104.5	76.8	92.3
595	Havre	101.4	75.0	89.7
596	Helena	102.7	75.0	90.5
597	Butte	103.1	77.5	91.8
598	Missoula	100.6	77.0	90.2
599	Kalispell	99.3	76.3	89.1
NEBRASKA				
680-681	Omaha	100.4	79.5	91.2
683-685	Lincoln	98.6	73.0	87.3
686	Columbus	96.8	70.6	85.3
687	Norfolk	98.6	74.0	87.8
688	Grand Island	98.7	77.8	89.5
689	Hastings	97.8	78.1	89.1
690	Mccook	97.6	69.6	85.3
691	North Platte	98.1	78.3	89.4
692	Valentine	100.0	66.6	85.3
693	Alliance	100.0	64.8	84.5
NEVADA				
889-891	Las Vegas	100.2	109.0	104.1
893	Ely	98.7	99.3	99.0
894-895	Reno	99.0	92.0	95.9
897	Carson City	98.9	91.8	95.8
898	Elko	97.6	84.5	91.8
NEW HAMPSHIRE				
030	Nashua	99.8	94.0	97.2
031	Manchester	100.6	94.0	97.7

STATE/ZIP	CITY	MAT.	INST.	TOTAL
NEW HAMPSHIRE (CONT'D)				
032-033	Concord	98.4	92.8	95.9
034	Keene	96.4	74.6	86.8
035	Littleton	96.0	77.7	88.0
036	Charleston	95.8	70.6	84.7
037	Claremont	94.9	70.7	84.2
038	Portsmouth	97.3	94.5	96.1
NEW JERSEY				
070-071	Newark	102.7	124.8	112.4
072	Elizabeth	100.1	124.8	111.0
073	Jersey City	99.0	124.7	110.3
074-075	Paterson	100.9	124.9	111.5
076	Hackensack	98.9	124.9	110.4
077	Long Branch	98.4	122.3	108.9
078	Dover	99.2	124.9	110.5
079	Summit	99.2	124.8	110.5
080,083	Vineland	97.1	123.4	108.7
081	Camden	99.0	122.8	109.5
082,084	Atlantic City	97.6	123.3	108.9
085-086	Trenton	99.5	122.5	109.6
087	Point Pleasant	98.9	121.6	108.9
088-089	New Brunswick	99.4	123.9	110.2
NEW MEXICO				
870-872	Albuquerque	99.1	73.5	87.9
873	Gallup	99.0	73.5	87.8
874	Farmington	99.7	73.5	88.2
875	Santa Fe	100.4	73.5	88.5
877	Las Vegas	97.6	73.5	87.0
878	Socorro	97.1	73.5	86.7
879	Truth/Consequences	97.1	70.2	85.2
880	Las Cruces	96.2	69.4	84.4
881	Clovis	97.7	73.4	87.0
882	Roswell	99.6	73.5	88.1
883	Carrizozo	99.7	73.5	88.2
884	Tucumcari	98.4	73.4	87.4
NEW YORK				
100-102	New York	104.2	167.0	131.9
103	Staten Island	100.0	164.8	128.6
104	Bronx	98.3	164.8	127.6
105	Mount Vernon	98.0	133.0	113.4
106	White Plains	98.3	133.0	113.6
107	Yonkers	102.9	133.1	116.2
108	New Rochelle	98.3	133.0	113.6
109	Suffern	98.1	121.5	108.4
110	Queens	101.1	164.8	129.1
111	Long Island City	102.5	164.8	130.0
112	Brooklyn	103.0	164.8	130.2
113	Flushing	102.4	164.8	129.9
114	Jamaica	100.7	164.8	129.0
115,117,118	Hicksville	101.0	145.4	120.5
116	Far Rockaway	102.6	164.8	130.0
119	Riverhead	101.7	145.1	120.9
120-122	Albany	97.0	99.8	98.2
123	Schenectady	97.6	99.0	98.2
124	Kingston	100.4	118.5	108.4
125-126	Poughkeepsie	99.6	128.6	112.4
127	Monticello	99.0	117.6	107.2
128	Glens Falls	92.2	94.9	93.4
129	Plattsburgh	96.8	88.6	93.2
130-132	Syracuse	98.7	96.2	97.6
133-135	Utica	96.7	93.3	95.2
136	Watertown	98.2	93.1	95.9
137-139	Binghamton	98.2	100.1	99.0
140-142	Buffalo	100.0	104.1	101.8
143	Niagara Falls	97.5	105.5	101.0
144-146	Rochester	100.1	96.0	98.3
147	Jamestown	96.5	87.9	92.7
148-149	Elmira	96.3	92.7	94.8
NORTH CAROLINA				
270,272-274	Greensboro	99.0	56.2	80.2
271	Winston-Salem	98.8	57.5	80.6
275-276	Raleigh	99.0	55.1	79.7
277	Durham	100.3	53.9	79.9
278	Rocky Mount	96.1	53.9	77.5
279	Elizabeth City	97.0	52.2	77.2
280	Gastonia	97.8	56.5	79.6
281-282	Charlotte	98.7	56.5	80.1
283	Fayetteville	100.2	59.1	82.1
284	Wilmington	96.7	56.5	79.0
285	Kinston	94.5	54.3	76.8

540

Location Factors — RJ1040-010 Building Systems

STATE/ZIP	CITY	MAT.	INST.	TOTAL
NORTH CAROLINA (CONT'D)				
286	Hickory	94.8	55.3	77.4
287-288	Asheville	97.1	55.1	78.6
289	Murphy	95.7	49.9	75.5
NORTH DAKOTA				
580-581	Fargo	101.9	65.8	86.0
582	Grand Forks	102.4	52.6	80.5
583	Devils Lake	101.3	56.9	81.7
584	Jamestown	101.4	46.0	77.0
585	Bismarck	101.0	64.4	84.9
586	Dickinson	102.3	58.7	83.1
587	Minot	102.4	71.6	88.9
588	Williston	100.7	58.8	82.3
OHIO				
430-432	Columbus	98.6	90.0	94.8
433	Marion	94.8	84.1	90.1
434-436	Toledo	98.8	97.5	98.2
437-438	Zanesville	95.3	84.8	90.7
439	Steubenville	96.6	92.0	94.6
440	Lorain	98.1	90.8	94.9
441	Cleveland	98.2	100.7	99.3
442-443	Akron	99.1	93.6	96.6
444-445	Youngstown	98.4	89.1	94.3
446-447	Canton	98.5	85.0	92.5
448-449	Mansfield	95.6	86.1	91.4
450	Hamilton	97.5	84.7	91.9
451-452	Cincinnati	97.9	85.1	92.3
453-454	Dayton	97.7	83.9	91.6
455	Springfield	97.6	84.8	91.9
456	Chillicothe	96.2	91.9	94.3
457	Athens	98.9	82.2	91.6
458	Lima	99.4	85.0	93.1
OKLAHOMA				
730-731	Oklahoma City	98.5	59.8	81.4
734	Ardmore	95.1	59.3	79.3
735	Lawton	97.7	60.6	81.3
736	Clinton	96.6	58.0	79.6
737	Enid	97.5	58.0	80.1
738	Woodward	95.3	58.0	78.9
739	Guymon	96.4	30.7	67.4
740-741	Tulsa	97.5	52.8	77.8
743	Miami	94.0	66.9	82.1
744	Muskogee	96.8	38.8	71.3
745	Mcalester	93.7	49.9	74.4
746	Ponca City	94.3	57.9	78.3
747	Durant	94.3	56.9	77.8
748	Shawnee	95.8	55.8	78.2
749	Poteau	93.5	60.6	79.0
OREGON				
970-972	Portland	98.4	101.2	99.6
973	Salem	98.9	100.4	99.5
974	Eugene	98.2	100.3	99.1
975	Medford	99.7	98.3	99.1
976	Klamath Falls	99.3	98.3	98.9
977	Bend	98.3	100.4	99.3
978	Pendleton	93.8	102.9	97.8
979	Vale	91.8	91.1	91.5
PENNSYLVANIA				
150-152	Pittsburgh	99.4	104.7	101.8
153	Washington	96.3	103.0	99.2
154	Uniontown	96.7	101.9	99.0
155	Bedford	97.5	92.8	95.5
156	Greensburg	97.5	101.2	99.2
157	Indiana	96.4	99.1	97.6
158	Dubois	97.9	97.2	97.6
159	Johnstown	97.5	97.4	97.4
160	Butler	92.0	101.5	96.2
161	New Castle	92.0	98.0	94.7
162	Kittanning	92.5	102.9	97.1
163	Oil City	92.1	96.1	93.8
164-165	Erie	94.2	95.2	94.6
166	Altoona	94.4	93.2	93.9
167	Bradford	95.2	96.3	95.7
168	State College	94.7	93.7	94.3
169	Wellsboro	95.8	93.4	94.7
170-171	Harrisburg	97.8	96.5	97.2
172	Chambersburg	95.2	90.6	93.2
173-174	York	96.0	96.4	96.2
175-176	Lancaster	94.1	90.0	92.3

STATE/ZIP	CITY	MAT.	INST.	TOTAL
PENNSYLVANIA (CONT'D)				
177	Williamsport	92.8	79.3	86.9
178	Sunbury	94.7	95.8	95.2
179	Pottsville	94.0	98.8	96.1
180	Lehigh Valley	95.3	113.4	103.3
181	Allentown	97.7	109.5	102.9
182	Hazleton	94.9	98.6	96.5
183	Stroudsburg	94.7	102.7	98.2
184-185	Scranton	98.4	99.9	99.1
186-187	Wilkes-Barre	94.6	98.5	96.3
188	Montrose	94.2	97.3	95.6
189	Doylestown	94.5	120.2	105.8
190-191	Philadelphia	99.2	132.2	113.7
193	Westchester	95.5	122.7	107.5
194	Norristown	94.7	128.4	109.5
195-196	Reading	97.1	101.1	98.8
PUERTO RICO				
009	San Juan	121.0	23.8	78.2
RHODE ISLAND				
028	Newport	98.7	116.7	106.6
029	Providence	100.1	116.7	107.4
SOUTH CAROLINA				
290-292	Columbia	97.3	48.8	75.9
293	Spartanburg	96.3	48.4	75.2
294	Charleston	97.7	56.7	79.6
295	Florence	96.0	48.7	75.2
296	Greenville	96.0	48.4	75.0
297	Rock Hill	95.3	46.6	73.8
298	Aiken	96.1	71.8	85.4
299	Beaufort	96.8	42.1	72.7
SOUTH DAKOTA				
570-571	Sioux Falls	100.9	57.1	81.6
572	Watertown	99.5	50.9	78.1
573	Mitchell	98.4	50.5	77.3
574	Aberdeen	101.3	51.7	79.4
575	Pierre	100.5	53.7	79.8
576	Mobridge	99.1	50.9	77.8
577	Rapid City	101.2	54.6	80.6
TENNESSEE				
370-372	Nashville	97.6	73.1	86.8
373-374	Chattanooga	98.9	65.5	84.2
375, 380-381	Memphis	97.4	70.0	85.3
376	Johnson City	98.4	56.0	79.8
377-379	Knoxville	95.5	60.8	80.2
382	Mckenzie	96.8	61.1	81.1
383	Jackson	98.9	63.2	83.2
384	Columbia	95.5	66.2	82.6
385	Cookeville	96.7	59.5	80.3
TEXAS				
750	Mckinney	98.5	49.7	77.0
751	Waxahackie	98.5	56.7	80.1
752-753	Dallas	99.5	66.5	85.0
754	Greenville	98.6	42.9	74.0
755	Texarkana	97.8	50.5	77.0
756	Longview	98.4	40.0	72.7
757	Tyler	98.6	54.1	79.0
758	Palestine	95.0	55.9	77.8
759	Lufkin	95.5	58.8	79.3
760-761	Fort Worth	97.1	62.9	82.0
762	Denton	97.3	47.3	75.2
763	Wichita Falls	97.9	54.1	78.6
764	Eastland	96.3	49.1	75.5
765	Temple	95.0	48.8	74.6
766-767	Waco	97.3	55.8	79.0
768	Brownwood	97.8	48.4	76.0
769	San Angelo	97.5	49.1	76.2
770-772	Houston	99.3	70.3	86.5
773	Huntsville	97.6	56.3	79.4
774	Wharton	98.5	53.4	78.6
775	Galveston	96.6	68.3	84.1
776-777	Beaumont	97.3	61.3	81.4
778	Bryan	94.5	61.6	80.0
779	Victoria	98.7	43.4	74.3
780	Laredo	96.7	53.2	77.5
781-782	San Antonio	97.0	65.6	83.2
783-784	Corpus Christi	99.7	51.7	78.6
785	Mc Allen	99.4	46.1	75.9
786-787	Austin	97.1	58.1	79.9

Location Factors — RJ1040-010 Building Systems

STATE/ZIP	CITY	MAT.	INST.	TOTAL
TEXAS (CONT'D)				
788	Del Rio	98.9	45.9	75.6
789	Giddings	95.8	55.4	78.0
790-791	Amarillo	98.6	60.1	81.7
792	Childress	97.5	56.5	79.4
793-794	Lubbock	99.7	56.2	80.6
795-796	Abilene	98.1	53.6	78.5
797	Midland	99.5	51.3	78.2
798-799,885	El Paso	97.2	50.6	76.7
UTAH				
840-841	Salt Lake City	101.9	67.5	86.7
842,844	Ogden	97.6	69.1	85.1
843	Logan	99.4	69.1	86.1
845	Price	99.8	65.5	84.7
846-847	Provo	100.0	67.4	85.6
VERMONT				
050	White River Jct.	97.3	67.7	84.3
051	Bellows Falls	96.0	81.8	89.7
052	Bennington	96.3	80.1	89.1
053	Brattleboro	96.6	86.2	92.0
054	Burlington	100.5	77.8	90.5
056	Montpelier	97.2	77.8	88.7
057	Rutland	98.8	77.8	89.6
058	St. Johnsbury	97.3	67.1	84.0
059	Guildhall	96.1	67.1	83.3
VIRGINIA				
220-221	Fairfax	99.3	84.9	93.0
222	Arlington	100.8	85.3	93.9
223	Alexandria	99.9	86.0	93.8
224-225	Fredericksburg	98.1	83.8	91.8
226	Winchester	98.6	82.7	91.6
227	Culpeper	98.6	81.6	91.1
228	Harrisonburg	98.8	69.3	85.8
229	Charlottesville	99.2	68.4	85.6
230-232	Richmond	100.1	70.1	86.9
233-235	Norfolk	100.8	69.4	87.0
236	Newport News	99.8	69.3	86.3
237	Portsmouth	99.2	66.7	84.9
238	Petersburg	99.0	70.1	86.3
239	Farmville	98.2	54.4	78.9
240-241	Roanoke	100.4	66.2	85.3
242	Bristol	97.9	56.7	79.8
243	Pulaski	97.6	57.3	79.8
244	Staunton	98.4	63.9	83.2
245	Lynchburg	98.5	67.4	84.8
246	Grundy	97.9	54.5	78.8
WASHINGTON				
980-981,987	Seattle	102.8	105.1	103.8
982	Everett	102.4	99.7	101.2
983-984	Tacoma	102.5	100.3	101.6
985	Olympia	100.1	100.3	100.2
986	Vancouver	103.5	94.1	99.4
988	Wenatchee	102.1	87.5	95.6
989	Yakima	102.7	93.6	98.7
990-992	Spokane	102.0	82.4	93.4
993	Richland	101.8	88.7	96.0
994	Clarkston	100.0	81.4	91.8
WEST VIRGINIA				
247-248	Bluefield	96.8	83.2	90.8
249	Lewisburg	98.3	83.2	91.7
250-253	Charleston	98.9	90.1	95.0
254	Martinsburg	98.2	80.3	90.3
255-257	Huntington	99.9	92.5	96.6
258-259	Beckley	96.7	87.2	92.5
260	Wheeling	100.0	90.3	95.7
261	Parkersburg	98.9	89.6	94.8
262	Buckhannon	98.1	90.3	94.7
263-264	Clarksburg	98.7	90.1	94.9
265	Morgantown	98.7	89.2	94.5
266	Gassaway	98.0	90.2	94.6
267	Romney	98.0	86.0	92.7
268	Petersburg	97.8	86.1	92.6
WISCONSIN				
530,532	Milwaukee	101.1	108.6	104.4
531	Kenosha	100.8	99.8	100.4
534	Racine	100.3	99.9	100.1
535	Beloit	100.2	94.8	97.8
537	Madison	99.9	96.0	98.2

STATE/ZIP	CITY	MAT.	INST.	TOTAL
WISCONSIN (CONT'D)				
538	Lancaster	97.7	90.3	94.4
539	Portage	96.3	93.4	95.0
540	New Richmond	97.0	93.0	95.2
541-543	Green Bay	101.5	92.6	97.6
544	Wausau	96.3	91.0	94.0
545	Rhinelander	99.5	89.2	94.9
546	La Crosse	98.0	92.4	95.5
547	Eau Claire	99.5	93.5	96.9
548	Superior	96.8	95.5	96.2
549	Oshkosh	96.9	81.4	90.1
WYOMING				
820	Cheyenne	99.4	66.7	85.0
821	Yellowstone Nat'l Park	96.7	65.9	83.2
822	Wheatland	97.8	65.5	83.6
823	Rawlins	99.3	68.6	85.8
824	Worland	97.4	64.9	83.0
825	Riverton	98.4	65.6	84.0
826	Casper	99.4	61.5	82.7
827	Newcastle	97.2	67.5	84.1
828	Sheridan	100.0	66.1	85.1
829-831	Rock Springs	101.2	65.6	85.5
CANADIAN FACTORS (reflect Canadian currency)				
ALBERTA				
	Calgary	127.9	97.4	114.5
	Edmonton	128.5	97.4	114.8
	Fort McMurray	125.2	96.2	112.4
	Lethbridge	119.7	95.6	109.1
	Lloydminster	114.7	92.2	104.8
	Medicine Hat	114.8	91.4	104.5
	Red Deer	115.3	91.4	104.8
BRITISH COLUMBIA				
	Kamloops	115.7	94.3	106.3
	Prince George	116.8	93.6	106.5
	Vancouver	126.8	91.9	111.4
	Victoria	116.8	90.0	105.0
MANITOBA				
	Brandon	114.9	78.9	99.0
	Portage la Prairie	114.7	77.2	98.2
	Winnipeg	127.3	72.3	103.1
NEW BRUNSWICK				
	Bathurst	112.9	70.0	94.0
	Dalhousie	113.0	70.2	94.1
	Fredericton	117.6	75.5	99.1
	Moncton	113.3	72.9	95.5
	Newcastle	113.0	70.2	94.1
	St. John	115.5	79.2	99.5
NEWFOUNDLAND				
	Corner Brook	118.2	69.4	96.7
	St Johns	121.7	81.9	104.2
NORTHWEST TERRITORIES				
	Yellowknife	123.9	87.6	107.9
NOVA SCOTIA				
	Bridgewater	114.3	76.6	97.7
	Dartmouth	115.9	76.6	98.6
	Halifax	119.5	82.2	103.0
	New Glasgow	114.0	76.6	97.6
	Sydney	111.7	76.6	96.3
	Truro	113.8	76.6	97.4
	Yarmouth	113.9	76.6	97.5
ONTARIO				
	Barrie	118.1	95.4	108.1
	Brantford	116.4	99.4	108.9
	Cornwall	116.5	95.7	107.3
	Hamilton	120.4	97.0	110.1
	Kingston	117.3	95.8	107.8
	Kitchener	112.6	93.5	104.2
	London	120.8	93.5	108.8
	North Bay	116.6	93.5	106.4
	Oshawa	115.8	96.8	107.4
	Ottawa	121.1	94.1	109.2
	Owen Sound	118.2	93.6	107.4
	Peterborough	116.4	95.5	107.2
	Sarnia	116.2	100.1	109.1

Location Factors — RJ1040-010 Building Systems

STATE/ZIP	CITY	MAT.	INST.	TOTAL
ONTARIO (CONT'D)				
	Sault Ste Marie	111.4	94.1	103.8
	St. Catharines	110.6	94.6	103.6
	Sudbury	111.1	93.2	103.2
	Thunder Bay	111.9	93.3	103.7
	Timmins	116.6	93.5	106.4
	Toronto	121.6	100.5	112.3
	Windsor	111.2	93.4	103.3
PRINCE EDWARD ISLAND				
	Charlottetown	119.0	64.5	95.0
	Summerside	115.7	64.5	93.2
QUEBEC				
	Cap-de-la-Madeleine	113.7	89.9	103.2
	Charlesbourg	113.7	89.9	103.2
	Chicoutimi	112.6	94.6	104.7
	Gatineau	113.3	89.7	102.9
	Granby	113.5	89.6	103.0
	Hull	113.4	89.7	102.9
	Joliette	113.8	89.9	103.3
	Laval	112.9	89.7	102.7
	Montreal	121.8	94.9	110.0
	Quebec	119.5	95.3	108.9
	Rimouski	113.1	94.6	104.9
	Rouyn-Noranda	113.2	89.7	102.9
	Saint Hyacinthe	112.6	89.7	102.5
	Sherbrooke	113.6	89.7	103.1
	Sorel	113.8	89.9	103.3
	St Jerome	113.3	89.7	102.9
	Trois Rivieres	114.0	89.9	103.4
SASKATCHEWAN				
	Moose Jaw	112.0	71.3	94.1
	Prince Albert	111.1	69.5	92.7
	Regina	117.6	90.1	105.5
	Saskatoon	112.0	90.1	102.4
YUKON				
	Whitehorse	118.4	70.3	97.2

Square Foot Costs — RK1010-010 Information

Square Foot and Cubic Foot Building Costs

The cost figures in Division K1010 were derived from more than 11,200 projects contained in the Means Data Bank of Construction Costs, and include the contractor's overhead and profit, but do not include architectural fees or land costs. The figures have been adjusted to January of the current year. New projects are added to our files each year, and outdated projects are discarded. For this reason, certain costs may not show a uniform annual progression. In no case are all subdivisions of a project listed.

These projects were located throughout the U.S. and reflect a tremendous variation in S.F. and C.F. costs. This is due to differences, not only in labor and material costs, but also in individual owner's requirements. For instance, a bank in a large city would have different features than one in a rural area. This is true of all the different types of buildings analyzed. Therefore, caution should be exercised when using Division K1010 costs. For example, for court houses, costs in the data bank are local court house costs and will not apply to the larger, more elaborate federal court houses. As a general rule, the projects in the 1/4 column do not include any site work or equipment, while the projects in the 3/4 column may include both equipment and site work. The median figures do not generally include site work.

None of the figures "go with" any others. All individual cost items were computed and tabulated separately. Thus the sum of the median figures for Plumbing, HVAC, and Electrical will not normally total up to the total Mechanical and Electrical costs arrived at by separate analysis and tabulation of the projects.

Each building was analyzed as to total and component costs and percentages. The figures were arranged in ascending order with the results tabulated as shown. The 1/4 column shows that 25% of the projects had lower costs, 75% higher. The 3/4 column shows that 75% of the projects had lower costs, 25% had higher. The median column shows that 50% of the projects had lower costs, 50% had higher.

There are two times when square foot costs are useful. The first is in the conceptual stage when no details are available. Then square foot costs make a useful starting point. The second is after the bids are in and the costs can be worked back into their appropriate units for information purposes. As soon as details become available in the project design, the square foot approach should be discontinued and the project priced as to its particular components. When more precision is required or for estimating the replacement cost of specific buildings, the current edition of **RSMeans Square Foot Costs** should be used.

In using the figures in Division K1010, it is recommended that the median column be used for preliminary figures if no additional information is available. The median figures, when multiplied by the total city construction cost index figures (see City Cost Indexes) and then multiplied by the project size modifier in RK1010-050, should present a fairly accurate base figure, which would then have to be adjusted in view of the estimator's experience, local economic conditions, code requirements, and the owner's particular requirements. There is no need to factor the percentage figures as these should remain constant from city to city. All tabulations mentioning air conditioning had at least partial air conditioning.

The editors of this book would greatly appreciate receiving cost figures on one or more of your recent projects, which would then be included in the averages for next year. All cost figures received will be kept confidential except that they will be averaged with other similar projects to arrive at S.F. and C.F. cost figures for next year's book. See the last page of the book for details and the discount available for submitting one or more of your projects.

Square Foot Costs — RK1010-030 Space Planning

Table K1010-031 Unit Gross Area Requirements

The figures in the table below indicate typical ranges in square feet as a function of the "occupant" unit. This table is best used in the preliminary design stages to help determine the probable size requirement for the total project. See RK1010-050 for the typical total size ranges for various types of buildings.

Building Type	Unit	Gross Area in S.F.		
		1/4	Median	3/4
Apartments	Unit	660	860	1,100
Auditorium & Play Theaters	Seat	18	25	38
Bowling Alleys	Lane		940	
Churches & Synagogues	Seat	20	28	39
Dormitories	Bed	200	230	275
Fraternity & Sorority Houses	Bed	220	315	370
Garages, Parking	Car	325	355	385
Hospitals	Bed	685	850	1,075
Hotels	Rental Unit	475	600	710
Housing for the Elderly	Unit	515	635	755
Housing, Public	Unit	700	875	1,030
Ice Skating Rinks	Total	27,000	30,000	36,000
Motels	Rental Unit	360	465	620
Nursing Homes	Bed	290	350	450
Restaurants	Seat	23	29	39
Schools, Elementary	Pupil	65	77	90
Junior High & Middle		85	110	129
Senior High		102	130	145
Vocational		110	135	195
Shooting Ranges	Point		450	
Theaters & Movies	Seat		15	

Square Foot Costs

RK1010-050 Project Size Modifier

RK1010-050 Square Foot Project Size Modifier

One factor that affects the S.F. cost of a particular building is the size. In general, for buildings built to the same specifications in the same locality, the larger building will have the lower S.F. cost. This is due mainly to the decreasing contribution of the exterior walls plus the economy of scale usually achievable in larger buildings. The Area Conversion Scale shown below will give a factor to convert costs for the typical size building to an adjusted cost for the particular project.

The Square Foot Base Size lists the median costs, most typical project size in our accumulated data, and the range in size of the projects.

The Size Factor for your project is determined by dividing your project area in S.F. by the typical project size for the particular Building Type. With this factor, enter the Area Conversion Scale at the appropriate Size Factor and determine the appropriate cost multiplier for your building size.

Example: Determine the cost per S.F. for a 100,000 S.F. mid-rise apartment building.

$$\frac{\text{Proposed building area} = 100{,}000 \text{ S.F.}}{\text{Typical size from below} = 50{,}000 \text{ S.F.}} = 2.00$$

Enter Area Conversion scale at 2.0, intersect curve, read horizontally the appropriate cost multiplier of .94. Size adjusted cost becomes .94 x $114.00 = $107.00 based on national average costs.

Note: For Size Factors less than .50, the Cost Multiplier is 1.1
For Size Factors greater than 3.5, the Cost Multiplier is .90

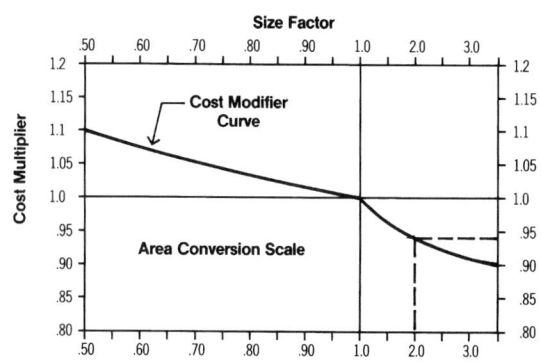

Square Foot Base Size							
Building Type	Median Cost per S.F.	Typical Size Gross S.F.	Typical Range Gross S.F.	Building Type	Median Cost per S.F.	Typical Size Gross S.F.	Typical Range Gross S.F.
Apartments, Low Rise	$ 90.00	21,000	9,700 - 37,200	Jails	$ 273.00	40,000	5,500 - 145,000
Apartments, Mid Rise	114.00	50,000	32,000 - 100,000	Libraries	178.00	12,000	7,000 - 31,000
Apartments, High Rise	123.00	145,000	95,000 - 600,000	Living, Assisted	146.00	32,300	23,500 - 50,300
Auditoriums	152.00	25,000	7,600 - 39,000	Medical Clinics	155.00	7,200	4,200 - 15,700
Auto Sales	112.00	20,000	10,800 - 28,600	Medical Offices	147.00	6,000	4,000 - 15,000
Banks	202.00	4,200	2,500 - 7,500	Motels	108.00	40,000	15,800 - 120,000
Churches	138.00	17,000	2,000 - 42,000	Nursing Homes	151.00	23,000	15,000 - 37,000
Clubs, Country	141.00	6,500	4,500 - 15,000	Offices, Low Rise	129.00	20,000	5,000 - 80,000
Clubs, Social	135.00	10,000	6,000 - 13,500	Offices, Mid Rise	127.00	120,000	20,000 - 300,000
Clubs, YMCA	158.00	28,300	12,800 - 39,400	Offices, High Rise	162.00	260,000	120,000 - 800,000
Colleges (Class)	160.00	50,000	15,000 - 150,000	Police Stations	202.00	10,500	4,000 - 19,000
Colleges (Science Lab)	257.00	45,600	16,600 - 80,000	Post Offices	150.00	12,400	6,800 - 30,000
College (Student Union)	190.00	33,400	16,000 - 85,000	Power Plants	1125.00	7,500	1,000 - 20,000
Community Center	144.00	9,400	5,300 - 16,700	Religious Education	128.00	9,000	6,000 - 12,000
Court Houses	197.00	32,400	17,800 - 106,000	Research	212.00	19,000	6,300 - 45,000
Dept. Stores	83.50	90,000	44,000 - 122,000	Restaurants	183.00	4,400	2,800 - 6,000
Dormitories, Low Rise	162.00	25,000	10,000 - 95,000	Retail Stores	89.00	7,200	4,000 - 17,600
Dormitories, Mid Rise	187.00	85,000	20,000 - 200,000	Schools, Elementary	135.00	41,000	24,500 - 55,000
Factories	81.00	26,400	12,900 - 50,000	Schools, Jr. High	138.00	92,000	52,000 - 119,000
Fire Stations	149.00	5,800	4,000 - 8,700	Schools, Sr. High	145.00	101,000	50,500 - 175,000
Fraternity Houses	139.00	12,500	8,200 - 14,800	Schools, Vocational	137.00	37,000	20,500 - 82,000
Funeral Homes	155.00	10,000	4,000 - 20,000	Sports Arenas	110.00	15,000	5,000 - 40,000
Garages, Commercial	99.50	9,300	5,000 - 13,600	Supermarkets	88.00	44,000	12,000 - 60,000
Garages, Municipal	126.00	8,300	4,500 - 12,600	Swimming Pools	206.00	20,000	10,000 - 32,000
Garages, Parking	53.50	163,000	76,400 - 225,300	Telephone Exchange	232.00	4,500	1,200 - 10,600
Gymnasiums	137.00	19,200	11,600 - 41,000	Theaters	132.00	10,500	8,800 - 17,500
Hospitals	246.00	55,000	27,200 - 125,000	Town Halls	146.00	10,800	4,800 - 23,400
House (Elderly)	123.00	37,000	21,000 - 66,000	Warehouses	61.50	25,000	8,000 - 72,000
Housing (Public)	113.00	36,000	14,400 - 74,400	Warehouse & Office	70.00	25,000	8,000 - 72,000
Ice Rinks	163.00	29,000	27,200 - 33,600				

K1010 | Square Foot Costs

K1010 | S.F. Costs

			UNIT	UNIT COSTS 1/4	UNIT COSTS MEDIAN	UNIT COSTS 3/4	% OF TOTAL 1/4	% OF TOTAL MEDIAN	% OF TOTAL 3/4	
01	0010	**APARTMENTS Low Rise (1 to 3 story)**	S.F.	71.50	90	119				01
	0020	Total project cost	C.F.	6.40	8.50	10.45				
	0100	Site work	S.F.	6.10	8.35	14.65	6.05%	10.55%	14.05%	
	0500	Masonry		1.40	3.26	5.65	1.54%	3.67%	6.35%	
	1500	Finishes		7.55	10.40	12.85	9.05%	10.75%	12.85%	
	1800	Equipment		2.34	3.54	5.25	2.73%	4.03%	5.95%	
	2720	Plumbing		5.55	7.15	9.10	6.65%	8.95%	10.05%	
	2770	Heating, ventilating, air conditioning		3.54	4.36	6.40	4.20%	5.60%	7.60%	
	2900	Electrical		4.14	5.50	7.45	5.20%	6.65%	8.40%	
	3100	Total: Mechanical & Electrical		14.30	18.20	22.50	15.90%	18.05%	23%	
	9000	Per apartment unit, total cost	Apt.	66,500	101,500	149,500				
	9500	Total: Mechanical & Electrical	"	12,600	19,800	25,800				
02	0010	**APARTMENTS Mid Rise (4 to 7 story)**	S.F.	94.50	114	141				02
	0020	Total project costs	C.F.	7.40	10.20	13.90				
	0100	Site work	S.F.	3.78	7.65	14.80	5.25%	6.70%	9.20%	
	0500	Masonry		6.30	8.65	11.85	5.10%	7.25%	10.50%	
	1500	Finishes		12.35	16.60	19.55	10.70%	13.50%	17.70%	
	1800	Equipment		2.75	4.11	5.40	2.54%	3.47%	4.31%	
	2500	Conveying equipment		2.14	2.64	3.18	2.05%	2.27%	2.69%	
	2720	Plumbing		5.55	8.90	9.40	5.70%	7.20%	8.95%	
	2900	Electrical		6.25	8.45	10.30	6.35%	7.20%	8.95%	
	3100	Total: Mechanical & Electrical		19.95	25	30.50	18.25%	21%	23%	
	9000	Per apartment unit, total cost	Apt.	107,000	126,000	209,000				
	9500	Total: Mechanical & Electrical	"	20,200	23,400	24,400				
03	0010	**APARTMENTS High Rise (8 to 24 story)**	S.F.	107	123	148				03
	0020	Total project costs	C.F.	10.40	12.10	15.50				
	0100	Site work	S.F.	3.89	6.30	8.80	2.58%	4.84%	6.15%	
	0500	Masonry		6.20	11.30	14.05	4.74%	9.65%	11.05%	
	1500	Finishes		11.90	14.85	17.55	9.75%	11.80%	13.70%	
	1800	Equipment		3.45	4.24	5.60	2.78%	3.49%	4.35%	
	2500	Conveying equipment		2.44	3.70	5.05	2.23%	2.78%	3.37%	
	2720	Plumbing		7.90	9.30	11.40	6.80%	7.20%	10.45%	
	2900	Electrical		7.35	9.30	12.60	6.45%	7.65%	8.80%	
	3100	Total: Mechanical & Electrical		22	28	34	17.95%	22.50%	24.50%	
	9000	Per apartment unit, total cost	Apt.	111,500	122,500	170,000				
	9500	Total: Mechanical & Electrical	"	24,100	27,500	29,100				
04	0010	**AUDITORIUMS**	S.F.	111	152	219				04
	0020	Total project costs	C.F.	6.95	9.70	13.90				
	2720	Plumbing	S.F.	6.60	9.75	11.65	5.85%	7.20%	8.70%	
	2900	Electrical		9	12.70	21.50	6.85%	9.45%	11.35%	
	3100	Total: Mechanical & Electrical		59.50	79.50	96.50	24.50%	28%	31%	
05	0010	**AUTOMOTIVE SALES**	S.F.	82.50	112	139				05
	0020	Total project costs	C.F.	5.45	6.50	8.45				
	2720	Plumbing	S.F.	3.76	6.55	7.10	2.89%	6.05%	6.50%	
	2770	Heating, ventilating, air conditioning		5.80	8.85	9.60	4.61%	10%	10.35%	
	2900	Electrical		6.50	10.20	14.15	7.25%	8.80%	12.15%	
	3100	Total: Mechanical & Electrical		21	29.50	35.50	18.90%	20.50%	22%	
06	0010	**BANKS**	S.F.	162	202	256				06
	0020	Total project costs	C.F.	11.55	15.70	20.50				
	0100	Site work	S.F.	18.55	28	41	7.90%	12.95%	16.95%	
	0500	Masonry		8.40	16.10	28.50	3.46%	7.65%	10.10%	
	1500	Finishes		15	22	27.50	5.95%	8.65%	11.70%	
	1800	Equipment		6.10	13.40	27	1.34%	5.55%	10.50%	
	2720	Plumbing		5.05	7.25	10.60	2.82%	3.90%	4.93%	
	2770	Heating, ventilating, air conditioning		9.65	12.85	17.15	4.86%	7.15%	8.50%	
	2900	Electrical		15.30	20.50	26.50	8.20%	10.20%	12.20%	
	3100	Total: Mechanical & Electrical		36.50	49.50	59	16.25%	19.45%	23%	
	3500	See also Divisions 11 16 00 & 11 17 00								

K1010 | Square Foot Costs

K1010 | S.F. Costs

				UNIT COSTS			% OF TOTAL		
			UNIT	1/4	MEDIAN	3/4	1/4	MEDIAN	3/4
13	0010	**CHURCHES**	S.F.	109	138	180			
	0020	Total project costs	C.F.	6.75	8.50	11.20			
	1800	Equipment	S.F.	1.30	3.11	6.60	.95%	2.11%	4.31%
	2720	Plumbing		4.24	5.95	8.75	3.51%	4.96%	6.25%
	2770	Heating, ventilating, air conditioning		9.90	12.90	18.30	7.50%	10%	12%
	2900	Electrical		9.15	12.60	17.10	7.30%	8.80%	10.95%
	3100	Total: Mechanical & Electrical	▼	28.50	38	50.50	18.30%	22%	25%
	3500	See also division 11040 (MF2004 11 91 00)							
15	0010	**CLUBS, COUNTRY**	S.F.	117	141	177			
	0020	Total project costs	C.F.	9.40	11.45	15.85			
	2720	Plumbing	S.F.	7.05	10.45	24	5.60%	7.90%	10%
	2900	Electrical		9.20	12.60	16.40	7%	8.95%	11%
	3100	Total: Mechanical & Electrical	▼	49	61	64.50	19%	26.50%	29.50%
17	0010	**CLUBS, SOCIAL Fraternal**	S.F.	98.50	135	179			
	0020	Total project costs	C.F.	5.80	8.85	10.55			
	2720	Plumbing	S.F.	5.85	7.30	11.05	5.60%	6.90%	8.55%
	2770	Heating, ventilating, air conditioning		8.45	10.20	13.15	8.20%	9.25%	14.40%
	2900	Electrical		7	11.50	13.20	5.95%	9.30%	10.55%
	3100	Total: Mechanical & Electrical	▼	36	39	50	21%	23%	23.50%
18	0010	**CLUBS, Y.M.C.A.**	S.F.	122	158	241			
	0020	Total project costs	C.F.	5.40	9	13.45			
	2720	Plumbing	S.F.	7.40	14.70	16.50	5.65%	7.60%	10.85%
	2900	Electrical		9.60	12.10	19.40	6.25%	8%	9.20%
	3100	Total: Mechanical & Electrical	▼	39	44.50	72.50	20.50%	21.50%	28.50%
19	0010	**COLLEGES Classrooms & Administration**	S.F.	117	160	208			
	0020	Total project costs	C.F.	8.30	12.30	19.35			
	0500	Masonry	S.F.	11.65	16.40	20	5.65%	8.05%	10.50%
	2720	Plumbing		6	12.30	23	5.10%	6.60%	8.95%
	2900	Electrical		9.60	14.95	20.50	7.70%	9.75%	12%
	3100	Total: Mechanical & Electrical	▼	40.50	55	74	23.50%	28%	32%
21	0010	**COLLEGES Science, Engineering, Laboratories**	S.F.	220	257	298			
	0020	Total project costs	C.F.	12.60	18.40	21			
	1800	Equipment	S.F.	12.25	27.50	30.50	2%	6.45%	12.65%
	2900	Electrical		18.15	26	39.50	7.10%	9.40%	12.10%
	3100	Total: Mechanical & Electrical	▼	67.50	80	124	28.50%	31.50%	41%
	3500	See also division 11600 (MF2004 11 53 00)							
23	0010	**COLLEGES Student Unions**	S.F.	140	190	230			
	0020	Total project costs	C.F.	7.80	10.25	12.70			
	3100	Total: Mechanical & Electrical	S.F.	52.50	57	67.50	23.50%	26%	29%
25	0010	**COMMUNITY CENTERS**	S.F.	116	144	194			
	0020	Total project costs	C.F.	7.55	10.80	14			
	1800	Equipment	S.F.	2.80	4.59	7.55	1.47%	3.01%	5.30%
	2720	Plumbing		5.50	9.60	13.10	4.85%	7%	8.95%
	2770	Heating, ventilating, air conditioning		8.80	12.80	18.30	6.80%	10.35%	12.90%
	2900	Electrical		9.80	12.95	19	7.20%	8.90%	10.40%
	3100	Total: Mechanical & Electrical	▼	32.50	41	59	19.05%	23.50%	31%
28	0010	**COURT HOUSES**	S.F.	167	197	271			
	0020	Total project costs	C.F.	12.80	15.30	19.30			
	2720	Plumbing	S.F.	7.95	11.10	12.55	5.95%	7.45%	8.20%
	2900	Electrical		17.75	19.70	29	8.90%	10.45%	11.85%
	3100	Total: Mechanical & Electrical	▼	49.50	65	71	22.50%	27.50%	30%
30	0010	**DEPARTMENT STORES**	S.F.	61.50	83.50	105			
	0020	Total project costs	C.F.	3.31	4.29	5.80			
	2720	Plumbing	S.F.	1.92	2.43	3.68	1.82%	4.21%	5.90%
	2770	Heating, ventilating, air conditioning		5.60	8.65	13.05	8.20%	9.10%	14.80%

K1010 | Square Foot Costs

K1010 | S.F. Costs

			UNIT	UNIT COSTS 1/4	MEDIAN	3/4	% OF TOTAL 1/4	MEDIAN	3/4	
30	2900	Electrical	S.F.	7.10	9.70	11.45	9.05%	12.15%	14.95%	30
	3100	Total: Mechanical & Electrical	↓	12.45	15.95	28	13.20%	21.50%	50%	
31	0010	DORMITORIES Low Rise (1 to 3 story)	S.F.	117	162	203				31
	0020	Total project costs	C.F.	6.60	10.70	16				
	2720	Plumbing	S.F.	7.05	9.40	11.90	8.05%	9%	9.65%	
	2770	Heating, ventilating, air conditioning		7.45	8.90	11.90	4.61%	8.05%	10%	
	2900	Electrical		7.75	11.80	16.15	6.40%	8.65%	9.50%	
	3100	Total: Mechanical & Electrical	↓	40.50	43	67.50	22%	25%	27%	
	9000	Per bed, total cost	Bed	49,600	55,000	118,000				
32	0010	DORMITORIES Mid Rise (4 to 8 story)	S.F.	144	187	231				32
	0020	Total project costs	C.F.	15.85	17.40	21				
	2900	Electrical	S.F.	15.25	17.35	23.50	8.20%	10.20%	11.95%	
	3100	Total: Mechanical & Electrical	"	42.50	85	86.50	25%	30.50%	35.50%	
	9000	Per bed, total cost	Bed	20,500	46,700	270,000				
34	0010	FACTORIES	S.F.	54.50	81	125				34
	0020	Total project costs	C.F.	3.49	5.20	8.65				
	0100	Site work	S.F.	6.20	11.35	17.90	6.95%	11.45%	17.95%	
	2720	Plumbing		2.93	5.45	9.05	3.73%	6.05%	8.10%	
	2770	Heating, ventilating, air conditioning		5.70	8.20	11.05	5.25%	8.45%	11.35%	
	2900	Electrical		6.75	10.70	16.30	8.10%	10.50%	14.20%	
	3100	Total: Mechanical & Electrical	↓	19.30	29.50	39	21%	28.50%	35.50%	
36	0010	FIRE STATIONS	S.F.	108	149	199				36
	0020	Total project costs	C.F.	6.30	8.65	11.55				
	0500	Masonry	S.F.	15.85	28.50	37	8.10%	11.30%	15.70%	
	1140	Roofing		3.52	9.55	10.85	1.90%	4.94%	5.05%	
	1580	Painting		2.73	4.08	4.17	1.37%	1.57%	2.07%	
	1800	Equipment		1.34	2.57	4.75	.62%	1.63%	3.42%	
	2720	Plumbing		6.05	9.60	13.60	5.85%	7.35%	9.45%	
	2770	Heating, ventilating, air conditioning		6	9.70	14.95	5.15%	7.40%	9.40%	
	2900	Electrical		7.70	13.45	18.35	6.85%	8.60%	10.60%	
	3100	Total: Mechanical & Electrical	↓	39.50	50.50	57	18.40%	23%	26%	
37	0010	FRATERNITY HOUSES & Sorority Houses	S.F.	108	139	190				37
	0020	Total project costs	C.F.	10.75	11.20	13.50				
	2720	Plumbing	S.F.	8.15	9.35	17.10	6.80%	8%	10.85%	
	2900	Electrical		7.10	15.40	18.85	6.60%	9.90%	10.65%	
	3100	Total: Mechanical & Electrical	↓	8.70	27	32.50		15.10%	15.90%	
38	0010	FUNERAL HOMES	S.F.	114	155	282				38
	0020	Total project costs	C.F.	11.60	12.95	25				
	2900	Electrical	S.F.	5.05	9.20	10.10	3.58%	4.44%	5.95%	
	3100	Total: Mechanical & Electrical	↓	17.80	26	36.50	12.90%	12.90%	12.90%	
39	0010	GARAGES, COMMERCIAL (Service)	S.F.	64.50	99.50	138				39
	0020	Total project costs	C.F.	4.24	6.25	9.10				
	1800	Equipment	S.F.	3.63	8.15	12.70	2.21%	4.62%	6.80%	
	2720	Plumbing		4.46	6.90	12.55	5.45%	7.85%	10.65%	
	2730	Heating & ventilating		5.85	7.60	9.70	5.25%	6.85%	8.20%	
	2900	Electrical		6.10	9.35	13.45	7.15%	9.25%	10.85%	
	3100	Total: Mechanical & Electrical	↓	13.50	26	38	12.35%	17.40%	26%	
40	0010	GARAGES, MUNICIPAL (Repair)	S.F.	100	126	178				40
	0020	Total project costs	C.F.	5.90	7.50	12.90				
	0500	Masonry	S.F.	8.90	17.40	27	4.03%	9.15%	12.50%	
	2720	Plumbing		4.25	8.15	15.35	3.59%	6.70%	7.95%	
	2730	Heating & ventilating		7.25	10.55	20.50	6.15%	7.45%	13.50%	
	2900	Electrical		7	11	16	6.65%	8.15%	11.15%	
	3100	Total: Mechanical & Electrical	↓	32.50	45	65	21.50%	23%	28.50%	

K1010 | Square Foot Costs

K1010 | S.F. Costs

				UNIT COSTS			% OF TOTAL		
			UNIT	1/4	MEDIAN	3/4	1/4	MEDIAN	3/4
41	0010	**GARAGES, PARKING**	S.F.	36.50	53.50	92.50			
	0020	Total project costs	C.F.	3.46	4.70	6.85			
	2720	Plumbing	S.F.	1.04	1.62	2.50	1.72%	2.70%	3.85%
	2900	Electrical		2.02	2.48	3.90	4.33%	5.20%	6.30%
	3100	Total: Mechanical & Electrical	↓	4.14	5.75	7.20	7%	8.80%	9.70%
	3200								
	9000	Per car, total cost	Car	15,500	19,500	24,900			
43	0010	**GYMNASIUMS**	S.F.	103	137	185			
	0020	Total project costs	C.F.	5.10	6.95	8.50			
	1800	Equipment	S.F.	2.44	4.57	8.75	1.81%	3.30%	6.70%
	2720	Plumbing		6.50	7.55	9.95	4.65%	6.40%	7.75%
	2770	Heating, ventilating, air conditioning		7	10.60	21.50	5.15%	9.05%	11.10%
	2900	Electrical		8.15	10.75	14.30	6.75%	8.50%	10.35%
	3100	Total: Mechanical & Electrical	↓	29	38	46.50	19.60%	23.50%	29%
	3500	See also division 11480 (MF2004 11 67 00)							
46	0010	**HOSPITALS**	S.F.	196	246	335			
	0020	Total project costs	C.F.	14.90	18.55	26.50			
	1800	Equipment	S.F.	4.99	9.60	16.55	.80%	2.53%	4.80%
	2720	Plumbing		16.95	24	30.50	7.60%	9.10%	10.85%
	2770	Heating, ventilating, air conditioning		25	32	44.50	7.80%	12.95%	16.65%
	2900	Electrical		21.50	29	41.50	9.90%	11.75%	14.10%
	3100	Total: Mechanical & Electrical	↓	63.50	88.50	131	28%	33.50%	37%
	9000	Per bed or person, total cost	Bed	228,000	314,000	362,000			
	9900	See also division 11700 (MF2004 11 71 00)							
48	0010	**HOUSING For the Elderly**	S.F.	97	123	150			
	0020	Total project costs	C.F.	6.90	9.60	12.25			
	0100	Site work	S.F.	6.75	10.50	15.35	5.05%	7.90%	12.10%
	0500	Masonry		2.95	11	16.10	1.30%	6.05%	11%
	1800	Equipment		2.34	3.22	5.15	1.88%	3.23%	4.43%
	2510	Conveying systems		2.35	3.16	4.29	1.78%	2.20%	2.81%
	2720	Plumbing		7.20	9.15	11.55	8.15%	9.55%	10.50%
	2730	Heating, ventilating, air conditioning		3.69	5.20	7.80	3.30%	5.60%	7.25%
	2900	Electrical		7.20	9.80	12.55	7.30%	8.50%	10.25%
	3100	Total: Mechanical & Electrical	↓	26.50	31	39.50	18.10%	22.50%	29%
	9000	Per rental unit, total cost	Unit	90,000	105,500	117,000			
	9500	Total: Mechanical & Electrical	"	20,000	23,100	26,900			
50	0010	**HOUSING Public (Low Rise)**	S.F.	81.50	113	147			
	0020	Total project costs	C.F.	7.25	9.05	11.25			
	0100	Site work	S.F.	10.35	14.90	24	8.35%	11.75%	16.50%
	1800	Equipment		2.21	3.61	5.50	2.26%	3.03%	4.24%
	2720	Plumbing		5.85	7.75	9.80	7.15%	9.05%	11.60%
	2730	Heating, ventilating, air conditioning		2.95	5.70	6.25	4.26%	6.05%	6.45%
	2900	Electrical		4.92	7.35	10.15	5.10%	6.55%	8.25%
	3100	Total: Mechanical & Electrical	↓	23.50	30	33.50	14.50%	17.55%	26.50%
	9000	Per apartment, total cost	Apt.	89,500	101,500	127,500			
	9500	Total: Mechanical & Electrical	"	19,100	23,500	26,000			
51	0010	**ICE SKATING RINKS**	S.F.	69.50	163	178			
	0020	Total project costs	C.F.	5.10	5.25	6.05			
	2720	Plumbing	S.F.	2.60	4.87	4.98	3.12%	3.23%	5.65%
	2900	Electrical		7.45	11.40	12.10	6.30%	10.15%	15.05%
	3100	Total: Mechanical & Electrical	↓	31	31	31	18.95%	18.95%	18.95%
52	0010	**JAILS**	S.F.	211	273	355			
	0020	Total project costs	C.F.	19.10	26.50	31			
	1800	Equipment	S.F.	8.25	24.50	41.50	2.80%	6.05%	9.80%
	2720	Plumbing		20.50	27.50	36	7%	8.90%	13.35%
	2770	Heating, ventilating, air conditioning		19.10	25.50	49.50	7.50%	9.45%	17.75%
	2900	Electrical		22	30	37.50	8.80%	11.65%	15%

K1010 | Square Foot Costs

K1010 | S.F. Costs

				UNIT COSTS			% OF TOTAL			
			UNIT	1/4	MEDIAN	3/4	1/4	MEDIAN	3/4	
52	3100	Total: Mechanical & Electrical	S.F.	59.50	105	125	28%	30%	34%	52
53	0010	**LIBRARIES**	S.F.	136	178	232				53
	0020	Total project costs	C.F.	9.15	11.45	14.65				
	0500	Masonry	S.F.	10.70	18.50	31	5.65%	7.45%	11.60%	
	1800	Equipment		1.83	4.92	7.45	.28%	1.39%	4.07%	
	2720	Plumbing		4.94	7	9.75	3.38%	4.60%	5.70%	
	2770	Heating, ventilating, air conditioning		10.95	18.50	22.50	7.80%	10.95%	12.80%	
	2900	Electrical		13.75	18.15	24	8.40%	10.50%	12%	
	3100	Total: Mechanical & Electrical	↓	41.50	52.50	64	20.50%	23%	26.50%	
54	0010	**LIVING, ASSISTED**	S.F.	124	146	172				54
	0020	Total project costs	C.F.	10.40	12.15	13.80				
	0500	Masonry	S.F.	3.59	4.33	5.30	2.36%	3.16%	3.90%	
	1800	Equipment		2.74	3.21	4.19	2.06%	2.45%	2.87%	
	2720	Plumbing		10.35	13.90	14.35	6.05%	8.15%	10.60%	
	2770	Heating, ventilating, air conditioning		12.30	12.85	14.10	7.95%	9.35%	9.70%	
	2900	Electrical		12.10	13.40	15.35	8.95%	9.95%	10.65%	
	3100	Total: Mechanical & Electrical	↓	34	39.50	45.50	25%	28.50%	31.50%	
55	0010	**MEDICAL CLINICS**	S.F.	126	155	197				55
	0020	Total project costs	C.F.	9.25	11.95	15.85				
	1800	Equipment	S.F.	3.40	7.15	11.15	1.05%	2.94%	6.35%	
	2720	Plumbing		8.35	11.80	15.75	6.15%	8.40%	10.10%	
	2770	Heating, ventilating, air conditioning		10	13.10	19.20	6.65%	8.85%	11.35%	
	2900	Electrical		10.80	15.35	20	8.10%	10%	12.25%	
	3100	Total: Mechanical & Electrical	↓	34.50	47	64.50	22.50%	27%	33.50%	
	3500	See also division 11700 (MF2004 11 71 00)								
57	0010	**MEDICAL OFFICES**	S.F.	118	147	180				57
	0020	Total project costs	C.F.	8.85	11.95	16.20				
	1800	Equipment	S.F.	3.91	7.75	11	.68%	4.97%	7.05%	
	2720	Plumbing		6.55	10.10	13.60	5.60%	6.80%	8.50%	
	2770	Heating, ventilating, air conditioning		7.90	11.40	15.10	6.10%	8%	9.70%	
	2900	Electrical		9.50	13.80	19.25	7.60%	9.80%	11.70%	
	3100	Total: Mechanical & Electrical	↓	25.50	37	53.50	19.30%	23%	27.50%	
59	0010	**MOTELS**	S.F.	74.50	108	141				59
	0020	Total project costs	C.F.	6.65	8.90	14.55				
	2720	Plumbing	S.F.	7.60	9.65	11.50	9.45%	10.60%	12.55%	
	2770	Heating, ventilating, air conditioning		4.62	6.90	12.35	5.60%	5.60%	10%	
	2900	Electrical		7.05	8.95	11.15	7.45%	9.05%	10.45%	
	3100	Total: Mechanical & Electrical	↓	22.50	30	51.50	18.50%	24%	25.50%	
	5000										
	9000	Per rental unit, total cost	Unit	38,100	72,500	78,000				
	9500	Total: Mechanical & Electrical	"	7,400	11,200	13,000				
60	0010	**NURSING HOMES**	S.F.	117	151	188				60
	0020	Total project costs	C.F.	9.20	11.50	15.70				
	1800	Equipment	S.F.	3.69	4.89	8.15	2%	3.62%	4.99%	
	2720	Plumbing		10.05	15.20	18.35	8.75%	10.10%	12.70%	
	2770	Heating, ventilating, air conditioning		10.60	16.05	21.50	9.70%	11.45%	11.80%	
	2900	Electrical		11.60	14.55	19.75	9.40%	10.60%	12.50%	
	3100	Total: Mechanical & Electrical	↓	27.50	38.50	65	26%	28%	30%	
	9000	Per bed or person, total cost	Bed	52,000	65,000	84,000				
61	0010	**OFFICES Low Rise (1 to 4 story)**	S.F.	99	129	167				61
	0020	Total project costs	C.F.	7.05	9.70	12.80				
	0100	Site work	S.F.	7.95	13.85	20	6.20%	9.70%	13.55%	
	0500	Masonry		3.93	7.70	14.05	2.62%	5.45%	8.45%	
	1800	Equipment		1.05	2.06	5.60	.60%	1.50%	3.50%	
	2720	Plumbing		3.53	5.45	8	3.66%	4.50%	6.10%	
	2770	Heating, ventilating, air conditioning		7.80	10.90	15.90	7.20%	10.30%	11.70%	
	2900	Electrical		8.05	11.50	16.35	7.45%	9.65%	11.40%	

K1010 | Square Foot Costs

		K1010	S.F. Costs	UNIT	UNIT COSTS 1/4	UNIT COSTS MEDIAN	UNIT COSTS 3/4	% OF TOTAL 1/4	% OF TOTAL MEDIAN	% OF TOTAL 3/4	
61	3100		Total: Mechanical & Electrical	S.F.	22	30.50	46	18.15%	22%	27%	61
62	0010	OFFICES Mid Rise (5 to 10 story)		S.F.	105	127	172				62
	0020		Total project costs	C.F.	7.40	9.45	13.40				
	2720		Plumbing	S.F.	3.16	4.90	7.05	2.83%	3.74%	4.50%	
	2770		Heating, ventilating, air conditioning		7.95	11.35	18.10	7.65%	9.40%	11%	
	2900		Electrical		7.75	9.95	15.05	6.35%	7.80%	10%	
	3100		Total: Mechanical & Electrical		20	26	49	18.95%	21%	27.50%	
63	0010	OFFICES High Rise (11 to 20 story)		S.F.	128	162	199				63
	0020		Total project costs	C.F.	9	11.20	16.10				
	2900		Electrical	S.F.	7.80	9.50	14.15	5.80%	7.85%	10.50%	
	3100		Total: Mechanical & Electrical		25	33.50	57	16.90%	23.50%	34%	
64	0010	POLICE STATIONS		S.F.	154	202	255				64
	0020		Total project costs	C.F.	12.25	15.05	20.50				
	0500		Masonry	S.F.	14.45	25.50	32	6.70%	9.10%	11.35%	
	1800		Equipment		2.45	10.75	17.05	.98%	3.35%	6.70%	
	2720		Plumbing		8.60	17.15	21.50	5.65%	6.90%	10.75%	
	2770		Heating, ventilating, air conditioning		13.45	17.85	25	5.85%	10.55%	11.70%	
	2900		Electrical		16.85	24	32	9.80%	11.85%	14.80%	
	3100		Total: Mechanical & Electrical		55	66.50	90	25%	31.50%	32.50%	
65	0010	POST OFFICES		S.F.	122	150	191				65
	0020		Total project costs	C.F.	7.30	9.25	10.55				
	2720		Plumbing	S.F.	5.45	6.80	8.55	4.24%	5.30%	5.60%	
	2770		Heating, ventilating, air conditioning		8.55	10.55	11.75	6.65%	7.15%	9.35%	
	2900		Electrical		10.05	14.10	16.75	7.25%	9%	11%	
	3100		Total: Mechanical & Electrical		29	38	43	16.25%	18.80%	22%	
66	0010	POWER PLANTS		S.F.	845	1,125	2,050				66
	0020		Total project costs	C.F.	23	50.50	108				
	2900		Electrical	S.F.	59.50	126	189	9.30%	12.75%	21.50%	
	8100		Total: Mechanical & Electrical		148	485	1,075	32.50%	32.50%	52.50%	
67	0010	RELIGIOUS EDUCATION		S.F.	98	128	157				67
	0020		Total project costs	C.F.	5.45	7.80	9.75				
	2720		Plumbing	S.F.	4.09	5.80	8.15	4.40%	5.30%	7.10%	
	2770		Heating, ventilating, air conditioning		10.35	11.70	16.55	10.05%	11.45%	12.35%	
	2900		Electrical		7.75	10.95	14.50	7.60%	9.05%	10.35%	
	3100		Total: Mechanical & Electrical		31.50	41.50	50.50	22%	23%	27%	
69	0010	RESEARCH Laboratories & Facilities		S.F.	150	212	310				69
	0020		Total project costs	C.F.	10.90	21.50	26				
	1800		Equipment	S.F.	6.50	12.80	31	.94%	4.58%	9.10%	
	2720		Plumbing		14.70	18.80	30	6.15%	8.30%	10.80%	
	2770		Heating, ventilating, air conditioning		13.20	44.50	52.50	7.25%	16.50%	17.50%	
	2900		Electrical		17.65	28.50	46.50	9.55%	11.50%	15.40%	
	3100		Total: Mechanical & Electrical		56.50	97.50	140	29.50%	37%	41%	
70	0010	RESTAURANTS		S.F.	142	183	238				70
	0020		Total project costs	C.F.	11.95	15.70	20.50				
	1800		Equipment	S.F.	9.20	22.50	34	6.10%	13%	15.65%	
	2720		Plumbing		11.25	13.65	17.95	6.10%	8.15%	9%	
	2770		Heating, ventilating, air conditioning		14.30	19.75	24	9.20%	12%	12.40%	
	2900		Electrical		15	18.55	24	8.35%	10.55%	11.55%	
	3100		Total: Mechanical & Electrical		46.50	50	64.50	21%	25%	29.50%	
	9000		Per seat unit, total cost	Seat	5,225	6,950	8,225				
	9500		Total: Mechanical & Electrical	"	1,300	1,725	2,050				
72	0010	RETAIL STORES		S.F.	66.50	89	118				72
	0020		Total project costs	C.F.	4.50	6.40	8.90				
	2720		Plumbing	S.F.	2.41	4.02	6.85	3.26%	4.60%	6.80%	
	2770		Heating, ventilating, air conditioning		5.20	7.10	10.70	6.75%	8.75%	10.15%	
	2900		Electrical		6	8.15	11.80	7.25%	9.90%	11.60%	
	3100		Total: Mechanical & Electrical		15.90	20.50	28	17.05%	21%	23.50%	

K1010 | Square Foot Costs

K1010 | S.F. Costs

			UNIT	UNIT COSTS 1/4	MEDIAN	3/4	% OF TOTAL 1/4	MEDIAN	3/4	
74	0010	**SCHOOLS Elementary**	S.F.	109	135	165				74
	0020	Total project costs	C.F.	7.10	9.10	11.70				
	0500	Masonry	S.F.	9.40	16.55	24.50	5.25%	10.65%	14.05%	
	1800	Equipment		2.96	4.91	9.25	1.80%	3.13%	4.61%	
	2720	Plumbing		6.25	8.80	11.75	5.70%	7.15%	9.35%	
	2730	Heating, ventilating, air conditioning		9.35	14.85	21	8.15%	10.80%	14.90%	
	2900	Electrical		10.25	13.65	17.40	8.50%	10.05%	11.80%	
	3100	Total: Mechanical & Electrical		37	46	57.50	25%	27.50%	30%	
	9000	Per pupil, total cost	Ea.	12,400	18,500	41,400				
	9500	Total: Mechanical & Electrical	"	3,500	4,450	11,200				
76	0010	**SCHOOLS Junior High & Middle**	S.F.	112	138	169				76
	0020	Total project costs	C.F.	7.10	9.15	10.30				
	0500	Masonry	S.F.	13.90	17.95	22	7.55%	11.10%	14.30%	
	1800	Equipment		3.57	5.75	8.45	1.79%	3.03%	4.80%	
	2720	Plumbing		6.50	8	9.95	5.30%	6.80%	7.25%	
	2770	Heating, ventilating, air conditioning		13	15.80	28	8.90%	11.55%	14.20%	
	2900	Electrical		11.10	13.90	17.75	7.90%	9.40%	10.60%	
	3100	Total: Mechanical & Electrical		35.50	45.50	57	23%	25.50%	29.50%	
	9000	Per pupil, total cost	Ea.	14,200	18,600	25,000				
78	0010	**SCHOOLS Senior High**	S.F.	117	145	181				78
	0020	Total project costs	C.F.	7	10	16.65				
	1800	Equipment	S.F.	3.10	7.20	10	1.88%	2.67%	4.30%	
	2720	Plumbing		6.50	9.80	17.90	5.60%	6.90%	8.30%	
	2770	Heating, ventilating, air conditioning		13.30	15.25	23	8.95%	11.60%	15%	
	2900	Electrical		11.75	15.65	23	8.70%	10.35%	12.50%	
	3100	Total: Mechanical & Electrical		39.50	46	75	24%	26.50%	28.50%	
	9000	Per pupil, total cost	Ea.	11,000	22,300	27,900				
80	0010	**SCHOOLS Vocational**	S.F.	94	137	169				80
	0020	Total project costs	C.F.	5.85	8.40	11.60				
	0500	Masonry	S.F.	5.50	13.65	21	3.20%	4.61%	10.95%	
	1800	Equipment		2.94	7.30	10.10	1.24%	3.10%	4.26%	
	2720	Plumbing		6	8.95	13.20	5.40%	6.90%	8.55%	
	2770	Heating, ventilating, air conditioning		8.40	15.65	26	8.60%	11.90%	14.65%	
	2900	Electrical		9.80	13.40	18.30	8.45%	11%	13.20%	
	3100	Total: Mechanical & Electrical		37	37.50	64.50	27.50%	29.50%	31%	
	9000	Per pupil, total cost	Ea.	13,100	35,100	52,500				
83	0010	**SPORTS ARENAS**	S.F.	82	110	169				83
	0020	Total project costs	C.F.	4.46	8	10.30				
	2720	Plumbing	S.F.	4.77	7.25	15.25	4.35%	6.35%	9.40%	
	2770	Heating, ventilating, air conditioning		10.25	12.15	16.85	8.80%	10.20%	13.55%	
	2900	Electrical		8.55	11.65	15	7.65%	9.75%	11.90%	
	3100	Total: Mechanical & Electrical		21.50	38	51	18.05%	21.50%	25%	
85	0010	**SUPERMARKETS**	S.F.	76	88	103				85
	0020	Total project costs	C.F.	4.23	5.10	7.75				
	2720	Plumbing	S.F.	4.24	5.35	6.20	5.40%	6%	7.45%	
	2770	Heating, ventilating, air conditioning		6.25	8.30	10.10	8.60%	8.65%	9.60%	
	2900	Electrical		9.50	10.85	12.90	10.40%	12.45%	13.60%	
	3100	Total: Mechanical & Electrical		24.50	26	34.50	23.50%	27.50%	28.50%	
86	0010	**SWIMMING POOLS**	S.F.	123	206	440				86
	0020	Total project costs	C.F.	9.85	12.30	13.40				
	2720	Plumbing	S.F.	11.35	13	17.45	4.80%	9.70%	20.50%	
	2900	Electrical		9.25	15	22	5.75%	6.95%	7.60%	
	3100	Total: Mechanical & Electrical		22.50	57.50	77	11.15%	14.10%	23.50%	
87	0010	**TELEPHONE EXCHANGES**	S.F.	163	232	305				87
	0020	Total project costs	C.F.	10.15	16.30	22.50				
	2720	Plumbing	S.F.	6.90	10.65	15.60	4.52%	5.80%	6.90%	
	2770	Heating, ventilating, air conditioning		16	32	40	11.80%	16.05%	18.40%	

K1010 | Square Foot Costs

K1010 | S.F. Costs

			UNIT	UNIT COSTS			% OF TOTAL		
				1/4	MEDIAN	3/4	1/4	MEDIAN	3/4
87	2900	Electrical	S.F.	16.65	26.50	46.50	10.90%	14%	17.85%
	3100	Total: Mechanical & Electrical	↓	49	93	132	29.50%	33.50%	44.50%
91	0010	**THEATERS**	S.F.	103	132	194			
	0020	Total project costs	C.F.	4.74	7	10.30			
	2720	Plumbing	S.F.	3.42	3.71	15.15	2.92%	4.70%	6.80%
	2770	Heating, ventilating, air conditioning		9.95	12.10	14.95	8%	12.25%	13.40%
	2900	Electrical		9	12.10	24.50	8.05%	9.35%	12.25%
	3100	Total: Mechanical & Electrical	↓	23	31	36.50	21.50%	25.50%	27.50%
94	0010	**TOWN HALLS City Halls & Municipal Buildings**	S.F.	115	146	190			
	0020	Total project costs	C.F.	10.45	12.05	17.65			
	2720	Plumbing	S.F.	4.79	8.95	15.70	4.31%	5.95%	7.95%
	2770	Heating, ventilating, air conditioning		8.65	17.20	25	7.05%	9.05%	13.45%
	2900	Electrical		10.90	15	20.50	8.05%	9.45%	11.65%
	3100	Total: Mechanical & Electrical	↓	38	48	73	22%	26.50%	31%
97	0010	**WAREHOUSES & Storage Buildings**	S.F.	42.50	61.50	91			
	0020	Total project costs	C.F.	2.24	3.51	5.80			
	0100	Site work	S.F.	4.42	8.75	13.20	6.05%	12.95%	19.55%
	0500	Masonry		2.53	6.10	13.10	3.60%	7.35%	12%
	1800	Equipment		.69	1.48	8.30	.72%	1.69%	5.55%
	2720	Plumbing		1.43	2.56	4.78	2.90%	4.80%	6.55%
	2730	Heating, ventilating, air conditioning		1.63	4.59	6.15	2.41%	5%	8.90%
	2900	Electrical		2.53	4.77	7.80	5.15%	7.20%	10.05%
	3100	Total: Mechanical & Electrical	↓	7.05	10.85	21.50	13.30%	18.90%	26%
99	0010	**WAREHOUSE & OFFICES Combination**	S.F.	52.50	70	96.50			
	0020	Total project costs	C.F.	2.69	3.91	5.75			
	1800	Equipment	S.F.	.91	1.76	2.62	.52%	1.19%	2.40%
	2720	Plumbing		2.03	3.53	5.25	3.74%	4.76%	6.30%
	2770	Heating, ventilating, air conditioning		3.21	5	7	5%	5.65%	10.05%
	2900	Electrical		3.59	5.25	8.30	5.85%	8%	10%
	3100	Total: Mechanical & Electrical	↓	10	15.15	24	14.55%	19.95%	24.50%

Reference Aids — RL1010-100 Live Loads

L1010-101 Minimum Design Live Loads in Pounds per S.F. for Various Building Codes

Occupancy	Description	Minimum Live Loads, Pounds per S.F. — IBC
Access Floors	Office use	50
	Computer use	100
Armories		150
Assembly	Fixed seats	60
	Movable seats	100
	Platforms or stage floors	125
Commercial & Industrial	Light manufacturing	125
	Heavy manufacturing	250
	Light storage	125
	Heavy storage	250
	Stores, retail, first floor	100
	Stores, retail, upper floors	75
	Stores, wholesale	125
Dance halls	Ballrooms	100
Dining rooms	Restaurants	100
Fire escapes	Other than below	100
	Single family residential	40
Garages	Passenger cars only	40
Gymnasiums	Main floors and balconies	100
Hospitals	Operating rooms, laboratories	60
	Private room	40
	Corridors, above first floor	80
Libraries	Reading rooms	60
	Stack rooms	150
	Corridors, above first floor	80
Marquees		75
Office Buildings	Offices	50
	Lobbies	100
	Corridors, above first floor	80
Residential	Multi family private apartments	40
	Multi family, public rooms	100
	Multi family, corridors	100
	Dwellings, first floor	40
	Dwellings, second floor & habitable attics	30
	Dwellings, uninhabitable attics	20
	Hotels, guest rooms	40
	Hotels, public rooms	100
	Hotels, corridors serving public rooms	100
	Hotels, corridors	100
Schools	Classrooms	40
	Corridors	100
Sidewalks	Driveways, etc. subject to trucking	250
Stairs	Exits	100
Theaters	Aisles, corridors and lobbies	See Assembly
	Orchestra floors	↓
	Balconies	↓
	Stage floors	125
Yards	Terrace, pedestrian	100

UBC = Uniform Building Code, International Conference of Building Officials, 2006 Edition

Reference Aids — RL1010-200 Dead Load, Material Weights

Table L1010-201 Design Weight Per S.F. for Walls and Partitions

Type	Wall Thickness	Description	Weight Per S.F.	Type	Wall Thickness	Description	Weight Per S.F.
Brick	4"	Clay brick, high absorption	34 lb.	Clay tile	2"	Split terra cotta furring	10 lb.
		Clay brick, medium absorption	39			Non load bearing clay tile	11
		Clay brick, low absorption	46		3"	Split terra cotta furring	12
		Sand-lime brick	38			Non load bearing clay tile	18
		Concrete brick, heavy aggregate	46		4"	Non load bearing clay tile	20
		Concrete brick, light aggregate	33			Load bearing clay tile	24
	8"	Clay brick, high absorption	69		6"	Non load bearing clay tile	30
		Clay brick, medium absorption	79			Load bearing clay tile	36
		Clay brick, low absorption	89		8"	Non load bearing clay tile	36
		Sand-lime brick	74			Load bearing clay tile	42
		Concrete brick, heavy aggregate	89		12"	Non load bearing clay tile	46
		Concrete brick, light aggregate	68			Load bearing clay tile	58
	12"	Common brick	120	Gypsum block	2"	Hollow gypsum block	9.5
		Pressed brick	130			Solid gypsum block	12
		Sand-lime brick	105		3"	Hollow gypsum block	10
		Concrete brick, heavy aggregate	130			Solid gypsum block	18
		Concrete brick, light aggregate	98		4"	Hollow gypsum block	15
	16"	Clay brick, high absorption	134			Solid gypsum block	24
		Clay brick, medium absorption	155		5"	Hollow gypsum block	18
		Clay brick, low absorption	173		6"	Hollow gypsum block	24
		Sand-lime brick	138	Structural facing tile	2"	Facing tile	15
		Concrete brick, heavy aggregate	174		4"	Facing tile	25
		Concrete brick, light aggregate	130		6"	Facing tile	38
Concrete block	4"	Solid conc. block, stone aggregate	45	Glass	4"	Glass block	18
		Solid conc. block, lightweight	34		1"	Structural glass	15
		Hollow conc. block, stone aggregate	30	Plaster	1"	Gypsum plaster (1 side)	5
		Hollow conc. block, lightweight	20			Cement plaster (1 side)	10
	6"	Solid conc. block, stone aggregate	50			Gypsum plaster on lath	8
		Solid conc. block, lightweight	37			Cement plaster on lath	13
		Hollow conc. block, stone aggregate	42	Plaster partition (2 finished faces)	2"	Solid gypsum on metal lath	18
		Hollow conc. block, lightweight	30			Solid cement on metal lath	25
	8"	Solid conc. block, stone aggregate	67			Solid gypsum on gypsum lath	18
		Solid conc. block, lightweight	48			Gypsum on lath & metal studs	18
		Hollow conc. block, stone aggregate	55		3"	Gypsum on lath & metal studs	19
		Hollow conc. block, lightweight	38		4"	Gypsum on lath & metal studs	20
	10"	Solid conc. block, stone aggregate	84		6"	Gypsum on lath & wood studs	18
		Solid conc. block, lightweight	62	Concrete	6"	Reinf concrete, stone aggregate	75
		Hollow conc. block, stone aggregate	55			Reinf. concrete, lightweight	36-60
		Hollow conc. block, lightweight	38		8"	Reinf. concrete, stone aggregate	100
	12"	Solid conc. block, stone aggregate	108			Reinf. concrete, lightweight	48-80
		Solid conc. block, lightweight	72		10"	Reinf. concrete, stone aggregate	125
		Hollow conc. block, stone aggregate	85			Reinf. concrete, lightweight	60-100
		Hollow conc. block, lightweight	55		12"	Reinf concrete stone aggregate	150
Drywall	6"	Drywall on wood studs	10			Reinf. concrete, lightweight	72-120

Table L1010-202 Design Weight per S.F. for Roof Coverings

Type		Description	Weight lb. Per S.F.	Type	Wall Thickness	Description	Weight lb. Per S.F.
Sheathing	Gypsum	1" thick	4	Metal	Aluminum	Corr. & ribbed, .024" to .040"	.4-.8
	Wood	¾" thick	3		Copper	or tin	1.5-2.5
Insulation	per 1"	Loose	.5		Steel	Corrugated, 29 ga. to 12 ga.	.6-5.0
		Poured in place	2	Shingles	Asphalt	Strip shingles	1.7-2.8
		Rigid	1.5		Clay	Tile	8-16
Built-up	Tar &	3 ply felt	5.5		Slate	¼" thick	9.5
	gravel	5 ply felt	6.5		Wood		2-3

Reference Aids — RL1010-200 Dead Load, Material Weights

Table L1010-203 Design Weight per Square Foot for Floor Fills and Finishes

Type		Description	Weight lb. per S.F.	Type		Description	Weight lb. per S.F.
Floor fill	per 1"	Cinder fill	5	Wood	Single 7/8"	On sleepers, light concrete fill	16
		Cinder concrete	9			On sleepers, stone concrete fill	25
		Lightweight concrete	3-9		Double 7/8"	On sleepers, light concrete fill	19
		Stone concrete	12			On sleepers, stone concrete fill	28
		Sand	8		3"	Wood block on mastic, no fill	15
		Gypsum	6			Wood block on ½" mortar	16
Terrazzo	1"	Terrazzo, 2" stone concrete	25		per 1"	Hardwood flooring (25/32")	4
Marble	and mortar	on stone concrete fill	33			Underlayment (Plywood per 1")	3
Resilient	1/16"-1/4"	Linoleum, asphalt, vinyl tile	2	Asphalt	1-1/2"	Mastic flooring	18
Tile	3/4"	Ceramic or quarry	10		2"	Block on ½" mortar	30

Table L1010-204 Design Weight Per Cubic Foot for Miscellaneous Materials

Type		Description	Weight lb. per C.F.	Type		Description	Weight lb. per C.F.
Bituminous	Coal, piled	Anthracite	47-58	Masonry	Ashlar	Granite	165
		Bituminous	40-54			Limestone, crystalline	165
		Peat, turf, dry	47			Limestone, oolitic	135
		Coke	75			Marble	173
	Petroleum	Unrefined	54			Sandstone	144
		Refined	50		Rubble, in mortar	Granite	155
		Gasoline	42			Limestone, crystalline	147
	Pitch		69			Limestone, oolitic	138
	Tar	Bituminous	75			Marble	156
Concrete	Plain	Stone aggregate	144			Sandstone & Bluestone	130
		Slag aggregate	132		Brick	Pressed	140
		Expanded slag aggregate	100			Common	120
		Haydite (burned clay agg.)	90			Soft	100
		Vermiculite & perlite, load bearing	70-105		Cement	Portland, loose	90
		Vermiculite & perlite, non load bear	35-50			Portland set	183
	Reinforced	Stone aggregate	150		Lime	Gypsum, loose	53-64
		Slag aggregate	138		Mortar	Set	103
		Lightweight aggregates	30-120	Metals	Aluminum	Cast, hammered	165
Earth	Clay	Dry	63		Brass	Cast, rolled	534
		Damp, plastic	110		Bronze	7.9 to 14% Sn	509
		and gravel, dry	100		Copper	Cast, rolled	556
	Dry	Loose	76		Iron	Cast, pig	450
		Packed	95			Wrought	480
	Moist	Loose	78		Lead		710
		Packed	96		Monel		556
	Mud	Flowing	108		Steel	Rolled	490
		Packed	115		Tin	Cast, hammered	459
	Riprap	Limestone	80-85		Zinc	Cast rolled	440
		Sandstone	90	Timber	Cedar	White or red	24.2
		Shale	105		Fir	Douglas	23.7
	Sand & gravel	Dry, loose	90-105			Eastern	25
		Dry, packed	100-120		Maple	Hard	44.5
		Wet	118-120			White	33
Gases	Air	0C., 760 mm.	.0807		Oak	Red or Black	47.3
	Gas	Natural	.0385			White	47.3
Liquids	Alcohol	100%	49		Pine	White	26
	Water	4°C., maximum density	62.5			Yellow, long leaf	44
		Ice	56			Yellow, short leaf	38
		Snow, fresh fallen	8		Redwood	California	26
		Sea water	64		Spruce	White or black	27

Reference Aids — RL1010-220 Dead Load, System Weights

Table L1010-225 Design Weight Per S.F. for Structural Floor and Roof Systems

Type Slab		Description	Weight in Pounds per S.F. — Slab Depth in Inches												
Concrete Slab	Reinforced	Stone aggregate	1"	12.5	2"	25	3"	37.5	4"	50	5"	62.5	6"	75
		Lightweight sand aggregate		9.5		19		28.5		38		47.5		57
		All lightweight aggregate		9.0		18		27.0		36		45.0		54
	Plain, nonreinforced	Stone aggregate		12.0		24		36.0		48		60.0		72
		Lightweight sand aggregate		9.0		18		27.0		36		45.0		54
		All lightweight aggregate	↓	8.5	↓	17	↓	25.5	↓	34	↓	42.5	↓	51
Concrete Waffle	19" x 19"	5" wide ribs @ 24" O.C.	6+3	77	8+3	92	10+3	100	12+3	118				
			6+4½	96	8+4½	110	10+4½	119	12+4½	136				
	30" x 30"	6" wide ribs @ 36" O.C.	8+3	83	10+3	95	12+3	109	14+3	118	16+3	130	20+3	155
			8+4½	101	10+4½	113	12+4½	126	14+4½	137	16+4½	149	20+4½	173
Concrete Joist	20" wide form	5" wide rib	8+3	60	10+3	67	12+3	74	14+3	81				
		6" wide rib	↓	63	↓	70	↓	78	↓	86	16+3	94	20+3	111
		7" wide rib									↓	99	↓	118
		5" wide rib	8+4½	79	10+4½	85	12+4½	92	14+4½	99				
		6" wide rib	↓	82	↓	89	↓	97	↓	104	16+4½	113	20+4½	130
		7" wide rib									↓	118	↓	136
	30" wide form	5" wide rib	8+3	54	10+3	58	12+3	63	14+3	68				
		6" wide rib	↓	56	↓	61	↓	67	↓	72	16+3	78	20+3	91
		7" wide rib									↓	83	↓	96
		5" wide rib	8+4½	72	10+4½	77	12+4½	82	14+4½	87				
		6" wide rib	↓	75	↓	80	↓	85	↓	91	16+4½	97	20+4½	109
		7" wide rib									↓	101	↓	115
Wood Joists	Incl. Subfloor	12" O.C.	2x6	6	2x8	6	2x10	7	2x12	8	3x8	8	3x12	11
		16" O.C.	↓	5	↓	6	↓	6	↓	7	↓	7	↓	9

Table L1010-226 Superimposed Dead Load Ranges

Component	Load Range (PSF)
Ceiling	5-10
Partitions	20-30
Mechanical	4-8

Reference Aids — RL1010-300 Wind Loads

Table L1010-301 Design Loads for Structures for Wind Load

Wind Loads: Structures are designed to resist the wind force from any direction. Usually 2/3 is assumed to act on the windward side, 1/3 on the leeward side.

For more than 1/3 openings, add 10 psf for internal wind pressure or 5 psf for suction, whichever is critical.

For buildings and structures, use psf values from Table L1010-301.

For glass over 4 S.F. use values in Table L1010-302 after determining 30′ wind velocity from Table L1010-303.

Type Structure	Height Above Grade	Horizontal Load in Lb. per S.F.
Buildings	Up to 50 ft.	15
	50 to 100 ft.	20
	Over 100 ft.	20 + .025 per ft.
Ground signs & towers	Up to 50 ft.	15
	Over 50 ft.	20
Roof structures		30
Glass	See Table below	

Table L1010-302 Design Wind Load in PSF for Glass at Various Elevations (BOCA Code)

Height From Grade	Vel.	PSF	Vel.	PSF	Vel.	PSF	Vel.	PSF	Vel.	PSF	Vel.	PSF	Vel.	PSF	Vel.	PSF	Vel.	PSF	Vel.	PSF	Vel.	PSF
To 10 ft.	42	6	46	7	49	8	52	9	55	10	59	11	62	12	66	14	69	15	76	19	83	22
10-20	52	9	58	11	61	11	65	14	70	16	74	18	79	20	83	22	87	24	96	30	105	35
20-30*	60	12	67	14	70	16	75	18	80	20	85	23	90	26	95	29	100	32	110	39	120	46
30-60	66	14	74	18	77	19	83	22	88	25	94	28	99	31	104	35	110	39	121	47	132	56
60-120	73	17	82	21	85	12	92	27	98	31	104	35	110	39	116	43	122	48	134	57	146	68
120-140	81	21	91	26	95	29	101	33	108	37	115	42	122	48	128	52	135	48	149	71	162	84
240-480	90	26	100	32	104	35	112	40	119	45	127	51	134	57	142	65	149	71	164	86	179	102
480-960	98	31	110	39	115	42	123	49	131	55	139	62	148	70	156	78	164	86	180	104	197	124
Over 960	98	31	110	39	115	42	123	49	131	55	139	62	148	70	156	78	164	86	180	104	197	124

*Determine appropriate wind at 30′ elevation Fig. L1010-303 below.

Table L1010-303 Design Wind Velocity at 30 Ft. Above Ground

SPEEDS ARE FOR NORMAL EXPOSURE WHERE SURFACE FRICTION IS RELATIVELY UNIFORM FOR A FETCH OF ABOUT 25 MILES. IF THE EXPOSURE IS ELEVATED, SUBJECT TO CHANNELING, OR OTHER SPECIAL CONDITIONS AFFECTING THE EXTREME WIND SPEEDS, ADJUSTMENTS MUST BE MADE TO THE MAP VALUES.

Reference Aids   RL1010-400   Snow Loads

Table L1010-401   Snow Load in Pounds Per Square Foot on the Ground

Table L1010-402   Snow Load in Pounds Per Square Foot on the Ground

Reference Aids

RL1010-400 Snow Loads

Table L1010-403 Snow Loads

To convert the ground snow loads on the previous page to roof snow loads, the ground snow loads should be multiplied by the following factors depending upon the roof characteristics.

Note, in all cases $\frac{\alpha - 30}{50}$ is valid only for > 30 degrees.

Description	Sketch	Formula	Angle	Conversion Factor = C_f Sheltered	Exposed
Simple flat and shed roofs	Load Diagram	For $\alpha > 30°$ $C_f = 0.8 - \frac{\alpha - 30}{50}$	0° to 30°	0.8	0.6
			40°	0.6	0.45
			50°	0.4	0.3
			60°	0.2	0.15
			70° to 90°	0	0
				Case I	Case II
Simple gable and hip roofs	Case I Load Diagram; Case II Load Diagram	$C_f = 0.8 - \frac{\alpha - 30}{50}$ $C_f = 1.25 \left(0.8 - \frac{\alpha - 30}{50}\right)$	10°	0.8	—
			20°	0.8	—
			30°	0.8	1.0
			40°	0.6	0.75
			50°	0.4	0.5
			60°	0.2	0.25
Valley areas of Two span roofs	Case I; Case II	$\beta = \frac{\alpha_1 + \alpha_2}{2}$ $C_f = 0.8 - \frac{\alpha - 30}{50}$	$\beta \leq 10°$ use Case I only $\beta > 10°$ $\beta < 20°$ use Case I & II $\beta \geq 20°$ use Case I, II & III		
Lower level of multi level roofs (or on an adjacent building not more then 15 ft. away)	$W = 2h$	$C_f = 15 \frac{h}{g}$ h = difference in roof height in feet g = ground snow load in psf w = width of drift For h < 5, w = 10 h > 15, w = 30	When $15 \frac{h}{g} < .8$, use 0.8 When $15 \frac{h}{g} > 3.0$, use 3.0		

Summary of above:

1. For flat roofs or roofs up to 30°, use 0.8 x ground snow load for the roof snow load.
2. For roof pitches in excess of 30°, conversion factor becomes lower than 0.8.
3. For exposed roofs there is a further 25% reduction of conversion factor.
4. For steep roofs a more highly loaded half span must be considered.
5. For shallow roof valleys conversion factor is 0.8.
6. For moderate roof valleys, conversion factor is 1.0 for half the span.
7. For steep roof valleys, conversion factor is 1.5 for one quarter of the span.
8. For roofs adjoining vertical surfaces the conversion factor is up to 3.0 for part of the span.
9. If snow load is less than 30 psf, use water load on roof for clogged drain condition.

Reference Aids — RL1020-100 Area Requirements

Table L1020-101 Floor Area Ratios

Table below lists commonly used gross to net area and net to gross area ratios expressed in % for various building types.

Building Type	Gross to Net Ratio	Net to Gross Ratio	Building Type	Gross to Net Ratio	Net to Gross Ratio
Apartment	156	64	School Buildings (campus type)		
Bank	140	72	Administrative	150	67
Church	142	70	Auditorium	142	70
Courthouse	162	61	Biology	161	62
Department Store	123	81	Chemistry	170	59
Garage	118	85	Classroom	152	66
Hospital	183	55	Dining Hall	138	72
Hotel	158	63	Dormitory	154	65
Laboratory	171	58	Engineering	164	61
Library	132	76	Fraternity	160	63
Office	135	75	Gymnasium	142	70
Restaurant	141	70	Science	167	60
Warehouse	108	93	Service	120	83
			Student Union	172	59

The gross area of a building is the total floor area based on outside dimensions.

The net area of a building is the usable floor area for the function intended and excludes such items as stairways, corridors, and mechanical rooms.

In the case of a commercial building, it might be considered as the "leasable area."

Reference Aids — RL1020-200 Partition/Door Guidelines

Table L1020-201 Partition/Door Density

Building Type		Stories	Partition/Density	Doors	Description of Partition
Apartments		1 story	9 SF/LF	90 SF/door	Plaster, wood doors & trim
		2 story	8 SF/LF	80 SF/door	Drywall, wood studs, wood doors & trim
		3 story	9 SF/LF	90 SF/door	Plaster, wood studs, wood doors & trim
		5 story	9 SF/LF	90 SF/door	Plaster, metal studs, wood doors & trim
		6-15 story	8 SF/LF	80 SF/door	Drywall, metal studs, wood doors & trim
Bakery		1 story	50 SF/LF	500 SF/door	Conc. block, paint, door & drywall, wood studs
		2 story	50 SF/LF	500 SF/door	Conc. block, paint, door & drywall, wood studs
Bank		1 story	20 SF/LF	200 SF/door	Plaster, wood studs, wood doors & trim
		2-4 story	15 SF/LF	150 SF/door	Plaster, metal studs, wood doors & trim
Bottling Plant		1 story	50 SF/LF	500 SF/door	Conc. block, drywall, metal studs, wood trim
Bowling Alley		1 story	50 SF/LF	500 SF/door	Conc. block, wood & metal doors, wood trim
Bus Terminal		1 story	15 SF/LF	150 SF/door	Conc. block, ceramic tile, wood trim
Cannery		1 story	100 SF/LF	1000 SF/door	Drywall on metal studs
Car Wash		1 story	18 SF/LF	180 SF/door	Concrete block, painted & hollow metal door
Dairy Plant		1 story	30 SF/LF	300 SF/door	Concrete block, glazed tile, insulated cooler doors
Department Store		1 story	60 SF/LF	600 SF/door	Drywall, metal studs, wood doors & trim
		2-5 story	60 SF/LF	600 SF/door	30% concrete block, 70% drywall, wood studs
Dormitory		2 story	9 SF/LF	90 SF/door	Plaster, concrete block, wood doors & trim
		3-5 story	9 SF/LF	90 SF/door	Plaster, concrete block, wood doors & trim
		6-15 story	9 SF/LF	90 SF/door	Plaster, concrete block, wood doors & trim
Funeral Home		1 story	15 SF/LF	150 SF/door	Plaster on concrete block & wood studs, paneling
		2 story	14 SF/LF	140 SF/door	Plaster, wood studs, paneling & wood doors
Garage Sales & Service		1 story	30 SF/LF	300 SF/door	50% conc. block, 50% drywall, wood studs
Hotel		3-8 story	9 SF/LF	90 SF/door	Plaster, conc. block, wood doors & trim
		9-15 story	9 SF/LF	90 SF/door	Plaster, conc. block, wood doors & trim
Laundromat		1 story	25 SF/LF	250 SF/door	Drywall, wood studs, wood doors & trim
Medical Clinic		1 story	6 SF/LF	60 SF/door	Drywall, wood studs, wood doors & trim
		2-4 story	6 SF/LF	60 SF/door	Drywall, metal studs, wood doors & trim
Motel		1 story	7 SF/LF	70 SF/door	Drywall, wood studs, wood doors & trim
		2-3 story	7 SF/LF	70 SF/door	Concrete block, drywall on metal studs, wood paneling
Movie Theater	200-600 seats	1 story	18 SF/LF	180 SF/door	Concrete block, wood, metal, vinyl trim
	601-1400 seats		20 SF/LF	200 SF/door	Concrete block, wood, metal, vinyl trim
	1401-22000 seats		25 SF/LF	250 SF/door	Concrete block, wood, metal, vinyl trim
Nursing Home		1 story	8 SF/LF	80 SF/door	Drywall, metal studs, wood doors & trim
		2-4 story	8 SF/LF	80 SF/door	Drywall, metal studs, wood doors & trim
Office		1 story	20 SF/LF	200-500 SF/door	30% concrete block, 70% drywall on wood studs
		2 story	20 SF/LF	200-500 SF/door	30% concrete block, 70% drywall on metal studs
		3-5 story	20 SF/LF	200-500 SF/door	30% concrete block, 70% movable partitions
		6-10 story	20 SF/LF	200-500 SF/door	30% concrete block, 70% movable partitions
		11-20 story	20 SF/LF	200-500 SF/door	30% concrete block, 70% movable partitions
Parking Ramp (Open)		2-8 story	60 SF/LF	600 SF/door	Stair and elevator enclosures only
Parking garage		2-8 story	60 SF/LF	600 SF/door	Stair and elevator enclosures only
Pre-Engineered	Steel	1 story	0		
	Store	1 story	60 SF/LF	600 SF/door	Drywall on metal studs, wood doors & trim
	Office	1 story	15 SF/LF	150 SF/door	Concrete block, movable wood partitions
	Shop	1 story	15 SF/LF	150 SF/door	Movable wood partitions
	Warehouse	1 story	0		
Radio & TV Broadcasting		1 story	25 SF/LF	250 SF/door	Concrete block, metal and wood doors
& TV Transmitter		1 story	40 SF/LF	400 SF/door	Concrete block, metal and wood doors
Self Service Restaurant		1 story	15 SF/LF	150 SF/door	Concrete block, wood and aluminum trim
Cafe & Drive-in Restaurant		1 story	18 SF/LF	180 SF/door	Drywall, metal studs, ceramic & plastic trim
Restaurant with seating		1 story	25 SF/LF	250 SF/door	Concrete block, paneling, wood studs & trim
Supper Club		1 story	25 SF/LF	250 SF/door	Concrete block, paneling, wood studs & trim
Bar or Lounge		1 story	24 SF/LF	240 SF/door	Plaster or gypsum lath, wooded studs
Retail Store or Shop		1 story	60 SF/LF	600 SF/door	Drywall metal studs, wood doors & trim
Service Station	Masonry	1 story	15 SF/LF	150 SF/door	Concrete block, paint, door & drywall, wood studs
	Metal panel	1 story	15 SF/LF	150 SF/door	Concrete block, paint, door & drywall, wood studs
	Frame	1 story	15 SF/LF	150 SF/door	Drywall, wood studs, wood doors & trim
Shopping Center	(strip)	1 story	30 SF/LF	300 SF/door	Drywall, metal studs, wood doors & trim
	(group)	1 story	40 SF/LF	400 SF/door	50% concrete block, 50% drywall, wood studs
		2 story	40 SF/LF	400 SF/door	50% concrete block, 50% drywall, wood studs
Small Food Store		1 story	30 SF/LF	300 SF/door	Concrete block drywall, wood studs, wood trim
Store/Apt. above	Masonry	2 story	10 SF/LF	100 SF/door	Plaster, metal studs, wood doors & trim
	Frame	2 story	10 SF/LF	100 SF/door	Plaster, metal studs, wood doors & trim
	Frame	3 story	10 SF/LF	100 SF/door	Plaster, metal studs, wood doors & trim
Supermarkets		1 story	40 SF/LF	400 SF/door	Concrete block, paint, drywall & porcelain panel
Truck Terminal		1 story	0		
Warehouse		1 story	0		

Reference Aids RL1020-300 Building Code Requirements

Table L1020-301 Occupancy Determinations

Description		S.F. Required per Person
		IBC
Assembly Areas	Fixed seats**	# fixed seats
	Without fixed seats	
	Concentrated (chairs only)	7 net
	Standing Space	5 net
	Unconcentrated (tables and chairs)	15 net
	Gaming floors (Keno, slots, etc.)	11 gross
Educational	Classroom area	20 net
	Shops and other vocational room areas	50 net
Institutional	Inpatient treatment areas	240 gross
	Outpatient areas	100 gross
	Sleeping Areas	120 gross
Mercantile	Areas on other floors	60 gross
	Basement and grade floor areas	30 gross
	Storage, stock, shipping areas	300 gross
Business Areas		100 gross

IBC = *International Building Code*, 2006 Edition

** The occupancy load for assembly area with fixed seats shall be determined by the number of fixed seats installed.

Table L1020-302 Length of Exitway Access Travel (ft.)

Use Group	Without Fire Suppression System	With Fire Suppression System
Assembly	200	250
Business	200	300
Factory and industrial	300	400
High hazard	—	75
Institutional	200	250
Mercantile	200	250
Residential	200	250
Storage, low hazard	300	400
Storage, moderate hazard	200	250

Note: The maximum length of exitway access travel in unlimited area buildings shall be 400'.

Table L1020-303 Capacity Per Unit Egress Width*

Use Group	Without Fire Suppression System (Inches per Person)*		With Fire Suppression System (Inches per Person)*	
	Stairways	Doors, Ramps and Corridors	Stairways	Doors, Ramps and Corridors
Assembly, Business, Educational, Factory Industrial, Mercantile, Residential, Storage	0.3	0.2	0.2	0.15
Institutional—1	0.3	0.2	0.2	0.15
Institutional—2	—	0.7	0.3	0.2
Institutional—3	0.3	0.2	0.2	0.15
High Hazard	0.7	0.4	0.3	0.2

* 1" = 25.4 mm

Reference Aids — RL1030-100 "U" Values/Type A & B Buildings

Table L1030-101 "U" Values for Type "A" Buildings

Type A buildings shall include:

- A1 Detached one and two family dwellings
- A2 All other residential buildings, three stories or less, including but not limited to: multi-family dwellings, hotels and motels.

Table L1030-102 "U" Values for Type "B" Buildings

For all buildings Not Classified Type "A"

Reference Aids — RL1030-200 "U" Values for Walls/Roofs

Table L1030-201 Combinations of Wall and Single-Glazed Openings (For Use With ASHRAE 90-75)

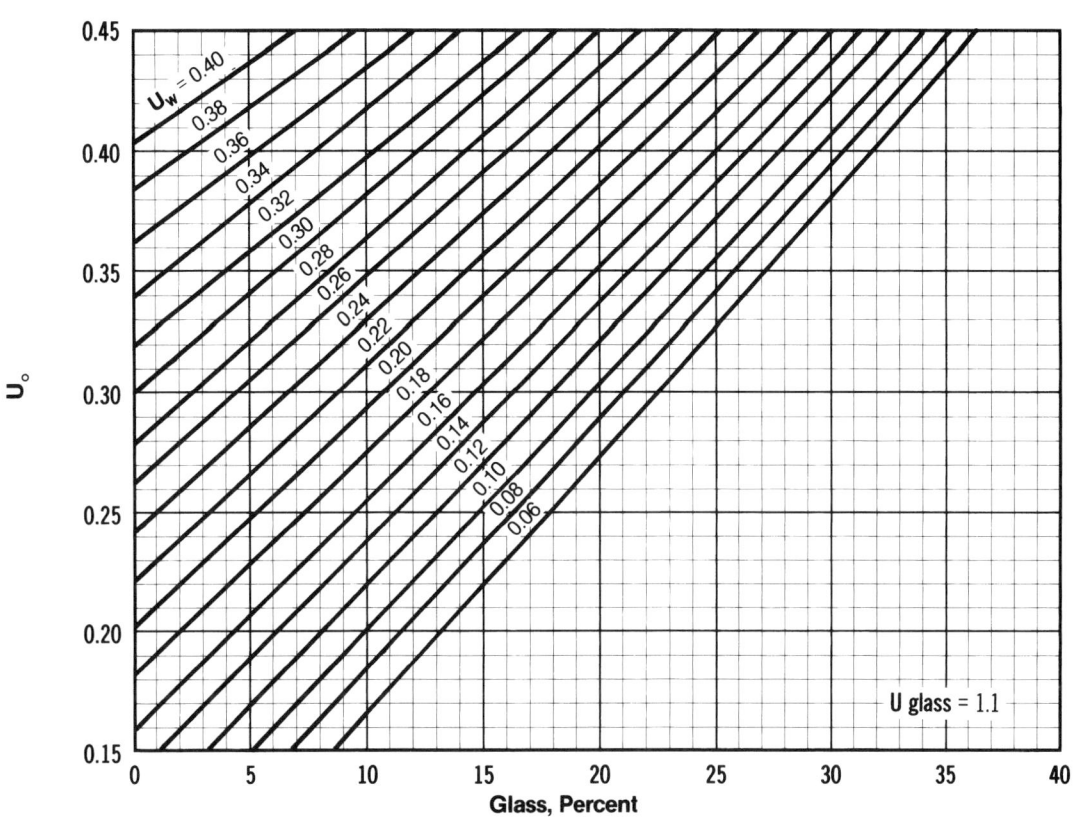

Table L1030-202 Combinations of Wall and Double-Glazed Openings (For Use With ASHRAE 90-75)

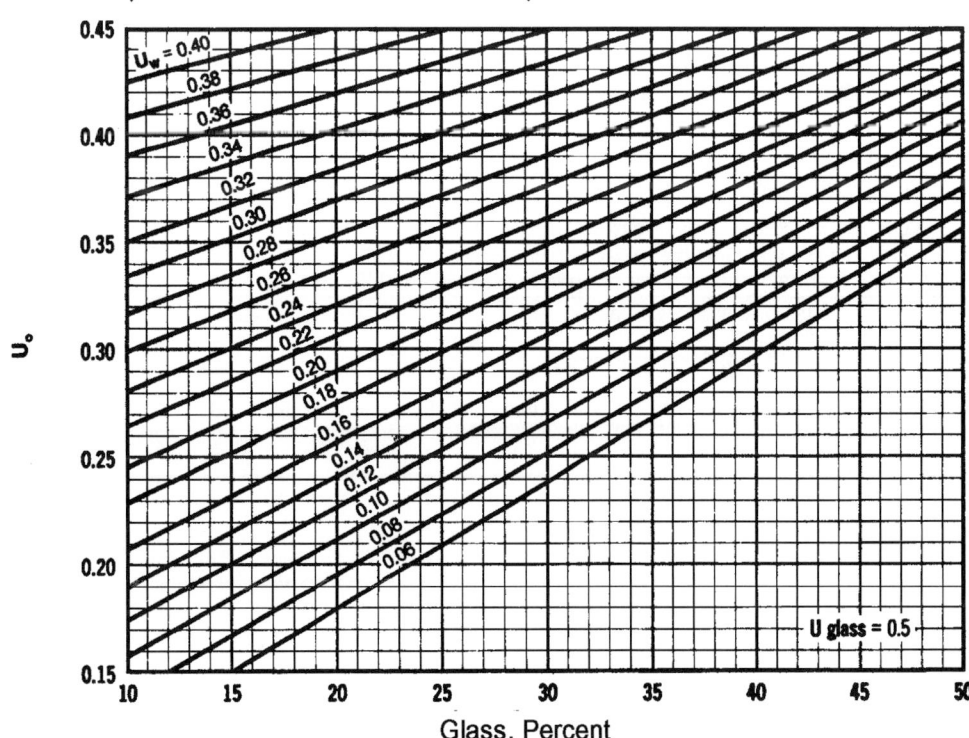

Reference Aids — RL1030-200 "U" Values for Walls/Roofs

Table L1030-203 Influence of Wall and Roof Weight on "U" Value Correction for Heating Design

Effective "U" for walls: $U_w = U_{w_{ss}} \times M$ (similarly, U_r for roofs)

where U_w = effective thermal transmittance of opaque wall area BTU/h × Ft.^{2} × °F
$U_{w_{ss}}$ = steady state thermal transmittance of opaque wall area BTU/h × Ft.^{2} × °F
(steady state "U" value)
M = weight correction factor

Example: $U_{w_{ss}} = 0.20$ with wall weight = 120 psf in Providence, R.I. (6000 degree days)
Enter chart on bottom at 6000, go up to 120 psf curve, read to the left .88 M = .88
and $U_w = 0.20 \times 0.88 = 0.176$

Reference Aids — RL1030-200 "U" Values for Walls/Roofs

Table L1030-204 "U" Values for Type "A" Buildings

One and two family dwellings and residential buildings of three stories or less shall have an overall U value as follows:

A. For roof assemblies in which the finished interior surface is essentially the underside of the roof deck, such as exposed concrete slabs, joist slabs, cored slabs, or cathedral ceilings or wood-beam construction.

$U_{or} = 0.08$ Btu/hr./ft.^{2}/° F. (all degree days)

B. For roof-ceiling assemblies, such as roofs with finished ceilings attached to or suspended below the roof deck.

$U_{or} = 0.05$ for below 8,000 degree days
$= 0.04$ for 8,000 degree days and greater.

Table L1030-205 "U" Values for Type "B" Buildings*

For all buildings not classified type "A".

*Minimum requirements for thermal design, ASHRAE Standard 90-75

Reference Aids — RL1030-300 "U" & "R" Values for Floors/Slabs

Table L1030-301 "U" Values for Floors Over Unheated Spaces

Table L1030-302 "R" Values for Slabs on Grade

Table L1030-303 Insulation Requirements for Slabs on Grade for all Buildings

Section

Section (alternate)

Reference Aids — RL1030-400 Thermal Properties of Materials

Table L1030-401 Resistances ("R") of Building and Insulating Materials

Material	Wt./Lbs. per C.F.	R per Inch	R Listed Size
Air Spaces and Surfaces			
Enclosed non-reflective spaces, E=0.82			
50° F mean temp., 30°/10° F diff.			
.5"			.90/.91
.75"			.94/1.01
1.50"			.90/1.02
3.50"			.91/1.01
Inside vert. surface (still air)			0.68
Outside vert. surface (15 mph wind)			0.17
Building Boards			
Asbestos cement, 0.25" thick	120		0.06
Gypsum or plaster, 0.5" thick	50		0.45
Hardboard regular	50	1.37	
Tempered	63	1.00	
Laminated paper	30	2.00	
Particle board	37	1.85	
	50	1.06	
	63	0.85	
Plywood (Douglas Fir), 0.5" thick	34		0.62
Shingle backer, .375" thick	18		0.94
Sound deadening board, 0.5" thick	15		1.35
Tile and lay-in panels, plain or			
acoustical, 0.5" thick	18		1.25
Vegetable fiber, 0.5" thick	18		1.32
	25		1.14
Wood, hardwoods	48	0.91	
Softwoods	32	1.25	
Flooring Carpet with fibrous pad			2.08
With rubber pad			1.23
Cork tile, 1/8" thick			0.28
Terrazzo			0.08
Tile, resilient			0.05
Wood, hardwood, 0.75" thick			0.68
Subfloor, 0.75" thick			0.94
Glass			
Insulation, 0.50" air space			2.04
Single glass			0.91
Insulation Blanket or Batt, mineral, glass			
or rock fiber, approximate thickness			
3.0" to 3.5" thick			11
3.5" to 4.0" thick			13
6.0" to 6.5" thick			19
6.5" to 7.0" thick			22
8.5" to 9.0" thick			30
Boards			
Cellular glass	8.5	2.63	
Fiberboard, wet felted			
Acoustical tile	21	2.70	
Roof insulation	17	2.94	
Fiberboard, wet molded			
Acoustical tile	23	2.38	
Mineral fiber with resin binder	15	3.45	
Polystyrene, extruded,			
cut cell surface	1.8	4.00	
smooth skin surface	2.2	5.00	
	3.5	5.26	
Bead boards	1.0	3.57	
Polyurethane	1.5	6.25	
Wood or cane fiberboard, 0.5" thick			1.25

Material	Wt./Lbs. per C.F.	R per Inch	R Listed Size
Insulation Loose Fill			
Cellulose	2.3	3.13	
	3.2	3.70	
Mineral fiber, 3.75" to 5" thick	2-5		11
6.5" to 8.75" thick			19
7.5" to 10" thick			22
10.25" to 13.75" thick			30
Perlite	5-8	2.70	
Vermiculite	4-6	2.27	
Wood fiber	2-3.5	3.33	
Masonry Brick, Common	120	0.20	
Face	130	0.11	
Cement mortar	116	0.20	
Clay tile, hollow			
1 cell wide, 3" width			0.80
4" width			1.11
2 cells wide, 6" width			1.52
8" width			1.85
10" width			2.22
3 cells wide, 12" width			2.50
Concrete, gypsum fiber	51	0.60	
Lightweight	120	0.19	
	80	0.40	
	40	0.86	
Perlite	40	1.08	
Sand and gravel or stone	140	0.08	
Concrete block, lightweight			
3 cell units, 4"-15 lbs. ea.			1.68
6"-23 lbs. ea.			1.83
8"-28 lbs. ea.			2.12
12"-40 lbs. ea.			2.62
Sand and gravel aggregates,			
4"-20 lbs. ea.			1.17
6"-33 lbs. ea.			1.29
8"-38 lbs. ea.			1.46
12"-56 lbs. ea.			1.81
Plastering Cement Plaster,			
Sand aggregate	116	0.20	
Gypsum plaster, Perlite aggregate	45	0.67	
Sand aggregate	105	0.18	
Vermiculite aggregate	45	0.59	
Roofing			
Asphalt, felt, 15 lb.			0.06
Rolled roofing	70		0.15
Shingles	70		0.44
Built-up roofing .375" thick	70		0.33
Cement shingles	120		0.21
Vapor-permeable felt			0.06
Vapor seal, 2 layers of			
mopped 15 lb. felt			0.12
Wood, shingles 16"-7.5" exposure			0.87
Siding			
Aluminum or steel (hollow backed)			
oversheathing			0.61
With .375" insulating backer board			1.82
Foil backed			2.96
Wood siding, beveled, ½" x 8"			0.81

Reference Aids — RL1030-500 Weather Information

Table L1030-501 Weather Data and Design Conditions

City	Latitude (1) °	Latitude (1) '	Winter Temperatures (1) Med. of Annual Extremes	Winter Temperatures (1) 99%	Winter Temperatures (1) 97½%	Winter Degree Days (2)	Summer (Design Dry Bulb) Temperatures and Relative Humidity 1%	Summer (Design Dry Bulb) Temperatures and Relative Humidity 2½%	Summer (Design Dry Bulb) Temperatures and Relative Humidity 5%
UNITED STATES									
Albuquerque, NM	35	0	5.1	12	16	4,400	96/61	94/61	92/61
Atlanta, GA	33	4	11.9	17	22	3,000	94/74	92/74	90/73
Baltimore, MD	39	2	7	14	17	4,600	94/75	91/75	89/74
Birmingham, AL	33	3	13	17	21	2,600	96/74	94/75	92/74
Bismarck, ND	46	5	-32	-23	-19	8,800	95/68	91/68	88/67
Boise, ID	43	3	1	3	10	5,800	96/65	94/64	91/64
Boston, MA	42	2	-1	6	9	5,600	91/73	88/71	85/70
Burlington, VT	44	3	-17	-12	-7	8,200	88/72	85/70	82/69
Charleston, WV	38	2	3	7	11	4,400	92/74	90/73	87/72
Charlotte, NC	35	1	13	18	22	3,200	95/74	93/74	91/74
Casper, WY	42	5	-21	-11	-5	7,400	92/58	90/57	87/57
Chicago, IL	41	5	-8	-3	2	6,600	94/75	91/74	88/73
Cincinnati, OH	39	1	0	1	6	4,400	92/73	90/72	88/72
Cleveland, OH	41	2	-3	1	5	6,400	91/73	88/72	86/71
Columbia, SC	34	0	16	20	24	2,400	97/76	95/75	93/75
Dallas, TX	32	5	14	18	22	2,400	102/75	100/75	97/75
Denver, CO	39	5	-10	-5	1	6,200	93/59	91/59	89/59
Des Moines, IA	41	3	-14	-10	-5	6,600	94/75	91/74	88/73
Detroit, MI	42	2	-3	3	6	6,200	91/73	88/72	86/71
Great Falls, MT	47	3	-25	-21	-15	7,800	91/60	88/60	85/59
Hartford, CT	41	5	-4	3	7	6,200	91/74	88/73	85/72
Houston, TX	29	5	24	28	33	1,400	97/77	95/77	93/77
Indianapolis, IN	39	4	-7	-2	2	5,600	92/74	90/74	87/73
Jackson, MS	32	2	16	21	25	2,200	97/76	95/76	93/76
Kansas City, MO	39	1	-4	2	6	4,800	99/75	96/74	93/74
Las Vegas, NV	36	1	18	25	28	2,800	108/66	106/65	104/65
Lexington, KY	38	0	-1	3	8	4,600	93/73	91/73	88/72
Little Rock, AR	34	4	11	15	20	3,200	99/76	96/77	94/77
Los Angeles, CA	34	0	36	41	43	2,000	93/70	89/70	86/69
Memphis, TN	35	0	10	13	18	3,200	98/77	95/76	93/76
Miami, FL	25	5	39	44	47	200	91/77	90/77	89/77
Milwaukee, WI	43	0	-11	-8	-4	7,600	90/74	87/73	84/71
Minneapolis, MN	44	5	-22	-16	-12	8,400	92/75	89/73	86/71
New Orleans, LA	30	0	28	29	33	1,400	93/78	92/77	90/77
New York, NY	40	5	6	11	15	5,000	92/74	89/73	87/72
Norfolk, VA	36	5	15	20	22	3,400	93/77	91/76	89/76
Oklahoma City, OK	35	2	4	9	13	3,200	100/74	97/74	95/73
Omaha, NE	41	2	-13	-8	-3	6,600	94/76	91/75	88/74
Philadelphia, PA	39	5	6	10	14	4,400	93/75	90/74	87/72
Phoenix, AZ	33	3	27	31	34	1,800	109/71	107/71	105/71
Pittsburgh, PA	40	3	-1	3	7	6,000	91/72	88/71	86/70
Portland, ME	43	4	-10	-6	-1	7,600	87/72	84/71	81/69
Portland, OR	45	4	18	17	23	4,600	89/68	85/67	81/65
Portsmouth, NH	43	1	-8	-2	2	7,200	89/73	85/71	83/70
Providence, RI	41	4	-1	5	9	6,000	89/73	86/72	83/70
Rochester, NY	43	1	-5	1	5	6,800	91/73	88/71	85/70
Salt Lake City, UT	40	5	0	3	8	6,000	97/62	95/62	92/61
San Francisco, CA	37	5	36	38	40	3,000	74/63	71/62	69/61
Seattle, WA	47	4	22	22	27	5,200	85/68	82/66	78/65
Sioux Falls, SD	43	4	-21	-15	-11	7,800	94/73	91/72	88/71
St. Louis, MO	38	4	-3	3	8	5,000	98/75	94/75	91/75
Tampa, FL	28	0	32	36	40	680	92/77	91/77	90/76
Trenton, NJ	40	1	4	11	14	5,000	91/75	88/74	85/73
Washington, DC	38	5	7	14	17	4,200	93/75	91/74	89/74
Wichita, KS	37	4	-3	3	7	4,600	101/72	98/73	96/73
Wilmington, DE	39	4	5	10	14	5,000	92/74	89/74	87/73
ALASKA									
Anchorage	61	1	-29	-23	-18	10,800	71/59	68/58	66/56
Fairbanks	64	5	-59	-51	-47	14,280	82/62	78/60	75/59
CANADA									
Edmonton, Alta.	53	3	-30	-29	-25	11,000	85/66	82/65	79/63
Halifax, N.S.	44	4	-4	1	5	8,000	79/66	76/65	74/64
Montreal, Que.	45	3	-20	-16	-10	9,000	88/73	85/72	83/71
Saskatoon, Sask.	52	1	-35	-35	-31	11,000	89/68	86/66	83/65
St. John, Nwf.	47	4	1	3	7	8,600	77/66	75/65	73/64
Saint John, N.B.	45	2	-15	-12	-8	8,200	80/67	77/65	75/64
Toronto, Ont.	43	4	-10	-5	-1	7,000	90/73	87/72	85/71
Vancouver, B.C.	49	1	13	15	19	6,000	79/67	77/66	74/65
Winnipeg, Man.	49	5	-31	-30	-27	10,800	89/73	86/71	84/70

(1) Handbook of Fundamentals, ASHRAE, Inc., NY 1989
(2) Local Climatological Annual Survey, USDC Env. Science Services Administration, Asheville, NC

Reference Aids — RL1030-500 Weather Information

Table L1030-502 Maximum Depth of Frost Penetration in Inches

THIS MAP IS REASONABLY ACCURATE FOR MOST PARTS OF THE UNITED STATES BUT IS NECESSARILY HIGHLY GENERALIZED, AND CONSEQUENTLY NOT TOO ACCURATE IN MOUNTAINOUS REGIONS, PARTICULARLY IN THE ROCKIES.

Reference Aids — RL1040-100 Construction Classification

Table L1040-101 Fire-Resisting Ratings of Structural Elements (in hours)

Description of the Structural Element	No. 1 Fireproof		No. 2 Non Combustible			No. 3 Exterior Masonry Wall			No. 4 Frame	
			Protected		Unprotected	Heavy Timber	Ordinary		Protected	Unprotected
							Protected	Unprotected		
	1A	1B	2A	2B	2C	3A	3B	3C	4A	4B
Exterior, Bearing Walls	4	3	2	1½	1	2	2	2	1	1
Nonbearing Walls	2	2	1½	1	1	2	2	2	1	1
Interior Bearing Walls and Partitions	4	3	2	1	0	2	1	0	1	0
Fire Walls and Party Walls	4	3	2	2	2	2	2	2	2	2
Fire Enclosure of Exitways, Exit Hallways and Stairways	2	2	2	2	2	2	2	2	1	1
Shafts other than Exitways, Hallways and Stairways	2	2	2	2	2	2	2	2	1	1
Exitway access corridors and Vertical separation of tenant space	1	1	1	1	0	1	1	0	1	0
Columns, girders, trusses (other than roof trusses) and framing: Supporting more than one floor	4	3	2	1	0	—	1	0	1	0
Supporting one floor only	3	2	1½	1	0	—	1	0	1	0
Structural members supporting wall	3	2	1½	1	0	1	1	0	1	0
Floor construction including beams	3	2	1½	1	0	—	1	0	1	0
Roof construction including beams, trusses and framing arches and roof deck 15' or less in height to lowest member	2	1½	1	1	0	—	1	0	1	0

Note:
a. Codes include special requirements and exceptions that are not included in the table above.
b. Each type of construction has been divided into sub-types which vary according to the degree of fire resistance required. Sub-types (A) requirements are more severe than those for sub-types (B).
c. Protected construction means all structural members are chemically treated, covered or protected so that the unit has the required fire resistance.

Type No. 1, Fireproof Construction — Buildings and structures of fireproof construction are those in which the walls, partitions, structural elements, floors, ceilings, and roofs, and the exitways are protected with approved noncombustible materials to afford the fire-resistance rating specified in Table L1040-101; except as otherwise specifically regulated. Fire-resistant treated wood may be used as specified.

Type No. 2, Noncombustible Construction — Buildings and structures of noncombustible construction are those in which the walls, partitions, structural elements, floors, ceilings, roofs and the exitways are approved noncombustible materials meeting the fire-resistance rating requirements specified in Table L1040-101; except as modified by the fire limit restrictions. Fire-retardant treated wood may be used as specified.

Type No. 3, Exterior Masonry Wall Construction — Buildings and structures of exterior masonry wall construction are those in which the exterior, fire and party walls are masonry or other approved noncombustible materials of the required fire-resistance rating and structural properties. The floors, roofs, and interior framing are wholly or partly wood or metal or other approved construction. The fire and party walls are ground-supported; except that girders and their supports, carrying walls of masonry shall be protected to afford the same degree of fire-resistance rating of the supported walls. All structural elements have the required fire-resistance rating specified in Table L1040-101.

Type No. 4, Frame Construction — Buildings and structures of frame construction are those in which the exterior walls, bearing walls, partitions, floor and roof construction are wholly or partly of wood stud and joist assemblies with a minimum nominal dimension of two inches or of other approved combustible materials. Fire stops are required at all vertical and horizontal draft openings in which the structural elements have required fire-resistance ratings specified in Table L1040-101.

Reference Aids — RL1040-200 Fire Hazard Classification

Table L1040-201 Fire Grading of Building in Hours for Fire Walls and Fire Separation Walls

Fire Hazard for Fire Walls

The degree of fire hazard of buildings relating to their intended use is defined by "Fire Grading" the buildings. Such a grading system is listed in the Table L1040-201 below. This type of grading determines the requirements for fire walls and fire separation walls (exterior fire exposure). For mixed use occupancy, use the higher Fire Grading requirement of the components.

Building Type	Hours	Building Type	Hours
Businesses	2	Recreation Centers	2
Churches	1½	Residential Hotels	2
Factories	3	Residential, Multi-family Dwellings	1½
High Hazard*	4	Residential, One & Two Family Dwellings	¾
Industrial	3	Restaurants	2
Institutional, Incapacitated Occupants	2	Schools	1½
Institutional, Restrained Occupants	3	Storage, Low Hazard*	2
Lecture Halls	2	Storage, Moderate Hazard*	3
Mercantile	3	Terminals	2
Night Club	3	Theatres	3

Note: *The difference in "Fire Hazards" is determined by their occupancy and use.

High Hazard: Industrial and storage buildings in which the combustible contents might cause fires to be unusually intense or where explosives, combustible gases or flammable liquids are manufactured or stored.

Moderate Hazard: Mercantile buildings, industrial and storage buildings in which combustible contents might cause fires of moderate intensity.

Low Hazard: Business buildings that ordinarily do not burn rapidly.

Table L1040-202 Interior Finish Classification

Flame Spread for Interior Finishes

The flame spreadability of a material is the burning characteristic of the material relative to the fuel contributed by its combustion and the density of smoke developed. The flame spread classification of a material is based on a ten minute test on a scale of 0 to 100. Cement asbestos board is assigned a rating of 0 and select red oak flooring a rating of 100. The four classes are listed in Table L1040-202.

The flame spread ratings for interior finish walls and ceilings shall not be greater than the Class listed in the Table L1040-203 below.

Class of Material	Surface Burning Characteristics
I	0 to 25
II	26 to 75
III	76 to 200
IV	201 to 5000

Table L1040-203 Interior Finish Requirements by Class

Building Type	Vertical Exitways & Passageways	Corridors Providing Exitways	Rooms or Enclosed Spaces	Building Type	Vertical Exitways & Passageways	Corridors Providing Exitways	Rooms or Enclosed Space
Assembly Halls	I	I	II	Mercantile Walls	I	II	III
Businesses	I	II	III	Night Clubs	I	I	II
Churches	I	I	II	Residential Hotels	I	II	III
Factory	I	II	III	Residential, Multi-family	I	II	III
High Hazard	I	II	II	Residential, 1 & 2 Family	IV	IV	IV
Industrial	I	II	III	Restaurants	I	I	II
Institutional, Incapacitated	I	II	I	Storage, Low Hazard	I	II	III
Institutional, Restrained	I	I	I	Storage, Moderate Hazard	I	II	III
Mercantile Ceilings	I	II	II	Terminals	I	I	II

Reference Aids · RL1090-100 Conversion Factors

Description: This table is primarily for converting customary U.S. units in the left hand column to SI metric units in the right hand column. In addition, conversion factors for some commonly encountered Canadian and non-SI metric units are included.

Table L1090-101 Metric Conversion Factors

	If You Know		Multiply By		To Find
Length	Inches	x	25.4[a]	=	Millimeters
	Feet	x	0.3048[a]	=	Meters
	Yards	x	0.9144[a]	=	Meters
	Miles (statute)	x	1.609	=	Kilometers
Area	Square inches	x	645.2	=	Square millimeters
	Square feet	x	0.0929	=	Square meters
	Square yards	x	0.8361	=	Square meters
Volume (Capacity)	Cubic inches	x	16,387	=	Cubic millimeters
	Cubic feet	x	0.02832	=	Cubic meters
	Cubic yards	x	0.7646	=	Cubic meters
	Gallons (U.S. liquids)[b]	x	0.003785	=	Cubic meters[c]
	Gallons (Canadian liquid)[b]	x	0.004546	=	Cubic meters[c]
	Ounces (U.S. liquid)[b]	x	29.57	=	Milliliters[c,d]
	Quarts (U.S. liquid)[b]	x	0.9464	=	Liters[c,d]
	Gallons (U.S. liquid)[b]	x	3.785	=	Liters[c,d]
Force	Kilograms force[d]	x	9.807	=	Newtons
	Pounds force	x	4.448	=	Newtons
	Pounds force	x	0.4536	=	Kilograms force[d]
	Kips	x	4448	=	Newtons
	Kips	x	453.6	=	Kilograms force[d]
Pressure, Stress, Strength (Force per unit area)	Kilograms force per square centimeter[d]	x	0.09807	=	Megapascals
	Pounds force per square inch (psi)	x	0.006895	=	Megapascals
	Kips per square inch	x	6.895	=	Megapascals
	Pounds force per square inch (psi)	x	0.07031	=	Kilograms force per square centimeter[d]
	Pounds force per square foot	x	47.88	=	Pascals
	Pounds force per square foot	x	4.882	=	Kilograms force per square meter[d]
Bending Moment Or Torque	Inch-pounds force	x	0.01152	=	Meter-kilograms force[d]
	Inch-pounds force	x	0.1130	=	Newton-meters
	Foot-pounds force	x	0.1383	=	Meter-kilograms force[d]
	Foot-pounds force	x	1.356	=	Newton-meters
	Meter-kilograms force[d]	x	9.807	=	Newton-meters
Mass	Ounces (avoirdupois)	x	28.35	=	Grams
	Pounds (avoirdupois)	x	0.4536	=	Kilograms
	Tons (metric)	x	1000	=	Kilograms
	Tons, short (2000 pounds)	x	907.2	=	Kilograms
	Tons, short (2000 pounds)	x	0.9072	=	Megagrams[e]
Mass per Unit Volume	Pounds mass per cubic foot	x	16.02	=	Kilograms per cubic meter
	Pounds mass per cubic yard	x	0.5933	=	Kilograms per cubic meter
	Pounds mass per gallon (U.S. liquid)[b]	x	119.8	=	Kilograms per cubic meter
	Pounds mass per gallon (Canadian liquid)[b]	x	99.78	=	Kilograms per cubic meter
Temperature	Degrees Fahrenheit	(F−32)/1.8		=	Degrees Celsius
	Degrees Fahrenheit	(F+459.67)/1.8		=	Degrees Kelvin
	Degrees Celsius	C+273.15		=	Degrees Kelvin

[a] The factor given is exact
[b] One U.S. gallon = 0.8327 Canadian gallon
[c] 1 liter = 1000 milliliters = 1000 cubic centimeters
    1 cubic decimeter = 0.001 cubic meter
[d] Metric but not SI unit
[e] Called "tonne" in England and "metric ton" in other metric countries

Reference Aids — RL1090-200 Metric Equivalents

Table L1090-201 Metric Equivalents of Cement Content for Concrete Mixes

94 Pound Bags per Cubic Yard	Kilograms per Cubic Meter	94 Pound Bags per Cubic Yard	Kilograms per Cubic Meter
1.0	55.77	7.0	390.4
1.5	83.65	7.5	418.3
2.0	111.5	8.0	446.2
2.5	139.4	8.5	474.0
3.0	167.3	9.0	501.9
3.5	195.2	9.5	529.8
4.0	223.1	10.0	557.7
4.5	251.0	10.5	585.6
5.0	278.8	11.0	613.5
5.5	306.7	11.5	641.3
6.0	334.6	12.0	669.2
6.5	362.5	12.5	697.1

a. If you know the cement content in pounds per cubic yard, multiply by .5933 to obtain kilograms per cubic meter.

b. If you know the cement content in 94 pound bags per cubic yard, multiply by 55.77 to obtain kilograms per cubic meter.

Table L1090-202 Metric Equivalents of Common Concrete Strengths (to convert other psi values to megapascals, multiply by .006895)

U.S. Values psi	SI Value Megapascals	Non-SI Metric Value kgf/cm^2*
2000	14	140
2500	17	175
3000	21	210
3500	24	245
4000	28	280
4500	31	315
5000	34	350
6000	41	420
7000	48	490
8000	55	560
9000	62	630
10,000	69	705

* kilograms force per square centimeter

Reference Aids — RL1090-200 Metric Equivalents

Table L1090-203 Comparison of U.S. Customary Units and SI Units for Reinforcing Bars

		U.S. Customary Units					
		Nominal Dimensions[a]			Deformation Requirements, in.		
Bar Designation No.[b]	Nominal Weight, lb/ft	Diameter in.	Cross Sectional Area, in.^{2}	Perimeter in.	Maximum Average Spacing	Minimum Average Height	Maximum Gap (Chord of 12-1/2% of Nominal Perimeter)
3	0.376	0.375	0.11	1.178	0.262	0.015	0.143
4	0.668	0.500	0.20	1.571	0.350	0.020	0.191
5	1.043	0.625	0.31	1.963	0.437	0.028	0.239
6	1.502	0.750	0.44	2.356	0.525	0.038	0.286
7	2.044	0.875	0.60	2.749	0.612	0.044	0.334
8	2.670	1.000	0.79	3.142	0.700	0.050	0.383
9	3.400	1.128	1.00	3.544	0.790	0.056	0.431
10	4.303	1.270	1.27	3.990	0.889	0.064	0.487
11	5.313	1.410	1.56	4.430	0.987	0.071	0.540
14	7.65	1.693	2.25	5.32	1.185	0.085	0.648
18	13.60	2.257	4.00	7.09	1.58	0.102	0.864

		SI UNITS					
		Nominal Dimensions[a]			Deformation Requirements, mm		
Bar Designation No.[b]	Nominal Weight kg/m	Diameter, mm	Cross Sectional Area, cm^2	Perimeter, mm	Maximum Average Spacing	Minimum Average Height	Maximum Gap (Chord of 12-1/2% of Nominal Perimeter)
3	0.560	9.52	0.71	29.9	6.7	0.38	3.5
4	0.994	12.70	1.29	39.9	8.9	0.51	4.9
5	1.552	15.88	2.00	49.9	11.1	0.71	6.1
6	2.235	19.05	2.84	59.8	13.3	0.96	7.3
7	3.042	22.22	3.87	69.8	15.5	1.11	8.5
8	3.973	25.40	5.10	79.8	17.8	1.27	9.7
9	5.059	28.65	6.45	90.0	20.1	1.42	10.9
10	6.403	32.26	8.19	101.4	22.6	1.62	11.4
11	7.906	35.81	10.06	112.5	25.1	1.80	13.6
14	11.384	43.00	14.52	135.1	30.1	2.16	16.5
18	20.238	57.33	25.81	180.1	40.1	2.59	21.9

[a] Nominal dimensions of a deformed bar are equivalent to those of a plain round bar having the same weight per foot as the deformed bar.
[b] Bar numbers are based on the number of eighths of an inch included in the nominal diameter of the bars.

Table L1090-204 Metric Rebar Specification - ASTM A615-81

	Grade 300 (300 MPa* = **43,560 psi; +8.7% vs. Grade 40**)			
	Grade 400 (400 MPa* = **58,000 psi; –3.4% vs. Grade 60**)			
Bar No.	Diameter mm	Area mm^2	Equivalent in.^{2}	Comparison with U.S. Customary Bars
10	11.3	100	.16	Between #3 & #4
15	16.0	200	.31	#5 (.31 in.^{2})
20	19.5	300	.47	#6 (.44 in.^{2})
25	25.2	500	.78	#8 (.79 in.^{2})
30	29.9	700	1.09	#9 (1.00 in.^{2})
35	35.7	1000	1.55	#11 (1.56 in.^{2})
45	43.7	1500	2.33	#14 (2.25 in.^{2})
55	56.4	2500	3.88	#18 (4.00 in.^{2})

* MPa = megapascals   Grade 300 bars are furnished only in sizes 10 through 35

Estimating Form — Project Summary

PRELIMINARY ESTIMATE

PROJECT		TOTAL SITE AREA	
BUILDING TYPE		OWNER	
LOCATION		ARCHITECT	
DATE OF CONSTRUCTION		ESTIMATED CONSTRUCTION PERIOD	
BRIEF DESCRIPTION			

TYPE OF PLAN		TYPE OF CONSTRUCTION	
QUALITY		BUILDING CAPACITY	

Floor			**Wall Areas**			
Below Grade Levels			Foundation Walls	L.F.	Ht.	S.F.
Area		S.F.	Frost Walls	L.F.	Ht.	S.F.
Area		S.F.	Exterior Closure		Total	S.F.
Total Area		S.F.	Comment			
Ground Floor			Fenestration		%	S.F.
Area		S.F.			%	S.F.
Area		S.F.	Exterior Wall		%	S.F.
Total Area		S.F.			%	S.F.
Supported Levels			**Site Work**			
Area		S.F.	Parking		S.F. (For	Cars)
Area		S.F.	Access Roads		L.F. (X	Ft. Wide)
Area		S.F.	Sidewalk		L.F. (X	Ft. Wide)
Area		S.F.	Landscaping		S.F. (	% Unbuilt Site)
Area		S.F.	**Building Codes**			
Total Area		S.F.	City		County	
Miscellaneous			National		Other	
Area		S.F.	**Loading**			
Area		S.F.	Roof	psf	Ground Floor	psf
Area		S.F.	Supported Floors	psf	Corridor	psf
Area		S.F.	Balcony	psf	Partition, allow	psf
Total Area		S.F.	Miscellaneous			psf
Net Finished Area		S.F.	Live Load Reduction			
Net Floor Area		S.F.	Wind			
Gross Floor Area		S.F.	Earthquake		Zone	
Roof			Comment			
Total Area		S.F.	Soil Type			
Comments			Bearing Capacity			K.S.F.
			Frost Depth			Ft.
Volume			**Frame**			
Depth of Floor System			Type		Bay Spacing	
Minimum		In.	Foundation, Standard			
Maximum		In.	Special			
Foundation Wall Height		Ft.	Substructure			
Floor to Floor Height		Ft.	Comment			
Floor to Ceiling Height		Ft.	Superstructure, Vertical		Horizontal	
Subgrade Volume		C.F.	Fireproofing		☐ Columns	Hrs.
Above Grade Volume		C.F.	☐ Girders	Hrs.	☐ Beams	Hrs.
Total Building Volume		C.F.	☐ Floor	Hrs.	☐ None	

Estimating Form — Systems Costs

ASSEMBLY NUMBER	DESCRIPTION	QTY	UNIT	TOTAL COST UNIT	TOTAL COST TOTAL	COST PER S.F.
A10	Foundations					
A20	Basement Construction					
B10	Superstructure					
B20	Exterior Closure					
B30	Roofing					

Estimating Form — Systems Costs

ASSEMBLY NUMBER	DESCRIPTION	QTY	UNIT	TOTAL COST UNIT	TOTAL COST TOTAL	COST PER S.F.
C	**Interior Construction**					
D10	**Conveying**					
D20	**Plumbing**					

Estimating Form — Systems Costs

ASSEMBLY NUMBER	DESCRIPTION	QTY	UNIT	TOTAL COST UNIT	TOTAL COST TOTAL	COST PER S.F.
D30	**HVAC**					
D40	**Fire Protection**					
D50	**Electrical**					

Estimating Form — Systems Costs

ASSEMBLY NUMBER	DESCRIPTION	QTY	UNIT	TOTAL COST UNIT	TOTAL COST TOTAL	COST PER S.F.
E	**Equipment & Furnishings**					
F	**Special Construction**					
G	**Sitework**					

Estimating Form — Systems Costs Summary

Preliminary Estimate Cost Summary

PROJECT	TOTAL AREA	SHEET NO.
LOCATION	TOTAL VOLUME	ESTIMATE NO.
ARCHITECT	COST PER S.F.	DATE
OWNER	COST PER C.F.	NO OF STORIES
QUANTITIES BY	EXTENSIONS BY	CHECKED BY

DIV	DESCRIPTION	SUBTOTAL COST	COST/S.F.	PERCENTAGE
A	SUBSTRUCTURE			
B10	SHELL: SUPERSTRUCTURE			
B20	SHELL: EXTERIOR CLOSURE			
B30	SHELL: ROOFING			
C	INTERIOR CONSTRUCTION			
D10	SERVICES: CONVEYING			
D20	SERVICES: PLUMBING			
D30	SERVICES: HVAC			
D40	SERVICES: FIRE PROTECTION			
D50	SERVICES: ELECTRICAL			
E	EQUIPMENT & FURNISHINGS			
F	SPECIAL CONSTRUCTION			
G	SITEWORK			

BUILDING SUBTOTAL $ _____ -

Sales Tax ___ % x Subtotal /2 $ _____ -

General Conditions ___ % x Subtotal $ _____ -

                                               Subtotal "A" $ _____ -

Overhead ___ % x Subtotal "A" $ _____ -

                                               Subtotal "B" $ _____ -

Profit ___ % x Subtotal "B" $ _____ -

                                               Subtotal "C" $ _____ -

Location Factor ___ % x Subtotal "C"   Localized Cost $ _____ -
   (Boston, MA)

Architects Fee ___ x Localized Cost = $ _____ -
Contingency ___ x Localized Cost = $ _____ -

                                               Project Total Cost $ _____ -

Square Foot Cost $ _____ / _____ S.F. =   **S.F. Cost** $ _____ -
Cubic Foot Cost $ _____ / _____ C.F. =   **C.F. Cost** $ _____ -

Abbreviations

A	Area Square Feet; Ampere	Brk., brk	Brick	Csc	Cosecant
AAFES	Army and Air Force Exchange Service	brkt	Bracket	C.S.F.	Hundred Square Feet
		Brng.	Bearing	CSI	Construction Specifications Institute
ABS	Acrylonitrile Butadiene Stryrene; Asbestos Bonded Steel	Brs.	Brass		
		Brz.	Bronze	CT	Current Transformer
A.C., AC	Alternating Current; Air-Conditioning; Asbestos Cement; Plywood Grade A & C	Bsn.	Basin	CTS	Copper Tube Size
		Btr.	Better	Cu	Copper, Cubic
		Btu	British Thermal Unit	Cu. Ft.	Cubic Foot
		BTUH	BTU per Hour	cw	Continuous Wave
ACI	American Concrete Institute	Bu.	bushels	C.W.	Cool White; Cold Water
ACR	Air Conditioning Refrigeration	BUR	Built-up Roofing	Cwt.	100 Pounds
ADA	Americans with Disabilities Act	BX	Interlocked Armored Cable	C.W.X.	Cool White Deluxe
AD	Plywood, Grade A & D	°C	degree centegrade	C.Y.	Cubic Yard (27 cubic feet)
Addit.	Additional	c	Conductivity, Copper Sweat	C.Y./Hr.	Cubic Yard per Hour
Adj.	Adjustable	C	Hundred; Centigrade	Cyl.	Cylinder
af	Audio-frequency	C/C	Center to Center, Cedar on Cedar	d	Penny (nail size)
AFUE	Annual Fuel Utilization Efficiency	C-C	Center to Center	D	Deep; Depth; Discharge
AGA	American Gas Association	Cab	Cabinet	Dis., Disch.	Discharge
Agg.	Aggregate	Cair.	Air Tool Laborer	Db	Decibel
A.H., Ah	Ampere Hours	Cal.	caliper	Dbl.	Double
A hr.	Ampere-hour	Calc	Calculated	DC	Direct Current
A.H.U., AHU	Air Handling Unit	Cap.	Capacity	DDC	Direct Digital Control
A.I.A.	American Institute of Architects	Carp.	Carpenter	Demob.	Demobilization
AIC	Ampere Interrupting Capacity	C.B.	Circuit Breaker	d.f.t.	Dry Film Thickness
Allow.	Allowance	C.C.A.	Chromate Copper Arsenate	d.f.u.	Drainage Fixture Units
alt., alt	Alternate	C.C.F.	Hundred Cubic Feet	D.H.	Double Hung
Alum.	Aluminum	cd	Candela	DHW	Domestic Hot Water
a.m.	Ante Meridiem	cd/sf	Candela per Square Foot	DI	Ductile Iron
Amp.	Ampere	CD	Grade of Plywood Face & Back	Diag.	Diagonal
Anod.	Anodized	CDX	Plywood, Grade C & D, exterior glue	Diam., Dia	Diameter
ANSI	American National Standards Institute			Distrib.	Distribution
		Cefi.	Cement Finisher	Div.	Division
APA	American Plywood Association	Cem.	Cement	Dk.	Deck
Approx.	Approximate	CF	Hundred Feet	D.L.	Dead Load; Diesel
Apt.	Apartment	C.F.	Cubic Feet	DLH	Deep Long Span Bar Joist
Asb.	Asbestos	CFM	Cubic Feet per Minute	dlx	Deluxe
A.S.B.C.	American Standard Building Code	c.g.	Center of Gravity	Do.	Ditto
Asbe.	Asbestos Worker	CHW	Chilled Water; Commercial Hot Water	DOP	Dioctyl Phthalate Penetration Test (Air Filters)
ASCE	American Society of Civil Engineers				
A.S.H.R.A.E.	American Society of Heating, Refrig. & AC Engineers	C.I., CI	Cast Iron	Dp., dp	Depth
		C.I.P., CIP	Cast in Place	D.P.S.T.	Double Pole, Single Throw
ASME	American Society of Mechanical Engineers	Circ.	Circuit	Dr.	Drive
		C.L.	Carload Lot	DR	Dimension Ratio
ASTM	American Society for Testing and Materials	CL	Chain Link	Drink.	Drinking
		Clab.	Common Laborer	D.S.	Double Strength
Attchmt.	Attachment	Clam	Common maintenance laborer	D.S.A.	Double Strength A Grade
Avg., Ave.	Average	C.L.F.	Hundred Linear Feet	D.S.B.	Double Strength B Grade
AWG	American Wire Gauge	CLF	Current Limiting Fuse	Dty.	Duty
AWWA	American Water Works Assoc.	CLP	Cross Linked Polyethylene	DWV	Drain Waste Vent
Bbl.	Barrel	cm	Centimeter	DX	Deluxe White, Direct Expansion
B&B, BB	Grade B and Better; Balled & Burlapped	CMP	Corr. Metal Pipe	dyn	Dyne
		CMU	Concrete Masonry Unit	e	Eccentricity
B&S	Bell and Spigot	CN	Change Notice	E	Equipment Only; East; emissivity
B.&W.	Black and White	Col.	Column	Ea.	Each
b.c.c.	Body-centered Cubic	CO$_2$	Carbon Dioxide	EB	Encased Burial
B.C.Y.	Bank Cubic Yards	Comb.	Combination	Econ.	Economy
BE	Bevel End	comm.	Commercial, Communication	E.C.Y	Embankment Cubic Yards
B.F.	Board Feet	Compr.	Compressor	EDP	Electronic Data Processing
Bg. cem.	Bag of Cement	Conc.	Concrete	EIFS	Exterior Insulation Finish System
BHP	Boiler Horsepower; Brake Horsepower	Cont., cont	Continuous; Continued, Container	E.D.R.	Equiv. Direct Radiation
		Corr.	Corrugated	Eq.	Equation
B.I.	Black Iron	Cos	Cosine	EL	elevation
bidir.	bidirectional	Cot	Cotangent	Elec.	Electrician; Electrical
Bit., Bitum.	Bituminous	Cov.	Cover	Elev.	Elevator; Elevating
Bit., Conc.	Bituminous Concrete	C/P	Cedar on Paneling	EMT	Electrical Metallic Conduit; Thin Wall Conduit
Bk.	Backed	CPA	Control Point Adjustment		
Bkrs.	Breakers	Cplg.	Coupling	Eng.	Engine, Engineered
Bldg., bldg	Building	CPM	Critical Path Method	EPDM	Ethylene Propylene Diene Monomer
Blk.	Block	CPVC	Chlorinated Polyvinyl Chloride		
Bm.	Beam	C.Pr.	Hundred Pair	EPS	Expanded Polystyrene
Boil.	Boilermaker	CRC	Cold Rolled Channel	Eqhv.	Equip. Oper., Heavy
bpm	Blows per Minute	Creos.	Creosote	Eqlt.	Equip. Oper., Light
BR	Bedroom	Crpt.	Carpet & Linoleum Layer	Eqmd.	Equip. Oper., Medium
Brg.	Bearing	CRT	Cathode-ray Tube	Eqmm.	Equip. Oper., Master Mechanic
Brhe.	Bricklayer Helper	CS	Carbon Steel, Constant Shear Bar Joist	Eqol.	Equip. Oper., Oilers
Bric.	Bricklayer			Equip.	Equipment
				ERW	Electric Resistance Welded

Abbreviations

E.S.	Energy Saver	H	High Henry	Lath.	Lather
Est.	Estimated	HC	High Capacity	Lav.	Lavatory
esu	Electrostatic Units	H.D., HD	Heavy Duty; High Density	lb.; #	Pound
E.W.	Each Way	H.D.O.	High Density Overlaid	L.B., LB	Load Bearing; L Conduit Body
EWT	Entering Water Temperature	HDPE	High density polyethelene plastic	L. & E.	Labor & Equipment
Excav.	Excavation	Hdr.	Header	lb./hr.	Pounds per Hour
excl	Excluding	Hdwe.	Hardware	lb./L.F.	Pounds per Linear Foot
Exp., exp	Expansion, Exposure	H.I.D., HID	High Intensity Discharge	lbf/sq.in.	Pound-force per Square Inch
Ext., ext	Exterior; Extension	Help.	Helper Average	L.C.L.	Less than Carload Lot
Extru.	Extrusion	HEPA	High Efficiency Particulate Air Filter	L.C.Y.	Loose Cubic Yard
f.	Fiber stress			Ld.	Load
F	Fahrenheit; Female; Fill	Hg	Mercury	LE	Lead Equivalent
Fab., fab	Fabricated; fabric	HIC	High Interrupting Capacity	LED	Light Emitting Diode
FBGS	Fiberglass	HM	Hollow Metal	L.F.	Linear Foot
F.C.	Footcandles	HMWPE	high molecular weight polyethylene	L.F. Nose	Linear Foot of Stair Nosing
f.c.c.	Face-centered Cubic			L.F. Rsr	Linear Foot of Stair Riser
f'c.	Compressive Stress in Concrete; Extreme Compressive Stress	HO	High Output	Lg.	Long; Length; Large
		Horiz.	Horizontal	L & H	Light and Heat
F.E.	Front End	H.P., HP	Horsepower; High Pressure	LH	Long Span Bar Joist
FEP	Fluorinated Ethylene Propylene (Teflon)	H.P.F.	High Power Factor	L.H.	Labor Hours
		Hr.	Hour	L.L., LL	Live Load
F.G.	Flat Grain	Hrs./Day	Hours per Day	L.L.D.	Lamp Lumen Depreciation
F.H.A.	Federal Housing Administration	HSC	High Short Circuit	lm	Lumen
Fig.	Figure	Ht.	Height	lm/sf	Lumen per Square Foot
Fin.	Finished	Htg.	Heating	lm/W	Lumen per Watt
FIPS	Female Iron Pipe Size	Htrs.	Heaters	LOA	Length Over All
Fixt.	Fixture	HVAC	Heating, Ventilation & Air-Conditioning	log	Logarithm
FJP	Finger jointed and primed			L-O-L	Lateralolet
Fl. Oz.	Fluid Ounces	Hvy.	Heavy	long.	longitude
Flr.	Floor	HW	Hot Water	L.P., LP	Liquefied Petroleum; Low Pressure
FM	Frequency Modulation; Factory Mutual	Hyd.; Hydr.	Hydraulic	L.P.F.	Low Power Factor
		Hz	Hertz (cycles)	LR	Long Radius
Fmg.	Framing	I.	Moment of Inertia	L.S.	Lump Sum
FM/UL	Factory Mutual/Underwriters Labs	IBC	International Building Code	Lt.	Light
Fdn.	Foundation	I.C.	Interrupting Capacity	Lt. Ga.	Light Gauge
FNPT	Female National Pipe Thread	ID	Inside Diameter	L.T.L.	Less than Truckload Lot
Fori.	Foreman, Inside	I.D.	Inside Dimension; Identification	Lt. Wt.	Lightweight
Foro.	Foreman, Outside	I.F.	Inside Frosted	L.V.	Low Voltage
Fount.	Fountain	I.M.C.	Intermediate Metal Conduit	M	Thousand; Material; Male; Light Wall Copper Tubing
fpm	Feet per Minute	In.	Inch		
FPT	Female Pipe Thread	Incan.	Incandescent	M^2CA	Meters Squared Contact Area
Fr	Frame	Incl.	Included; Including	m/hr.; M.H.	Man-hour
F.R.	Fire Rating	Int.	Interior	mA	Milliampere
FRK	Foil Reinforced Kraft	Inst.	Installation	Mach.	Machine
FSK	Foil/scrim/kraft	Insul., insul	Insulation/Insulated	Mag. Str.	Magnetic Starter
FRP	Fiberglass Reinforced Plastic	I.P.	Iron Pipe	Maint.	Maintenance
FS	Forged Steel	I.P.S., IPS	Iron Pipe Size	Marb.	Marble Setter
FSC	Cast Body; Cast Switch Box	IPT	Iron Pipe Threaded	Mat; Mat'l.	Material
Ft., ft	Foot; Feet	I.W.	Indirect Waste	Max.	Maximum
Ftng.	Fitting	J	Joule	MBF	Thousand Board Feet
Ftg.	Footing	J.I.C.	Joint Industrial Council	MBH	Thousand BTU's per hr.
Ft lb.	Foot Pound	K	Thousand; Thousand Pounds; Heavy Wall Copper Tubing, Kelvin	MC	Metal Clad Cable
Furn.	Furniture			MCC	Motor Control Center
FVNR	Full Voltage Non-Reversing	K.A.H.	Thousand Amp. Hours	M.C.F.	Thousand Cubic Feet
FVR	Full Voltage Reversing	kcmil	Thousand Circular Mils	MCFM	Thousand Cubic Feet per Minute
FXM	Female by Male	KD	Knock Down	M.C.M.	Thousand Circular Mils
Fy.	Minimum Yield Stress of Steel	K.D.A.T.	Kiln Dried After Treatment	MCP	Motor Circuit Protector
g	Gram	kg	Kilogram	MD	Medium Duty
G	Gauss	kG	Kilogauss	MDF	Medium-density fibreboard
Ga.	Gauge	kgf	Kilogram Force	M.D.O.	Medium Density Overlaid
Gal., gal.	Gallon	kHz	Kilohertz	Med.	Medium
gpm, GPM	Gallon per Minute	Kip	1000 Pounds	MF	Thousand Feet
Galv., galv	Galvanized	KJ	Kiljoule	M.F.B.M.	Thousand Feet Board Measure
GC/MS	Gas Chromatograph/Mass Spectrometer	K.L.	Effective Length Factor	Mfg.	Manufacturing
		K.L.F.	Kips per Linear Foot	Mfrs.	Manufacturers
Gen.	General	Km	Kilometer	mg	Milligram
GFI	Ground Fault Interrupter	KO	Knock Out	MGD	Million Gallons per Day
GFRC	Glass Fiber Reinforced Concrete	K.S.F.	Kips per Square Foot	MGPH	Thousand Gallons per Hour
Glaz.	Glazier	K.S.I.	Kips per Square Inch	MH, M.H.	Manhole; Metal Halide; Man-Hour
GPD	Gallons per Day	kV	Kilovolt	MHz	Megahertz
gpf	Gallon per flush	kVA	Kilovolt Ampere	Mi.	Mile
GPH	Gallons per Hour	kVAR	Kilovar (Reactance)	MI	Malleable Iron; Mineral Insulated
GPM	Gallons per Minute	KW	Kilowatt	MIPS	Male Iron Pipe Size
GR	Grade	KWh	Kilowatt-hour	mj	Mechanical Joint
Gran.	Granular	L	Labor Only; Length; Long; Medium Wall Copper Tubing	m	Meter
Grnd.	Ground			mm	Millimeter
GVW	Gross Vehicle Weight	Lab.	Labor	Mill.	Millwright
GWB	Gypsum wall board	lat	Latitude	Min., min.	Minimum, minute

Abbreviations

Misc.	Miscellaneous	PDCA	Painting and Decorating Contractors of America	SC	Screw Cover
ml	Milliliter, Mainline			SCFM	Standard Cubic Feet per Minute
M.L.F.	Thousand Linear Feet	P.E., PE	Professional Engineer; Porcelain Enamel; Polyethylene; Plain End	Scaf.	Scaffold
Mo.	Month			Sch., Sched.	Schedule
Mobil.	Mobilization			S.C.R.	Modular Brick
Mog.	Mogul Base	Perf.	Perforated	S.D.	Sound Deadening
MPH	Miles per Hour	PEX	Cross linked polyethylene	SDR	Standard Dimension Ratio
MPT	Male Pipe Thread	Ph.	Phase	S.E.	Surfaced Edge
MRGWB	Moisture Resistant Gypsum Wallboard	P.I.	Pressure Injected	Sel.	Select
		Pile.	Pile Driver	SER, SEU	Service Entrance Cable
MRT	Mile Round Trip	Pkg.	Package	S.F.	Square Foot
ms	Millisecond	Pl.	Plate	S.F.C.A.	Square Foot Contact Area
M.S.F.	Thousand Square Feet	Plah.	Plasterer Helper	S.F. Flr.	Square Foot of Floor
Mstz.	Mosaic & Terrazzo Worker	Plas.	Plasterer	S.F.G.	Square Foot of Ground
M.S.Y.	Thousand Square Yards	plf	Pounds Per Linear Foot	S.F. Hor.	Square Foot Horizontal
Mtd., mtd., mtd	Mounted	Pluh.	Plumbers Helper	SFR	Square Feet of Radiation
Mthe.	Mosaic & Terrazzo Helper	Plum.	Plumber	S.F. Shlf.	Square Foot of Shelf
Mtng.	Mounting	Ply.	Plywood	S4S	Surface 4 Sides
Mult.	Multi; Multiply	p.m.	Post Meridiem	Shee.	Sheet Metal Worker
M.V.A.	Million Volt Amperes	Pntd.	Painted	Sin.	Sine
M.V.A.R.	Million Volt Amperes Reactance	Pord.	Painter, Ordinary	Skwk.	Skilled Worker
MV	Megavolt	pp	Pages	SL	Saran Lined
MW	Megawatt	PP, PPL	Polypropylene	S.L.	Slimline
MXM	Male by Male	P.P.M.	Parts per Million	Sldr.	Solder
MYD	Thousand Yards	Pr.	Pair	SLH	Super Long Span Bar Joist
N	Natural; North	P.E.S.B.	Pre-engineered Steel Building	S.N.	Solid Neutral
nA	Nanoampere	Prefab.	Prefabricated	SO	Stranded with oil resistant inside insulation
NA	Not Available; Not Applicable	Prefin.	Prefinished		
N.B.C.	National Building Code	Prop.	Propelled	S-O-L	Socketolet
NC	Normally Closed	PSF, psf	Pounds per Square Foot	sp	Standpipe
NEMA	National Electrical Manufacturers Assoc.	PSI, psi	Pounds per Square Inch	S.P.	Static Pressure; Single Pole; Self-Propelled
		PSIG	Pounds per Square Inch Gauge		
NEHB	Bolted Circuit Breaker to 600V.	PSP	Plastic Sewer Pipe	Spri.	Sprinkler Installer
NFPA	National Fire Protection Association	Pspr.	Painter, Spray	spwg	Static Pressure Water Gauge
NLB	Non-Load-Bearing	Psst.	Painter, Structural Steel	S.P.D.T.	Single Pole, Double Throw
NM	Non-Metallic Cable	P.T.	Potential Transformer	SPF	Spruce Pine Fir
nm	Nanometer	P. & T.	Pressure & Temperature	S.P.S.T.	Single Pole, Single Throw
No.	Number	Ptd.	Painted	SPT	Standard Pipe Thread
NO	Normally Open	Ptns.	Partitions	Sq.	Square; 100 Square Feet
N.O.C.	Not Otherwise Classified	Pu	Ultimate Load	Sq. Hd.	Square Head
Nose.	Nosing	PVC	Polyvinyl Chloride	Sq. In.	Square Inch
NPT	National Pipe Thread	Pvmt.	Pavement	S.S.	Single Strength; Stainless Steel
NQOD	Combination Plug-on/Bolt on Circuit Breaker to 240V.	PRV	Pressure Relief Valve	S.S.B.	Single Strength B Grade
		Pwr.	Power	sst, ss	Stainless Steel
N.R.C., NRC	Noise Reduction Coefficient/ Nuclear Regulator Commission	Q	Quantity Heat Flow	Sswk.	Structural Steel Worker
		Qt.	Quart	Sswl.	Structural Steel Welder
N.R.S.	Non Rising Stem	Quan., Qty.	Quantity	St.; Stl.	Steel
ns	Nanosecond	Q.C.	Quick Coupling	STC	Sound Transmission Coefficient
nW	Nanowatt	r	Radius of Gyration	Std.	Standard
OB	Opposing Blade	R	Resistance	Stg.	Staging
OC	On Center	R.C.P.	Reinforced Concrete Pipe	STK	Select Tight Knot
OD	Outside Diameter	Rect.	Rectangle	STP	Standard Temperature & Pressure
O.D.	Outside Dimension	recpt.	receptacle	Stpi.	Steamfitter, Pipefitter
ODS	Overhead Distribution System	Reg.	Regular	Str.	Strength; Starter; Straight
O.G.	Ogee	Reinf.	Reinforced	Strd.	Stranded
O.H.	Overhead	Req'd.	Required	Struct.	Structural
O&P	Overhead and Profit	Res.	Resistant	Sty.	Story
Oper.	Operator	Resi.	Residential	Subj.	Subject
Opng.	Opening	RF	Radio Frequency	Subs.	Subcontractors
Orna.	Ornamental	RFID	Radio-frequency identification	Surf.	Surface
OSB	Oriented Strand Board	Rgh.	Rough	Sw.	Switch
OS&Y	Outside Screw and Yoke	RGS	Rigid Galvanized Steel	Swbd.	Switchboard
OSHA	Occupational Safety and Health Act	RHW	Rubber, Heat & Water Resistant; Residential Hot Water	S.Y.	Square Yard
				Syn.	Synthetic
Ovhd.	Overhead	rms	Root Mean Square	S.Y.P.	Southern Yellow Pine
OWG	Oil, Water or Gas	Rnd.	Round	Sys.	System
Oz.	Ounce	Rodm.	Rodman	t.	Thickness
P.	Pole; Applied Load; Projection	Rofc.	Roofer, Composition	T	Temperature; Ton
p.	Page	Rofp.	Roofer, Precast	Tan	Tangent
Pape.	Paperhanger	Rohe.	Roofer Helpers (Composition)	T.C.	Terra Cotta
P.A.P.R.	Powered Air Purifying Respirator	Rots.	Roofer, Tile & Slate	T & C	Threaded and Coupled
PAR	Parabolic Reflector	R.O.W.	Right of Way	T.D.	Temperature Difference
P.B., PB	Push Button	RPM	Revolutions per Minute	Tdd	Telecommunications Device for the Deaf
Pc., Pcs.	Piece, Pieces	R.S.	Rapid Start		
P.C.	Portland Cement; Power Connector	Rsr	Riser	T.E.M.	Transmission Electron Microscopy
P.C.F.	Pounds per Cubic Foot	RT	Round Trip	temp	Temperature, Tempered, Temporary
PCM	Phase Contrast Microscopy	S.	Suction; Single Entrance; South	TFFN	Nylon Jacketed Wire

Abbreviations

TFE	Tetrafluoroethylene (Teflon)	U.L., UL	Underwriters Laboratory	w/	With
T. & G.	Tongue & Groove; Tar & Gravel	Uld.	unloading	W.C., WC	Water Column; Water Closet
Th., Thk.	Thick	Unfin.	Unfinished	W.F.	Wide Flange
Thn.	Thin	UPS	Uninterruptible Power Supply	W.G.	Water Gauge
Thrded	Threaded	URD	Underground Residential Distribution	Wldg.	Welding
Tilf.	Tile Layer, Floor	US	United States	W. Mile	Wire Mile
Tilh.	Tile Layer, Helper	USGBC	U.S. Green Building Council	W-O-L	Weldolet
THHN	Nylon Jacketed Wire	USP	United States Primed	W.R.	Water Resistant
THW.	Insulated Strand Wire	UTMCD	Uniform Traffic Manual For Control Devices	Wrck.	Wrecker
THWN	Nylon Jacketed Wire			W.S.P.	Water, Steam, Petroleum
T.L., TL	Truckload	UTP	Unshielded Twisted Pair	WT., Wt.	Weight
T.M.	Track Mounted	V	Volt	WWF	Welded Wire Fabric
Tot.	Total	VA	Volt Amperes	XFER	Transfer
T-O-L	Threadolet	V.C.T.	Vinyl Composition Tile	XFMR	Transformer
tmpd	Tempered	VAV	Variable Air Volume	XHD	Extra Heavy Duty
T.S.	Trigger Start	VC	Veneer Core	XHHW,	Cross-Linked Polyethylene Wire
Tr.	Trade	VDC	Volts Direct Current	XLPE	Insulation
Transf.	Transformer	Vent.	Ventilation	XLP	Cross-linked Polyethylene
Trhv.	Truck Driver, Heavy	Vert.	Vertical	Xport	Transport
Trlr	Trailer	V.F.	Vinyl Faced	Y	Wye
Trlt.	Truck Driver, Light	V.G.	Vertical Grain	yd	Yard
TTY	Teletypewriter	VHF	Very High Frequency	yr	Year
TV	Television	VHO	Very High Output	Δ	Delta
T.W.	Thermoplastic Water Resistant Wire	Vib.	Vibrating	%	Percent
		VLF	Vertical Linear Foot	~	Approximately
UCI	Uniform Construction Index	VOC	Volitile Organic Compound	Ø	Phase; diameter
UF	Underground Feeder	Vol.	Volume	@	At
UGND	Underground Feeder	VRP	Vinyl Reinforced Polyester	#	Pound; Number
UHF	Ultra High Frequency	W	Wire; Watt; Wide; West	<	Less Than
U.I.	United Inch			>	Greater Than
				Z	zone

Index

A

Acoustical and plaster ceiling	259, 260, 261
ceiling	260
enclosure	416
Adjustment for air conditioning	487
Air conditioning and vent.	487, 488
conditioning requirements	486
cooled condensing unit	331
curtain	415
supported storage tank cover	414
supported structure	414
Air conditioner air cooled	330-332
chilled water	324, 332
direct expansion	327, 329-331
fan coil	324
self-contained	329, 330
single zone	327
split system	331
water cooled	329, 332
Air conditioning computer	332
drug store	328
power central	368
Aluminum and glass door	206
corrugated pipe	435
door	208, 209
flashing	218
folding	238
gravel stop	219
gutter	221
light pole	440
siding	194, 195
window wall	241, 242
windows	201, 202
Anechoic chamber	415
Appliance	408
Architectural equip.	404, 406, 408
fee	518
Asphalt shingle	216
Audio masking system	416
Audiometric room	415
Auditorium chair	411
Automotive equipment	407

B

Backhoe trenching	420, 422, 424
Balanced door	206, 210
Bank air conditioning	331
equipment	404
Baptistry	405
Barber equipment	404
Base flashing	218
Bath and toilet accessory	250, 251
Bathroom accessory curtain rod	250
five fixture	285
four fixture	284
system	283-285
three fixture	283
two fixture	282
Beam and double tee precast	76
and joist wood	99
and plank precast	74, 75
and slab C.I.P. one way	58, 59
and slab C.I.P. two way	60
concrete	49-54
double tee precast	71
L shaped precast	53, 54
laminated wood	100
rectangular precast	49, 50
steel fireproofing	102, 103
T shaped precast	51, 52
W shape and deck	92
wide flange and deck	93
wood	55
Belt conveyor	268
Bituminous parking lot	430
paving	428-430
waterproofing	8
Bleacher	409
Blind	411
Block lintel concrete	190
manhole/catch basin	436
partition	228
wall glass	189
wall miscellaneous	153
Boiler coal-fired hydronic	321
electric	318
fossil fuel hydronic	321
hot water and steam	319
hot water electric	318
hydronic fossil fuel	321
oil fired hydronic	321
steam electric	317
Booth for attendant	407
Bowling alley	417
alley air conditioning	329
Box culvert precast	434
Brick composite wall	
double wythe	169-172
composite wall triple wythe	172, 173
curtain wall	197
paving	431
veneer wall exterior	163-168
wall solid	157
wall solid double wythe	158, 159
wall solid reinforced	159, 160
wall solid single wythe	157, 158
Broiler without oven	408
Builder's risk insurance	519
Building air conditioning	324
air conditioning	324-332
code requirements	564
excavation and backfill	27
slab	24
subdrain	7
wall exterior	189
Built-up roofing	211
Bulb end pile	23
Bulletin board	250
Bus duct	503

C

Cabinet dormitory	412
household	251
school	251
Cable high voltage in conduit	354
Caisson	22
Canopy	123
Cap pile	5
Casement windows vinyl clad	202
Cast boxes	504
iron fitting	302-305
iron piping	300-302
iron stair	252, 253
Catch basin and manhole	436, 437
Cedar deck price	101
drop siding	194, 195
Ceiling acoustical	261
plaster and acoustical	260, 261
Cellular deck steel	122, 123
Cement content	575
fill stair	252
mortar	470, 472
Cementitious coating	8
Cementitious/metallic slurry	8
Central A.C. power system	368
Chalkboard	250
Checkout counter	404
Chilled wtr. cooling tower sys.	325
Chimney and vent	322
stainless steel insulated	323
Church equipment	405
Chute linen or refuse	268
mail	268
C.I. roof drain system	298
C.I.P. beam and slab one way	58, 59
beam and slab two way	60
flat plate	64
flat plate w/drop panel	63
flat slab w/drop panel	62
multi span joist slab	65, 66
pile	9, 10
round column	34, 36
slab one way	56, 57
square column	37
waffle slab	67
wall	28, 29
Circuit length	500
Closer	248
Coal tar roofing	211
Column C.I.P. square	39
footing	3, 4
precast	40
round C.I.P.	34, 36
round tied	34, 36
square C.I.P.	37
square tied	37
steel	42-45
steel fireproofing	104
wood	47, 48
Comfort station	414
Commercial air conditioner	324
air conditioning	325, 327-332
dish washer	408
electric radiation	402
electric service	355
fin tube heating	308
unit heater heating	309
water heater	287-289
Compactor	408
Composite block wall	185
brick wall	169
fill slab	94
supported slab	90, 91
Concrete beam	49-54
beam and double tee	71, 76
block nonreinforced wall	131
block partition	226, 228
block specialties	191
block wall lightweight	132, 133
block wall regular weight	131, 132
block walls	470
caisson	22
C.I.P. wall	124, 125
column	34, 35, 37, 38, 41, 457
fireproofing column	104, 105
flat plate	64
flat plate w/drop panel	63
flat slab w/drop panel	62
footing	2-4
Concrete, metal & wood stairs	474
Concrete panel tilt-up	129, 130
paving	428, 429
pile	9-11
pile cap	5
plank and beam	74, 75
plank precast	69, 70
precast flat	126
precast fluted window	127
precast ribbed	128
precast specialties	128
precast unit price	126, 128
slab	24
slab and beam	58-60
slab C.I.P.	56, 57
slab multispan joist	65, 66
slab waffle	67, 68
stair	253
stengths	575
wall	28, 29
Condenser chilled wtr. air cld.	324
Conductor high volt. shielded	354
Conductors	502
Conduit	502
encasement	503
weight	504
Confessional	405
Construction time	518
Continuous footing	2
Control board	250
tower	417
Cooler beverage	408
Cooling computer	332
Coping precast	128
Copper fitting	302-306
flashing	218, 220
gravel stop	219
roof edge	219
tubing piping	300-302
Correspondence lift	268
Corridor air conditioning	329, 331
Counter window post office	251
Covering and paint wall	254
and tile floor	256
Culvert box precast	434
Cup sink	275
Curb & berm	429
Curtain wall panel	205

D

Darkroom	415
equipment	408
Data communication	397
Deck and beam W shape	92
and beam wide flange	93
and joist on bearing wall	111
and steel joist	83-85
double span	121
metal concrete fill	96, 97
price wood	101
single span	121
steel cellular	122
steel long span	121
steel short span	119
Delicatessen case	404
Deluge sprinkler	339, 341
Dental equipment	406
Design wt. for flr. fills and fin.	556
wt. per S.F. for walls and part.	555
wind velocity	558
Detention equipment	406
Directory board	250
Disappearing stairway	416
Dish washer	408
Dispenser towel	250
Display case	404
Diving board	409
Dock bumper	407
Dome	414
Door aluminum	206
balanced	208, 209
fire	243
glass and aluminum	206
kalamein	243
lauan	243

587

Index

mall front 206	Factory air conditioning 324, 325	FM200 fire suppression 353	pile steel 15
mall front panel 243	Feeder in conduit w/neutral 359	system 353	Hangar 414, 417
overhead 208, 209	installation system 359	Food warmer counter 408	Hardware 248
paneled pine 243	Fence chain link 432	Footing column 3, 4	Hatch roof 223
prehung 208, 209	Fiberboard insulation 217	concrete 2, 4	Haul earth 27
revolving 206	Fiberglass/urethane 217	pressure injected 23	Heat loss 484
rolling 208, 209	Field assembly panel siding . 192, 193	spread 3, 4	Heater swimming pool solar 311
sliding 243	Fill slab composite 94	strip 2	terminal unit comm. bldg. 309
stainless steel and glass .. 208, 209	Film transport 405	Formed metal roofing 215	Heating coal hydronic 320
storm 208, 209	Fin tube radiation apt. bldg 307	Fossil fuel hydronic heating . 320, 321	commercial bldg. 309
swing 243	radiation commercial bldg. ... 308	Foundation 13, 15, 17, 18, 20, 22	electric baseboard 402
wood 243	Fire detection system 396	backfill 27	fin tube 321
Double oven 408	door 209, 247	beam 20	fin tube residential 307
span cellular deck 122, 123	extinguishing 333, 335-338,	dampproofing 8	hot water solar 293
tee and beam precast 76	340-348, 350-353	excavation 26, 27	hot water and steam 319
Douglas fir deck 101	protection classification 489	underdrain 7	hydronic fossil fuel 320, 321
Downspout 221, 222	resistant ceiling 260, 261	walls 28, 29, 453, 454	oil hydronic 320, 321
Drapery 411	suppression 353	waterproofing 8	passive solar 312, 313
Dripproof motor and starter . 375-377	wall 573	Framing aluminum tube 204	solar passive 315, 316
Drug store air cond. 324, 329, 331	Firecycle sprinkler 343	light gage steel 82	solar-energy 311
Dry cleaner 404	Fireplace prefabricated 416	steel floor 78, 79	systems 483
standpipe riser 350	Fireproofing steel beam 102, 103	steel joist 83, 84, 87, 88	terminal unit heater 320
wall partition 231	steel column 104, 105	Frost penetration 571	Heating/cooling system 326
Drywall ceiling 259	Fitting pipe 302-306	Frozen food chest 404	H.I.D. fixture high bay 383
partition 230, 231	Fittings gas vent 322	Fryer with submerger 408	fixture low bay 391
Ductwork 487	Fixture fluorescent installed 378	Furnishing 411, 412	High intensity discharge 512
Duranodic gravel stop 219	type/S.F. H.I.D. 383, 385, 387	cabinet 412	intensity fixture 386
D.W.V. fitting 302-305	type/S.F. incandescent 380	Furniture and accessory 412	voltage shielded conductor ... 354
	watt/S.F. fluorescent 382		Hinge 248
	watt/S.F. HID 384, 386, 388		Historical Cost Indexes 526
E	watt/S.F. incandescent 381	**G**	Hoist 407
Ecclesiastical equipment 405	Flagpole 432	Galvanized flashing 218	Hollow metal door 244, 245
Eggcrate ceiling 261	Flashing 220	gravel stop 219	Hook-up electric service 355
E.I.F.S. 197	Flat plate C.I.P. 64	roof edge 219	Hospital furniture 412
Electric circuit voltages 498-500	plate w/drop panel 62, 63	Garage cost 417	Hot water solar drain down 294
Electric baseboard radiation 401	precast concrete wall 126, 128	Garbage disposer 408	Hotel furnishing 412
board system 401	Flexible transparent curtain 416	Garden house 414	Hot water and steam boiler 319
boiler 317	Floor beam and double tee 76	Gas fired wtr. htr. comm. 288	commercial 287-289
boiler large hydronic 318	carpet 256	heating boiler 319	heating forced 307
boiler small hydronic 317	cellular deck steel 122, 123	vent 322	solar energy 291, 293, 295, 297
fire pump 352	composition 256	vent fitting galv. steel 322, 323	system 287-289
heating boiler 319	concrete topping 256	Gas service polyethylene pipe ... 438	H.P. value/cost determin. ... 507, 508
service 3 phase/4 wire ... 355, 356	flat plate 64	Generator 398	HV cable system 354
service installation 355	flat plate w/drop panel 62	diesel 398	Hydraulic elevator 264
systems 495-498, 501	framing steel 78-80	gas operated 398	
water heater commercial 287	framing steel joist 83-85, 87	system 398	
Elevator freight 264-266	joist and beam wood 99	weight 503	**I**
geared traction 265	joist wood 98	Geodesic 414	Ice skating rink 417
gearless traction 266	light gage steel 82	Glass and aluminum door 209	Identifying/visual aid specialty .. 250
hoistway sizes 475, 476	mat 411	fiber felts 211, 212	Incandescent fixture system 380
hospital 264-266	metal deck concrete fill 96, 97	fiber roofing 212	Incinerator 408
hydraulic 264	multi span joist slab 65, 66	panel 205	Incubator 406
passenger 264-266	plank and beam 74	Glazing solar direct gain 312, 313	Indirect gain solar wall 314
Equip athletic/rec./healing 409	precast beam and slab 75	solar indirect gain 314, 315	Insulating materials 569
Equipment fire 352	precast double tee 71, 72	Grab bar 250	Insulation roof deck rigid 217
fire hose 352	precast plank 69, 70	Grade beam 20	Insurance 519
food service 408	resilient 256	beams 451	Integrated ceiling 415
loading dock 407	slab and beam one way 58, 59	Grandstand 432	Intercom system 396
parking control 407	slab and beam two way 60	Granite paving 431	Interior finish 254
solid waste handling 408	slab C.I.P. 56	veneer wall 162	finish classification 573
theater and stage 405	slab composite fill 94, 95	Gravel stop aluminum 219	glazing 240
vehicular service 407	slab composite supported .. 90, 91	stop duranodic 219	window wall 241, 242
Escalator 267	steel deck 119, 121	Greenhouse 414	Interlocking paving 431
Estimate plumbing 286	systems 455	Ground face block wall 149	Intermediate rib steel deck 119
Excavation 514	systems cost 456	Group wash fountain 278	Ironer commercial 404
& backfill building 26, 27	tile and covering 256	Guardrail 432	
trench 420-422, 424	underlayment 256	Gutter roof 221	
Exterior lighting 439	W shape beam and deck 92	Gypsum board ceiling . 258, 260, 261	**J**
wall 157, 464	waffle slab 67	plaster ceiling 258, 260, 261	Joist and beam wood 99
wood siding 194, 195	wood 256	plaster partition 234, 237	and deck on bearing wall 111
	Flor and roof systems 557		steel on beam 87, 109
	Fluorescent fixture 378		steel on beam and wall 84, 85
F	fixture high bay 382	**H**	steel on beam wall 107
Fabricated cabinet & counter 251	fixture system 378	H pile 15	steel on bearing wall 83
	light by wattage 379		
	Fluted window precast concrete . 127		

Index

steel on girder 116-118
steel on girder and wall ... 113-115
wood 98
wood pitched 106

K

Key station on pedestal 407
Kiosk 414
Kips and in-kips 467-469
Kitchen equipment 408
   sink 273
kVA value/cost determination ... 506
KW value/cost determination 505

L

Lab equip. 406
Laboratory equipment 406
   sink 275
Laminated wood beam 100
Lamp 513
Lamphouse 405
Lamps 512
Laundry equipment 404
   sink 274
Laundry/dry cleaning 404
Lavatories 272
Letter slot post office 251
Library air conditioning 330
   equipment 405
Light gage steel floor 82
   pole 440
   switch 366
Limestone veneer wall 161
Lintel concrete block 190, 191
Live loads per S.F. 554
Locker 250
Lockset 248
L shaped precast beam 53, 54
Lube equipment 407

M

Magnetic starter 372
   starter & motor 375-377
Mail box 251
Maintenance equipment 408
Mall front door 208, 209
Malleable iron fitting 302-305
Manual starter & motor ... 376, 377
Marble sink 272
Masonry block lintels 464
   partition 226
Master clock system 396
   T.V. antenna system 397
Meat case 404
Median barrier 432
Medical center air cond. ... 329, 330
   equipment 406
Medicine cabinet 251
Membrane roofing 213
Mercantile equipment 404
Metal deck & joist on brg. wall .. 111
   deck concrete fill 97
   door 244, 245
   roofing formed 215
   shingle 216
   siding panel 192
   siding support 198, 199
Metric conversion factors 574
   rebar specs 576
Mineral fiber shingle 216

Mirror 251
Misc. power system 367
Miscellaneous power 367
Modular office 238
Motor and starter 375
   connections 374
   feeder system 371
   installation 369
Motors 503
Movie equipment 405
Moving walk 267
Mullion precast concrete 127
Multispan joist slab C.I.P. 65
Music practice room 415

N

Narrow rib steel deck 119
Neoprene flashing 220
Night depository 404

O

Occupancy determinations 563
Office furniture 412
Oil fired wtr. htr. comm. 289
Organic felts 211
Overhead door 206
Overtime 523

P

Paint and covering wall 254, 255
   cabinets 254
   casework 254
   ceiling 254
   miscellaneous metal 254
   structural steel 254
   trim 255
   woodwork 254
Pan tread stair 252, 253
Panel concrete tilt-up 129, 130
   sandwich 205
   solar indirect gain 315
   spandrel glass 205
Panelboard 361, 362
   installation system 361
Panic device 248
Parking equipment 407
   lot gravel base 430
   lots 430, 515, 516
Part bin 250
Partition accordion 238
   acoustical 238
   bathroom 249
   commercial 238
   drywall 233
   folding 238
   folding accordion 239
   gypsum board 230, 233
   handicap 249
   hospital 251
   industrial 238
   marble 249
   metal 249
   movable 238
   office 239
   plastic laminate 249
   porcelain enamel 249
   shower stall 249
   steel folding 238
   toilet 249
   urinal 249

   vinyl folding 238
   wood folding 238
Partition/door density 562, 563
Passive solar indirect gain panel . 315
   solar indirect gain wall 314
   solar sunspace 316
Patient wall system 412
Patio door 209
Paving roadway 428, 429
Pedestal access floor 415
   pile 23
Pedestrian traffic control 412
Perforated H.D.P.E. pipe 31
Perimeter drain 7
Perlite plaster 260, 261
   plaster ceiling 258-261
Photovoltaic power 399, 400
   system grid connected 400
Pile 13, 17, 18, 23
   caisson 22
   cap 5
   caps, piles and caissons 448
   cast in place 9, 10
   C.I.P. concrete 9, 10
   concrete 9, 10
   concrete filled pipe 13, 14
   precast concrete 11
   prestressed concrete 11
   steel H 15
   steel pipe 13, 14
   steel shell 10
   step-tapered steel 17
   treated wood 18
Piles caps, piles and caissons ... 450
Piles, general 449
Pipe and fittings 477
   bedding 426, 427
   bituminous coated metal 434
   bituminous fiber 434
   bituminous fiber perforated 31
   concrete nonreinforced 434
   concrete porous wall 31
   concrete reinforced 434
   copper in trench 433
   corrugated metal coated 434
   corrugated metal plain 434
   corrugated oval arch coated .. 435
   corrugated oval arch plain ... 435
   ductile iron in trench 433
   fitting 302-306
   in trench 31, 433-435, 438
   iron ductile mech joint ... 433
   iron ductile Tyton joint ... 433
   metal perforated coated 31
   polyethylene gas service 438
   polyethylene in trench 433
   polyvinyl chloride 435
   PVC in trench 433
   steel gas service 438
   steel tarred & wrapped 438
Piping 300, 301
   drainage & sewage 434
   excavation 420, 422, 424
   gas service 438
   water distribution 433
Plain & reinf. slab on grade ... 24, 25
Plank and beam precast 74, 75
   precast 69
Plaque 250
Plaster and acoustical ceiling . 258-261
   ceiling 258
   partition 234, 235, 237
Plasterboard 258, 260, 261
   ceiling 260, 261
Plaza brick & tile 431
Plumbing approx. for quick est. .. 477

   fixture 273, 280
   fixture requirements ... 478, 479
   system estimate 286
Pneumatic tube system ... 268, 404
Porcelain enamel sink 272
Portable booth 415
Precast beam and double tee .. 76
   beam and plank 74, 75
   beam L shaped 53, 54
   beam T shaped 51, 52
   concrete column 40, 41
   concrete flat 126, 128
   concrete fluted window ... 127
   concrete mullion 127, 128
   concrete pile 11
   concrete ribbed 128
   concrete specialties 128
   concrete unit price .. 127, 128
   double tee beam 71, 72
   paving 431
   plank 69, 70
   rectangular beam 49, 50
   window sills 128
Pre-engineered steel building ... 471
   structure 414
Preformed metal roofing 214
Prehung door 206
Pressure injected footing 23
Prestressed concrete pile 11
Produce case 404
Project size modifier 545
Projection screen 405
Projector mechanism 405
Pump gasoline 407
Push 248
PVC fitting 302-305
   roof drain system 298
   SDR 35 pipe 435
   tubing piping 300-302

R

R values 568
Rabbetted cdr. bev. siding .. 194, 195
Radiation electric 401
   electric baseboard 402
   protection 416
Radio tower 414
Rafter wood/flat or pitched 106
Range restaurant 408
Receptacle 363
   floor 364
Rectangular precast beam 49
Red cedar shingle siding ... 194, 195
Redwood shiplap siding ... 194, 195
Refrigerated food case 404
Refrigerator 408
   walk in 408
Reinforced beam 20
   block wall miscellaneous . 155, 156
   concrete block lightwght . 134-136
   ground face block wall 150-152
   split face block wall ... 145, 146
   split face lightweight 146-148
   split ribbed block wall .. 139, 140
   split ribbed lightweight ... 140, 141
Residential electric radiation 401
   equipment 408
   fin tube heating 307
Restaurant air conditioning 328
   furniture 412
Revolving door 209
Ribbed precast concrete ... 127, 128
Rigid insulation 217
Riser standpipe 348, 351

Index

Roadway 428, 429
   pavement 428, 429
Roll roofing 212
   up grille 243
Rolling door 206
   grille 208
   overhead door 243
Roof deck and joist 111
   drain system 299
   gravel stop 221
   joist steel 107-110, 113-118
   joist wood 106
Roofing formed metal 215
   gravel stop 222
Rooftop air conditioner 327, 328
   multizone unit 328
   single zone unit 327
Round tied column 34, 35
Rubber flashing 218

S

Safe 404
Safety switch 373
Sales tax 523
Sand backfill 426
Sandwich panel 205
   panel siding 193
Sauna 409
Sawn cedar batten siding ... 194, 195
Scaffolding 524
Scale 407
School air conditioning 328
   equipment 409
Scoreboard 409
Sealing foundation 8
Seating 411
   fixed multiple 411
   lecture hall 411
Security gate 404
Security/vault 404
Self-contained air cooled unit . 330
   water cooled unit 329
Service entrance electric 355
   sink 276
Sewage piping 434
Shade 411
Sheet metal edge 219
Shelter 414
Shelving metal 250
Shielding lead 416
   lead X-ray 416
Shingle and tile roofing 216
Shooting range 409
Shower system 279
Shredder 408
Sidewalk bituminous 431
Siding aluminum 194, 195
   aluminum panel 192
   cedar drop 195
   field assembly panel 192
   metal panel 192
   rabbetted cedar bev. 195
   red cedar shingle 195
   redwood shiplap 195
   sandwich panel 192
   sawn cedar batten 195
   steel panel 192
   support metal 198
   vertical T.&G. redwood 195
   wood exterior 194, 195
Sign exit 250
Single oven 408
   ply membrane 213
   span cellular deck 122

Site preparation 514
   utility 426
Sitework 26, 27
   manhole 436
Skylight 223
Slab and beam C.I.P. one way 58
   and beam C.I.P. two way 60
   C.I.P. one way 56, 57
   composite fill 94
   composite supported 90, 91
   concrete 24
   multispan joist C.I.P. 65, 66
   on grade 24, 452, 568
   waffle C.I.P. 67, 68
Slate roofing 216
Sliding patio door 207
Smoke hatch 223
Snow loads 559, 560
Solar closed loop add-on
   hot water 291
   closed loop hot water system . 297
   collectors 482
   direct gain glazing 312, 313
   direct gain glazing door 312
   direct gain glazing window ... 313
   direct gain sunspace 316
   drain back hot water 292
   draindown hot-water 295
   energy system 291, 297
   indirect gain panel 315
   passive heating 312-315
   swimming pool heater 311
Solar energy pool heater 311
Solid core door 208, 209
Sound absorbing panel 416
   system 396, 405
Sound/vibration/seismic
construct 416
Spandrel glass panel 205
Special construction ... 409, 415, 416
Specialties precast concrete ... 128
Specialty item 250, 251
Split face block reg. wt. ... 142, 143
   face block wall 142
   face block wall lightweight 143-145
   ribbed block wall 137
   ribbed block wall ltwt. ... 138, 139
Sport court 409
Spread footing 3, 4
   footings 442-445
Sprinkler deluge system ... 340, 341
   dry pipe 333
   dry pipe system 334
   firecycle system 343, 344
   preaction system 337
   wet pipe system 345, 346, 347
Square foot building costs 544
   tied column 37, 38
Stage equipment 405
Stainless steel and glass door . 208
   steel sink 272, 273
Stair 252, 253
Standpipe equipment 352
   systems 490
Starter and motor 375-377
Station scrub 406
Steam and hot water boiler 319
   bath heater 410
   heating commercial 308
Steel beam fireproofing 102, 103
   column 42-46
   column fireproofing 105
   columns 458
   deck cellular 122, 123
   deck long span 121
   deck short span 119, 120

door 208, 209, 244, 245
floor light gage 82
framing floor 78, 92, 93
framing/floor 78
galv. gutter 221
grate stair 253
H pile 15
joist on beam 87, 88, 109, 110
joist on beam and wall .. 84, 85, 107, 108
joist on bearing wall 83
joist on girder 116-118
joist on girder and wall ... 113-115
light pole 440
oval arch corrugated pipe 435
pile step-tapered 17
pipe pile 13, 14
roof drain system 298, 299
shell pile 9, 10
sink 272, 273
step-tapered pile 17
tubing piping 300-302
window 201
Steeple 405
Step tapered steel pile 17
Sterilizer 406
Stone faced wall 162
   veneer wall 161, 162
Storage specialties 250
Store air conditioning 329
Storm door 206
   drainage system 299
Street sign 250
Strip footings 2, 446, 447
Structural wall support 198
Stucco wall 196
Stud wall 230, 231, 234-236
   wall wood 194, 195
Subdrainage piping 31
Substructure 24
Sunspace solar direct gain ... 316
Supported composite slab 90
Surgery table 406
Suspension system 258, 260, 261
   system ceiling 260, 261
Swimming pool 409
   pool enclosure 414
Switchgear 360
Swithgear installation system . 360
System baseboard electric 401
   baseboard heat commercial ... 402
   boiler hot water 318
   boiler oil fired 321
   boiler steam 317
   cable in conduit w/neutral .. 354
   comm. and alarm 396, 397
   deluge sprinkler 340, 341
   drinking fountain 280
   dry pipe sprinkler 334, 335
   electric feeder 359
   electric service 355
   fin-tube radiation 321
   fire detection 396
   fire sprinkler 333, 335-347
   firecycle sprinkler 342-344
   fluorescent fixture 378
   fluorescent watt/S.F. 379
   FM200 fire suppression 353
   heating hydronic 321
   high bay H.I.D. fixture 383
   high intensity fixture .. 382-394
   hydronic fossil fuel 320, 321
   incandescent fixture 381
   incandescent fixture/S.F. ... 380
   intercom 396
   light pole 440

master clock 396
plumbing estimate 286
preaction sprinkler 337, 338
receptacle 364, 365
receptacle watt/S.F. 363
riser dry standpipe 351
sitework catch basin 436
sitework trenching 420, 422, 424
solar energy 311
standpipe riser dry 350, 351
standpipe riser wet 348
storm drainage 298
switch wall 365
telephone underfloor 395
toggleswitch 365
underfloor duct 364
wall concrete block 131
wall exterior block 180, 185
wall ext. brick ... 157, 163, 169, 174
wall exterior glass block 189
wall exterior stone 161
wall ground face block 149
wall miscellaneous block 153
wall split face block 142
wall split ribbed block 137
wall switch 366
wet pipe sprinkler 346, 347

T

Table exhibit 404
Tank liquid & gas storage 417
Tavern air cond. .. 324, 325, 328, 329
Taxes 523
Tee double and beam precast 76
   double precast 71
Telephone enclosure 250
   systems 395
Teller automated 404
   window 404
Tension structure 414
Texture 1-11 siding 194, 195
Thermal solar wall 314
Thermosypon solar panel 315
THW wire 503
Tile and covering floor ... 256, 257
   ceiling 258, 260, 261
   partition 229
Tilt-up concrete panel 129, 130
Toilet tissue dispenser 251
Towel bar 251
Traction geared elevator 265
   gearless elevator 266
Traffic detector 407
Transmission of heat 485
Treated wood pile 18
Trench backfill 426, 427
Trenching 421
   common earth 420, 421
   loam & sandy clay 422, 423
   sand & gravel 424, 425
Triple span cellular deck . 122, 123
T shaped precast beam 51, 52
Tub enclosure 249
Tubing copper type K 433
Tubular aluminum framing 204
Turnstile 251

U

U & R factors 465, 466, 472, 473
Underfloor duct telephone sys. . 395
Underground elec. service . 357, 358
   piping 426

Index

Unit gross area requirements 544
Urinal screen 249
   system 271
U. S. customary and SI units for
reinforcing bars 576
Utility excavation 420, 422, 424
   structure 436

V

Vacuum cleaning 408
Vanity top lavatory 272
Veneer brick wall exterior 163
   stone wall 161
   wall stone 162
Vent and chimney 322
Ventilation 487
Vertical T. and G. redwood
siding 194, 195
Vinyl steel gutter 221
Vitreous china lavatory 272
Vitreous-china service sink 276
Vocational shop equipment 409

W

W shape beam and deck 92
Waffle slab C.I.P. 67, 68
Wall block face cavity 180-182
   block face composite 185-188
   block face concrete backup ... 180
   block face w/insul cavity .. 182-184
   block faced concrete masonry . 185
   brick face back up w/cavity ... 177
      , 178, 179
   brick face cavity 174, 175
   brick faced concrete masonry . 169
   cast in place 29
   cast in place concrete 124, 125
   C.I.P. concrete 124, 125
   concrete 28
   foundation 29
   glass block 189
   indirect gain solar 314
   precast concrete 126, 128
   stone faced metal stud ... 161, 162
   stone faced wood stud ... 161, 162
   structural support 198, 199
   stucco 196
   switch installed/ea 365
   tilt-up panel concrete 129, 130
   wood stud 194, 195
Wallpaper wall 254
Walls, exterior 464
Wardrobe unit 412
Washer comm. coin operated ... 404
   residential 404
Waste handling 408
Water bubbler 280
   cooler 281, 480
   piping polyethylene 433
Water closet 270
   battery mount 270
   system 269
Water-heater electric 287
   gas fired 288
   oil fired 289
   solar energy 293, 295, 297
Waterproofing foundation 8
Weather data and design cond. .. 570
Weatherstripping 248
Weighted averages, sys. division . 525
Wet standpipe riser 348
White spruce deck 101

Wide rib steel deck 119
Wind load 558, 559, 561
Window sills precast 128
   treatment 411
   wall 204
Windows vinyl clad 202, 203
   vinyl clad double hung 203
Wire 501, 502
Wood beam 55
   beam and joist 99
   beams 459
   column 47
   deck 101
   deck price 101
   door 208, 209, 245-247
   floor joists 460
   folding panel 411
   joist 98
   joist pitched 106
   joists and rafters 462, 463
   laminated beam 100
   partition 236
   pile treated 18
   shingle 216
   siding exterior 194, 195
   stud wall 195
   window 200, 201
Woodburning stove 416
Wood decking 461
Workers' compensation 520-522

X

X-ray 406
   unit 406

Notes

Notes

Notes

Notes

Notes

Notes

Notes

Notes

Division Notes

	CREW	DAILY OUTPUT	LABOR-HOURS	UNIT	2012 BARE COSTS				TOTAL INCL O&P
					MAT.	LABOR	EQUIP.	TOTAL	

Division Notes

	CREW	DAILY OUTPUT	LABOR-HOURS	UNIT	2012 BARE COSTS				TOTAL INCL O&P
					MAT.	LABOR	EQUIP.	TOTAL	

Division Notes

Division Notes

Division Notes

Reed Construction Data/RSMeans—a tradition of excellence in construction cost information and services since 1942

Table of Contents
Annual Cost Guides
Seminars
New Titles
Reference Books
Order Form

For more information visit the RSMeans website at **www.rsmeans.com** Unit prices according to the latest MasterFormat!

Book Selection Guide

The following table provides definitive information on the content of each cost data publication. The number of lines of data provided in each unit price or assemblies division, as well as the number of reference tables and crews, is listed for each book. The presence of other elements such as equipment rental costs, historical cost indexes, city cost indexes, square foot models, or cross-referenced indexes is also indicated. You can use the table to help select the RSMeans book that has the quantity and type of information you most need in your work.

Unit Cost Divisions	Building Construction	Mechanical	Electrical	Commercial Renovation	Square Foot	Site Work Landsc.	Green Building	Interior	Concrete Masonry	Open Shop	Heavy Construction	Light Commercial	Facilities Construction	Plumbing	Residential
1	573	395	410	499		517	124	319	463	572	507	233	1038	411	176
2	755	281	87	706		1005	180	366	223	751	744	447	1197	291	258
3	1678	346	223	1053		1474	1004	323	2037	1678	1680	469	1774	315	376
4	900	22	0	698		701	183	573	1135	876	591	490	1147	0	405
5	1844	159	156	1040		809	1777	1047	712	1835	1031	924	1845	204	720
6	2294	67	69	1881		411	534	1421	300	2273	383	1968	1908	22	2480
7	1467	215	125	1501		563	776	572	497	1464	26	1178	1529	224	942
8	2122	98	48	2351		299	1157	1768	657	2066	0	1790	2581	0	1410
9	1927	72	26	1720		313	449	2012	389	1868	15	1639	2171	54	1416
10	987	17	10	610		215	27	807	156	987	29	497	1073	230	221
11	1022	208	165	486		126	54	878	28	1007	0	231	1042	169	110
12	563	0	2	325		248	150	1698	12	552	19	371	1748	23	324
13	727	117	112	246		346	128	251	66	726	253	82	754	71	80
14	277	36	0	225		28	0	259	0	276	0	12	296	18	6
21	78	0	16	30		0	0	248	0	78	0	60	429	435	224
22	1140	7318	154	1105		1536	1100	605	20	1102	1659	830	7152	9029	659
23	1211	7086	599	846		158	940	695	38	1132	111	752	5176	1860	416
26	1344	499	10087	1019		816	644	1132	55	1329	562	1218	10,011	444	618
27	72	0	266	34		13	0	60	0	72	39	52	268	0	4
28	96	58	124	71		0	21	78	0	99	0	40	139	44	25
31	1482	735	610	803		3207	290	7	1208	1426	3284	600	1545	660	611
32	770	54	0	836		4341	341	362	274	741	1796	381	1647	166	428
33	509	1014	461	207		2027	41	0	231	508	1937	119	1557	1205	117
34	109	0	20	4		192	0	0	31	64	170	0	129	0	0
35	18	0	0	0		327	0	0	0	18	442	0	83	0	0
41	52	0	0	24		7	0	22	0	52	30	0	59	14	0
44	75	83	0	0		0	0	0	0	0	0	0	75	75	0
46	23	16	0	0		274	181	0	0	23	264	0	33	33	0
48	13	0	28	0		4	28	0	0	13	4	28	13	0	26
Totals	24128	18896	13798	18320		19957	10,129	15503	8532	23588	15576	14411	48419	15997	12052

Assem Div	Building Construction	Mechanical	Electrical	Commercial Renovation	Square Foot	Site Work Landscape	Assemblies	Green Building	Interior	Concrete Masonry	Heavy Construction	Light Commercial	Facilities Construction	Plumbing	Asm Div	Residential
A		15	0	188	150	577	598	0	0	536	571	154	24	0	1	368
B		0	0	848	2483	0	5643	56	329	1966	368	2074	174	0	2	211
C		0	0	647	863	0	1233	0	1558	146	0	763	245	0	3	588
D		1067	886	712	1864	72	2464	265	826	0	0	1348	1084	1083	4	839
E		0	0	86	260	0	298	0	5	0	0	257	5	0	5	392
F		0	0	0	115	0	115	0	0	0	0	115	3	0	6	357
G		527	447	318	202	3456	668	0	0	535	1431	200	293	707	7	296
															8	760
															9	80
															10	0
															11	0
															12	0
Totals		1609	1333	2799	5937	4105	11019	321	2718	3183	2370	4911	1828	1790		3891

Reference Section	Building Construction Costs	Mechanical	Electrical	Commercial Renovation	Square Foot	Site Work Landscape	Assem.	Green Building	Interior	Concrete Masonry	Open Shop	Heavy Construction	Light Commercial	Facilities Construction	Plumbing	Resi.
Reference Tables	yes	yes	yes	yes	no	yes	yes	yes	yes	yes	yes	yes	yes	yes	yes	yes
Models					109								48			28
Crews	558	558	558	536		558		558	558	558	534	558	534	536	558	534
Equipment Rental Costs	yes	yes	yes	yes		yes		yes	yes	yes	yes	yes	yes	yes	yes	yes
Historical Cost Indexes	yes	yes	yes	yes	yes	yes	yes	yes	yes	yes	yes	yes	yes	yes	yes	no
City Cost Indexes	yes	yes	yes	yes	yes	yes	yes	yes	yes	yes	yes	yes	yes	yes	yes	yes

Annual Cost Guides

Cost Guides also available in CD and at rsmeansonline.com

For more information visit the RSMeans website at **www.rsmeans.com**

Unit prices according to the latest MasterFormat!

RSMeans Building Construction Cost Data 2013

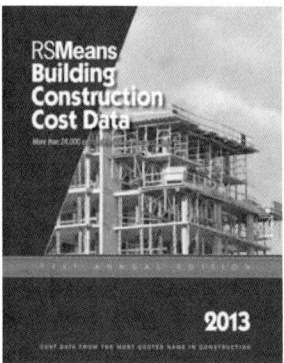

Offers you unchallenged unit price reliability in an easy-to-use format. Whether used for verifying complete, finished estimates or for periodic checks, it supplies more cost facts better and faster than any comparable source. Over 23,000 unit prices have been updated for 2013. The City Cost Indexes and Location Factors cover over 930 areas, for indexing to any project location in North America. Order and get *RSMeans Quarterly Update Service* FREE. You'll have year-long access to the RSMeans Estimating **HOTLINE** FREE with your subscription. Expert assistance when using RSMeans data is just a phone call away.

$186.95 per copy | Available Sept. 2012 | Catalog no. 60013

RSMeans Green Building Cost Data 2013

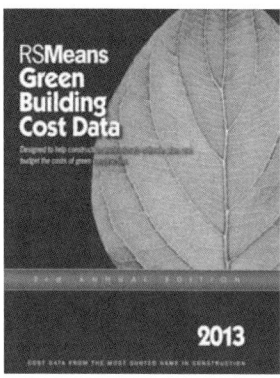

Estimate, plan, and budget the costs of green building for both new commercial construction and renovation work with this first edition of RSMeans Green Building Cost Data. More than 9,000 unit costs for a wide array of green building products plus assemblies costs. Easily identified cross references to LEED and Green Globes building rating systems criteria.

$155.95 per copy | Available Nov. 2012 | Catalog no. 60553

RSMeans Mechanical Cost Data 2013

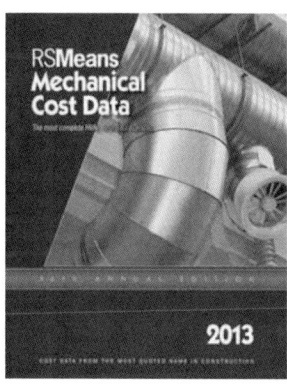

Total unit and systems price guidance for mechanical construction... materials, parts, fittings, and complete labor cost information. Includes prices for piping, heating, air conditioning, ventilation, and all related construction.

Plus new 2013 unit costs for:

- Thousands of installed HVAC/ controls assemblies components
- "On-site" Location Factors for over 930 cities and towns in the U.S. and Canada
- Crews, labor, and equipment

$183.95 per copy | Available Oct. 2012 | Catalog no. 60023

RSMeans Facilities Construction Cost Data 2013

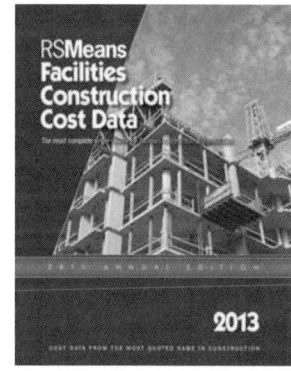

For the maintenance and construction of commercial, industrial, municipal, and institutional properties. Costs are shown for new and remodeling construction and are broken down into materials, labor, equipment, and overhead and profit. Special emphasis is given to sections on mechanical, electrical, furnishings, site work, building maintenance, finish work, and demolition.

More than 47,000 unit costs, plus assemblies costs and a comprehensive Reference Section are included.

$476.95 per copy | Available Nov. 2012 | Catalog no. 60203

RSMeans Square Foot Costs 2013

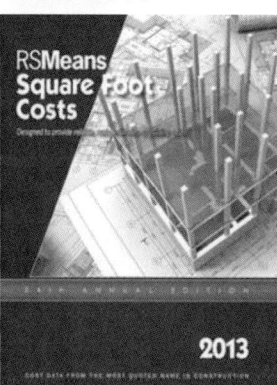

Accurate and Easy To Use

- **Updated price information** based on nationwide figures from suppliers, estimators, labor experts, and contractors
- "How-to-Use" sections, with **clear examples** of commercial, residential, industrial, and institutional structures
- Realistic graphics, offering true-to-life illustrations of building projects
- Extensive information on using square foot cost data, including sample estimates and alternate pricing methods

$197.95 per copy | Available Oct. 2012 | Catalog no. 60053

RSMeans Commercial Renovation Cost Data 2013

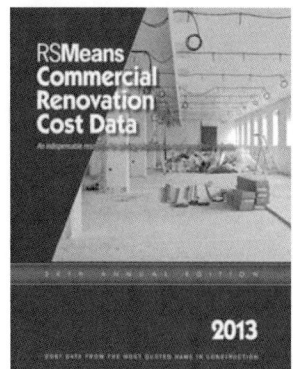

Commercial/Multifamily Residential

Use this valuable tool to estimate commercial and multifamily residential renovation and remodeling.

Includes: New costs for hundreds of unique methods, materials, and conditions that only come up in repair and remodeling, PLUS:

- Unit costs for more than 17,000 construction components
- Installed costs for more than 2,000 assemblies
- More than 930 "on-site" localization factors for the U.S. and Canada

$152.95 per copy | Available Oct. 2012 | Catalog no. 60043

Annual Cost Guides

Cost Guides also available in CD and at rsmeansonline.com

For more information visit the RSMeans website at **www.rsmeans.com**

Unit prices according to the latest MasterFormat!

RSMeans Electrical Cost Data 2013

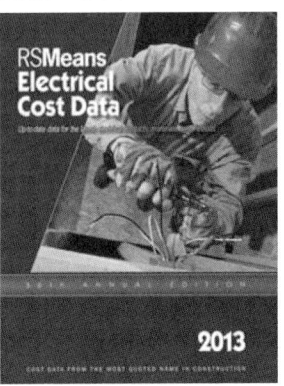

Pricing information for every part of electrical cost planning. More than 13,000 unit and systems costs with design tables; clear specifications and drawings; engineering guides; illustrated estimating procedures; complete labor-hour and materials costs for better scheduling and procurement; and the latest electrical products and construction methods.

- A variety of special electrical systems including cathodic protection
- Costs for maintenance, demolition, HVAC/mechanical, specialties, equipment, and more

$188.95 per copy | Available Oct. 2012 | Catalog no. 60033

RSMeans Electrical Change Order Cost Data 2013

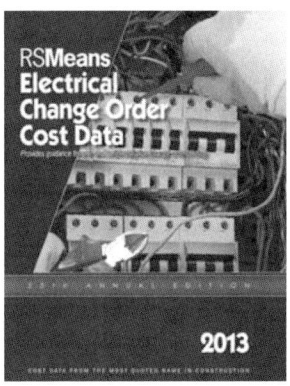

RSMeans Electrical Change Order Cost Data provides you with electrical unit prices exclusively for pricing change orders—based on the recent, direct experience of contractors and suppliers. Analyze and check your own change order estimates against the experience others have had doing the same work. It also covers productivity analysis and change order cost justifications. With useful information for calculating the effects of change orders and dealing with their administration.

$183.95 per copy | Available Dec. 2012 | Catalog no. 60233

RSMeans Assemblies Cost Data 2013

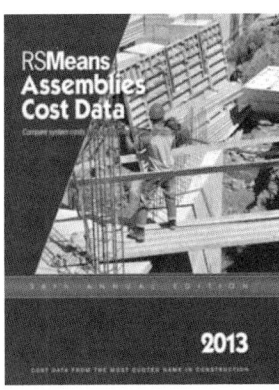

RSMeans Assemblies Cost Data takes the guesswork out of preliminary or conceptual estimates. Now you don't have to try to calculate the assembled cost by working up individual component costs. We've done all the work for you.

Presents detailed illustrations, descriptions, specifications, and costs for every conceivable building assembly—over 350 types in all—arranged in the easy-to-use UNIFORMAT II system. Each illustrated "assembled" cost includes a complete grouping of materials and associated installation costs, including the installing contractor's overhead and profit.

$305.95 per copy | Available Sept. 2012 | Catalog no. 60063

RSMeans Open Shop Building Construction Cost Data 2013

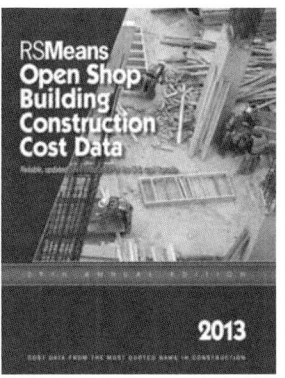

The latest costs for accurate budgeting and estimating of new commercial and residential construction... renovation work... change orders... cost engineering.

RSMeans Open Shop "BCCD" will assist you to:

- Develop benchmark prices for change orders
- Plug gaps in preliminary estimates and budgets
- Estimate complex projects
- Substantiate invoices on contracts
- Price ADA-related renovations

$160.95 per copy | Available Dec. 2012 | Catalog no. 60153

RSMeans Residential Cost Data 2013

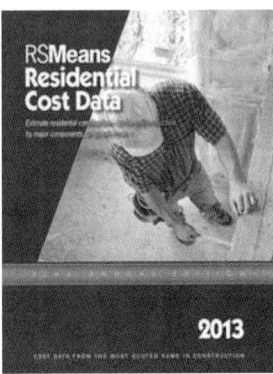

Contains square foot costs for 28 basic home models with the look of today, plus hundreds of custom additions and modifications you can quote right off the page. Includes costs for more than 700 residential systems. Complete with blank estimating forms, sample estimates, and step-by-step instructions.

Now contains line items for cultured stone and brick, PVC trim lumber, and TPO roofing.

$134.95 per copy | Available Oct. 2012 | Catalog no. 60173

RSMeans Site Work & Landscape Cost Data 2013

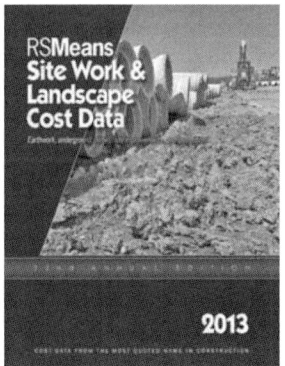

Includes unit and assemblies costs for earthwork, sewerage, piped utilities, site improvements, drainage, paving, trees and shrubs, street openings/repairs, underground tanks, and more. Contains 78 types of assemblies costs for accurate conceptual estimates.

Includes:

- Estimating for infrastructure improvements
- Environmentally-oriented construction
- ADA-mandated handicapped access
- Hazardous waste line items

$181.95 per copy | Available Dec. 2012 | Catalog no. 60283

Annual Cost Guides

Cost Guides also available in CD and at rsmeansonline.com

For more information visit the RSMeans website at **www.rsmeans.com**

Unit prices according to the latest MasterFormat!

RSMeans Facilities Maintenance & Repair Cost Data 2013

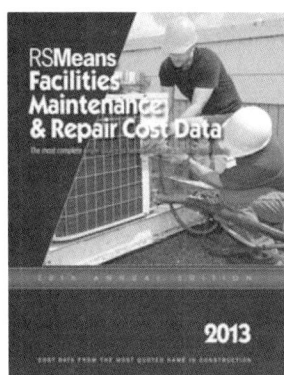

RSMeans Facilities Maintenance & Repair Cost Data gives you a complete system to manage and plan your facility repair and maintenance costs and budget efficiently. Guidelines for auditing a facility and developing an annual maintenance plan. Budgeting is included, along with reference tables on cost and management, and information on frequency and productivity of maintenance operations.

The only nationally recognized source of maintenance and repair costs. Developed in cooperation with the Civil Engineering Research Laboratory (CERL) of the Army Corps of Engineers.

$417.95 per copy | Available Nov. 2012 | Catalog no. 60303

RSMeans Light Commercial Cost Data 2013

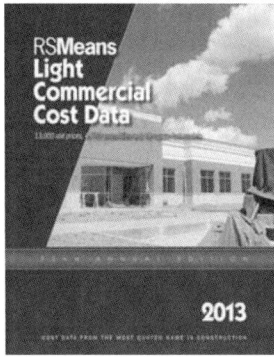

Specifically addresses the light commercial market, which is a specialized niche in the construction industry. Aids you, the owner/designer/contractor, in preparing all types of estimates—from budgets to detailed bids. Includes new advances in methods and materials.

Assemblies Section allows you to evaluate alternatives in the early stages of design/planning.

Over 13,000 unit costs ensure that you have the prices you need. . . when you need them.

$139.95 per copy | Available Nov. 2012 | Catalog no. 60183

RSMeans Concrete & Masonry Cost Data 2013

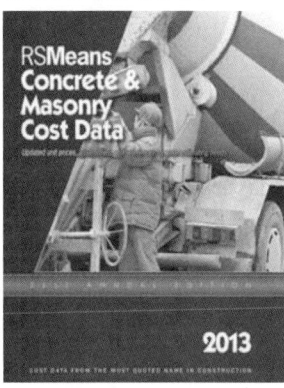

Provides you with cost facts for virtually all concrete/masonry estimating needs, from complicated formwork to various sizes and face finishes of brick and block—all in great detail. The comprehensive Unit Price Section contains more than 8,000 selected entries. Also contains an Assemblies [Cost] Section, and a detailed Reference Section that supplements the cost data.

$171.95 per copy | Available Dec. 2012 | Catalog no. 60113

RSMeans Labor Rates for the Construction Industry 2013

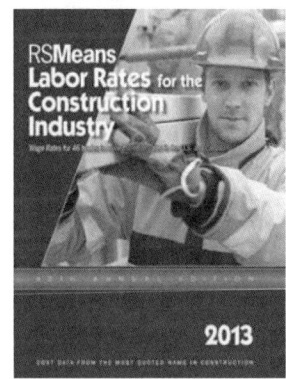

Complete information for estimating labor costs, making comparisons, and negotiating wage rates by trade for more than 300 U.S. and Canadian cities. With 46 construction trades listed by local union number in each city, and historical wage rates included for comparison. Each city chart lists the county and is alphabetically arranged with handy visual flip tabs for quick reference.

$415.95 per copy | Available Dec. 2012 | Catalog no. 60123

RSMeans Construction Cost Indexes 2013

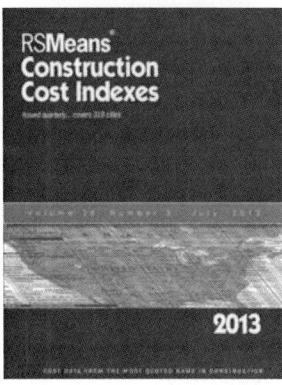

What materials and labor costs will change unexpectedly this year? By how much?

- Breakdowns for 318 major cities
- National averages for 30 key cities
- Expanded five major city indexes
- Historical construction cost indexes

$362.00 per year (subscription) | Catalog no. 50143
$90.50 individual quarters | Catalog no. 60143 A,B,C,D

RSMeans Interior Cost Data 2013

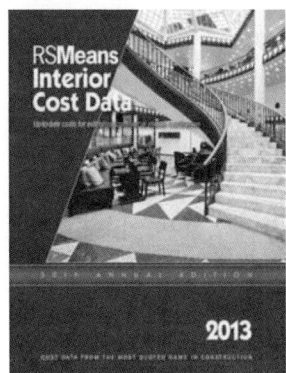

Provides you with prices and guidance needed to make accurate interior work estimates. Contains costs on materials, equipment, hardware, custom installations, furnishings, and labor costs. . . for new and remodel commercial and industrial interior construction, including updated information on office furnishings, and reference information.

$192.95 per copy | Available Nov. 2012 | Catalog no. 60093

Annual Cost Guides

Cost Guides also available in CD and at rsmeansonline.com

For more information visit the RSMeans website at **www.rsmeans.com**

Unit prices according to the latest MasterFormat!

RSMeans Heavy Construction Cost Data 2013

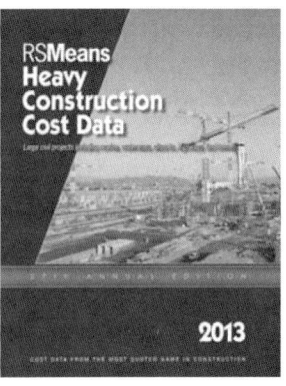

A comprehensive guide to heavy construction costs. Includes costs for highly specialized projects such as tunnels, dams, highways, airports, and waterways. Information on labor rates, equipment, and materials costs is included. Features unit price costs, systems costs, and numerous reference tables for costs and design.

$186.95 per copy | Available Dec. 2012 | Catalog no. 60163

RSMeans Plumbing Cost Data 2013

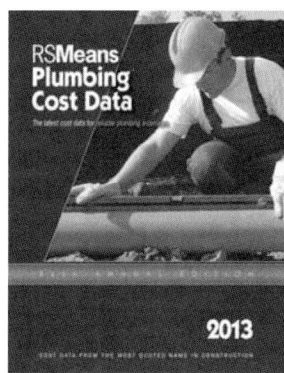

Comprehensive unit prices and assemblies for plumbing, irrigation systems, commercial and residential fire protection, point-of-use water heaters, and the latest approved materials. This publication and its companion, *RSMeans Mechanical Cost Data*, provide full-range cost estimating coverage for all the mechanical trades.

Now contains updated costs for potable water and radiant heat systems and high efficiency fixtures.

$188.95 per copy | Available Oct. 2012 | Catalog no. 60213

2013 RSMeans Seminar Schedule

Note: call for exact dates and details.

Location	Dates	Location	Dates
Las Vegas, NV	March	Las Vegas, NV	October
Washington, DC	April	Orlando, FL	November
San Francisco, CA	June	San Diego, CA	December
Washington, DC	September		
Dallas, TX	September		

1-800-334-3509, Press 1

2013 Regional Seminar Schedule

Note: call for exact dates and details.

Location	Dates	Location	Dates
Seattle, WA	January and August	Columbus, GA	June
Dallas/Ft. Worth, TX	January	El Segundo, CA	August
Austin, TX	February	Jacksonville FL	September
Oklahoma City, OK	February	St. Louis, MO	October
Anchorage, AK	March and September	Houston, TX	October
Phoenix, AZ	March	Vancouver, BC	November
San Bernardino, CA	April	Baltimore, MD	November
Kansas City, MO	April	San Antonio, TX	December
Toronto	May	Raleigh, NC	December
Denver, CO	May		
Bethesda, MD	June		

1-800-334-3509, Press 1

Professional Development

For more information visit the RSMeans website at **www.rsmeans.com**

Registration Information

Register early... Save up to $100! Register 30 days before the start date of a seminar and save $100 off your total fee. *Note: This discount can be applied only once per order. It cannot be applied to team discount registrations or any other special offer.*

How to register Register by phone today! The RSMeans toll-free number for making reservations is **1-800-334-3509, Press 1.**

Individual seminar registration fee - $935. *Means CostWorks*® training registration fee - $375. To register by mail, complete the registration form and return, with your full fee, to: RSMeans Seminars, 700 Longwater Drive, Norwell, MA 02061.

Government pricing All federal government employees save off the regular seminar price. Other promotional discounts cannot be combined with the government discount.

Team discount program for two to four seminar registrations. Call for pricing: 1-800-334-3509, Press 1

Multiple course discounts When signing up for two or more courses, call for pricing.

Refund policy Cancellations will be accepted up to ten business days prior to the seminar start. There are no refunds for cancellations received later than ten working days prior to the first day of the seminar. A $150 processing fee will be applied for all cancellations. Written notice of cancellation is required. Substitutions can be made at any time before the session starts. **No-shows are subject to the full seminar fee.**

AACE approved courses Many seminars described and offered here have been approved for 14 hours (1.4 recertification credits) of credit by the AACE International Certification Board toward meeting the continuing education requirements for recertification as a Certified Cost Engineer/Certified Cost Consultant.

AIA Continuing Education We are registered with the AIA Continuing Education System (AIA/CES) and are committed to developing quality learning activities in accordance with the CES criteria. Many seminars meet the AIA/CES criteria for Quality Level 2. AIA members may receive (14) learning units (LUs) for each two-day RSMeans course.

Daily course schedule The first day of each seminar session begins at 8:30 a.m. and ends at 4:30 p.m. The second day begins at 8:00 a.m. and ends at 4:00 p.m. Participants are urged to bring a hand-held calculator, since many actual problems will be worked out in each session.

Continental breakfast Your registration includes the cost of a continental breakfast, and a morning and afternoon refreshment break. These informal segments allow you to discuss topics of mutual interest with other seminar attendees. (You are free to make your own lunch and dinner arrangements.)

Hotel/transportation arrangements RSMeans arranges to hold a block of rooms at most host hotels. To take advantage of special group rates when making your reservation, be sure to mention that you are attending the RSMeans seminar. You are, of course, free to stay at the lodging place of your choice. (**Hotel reservations and transportation arrangements should be made directly by seminar attendees.**)

Important Class sizes are limited, so please register as soon as possible.

Note: Pricing subject to change.

Registration Form

ADDS-1000

Call **1-800-334-3509, Press 1** to register or FAX this form **1-800-632-6732**. Visit our website: **www.rsmeans.com**

Please register the following people for the RSMeans construction seminars as shown here. We understand that we must make our own hotel reservations if overnight stays are necessary.

☐ Full payment of $_____ enclosed.

☐ Bill me.

Please print name of registrant(s).
(To appear on certificate of completion)

_____
_____
_____

P.O. #: _____
GOVERNMENT AGENCIES MUST SUPPLY PURCHASE ORDER NUMBER OR TRAINING FORM.

Firm name _____

Address _____

City/State/Zip _____

Telephone no. _____ Fax no. _____

E-mail address _____

Charge registration(s) to: ☐ MasterCard ☐ VISA ☐ American Express

Account no. _____ Exp. date _____

Cardholder's signature _____

Seminar name _____

Seminar City _____

Please mail check to: RSMeans Seminars, 700 Longwater Drive, Norwell, MA 02061 USA

Professional Development

For more information visit the RSMeans website at **www.rsmeans.com**

RSMeans Online™ Training

Construction estimating is vital to the decision-making process at each state of every project. Means CostWorks.com works the way you do. It's systematic, flexible and intuitive. In this one day class you will see how you can estimate any phase of any project faster and better.

Some of what you'll learn:
- Customizing Means CostWorks.com
- Making the most of RSMeans "Circle Reference" numbers
- How to integrate your cost data
- Generate reports, exporting estimates to MS Excel, sharing, collaborating and more

Also available: RSMeans Online™ training webinar

Facilities Construction Estimating

In this *two-day* course, professionals working in facilities management can get help with their daily challenges to establish budgets for all phases of a project.

Some of what you'll learn:
- Determining the full scope of a project
- Identifying the scope of risks and opportunities
- Creative solutions to estimating issues
- Organizing estimates for presentation and discussion
- Special techniques for repair/remodel and maintenance projects
- Negotiating project change orders

Who should attend: facility managers, engineers, contractors, facility tradespeople, planners, and project managers.

Scheduling with Microsoft Project

Two days of hands-on training gives you a look at the basics to putting it all together with MS Project to create a complete project schedule. Learn to work better and faster, manage changes and updates, enhance tracking, generate actionable reports, and boost control of your time and budget.

Some of what you'll learn:
- Defining task/activity relationships: link tasks, establish predecessor/successor relationships and create logs
- Baseline scheduling: the essential what/when
- Using MS Project to determine the critical path
- Integrate RSMeans data with MS Project
- The dollar-loaded schedule: enhance the reliability of this "must-know" technique for scheduling public projects.

Who should attend: contractors' project-management teams, architects and engineers, project owners and their representatives, and those interested in improving their project planning and scheduling and management skills.

Maintenance & Repair Estimating for Facilities

This *two-day* course teaches attendees how to plan, budget, and estimate the cost of ongoing and preventive maintenance and repair for existing buildings and grounds.

Some of what you'll learn:
- The most financially favorable maintenance, repair, and replacement scheduling and estimating
- Auditing and value engineering facilities
- Preventive planning and facilities upgrading
- Determining both in-house and contract-out service costs
- Annual, asset-protecting M&R plan

Who should attend: facility managers, maintenance supervisors, buildings and grounds superintendents, plant managers, planners, estimators, and others involved in facilities planning and budgeting.

Practical Project Management for Construction Professionals

In this *two-day* course, acquire the essential knowledge and develop the skills to effectively and efficiently execute the day-to-day responsibilities of the construction project manager.

Covers:
- General conditions of the construction contract
- Contract modifications: change orders and construction change directives
- Negotiations with subcontractors and vendors
- Effective writing: notification and communications
- Dispute resolution: claims and liens

Who should attend: architects, engineers, owner's representatives, project managers.

Mechanical & Electrical Estimating

This *two-day* course teaches attendees how to prepare more accurate and complete mechanical/electrical estimates, avoiding the pitfalls of omission and double-counting, while understanding the composition and rationale within the RSMeans mechanical/electrical database.

Some of what you'll learn:
- The unique way mechanical and electrical systems are interrelated
- M&E estimates—conceptual, planning, budgeting, and bidding stages
- Order of magnitude, square foot, assemblies, and unit price estimating
- Comparative cost analysis of equipment and design alternatives

Who should attend: architects, engineers, facilities managers, mechanical and electrical contractors, and others who need a highly reliable method for developing, understanding, and evaluating mechanical and electrical contracts.

Professional Development

For more information visit the RSMeans website at **www.rsmeans.com**

Unit Price Estimating

This interactive *two-day* seminar teaches attendees how to interpret project information and process it into final, detailed estimates with the greatest accuracy level.

The most important credential an estimator can take to the job is the ability to visualize construction and estimate accurately.

Some of what you'll learn:
- Interpreting the design in terms of cost
- The most detailed, time-tested methodology for accurate pricing
- Key cost drivers—material, labor, equipment, staging, and subcontracts
- Understanding direct and indirect costs for accurate job cost accounting and change order management

Who should attend: corporate and government estimators and purchasers, architects, engineers, and others who need to produce accurate project estimates.

RSMeans CostWorks® CD Training

This one-day course helps users become more familiar with the functionality of *RSMeans CostWork*s program. Each menu, icon, screen, and function found in the program is explained in depth. Time is devoted to hands-on estimating exercises.

Some of what you'll learn:
- Searching the database using all navigation methods
- Exporting RSMeans data to your preferred spreadsheet format
- Viewing crews, assembly components, and much more
- Automatically regionalizing the database

This training session requires you to bring a laptop computer to class.

When you register for this course you will receive an outline for your laptop requirements.

Also offering web training for CostWorks CD!

Facilities Est. Using RSMeans CostWorks® CD

Combines hands-on skill building with best estimating practices and real-life problems. Brings you up-to-date with key concepts, and provides tips, pointers, and guidelines to save time and avoid cost oversights and errors.

Some of what you'll learn:
- Estimating process concepts
- Customizing and adapting RSMeans cost data
- Establishing scope of work to account for all known variables
- Budget estimating: when, why, and how
- Site visits: what to look for—what you can't afford to overlook
- How to estimate repair and remodeling variables

This training session requires you to bring a laptop computer to class.

Who should attend: facility managers, architects, engineers, contractors, facility tradespeople, planners, project managers and anyone involved with JOC, SABRE, or IDIQ.

Conceptual Estimating Using RSMeans CostWorks® CD

This *two day* class uses the leading industry data and a powerful software package to develop highly accurate conceptual estimates for your construction projects. All attendees must bring a laptop computer loaded with the current year *Square Foot Model Costs* and the *Assemblies Cost Data* CostWorks titles.

Some of what you'll learn:
- Introduction to conceptual estimating
- Types of conceptual estimates
- Helpful hints
- Order of magnitude estimating
- Square foot estimating
- Assemblies estimating

Who should attend: architects, engineers, contractors, construction estimators, owner's representatives, and anyone looking for an electronic method for performing square foot estimating.

Assessing Scope of Work for Facility Construction Estimating

This *two-day* practical training program addresses the vital importance of understanding the SCOPE of projects in order to produce accurate cost estimates for facility repair and remodeling.

Some of what you'll learn:
- Discussions of site visits, plans/specs, record drawings of facilities, and site-specific lists
- Review of CSI divisions, including means, methods, materials, and the challenges of scoping each topic
- Exercises in SCOPE identification and SCOPE writing for accurate estimating of projects
- Hands-on exercises that require SCOPE, take-off, and pricing

Who should attend: corporate and government estimators, planners, facility managers, and others who need to produce accurate project estimates.

Unit Price Est. Using RSMeans CostWorks® CD

Step-by-step instruction and practice problems to identify and track key cost drivers—material, labor, equipment, staging, and subcontractors—for each specific task. Learn the most detailed, time-tested methodology for accurately "pricing" these variables, their impact on each other and on total cost.

Some of what you'll learn:
- Unit price cost estimating
- Order of magnitude, square foot, and assemblies estimating
- Quantity takeoff
- Direct and indirect construction costs
- Development of contractor's bill rates
- How to use *RSMeans Building Construction Cost Data*

This training session requires you to bring a laptop computer to class.

Who should attend: architects, engineers, corporate and government estimators, facility managers, and government procurement staff.

New Titles - Reference Books

For more information visit the RSMeans website at **www.rsmeans.com**

Complete Book of Framing, 2nd Edition

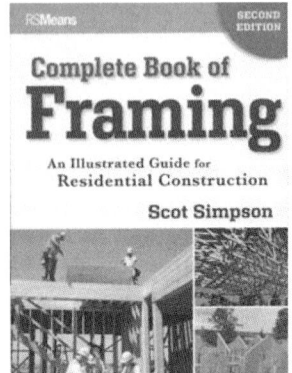

by Scot Simpson

This updated, easy-to-learn guide to rough carpentry and framing is written by an expert with more than thirty years of framing experience. Starting with the basics, this book begins with types of lumber, nails, and what tools are needed, followed by detailed, fully illustrated steps for framing each building element. Framer-Friendly Tips throughout the book show how to get a task done right—and more easily.

$29.95 per copy | 352 pages, softcover | Catalog no. 67353A

Estimating Building Costs, 2nd Edition

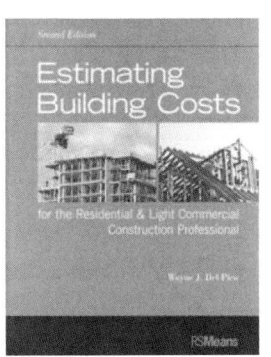

For the Residential & Light Commercial Construction Professional
by Wayne J. DelPico

This book guides readers through the entire estimating process, explaining in detail how to put together a reliable estimate that can be used not only for budgeting, but also for developing a schedule, managing a project, dealing with contingencies, and ultimately making a profit.

Completely revised and updated to reflect the new CSI MasterFormat 2010 system, this practical guide describes estimating techniques for each building system and how to apply them according to the latest industry standards.

$65.00 per copy | 398 pages, softcover | Catalog no. 67343A

Risk Management for Design and Construction

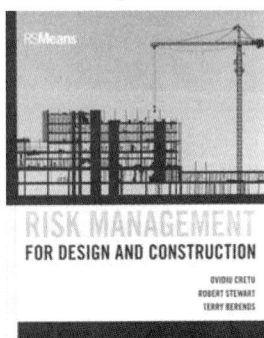

This book introduces risk as a central pillar of project management and shows how a project manager can be prepared for dealing with uncertainty. Written by experts in the field, *Risk Management for Design and Construction* uses clear, straightforward terminology to demystify the concepts of project uncertainty and risk.

Highlights include:

- Integrated cost and schedule risk analysis
- An introduction to a ready-to-use system of analyzing a project's risks and tools to proactively manage risks
- A methodology that was developed and used by the Washington State Department of Transportation
- Case studies and examples on the proper application of principles
- Information about combining value analysis with risk analysis

$125.00 per copy | Over 250 pages, softcover | Catalog no. 67359

How to Estimate with Means Data & CostWorks, 4th Edition

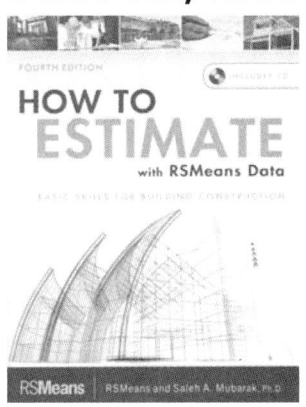

by RSMeans and Saleh A. Mubarak, Ph.D.

This step-by-step guide takes you through all the major construction items with extensive coverage of site work, concrete and masonry, wood and metal framing, doors and windows, and other divisions. The only construction cost estimating handbook that uses the most popular source of construction data, RSMeans, this indispensible guide features access to the instructional version of CostWorks in electronic form, enabling you to practice techniques to solve real-world estimating problems.

$70.00 per copy | 292 pages, softcover | Includes CostWorks CD
Catalog no. 67324C

How Your House Works, 2nd Edition

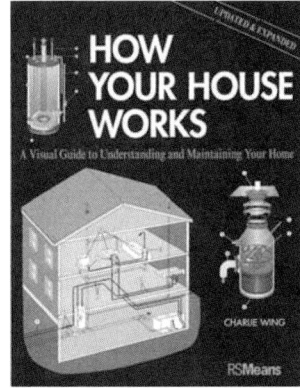

by Charlie Wing

Knowledge of your home's systems helps you control repair and construction costs and makes sure the correct elements are being installed or replaced. This book uncovers the mysteries behind just about every major appliance and building element in your house. See-through, cross-section drawings in full color show you exactly how these things should be put together and how they function, including what to check if they don't work. It just might save you having to call in a professional.

$22.95 per copy | 192 pages, softcover | Catalog no. 67351A

RSMeans Cost Data, Student Edition

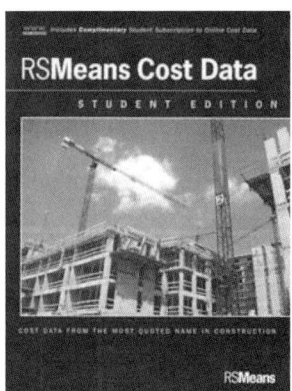

This book provides a thorough introduction to cost estimating in a self-contained print and online package. With clear explanations and a hands-on, example-driven approach, it is the ideal reference for students and new professionals.

Features include:

- Commercial and residential construction cost data in print and online formats
- Complete how-to guidance on the essentials of cost estimating
- A supplemental website with plans, problem sets, and a full sample estimate

$99.00 per copy | 512 pages, softcover | Catalog no. 67363

Reference Books

For more information visit the RSMeans website at **www.rsmeans.com**

Unit prices according to the latest MasterFormat!

Value Engineering: Practical Applications

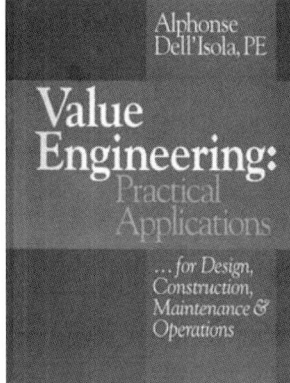

For Design, Construction, Maintenance & Operations
by Alphonse Dell'Isola, PE

A tool for immediate application—for engineers, architects, facility managers, owners, and contractors. Includes making the case for VE—the management briefing; integrating VE into planning, budgeting, and design; conducting life cycle costing; using VE methodology in design review and consultant selection; case studies; VE workbook; and a life cycle costing program on disk.

$79.95 per copy | Over 450 pages, illustrated, softcover | Catalog no. 67319A

Facilities Operations & Engineering Reference

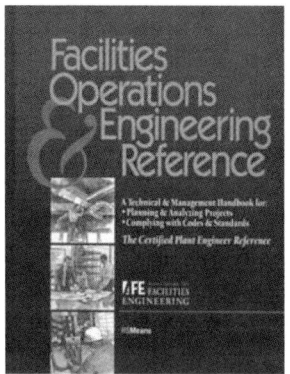

by the Association for Facilities Engineering and RSMeans

An all-in-one technical reference for planning and managing facility projects and solving day-to-day operations problems. Selected as the official certified plant engineer reference, this handbook covers financial analysis, maintenance, HVAC and energy efficiency, and more.

$109.95 per copy | Over 700 pages, illustrated, hardcover | Catalog no. 67318

The Building Professional's Guide to Contract Documents

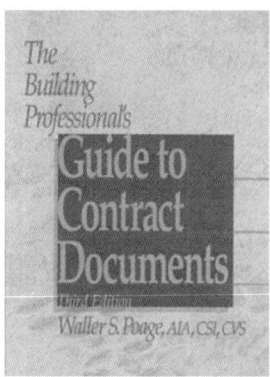

3rd Edition
by Waller S. Poage, AIA, CSI, CVS

A comprehensive reference for owners, design professionals, contractors, and students

- Structure your documents for maximum efficiency.
- Effectively communicate construction requirements.
- Understand the roles and responsibilities of construction professionals.
- Improve methods of project delivery.

$70.00 per copy | 400 pages Diagrams and construction forms, hardcover
Catalog no. 67261A

Green Home Improvement

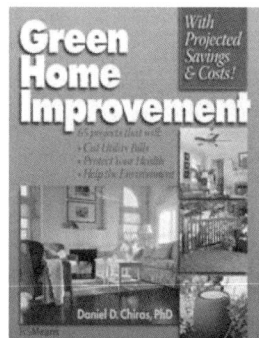

by Daniel D. Chiras, PhD

With energy costs rising and environmental awareness increasing, people are looking to make their homes greener. This book, with 65 projects and actual costs and projected savings help homeowners prioritize their green improvements.

Projects range from simple water savers that cost only a few dollars, to bigger-ticket items such as HVAC systems. With color photos and cost estimates, each project compares options and describes the work involved, the benefits, and the savings.

$34.95 | 320 pages, illustrated, softcover | Catalog no. 67355

Cost Planning & Estimating for Facilities Maintenance

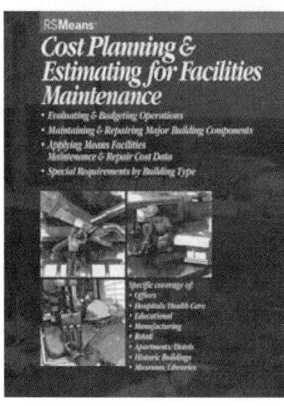

In this unique book, a team of facilities management authorities shares their expertise on:

- Evaluating and budgeting maintenance operations
- Maintaining and repairing key building components
- Applying *RSMeans Facilities Maintenance & Repair Cost Data* to your estimating

Covers special maintenance requirements of the ten major building types

$89.95 per copy | Over 475 pages, hardcover | Catalog no. 67314

Life Cycle Costing for Facilities

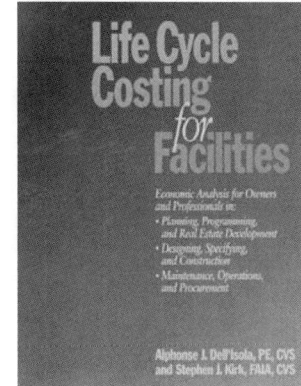

by Alphonse Dell'Isola and Dr. Steven Kirk

Guidance for achieving higher quality design and construction projects at lower costs! Cost-cutting efforts often sacrifice quality to yield the cheapest product. Life cycle costing enables building designers and owners to achieve both. The authors of this book show how LCC can work for a variety of projects — from roads to HVAC upgrades to different types of buildings.

$99.95 per copy | 396 pages, hardcover | Catalog no. 67341

Reference Books

For more information visit the RSMeans website at **www.rsmeans.com**

Unit prices according to the latest MasterFormat!

Interior Home Improvement Costs

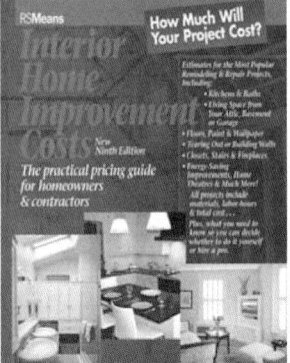

9th Edition

Updated estimates for the most popular remodeling and repair projects—from small, do-it-yourself jobs to major renovations and new construction. Includes: kitchens & baths; new living space from your attic, basement, or garage; new floors, paint, and wallpaper; tearing out or building new walls; closets, stairs, and fireplaces; new energy-saving improvements, home theaters, and more!

$24.95 per copy | 250 pages, illustrated, softcover | Catalog no. 67308E

Exterior Home Improvement Costs

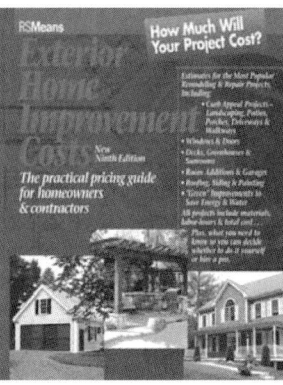

9th Edition

Updated estimates for the most popular remodeling and repair projects—from small, do-it-yourself jobs, to major renovations and new construction. Includes: curb appeal projects—landscaping, patios, porches, driveways, and walkways; new windows and doors; decks, greenhouses, and sunrooms; room additions and garages; roofing, siding, and painting; "green" improvements to save energy & water.

$24.95 per copy | Over 275 pages, illustrated, softcover | Catalog no. 67309E

Builder's Essentials: Plan Reading & Material Takeoff

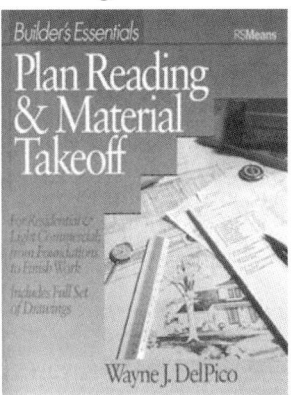

For Residential and Light Commercial Construction
by Wayne J. DelPico

A valuable tool for understanding plans and specs, and accurately calculating material quantities. Step-by-step instructions and takeoff procedures based on a full set of working drawings.

$35.95 per copy | Over 420 pages, softcover | Catalog no. 67307

Means Unit Price Estimating Methods

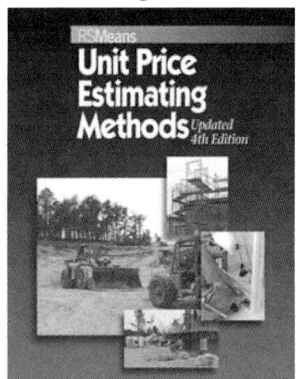

New 4th Edition

This new edition includes up-to-date cost data and estimating examples, updated to reflect changes to the CSI numbering system and new features of RSMeans cost data. It describes the most productive, universally accepted ways to estimate, and uses checklists and forms to illustrate shortcuts and timesavers. A model estimate demonstrates procedures. A new chapter explores computer estimating alternatives.

$65.00 per copy | Over 350 pages, illustrated, softcover | Catalog no. 67303B

Concrete Repair and Maintenance Illustrated

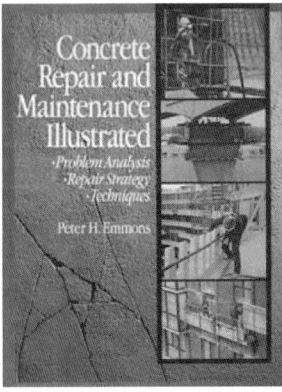

by Peter Emmons

Hundreds of illustrations show users how to analyze, repair, clean, and maintain concrete structures for optimal performance and cost effectiveness. From parking garages to roads and bridges to structural concrete, this comprehensive book describes the causes, effects, and remedies for concrete wear and failure. Invaluable for planning jobs, selecting materials, and training employees, this book is a must-have for concrete specialists, general contractors, facility managers, civil and structural engineers, and architects.

$69.95 per copy | 300 pages, illustrated, softcover | Catalog no. 67146

Total Productive Facilities Management

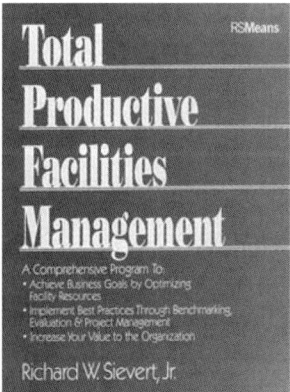

by Richard W. Sievert, Jr.

Today, facilities are viewed as strategic resources... elevating the facility manager to the role of asset manager supporting the organization's overall business goals. Now, Richard Sievert Jr., in this well-articulated guidebook, sets forth a new operational standard for the facility manager's emerging role... a comprehensive program for managing facilities as a true profit center.

$79.95 per copy | 275 pages, softcover | Catalog no. 67321

Reference Books

For more information visit the RSMeans website at **www.rsmeans.com**

Unit prices according to the latest MasterFormat!

Means Illustrated Construction Dictionary

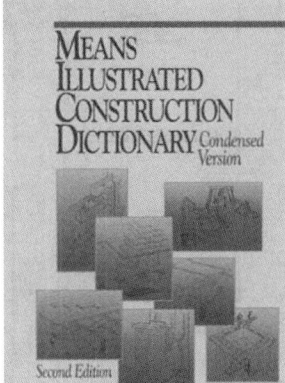

Condensed, 2nd Edition

The best portable dictionary for office or field use—an essential tool for contractors, architects, insurance and real estate personnel, facility managers, homeowners, and anyone who needs quick, clear definitions for construction terms. The second edition has been further enhanced with updates and hundreds of new terms and illustrations . . . in keeping with the most recent developments in the construction industry.

Now with a quick-reference Spanish section. Includes tools and equipment, materials, tasks, and more!

$59.95 per copy | Over 500 pages, softcover | Catalog no. 67282A

Facilities Planning & Relocation

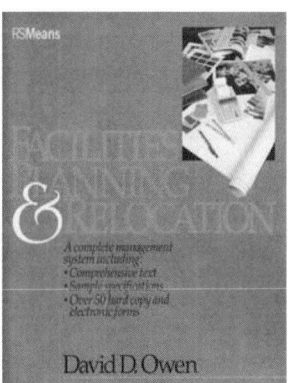

by David D. Owen

A complete system for planning space needs and managing relocations. Includes step-by-step manual, over 50 forms, and extensive reference section on materials and furnishings.

New lower price and user-friendly format.

$89.95 per copy | Over 450 pages, softcover | Catalog no. 67301

Electrical Estimating Methods

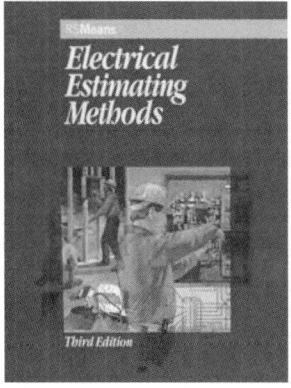

3rd Edition

Expanded edition includes sample estimates and cost information in keeping with the latest version of the CSI MasterFormat and UNIFORMAT II. Complete coverage of fiber optic and uninterruptible power supply electrical systems, broken down by components, and explained in detail. Includes a new chapter on computerized estimating methods. A practical companion to *RSMeans Electrical Cost Data.*

$64.95 per copy | Over 325 pages, hardcover | Catalog no. 67230B

Square Foot & UNIFORMAT Assemblies Estimating Methods

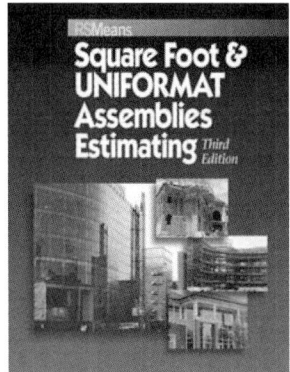

3rd Edition

Develop realistic square foot and assemblies costs for budgeting and construction funding. The new edition features updated guidance on square foot and assemblies estimating using UNIFORMAT II. An essential reference for anyone who performs conceptual estimates.

$69.95 per copy | Over 300 pages, illustrated, softcover | Catalog no. 67145B

Mechanical Estimating Methods

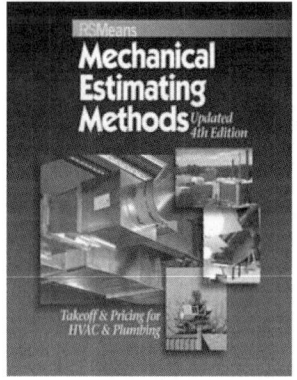

4th Edition

Completely updated, this guide assists you in making a review of plans, specs, and bid packages, with suggestions for takeoff procedures, listings, substitutions, and pre-bid scheduling for all components of HVAC. Includes suggestions for budgeting labor and equipment usage. Compares materials and construction methods to allow you to select the best option.

$64.95 per copy | Over 350 pages, illustrated, softcover | Catalog no. 67294B

Residential & Light Commercial Construction Standards

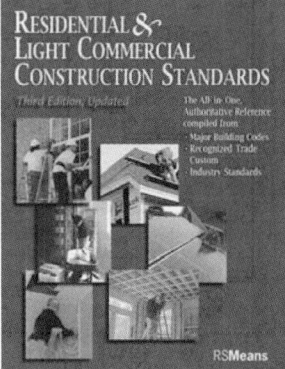

3rd Edition,
Updated by RSMeans and contributing authors

This book provides authoritative requirements and recommendations compiled from leading professional associations, industry publications, and building code organizations. This all-in-one reference helps establish a standard for workmanship, quickly resolve disputes, and avoid defect claims. Updated third edition includes new coverage of green building, seismic, hurricane, and mold-resistant construction.

$59.95 | Over 550 pages, illustrated, softcover | Catalog no. 67322B

Reference Books

For more information visit the RSMeans website at **www.rsmeans.com**

Unit prices according to the latest MasterFormat!

Construction Business Management

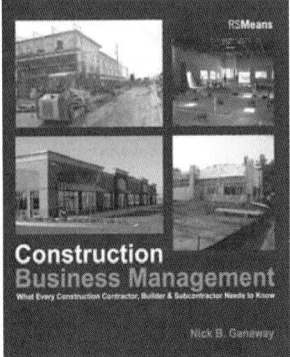

by Nick Ganaway

Only 43% of construction firms stay in business after four years. Make sure your company thrives with valuable guidance from a pro with 25 years of success as a commercial contractor. Find out what it takes to build all aspects of a business that is profitable, enjoyable, and enduring. With a bonus chapter on retail construction.

$49.95 per copy | 200 pages, softcover | Catalog no. 67352

Project Scheduling & Management for Construction

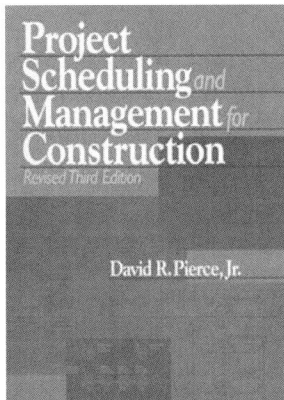

3rd Edition
by David R. Pierce, Jr.

A comprehensive yet easy-to-follow guide to construction project scheduling and control—from vital project management principles through the latest scheduling, tracking, and controlling techniques. The author is a leading authority on scheduling, with years of field and teaching experience at leading academic institutions. Spend a few hours with this book and come away with a solid understanding of this essential management topic.

$64.95 per copy | Over 300 pages, illustrated, hardcover | Catalog no. 67247B

The Practice of Cost Segregation Analysis

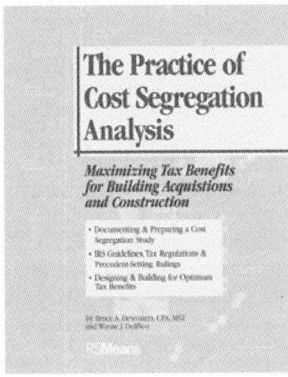

by Bruce A. Desrosiers and Wayne J. DelPico

This expert guide walks you through the practice of cost segregation analysis, which enables property owners to defer taxes and benefit from "accelerated cost recovery" through depreciation deductions on assets that are properly identified and classified.

With a glossary of terms, sample cost segregation estimates for various building types, key information resources, and updates via a dedicated website, this book is a critical resource for anyone involved in cost segregation analysis.

$99.95 per copy | Over 225 pages | Catalog no. 67345

Preventive Maintenance for Multi-Family Housing

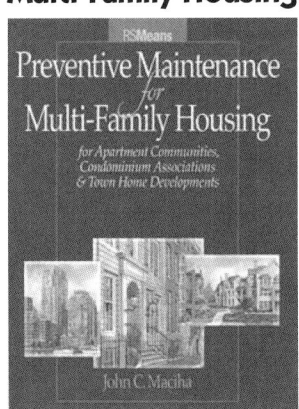

by John C. Maciha

Prepared by one of the nation's leading experts on multi-family housing.

This complete PM system for apartment and condominium communities features expert guidance, checklists for buildings and grounds maintenance tasks and their frequencies, a reusable wall chart to track maintenance, and a dedicated website featuring customizable electronic forms. A must-have for anyone involved with multi-family housing maintenance and upkeep.

$89.95 per copy | 225 pages | Catalog no. 67346

Green Building: Project Planning & Cost Estimating

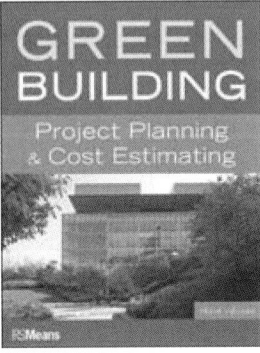

3rd Edition

Since the widely read first edition of this book, green building has gone from a growing trend to a major force in design and construction.

This new edition has been updated with the latest in green building technologies, design concepts, standards, and costs. Full-color with all new case studies—plus a new chapter on commercial real estate.

$99.95 per copy | Over 450 pages, softcover | Catalog no. 67338B

Job Order Contracting Expediting Construction Project Delivery

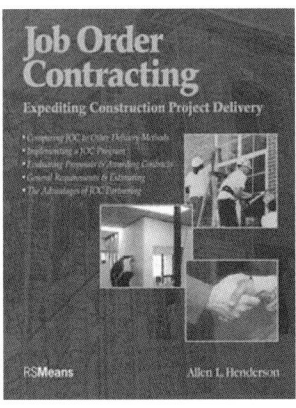

by Allen Henderson

Expert guidance to help you implement JOC—fast becoming the preferred project delivery method for repair and renovation, minor new construction, and maintenance projects in the public sector and in many states and municipalities. The author, a leading JOC expert and practitioner, shows how to:

• Establish a JOC program

• Evaluate proposals and award contracts

• Handle general requirements and estimating

• Partner for maximum benefits

$89.95 per copy | 192 pages, illustrated, hardcover | Catalog no. 67348

Reference Books

For more information visit the RSMeans website at **www.rsmeans.com**

Unit prices according to the latest MasterFormat!

Construction Supervision

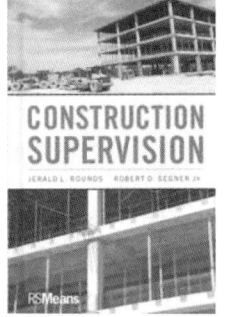

This brand new title inspires supervisory excellence with proven tactics and techniques applied by thousands of construction supervisors over the past decade. Recognizing the unique and critical role the supervisor plays in project success, the book's leadership guidelines carve out a practical blueprint for motivating work performance and increasing productivity through effective communication.

Features:

- A unique focus on field supervision and crew management
- Coverage of supervision from the foreman to the superintendent level
- An overview of technical skills whose mastery will build confidence and success for the supervisor
- A detailed view of "soft" management and communication skills

$90.00 per copy | Over 450 pages, softcover | Catalog no. 67358

Building & Renovating Schools

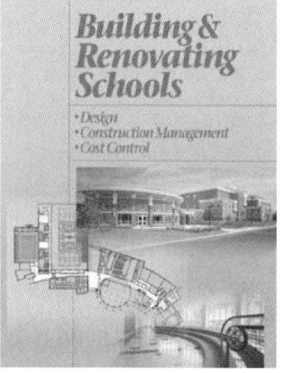

This all-inclusive guide covers every step of the school construction process—from initial planning, needs assessment, and design, right through moving into the new facility. A must-have resource for anyone concerned with new school construction or renovation. With square foot cost models for elementary, middle, and high school facilities, and real-life case studies of recently completed school projects.

The contributors to this book—architects, construction project managers, contractors, and estimators who specialize in school construction—provide start-to-finish, expert guidance on the process.

$99.95 per copy | Over 425 pages, hardcover | Catalog no. 67342

The Homeowner's Guide to Mold

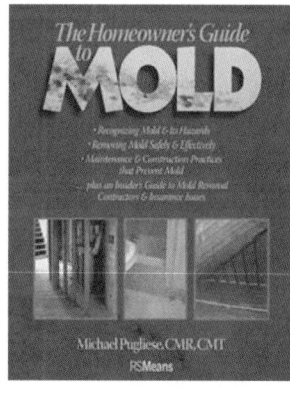

By Michael Pugliese

Expert guidance to protect your health and your home.

Mold, whether caused by leaks, humidity or flooding, is a real health and financial issue—for homeowners and contractors.

This full-color book explains:

- Construction and maintenance practices to prevent mold
- How to inspect for and remove mold
- Mold remediation procedures and costs
- What to do after a flood
- How to deal with insurance companies

$21.95 per copy | 144 pages, softcover | Catalog no. 67344

Universal Design Ideas for Style, Comfort & Safety

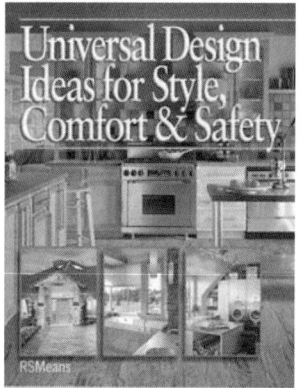

by RSMeans and Lexicon Consulting, Inc.

Incorporating universal design when building or remodeling helps people of any age and physical ability more fully and safely enjoy their living spaces. This book shows how universal design can be artfully blended into the most attractive homes. It discusses specialized products like adjustable countertops and chair lifts, as well as simple ways to enhance a home's safety and comfort. With color photos and expert guidance, every area of the home is covered. Includes budget estimates that give an idea how much projects will cost.

$21.95 | 160 pages, illustrated, softcover | Catalog no. 67354

Landscape Estimating Methods

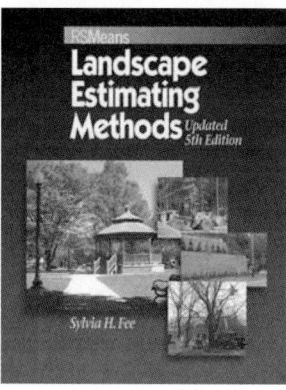

5th Edition

Answers questions about preparing competitive landscape construction estimates, with up-to-date cost estimates and the new MasterFormat classification system. Expanded and revised to address the latest materials and methods, including new coverage on approaches to green building. Includes:

- Step-by-step explanation of the estimating process
- Sample forms and worksheets that save time and prevent errors

$69.95 per copy | Over 350 pages, softcover | Catalog no. 67295C

Plumbing Estimating Methods

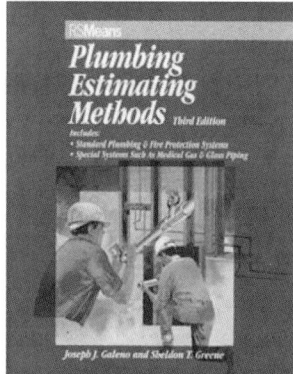

3rd Edition
by Joseph Galeno and Sheldon Greene

Updated and revised! This practical guide walks you through a plumbing estimate, from basic materials and installation methods through change order analysis. *Plumbing Estimating Methods* covers residential, commercial, industrial, and medical systems, and features sample takeoff and estimate forms and detailed illustrations of systems and components.

$59.95 per copy | 330+ pages, softcover | Catalog no. 67283B

Reference Books

For more information visit the RSMeans website at **www.rsmeans.com**

Unit prices according to the latest MasterFormat!

Understanding & Negotiating Construction Contracts

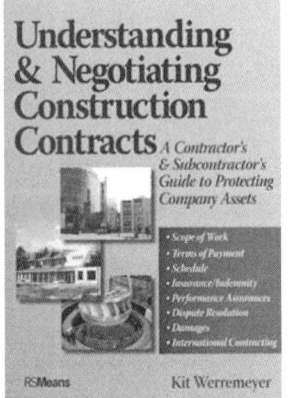

by Kit Werremeyer

Take advantage of the author's 30 years' experience in small-to-large (including international) construction projects. Learn how to identify, understand, and evaluate high risk terms and conditions typically found in all construction contracts—then negotiate to lower or eliminate the risk, improve terms of payment, and reduce exposure to claims and disputes. The author avoids "legalese" and gives real-life examples from actual projects.

$74.95 per copy | 300 pages, softcover | Catalog no. 67350

Means Illustrated Construction Dictionary

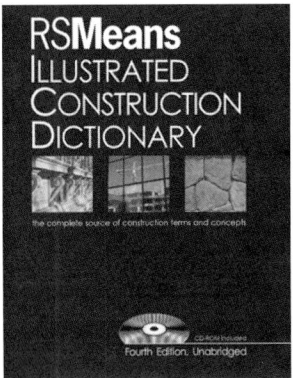

Unabridged 4th Edition, with CD-ROM

Long regarded as the industry's finest, *Means Illustrated Construction Dictionary* is now even better. With nearly 20,000 terms and more than 1,400 illustrations and photos, it is the clear choice for the most comprehensive and current information. The companion CD-ROM that comes with this new edition adds extra features such as: larger graphics and expanded definitions.

$99.95 per copy | Over 790 pages, illust., hardcover | Catalog no. 67292B

RSMeans Estimating Handbook

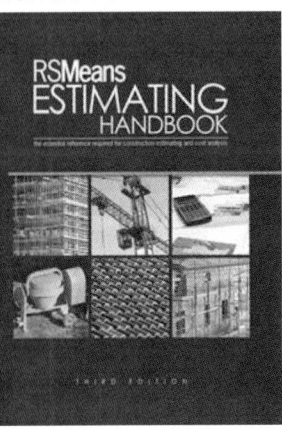

3rd Edition

Widely used in the industry for tasks ranging from routine estimates to special cost analysis projects, this handbook has been completely updated and reorganized with new and expanded technical information.

RSMeans Estimating Handbook will help construction professionals:

- Evaluate architectural plans and specifications
- Prepare accurate quantity takeoffs
- Compare design alternatives and costs
- Perform value engineering
- Double-check estimates and quotes
- Estimate change orders

$99.95 per copy | Over 900 pages, hardcover | Catalog No. 67276B

Spanish/English Construction Dictionary

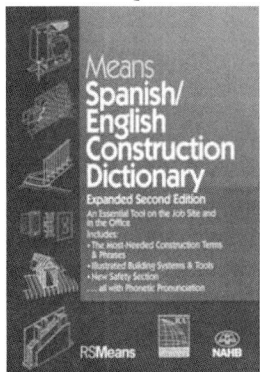

**2nd Edition
by RSMeans and the International Code Council**

This expanded edition features thousands of the most common words and useful phrases in the construction industry with easy-to-follow pronunciations in both Spanish and English. Over 800 new terms, phrases, and illustrations have been added. It also features a new stand-alone "Safety & Emergencies" section, with colored pages for quick access.

Unique to this dictionary are the systems illustrations showing the relationship of components in the most common building systems for all major trades.

$23.95 per copy | Over 400 pages | Catalog no. 67327A

The Gypsum Construction Handbook

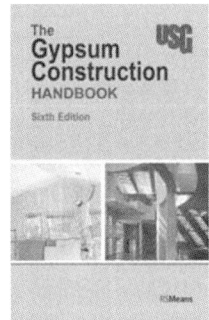

by USG

An invaluable reference of construction procedures for gypsum drywall, cement board, veneer plaster, and conventional plaster. This new edition includes the newest product developments, installation methods, fire- and sound-rated construction information, illustrated framing-to-finish application instructions, estimating and planning information, and more. Great for architects and engineers; contractors, builders, and dealers; apprentices and training programs; building inspectors and code officials; and anyone interested in gypsum construction.

Features information on tools and safety practices, a glossary of construction terms, and a list of important agencies and associations. Also available in Spanish.

$29.95 | Over 575 pages, illustrated, softcover | Catalog no. 67357A

2013 Order Form

ORDER TOLL FREE 1-800-334-3509
OR FAX 1-800-632-6732

For more information visit the RSMeans website at **www.rsmeans.com**

Qty.	Book no.	COST ESTIMATING BOOKS	Unit Price	Total
	60063	Assemblies Cost Data 2013	$305.95	
	60013	Building Construction Cost Data 2013	186.95	
	60113	Concrete & Masonry Cost Data 2013	171.95	
	50143	Construction Cost Indexes 2013 (subscription)	362.00	
	60143A	Construction Cost Index–January 2013	90.50	
	60143B	Construction Cost Index–April 2013	90.50	
	60143C	Construction Cost Index–July 2013	90.50	
	60143D	Construction Cost Index–October 2013	90.50	
	60343	Contr. Pricing Guide: Resid. R & R Costs 2013	39.95	
	60233	Electrical Change Order Cost Data 2013	183.95	
	60033	Electrical Cost Data 2013	188.95	
	60203	Facilities Construction Cost Data 2013	476.95	
	60303	Facilities Maintenance & Repair Cost Data 2013	417.95	
	60553	Green Building Cost Data 2013	155.95	
	60163	Heavy Construction Cost Data 2013	186.95	
	60093	Interior Cost Data 2013	192.95	
	60123	Labor Rates for the Const. Industry 2013	415.95	
	60183	Light Commercial Cost Data 2013	139.95	
	60023	Mechanical Cost Data 2013	183.95	
	60153	Open Shop Building Const. Cost Data 2013	160.95	
	60213	Plumbing Cost Data 2013	188.95	
	60043	Commercial Renovation Cost Data 2013	152.95	
	60173	Residential Cost Data 2013	134.95	
	60283	Site Work & Landscape Cost Data 2013	181.95	
	60053	Square Foot Costs 2013	197.95	
	62012	Yardsticks for Costing (2012)	176.95	
	62013	Yardsticks for Costing (2013)	185.95	
		REFERENCE BOOKS		
	67329	Bldrs Essentials: Best Bus. Practices for Bldrs	29.95	
	67298A	Bldrs Essentials: Framing/Carpentry 2nd Ed.	24.95	
	67298AS	Bldrs Essentials: Framing/Carpentry Spanish	24.95	
	67307	Bldrs Essentials: Plan Reading & Takeoff	35.95	
	67342	Building & Renovating Schools	99.95	
	67261A	The Building Prof. Guide to Contract Documents	70.00	
	67339	Building Security: Strategies & Costs	89.95	
	67353A	Complete Book of Framing, 2nd Ed.	29.95	
	67146	Concrete Repair & Maintenance Illustrated	69.95	
	67352	Construction Business Management	49.95	
	67358	Construction Supervision	90.00	
	67363	Cost Data Student Edition	99.00	
	67314	Cost Planning & Est. for Facil. Maint.	89.95	
	67328A	Designing & Building with the IBC, 2nd Ed.	99.95	
	67230B	Electrical Estimating Methods, 3rd Ed.	64.95	
	67343A	Est. Bldg. Costs for Resi. & Lt. Comm., 2nd Ed.	65.00	
	67276B	Estimating Handbook, 3rd Ed.	99.95	
	67318	Facilities Operations & Engineering Reference	109.95	
	67301	Facilities Planning & Relocation	89.95	
	67338B	Green Building: Proj. Planning & Cost Est., 3rd Ed.	99.95	
	67355	Green Home Improvement	34.95	
	67357A	The Gypsum Construction Handbook	29.95	
	67357S	The Gypsum Construction Handbook (Spanish)	29.95	

Qty.	Book no.	REFERENCE BOOKS (Cont.)	Unit Price	Total
	67148	Heavy Construction Handbook	99.95	
	67308E	Home Improvement Costs–Int. Projects, 9th Ed.	24.95	
	67309E	Home Improvement Costs–Ext. Projects, 9th Ed.	24.95	
	67344	Homeowner's Guide to Mold	21.95	
	67324C	How to Est.w/Means Data & CostWorks, 4th Ed.	70.00	
	67351A	How Your House Works, 2nd Ed.	22.95	
	67282A	Illustrated Const. Dictionary, Condensed, 2nd Ed.	59.95	
	67292B	Illustrated Const. Dictionary, w/CD-ROM, 4th Ed.	99.95	
	67348	Job Order Contracting	89.95	
	67295C	Landscape Estimating Methods, 5th Ed.	69.95	
	67341	Life Cycle Costing for Facilities	99.95	
	67294B	Mechanical Estimating Methods, 4th Ed.	64.95	
	67283B	Plumbing Estimating Methods, 3rd Ed.	59.95	
	67345	Practice of Cost Segregation Analysis	99.95	
	67346	Preventive Maint. for Multi-Family Housing	89.95	
	67326	Preventive Maint. Guidelines for School Facil.	149.95	
	67247B	Project Scheduling & Manag. for Constr., 3rd Ed.	64.95	
	67322B	Resi. & Light Commercial Const. Stds., 3rd Ed.	59.95	
	67359	Risk Management for Design & Construction	125.00	
	67327A	Spanish/English Construction Dictionary, 2nd Ed.	23.95	
	67145B	Sq. Ft. & Assem. Estimating Methods, 3rd Ed.	69.95	
	67321	Total Productive Facilities Management	79.95	
	67350	Understanding and Negotiating Const. Contracts	74.95	
	67303B	Unit Price Estimating Methods, 4th Ed.	65.00	
	67354	Universal Design	21.95	
	67319A	Value Engineering: Practical Applications	79.95	

MA residents add 6.25% state sales tax ____

Shipping & Handling** ____

Total (U.S. Funds)* ____

Prices are subject to change and are for U.S. delivery only. *Canadian customers may call for current prices. **Shipping & handling charges: Add 7% of total order for check and credit card payments. Add 9% of total order for invoiced orders.

Send order to: ADDV-1000

Name (please print) _____

Company _____

☐ Company
☐ Home Address _____

_____

City/State/Zip _____

Phone # _____ P.O. # _____

(Must accompany all orders being billed)

Mail to: RSMeans, 700 Longwater Drive, Norwell, MA 02061

RSMeans Project Cost Report

By filling out this report, your project data will contribute to the database that supports the RSMeans® Project Cost Square Foot Data. When you fill out this form, RSMeans will provide a $30 discount off one of the RSMeans products advertised in the preceding pages. Please complete the form including all items where you have cost data, and all the items marked (☑).

$30.00 Discount per product for each report you submit

Project Description (NEW construction only, please)

☑ Building Use (Office School...) _____

☑ Address (City, State) _____

_____

☑ Total Building Area (SF) _____

☑ Ground Floor (SF) _____

☑ Frame (Wood, Steel...) _____

☑ Exterior Wall (Brick, Tilt-up...) _____

☑ Basement: (check one) ☐ Full ☐ Partial ☐ None

☑ Number of Stories _____

☑ Floor-to-Floor Height _____

☑ Volume (C.F.) _____

% Air Conditioned _____ Tons _____

Total Project Cost $ _____

Owner _____

Architect _____

General Contractor _____

☑ Bid Date _____

☑ Typical Bay Size _____

☑ Occupant Capacity _____

☑ Labor Force: _____ % Union _____ % Non-Union

☑ Project Description (Circle one number in each line.)

   1. Economy 2. Average 3. Custom 4. Luxury

   1. Square 2. Rectangular 3. Irregular 4. Very Irregular

Comments _____

_____

A	☑	General Conditions	$
B	☑	Site Work	$
C	☑	Concrete	$
D	☑	Masonry	$
E	☑	Metals	$
F	☑	Wood & Plastics	$
G	☑	Thermal & Moisture Protection	$
GR		Roofing & Flashing	$
H	☑	Doors and Windows	$
J	☑	Finishes	$
JP		Painting & Wall Covering	$

K	☑	Specialties	$
L	☑	Equipment	$
M	☑	Furnishings	$
N	☑	SpecialConstruction	$
P	☑	ConveyingSystems	$
Q	☑	Mechanical	$
QP		Plumbing	$
QB		HVAC	$
R	☑	Electrical	$
S	☑	Mech./Elec. Combined	$

Please specify the RSMeans product you wish to receive.
Complete the address information.

Product Name _____

Product Number _____

Your Name _____

Title _____

Company _____

   ☐ Company
   ☐ Home   Street Address _____

City, State, Zip _____

Email Address _____

Return by mail or fax 888-492-6770.

Method of Payment:

Credit Card # _____

Expiration Date _____

Check _____

Purchase Order _____

Phone Number _____
(if you would prefer a sales call)

RSMeans
a division of Reed Construction Data
Square Foot Costs Department
700 Longwater Drive
Norwell, MA 02061